FLORA

OF

SHROPSHIRE.

A

FLORA

OF

SHROPSHIRE.

BY

W. A. LEIGHTON, B.A.

FELLOW OF THE BOTANICAL SOCIETY OF EDINBURGH, HONORARY CORRESPONDING
MEMBER OF THE WORCESTERSHIRE NATURAL HISTORY SOCIETY, &c.

THE SUNNE YRISEN HATHE,
THE BIRDES BIN SINGEN CLERE,
THE LARKE WITH CHEERIE LAYE
AWAKES THE BLUSHINGE MORNE,
UP, UP, MIE LOVE, NOR LONGER STAYE,
BUT THRO' THE VERDAUNTE MEADES LET'S STRAYE,—
OR BY THE BABBLINGE BROOKE,
OR MID THE FORESTE DANKE,
AND GATHER AS WE GO
THE GEMMIE FLOWERS THAT GROWE,
NOWE ALL BESPRENTE WITH DEWE.

OLD MS.

EP Publishing Limited
1978

This is a reprint of the 1841 edition, published by John van Voorst, London and John Davies, Shrewsbury, together with a reprint of *The Filices, Lycopodiaceae, Marsileaceae and Equisetaceae of Shropshire* by William Phillips (Shropshire Archaeological Society, 1877) which has been inserted following p. 573.

Republished 1978 by
EP Publishing Limited
East Ardsley, Wakefield
West Yorkshire, England

ISBN 0 7158 1336 6

British Library Cataloguing in Publication Data
Leighton, William Allport
  A flora of Shropshire.
  1. Botany – England – Shropshire
  I. Title   II. Phillips, William, fl.1877. The
  Filices, Lycopodiaceae, Marsileaceae and
  Equisetaceae of Shropshire
  581.9'424'5   QK306
  ISBN 0–7158–1336–6

Please address all enquiries to EP Publishing Limited
(address as above)

Printed in Great Britain by
The Scolar Press Limited
Ilkley, West Yorkshire

" Quanquam multas observaverim plantas et sedulo quidem, tamen non confido me semper veritatem invenisse."—LINK *Elem. Philos. Bot.*

PREFACE.

THE Botany of Shropshire has received little illustration except in the County Lists of Plants in Camden's Britannia, the Botanist's Guides, and the Agricultural Survey, and the few scanty notices scattered throughout the works of How, Ray, Purton, Withering, and Smith. No regular Flora of the County is in existence, nor have those whose attention has been heretofore directed to the subject transmitted any record of their investigations, with the single exception of the late Rev. Edward Williams, M.A., Minister of Battlefield and Uffington. This gentleman in the prosecution of his arduous researches into the History and Antiquities of this County, gradually compiled a MS. Catalogue of the Plants of Shropshire with their localities, which, with his other voluminous and valuable MSS., remains unpublished in the possession of the Right Hon. Lord Berwick. In this Catalogue are enumerated 715 species of Flowering plants and 300 species of the Cryptogamic tribes, (Filices; Musci; Hepaticæ; Lichenes; Characeæ; Algæ.) Had the writer been earlier apprised of the existence of this Catalogue, he would have deemed it incumbent upon him to have made it the basis of the present Flora, inasmuch as the greater portion of its contents was derived from Mr. Williams's personal observations, and consequently the most implicit reliance could be placed on its information. The present work however had proceeded nearly to completion, before its author had an opportunity of consulting the MS. Catalogue, the localities of which he has since in numerous instances personally verified, and considers it a delightful duty to bear his humble testimony to Mr. Williams's uniform fidelity and accuracy. In several cases indeed, it would notwithstanding have been desirable to have had access to the Herbarium of Mr. Williams, to clear up with greater satisfaction some dubious points, but owing to this collection having been most unfortunately separated from the MS., and removed from the County, his wishes in this respect have failed to be gratified.

With a view to supply the deficiency of a Flora of Shropshire, and to afford to other investigators a nucleus whereon to concentrate their discoveries, the present work was projected, and was at first intended to appear as a mere Catalogue of Plants with localities and observations, but the author, yielding to repeated solicitations, subsequently changed his entire plans, and has been induced to present it in the more extended form of a descriptive Flora. To render this account of the vegetable productions of his native County more complete and comprehensive than could be effected by his single and unaided efforts, circulars containing queries and heads of enquiry were distributed, to which many friends of science resident within the district most cheerfully and liberally responded. The result has been that 876 species of Flowering plants are now ascertained and described as natives of Shropshire. That the subject is yet exhausted the author is neither arrogant nor sanguine enough to imagine, and therefore, with all proper respect still urges his correspondents to renewed researches, which if crowned with success, and communicated to him, he proposes to embody with any additions of his own detection, in a Supplement to be published at some future period. Under these circumstances he judges it advisable here to reprint his heads of enquiry, to direct their attention to the most requisite objects of research, namely,—

Lists of the species and more remarkable varieties of the Plants (rare or common) growing wild in any portion of the County,—the general distribution of the commoner species, and the exact localities in which the rarer ones may be found.

The relative degree of scarcity or abundance of species in particular districts.

The periods of their flowering.

The dates, (if carefully noted) of the first flowering of any of the more generally diffused wild plants.

The soils and situations affected by each species.

The possible or probable introduction of particular species by human agency.

Changes that have occurred in the comparative scarcity or abundance of species.

Alterations in character, size, or general habit and appearance, resulting from differences in situation, soil, season, or botanical cultivation.

The wild, or apparently wild localities of our native trees, with anecdotes and memoranda of any remarkable for size, beauty, or connection with the real or legendary history of the neighbourhood.

Results of the examination of the woody concentric zones of trees with a view to ascertain their probable age.

The highest and lowest places at which species occur, whether in absolute height, in comparison with the appearance or cessation of other species, or in relation to the parts of particular mountains, as at the base, middle, or summit.

Mean winter and summer temperature of the air and water at the highest and lowest elevations at which species occur.

The hygrometrical condition of the air.

Any other information of an historical, œconomical, and philosophical nature, tending to illustrate the Flora of Shropshire or the science of Botany.

The present volume must however be considered as strictly complete in itself, as fulfilling all that was contemplated in the primary design of the author; although it is his full intention, should health and opportunities be continued to him, to investigate and describe the cryptogamic tribes in a subsequent and distinct work, for which he earnestly solicits information and specimens.

To those gentlemen and friends who have aided his present researches, he should be wanting both in respect and courtesy, did he not here present his sincere and heartfelt thanks for the valuable particulars which they have severally communicated, and for the undiminished interest which they have manifested for the successful completion of his labours.

To the Right Honourable LORD BERWICK he feels grateful for the opportunity of consulting the accurate unpublished MS. Catalogue of the late Rev. Edward Williams, and for the liberality displayed in his Lordship's unreserved permission to make use of its valuable contents.

To his excellent friend CHARLES CARDALE BABINGTON, ESQ., M.A., F.L.S., F.G.S., &c., of St. John's College, Cambridge, he owes unnumbered obligations for the unremitting kindness with which throughout the progress of the work he has examined herbaria in doubtful cases, consulted foreign authors, and communicated important hints and information, (always it is hoped incorporated with due acknowledgment,) of which his residence in the University of Cambridge has afforded ample opportunities, and his extended researches and acute botanical talents have so fully qualified him to supply.

To TH. FR. L. NEES AB ESENBECK, PH. and M.D., Professor of Botany in the University of Bonn, one of the learned authors of the "Rubi Germanici;" JOHN LINDLEY, ESQ., Ph.D, F.R.S., L.S., and G.S., Professor of Botany in the University of London; and WILLIAM BORRER, ESQ., F.L.S., he is indebted for the determination of the species of the difficult genus *Rubus;* and also to the latter eminent Botanist for additional assistance in the equally complex *Salices,* and the genus *Cerasus;* to DAVID DON, ESQ., F. and Lib. L.S., Professor of Botany in King's College, London; and to ROBERT GRAHAM, ESQ., M.D., F.R.S. Ed. F.L.S., Regius Professor of Botany in the University of Edinburgh, for their valuable aid in the genus *Quercus;* and to WILLIAM ARNOLD BROMFIELD, ESQ., M.D., F.L.S., of Ryde, Isle of Wight; and to EDWARD FORSTER, ESQ., F.L.S., Woodford, Essex, in the genus *Cerasus.*

To MISS MC GHIE, of Ludlow; the Rev. WATIES CORBETT, M.A., Chancellor of Hereford; Mr. F. DICKINSON, of Birmingham; Mr. HENRY

SPARE, of Ludlow; W. P. BROOKES, ESQ., of Much Wenlock; Mr. EDWARD ELSMERE, JUNR., of Astley; the Rev. ANDREW BLOXAM, M.A., of Atherstone, Warwickshire; GEORGE LLOYD, ESQ., M.D., of Leamington; T. C. EYTON, ESQ., F.L.S., F.Z.S., of Donnerville House, near Wellington; EDWIN LEES, ESQ., F.L.S., Forthampton, near Tewkesbury; Mr. GEORGE JORDEN, of Bewdley; and to his valued friends J. E. BOWMAN, ESQ., F.L.S., Manchester; J. F. M. DOVASTON, ESQ., M.A., Westfelton; and the Rev. THOMAS SALWEY, M.A., Vicar of Oswestry; his acknowledgements are due for very comprehensive lists of plants observed either in their immediate neighbourhoods or in other portions of the County: as also to MISS H. MOSELEY, Buildwas; ARTHUR AIKIN, ESQ., F.R.S., London; H. BIDWELL, ESQ., Wellington; J. S. BALY, ESQ.; T. BODENHAM, ESQ., Shrewsbury; the Rev. R. W. CROTCH, M.A., Taunton; the late THOMAS DU GARD, ESQ., M.D., F.G.S., Shrewsbury; R. G. HIGGINS, ESQ., Newport; the Rev. S. P. MANSEL, M.A., Church Stretton; Mr. W. G. PERRY, Warwick; T. W. WILSON, ESQ., M.D., formerly of Shrewsbury; and the late W. W. WATKINS, ESQ., Shotton; for communications of many localities.

In the preparation of the work the invaluable Floras of Smith and Hooker have been taken as text-books,* and the Shropshire plants have been compared with their descriptions, and also, when requisite, with the works of Koch, Fries, Reichenbach, Bluff and Fingerhuth, and other continental Botanists. The descriptions have been, so far as practicable, invariably drawn up from careful examination of living or dried specimens, at the same time adopting such portions of the information conveyed by the above authors as comparison with nature dictated, and rejecting it where apparently new and better characters displayed themselves. The authenticity of the localities rests on the authority of those persons whose names are appended to them, except in those instances in which the author has seen a specimen from the locality, which circumstance is indicated by a mark of admiration. (!) The remaining localities have been ascertained by personal observation, and for these the author is alone responsible. That inaccuracies do not exist cannot be for a moment supposed possible, their detection will be therefore always thankfully appreciated.

The etchings are not inserted as *embellishments;* their attempted accuracy will, it is hoped, render them not altogether useless. Relative proportion has been disregarded, except in those relating to the genera *Carex* and *Blysmus,* which have been prepared by viewing, with a low power of the microscope, the objects placed on a micrometer of $\frac{1}{100}$ of an inch (the squares being rendered visible by rubbing over with the scrapings

* The editions used were Smith's English Flora, 2nd ed. 1828, and Hooker's British Flora, 3rd ed. 1835, unless otherwise expressed.

of a black lead pencil,) and tracing their forms carefully on correspondent squares drawn on paper of the size indicated by the scale marked on the plates.

The author candidly confesses his inability to enter into any statement of the geological features of the County, a deficiency happily of little importance in consequence of the recent publication of Mr. Murchison's "Silurian Regions," in which the geology of the whole County is worked out with faithful accuracy of detail. In the Index of Localities, however, are inserted from Mr. Murchison's work, the geological formations of the places mentioned, with their elevation in feet above the level of the sea; and by references to localities in their immediate vicinities, the Flora of each particular portion of the County has been endeavoured to be exhibited.

Towards the elucidation of the question of the geographical distribution of Plants, the following important data have been kindly communicated by T. O. WARD, ESQ., M.D., Shrewsbury:—Latitude of Shrewsbury taken from St. Chad's Church, 52°, 42,' 40." Elevation of Shrewsbury Canal above 6 ft. 10 in. below Old Dock Sill, Liverpool, at the following places, viz., Ellesmere, Chirk, and Llangollen, 309 ft. 9 in.; Llanymynech, 261 ft. 9 in.; Norbury Junction, 346 ft. 5 in.; Wappensall Junction, 206 ft. 3 in.; Shrewsbury, 192 ft. 7 in.; River Severn at Shrewsbury, 175 ft.; River Severn at Bewdley, 120 ft. Temperature of the air derived from daily observations during 1832-3-4:—Mean annual temperature, 50°, 97' by observation; 51°, 9', 72" by calculation of latitude. Difference of hottest and coldest months, 26°. 07:—Mean temperature of seasons— Winter, 40.5; Spring, 51.63; Summer, 61.66; Autumn, 51.74. Mean difference of temperature of day and night, 32.8. Mean annual range, 61.73. Pressure—mean height of Barometer at 240 ft. above the sea, 29.72. Dew-point—mean maximum, 66.8; mean minimum, 25.5. Days with rain, 169; days without rain, 196. Winds from S.E. to W. inclusive, 245.46 days; winds from N.W. to E. inclusive, 120. 3 days. For the following estimate in inches of the rain I am indebted to the Rev. C. A. A. LLOYD, M.A., Rector of Whittington, near Oswestry:— 1836=26 in.; 1837=16½ in.; 1838=25 in.; 1839=28½ in; 1840, January to August 28=16¼ in.

That much apology is needful for the unusual length of time which has been expended in the printing the author cannot but be sensible, at the same time he feels himself wholly undeserving of blame inasmuch as the delay has proceeded from causes over which he could exercise no controul. This inconvenience he would hope has been doubly compensated by the constant opportunities thus afforded to him of incorporating additional information and discoveries, of revising his previous labours, and rendering the details generally more ample and accurate.

From these investigations fame or pecuniary profit can scarcely be expected, nor are they looked for in the present instance. The pursuit has brought with it its own peculiar and abundant reward; and happy will the author esteem himself if he shall be found to have added only one small pebble, coarse and rugged though it be, to the vast and incessantly accumulating cairn of human knowledge. The retrospect of his past labours brings with it many fond associations of delighted intercourse and rambles with friends bright, amiable, and gentle, amid the beauteous or majestic scenes of Nature; whilst from experience he can confidently foresee that any future exertions in the same fields will be accompanied with healthful freedom and exercise,— will prove a never-failing solace against those "secret woes the world has never known," and a constant resource against the empty haughtiness, the biting sarcasms, or the trickful chicanery of the world and its votaries,—will open to him ten thousand sources of intellectual gratification,—of silent though enraptured adoration of his Creator's benevolence and wisdom,—and will impel him perpetually to see and to feel that there is "good in every thing."

With the most ardent wishes that to all who may engage in these pleasing and not unimportant studies, they may bring similar and ten-fold delights, he, in the words of Davie Gellatley's song, kindly bids them—

> " Hie away, hie away,
> Over bank and over brae,
> Where the copsewood is the greenest,
> Where the fountains glisten sheenest,
> Where the lady-fern grows strongest,
> Where the morning dew lies longest,
> Where the black-cock sweetest sips it,
> Where the fairy latest trips it :
> Hie to haunts right seldom seen,
> Lovely, lonesome, cool and green,
> Over bank and over brae,
> Hie away, hie away."

W. A. LEIGHTON.

Shrewsbury, September, 1840.

CLASS I.

MONANDRIA.

—

> " Thy desire, which tends to know
> The works of God, thereby to glorify
> The great Workmaster, leads to no excess
> That reaches blame, but rather merits praise
> The more it seems excess; * * *
> * * * * * * *
> For wonderful indeed are all His works,
> Pleasant to know, and worthiest to be all
> Had in remembrance always with delight."
>
> MILTON.

FLORA OF SHROPSHIRE.

—

CLASS I.

MONANDRIA. 1 *Stamen.*

ORDER I. MONOGYNIA. 1 *Style.*

1. HIPPURIS. *Perianth* single, superior, forming an indistinct rim to the germen. *Fruit* a small one-seeded *nut.*—*Nat. Ord.* HALORAGEÆ, *Br.*—Name from ιππος, a horse, and ουρα, a tail.

(See *Centranthus ruber*, in CL. III. *Alchemilla arvensis*, in CL. IV.)

(ORD. II. DIGYNIA. 2 *Styles.* See *Callitriche* in CL. XXI.)

MONANDRIA—MONOGYNIA.

1. HIPPURIS. *Linn.* Mare's-Tail.

1. H. *vulgaris*, Linn. *Common Mare's-Tail.* Leaves linear, 6—8 or 10 in a whorl. *E. Bot. t.* 763. *E. Fl. v. i. p.* 4. *Hook. Br. Fl. p.* 2.

Ditches; rare. *Fl.* June, July. ♃.

In the ditches which intersect the moors between Kinnersley and Crudgington; *H. Bidwell, Esq.* Ditch on the right-hand side of the road between Kinnersley and Butterey; *Rev. E. Williams's MSS.*

Stem erect, simple, jointed. *Whorls* of about 8 *leaves* which are callous at the point. *Flowers* at the base of each of the upper leaves, not unfrequently destitute of stamen. *Germen* oval, inferior; within its minute rim or border, at the summit which constitutes the calyx, is situated the *stamen,* with its large 2-lobed *anther*; when young, having the style passing between the two lobes. *Seeds* fixed to the top of the cell of the *pericarp* and thus invested.

CLASS II.

DIANDRIA.

"Nature never did betray
The heart that loved her; 'tis her privilege
Through all the years of this our life, to lead
From joy to joy: for she can so inform
The mind that is within us, so impress
With quietness and beauty, and so feed
With lofty thoughts, that neither evil tongues,
Rash judgments, nor the sneers of selfish men,
Nor greetings where no kindness is, nor all
The dreary intercourse of daily life,
Shall e'er prevail against us, or disturb
Our cheerful faith, that all which we behold
Is full of blessings."

WORDSWORTH.

CLASS II.

DIANDRIA. 2 *Stamens.*

ORD. I. MONOGYNIA. 1 *Style.*

* *Perianth double, inferior, monopetalous, regular.*

1. LIGUSTRUM. *Calyx* of four minute teeth. *Corolla* 4-cleft, funnelshaped. *Berry* (drupe) 2-celled, cells 2-seeded.—*Nat. Ord.* JASMINACEÆ, *DC.*—Name from *ligo, to bind;* on account of the use sometimes made of its long pliant branches.

** *Perianth double, inferior, monopetalous, irregular. Seeds inclosed in a distinct pericarp. (Angiospermous.)*

2. VERONICA. *Calyx* 4-cleft. *Corolla* 4-cleft, rotate, lower segment narrower, upper broadest. *Capsule* 2-celled. *Seeds* numerous.—*Nat. Ord.* SCROPHULARINEÆ, *Juss.*—Name, a barbarous compound of Greek and Latin; *vero, in the truth, and νικη, victory:* —*Victory in the truth.*

3. PINGUICULA. *Calyx* 2-lipped, upper lip of 3, lower of 1 bifid reflexed segment. *Corolla* ringent, 5-cleft, spurred. *Germen* globose. *Stigma* large, of two unequal plates or lobes. *Capsule* 1-celled. *Seeds* numerous, attached to a central receptacle.—*Nat. Ord.* LENTIBULARIÆ, *Rich.*—Name from *pinguis, fat;* from the texture and appearance of the leaves.

4. UTRICULARIA. *Calyx* 2-leaved, equal, undivided. *Corolla* personate; upper lip obtuse, erect; lower larger and spurred; palate cordate, prominent. *Stigma* 2-lipped. *Capsule* globose, 1-celled. *Seeds* fixed to a central receptacle.—*Nat. Ord.* LENTIBULARIÆ, *Rich.*—Name from *utriculus, a little bladder.*

*** *Perianth double, inferior, monopetalous, irregular. Seeds 4, apparently naked, closely covered by the pericarp. (Gymnospermous.)*

5. LYCOPUS. *Calyx* tubular, 5-cleft, permanent. *Corolla* tubular, **limb** 4-cleft, obtuse, upper segments broader and notched. *Stamens* simple, distant.—*Nat. Ord.* LABIATÆ, *Juss.*—Name from

λυκος, a *wolf* and πους, a *foot*; from a fancied resemblance in the cut leaves of the plant, to a wolf's paw :—*der wolfsfuss*, in German. In English, *Gipsy-wort*, from the plant being used by gipsies in staining their skin.

6. SALVIA. *Calyx* tubular, ribbed, unequally 2-lipped. *Corolla* labiate, *tube* compressed, dilated upwards. *Filaments* with 2 divaricating branches, one only bearing a perfect single cell of an *anther.*—*Nat. Ord.* LABIATÆ, *Juss.*—Name from *salvo, to save* or *heal*; from the balsamic, sanative qualities of the plant.

**** *Perianth double, superior.*

7. CIRCÆA. *Calyx* 2-leaved, united into a short tube at the base, segments reflexed. *Corolla* of 2 petals. *Capsule* 2-celled, cells 1-seeded.—*Nat. Ord.* ONAGRARIÆ, *Juss.*—Name from the enchantress *Circe*, from its rivetting the attention by the beauty of its appearance and the dissimilarity of habit from the other plants growing in the damp shady places which it inhabits.

***** *Perianth single or none.*

8 FRAXINUS. *Calyx* 0, or 4-cleft. *Corolla* 0, or of 4 petals. *Capsule* 2-celled, compressed and foliaceous at the extremity. (a *Samara.) Seeds* one in each cell, pendulous. (Flowers sometimes without stamens.)—*Nat. Ord.* JASMINACEÆ, *DC.*—Name from φραξις, *a separation;* from the facility with which the wood splits.

9. LEMNA. *Perianth* single, monophyllous, membranaceous, urceolate, included in the marginal fissure of the frond. *Fruit* utricular, indehiscent, 1-celled, 1 or more seeded. *Seeds* erect or horizontal. *Fronds* without distinct stem or leaves, floating on the surface of the water, and increasing, not only by seeds, but, far more abundantly, by *gemmæ* or *buds*, concealed in lateral clefts of the parent frond, which growing out, on two opposite sides, into new plants, and these being similarly proliferous while still attached to their parent, present a most curious appearance.—*Nat. Ord.* AROIDEÆ, *DC.*—Name, λεμμα of the Greeks, from λεπω, *to pull away a scale;* probably in allusion to its mode of increase.

(See *Cladium* and *Anthoxanthum* in CL. III.)

DIANDRIA. MONOGYNIA.

1. LIGUSTRUM. *Linn.* Privet.

1. L. *vulgare*, Linn. *Privet.* Leaves opposite, elliptico-lanceolate, entire, glabrous; panicles terminal, compound, dense. *E. Bot. t.* 764. *E. Fl.* v. 1. *p.* 13. *Hook. Br. Fl. p.* 4.

Woods and thickets, in the limestone districts; rare. *Fl.* June, July. ♄. Benthall Edge; *Mr. F. Dickinson.*
A bushy *shrub*, with straight branches filled with pith; wood hard; bark smooth and bitter. *Leaves* shortly petiolate, obtuse, similar to myrtle, but of a duller hue, nearly evergreen. *Flowers* white, scented. *Berries* globose, black, nauseous and bitter.
Valuable for ornamental fences, as the plant bears clipping.

2. VERONICA. *Linn.* Speedwell.

* *Spikes or racemes terminal.*

a. *Root perennial.*

1. V. *serpyllifolia*, Linn. *Thyme-leaved Speedwell.* Raceme somewhat spiked, many-flowered; leaves elliptical, obtuse, slightly crenate, 3-ribbed, glabrous; capsule inversely reniform, compressed, as long as the style, smooth, with a glanduloso-pilose keel. *E. Bot. t.* 1075. *E. Fl.* v. i. *p.* 20. *Hook. Br. Fl. p.* 5.

Meadows and pastures; abundant. *Fl.* May—July. ♃.
Stem more or less procumbent and branched near the crown, throwing out long fibrous roots from the lower joints, pubescent. *Leaves* of a pale shining green, in pairs opposite, obtusely elliptical, 3-ribbed, slightly crenate, almost sessile, the lowermost on short tapering footstalks, somewhat obovate, the uppermost passing gradually into bracteas. *Racemes* solitary, erect. *Flowers* pale blue or white with dark streaks, each on a short pedicel with a bractea at the base. *Calyx* monosepalous, deeply 4-cleft, the segments elliptical, with 3 principal ribs, and a few erect teeth on the edges. *Capsule* inversely reniform, as long as the persistent style, glanduloso-pilose at the keel. *Seeds* numerous, obovate, smooth.

b. *Root annual.*

2. V. *arvensis*, Linn. *Wall Speedwell.* Raceme somewhat spiked, many-flowered; leaves cordato-ovate, deeply serrated, 5-ribbed, hairy; stems ascending; capsule obcordate, compressed, smooth, with a glanduloso-pilose keel, cells many-seeded. *E. Bot. t.* 734. *E. Fl.* v. i. *p.* 24. *Hook. Br. Fl. p.* 8.

Walls, dry banks, and gravelly ground; frequent. *Fl.* April—June. ☉.
Whole plant soft and hairy. *Root* fibrous. *Stem* erect, slender, pale green, 3—6 inches high, generally simple, occasionally branched in the lower part. *Lower leaves* shortly petiolate, upper ones sessile and gradually diminishing into lanceolate entire bracteas. *Flowers* pale blue, light in the centre, on short bracteated pedicels. *Calyx* monosepalous, deeply 4-cleft, segments lanceolate, unequal. *Seeds* 12 or 14 in each cell, flat, smooth.

** *Racemes axillary.*

a. *Root perennial.*

3. V. *scutellata*, Linn. *Marsh Speedwell.* Racemes alternate, pedicels divaricated, reflexed in fruit; leaves linear, dentate; capsule of two flattened orbicular lobes; stem tetragonous, angles winged. *E. Bot. t.* 782. *E. Fl.* v. i. *p.* 21. *Hook. Br. Fl. p.* 6.

Wet places and sides of pools; not very common. *Fl.* July, Aug. ♃.
•Ellesmere Mere; *J. E. Bowman, Esq.* Oakley Park, near Ludlow; *Mr. H. Spare.* Pools on the Clee Hills; *Rev. T. Salwey.* Near Ness; *T. C. Eyton, Esq.!* Broadmoor, near Hales Owen; *Withering.* In a bog 1½ mile from Shrewsbury, on Wenlock road; *Mr. F. Dickinson.* Forest of Wyre; *Mr. G. Jorden.*
Bomere Pool, and Hancott Pool, near Shrewsbury.
Root fibrous, proceeding from the lower joints. *Stem* trailing, at length erect, weak, tetragonous, the angles winged, smooth. *Leaves* in pairs, opposite, linear, dentate, semi-amplexicaul. *Racemes* alternate, axillary, considerably longer than the leaves; pedicels alternate, divaricated, reflexed in fruit, each with a small linear bractea. *Flowers* flesh-coloured, with darker bluish veins. *Calyx* monosepalous, 4-cleft, segments lanceolate, acute, shorter than the fruit, each with 4 veins. *Capsule* of 2 flattened orbicular lobes, wrinkled and veined. *Seeds* numerous, orbicular, rugged or minutely dotted.
A variety with the whole herbage hairy occurs on some boggy ground north of Bomere Pool, and at the east end of Blackmere.

4. V. *Anagallis*, Linn. *Water Speedwell.* Racemes opposite; leaves lanceolate, serrated; stem erect. *E. Bot. t.* 781. *E. Fl. v. i. p.* 21. *Hook. Br. Fl. p.* 6.

Ditches and watery places; not very common. *Fl.* July, Aug. ♃.
Burway Meadows, near Ludlow; *Mr. J. S. Baly!* Shotton; *Dr. Wilson!* Sutton; Shelton Rough; Uffington, near Shrewsbury.
Root creeping. *Stem* round, hollow, smooth, erect, very variable in height according to situation. *Leaves* in pairs, opposite, semi-amplexicaul, lanceolate, coarsely serrated, with 3 principal ribs. *Racemes* opposite, axillary, considerably longer than the leaves; pedicels alternate, shorter than the linear bracteas, divaricated, never reflexed, occasionally pubescent. *Flowers* pale blue. *Calyx* monosepalous, deeply 4-cleft, segments lanceolate, longer than the fruit, single ribbed. *Capsule* broadly ovate, notched, tipped with the very short style. *Seeds* numerous, elliptical.

5. V. *Beccabunga*, Linn. *Brooklime.* Racemes opposite; leaves elliptical, obtuse, subserrated, glabrous; stem procumbent at the base and rooting. *E. Bot. t.* 655. *E. Fl. v. i. p.* 20. *Hook. Br. Fl. p.* 6.

Rivulets, ditches, and wet places; very common. *Fl.* Summer months. ♃.
Whole plant glabrous and succulent. *Stem* procumbent, rooting from the lower joints. *Leaves* opposite, shortly petiolate. *Racemes* of numerous bright blue *flowers*, on erect bracteated pedicels. *Calyx-segments* ovate, acute, 3—5 ribbed. *Capsule* roundish, tumid, scarcely reticulated. *Seeds* small, numerous.

6. V. *officinalis*, Linn. *Common Speedwell.* Racemes axillary, spicate; leaves elliptical, serrated, rough with pubescence; stem procumbent, very downy; capsule obcordate, truncated, notched. *E. Bot. t.* 765. *E. Fl. v. i. p.* 22. *Hook. Br. Fl. p.* 6.

Abundant in woods and dry hilly pastures.—*Fl.* May—Aug. ♃.
Near Dowles Brook, Wyre Forest; *Mr. W. G. Perry.* Hawkestone; *Mr. J. S. Baly.* Cox Wood, near Coalbrookdale; *Mr. F. Dickinson.*
Whitecliff Coppice, near Ludlow. Haughmond Hill and Bickley Coppice, near Shrewsbury.
Stem prostrate, creeping, round, covered, as is the whole herbage, with short, spreading, delicately jointed hairs. *Leaves* in pairs, opposite, elliptical, pointed, serrated, shortly petiolate, rather rigid, rough on both sides with jointed pubescence. *Racemes* alternate, axillary, erect; *pedicels* alternate, very short. *Bracteas* lanceolate, longer than the pedicels. *Flowers* light blue with dark streaks. *Calyx* monosepalous, 4-cleft, segments lanceolate, shorter than the fruit, 3-ribbed, densely covered with jointed glandular hairs. *Capsule* obcordate, truncate, slightly notched, covered with jointed glandular hairs, crowned with the long persistent style and capitate stigma. *Seeds* orbicular, flat.
Leaves astringent and bitter, occasionally used as a medicinal tea.

7. V. *montana*, Linn. *Mountain Speedwell.* Racemes axillary, alternate, lax, few-flowered; leaves broadly ovate, petiolate, serrated; stem procumbent, hairy all round; capsule compressed, orbicular, two-lobed, notched, edges denticulate, ciliato-glandulose, much larger than the calyx. *E. Bot. t.* 766. *E. Fl. v. i. p.* 23. *Hook. B. Fl. p.* 7.

Moist woods; not common. *Fl.* May, June. ♃.
In a moist woody place near Hord's Park; *Purton's Midl. Fl.*—Woods about Coalbrookdale, plentifully; *Turn. & Dillw. Bot. Guide.*—Ashford Coppices near Ludlow; *Miss Mc Ghie.* Upper Lumhole Pool, near Coalbrookdale; Old Brook Wood, near Coalbrookdale; *Mr. F. Dickinson!* Forest of Wyre; *Mr. G. Jorden.*
Whitecliff Coppice near Ludlow. Almond Park and Shelton Wood, near Shrewsbury.
Stems procumbent, creeping, covered all round with short, spreading, jointed, glandular hairs. *Leaves* in pairs, opposite, broadly ovate, petiolate, serrated, clothed on both sides with jointed hairs, most abundant on the veins. *Racemes* axillary, alternate, occasionally straggling, few-flowered, hairy; *pedicels* longer than the linear *bracteas.* *Flowers* pale blue. *Calyx* monosepalous, 4-cleft, segments obovate, acute, with 3 principal veins, margins ciliated with jointed glandulose hairs. *Capsule* orbicular, compressed, two-lobed, slightly notched, denticulate and ciliato-glandulose at the edges, larger than the calyx, and crowned with the long persistent style.
This species can never be confounded with V. *Chamædrys*, if attention is directed to the form of the capsule and the pubescence of the stem.

8. V. *Chamædrys*, Linn. *Germander Speedwell.* Racemes axillary, alternate, elongated, many-flowered; leaves cordato-ovate, nearly sessile, inciso-serrate; stem procumbent, bifariously hairy; capsule flat, obcordate, two-lobed, deeply notched, hairy, much smaller than the calyx. *E. Bot. t.* 623. *E. Fl. v. i. p.* 23. *Hook. Br. Fl. p.* 7.

Woods, pastures, and hedge-banks; common. *Fl.* May, June. ♃.
Stem procumbent, creeping, with a line of jointed hairs on its opposite sides, the hairy lines taking different sides above and below each pair of leaves, or decussate. *Leaves* in pairs, opposite, nearly sessile, cordato-ovate, inciso-serrate, hairy on the under surface, the upper surface smooth, except near the margins. *Racemes* axillary, alternate, many-flowered, glanduloso-pilose; *pedicels* erect, glanduloso-pilose, longer than the bracteas. *Bracteas* ovato-lanceolate, acute,

margins and under side ciliato-glandulose. *Flowers* large and bright blue; segments of the corolla (except the upper one, to which the filaments are attached,) marked with a hairy line immediately above the summit of the tube. *Calyx* monosepalous, 4-cleft, segments lanceolate, acute, with 3 principal ribs, hairy, margins ciliated. *Capsule* obcordate, two-lobed, deeply notched, hairy, smaller than the calyx, and crowned with the long style.

*** *Flowers axillary, solitary. Root annual.*

9. V. ***hederifolia***, Linn. *Ivy-leaved Speedwell.* Stem procumbent; leaves alternate, petiolate, cordate, with 5—7 lobes; calyx-segments cordate, ciliated; capsule of 2 turgid lobes, seeds 2 in each cell, cupped, transversely wrinkled. *E. Bot. t.* 784. *E. Fl. v. i. p.* 25. *Hook. Br. Fl. p.* 7.

Corn fields and cultivated ground; very common. *Fl.* Apr.—June. ⊙.

Root fibrous. *Stem* procumbent, bifariously hairy. *Leaves* alternate, petiolate, cordate, with 5—7 lobes, the central one largest, pointed, fleshy and minutely dotted, with 3 principal ribs, and a few scattered jointed hairs on both surfaces. *Flowers* axillary, solitary, pale blue, streaked; *peduncles* scarcely so long as the leaf, recurved in fruit. *Calyx-segments* cordate, acute, longer than the fruit, with 3 principal veins, margin ciliated with long jointed hairs. *Capsule* of two turgid, rounded, glabrous lobes, crowned with the short style. *Seeds* 2 in each cell, large, oval, gibbous, transversely wrinkled, hollow or cupped on the under side, and peltate.

10. V. ***agrestis***, Linn. *Green procumbent Field Speedwell.* Stem procumbent; leaves petiolate, cordato-ovate, inciso-serrate, longer than the flower-stalk; calyx-segments oblong, obtuse; capsule of two turgid, keeled, minutely dotted and veined lobes, seeds 6 in each cell. *E. Bot. Suppl. t.* 2603. *E. Fl. v. p.* 24. *Hook. Br. Fl. p.* 8.

Corn fields and cultivated ground; common. *Fl.* Apr.—Sept. ⊙.

Root fibrous. *Stem* procumbent, branched and spreading at the base, round, pubescent. *Leaves*, the lower ones in pairs, opposite, upper ones alternate, all petiolate, cordato-ovate, inciso-serrate, with a few scattered hairs on both surfaces and three principal ribs. *Flowers* small, bright blue; *peduncles* scarcely so long as the leaf, recurved in fruit, slightly pubescent. *Calyx* larger than the fruit, segments oblong, obtuse, with 3 prominent and principal ribs, from whence smaller ones ramify, margins ciliated with jointed hairs, and a few scattered ones on the keel of the principal rib. *Capsule* of two turgid, keeled, minutely dotted and veined lobes, keels ciliated, crowned with the short style and capitate stigma. *Seeds* about six in each cell, ovate, transversely wrinkled, cupped and peltate on the under side.

11. V. ***polita***, Fries. *Grey procumbent Field Speedwell.* Stem procumbent; leaves petiolate, cordato-ovate, inciso-serrate, shorter than the flower-stalk; calyx-segments broadly ovate, acute; capsule of two turgid, keeled, densely pubescent lobes, seeds about 18. *V. agrestis E. Bot. t.* 783. *Hook. Br. Fl. p.* 8.

Corn fields and cultivated land with the preceeding; rare. *Fl.* throughout the summer. ⊙.

Corn fields near Coalbrookdale; *Mr. F. Dickinson.*

Cultivated ground near Shrewsbury.—Halford near Craven Arms.

Root fibrous. *Stem* procumbent, branched and spreading at the base, hairy. *Leaves*, lowermost opposite, upper ones alternate, all petiolate, cordato-ovate, inciso-serrate, hairy on both surfaces, with five principal ribs. *Flowers* small, bright

blue; *peduncles* pubescent, longer than the leaves, recurved in fruit. *Calyx* longer than the fruit, segments broadly ovate, acute, with 3 principal prominent ribs, from whence minor ramifications proceed, margins and keel of the principal rib ciliated with jointed hairs. *Capsule* of two turgid, keeled lobes, crowned with the very short style and capitate stigma, and covered with a short dense pubescence. *Seeds* about 18, (10 in one cell and 8 in the other,) ovate, transversely wrinkled, cupped, and peltate on the under side.

V. agrestis, polita, and *Buxbaumii* are all nearly allied species and similar in general habit and appearance, but may be at once accurately distinguished by their capsules. *V. polita* differs from *V. agrestis* in its capsule being covered with a short dense pubescence, that of *V. agrestis* being minutely dotted and veined, the keel only ciliated; whilst both are distinguished from *V. Buxbaumii* by the capsule of the latter being obcordate, its breadth nearly double its length, and its lobes divaricated, sharply keeled, smooth and veiny.

3. PINGUICULA. *Linn.* Butterwort.

1. P. ***vulgaris***, Linn. *Common Butterwort.* Spur subulato-cylindrical, acute, as long as the veinless limb of the corolla, whose segments are very unequal, rounded, even, and all entire; capsule ovate. *E. Bot. t.* 70. *E. Fl. v. i. p.* 28. *Hook. Br. Fl. p.* 9.

Bogs and moist heaths; not very common. *Fl.* May, June. ♃.

Rowley, near Much Wenlock; *W. P. Brookes, Esq!* Meadow near Walford; *T. C. Eyton, Esq!* Near Westfelton; *J. F. M. Dovaston, Esq.* Near Oswestry; *Rev. T. Salwey.* Bog at Underton, near Bridgnorth; *Purton's Mid. Fl.* Meadows near Downton Hall, near Ludlow; *Mr. H. Spare.* In boggy ground at the north-eastern base of the Wrekin, but not very plentiful; *Edwin Lees, Esq.* Bog south-western side of Wrekin; *Mr. F. Dickinson.* Clee Hill, Aston, and Vinalls, near Ludlow; *Miss Mc Ghie.* Oreton Common, plentiful; *Mr. G. Jorden.*

Boggy ground at the base of the Stiperstones Hill.

Root fibrous. *Leaves* all radical, ovato-oblong, pale green, fleshy, covered with minute raised crystalline points, margins involute. *Scapes* 3—9 inches high, single-flowered. *Calyx-segments* oblong, obtuse. *Flowers* purple, very handsome, drooping; palate covered with white, compactly jointed hairs. *Anthers* 1-celled, vertical, placed just beneath the large horizontal plate or lobe of the stigma. *Style* short. *Capsule* ovate, one-celled, bursting half-way into 2 valves. *Seeds* numerous, oblong, rough.

The expressed juice of the leaves may be used as a substitute for rennet in the manufacture of cheese.

On the gradual decay of the leaves in autumn, small, round, leafy buds or hybernacula are formed, which survive the winter, and are capable of developing new plants in the spring; a beneficent provision of Nature for the preservation of a tribe of plants which are re-produced but scantily by seed, and which, by reason of the delicate and succulent nature of their herbage, would otherwise be unable to survive the winter's cold in their chilled and marshy habitats.

4. UTRICULARIA. *Linn.* Bladderwort.

1. U. ***vulgaris***, Linn. *Greater Bladderwort.* Spur conical, upper lip entire, as long as the projecting palate, reflexed at the sides; leaves pinnato-multifid, bristly. *E. Fl. v. i. p.* 30. *Hook. Br. Fl. p.* 18.

Ditches and pools; rare. *Fl.* June, July. ♃.

In a small pond on the Sharpstones, near Sutton; *Arthur Aikin, Esq.* In a deep ditch on the north margin of Croesmere Mere; *J. E. Bowman, Esq!* Hancott Pool, near Shrewsbury; *Mr. F. Dickinson.* Naturalized in a pond in "Fairy-land," near Westfelton, by *J. F. M. Dovaston, Esq.!*

Roots much branched. *Shoots* or *runners* long, floating horizontally in the water. *Leaves* vascular, placed at regular intervals, capillaceo-multifid, bristly at the margin, bearing numerous crested vesicles of a green-purple or pink colour, having an orifice closed by an elastic valve, opening inwards and of much thinner texture than the bladder, to which it is attached where the crest is placed, margin armed with a few long spines. *Scape* erect. *Flowers* 6—8, bright yellow, somewhat corymbose, lower *lip* convex, much larger and broader than the upper one, with a projecting palate closing the mouth. *Spur* short, deflexed. *Filaments* curved, thick. *Stigma* large.

When it becomes necessary for the plant to rise to the surface of the water for the expansion of its blossoms, the vesicles on the leaves, which before contained water, are by some unknown means now filled with air, which, after fecundation, again gives place to the water, and the plant descends to ripen its seeds at the bottom.

2. U. ***minor***, Linn. *Lesser Bladderwort.* Spur extremely short, obtuse, keeled; upper lip emarginate, as long as the palate; leaves subtripartite, segments linear, dichotomous. *E. Bot. t.* 254. *E. Fl. v. i. p.* 31. *Hook. Br. Fl. p.* 11.

Ditches and pools; rare. *Fl.* June, July. ♃.

Whixall Moss; *J. E. Bowman, Esq!—Rev. A. Bloxam.*

Ditches on the north side of Bomere Pool.

Roots fibrous. *Shoots* or *runners* long, floating. *Leaves* alternate, subtripartite, segments linear, dichotomous, smooth. *Vesicles* attached to the leaves. *Scape* erect, short. *Flowers* about 3, small, pale yellow; lower *lip* almost plane, palate scarcely closing the mouth, not projecting beyond the lip. *Spur* scarcely any.

5. LYCOPUS. *Linn.* Gipsy-wort.

1. L. ***Europæus***, Linn. *Common Gipsy-wort, or Water Hore-hound.* Leaves deeply and irregularly pinnatifido-serrate. *E. Bot. t.* 1105. *E. Fl. v. i. p.* 34. *Hook. Br. Fl. p.* 11.

Ditches and margins of pools and rivers; not uncommon. *Fl.* June—Aug. ♃.

Buildwas Park; *W. P. Brookes, Esq.* Ruckley Wood; *Dr. G. Lloyd.* Walford; *T. C. Eyton, Esq!* Near Oswestry; *Rev. T. Salwey.* Oakley Park, near Ludlow; *Mr. H. Spare.* Near Ludlow; *Miss Mc Ghie.* Astley, near Shrewsbury; *Mr. E. Elsmere, junr.!* White Sitch Pool; *Mr. F. Dickinson.* Kinlet; *Mr. G. Jorden.*

Canal near Shrewsbury; Berrington, Hancott, and Bomere Pools.

Root creeping. *Stem* 2 feet high, erect, 4-sided, furrowed, pubescent, chiefly so at the angles. *Leaves* opposite, nearly sessile, ovato-lanceolate, sinuato-serrate, lower ones almost pinnatifid, upper surface rough with minute glandular points and a few scattered jointed hairs, under surface densely covered with a glandular mealiness, in which probably resides the dye of the plant, ribs copiously clothed with jointed appressed pubescence. *Flowers* small, sessile, in dense whorls in the axils of the upper leaves, whitish with purple dots, hairy within, and a few scattered hairs and glands on the outside. *Calyx* monosepalous, hairy, deeply 5-cleft, segments subulate, with a hairy acute keel and a strong nerve on each side, margins with erect teeth. *Seeds* 4, quadrangular, obtuse, in the bottom of the permanent calyx.

6. SALVIA. *Linn.* Sage or Clary.*

1. S. ***Verbenaca***, Linn. *Wild English Clary or Sage.* Lower

* Mr. H. Spare has communicated "Oakley Park, near Ludlow," as the habitat of "*Salvia pratensis*," Linn., but as that species is one of extremely rare occurrence, and has been too often confounded with the commoner one, *S. Verbenaca*, I have felt unwilling to insert a description without the examination of specimens from that locality.

leaves cordato-oblong, obtuse, sinuated and serrated, petiolate, upper ones acute, sessile, semiamplexicaul; bracteas cordate, sharply acuminate; corolla contracted, scarcely longer than the calyx. *E. Bot. t.* 154. *E. Fl v. i. p.* 35. *Hook. Br. Fl. p.* 11.

Dry pastures and hedge-banks; rare. *Fl.* June, July. ♃.

Hedge-bank of St. Giles's church-yard, Shrewsbury.

Root woody. *Stem* more or less decumbent, four-sided, furrowed, clothed with spreading glandular hairs. *Herbage* aromatic. *Leaves* greyish-green, lower ones cordato-oblong, obtuse, sinuated and serrated, petiolate, upper ones cordato-ovate, acute, sessile, semiamplexicaul, less lobed but more serrated, wrinkled, both surfaces covered with a minute glandular mealiness, hairy chiefly at the ribs and margins. *Bracteas* 2 under each whorl, cordato-acuminate, entire, ciliated and hairy with long jointed hairs. *Calyx* tubular, with 13 ribs, unequally 2-lipped, ciliated, permanent, upper lip 3-cleft, segments mucronate, hispid, lower lip broadly obovate, smooth, with a minute point above the central rib, hairy only at the base and ribs, reflexed in fruit. *Flowers* whorled at regular intervals. *Corolla* contracted, smaller than the calyx, purple, upper *lip* concave, compressed, lower broad, 3-lobed, middle lobe largest, cloven. *Seeds* 4 in the bottom of the dry converging calyx, oval, minutely dotted.

The seeds when steeped in water speedily give out a brownish mucilage, which on examination under a high power of the microscope, appears to be composed of delicate cylindrical fibres, containing masses of brown granular substance.

7. CIRCÆA. *Linn.* Enchanter's Nightshade.

1. C. ***Lutetiana***, Linn. *Common Enchanter's Nightshade.* Stem erect, pubescent; leaves ovato-cordate, acuminate, dentate, opaque, longer than the petiole. *E. Bot. t.* 1056. *E. Fl. v. i. p.* 15. *Hook. Br. Fl. p.* 12.

Moist shady woods and coppices; common. *Fl.* June, July. ♃.

Root creeping. *Stem* erect, more or less branched, round, swelling at the joints, clothed with short, curved, deflexed pubescence. *Leaves* in pairs, opposite, ovato-cordate, acuminate, dentate, pubescent, longer than the petioles, upper leaves ovate, acuminate. *Raceme* more or less branched, covered with short, glandular, spreading hairs. *Corolla* white or rose-coloured, *petals* 2, obcordate, spreading, alternate with the sepals. *Pedicels* erect, reflexed in fruit. *Stamens* 2, alternate with the petals, inserted into the calyx. *Calyx* tubular at the base, the tube filled with a large cup-shaped disk which projects beyond it, *sepals* 2, ovate, acute, coloured, reflexed. *Capsule* obovate, covered with glandular, hooked bristles, 2-celled. *Seeds* one in each cell, obovate, flat on the inside.

8. FRAXINUS. *Linn.* Ash.

1. F. ***excelsior***, Linn. *Common Ash.* Leaves pinnated, leaflets ovato-lanceolate, acuminate, serrated; flowers without either calyx or corolla. *E. Bot. t.* 1692. *E. Fl. v. i. p.* 14. *Hook. Br. Fl. p.* 12.

Woods and hedges; frequent. *Fl.* April, May. ♄.

One of the noblest of our trees, graceful and elegant in its form and appearance, particularly in old individuals whose lower branches become pendent, with their extremities curving upwards. "Fraxinus in silvis pulcherrima." Timber highly valuable for implements of husbandry, and other wheelwright's works.

Bark smooth, grey. *Wood* tough and white. *Buds* dark brown or black, downy. *Leaves* pinnate, *leaflets* in 4—8 pairs with an odd terminal one, nearly sessile, pubescent on the under side at the base and angles of the veins. *Flowers* in axillary clusters, appearing before the leaves, at the extremity of the last year's wood, small, brown, and very simple, without *calyx* or *corolla*. *Pistils* and *stamens*, often one of each, sometimes, however, produced on separate trees. *Capsule* glittering with rusty meal.

9. LEMNA. *Linn.* Duckweed.

1. L. *trisulca*, Linn. *Ivy-leaved Duckweed.* Fronds thin, pellucid, elliptico-lanceolate, caudate at one extremity, at the other serrated ; roots solitary. *E. Bot. t.* 926. *E. Fl. v. i. p.* 32. *Hook. Br. Fl. p.* 13.

Clear stagnant waters ; not unfrequent.
Pond near Uffington.—Canal near Shrewsbury.
Root solitary, filiform, pendulous from the centre of the under surface of the frond, tipped at the extremity with a small sheath. *Fronds* ½–¾ of an inch in length, reticulated, 3-nerved, the central nerve proceeding from the caudate extremity, proliferous laterally at right angles, the young fronds apparently proceeding from the centre where the root is attached, and passing out between the laminæ of the parent frond. *Flowers* minute, white with yellow anthers. Several flowers want the pistil, none the stamens, which latter are both at one side of the germen. *Seed* single, very hard. *Embryo* simple, horizontal.

2. L. *minor*, Linn. *Lesser Duckweed.* Fronds opaque, nearly ovate, compressed ; roots solitary. *E. Bot. t.* 1095. *E. Fl. v. i. p.* 32. *Hook. Br. Fl. p.* 13.

Stagnant waters ; most abundant. *Fl.* July. ☉.
Frond a line, or a line and a half long, thick, succulent, slightly convex beneath, proliferous. *Flowers* similar to the last. *Capsule* single seeded. *Seed* horizontal, with its hilum directed towards the narrow end of the frond.

3. L. *polyrrhiza*, Linn. *Greater Duckweed.* Fronds opaque, obovato-rotundate, compressed ; roots numerous, clustered. *E. Bot. t.* 2458. *E. Fl. v. i. p.* 33. *Hook. Br. Fl. p.* 13.

Stagnant waters. *Fl.* unknown in Britain. ☉.
Pool near Hanmer Hall ; *Mr. F. Dickinson.* Ellesmere ; *Rev. A. Bloxam.* Canal near Shrewsbury.
Roots numerous, filiform, proceeding from the base of the frond. *Frond* half an inch long and nearly as broad, succulent, firm, upper surface green, minutely tuberculated, margin purple, under surface slightly convex, deep purple, proliferous from the base or part where the roots are fixed.
The fronds of this species appear under the microscope to be composed of large purple meshes, the interstices of which are filled with very minute reticulations.

CLASS III.

TRIANDRIA.

—

" Let the earth
Put forth the verdant grass, herb yielding seed,
And fruit-tree yielding fruit after her kind,
Whose seed is in herself upon the earth.
He scarce had said, when the bare earth, till then
Desert and bare, unsightly, unadorned,
Brought forth the tender grass, whose verdure clad
Her universal face with pleasant green ;
Then herbs of every leaf, that sudden flowered,
Opening their various colours, and made gay
Her bosom, smelling sweet."
                                        MILTON.

CLASS III.

TRIANDRIA. 3 *Stamens.*

ORDER I. MONOGYNIA. 1 *Style.*

* *Flowers superior.*

1. VALERIANA. *Calyx* a thickened margin to the top of the germen, at length unfolding into a feathery *pappus*. *Corolla* monopetalous, tubular, gibbous at the base, irregularly 5-lobed. *Fruit* (Achenium) 1-celled, crowned with the feathery *pappus.*— *Nat. Ord.* VALERIANEÆ, *DC.*—Name from *valeo, to be powerful ;* in allusion to the medicinal effects.

2. CENTRANTHUS. *Calyx* a thickened margin to the top of the germen, at length unfolding into a feathery *pappus*. *Corolla* monopetalous, tubular, spurred at the base, regularly 5-lobed. *Fruit* (Achenium) 1-celled, crowned with the feathery *pappus.*—*Nat. Ord.* VALERIANEÆ, *DC.*—Name from κεντρον, *a spur,* and ανθος, *a flower ;* in allusion to the basal spur of the corolla.

3. VALERIANELLA. *Calyx* small, erect, unequally toothed, crowning the fruit. *Corolla* monopetalous, funnel-shaped, regularly 5-lobed. *Fruit* (Achenium) 3-celled, 2 cells abortive, crowned with the teeth of the persistent calyx.—*Nat. Ord.* VALERIANEÆ, *DC.*— Name, a diminutive of *Valeriana.*

4. CROCUS. *Perianth* infundibuliform, coloured ; *tube* very long ; *limb* cut into six equal erect segments. *Stigma* 3-lobed, dilated, involute. *Capsule* ovate, triquetrous, 3-celled, 3-valved.—*Nat. Ord.* IRIDEÆ, *Juss.*—Name from κροκη, *a thread,* or *filament ;* in allusion to the appearance of the *Saffron* of commerce, which is the dried stigmas of *C. sativus.*

5. IRIS. *Perianth* single, petaloid, 6-cleft, *exterior segments* larger and reflexed, *interior* ones erect. *Stigmas* 3, petaloid, covering the *stamens*. *Capsule* oblong, angular, 3-celled, 3-valved.—*Nat. Ord.* IRIDEÆ, *Juss.*—Name in allusion to the beauty and variety of the colours of the flowers.

** *Flowers inferior. Seeds* 3.

6. MONTIA. *Calyx* of 2 leaves. *Corolla* of 5 irregular *petals,* united at the base into one. *Capsule* 3-valved, 3-seeded.—*Nat. Ord.* PORTULACEÆ, *Juss.*—Named in honour of *Joseph de Monti,* a Professor of Botany and Natural History at Bologna.

c

*** *Flowers inferior, glumaceous, dry and chaffy.*[1] *Seed one.*

A. *Glumes imbricated on all sides.*

† *Perigynium none.*

7. ISOLEPIS. *Stems* simple, leafy. *Spikes* fasciculate, sublateral, 1, 2, or more, bracteated. *Glumes* uniform, all fertile. *Styles* filiform, not thickened at the base. *Stigmas* 3. *Fruit* (Caryopsis) trigonous, mucronate.—*Nat. Ord.* CYPERACEÆ, *Juss.*—Name from ισος, *equal,* and λεπις, *a scale;* in allusion to the uniform size of the glumes.

8. ELEOGITON. *Stems* branched, leafy. *Spikes* terminal, solitary, naked. *Glumes* imbricated in 4 rows, uniform, all fertile. *Style* short, not thickened at the base. *Stigmas* 2. *Fruit* (Caryopsis) compressed, plano-convex, mucronate.—*Nat. Ord.* CYPERACEÆ, *Juss.*—Name from ελος, ελεος, *a marsh,* and γειτων, *a neighbour;* in allusion to the places of its growth.

†† *Perigynium of numerous setæ or bristles.*

9. ERIOPHORUM. *Stems* simple, angular or round, leafy. *Spikes* solitary and terminal, or fasciculate and bracteated. *Glumes* all nearly equal, lower ones sometimes sterile. *Perigynium* of numerous, very long, silky hairs, in fascicles at the base of the fruit. *Stigmas* 3. *Fruit* (Caryopsis) trigonous, mucronate.—*Nat. Ord.* CYPERACEÆ, *Juss.*—Name from εριον, *wool,* and φερω, *to bear;* in allusion to the nature of the perigynium.

10. SCIRPUS. *Stems* simple, leafy. *Spikes* numerous, fasciculate or subcymose, bracteated. *Glumes* nearly equal, all fertile. *Perigynium* of 6 persistent, scabrous bristles. *Stigmas* 2 or 3. *Fruit* (Caryopsis) plano-convex, or obtusely trigonous, mucronate.—*Nat. Ord.* CYPERACEÆ, *Juss.*—Name from *Cirs,* Celtic, *a cord;* in allusion to the use formerly made of the stems.

[1] This groupe, together with the genus *Carex* in the 21st Class, constitute the Natural Order CYPERACEÆ, in which the inflorescence is so peculiar and so dissimilar from that of other plants, that it seems advisable to add an explanation of the terms here used in describing the structure of the flowers. These are collected into small *spikes,* composed of numerous minute chaffy scales, here called *glumes,* which are in reality true bracteas, imbricated, or arranged one over the other, similar to the tiles on the roof of a house. Within each of these glumes will be found the flower, whose outer covering consists of *hairs* or *bristles,* partly arranged between the glumes and the stamens, and partly between the stamens and fruit. These bristles have been considered as the true perianth, and are here styled *perigynium.* In some few genera the perigynium is absent, and in the genus *Carex* it is formed of two bracteas, which uniting at their edges constitute a membranous urceolate body, having an aperture at the apex for the transmission of the pistil. Within the perigynium are situated the organs of reproduction.

---

11. ELEOCHARIS. *Stems* simple, with leafless sheaths at the base. *Spikes* terminal, solitary, naked. *Glumes* nearly equal, all fertile. *Perigynium* of 6 persistent, rough bristles. *Styles* bulbous at the base. *Stigmas* 2 or 3. *Fruit* (Caryopsis) compressed, obtusely triquetrous, crowned with the persistent, bulbous base of the style.—*Nat. Ord.* CYPERACEÆ, *Juss.*—Name from ελος, ελεος, *a marsh,* and χαιρω, *to delight in;* in allusion to its place of growth.

12. SCIRPIDIUM. *Stems* simple, with leafless sheaths at the base. *Spikes* terminal, solitary, naked. *Glumes* nearly equal, all fertile. *Perigynium* of 3 deciduous, scabrous, short bristles. *Style* bulbous at the base. *Stigmas* 3. *Fruit* (Caryopsis) compressed, crowned with the persistent, thickened base of the style.—*Nat. Ord.* CYPERACEÆ, *Juss.*—Name, a diminutive of *Scirpus.*

13. BÆOTHRYON. *Stems* simple, with leafless or subfoliaceous sheaths at the base. *Spikes* terminal, solitary, naked. *Glumes* unequal, outermost the largest, all fertile. *Perigynium* of 6 persistent, scabrous bristles. *Style* filiform, not bulbous at the base. *Stigmas* 3. *Fruit* (Caryopsis) trigonous, obtusely mucronate.—*Nat. Ord.* CYPERACEÆ, *Juss.*—Name from βαιος, *small,* and θρυον, *a rush;* in allusion to its comparatively small size.

B. *Glumes distichous.*

† *Perigynium none.*

14. CLADIUM. *Stems* leafy. *Spikes* 1-flowered, fasciculate, subcymose, lateral and terminal, bracteated. *Glumes* subdistichous, lower ones smaller sterile, upper ones fertile. *Perigynium* 0. *Stamens* 2—3. *Style* attenuated. *Stigmas* 2—3. *Fruit* (Drupa) a nut with a thick fleshy epicarp, ovato-acuminate, mucronate.—*Nat. Ord.* CYPERACEÆ, *Juss.*—Name from κλαδος, *a branch;* in allusion to the branchy appearance of the inflorescence.

†† *Perigynium of numerous setæ or bristles.*

15. RHYNCHOSPORA. *Stems* leafy. *Spikes* 2 or few-flowered, compressed, fasciculate, terminal or axillary, bracteated. *Glumes* distichous, lower ones smaller, sterile. *Perigynium* of 6 scabrous, persistent bristles. *Styles* dilated at the base. *Stigmas* 2. *Fruit* (Caryopsis) compressed, convex on both sides, crowned with the dilated base of the style.—*Nat. Ord.* CYPERACEÆ, *Juss.*—Name from ρυγκος, *a beak,* and σπορα, *a seed;* in allusion to the long style crowning the germen.

(See *Nardus* in the next Order.)

ORDER II. DIGYNIA. 2 *Styles.*

* *Flowers spiked.*[1] *Spikelets sessile, or nearly so, on a common, toothed, flexuose rachis.*

† *Inferior valve of the paleæ, awnless.*

16. LOLIUM. *Spike* distichous. *Spikelets* alternate, compressed, many-flowered. *Glumes* 2, one often deficient, opposite to the channelled rachis. *Paleæ* 2, *inferior valve* awnless, or with an *awn* below the extremity. *Scales* acuminate, with a lateral tooth. Name from λαιον, *corn,* and ολοος, *pernicious;* in allusion to the poisonous qualities attributed to *Lolium temulentum.*

†† *Inferior valve of the paleæ, awned.*

17. NARDUS. *Spike* simple, filiform. *Spikelets* unilateral, sessile, alternate, single-flowered. *Glumes* 0. *Paleæ* 2, *inferior valve* tapering into a long, rigid, setigerous *awn,* and enveloping the *superior valve.* *Style* 1. *Scales* 0. Name from ναρδος, an old word for a kind of *withy* or *osier;* in allusion to the willow-like appearance of its unilateral spikes; or perhaps from ναρδος, *frankincense;* in allusion to the odoriferous oil which may probably reside in the glands on the paleæ.

18. AGROPYRUM. *Spike* distichous. *Spikelets* compressed, sessile, transverse, their sides directed to the rachis, many-flowered,

[1] The whole of the remaining plants in this Class constitute the true GRASSES or GRAMINEÆ, *Juss.* Their flowers, like those of the CYPERACEÆ, being dissimilar from those of other flowering plants, have had peculiar names applied to their different parts. The inflorescence consists of *spikelets,* either sessile on a common rachis or stalk, or borne at the extremity of a branched peduncle. At the base of each spikelet are two bracteas, here called *glumes,* one attached to the rachis a little above the base of the other; the lower one being termed the *inferior glume valve,* and the upper, the *superior.* Above the glumes appear 1, 2, 3, or more flowers, sitting in notches of the rachis, each consisting of two bracteas, here called *paleæ.* Of these one is usually larger, and is furnished with a mibrib, which quitting it a little below the apex, is elongated into a rough bristle or *awn;* this is termed the *inferior palea,* or the *inferior valve of the paleæ.* The other bractea or *superior palea,* as it is called, faces the first, and has its back towards the rachis, is bifid at the apex, and has the edges membranous, inflexed and ribbed. Within the paleæ are the stamens, and again between them and the ovary bearing the pistils, are two minute colourless fleshy scales, sometimes united or connate, called *squamulæ* or *scales.* The parts called glumes are termed by Smith and Hooker, the *calyx;* the paleæ, the *corolla,* and its pieces *glumes* or *valves;* the *scales* are the *nectary* of Linnæus and Smith. The stem is generally hollow and jointed, and called a *culm.* From each joint proceeds a *leaf,* which clasping the culm with its dilated petiole for some distance, receives the name of *sheath,* and is more or less split up on one side. At the top of the sheath, just where it divaricates or expands into the blade, is a small projecting membrane called a *ligule;* the *stipula* of Smith.

---

fertile. *Glumes* 2, nearly equal, lanceolate, acuminate, (not ventricose.) *Paleæ* 2, valves lanceolate, *inferior* acuminate, or awned at the extremity, *superior* retuse. *Scales* 2, acuminate, laterally toothed, ciliated at the apex. (*Root perennial.*)—Name from αγρος, *a field,* and πυρος, *wheat;* from its similarity to the true wheat.

19. BRACHYPODIUM. *Spikes* subdistichous. *Spikelets* cylindrical-compressed, shortly pedicellate, transverse, their sides directed to the rachis, many-flowered, uppermost sterile. *Glumes* 2, lanceolate, unequal. *Paleæ* 2, *inferior valve* awned at the extremity, *superior* retuse. *Scales* 2, ovate, obliquely acuminate, entire, ciliated, tumid at the base.—Name from βραχυς, *short,* and πους, *a foot;* in allusion to the short footstalks of the spikelets.

20. HORDEUM. *Spikelets* single-flowered, ternate, central one perfect, lateral ones mostly imperfect, with a bristle or abortive floret at the back of the superior valve. *Glumes* 2, lateral, subulato-aristate. *Paleæ* 2, *inferior valve* with a terminal *awn.* *Scales* 2, entire, hairy. *Fruit* invested with the glumes.—Name of dubious origin.

[Lolium temulentum.]

** *Flowers panicled.* *Panicle compact, so as to appear spiked.* (*Spicæform.*)

† *Inferior valve of the paleæ, awnless.*

21. PHALARIS. *Panicle* compact. *Spikelets* 3-flowered, two florets sterile, reduced to glabrous scales, one only fertile. *Glumes* 2, nearly equal, navicular, with a broad, winged, membranous keel, obtuse. *Paleæ* 2, navicular, unequal, obtuse, *inferior valve* larger, and enveloping the *superior* one. *Scales* 2, entire, minute, glabrous. *Fruit* laterally compressed, glabrous, invested with the hardened paleæ.—Name from φαλαρις, an old name for the plant derived from φαλος, *shining;* in allusion to the shining coat of the seed.

22. PHLEUM. *Panicle* compact. *Spikelets* 1-flowered. *Glumes* 2, equal, compressed, keeled, mucronato-aristate. *Paleæ* 2, unequal, awnless, *inferior valve* enveloping the *superior* one. *Scales* 2, 2-lobed, denticulate. *Seed* free.—Name from φλεος, or φλεως, formerly applied to the *Reed-mace,* (*Typha,*) to which this grass bears a distant resemblance.

†† *Inferior valve of the paleæ, awned.*

a. *awn from the apex.*

23. CYNOSURUS. *Panicle* compact, spicæform. *Spikelets* 2—5-flowered, distichous, with pectinated bracteas or abortive spikelets.

Glumes 2, nearly equal, awned. *Paleæ* 2, linear-lanceolate, *inferior valve* awned, or mucronate at the apex. *Scales* 2, acuminate, laterally toothed.—Name from κυων, *a dog,* and ουρα, *a tail;* from the shape of the inflorescence.

24. SESLERIA. *Panicle* simple, spicæform. *Spikelets* 2—6-flowered, distichous. *Glumes* 2, nearly equal, mucronate. *Paleæ* 2, keeled, *inferior valve* denticulate, and awned at the apex, *superior* emarginate. *Scales* 2, long, dentate.—Named from *Leonard Sesler,* an Italian Physician and Botanist.

b. *awn from near to the base.*

25. ANTHOXANTHUM. *Panicle* compact, spicæform. *Spikelets* 3-flowered, 2 outer or lateral florets sterile, univalved, awned from near to the base, central or superior floret fertile, awnless. *Glumes* 2, unequal. *Paleæ* 2, membranaceous, obtuse. *Scales* 0.—Name from ανθος, *a flower,* and ξανθος, *yellow;* from the yellow colour of the inflorescence, especially when old.

26. ALOPECURUS. *Panicle* spicæform, cylindrical. *Spikelets* 1-flowered. *Glumes* nearly equal, compressed, keeled, connate at the base. *Paleæ* 1, awned from near to the base. *Scales* 0.—Name from αλωπηξ, *a fox,* and ουρα, *a tail;* in allusion to the form of the inflorescence.

*** *Flowers panicled. Panicle branched, diffuse.*

† *Inferior valve of the paleæ, awnless.*

27. BRIZA. *Panicle* lax. *Spikelets* many-flowered, compressed, bifariously imbricated. *Glumes* 2, nearly equal, ventricose. *Paleæ* 2, awnless, *inferior valve* ventricose, *superior* very small and flat. *Scales* 2, glabrous, entire or slightly lobed.—Name from βριθω, *to balance;* in allusion to the delicate manner in which the spikelets are suspended.

28. MELICA. *Panicle* subracemose. *Spikelets* 3—5-flowered, upper florets sterile, imperfect and involute. *Glumes* 2, nearly equal, about as long as the florets. *Paleæ* 2, unequal, awnless. *Scales* 2, gibbous, cuspidate. *Fruit* free, covered by the cartilaginous paleæ.—Name from *mel, honey;* on account of the sweet flavour of the stem of a nearly allied plant, *Sorghum vulgare,* to which the name of *Melica* or *Melliga* is given in Italy, and applied by Linnæus to this genus.

29. GLYCERIA. *Panicle* loose. *Spikelets* linear, compressed, 2 or many-flowered. *Glumes* 2, unequal, obtuse, shorter than the

florets. *Paleæ* 2, nearly equal, obtuse, awnless. *Scales* 2, truncate, sometimes connate.—Name from γλυκερος, *sweet;* in allusion to the sweetness of the seeds of *Glyceria fluitans,* which constitute the *Manna-seeds* of commerce.

30. CATABROSA. *Panicle* loose, spreading. *Spikelets* 2—3-flowered, often with a 4th imperfect floret. *Glumes* 2, unequal, membranous, truncate, much shorter than the florets. *Paleæ* 2, nearly equal, coriaceous, membranous only at the extremity, ribbed, truncate, erose, awnless.—Name from καταβρωσις, *a gnawing;* from the erose extremity of the valves.

31. POA. *Panicle* loose. *Spikelets* 3 or many-flowered, florets articulated with their rachis. *Glumes* 2, nearly equal, a little shorter than the lower florets. *Paleæ* 2, nearly equal, awnless, *inferior valve* 5-ribbed, hairy or webbed at the base. *Scales* 2, oval, acute, gibbous at the base.—Name from ποα, *grass* or *pasturage,* from παω, *to feed;* in allusion to the excellent pasturage for cattle afforded by this genus.

32. SCLEROCHLOA. *Panicle* contracted, rigid, pedicels articulated with the spikelets. *Spikelets* many-flowered, cylindrical, compressed. *Glumes* 2, shorter than the lower florets, obtuse. *Paleæ* 2, awnless or mucronate. *Scales* 2, emarginate.—Name from σκληρος, *hard* or *rigid,* and χλοα, *grass;* in allusion to the rigidity of its culms.

33. MILIUM. *Panicle* spreading. *Spikelets* 1-flowered. *Glumes* 2, equal, flattish, rather acute, awnless, longer than the paleæ. *Paleæ* 2, nearly equal, awnless. *Fruit* invested with the permanent hardened paleæ. *Scales* 2, acute, with a marginal tooth.—Name from *mille, a thousand,* on account of its fertility, or from *mil,* Celtic, *a stone;* in allusion to the hardness of its fruit.

34. MOLINIA. *Panicle* narrow. *Spikelets* conical, 2—5-flowered, upper floret abortive. *Glumes* 2, membranaceous, unequal, much shorter than the florets. *Paleæ* 2, nearly equal, awnless. *Scales* 2, short, truncate. *Fruit* free, covered by the cartilaginous paleæ.—Name in honour of *Giovanni Ignatio Molinia,* author of an account of the plants of Chili in 1782.

35. BALDINGERA. *Panicle* branched, diffuse. *Spikelets* 3-flowered, two florets sterile reduced to hairy scales, one only fertile. *Glumes* 2, nearly equal, keeled, but not winged at the back, acute. *Paleæ* 2, keeled, unequal, obtuse, *inferior valve* larger, and enveloping the *superior* one. *Scales* 2, entire, minute, glabrous. *Fruit* laterally compressed, glabrous, invested with the hardened paleæ.

36. PHRAGMITES. *Panicle* very much branched, diffuse. *Spikelets* 3—6-flowered, florets distichous, beardless, lower floret with stamens only, rachis with long silky hairs. *Glumes* 2, unequal, shorter than the florets. *Paleæ* 2, *inferior valve* very long and acuminate. *Scales* 2, very large, obtuse.—Name from φρασσω, to *hedge* or *fence;* in allusion to the tall, erect, and rigid appearance of groups of the plant, or perhaps to the uses to which the culms are appropriated.

[Agrostis vulgaris and alba.]

†† *Inferior valve of the paleæ, awned.*

a. *awn at the apex.*

37. DACTYLIS. *Panicle* lax, secondary branches contracted, short, and very dense, subsecund. *Spikelets* 2—7-flowered, compressed. *Glumes* 2, unequal, keeled. *Paleæ* 2, unequal, *inferior valve* lanceolate, keeled, more or less awned. *Fruit* enclosed in the paleæ. *Scales* 2, bifid, glabrous.—Name from δακτυλος, *a finger.*

38. FESTUCA. *Panicle* branched, loose, more or less spreading, filiform. *Spikelets* many-flowered, florets deciduous. *Glumes* 2, unequal, or nearly equal, acute. *Paleæ* 2, *inferior valve* mucronate or awned at the apex, awn shorter than the paleæ. *Scales* 2, tumid at the base, apex bidentate. *Styles* terminal. *Sheaths* entirely divided.—Name from *fest,* Celtic, *food* or *pasturage.*

39. VULPIA. *Panicle* racemose or contracted, subsecund, with very thick pedicels. *Spikelets* many-flowered, florets deciduous. *Glumes* 2, very unequal, *inferior valve* much smaller, or wholly deficient, very acuminate. *Paleæ* 2, *inferior valve* awned at the apex. *Awn* as long as, or longer than the paleæ. *Scales* 2, tumid at the base, ovato-acuminate, laterally toothed.

40. DANTHONIA. *Panicle* racemose. *Spikelets* many-flowered. *Glumes* 2, nearly equal, awnless, often equalling, or exceeding in length the florets. *Paleæ* 2, *inferior valve* bifid, with a short straight *awn* proceeding from between the segments. *Scales* 2, entire.

b. *awn from below the apex.*

41. BROMUS. *Panicle* more or less branched and spreading. *Spikelets* many-flowered, florets distichous. *Glumes* 2, unequal, shorter than the florets. *Paleæ* 2, very unequal, *inferior valve* awned below the bifid extremity, *superior* ciliated on the marginal ribs with remote bristles. *Scales* 2, entire. *Styles* lateral. *Sheaths* slightly divided at the summit only.—Name from βρωμος, given by the Greeks to a kind of *oat,* and that again from βρωμα, *food.*

42. HOLCUS. *Panicle* lax. *Spikelets* 2-flowered, florets remote, inferior one perfect and awnless, upper one with stamens only, and awned from below the apex. *Glumes* 2, nearly equal, awnless, longer than the florets. *Paleæ* 2, nearly equal. *Scales* 2, obliquely acuminate.—Name ολκος, from ελκω, *to extract;* in allusion to its imaginary property of extracting thorns from the flesh.

43. AGROSTIS. *Panicle* loose. *Spikelets* 1-flowered. *Glumes* 2, unequal, longer than the paleæ, awnless. *Paleæ* 2, unequal, *superior valve* sometimes wanting, *inferior* with or without an *awn* from below the apex. *Scales* 2, nearly entire, glabrous. *Seed* free.—Name from αγρος, *a field;* given by the Greeks to grasses generally, from their abundance in such situations.

c. *awn from about the middle.*

44. CALAMAGROSTIS. *Panicle* loose. *Spikelets* 1-flowered. *Glumes* 2, much longer than the floret. *Paleæ* 2, surrounded at the base by long hairs, *inferior valve* awned. *Scales* 2, entire, glabrous. —Name compounded of καλαμος, *a reed,* and αγροστις, a genus of grasses.

45. AVENA. *Panicle* lax. *Spikelets* 2 or more flowered, florets remote, upper one imperfect. *Glumes* 2, nearly equal. *Paleæ* 2, lanceolate, nearly equal, firmly enclosing the seed, *inferior valve* bicuspidate, bearing a twisted dorsal *awn.*—Name of doubtful origin.

d. *awn from near to the base.*

46. AIRA. *Panicle* loose. *Spikelets* 2—3 flowered, two florets only fertile. *Glumes* 2, nearly equal. *Paleæ* 2, unequal, membranaceous and thin, *inferior valve* with a straight or twisted *awn* proceeding from above the base. *Scales* 2, entire, glabrous. *Fruit* free.—Name from αιρω, to *destroy;* in allusion to the injurious effects of *Lolium temulentum,* to which the name was anciently applied.

47. ARRHENATHERUM. *Panicle* lax. *Spikelets* 2-flowered, with the rudiments of a third floret; lowermost floret with stamens only, and a long twisted *awn* from above the base, upper one fertile, with a short straight bristle below the point. *Glumes* 2, unequal. *Paleæ* 2. *Scales* 2, very long, lanceolato-linear, entire, glabrous.—Name from αρρην, *male,* and αθηρ, *an awn;* in allusion to the long *awn* of the imperfect floret.—This genus has the habit of *Avena,* but differs in the number and structure of its flowers.

TRIANDRIA—MONOGYNIA.

1. VALERIANA. *Linn.* Valerian.

1. V. *dioica*, Linn. *Small Marsh Valerian.* Flowers diœcious; radical leaves ovate, petiolate, cauline ones lyrato-pinnatifid. *E. Bot. t.* 628. *E. Fl. v. i. p.* 43. *Hook. Br. Fl. p.* 23.

Marshy meadows; frequent. *Fl.* May, June. ♃.

Boggy meadow, near Titterstone Clee Hills; *Mr. J. S. Baly!* Eyton and Walford; *T. C. Eyton, Esq.!* Oakley Park; *Mr. H. Spare.* Near Park Brook, Wyre Forest; *Mr. W. G. Perry.* Wyre Forest; *Mr. G. Jorden.* Rowley, near Harley; *W. P. Brookes, Esq.*

Moist places about Shrewsbury, generally.

Root creeping. *Stem* erect, 6—12 inches high, fertile plant most robust. *Leaves* in pairs, opposite, smooth, margins bristly or ciliated, radical ones entire on long dilated petioles, those of the stem sessile, with 2—6 pairs of acute leaflets, and a larger terminal 3-cleft one. *Flowers* flesh-coloured, in dense bracteated terminal corymbs, those of the fertile plant denser and smaller. *Stamens* and *pistils* occasionally present, but rarely perfect on the same plant. *Stigma* 2—3 fid.

2. V. *officinalis*, Linn. *Great Wild Valerian.* Flowers hermaphrodite; leaves all decursively pinnated, leaflets lanceolate, nearly uniform, dentato-serrate. *E. Bot. t.* 698. *E. Fl. v. i. p.* 43. *Hook. Br. Fl. p.* 23.

Sides of rivers and water courses; frequent. *Fl.* June, July. ♃.

Near Dowles Brook, Wyre Forest; *Mr. W. G. Perry.* About Tong and Albrighton; *Dr. G. Lloyd.* Astley, near Shrewsbury; *Mr. E. Elsmere, junr.!* Eyton and Walford; *T. C. Eyton, Esq.!* Oakley Park, near Ludlow; *Mr. H. Spare.* Near Ludlow; *Miss Mc Ghie.* Banks of the Severn, near Bewdley; *Mr. G. Jorden.* Cox Wood, Coalbrookdale; *Mr. F. Dickinson.*

Near Shrewsbury, generally.

Root tuberous, somewhat creeping, warm, aromatic, and employed in medicine. *Stem* 2—4 feet high, furrowed, slightly hairy. *Leaves* in pairs, opposite, decursively pinnate, with a terminal leaflet, petiolate, connate at the base, *petioles* covered with deflexed hairs, *leaflets* lanceolate, under surface hairy on the veins, upper surface with a few scattered hairs, and a row of short dense bristles on the margins. *Panicle* corymbose, trichotomously branched, with a pair of lanceolate, acuminate *bracteas* at the base of each branch. *Flowers* pale flesh-coloured. *Seeds* ovato-oblong, compressed, smooth, margins strongly nerved, with 3 prominent ribs on the one side and one on the other, somewhat gibbous at the apex on the 3-ribbed side. *Pappus* of 11 feathered rays, connected at the base by an irregular membrane.

2. CENTRANTHUS. *DC.* Spur-flower.

1. C. *ruber*, DC. *Red Spur-flower.* Corolla tubular, spur longer than the germen; leaves ovato-lanceolate. *Valeriana rubra,* Linn. *E. Bot. t.* 1531. *E. Fl. v. i. p.* 42. *Hook. Br. Fl. p.* 22.

Old walls; rare. *Fl.* June—September. ♃.

Walls of Ludlow Castle; *Mr. J. S. Baly!*

Wenlock Abbey.

Root somewhat woody, spreading. *Stem* erect, leafy, smooth, somewhat glaucous. *Leaves* in pairs, opposite, ovato-lanceolate, obtuse, sinuato-dentate,

Pl. 1. to Face P. 27.

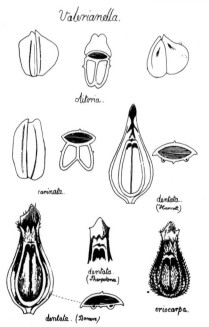

Valerianella.

olitoria.

carinata.

dentata.
(*Hancroft*)

dentata.
(*Sharpstone*)

dentata. (*Bomere*)

eriocarpa.

petiolate, upper ones sessile, and more acuminated. *Flowers* deep rose-colour, erect, arranged in numerous unilateral, cymose spikes. *Stamen* single. *Seed* ovate, narrowed upwards, slightly scabrous, margins strongly nerved, 3-ribbed and gibbous at the apex on one side, 1-ribbed on the other.

3. VALERIANELLA. *Moënch.* Corn-Sallad.

1. V. *olitoria*, Moënch. *Common Corn-Sallad or Lamb's Lettuce.* Capsule subglobose, laterally compressed, inflated, slightly pubescent, furrowed in front and on the back, the two sterile cells larger than the fertile one which is backed by a mass of cellular matter, crowned with the three obscure teeth of the calyx; flowers capitate. *Valeriana Locusta,* Linn. *E. Bot. t.* 811. *Fedia olitoria,* Vahl. *E. Fl. v. i. p.* 45. *Hook·Br. Fl. p.* 23.

Hedge-banks and corn-fields; abundant. *Fl.* April—June. ☉.

Root fibrous. *Stem* dichotomous, more or less rough with a rib armed with deflexed bristles continued down the stem from the edges and keels of the leaves. Radical *leaves* spathulate, those of the stem oblong, obtuse, entire or a little toothed, semi-amplexicaul, keels and the margins near the bases ciliated with short bristles continued down the ribs of the stem. *Flowers* pale blue, in terminal compact heads, with a kind of *involucre* at their base, formed of numerous, crowded, linear-oblong, often divided, ciliated *bracteas.* *Capsule* 3-lobed, obscurely furrowed both on the back and in front, somewhat wrinkled or pitted especially the larger lobe, pubescent, 3-celled, 2 of the cells parallel, oblong, single-ribbed on each side and abortive, the fertile one elliptical, transverse, backed by a mass of cellular matter, forming the larger lobe of the capsule. *Seed* solitary in the fertile cell, oval, smooth.

2. V. *carinata*, Lois. *Carinated Corn-Sallad.* Capsule oblong, subtetragonous, punctato-striate, slightly pubescent, deeply furrowed in front, the two sterile cells nearly equal to the fertile one, crowned with the straight single tooth of the calyx; flowers capitate. *Fedia carinata,* Stev. *Hook. Br. Fl.* addenda to 3rd edition. *E. Bot. Suppl. t.* 2810.

Hedge-bank close to the Long Lane Quarry, near Cheney Longville. *Fl.* June, July. ☉.

Root small. *Stem* 3—12 inches high, dichotomously branched from the base, rough with ribs armed with rigid deflexed bristles, proceeding from the edges and keels of the leaves. Radical *leaves* spathulate, those of the stem oblong, obtuse, entire or slightly toothed, semi-amplexicaul, margins ·ciliated with short bristles. *Flowers* pale blue, in terminal compact heads, with a kind of *involucre* at their base, formed of numerous, crowded, oblong *bracteas,* ciliated and membranaceous at their margins. *Capsule* oblong, 3-lobed, deeply furrowed in front, punctato-striate, pubescent, 3-celled, equal, two of the cells somewhat divergent, single-ribbed on each side and abortive, the fertile one transverse, with a longitudinal rib at the back, and terminated in a short obtuse tooth. *Seed* solitary in the fertile cell, oval, smooth.

Closely allied to *V. olitoria,* with which it agrees in its habit and general appearance, but essentially distinguished by the different form of its capsule.

3. V. *dentata*, DC. *Smooth narrow-fruited Corn-Sallad.* Capsule obpyriform, dorsally compressed, punctato-striate, acuminate, the two sterile cells reduced to hollow ribs, crowned with the prominent, cup-shaped, oblique, unequally 4-toothed calyx; flowers corymbose,

a sessile flower in the forks. *Valeriana dentata, Willd.* E. Bot. t. 1370. *Fedia dentata, Vahl.* E. Fl. v. i. p. 45. Hook. Br. Fl. p. 23.

Hedge-banks and corn-fields; not common. *Fl.* June, July. ☉.

Corn-fields near Little Wenlock; *Mr. F. Dickinson.* Astley; *Mr. E. Elsmere, junr.!*

Fields near Bomere Pool. Hedge-banks at Cross Hill, near Shrewsbury. Fields, Hill Top, Wenlock Edge. Corn-fields near Sharpstones Hill.

Root small. *Stem* rough as in the preceding, but much more dichotomously branched. *Leaves* narrower, linear tongue-shaped, upper ones with about 2 or 3 distinct teeth at the base on each side. *Flowers* flesh-coloured, in small terminal corymbs. *Bracteas* two at the base of each ramification of the inflorescence, linear, entire or with a membranaceous, subserrated, or ciliated margin. *Fruit* obpyriform, distinctly margined, convex and single-ribbed on the back, where the fertile cell is situated, nearly plane in front, where are the two abortive cells reduced to two projecting hollow ribs or lines curving below and uniting above the base, and terminated upward in two small subulate teeth; between them is a single rib surmounted by a similar tooth. The perfect cell is lengthened into a large broad and sharp tooth, with two smaller teeth at the base, one on each side. The surface of the fruit is minutely punctata-striate, the margins of the calyx-teeth ciliated, with a few scattered hairs on the back of the larger tooth. The mouth of the calyx is in some specimens open and the calyx-teeth divergent, whilst in others it is more closed by the inflexed convergence of the calyx-teeth.

Var. β. eriocarpa. Capsule clothed with spreading incurved rigid hairs, cup of the calyx large and open.

Three or four specimens of this plant occurred in a corn-field near the Sharpstones Hill, in August 1837, growing intermixed with the smooth-fruited state of *V. dentata.*

4. CROCUS. *Linn.* Crocus.

1. C. *vernus,* Willd. *Purple Spring Crocus.* Stigma within the flower, erect, cut into three jagged, wedge-shaped lobes; tube of the corolla hairy at the mouth. E. Bot. t. 344. E. Fl. v. i. p. 46. Hook. Br. Fl. p. 24.

Meadows and fields; naturalized. *Fl.* March. ♃.

" Common in meadows near Ludlow, supposed to be the sites of old cottages;" *Miss Mc Ghie.*

Three specimens, collected (1835) in Dorset's Barn Fields, Shrewsbury.

Bulb solid, external coat of numerous parallel fibres. *Leaves* all radical, enclosed in several imbricated membranous sheaths, linear, obtuse, broadly keeled, margins revolute, smooth, with a white central stripe on the upper surface. *Flowers* purple, from a loose, clasping, acuminated *spatha*, *petals* elliptical oblong, twice as long as the *pistils* and *stamens*. Mouth of the *tube* with numerous pellucid glandular hairs.

The bulb of the Crocus appears to consist, in reality, of the base of the stem much swollen, enveloped by several swollen and modified leaves closely agglutinated, and concentrically overlapping each other, and supporting in their axils, a series of embryonic bulbs or buds spirally arranged. On tracing these concentric leaves throughout the bulb to its summit, it will be found that the shoot or shoots destined to produce flowers, &c. in the present year, are one or more of these embryonic bulbs more highly developed than the rest. In these shoots also the same concentric arrangement of the leaves exists.

5. IRIS. *Linn.* Iris or Fleur de Luce.

1. I. *Pseud-acorus,* Linn. *Yellow Water-Iris or Corn-flag.*

Leaves sword-shaped; perianth beardless, inner segments smaller than the stigma; stem round; seeds angular. E. Bot. t. 578. E. Fl. v. i. p. 48. Hook. Br. Fl. v. i. p. 26.

Watery places, wet meadows, margins of pools and rivulets; frequent. *Fl.* May—July. ♃.

Astley; *Mr. E. Elsmere, junr.!* Near Oswestry; *Rev. T. Salwey.* Eyton and Walford; *T. C. Eyton, Esq.* Oakley Park, near Ludlow; *Mr. H. Spare.* Near Ludlow; *Miss Mc Ghie.* Buildwas; *Mr. F. Dickinson.*

Near Shrewsbury, and other parts of the county, generally.

Root large, horizontal, depressed, sending long fibres from its lower part. *Stem* erect, somewhat zig-zag, round, smooth. *Leaves* erect, in 2 opposite rows, equitant, ribbed. *Bracteas* thin and membranous at the edges. *Flowers* erect, large, handsome, bright yellow, 3 large segments placed opposite to and underneath the stigmas, smaller ones erect, narrow, all united by a firm thick base. *Stigmas* 2-lipped, upper lip cloven, erect, lower minute, with a cleft between them to receive the pollen. *Capsule* angular, 3-celled, 3-valved. *Seeds* numerous, 2-ranked, globular or angular from pressure.

The dried root is used in the preparation of ink, black dyes, &c.

2. I. *fœtidissima,* Linn. *Stinking Iris.* Leaves sword-shaped; perianth beardless, inner segments spreading about as large as the stigma; stem one-angled; seeds globose. E. Bot. t. 596. E. Fl. v. i. p. 50. Hook. Br. Fl. p. 26.

Woods, thickets, and pastures; very rare. *Fl.* May. ♃.

"On a warm bank near Shortwood, in the neighbourhood of Ludlow;" *Miss Mc Ghie.* "One large root near a small plantation of fir-trees, between Sharpstones Hill and Sutton;" *Mr. F. Dickinson.*

Stem 1—2 feet high. *Leaves* emitting, when bruised, a peculiar disagreeable smell. *Flowers* smaller than the last, dull livid purple, streaked with darker veins. *Capsule* 3-celled. *Seeds* numerous, smooth, orange.

6. MONTIA. *Linn.* Blinks.

1. M. *fontana,* Linn. *Water Blinks or Water Chickweed.* Leaves opposite, spathulate, entire; flowers axillary or terminal; capsule roundish, 1-celled, seeds 3, subreniform, dotted. E. Bot. t. 1206. E. Fl. v. i. p. 187. Hook. Br. Fl. p. 59.

Springy and wet places, and on the margins of pools; not very common. *Fl.* March—June. ☉.

Ellesmere; *Rev. A. Bloxam.* Hawkestone Heath; *Mr. F. Dickinson.* Pulley Common, Bomere Pool (east side), Sharpstones Hill, Haughmond Hill, Lyth Hill, all near Shrewsbury.

Whole plant much branched, smooth and succulent, varying considerably in size. *Stem* prostrate and rooting. *Peduncles* nearly terminal, often forked from the axils of the upper leaves. *Flowers* white, at first drooping. *Stamens* short, inserted on the corolla. *Capsule* surrounded with the persistent calyx. *Seeds* black, concentrically dotted.

7. ISOLEPIS. *Br.* Moor-Rush.

1. I. *setacea,* Br. *Bristle-stalked Moor-rush.* Stems setaceous, compressed, with 1 or 2 leaves at the base; spikelets 1 or 2, sessile; bractea erect, leafy, much shorter than the stem; fruit rotundo-obovate, trigonous, longitudinally ribbed and marked transversely

---

with minute parallel striæ. *Scirpus setaceus, Linn.* E. Bot. t. 1693. E. Fl. v. i. p. 59. Hook. Br. Fl. p. 28.

Moist gravelly places; rare. *Fl.* July, August. ♃.

Bog below the south-west side of the Wrekin; *Mr. F. Dickinson.* Near Park Brook, Wyre Forest; *Mr. W. G. Perry.* Shawbury Heath.

Root fibrous, with numerous creeping scyons. *Stems* tufted, 2—6 inches high, very slender. *Leaves* setaceous, chiefly at the base. *Spikelets* ovate. *Glumes* obtuse, pinkish brown, margins and keels green. *Stamens* 2. *Stigmas* 3. *Bractea* leafy, dilated at the base, with membranous edges, obtuse, with a smaller deciduous one opposite, similar in colour and texture to the glumes, but larger. *Fruit* brown, slightly pointed, with prominent longitudinal ribs, and marked transversely with minute parallel striæ.

8. ELEOGITON. *Link.* Water-Rush.

1. E. *fluitans,* Link. *Floating Water-Rush.* Stem, or rather rhizoma, floating, compressed, branched, leafy, pliant; leaves fasciculate; flower-stalks alternate, with a sheathing leaf at the base; spikes terminal, ovate, few-flowered; fruit obovate, plano-convex, minutely pitted. *Scirpus fluitans, Linn.* E. Bot. t. 216. E. Fl. v. i. p. 57. *Eleocharis fluitans, Hook.* Hook. Br. Fl. p. 31.

Sides of pools which are sometimes dried up; not uncommon. *Fl.* June, July. ♃.

Bomere Pool, near Shrewsbury; *Rev. Edw. Williams.* Side of pit near Astley; *Dr. T. W. Wilson, and Mr. Edw. Elsmere, junr.!*

Hancott Pool, near Shrewsbury.

Root fibrous. *Stem* slender, zig-zag, trailing and throwing out fibres from the joints, branched alternately. *Leaves* awl-shaped, 3-nerved, nerves parallel, connected at intervals by transverse ones, keeled, spreading at nearly right angles with their membranous, 3-nerved, sheaths. *Flower-stalks* 2 or 3 inches long, compressed, contracted at the top. *Spikes* small. *Glumes* obtuse, pale-green. *Stigmas* 2, long and feathery. *Style* short. *Fruit* obovate, plano-convex, very minutely pitted, white, crowned with the base of the style.

9. ERIOPHORUM. *Linn.* Cotton-grass.

* Spike solitary, terminal.

1. E. *vaginatum,* Linn. *Hare's-tail Cotton-grass.* Stem triquetrous above, round below, invested with numerous reticulated sheaths, the lower ones elongated into long setaceous leaves, upper ones leafless, obtuse, inflated; spike ovate; fruit obovate, triquetrous, covered with minutely elevated points. E. Bot. t. 873. E. Fl. v. i. p. 66. Hook. Br. Fl. v. i. p. 31.

Boggy ground; not unfrequent. *Fl.* March—May. ♃.

Clee Hills; *Mr. G. Jorden.* Ellardine Moss; *Mr. F. Dickinson.* Near Ellesmere, abundantly; *Dr. Evans in Bot. Guide.* Hancott Bog; *Mr. A. Aikin in Bot. Guide.* Bog near Ellesmere; *Rev. A. Bloxam.* Felton Farm, near Ludlow; *Mr. Spare.*

Bomere Pool, near Shrewsbury. Knockin Heath. Vownog Bog, near Westfelton.

Root tufted, creeping. *Stem* jointed, smooth, erect, with several inflated, strongly reticulated sheaths in the lower part, 1 or 2 of them elongated

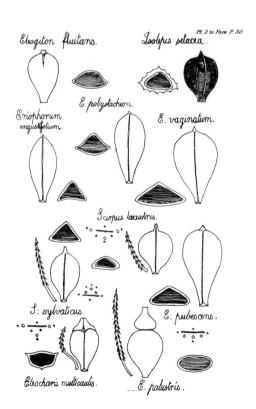

Pl. 2 to Face P. 30.

Elegiton fluitans.   Isolepis setacea.

Eriophorum angustifolium.   E. polystachion.   E. vaginatum.

Scirpus lacustris.

S. sylvaticus.   E. pubescens.

Eleocharis multicaulis.   E. palustris.

into leaves. *Leaves* proceeding from the lower portion of the stem, numerous, erect, slender, triangular, striated, sharp-pointed, nearly as tall as the stem. *Spike* silvery grey in flower. *Glumes* with long points, thin, membranous, shining, single-ribbed. *Fruit* brown, flatter on one side, rough with longitudinal rows of minutely elevated points, and crowned with the short base of the style.

** *Spikes many, fasciculate.*

2. E. *polystachion,* Linn. *Broad-leaved Cotton-grass.* Stem obscurely angular; leaves lanceolate, flat, contracted above the middle into a triangular scabrous point; stalks of the spikes smooth; fruit linear-obovate, triquetrous, hairs thrice the length of the spikes. *E. Bot. t.* 563. *E. Fl. v. i, p.* 64. *Hook. Br. Fl. v. i. p.* 32.

Bogs; rare. *Fl.* April—June. ♃.
In Shropshire; *Rev. Edward Williams in E. Fl. l. c.* Bog below the Wrekin, on the south side; *Mr. F. Dickinson.* Coreley; *Rev. Waties Corbet.* Astley; *Mr. E. Elsmere, junr.*
Root tufted, creeping. *Stem* 1—2 feet high, jointed, striated, smooth, leafy. *Leaves* numerous, smooth, with long striated slightly inflated sheaths, broad, flat, striated with a narrow acute keel, suddenly contracted above the middle into a long triangular obtuse point, edges and keels scabrous. *Spikes* several, sessile and stalked, ovate, grey, stalks striated, smooth, pendulous after flowering. *Bracteas* sheathing, leafy, triangular and obtuse at the apex, edges scabrous. *Glumes* filmy, blunt, slightly keeled, single-ribbed. *Stigmas* 3, downy.

3. E. *angustifolium,* Roth. *Common Cotton-grass.* Stem nearly round; leaves linear-lanceolate, complicate at the base, flat in the centre, and suddenly contracted above the middle into a triangular scabrous point; stalks of the spikes smooth; fruit elliptical acuminate, triquetrous; hairs four times the length of the spikes. *E. Bot. t.* 564. *E. Fl. v. i. p.* 69. *Hook. Br. Fl. v. i. p.* 32.

Bogs; not uncommon. *Fl.* April. ♃.
Aqualate Meer, near Newport; *Withering.* Near Ludlow; *Miss Mc Ghie.* Bog near Ellesmere; *Rev. A. Bloxam.* Astley, near Shrewsbury; *Mr. E. Elsmere.* Moss near Walford; *T. C. Eyton, Esq.* Titterstone Clee Hills; *Mr. J. S. Baly.* Marshy ground near Madeley, and at Rowley, near Harley; *W. P. Brookes, Esq.* Ellardine Moss, near Hawkstone; *Mr. F. Dickinson.* Wyre Forest; *Mr. G. Jorden.* Berrington and Bomere Pools, near Shrewsbury. Stiperstones Hill. Moor beyond Haughmond Hill. Shawbury Heath.
Root creeping. *Stem* 3 feet high, jointed, striated, smooth, leafy. *Leaves* very long, edges and keels scabrous. *Spikes* numerous, sessile and stalked, stalks striated, smooth. *Bracteas* long, sheathing, leafy, triangular at the apex, edges scabrous. *Glumes* filmy, single-ribbed. *Stigmas* 3.
There exists great difficulty in satisfactorily distinguishing between *E. polystachion,* and *angustifolium,* which are considered by many Botanists as forms of the same plant. I have never gathered any plant in Shropshire which I could positively refer to *E. polystachyon,* though it appears to occur in this county on the authority of an accurate Botanist, the late Rev. Edw. Williams, as recorded in English Flora; nor have I had opportunities of identifying the plant intended in that work. The above description was drawn up from a dried specimen sent from Gresford, Denbighshire, by my friend, Mr. J. E. Bowman, and that of *E. angustifolium,* from a comparison of numerous specimens from Shropshire and other parts, all of which correspond generally with specimens from the *Hort. Sic. Brit.* of the late Mr. Dickson. Whether they be really distinct species I must not presume to affirm, but would merely state, that in the above specimens there certainly exists

a difference in the shape of the fruit, that of *polystachyon* being linear-obovate, of *angustifolium* elliptical acuminate, and both triquetrous. These differences agree generally with the descriptions of De Candolle, who says of *polystachyon,* "seminibus trigonis oblanceolatis basi attenuatis obtusis," and of *angustifolium,* "seminibus subacutis lanceolatis acute triquetris." The same appearances seem indicated in Scheuchzer's *Agrostographia, p.* 307, where *polystachyon* is stated as possessing "semen triquetrum, fuscum aut castanei coloris, alterâ parte planum feré, e basi angustata sensim dilatatum, ut, propè rotundatum feré mucronem, duos circiter lineæ trientes latum evadat." In *p.* 310 of the same work the fruit of *E. angustifolium* is described to be "fuscum utrinque mucronatum ex triquetro teretiusculum."

4. E. *pubescens,* Sm. *Downy-stalked Cotton-grass.* Stem nearly triquetrous; leaves lanceolate, flat, keeled, suddenly contracted above the middle into a triquetrous scabrous point; stalks of the spikes rough, with short erect appressed bristles; fruit obpyriform, triquetrous, punctato-striate; hairs twice the length of the spike. *E. Bot. Suppl. t.* 2633. *E. Fl. v. i. p.* 69. *Hook. Br. Fl. p.* 32.

Bogs and marshes; rare. *Fl.* April, June. ♃.
Bog in Wyre Forest, also near Park Brook, Wyre Forest; *Mr. W. G. Perry.* Bog below the Wrekin, south-west side; *Mr. F. Dickinson!*
Root fibrous. *Stem* 1—2 feet high, smooth, striated, leafy. *Leaves* short and broad, with tight clasping sheaths, scabrous at the margins and keels. *Spikes* 5—7, their stalks somewhat angular, compressed and striated, clothed with short erect appressed bristles. *Bracteas* short, sheathing, dilated, membranous and striated at the base, apex triquetrous. *Glumes* membranous, dark, single-ribbed. *Stigmas* 3. Hairs very white and silky.

10. SCIRPUS. *Linn.* Club-rush.

1. S. *lacustris,* Linn. *Lake Club-rush or Bull-rush.* Stem round, naked above, leafy below; panicle compound, spikelets fasciculato-aggregate; involucre of 2 leaves; glumes mucronate, emarginate, fringed, glabrous; stigmas 3; anthers bearded at the apex; fruit rotundo-obovate, obtusely trigonous, punctato-striate; bristles 6, with deflexed teeth. *E. Bot. t.* 666. *E. Fl. v. i. p.* 57. *Hook. Br. Fl. p.* 27.

Margins of lakes and pools; not common. *Fl.* July, August. ♃.
Snowdon Pool; *Mr. F. Dickinson.*
Hancott Pool and Mare Pool, near Shrewsbury.
Root extensively creeping. *Stem* 4—6 feet high, soft, spongy, smooth. *Leaves* 1 or 2 at the base, with long sheaths. *Panicles* various in luxuriance, peduncles triquetrous, angles scabrous. *Spikelets* numerous, oblong, clustered at the extremities of the peduncles. *Glumes* brown. *Filaments* extended beyond the anther, and terminating in a downy tuft. *Stigmas* 3, pubescent. *Fruit* pale yellow-green, rotundo-obovate, obtusely trigonous, punctato-striate, tipped with a short thick blunt mucro. *Bristles* 6, with deflexed teeth, about as long as the fruit, arranged on the outside of the stamens.

2. S. *sylvaticus,* Linn. *Wood Club-rush.* Stem triangular, leafy; panicle terminal, cymose, decompound; spikelets fasciculato-aggregate; involucre of many foliaceous leaflets; glumes obtuse, slightly fringed; fruit obovate, obtusely trigonous, smooth; bristles 6, with deflexed teeth. *E. Bot. t.* 919. *E. Fl. v. i. p.* 62. *Hook. Br. Fl. p.* 29.

Moist woods and banks of rivers; not common. *Fl.* July. ♃.
Cox Wood Pool, near Coalbrookdale; *Mr. F. Dickinson.* Wyre Forest; *Mr. W. G. Perry.* Seven miles from Ludlow, on the Tenbury road; *Mr. A. Aikin.* Sides of a pool in Coalbrookdale; *Turn. & Dillw. Bot. Guide.* Banks of the Severn, near Bridgnorth; *Rev. W. R. Crotch.* In a field close to road leading from Ludlow to Tenbury, about 7 miles from the former place; *Plym. Agric. Surv.*
Near the confluence of the rivers Teme and Corve, Ludlow. Shelton Wood, near Shrewsbury.
Root extensively creeping. *Stem* 3 feet high, smooth, jointed. *Leaves* with elongated sheaths, broadly linear, flat, rough at the edges and keels. *Panicle* of innumerable small, dark-green, ovate spikelets, clustered on the extremities of spreading peduncles furrowed with rough angles. *Stamens* mucronate. *Fruit* pale tawny, with a short beak. *Bristles* 6, longer than the germen, placed on the outside of the stamens.

11. ELEOCHARIS. *Br.* Spike-rush.

1. E. *palustris,* Br. *Creeping Spike-rush.* Stem rounded; root much creeping; stigmas 2; fruit roundish-obovate, plano-convex, shorter than the 4 bristles, crowned with the broadly-ovate or battledore-shaped jointed base of the style; glumes acute, outer one smaller than the rest. *E. Fl. v. i. p.* 14. *Hook. Br. Fl. p.* 29. *Scirpus palustris,* Linn. *E. Bot. t.* 131.

Wet marshy places and sides of pools; not frequent. *Fl.* May—July. ♃.
Pastures near Marn Wood, Coalbrookdale; *Mr. F. Dickinson.* Oakley Park, near Ludlow; *Mr. Spare.* Mae-bury; *Rev. T. Salwey.*
Shawbury Heath. Hancott and Berrington Pools, near Shrewsbury.
Root extensively creeping, black and shining as well as the exterior sheaths of the stem. *Stems* several, erect, smooth, without central pith, and consisting of large membranous tubes, surrounded by smaller ones. *Sheaths* tight, entire and cylindrical. *Leaves* none. *Spike* ovato-oblong, acute, about half an inch long. *Glumes* brown, ovate, acute, keeled, membranous at the edges. *Bristles* 4, occasionally 5 or 6, longer than the germen and much narrower than the dilated membranous *filaments,* clothed with deflexed teeth except at the base, which is slightly dilated. *Stigmas* downy, spreading, as long as the *style.* Base of the style dilated, broadly ovate or battledore-shaped, contracted at its point of attachment with the germen. *Fruit* brown, striated, roundish-obovate, tumid at each side, but most on that next the glume, crowned by the brown, wrinkled, compressed, permanent, unpolished base of the style.

2. E. *multicaulis,* Sm. *Many-stalked Spike-rush.* Stem round; root tufted, fibrous; stigmas 3; fruit obovate, triquetrous, as long as the 6 bristles, crowned with the broad triquetrous base of the style; glumes obtuse, outer one emarginate. *E. Fl. v. i. p.* 64. *Hook. Br. Fl. p.* 30. *Scirpus multicaulis,* Sm. *Eng. Bot. t.* 1187.

Marshy places; rare. *Fl.* July. ♃.
Shawbury Heath; *Mr. E. Elsmere, junr.!*
Root sending out rootstocks to the length of 2 or 3 inches, from which fibres proceed below and new shoots above. *Stems* several, spreading, ascending, round, smooth, with a stout central pith and membranous tubes of looser texture interposed between it and the external part. *Sheaths* 1 or more at the base, tight, entire, cylindrical, pale. *Leaves* none. *Spike* oblong, more acute and slender than the last. *Glumes* brown, ovate, obtuse, keeled, edges membranous. *Bristles* 6, about as long as the fruit, not dilated at the base; three of them placed between the

stamens and germen, and three immediately without the stamens, clothed in the upper part with deflexed teeth, naked below. *Style* triquetrous, dilated and thickened at the base, which is as broad or broader than the fruit, to which it is jointed. *Fruit* tawny, shining, striated, obovate, triquetrous, flat on one side, keeled on the other, crowned by the broad indurated base of the brownish green style.

12. SCIRPIDIUM. *Nees ab. Esenb.* Needle-rush.

1. S. *aciculare*, N. ab. E. *Small Needle-rush.* Root creeping; stem setaceous, compressed, grooved; sheaths leafless; spikes ovate, acute; glumes equal, acute; fruit obovate-oblong, compressed, longitudinally ribbed, and marked with fine transverse lines. *Scirpus acicularis, Linn.* E. *Bot. t.* 749. *Eleocharis acicularis, Roem et Sch.* E. *Fl. v. i. p.* 65. Hook. *Br. Fl. p.* 31.

    Sides of lakes and wet sandy and marshy places; rare. *Fl.* July, Aug. ♃.
Ellesmere Mere, between the House of Industry and Otley; *J. E. Bowman, Esq.—Rev. A. Bloxam!* White Mere, near Ellesmere; *Mr. F. Dickinson.*
    Root fibrous, with filiform runners. *Stems* numerous, many barren and leaf-like, erect, smooth, with a tight entire sheath at the base. *Leaves* none. *Spikes* minute, 5—6 flowered. *Glumes* brown, acute, keeled, edges membranous. *Bristles* short, with deflexed teeth, deciduous. *Style* short. *Fruit* pale, crowned with the short turbinated base of the style.

13. BÆOTHRYON. *Nees ab Esenb.* Small Turf-rush.

1. B. *Halleri*, N. ab. E. *Haller's Turf-rush.* Stem round; sheaths leafless; two outer glumes obtuse, larger than the rest, but shorter than the spike; fruit obovate, obtusely triangular, as long again as the bristles, crowned with the narrow, triquetrous, (not jointed nor dilated), base of the style, reticulato-striated; bristles 6, with deflexed teeth. *Scirpus pauciflorus, Lightf.* E. *Bot. t.* 1029. E. *Fl. v. i. p.* 56. *Eleocharis pauciflora, Link.* Hook. *Br. Fl. p.* 30.

    Boggy ground; rare. *Fl.* July, August. ♃.
    South base of Broom Hill, near Pulverbatch; *Mr. T. Bodenham!* Bog below the Wrekin, south-west side; *Mr. F. Dickinson.* Bomere Pool, near Shrewsbury; *Rev. E. Williams.*
    Shawbury Heath.
    Root fibrous, sending out jointed runners. *Stems* numerous, many of them barren, erect, smooth, with 1 or more tight cylindrical sheaths at the base. *Leaves* none. *Spike* small, ovate, acute. *Glumes* except the two outer ones, acute, keeled, membranous at the edges. *Bristles* 6, about half the length of the fruit, not dilated at the base, clothed with deflexed teeth, 3 of them placed between the stamens and germen, and 3 immediately behind the stamens. *Stigmas* downy. *Style* long, triquetrous. *Fruit* pale-greyish.

2. B. *cæspitosum*, N. ab. E. *Scaly-stalked Turf-rush.* Stem slightly compressed; sheaths with short subulate leaves; two outer-most glumes longer than the very small spikes, terminating in long rigid points; fruit obovate-oblong, triquetrous, shorter than the bristles, crowned with the long, narrow, triquetrous, (not jointed nor dilated), base of the style; bristles 6, with erect teeth. *Scirpus cæspitosus, Linn.* E. *Bot. t.* 1029. E. *Fl. v. i. p.* 35. *Eleocharis cæspitosa, Link.* Hook. *Br. Fl. p.* 30.

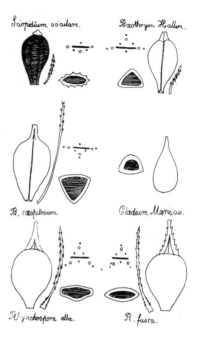

Pl. 3. *to Face P.* 34.

Scirpidium aciculare.   *Bæothryon Halleri.*

B. cæspitosum   *Cladium Mariscus.*

Rhynchospora alba.   *R. fusca.*

    Boggy ground; not frequent. *Fl.* June, July. ♃.
    Whixall Moss; *Rev. A. Bloxam.* Shawbury Heath; *Mr. E. Elsmere.*
    Vownog Bog, near Westfelton.
    Root fibrous, tufted. *Stems* numerous, many of them barren, erect, their bases invested with numerous, shining, pale-brown sheaths, elongated into short rigid subulate leaves. Outer *glume* ovate, rigid, 3-nerved, the rest ovate, acute, membranous at the edges. *Bristles* longer than the fruit, with a few erect teeth near the apex, which is forked, naked below, placed as in the last species. *Stigmas* downy. *Fruit* pale-yellow.

14. CLADIUM. *Br.* Twig-rush.

1. C. *Mariscus*, Br. *Prickly Twig-rush.* Stem round, smooth; corymbs crowded, spikes fasciculate; leaves serrated at the margins and keels with strong, erect, bristly teeth. E. *Fl. v. i. p.* 36. Hook. *Br. Fl. p.* 14. *Schœnus Mariscus, Linn.* E. *Bot. t.* 950.

    Boggy and fenny places; rare. *Fl.* July, August. ♃.
    South-west margin of Croesmere Mere; *J. E. Bowman, Esq.!* Oakley Park, near Ludlow; *Mr. H. Spare.*
    Root long and creeping. *Stem* 3—5 feet high, erect, polished, jointed, obso-letely angular upwards. *Leaves* sheathing, linear, keeled, triangular at the apex, margins and keels strongly serrated. *Panicle* erect, much divided, leafy, peduncles compressed, flat on the upper sides, rounded beneath, smooth, bracteated. *Spikelets* ovate, 6—12 in a dense head. *Glumes* numerous, imbricated, brown, inner ones the longest, the two or three innermost floriferous, of which one or two bears a coated nut almost as large as the spikelet. *Stigmas* generally 2, sometimes cloven, downy.

15. RHYNCHOSPORA. *Vahl.* Beak-rush.

1. R. *alba*, Vahl. *White Beak-rush.* Spikelets in a compact corymb shorter than the outer bracteas; stamens 2; filaments scarcely broader than the bristles; bristles 9—12, with reflexed teeth; fruit obovate, compressed, distinctly margined, minutely reticulated; style without teeth; leaves narrow-linear. E. *Fl. v. i. p.* 52. Hook. *Br. Fl. v. i. p.* 27. *Schœnus albus, Linn.* E. *Bot. t.* 985.

    Wet turfy bogs; not common. *Fl.* June—August. ♃.
    Whixall Moss, abundant; *J. E. Bowman, Esq.* Bog near Ellesmere; *Rev. A. Bloxam.*
    Bomere Pool (west side), and Shomere Moss, near Shrewsbury. Twyford Vownog, near Westfelton.
    Spikelets of *flowers* white or whitish, collected so as to form a level surface at the top, 2-flowered, both fertile. *Bristles* with reflexed teeth, and a few erect ciliæ at the base, both in this species and in *R. fusca,* much longer than the germen. *Fruit* greenish, tapering at the base into a short stalk. *Style* persistent, in texture and colour like the fruit, dilated at the base, which is not articulated, nor so broad as the seed. *Culm* leafy, triangular. *Leaves* keeled, their edges and keels, as well as those of the bracteas, and also the angles of the flower-stems, scabrous. *Root* creeping.
    At Bomere Pool, and on Twyford Vownog, near Westfelton, two varieties grow intermixed with the above state of the plant, and in nearly equal abundance, one having the corymb as long as the outer bracteas, and the other having the spikelets in a somewhat oval head, shorter than the outer bracteas.

The Natural Order CYPERACEÆ has been greatly elucidated by Mr. Brown, who considers the bristles, which in several of the genera accompany the fruit as the true perianth. Sir J. E. Smith (*E. Fl. v. i. p.* 50) objects to this view of the subject, inasmuch as the bristles are situated between the stamens and germen, and cannot therefore be either calyx or corolla, but are rather an appendage to the germen and seed. Dr. Hooker (*Br. Fl.* p. 15) states, that Mr. W. Wilson has proved that the bristles are not placed immediately at the base of the germen, between it and the stamens, but on the outside of the latter, and therefore concludes Mr. Brown's opinion to be correct.

I have carefully examined the relative position of the bristles, germen, and stamens, in *R. alba and fusca*, and find that the bristles are apparently arranged in whorls of three, and that the stamens are situated immediately before the exterior whorl, between that and the other whorls. This situation of the bristles exterior to the stamens is observable also in *Scirpus lacustris and sylvaticus, Eleocharis multicaulis, Bæothryon Halleri and cæspitosum.*

2. *R. fusca*, Sm. *Brown Beak-rush.* Spikelets in an oval head, considerably shorter than the broad outer bracteas; stamens 3; filaments dilated, membranous, much broader than the bristles; bristles 6, with erect teeth; fruit obovate, somewhat turbinate, compressed, obscurely margined, minutely reticulated; style with a row of erect teeth on each margin; leaves almost filiform. *E. Fl. v. i. p.* 53. *Hook. Br. Fl. v. i. p.* 27. *Schœnus fuscus, Linn. E. Bot. t.* 1575.

Bogs; very rare. *Fl.* July, August. ♃.

Bomere Pool; *Rev. Edward Williams.*

Spikelets oval, rich brown, longer than in *R. alba*, 2-flowered, one of which is abortive. *Bristles* 6. *Fruit* brownish, tapering at the base into a very short stalk, much shorter than in *R. alba*. *Style* dilated at the base as in *R. alba*, persistent, thin, pellucid, greenish, thickened at the outer margins, which are each clothed with a row of erect teeth. *Culm* leafy, compressed. *Leaves* narrow, almost filiform, edges and keels somewhat scabrous. *Bracteas* broad with about 7 nerves, edges and keels scabrous. *Root* creeping.

In vain have I repeatedly and most carefully searched the above locality for this very rare plant.

TRIANDRIA—DIGYNIA.

16. LOLIUM. *Linn.* Darnel.

1. L. *perenne*, Linn. *Perennial Darnel or Rye-grass.* Spikelets longer than the glume; florets linear oblong compressed awnless; root perennial. *E. Bot. t.* 315. *E. Fl. v. i. p.* 173. *Hook. Br. Fl.* 57.

Way-sides, pastures and waste places; frequent. *Fl.* June, July. ♃.

Root fibrous. *Culms* several, 1—2 feet high, round, smooth, rigid, striated, leafy, with purplish tumid joints, the lowermost bent. *Leaves* dark green, linear, pointed, flat, smooth, striated. *Sheaths* nearly as long as the leaves, compressed, striated, smooth. *Ligule* short, entire, acutely auricled at each side. *Spike* nearly erect, *rachis* smooth, channelled alternately at each side to receive the 6-flowered spikelets. *Glume* linear-lanceolate, smooth, 5-ribbed. *Paleæ* nearly equal, lanceolate, *inferior valve* 5-ribbed, exterior ribs very strong, slightly apiculate and cloven, *superior valve* with 2 ciliated marginal ribs, edges incurved, apex entire, obtuse.

This plant has the aspect of *Triticum repens*, from which, it is at once essentially distinguished by the single-valved glume, and the relative position of the spikelets and rachis. Highly valuable to the agriculturist for early crops and for forming with clover artificial pasture.

2. L. *temulentum*, Linn. *Bearded Darnel.* Spikes equal in length with the glumes; florets lanceolate awned; root annual. *E. Bot. t.* 124. *E. Fl. v. i. p.* 174. *Hook. Br. Fl. p.* 57.

Corn-fields. *Fl.* July. ⊙.

Root fibrous. *Stem* 1—2 feet high, erect, round, smooth, striated, rough in the upper part. *Leaves* linear with a tapering point, striated and roughish on the upper surface. *Sheaths* long, striated, auricled at the summit. *Ligule* short, obtuse, mostly torn. *Spikes* long, *rachis* notched, angular, wavy and rough. *Spikelets* alternate, compressed, lower ones with 2 glumes, *outer glume* as long or longer than the spikelets, linear-lanceolate, smooth, many-ribbed, *inner* smaller and membranous. *Paleæ* equal, *inferior valve* broadly ovate, emarginate, with about 9 ribs, the middle one the strongest, and terminating in a roughish awn of greater or less length and rigidity, proceeding from a little below the emarginate or cloven apex, *superior valve* membranous, emarginate, with two rough lateral ribs and incurved edges.

β. arvense. florets with or without short soft imperfect awns. *Lolium arvense, With. E. Bot. t.* 1125. *E. Fl. v. i. p.* 175. *Hook. Br. Fl. p.* 57.

Corn-fields. *Fl.* July. ⊙.

Welbatch, near Shrewsbury; *Mr. T. Bodenham!* Hadnall, near Shrewsbury; *Mr. E. Elsmere, junr.!* Near Ludlow; *Miss Mc Ghie.* Huck's Barn, near Ludlow; *Rev. T. Salwey.*

3. L. *festucaceum*, Link. *Spiked Darnel.* Spikelets many-flowered, compressed, much longer than the glumes, remote; florets cylindrical, awnless, inferior palea obtuse; root perennial. *Festuca loliacea, Huds. E. Bot. t.* 1821. *E. Fl. v. i. p.* 147. *Hook. Br. Fl. p.* 50.

Moist pastures and meadows; not unfrequent. *Fl.* June, July. ♃.

Welbatch, near Shrewsbury; *Mr. T. Bodenham!* Fields near Shrewsbury; *Turn. & Dilw. Bot. Guide.* Hadnall, near Shrewsbury; *Mr. E. Elsmere, junr.!* Fields near Little Wenlock. Steeraway Lime Works; *Mr. F. Dickinson.*

Root fibrous. *Culms* several, erect, 2 feet high, simple, leafy, round, very smooth, jointed. *Leaves* linear, narrow, flat, smooth, acute, ribbed, with a few scattered hairs on the upper surface. *Sheaths* long, ribbed, very smooth. *Ligule* short, clasping the stem, with a small acute auricle on each side. *Racemes* 2—5 inches long, *rachis* flexuose, angular, rough, *spikelets* nearly sessile, especially the upper ones, alternate, smooth, pale, erect, compressed, 8—10-flowered. *Glumes* very unequal, *outer one* lanceolate, acuminate, 7-ribbed, *inner* single-ribbed. *Inferior palea* ovato-lanceolate, obtuse and diaphanous at the apex, with 5 slightly scabrous ribs, *superior valve* covered with minute resinoso-glandulose points, bifid, marginal ribs serrated.

17. NARDUS. *Linn.* Mat-grass.

1. N. *stricta*, Linn. *Mat-grass.* Spike erect, slender, the florets all pointing one way. *E. Bot. t.* 290. *E. Fl. v. i. p.* 70. *Hook. Br. Fl. p.* 32.

Moors and heaths; not uncommon. *Fl. June.* ♃.

Titterstone Clee Hills; *Mr. J. S. Baly!* Oakley Park, near Ludlow; *Mr. Spare.* Bayston Hill, near Shrewsbury; *Mr. T. Bodenham!* Lilleshall Mill-pool Dam; *Withering.*

Lyth Hill and Haughmond Hill, near Shrewsbury. Shawbury Heath. Stiperstones Hill. Caer Caradoc.

Root fibrous, downy. *Stems* copiously tufted, erect, jointed, striated, rough with scattered appressed bristles, the base invested with the numerous long pale sheaths of the leaves. *Leaves* rigid, setaceous, furrowed, furrows with appressed bristles. *Spike* long, erect, grooved in two rows for the insertion of the unilateral, alternate, sessile florets, each of which has two teeth in front at its base, (obsolete glumes). *Paleæ* lanceolate, *outer valve* coriaceous, 3-nerved, tapering into a long, rigid, setigerous *awn*, covered with resinoso-glandulose tubercles, margins bristly, *inner one* membranous, smooth, awnless. *Stamens* 3. *Style* and *stigma* single.

18. AGROPYRUM. *Gærtn.* Couch-grass.

1. A. *repens*, Br. *Creeping Couch-grass.* Spikelets of about 5, more or less awned flowers; glumes lanceolate, 5-ribbed; leaves plane; root creeping. *Triticum repens, Linn. E. Bot. t.* 909. *E. Fl. v. i. p.* 183. *Hook. Br. Fl. p.* 56.

Fields and waste places; too frequent. *Fl.* Summer months. ♃.

Root extensively creeping and most difficult of extirpation. *Stem* 2 feet high, erect. *Leaves* of a dull glaucous green, linear, flat, ribbed, upper surface clothed with long soft hairs, rough on the ribs of the under surface and at the edges. *Sheaths* long, ribbed, smooth, tight and united below, open or divided in the upper portion for about one-third of their length, and membranous at the margins. *Ligule* short, truncate. *Spikes* 3—4 inches long. *Rachis* flattish, furrowed, with bristly edges. *Spikelets* about 5-flowered. *Glumes* lanceolate, glandulose and scabrous, 5-ribbed, mid-rib elongated into a short rough awn, edges membranous. *Paleæ* 2, *inferior valve* lanceolate, glandulose, 5-ribbed, mid-rib terminating in a short rough awn, *superior valve* flat, glandulose, retuse, ciliated, with 2 marginal fringed ribs and incurved edges. *Scales* 2, irregularly dentate, and tumid near the base, apex somewhat acute, ciliated.

The pest of cultivation being most difficult to extirpate by reason of the excessive brittleness of the numerously jointed and creeping under-ground stems. These contain a large proportion of nutritive matter, and are collected on the continent as a wholesome food for horses and other cattle. When dried, they are said to afford a good flour, which has been used in the preparation of bread during famine, or times of scarcity. They are also mentioned as a substitute for malt in the brewing of beer.

β. aristatum. Glumes awned. *T. repens β. E. Fl. v. i. p.* 183.

Way-sides and wastes; not unfrequent.

Cross Hill, near Shrewsbury; and in the neighbourhood of Shrewsbury generally.

The var *β.* appears to differ only in the glumes and paleæ having awns about half their length.

2. A. *caninum*, R. & Sch. *Fibrous-rooted Couch-grass.* Spikelets of 5 awned flowers, glumes lanceolate, 3—4-ribbed; leaves plane; root fibrous. *Triticum caninum, Huds. E. Bot. t.* 1327. *E. Fl. v. i. p.* 184. *Hook. Br. Fl. p.* 56. *Elymus caninus, Linn.*

Woods and hedges; not common. *Fl.* July. ♃.

Woods and hedges about Madeley; *W. P. Brookes, Esq.!* In the Dingle, Coalbrookdale; *Mr. F. Dickinson.*

Root simple, fibrous. *Stem* erect, 2—3 feet high, round, smooth, striated, occasionally downy in the lower part. *Leaves* flat, linear-lanceolate, scabrous on both sides, hairy above. *Sheaths* close, striated, smooth, occasionally hairy, entirely divided, edges membranous. *Ligule* very short, scarcely perceptible. *Spikes* 3—4 inches long, *rachis* bristly at the angles. *Spikelets* 5-flowered. *Glumes* unequal, lanceolate, acuminate, scabrous, 3—4-ribbed, mid-rib elongated into an awn. *Paleæ* 2, *inferior valve* concave, 5-ribbed, the mid-rib elongated into a long slender rough awn; *superior valve* flat, with 2 marginal fringed ribs and incurved edges.

Very nearly allied to the preceding species, from which it is chiefly distinguished by its fibrous root.

19. BRACHYPODIUM. *Beauv.* False-Brome-grass.

1. B. *sylvaticum*, Beauv. *Slender False Brome-grass.* Spike drooping, spikelets cylindrical, secund, hairy, awns longer than the florets. *Hook. Br. Fl. p.* 57. *Bromus sylvaticus, Poll. E. Bot. t.* 729. *Festuca sylvatica, Huds. E. Fl. v. i. p.* 149.

Dry copses, thickets and hedges; frequent. *Fl.* July. ♃.

Captain's Coppice, Coalbrookdale; *Mr. F. Dickinson.* Welbach, near Shrewsbury; *Mr. T. Bodenham!*

Hedge-banks, near Berrington, Bomere Pool, Sutton Spa, Shelton Rough, Bicton. Between Shrewsbury and Uffington. Woods at the foot of Caer Caradoc, and hedge-banks near Longnor.

Root fibrous, tufted. *Culms* 2—3 feet high, round, striated, smooth above, hairy below, densely so at the joints, leafy. *Leaves* broadly linear-lanceolate, very hairy, with a strong central mid-rib, and about 3 parallel nerves, the intermediate interstices finely striated and rough. *Sheaths* long, ribbed, close below, divided above for about one-third of their length, covered with deflexed hairs. *Ligule* short, blunt, notched or torn. *Spike* simple, 3—4 inches long. *Spikelets* alternate, remote, nearly sessile, slender, compressed, about 9-flowered, *rachis* wavy, angular, scarcely rough, partial peduncles pubescent. *Glumes* unequal, lanceolate, subaristate, hairy and glandulose, with 7 rough ribs, edges membranous, fringed. *Inferior valve* of the *paleæ* linear-lanceolate, hairy, 7-ribbed, ribs rough, the central one elongated into a rough awn, as long or longer than the florets; *superior valve* linear, flat, retuse, inflexed at the strongly fringed ribs. *Scales* 2, acute, fringed, tumid at the base.

20. HORDEUM. *Linn.* Barley.

1. H. *pratense*, Huds. *Meadow Barley.* Glumes all setaceous and scabrous. *E. Bot. t.* 409. *E. Fl. v. i. p.* 180. *Hook. Br. Fl. v. i. p.* 55.

Moist meadows and pastures; not common. *Fl.* July. ⊙. (Hooker) or ♃. (Smith).

Meadows near the New Park, Shrewsbury; *Mr. E. Elsmere, junr.!* Oakley Park, near Ludlow; *Mr. Spare.*

Root fibrous. *Culm* slender, about one foot high, erect, rigid, round, striated, smooth and naked above. *Leaves* narrow, roughish, occasionally hairy as well as their *sheaths* which are close, not swelling, with a scarcely perceptible *ligule*. *Spike* about 2 inches long, *rachis* alternately toothed on each side, flattened, edges bristly, *central spikelet* sessile and perfect, occasionally barren, *lateral* ones barren, on a rough peduncle. At the back of the superior valve of the paleæ of all the florets, whether fertile or barren, is a rough peduncle of another floret. *Glumes* narrow and equal. *Inferior valve of paleæ* ovate, very acuminate, smooth on the exterior, hairy in the interior near the apex, 5-ribbed; *superior valve* lanceolate, apex entire, with 2 smooth ribs or nerves subtending a glandulose furrow in which lies the rudimentary peduncle, edges membranous. *Awn* of central floret as long or longer than the valve, that of the lateral spikelets somewhat shorter. *Seed* oblong, channelled on the upper side, blunt and downy at the summit. *Scales* elongated, acute, fringed.

2. H. *murinum,* Linn. *Wall Barley.* Glumes of the intermediate floret linear-lanceolate, ciliated, the exterior glume of the lateral florets setaceous and scabrous. *E. Bot. t.* 1971. *E. Fl. v. i. p.* 179. *Hook. Br. Fl. v. i. p.* 55.

Waste ground and by way-sides; not common. *Fl.* June—August. ⊙.

Coalbrookdale; *Mr. F. Dickinson.*

Shrewsbury and Ludlow Castles.

Root fibrous. *Culms* several, spreading and decumbent at the base, then ascending, slender, leafy, smooth, jointed. *Leaves* short, linear, acute, flat, ribbed, rough, hairy principally at the margins on the upper side. *Sheaths* long, rather lax, entirely divided, strongly ribbed, upper ones smooth, lower ones with short deflexed pubescence. *Ligule* short, obtuse. *Spike* 2 or 3 inches long, cylindrical, very dense and uniform, 2-ranked, brittle, *rachis* alternately toothed, edges bristly, *central spikelet* sessile, fertile, *lateral ones* on rough peduncles, barren. In all the florets, whether barren or fertile, there is a rudimentary peduncle at the back of the superior valve of the paleæ. *Glumes* of central floret linear-lanceolate, scabrous at the back, fringed with spreading hairs, *superior* one of the lateral florets similar but somewhat narrower, and smooth on the back, *inferior* one setaceous and scabrous. *Paleæ* 2, *inferior valve* concave, ovato-lanceolate, acuminate, near the apex slightly downy both outside and inside, 5-ribbed, *superior* similar to that of the last species but more obtuse. *Awn* of central floret thrice as long as the paleæ, of the lateral ones about twice the length. *Scale* as in the last.

21. PHALARIS. *Linn.* Canary-grass.

1. P. *canariensis,* Linn. *Cultivated Canary-grass.* Panicle spicæform, ovate; glumes boat-shaped, entire at the point, accompanied by the single glabrous valves of 2 other florets; paleæ clothed with appressed pubescence. *E. Bot. t.* 1310. *E. Fl. v. i. p.* 75. *Hook. Br. Fl. p.* 34.

Cultivated and waste-ground; doubtless naturalized. *Fl.* July. ⊙.

Oakley Park, near Ludlow; *Mr. Spare.*

Root fibrous. *Culm* 1—2 feet high, erect, leafy, striated, rough, with brown joints. *Leaves* broad, striated, rough. *Sheaths* long, tumid in the upper part, entirely divided. *Ligule* large, laciniate. *Panicle* very compact, erect, elegantly variegated with green and white. *Glumes* membranous, navicular, compressed, larger than the paleæ, with a single green rib on each side near the margin, and a broader green keel which is dilated into a winged somewhat bristly membrane. *Paleæ* unequal, *inferior valve* ovato-acuminate, obscurely nerved, covered with appressed pubescence, bifid at the apex, *superior valve* smaller, truncate and ciliated at the extremity. *Sterile florets* consisting of narrow, single, membranous scales, nearly glabrous or only slightly hairy, apex ciliated. *Seed* polished and shining, (the *Canary-seed* of the shops.)

22. PHLEUM. *Linn.* Cat's-tail-grass.

1. P. *pratense,* Linn. *Common Cat's-tail-grass, Timothy-grass.* Panicle spicæform, cylindrical; glumes compressed, truncate, mucronato-aristate, ciliated at the keel, longer than the awn. *E. Bot. t.* 1076. *E. Fl. v. i. p.* 76. *Hook. Br. Fl. p.* 35.

Meadows and pastures; very common. *Fl.* June. ♃.

Root somewhat creeping. *Culm* erect, knotty, round, striated, leafy except near the top. *Leaves* flat, short, rough. *Sheaths* long, close, striated, smooth, entirely divided. *Ligule* obtuse, torn. *Panicle* erect, cylindrical, obtuse, 2—6 inches long, flowers crowded, footstalks short sub-divided. *Glumes* green or purplish, compressed, keeled with a dorsal green nerve running out into a spreading serrated awn scarcely half so long as the valve, single-ribbed on each side close to the keel, margins dilated, white, membranous, fringed, covered with pale glandular protuberances. *Valves* of the paleæ small, unequal, *outer* one crenated, enveloping the *inner,* smooth, ribbed, keeled, hairy at the keel.

Var. β. nodosum. Whole plant smaller; root tuberous. *Phleum nodosum, Willd. P. pratense β. and γ. E. Fl. v. i. p.* 76.

In barren dry ground; common. *Fl.* June. ♃.

Differs only in the root being tuberous, and the culm partly decumbent and bent at the lower joints.

23. CYNOSURUS. *Linn.* Dog's-tail-grass.

1. C. *cristatus,* Linn. *Crested Dog's-tail-grass.* Panicle coarctate, spicate, linear; florets with a very short awn. *E. Bot. t.* 316. *E. Fl. v.* 1. *p.* 137. *Hook. Br. Fl. p.* 48.

Dry pastures; common. *Fl.* July. ♃.

Root tufted, with long simple fibres. *Culms* several, slender, erect, round, jointed, striated, smooth, leafy. *Leaves* narrow, linear, acuminate, ribbed, smooth. *Sheaths* long, ribbed, smooth, entirely divided. *Ligule* short and abrupt. *Panicle* erect, rigid, secund, *rachis* angular, wavy, and rough. *Involucres* beautifully pectinated, one at the base of each spikelet, their divisions linear, acute, greenish, subglumaceous, a little curved, covered with glandulose dots, edges serrated with bristly teeth. *Spikelets* 3—5 flowered, on short peduncles. *Glumes* linear-lanceolate, nearly equal, membranous, rough at the keel, as long as the florets. *Inferior valve of the paleæ* lanceolate, scabrous, especially at the edges, keel, and near the apex, obscurely nerved; *superior* white, bifid, pubescent on the marginal ribs, edges membranous, incurved. *Seed* elliptic-oblong, acute, filling the valves of the corolla.

24. SESLERIA. *Linn.* Moor-grass.

1. S. *cærulea,* Scop. *Blue Moor-grass.* Panicle spicæform, ovate, bracteated; spikelets 2—3-flowered, inferior palea 3—5-toothed, mucronate or shortly aristate; leaves linear; culm with striated sheaths in the lower part. *E. Bot. t.* 1613. *E. Fl. v. i. p.* 114. *Hook. Br. Fl. p.* 42.

Limestone districts; rare. *Fl.* April—June. ♃.

Oakley Park, near Ludlow; *Mr. Spare.*

Root tufted. *Culm* 6—18 inches high, smooth, striated, clothed below with tight sheaths, surmounted with partially developed leaves. *Leaves* numerous, linear, obtuse, smooth, striated, of a glaucous green, edges and keel rough. *Sheaths* short and close, without *ligules. Inflorescence* close, a short, ovate, shining, bluish-grey spiked panicle, the lower spikelets accompanied by thin membranous jagged or ciliated bracteas at their base. *Spikelets* generally in pairs, oblongo-ovate. *Glumes* ovato-lanceolate, 3-toothed, middle tooth mucronate or lengthened into a short rough awn, slightly pubescent, shorter than the florets. *Paleæ* oblongo-ovate, *inferior valve* the larger, 3—5-nerved, ciliated or jagged with about 3—5 teeth, the middle one lengthened into a short awn, *superior valve* with 2 rough ribs, bifid at the apex, edges membranous and incurved. *Anthers* large, yellow, tipped with purple.

25. ANTHOXANTHUM. *Linn.* Vernal-grass.

1. A. *odoratum,* Linn. *Sweet-scented Vernal-grass.* Panicle spicæform, ovato-oblong; spikelets upon short partial stalks, 3-flowered. *E. Bot. t.* 647. *E. Fl. v. i. p.* 37. *Hook. Br. Fl. p.* 14.

Meadows and pastures; very common. *Fl.* May, June. ♃.

Root fibrous. *Culm* a foot high, slender, smooth, rigid, with 1 or 2 joints. *Leaves* short, flat, linear, slightly hairy, striated, with a triple row of bristles on each stria. *Glumes* very unequal, acuminate, strongly keeled, the rigid keels projecting into very short awns, margins ciliated, *inferior valve* smaller, single-ribbed, *superior* compressed, with a strong rib on each side the keel, slightly hairy, and covered with minute, round, resinoso-glandulose bodies *Florets* 3 in each spikelet, two outer ones sterile, each consisting of a single, ovate, obtuse, bifid, hairy, resinoso-glandulose valve, densely ciliated at the margin, awned. *Awn* of the *superior* floret long, proceeding from above the base of the valve, lower portion as far as the apex of the valve, contorted, striated, upper slightly divergent, rough with erect teeth; *awn* of the *inferior floret* proceeding from above the middle, triquetrous, edges toothed. *Central floret* fertile. *Paleæ* short, unequal, awnless, *inferior valve* broadly ovate, obtuse, compressed, *superior* narrow, acute. *Stamens* 2. *Anthers* oblong, forked at each end. *Stigmas* 2, long and feathery. *Germen* spurred at the base.

The grateful and peculiar fragrance of new-made hay is attributed to the drying of this grass, but a microscopic examination of the glumes and palea of most, if not of all grasses, will shew them to be more or less covered with innumerable, minute, resinoso-glandulose dots, probably containing essential oil which diffuses its odour by evaporation.

26. ALOPECURUS. *Linn.* Fox-tail-grass.

1. A. *pratensis,* Linn. *Meadow Fox-tail-grass.* Culm erect, striated, smooth; panicle spicæform, cylindrical, obtuse; glumes lanceolate, acute, hairy, connate at the base, keels ciliated; palea as long as the glumes, acute, awn inserted near to the base of the palea, and twice its length. *E. Bot. t.* 759. *E. Fl. v. i. p.* 80. *Hook. Br. Fl. p.* 33.

Meadows and pastures; common. *Fl.* May, June. ♃.

Root fibrous. *Culm* 1—2 feet high, smooth, leafy, often prostrate to the first joint at the base. *Leaves* flat, somewhat glaucous, nearly smooth, upper ones with long, smooth, furrowed, slightly swelling, entirely divided *sheaths. Ligule* short, obtuse. *Panicle* of a yellow-green colour with silvery hairs, *rachis* glabrous, *peduncles* very short, crowded, somewhat branched, bearing 3—5 florets. *Glumes* connate, remarkably compressed, hairy and covered with minute pale glandular dots, strongly keeled and ciliated with long soft hairs, with a green rib on each side. *Paleæ* folded, as long as the glumes, acute, with 5 green ribs, pubescent at the apex, and minutely glandular. *Awn* proceeding from below the middle near to the base, twice the length of the palea, somewhat geniculate, contorted, striated and smooth in the portion included by the glume, pale and rough with erect bristles above. *Anthers* yellow.

2. A. *geniculatus,* Linn. *Floating Fox-tail-grass.* Culm ascending, bent at the joints, striated, smooth; panicle spicæform, cylindrical, obtuse; glumes ovate, obtuse, hairy, scarcely connate at the base, keel hairy; palea shorter than the glumes, obtuse, awn inserted near to the base of the palea, and twice its length. *E. Bot. t.* 1250. *E. Fl. v. i. p.* 83. *Hook. Br. Fl. p.* 33.

In pools and wet marshy places, sometimes on dry ground; not unfrequent. *Fl.* June—August. ♃.

Welbach Coppice, near Shrewsbury; *Mr. T. Bodenham!* Hadnall; *Mr. E. Elsmere, junr.!* Oakley Park, near Ludlow; *Mr. Spare.*

Hancott Pool, Cross Hill, and road-side near Berrington, all near Shrewsbury.

Root fibrous. *Culms* 8 inches to 2 feet high, spreading and prostrate in the lower part, frequently rooting from the joints, leafy, smooth. *Leaves* short, flat, striated and scabrous on the upper surface, *sheaths* compressed, smooth, furrowed, ventricose, entirely divided. *Ligule* oblong, obtuse, very thin. *Panicle* of a purplish-green silvery appearance, *florets* half the size of those of *A. pratensis, rachis* glabrous, *peduncles* very short, crowded, branched, 3—5-flowered. *Glumes* nearly equal, compressed, scarcely connate, hairy and covered with minute pale glandular dots, keeled and ciliated with long soft hairs, obtuse, with a green rib on each side. *Paleæ* folded, shorter than the glume, abrupt, irregularly notched, smooth, minutely glandular, 3-ribbed. *Awn* from below the middle near to the base, twice the length of the palea, somewhat geniculate, contorted, striated, pale and smooth in the portion included by the glume, purplish and rough with erect bristles above. *Anthers* violet.

3. A. *agrestis,* Linn. *Slender Fox-tail-grass.* Culm erect, striated, roughish; panicle spicæform, cylindrical, acuminate; glumes lanceolate, acute, scabrous, connate as far as to the middle, keels dilated, ciliato-scabrous; palea a little shorter than the glumes, slightly obtuse, awn inserted near to the base of the palea, and twice its length. *E. Bot. t.* 848. *E. Fl. v.* 1. *p.* 81. *Hook. Br. Fl. p.* 33.

Fields and way-sides; not common. *Fl.* June, July. ⊙.

Oakley Park, near Ludlow; *Mr. Spare.* Field a little south of Ludlow, on the Leominster road; *J. E. Bowman, Esq.*

Root fibrous. *Culm* 1—2 feet high, erect, sometimes geniculate in the lower joints, smooth below, roughish above, leafy. *Leaves* rather narrow, striated and scabrous on the upper surface. *Sheaths* as long as the leaves, smooth except on the mid-rib, deeply furrowed, slightly ventricose, entirely divided. *Panicle* cylindrical, slender, long. *Florets* loosely imbricated, as large or larger than those of *A. pratensis, rachis* roughish, *peduncles* very short, simple, bearing generally a solitary floret. *Glumes* compressed, connate as far as to the middle, covered with minute, pale, scabrous points, sharply keeled, keel dilated, rigid, bristly in the upper part, ciliated with soft hairs below, acute, with a prominent rigid green rib on each side, scabrous or bristly above, hairy below, margins serrulate. *Paleæ* folded, scarcely shorter than the glumes, acute, smooth, 5-ribbed, 3 on one side the keel, and 2 on the other, slightly scabrous above. *Awn* from below the middle near to the base, twice the length of the palea, somewhat geniculate, smooth, contorted and striated in the included part, rough with erect bristles above. *Anthers* yellow.

27. BRIZA. *Linn.* Quaking-grass.

1. B. *media,* Linn. *Common Quaking-grass.* Spikelets broadly ovate, about 7-flowered; glumes shorter than the paleæ; ligule short, truncate. *E. Bot. t.* 340. *E. Fl. v.* 1. *p.* 133. *Hook. Br. Fl. p.* 47.

Meadows and pastures; frequent. *Fl.* June. ♃.

Root fibrous, tufted. *Culms* erect, slender, smooth, jointed, chiefly leafy at the base. *Leaves* short, linear-acuminate, with rough ribs and edges. *Sheaths* long, ribbed, smooth. *Ligule* short, truncate. *Panicle* very much branched, slender, *peduncles* divaricating in pairs, angular and roughish, purple, *pedicels* capillary and flexuose. *Spikelets* tremulous, very smooth, shining, purple on the back, whitish green at the edges. *Glumes* very concave, ovate, contracted at the base, slightly compressed, 3-nerved. *Inferior palea* similar to the glumes, compressed, and membranous at the margins, cordate or auricled at the base, with 3 strong principal ribs, and a smaller intermediate one on each side the mid-rib, the lateral ribs branching into three; *superior valve* obovate, flat, 2-ribbed, apex retuse, edges and apex finely ciliated.

28. MELICA. *Linn.* Melic-grass.

1. M. *nutans*, Linn. *Mountain Melic-grass.* Panicle nearly simple, racemed, secund; spikelets drooping, ovate, with 2 perfect flowers. *E. Bot. t.* 1059. *E. Fl. v. i. p.* 113. *Hook. Br. Fl. p.* 41.

Woods; not common. *Fl.* May, June. ♃.

In the Shropshire part of Wyre Forest, near Bewdley; *E. Lees, Esq.! Rev. A. Bloxam!*

Root creeping. *Culm* 1—2 feet high, erect, simple, jointed, tetragonous, furrowed, scabrous, leafy. *Leaves* linear-lanceolate, with fine rough ribs and edges. *Sheaths* nearly as long as the leaves, with strong scabrous ribs. *Ligule* very short. *Flowers* elegant, larger and more pendulous than in the next species. *Glumes* ovate, obtuse, convex, keeled, with 2 nerves on each side, deep purple-brown, margin pale, minutely glandular. *Paleæ* cartilaginous, unequal, *inferior valve* large, strongly and numerously nerved, covered with short glandular bristly pubescence; *superior* flat, with 2 marginal pubescent nerves and incurved membranous edges.

Between the two perfect florets is a pedicel, bearing a single involute membranous valve enclosing another pedicel bearing a similar but smaller valve, which again envelopes a still smaller pedicel and valve, without pistil or stamens in any of them.

2. M. *uniflora*, Linn. *Wood Melic-grass.* Panicle branched, slightly drooping; spikelets erect, ovate, with only one perfect floret. *E. Bot. t.* 1058. *E. Fl. v. i. p.* 112. *Hook. Br. Fl. p.* 41.

Shady woods; not unfrequent. *Fl.* May, June. ♃.

Welbach Coppice, near Shrewsbury; *Mr. T. Bodenham!* Benthal Edge; *Dr. G. Lloyd.* Near Ludlow; *Plymley.* Foot of the High Rock, Bridgnorth; *Rev. W. R. Crotch.* Woods, Coalbrookdale; *Mr. F. Dickinson.*

Near Cheney Longville. Ebury, Bayston Hill, near Shrewsbury.

Root creeping. *Culm* slender, unbranched, jointed, leafy, angular, smooth. *Leaves* linear-lanceolate, broader than in the last species, upper surface hairy, under one finely ribbed, ribs and edges rough. *Sheaths* shorter than the leaves, ribbed, slightly hairy, close and tight, entirely undivided. *Ligule* short, truncate, usually elongated into a point opposite to the leaf. *Panicle* drooping to one side, branches capillary, angular, rough. *Flowers* upright, tremulous, elegant, variegated with green white and reddish brown. *Glumes* unequal, ovate, acute, covered with a slight glandular pubescence, with a rough keel and 2 ribs on either side, *inferior* glume with one rib only on each side. *Paleæ* unequal, cartilaginous, clothed with minute glandular points, *inferior valve* larger, strongly nerved, *superior* flat, with two marginal densely downy nerves and incurved edges. *Imperfect florets* as in the last species.

29. GLYCERIA. *Br.* Sweet-grass.

1. G. *fluitans*, Br. *Floating Sweet-grass.* Panicle sub-secund, slightly branched; spikelets linear, appressed; florets 7—11, obtuse, 7-ribbed, with short intermediate ribs at the base; root creeping. *E. Fl. v.* 1. *p.* 117. *Poa fluitans, Scop. E. Bot. t.* 1520. *Hook. Br. Fl. p.* 43. *Festuca, Linn.*

In stagnant waters and slow streams; frequent. *Fl.* June—August. ♃.

Field near the Tanyard, Much Wenlock; *W. P. Brookes, Esq.* Hadnall; *Mr. E. Elsmere, jun.!* Coalbrookdale; *Mr. F. Dickinson.*

Berrington and Hancott pools; pit near the Flash, all near Shrewsbury.

Root long and creeping, or partly floating. *Culm* ascending, round, striated, leafy, smooth, hollow, partly decumbent on the surface of the water, as are many of the long linear-lanceolate, acute, flat *leaves*, which are ribbed and smooth except at the edges and mid-rib. *Sheaths* long, compressed, smooth. *Ligule* oblong, pointed, often torn, decumbent. *Panicle* sub-secund, very long and slender, sparingly branched, branches roughish, nearly erect, divaricated in flower. *Glumes* very unequal, small, membranous, ovate, obtuse, smooth, with a green keel. *Paleæ* ovato-oblong, thrice as long as the glumes, *inferior valve* cylindrical, obtuse, scabrous and bristly, membranous and notched at the summit, *superior valve* flat, bifid, with 2 marginal ribs and inflexed edges. *Scale* of 1 thick piece.

2. G. *spectabilis*, Mert. & Koch. *Reed Sweet-grass.* Panicle erect, very much branched, spreading; spikelets linear; florets 5—9, obtuse, 7-ribbed; root creeping. *Poa aquatica, Linn. E. Bot. t.* 1315. *Hook. Br. Fl. p.* 43. *Glyceria aquatica, Sm. E. Fl. v.* 1. *p.* 116.

Sides of rivers, ponds and ditches; not very common. *Fl.* July, August. ♃.

Oakley Park, near Ludlow; *Mr. H. Spare.*

Sides of Canal near Uffington.

Root jointed and creeping. *Culms* 3—6 feet high, erect, striated, smooth and leafy below, naked and rough above, slightly compressed. *Leaves* long, linear-lanceolate, broad, flat, single-ribbed, lower surface and edges rough. *Sheaths* close, smooth, and striated. *Ligule* obtuse, torn. *Panicle* nearly erect, 6—12 inches long, branches semiverticillate, repeatedly branched and spreading. *Glumes* unequal, membranous, smooth, *inferior valve* 1—, *superior* 3-ribbed. *Paleæ* obtuse, *inferior valve* 7-ribbed, covered with minute scabrous pubescence, *superior one* bifid at the apex, with 2 lateral ribs and membranous inflexed margins.

30. CATABROSA. *Beauv.* Whorl-grass.

1. C. *aquatica*, Beauv. *Water Whorl-grass.* Panicle erect, with whorled patent branches; spikelets linear; florets 2, obtuse, 3-ribbed; root creeping. *Aira aquatica, Linn. E. Bot. t.* 155. *E. Fl. v. i. p.* 102. *Catabrosa aquatica, Beauv. Hook. Br. Fl. p.* 39.

Banks of rivers, and pools of water. *Fl.* May, June. ♃.

Oakley Park, near Ludlow; *Mr. Spare.*

Root of whorled fibres proceeding from the joints of the long, branched, floating stems. *Culms* 1 foot or more in length, stout, smooth, striated and leafy. *Leaves* linear, short, broad, flat, obtuse, smooth. *Sheaths* loose, smooth, striated. *Ligule* broad, acute. *Panicle* erect, large, branched, smooth. *Glumes* thin and membranous, scarcely-ribbed, broadly oval, obtuse. *Paleæ* of a thick texture, brownish-green, 3-ribbed, white and membranous at the blunt erose extremity.

31. POA. *Linn.* Meadow-grass.

1. P. *annua*, Linn. *Annual Meadow-grass.* Panicle subsecund, divaricated; spikelets oblongo-ovate, 3—5-flowered; florets a little remote, 5-ribbed, destitute of web; culm ascending, compressed; root fibrous. *E. Bot. t.* 1141. *E. Fl. v.* 1. *p.* 127. *Hook. Br. Fl. p.* 46.

Meadows, pastures, road-sides, waste and cultivated ground; every where. *Fl.* at almost all seasons. ⊙.

Root fibrous. *Culms* numerous, pale, very smooth, compressed, leafy, jointed, branched and prostrate at the base, throwing out roots from the lower joints. *Leaves* distichous, linear, abruptly pointed, keeled, smooth except at the edges and keel near the point, flaccid, wrinkled above the middle. *Sheaths* long, compressed, smooth, divided about half-way. *Ligule* oblong, acute, in the lower leaves shorter and more obtuse. *Glumes* very unequal, shorter than the florets, ovato-lanceolate, rough at the back, membranous at the edges, larger one 3-nerved. *Inferior palea* ovato-lanceolate, acute, white and diaphanous at the margin, 5-ribbed, slightly hairy at the base, *superior* one flat, deeply bifid, 2-ribbed, ribs hairy, edges incurved.

A small peduncle extends beyond the upper floret.

Var. *β. villosa.*

Equally common, and flowering at the same time.

This only differs in the culms being more prostrate and spreading, throwing out fibres from the lower joints, and in the ribs of the superior valve of the paleæ being copiously covered with long soft white hairs.

2. P. *compressa*, Linn. *Flat-stemmed Meadow-grass.* Panicle secund, coarctate; spikelets ovato-oblong; florets 5—7, obtuse, connected by a web; culm compressed, decumbent at the base; root creeping. *E. Bot. t.* 365. *E. Fl. v. i. p.* 121. *Hook. Br. Fl. p.* 45.

On walls and in dry barren ground; not common. *Fl.* June, July. ♃.

On a wall at Wall-under-Haywood. Right-hand side of the road leading from Longville in the Dale to Wenlock Edge. Wenlock Abbey.

Root creeping. *Culm* 12—18 inches high, decumbent at the base, ascending, compressed, somewhat rigid, smooth, leafy. *Leaves* short, linear, acute, roughish on the upper surface. *Sheaths* as long as the leaves, compressed, smooth, divided for about three-fourths of their length, close at the base. *Ligule* short, truncate. *Panicle* erect, 1—2 inches long, coarctate, secund, branches few, short, angular and rough. *Glumes* nearly equal, ovate, acute, 3-ribbed, mid-rib rough, covered with minute glandular dots. *Inferior palea* ovate, obtuse, 4-ribbed, the keel and exterior lateral ribs furnished with a soft fine white web. *Superior valve* bifid at the apex, with 2 lateral roughish ribs and membranous inflexed margins.

3. P. *nemoralis*, Linn. *Wood Meadow-grass.* Panicle slender, slightly leaning one way, lax, attenuate; spikelets ovato-lanceolate, about 3-flowered; florets rather distant, slightly webbed; ligule short, truncate; culms sub-compressed; sheaths glabrous; root scarcely creeping. *E. Bot. t.* 1265. *E. Fl. v.* 1. *p.* 129. *Hook. Br. Fl. p.* 46.

Woods and thickets; not very common. *Fl.* June, July. ♃.

Benthal Edge; *Dr. G. Lloyd.*

Leaton Shelf, near Shrewsbury.

Root fibrous, scarcely creeping. *Whole plant* very slender and elegant. *Culms* several, erect, slightly compressed, smooth, striated, leafy, jointed. *Leaves* chiefly on the stem, narrow, linear, acute, with 3 principal nerves more or less rough. *Sheaths* entirely divided. *Panicle* half-whorled, branches almost erecto-patent, angular, wavy and rough. *Spikelets* 3-flowered, with the peduncle of a 4th floret. *Florets* with a hairy peduncle and a slightly connecting web. *Glumes* lanceolate, with a rough keel and a single rib on each side. *Paleæ* when seen under a magnifier, resinoso-glandulose. *Inferior valve* lanceolate, acute, 5-ribbed, pubescent on the keel and exterior lateral ribs, intermediate ones smooth and inconspicuous, *superior* bifid, with rough lateral ribs. *Scales* 2, acute, cloven.

4. P. *trivialis*, Linn. *Roughish Meadow-grass.* Panicle diffuse; spikelets oblongo-ovate, about 3-flowered; flowers acute, 5-nerved, connected by a web; culm, leaves, and sheaths roughish; ligule long, oblong, acute; root fibrous. *E. Bot. t.* 1072. *E. Fl. v. i. p.* 124. *Hook. Br. Fl. p.* 46.

Meadows and pastures; common. *Fl.* June, July. ♃.

Root tufted, fibrous. *Culms* several, erect, leafy, jointed, round, furrowed, roughish. *Leaves* linear, acute, rough. *Sheaths* as long as the leaves, furrowed, rough, entirely divided. *Ligule* acute, oblong, 1½ or 2 lines long. *Panicle* large, spreading, with half-whorled, horizontal, wavy, angular, rough, compound but very unequal branches. *Spikelets* sometimes only 2-flowered with the peduncle of a third, *lower florets* sessile, *upper one* pedunculate, with a few long very slender convoluted filaments proceeding from their bases and connecting them with the receptacle and with each other. *Glumes* lanceolate, acute, with rough keels, the larger with a single rough rib on each side of the keel. *Paleæ* lanceolate, acute, *inferior valve* concave, slightly compressed, keeled, smooth, with 2 lateral ribs at each side, membranous at the point, covered with minute resinoso-glandulose dots when viewed under a powerful lens, *superior valve* flat, bifid, with 2 rough marginal ribs and incurved edges.

5. P. *pratensis*, Linn. *Smooth-stalked Meadow-grass.* Panicle diffuse; spikelets oblongo-ovate, about 4-flowered; florets acute, 5-nerved, connected by a web; culm, leaves, and sheaths smooth; ligule short, truncate; root creeping. *E. Bot. t.* 1078. *E. Fl. v.* 1. *p.* 125. *Hook. Br. Fl. p.* 46.

Meadows and pastures; common. *Fl.* June, July. ♃.

Roots creeping with horizontal runners. *Whole plant* similar in general appearance to *P. trivialis*, though the culm, sheaths and leaves are smooth, and the florets broader, more ovate, obtuse, and membranous at the edges, with a more copious connecting web.

The most constant and invariable mark of distinction between *Poa trivialis* and *P. pratensis* exists in the form and size of the ligule; that of the former being oblong, acute and long, whilst that of the latter is very short, abrupt, and truncate.

32. SCLEROCHLOA. *Br.* Hard-grass.

1. S. *rigida*, Panzer. *Rigid Hard-grass.* Panicle lanceolate, disticho-secund, rigid; spikelets linear, acute; florets 5—11; inferior palea obsoletely 5-nerved, obtuse, subemarginate, and submucronate at the apex; root fibrous. *Poa rigida, Linn. E. Bot. t.* 1371. *Hook Br. Fl. p.* 44. *Glyceria rigida, Sm. E. Fl. v. i. p.* 120.

Dry barren soils chiefly in the limestone districts. *Fl.* June. ⊙.

Lincoln's Hill, Ironbridge; *Mr. F. Dickinson.* Benthal Edge; *Dr. G. Lloyd.* Near the Mill-pool, Lilleshall; *Withering.*

Whitecliff Coppice, near Ludlow, and all along the limestone ridge stretching from Ludlow to Wenlock.

Root fibrous, woolly. *Culm,* as is the whole plant, very rigid and wavy, slender, erect, 3—5 inches long, jointed, leafy. *Leaves* short, rigid, linear, setaceous, rough on the upper surface, more or less involute at the margins. *Sheaths* nearly as long as the leaves, compressed and keeled, strongly ribbed, smooth, divided for more than three-fourths of their length. *Ligule* oblong, jagged. *Rachis* angled, sometimes at once bearing the spikelets, but more usually throwing out branches. *Glumes* unequal, lanceolate, smaller 1—, larger 3-ribbed. *Paleæ* scabrous, *inferior valve* obsoletely 5-ribbed, the mid-rib or keel scabrous near the obtuse slightly emarginate apex, and terminating in a very minute mucro, membranous at the edges and extremity, *superior valve* flat, bifid, rough at the 2 lateral ribs, edges incurved.

33. MILIUM. *Linn.* Millet-grass.

1. M. *effusum*, Linn. *Spreading Millet-grass.* Panicle spreading, glabrous; branches subverticillate; leaves broad, lanceolate; ligule obtuse. *E. Bot. t.* 1106. *E. Fl. v.* 1. *p.* 87. *Hook. Br. Fl. p.* 36.

Moist shady places; not common. *Fl.* June. ♃.

Woods about Madeley; *W. P. Brookes, Esq.* Welbatch Coppice, near Shrewsbury; *Mr. T. Bodenham.!* Benthal Edge; *Dr. G. Lloyd.*

Borders of Hancott Pool, near Shrewsbury.

Root fibrous. *Culm* tall, erect, round, jointed, leafy, smooth. *Leaves* with a few scattered hairs on the upper surface, rough beneath, edges finely serrated. *Sheaths* long, strongly ribbed, smooth, divided nearly to their base. *Glumes* green, ovate, 3-ribbed, mid-rib and edges serrated, covered with pale glandular protuberances. *Paleæ* smooth and shining, nearly as large as and opposite the glumes, *inferior valve* obsoletely 3-ribbed, embracing the *superior* one which is emarginate and 2-ribbed, at length forming a hard shining coat to the seed.

34. MOLINIA. *Mœnch.* Molinia.

1. M. *cærulea*, Mœnch. *Purple Molinia.* Panicle erect, subcoarctate; spikelets erect, oblongo-cylindrical, with 2 perfect flowers; florets much longer than the calyx; ligule a tuft of hairs; stem nearly naked. *Melica cærulea, Linn. E. Bot. t.* 750. *E. Fl. v. i. p.* 113. *Hook. Br. Fl. p.* 41.

On the turfy, boggy margins of pools and wet moors; not common. *Fl.* August ♃.

Shawbury Heath; *Dr. T. W. Wilson!*

Bomere Pool, near Shrewsbury.

Root woody, with long stout fibres. *Culm* somewhat bulbous at the base, erect, 2—3 feet high, slightly compressed, finely striated and smooth, with one tumid joint a few inches above the base, from whence and from the base all the leaves proceed. *Leaves* long, linear, sharply accuminated, hairy on the upper surface, rough at the edges and ribs beneath. *Sheaths* long, striated, smooth, entirely divided. *Ligule* a tuft of hairs. *Panicle* dark purple, 2—8 inches long, branches numerous, compressed, short, angular and rough. *Lower floret* sessile, *second floret* elevated on a round pubescent pedicel with a small tuft of longer hairs at its summit. The *peduncle* is elongated beyond the second floret, and terminates in the rudiments of a *third floret*. *Glumes* lanceolate, acute, with rough keels. *Paleæ* slightly scabrous, keeled, with a nerve on each side, *inferior valve* lanceolate, acute, 3-ribbed, *superior* flat, ovate, obtuse, with 2 rough marginal ribs, edges incurved. *Anthers* large and purple.

35. BALDINGERA. *Gærtn.* Band-grass.

1. B. *arundinacea*, Dumort. *Reed Band-grass.* Panicle erect, branched, branches patent; flowers clustered, secund; imperfect florets small, hairy, ciliated valves. *Phalaris arundinacea, Linn. E. Bot. t.* 402, *and t.* 2160 *f.* 2. *E. Fl. v. i. p.* 75. *Hook. Br. Fl. p.* 34.

Sides of rivers, ponds and pools; common. *Fl.* July, August. ♃.

Welbach, near Shrewsbury; *Mr. T. Bodenham!* Hancott Pool, near Shrewsbury; *Mr. F. Dickinson.*

Banks of River Severn.

Root extensively creeping. *Culm* erect, reedy, jointed, leafy, hollow, smooth. *Leaves* flat, broad, lanceolate, taper-pointed, finely serrulate at the edges. *Sheaths* long, striated, smooth, entirely divided. *Ligule* short, bluntish, decurrent. *Panicle* 6—8 inches long, green or purplish, branches spreading, angular, rough. *Glumes* lanceolate, acute, covered with glandular protuberances, keeled, with a strong rib on each side, keels and edges finely serrated. *Paleæ* hairy, at length indurated smooth and shining, margins ciliated, valves unequal, *inferior* one the larger, ovate, acute, keeled, obscurely ribbed, *superior* one smaller, lanceolate, keeled. *Valve* of the imperfect florets very minute, linear, hairy, margins ciliated with hairs longer than the valves.

A variety with variegated leaves called "Ribband-grass" is frequent in gardens.

36. PHRAGMITES. *Trin.* Reed-grass.

1. Ph. *communis*, Trin. *Common Reed-grass.* Panicle diffuse, drooping, very much branched; stem and leaves glabrous; ligule a tuft of short hairs; root creeping. *Arundo Phragmites, Linn. E. Bot. t.* 401. *E. Fl. v. i. p.* 169. *Hook. Br. Fl. p.* 54.

Margins of pools; not unfrequent. *Fl.* July, August. ♃.

Croesmere, near Ellesmere; *Mr. F. Dickinson.* Bromfield, near Ludlow; *Miss Mc Ghie.* Oakley Park, near Ludlow; *Mr. Spare.* Wildmoors near Eyton; *T. C. Eyton, Esq.*

Berrington Pool, near Shrewsbury.

Root creeping. *Culm* erect, 6 feet or more in height, stout, round, smooth, striated and leafy. *Leaves* long, broad, lanceolate, very much acuminate, striated, smooth, somewhat glaucous, edges rough. *Sheaths* long, close, smooth, striated, entirely divided. *Ligule* a tuft of short hairs. *Panicle* very large, repeatedly compound, purple-brown, at length drooping, very graceful, branches semi-verticillate, much divided, angular and smooth, lower ones with a tuft of pale shining hairs at the base. *Florets* about 5, close, afterwards spreading, each on a short pedicel, lower one with stamens only and naked, upper ones fertile, their pedicels clothed with white shining silky hairs, which subsequently become elongated and spreading. *Glumes* very unequal, lanceolate, *inferior valve* 3-ribbed, *superior* twice its length, acuminate, obsoletely 3-ribbed. *Paleæ* unequal, *inferior valve* brownish-purple, very long and acuminate, 3-ribbed, twice or thrice the length of the *superior* one, which is pale and membranous, bifid at the apex, and with 2 rough lateral ribs.

The largest of our grasses, and used for the purposes of thatching, and reed-fences or protections. Reed-pens for sketching are also constructed from its culms. The spreading and creeping roots are of considerable service in securing the banks of streams from being washed away by the running water.

37. DACTYLIS. *Linn.* Cock's-foot-grass.

1. D. *glomerata*, Linn. *Rough Cock's-foot-grass.* Panicle distantly branched, flowers in densely crowded clusters, secund; spikelets about 3-flowered; corolla acuminate, somewhat awned. *E. Bot. t.* 335. *E. Fl. v. i. p.* 134. *Hook. Br. Fl. p.* 48.

Way-sides, meadows and hedges; abundant. *Fl.* June—August. ♃.

Root fibrous, tufted. *Culm* erect, round, jointed, ribbed, smooth, leafy chiefly in the lower part. *Leaves* broadly linear, flat, acuminate, ribbed, rough

and harsh, spreading. *Sheaths* long, ribbed, compressed, keeled, rough, divided nearly to the base. *Ligule* elongated, laciniate. *Panicle* alternately branched, branches angular, stiff, rough and spreading, the lower one longer more remote and spreading, each bearing a dense ovate cluster of 3 or 4-flowered spikelets. *Glumes* unequal, membranous, convex and broad on the one side, depressed and narrow on the other, shorter than the florets, lanceolate, acuminate, glandular, smooth except the more or less oblique scabrous keel. *Paleæ* sub-cartilaginous, *inferior valve* compressed, lanceolate, scabrous, 5-ribbed, keel ciliated, with a short rough *awn* at the point, *superior valve* bifid, with ciliated ribs and incurved edges.

38. FESTUCA. *Linn.* Fescue-grass.

1. F. *ovina*, Linn. *Sheep's Fescue-grass.* Panicle subcoarctate, erect, spikelets oblong, of about 4 or 5 flowers with short awns; culms square upwards; leaves setaceous. *E. Bot. t.* 585. *E. Fl. v. i. p.* 139. *Hook. Br. Fl. p.* 48.

Dry hilly pastures; frequent. *Fl.* June. ♃.

Root of many long black capillary fibres. *Culms* erect, slender, rigid, smooth, leafy below, square above. *Leaves* chiefly radical, very numerous, composing dense tufts, margins involute so as to appear setaceous, mostly short, smooth or slightly scabrous. *Sheaths* angular or furrowed, entirely divided. *Ligule* very short, with more or less of a polished knot-like tubercle on each side of the top of the sheath. *Panicle* small, close, slightly branched. *Branches* angular and rough. *Glumes* shorter than the paleæ, unequal, subglabrous, keeled, larger 3-ribbed, smaller single-ribbed. *Inferior palea* somewhat scabrous, more or less pubescent upwards, terminating in a short rough awn, 5-ribbed, edges slightly ciliated near the base, *superior* resinoso-glandulose, bifid, ribs somewhat scabrous, edges incurved. *Awn* variable in length, but not longer than half the length of the valve.

2. F. *duriuscula*, Linn. *Hard Fescue-grass.* Panicle spreading, spikelets oblong of about 6 florets with short awns; culms round; stem leaves nearly plane, radical ones subsetaceous; root fibrous. *E. Bot. t.* 470. *E. Fl. v. i. p.* 141. *Hook. Br. Fl. p.* 49.

Pastures and waste grounds; not common. *Fl.* June, July. ♃.

Lincoln's Hill, near Ironbridge; *Mr. F. Dickinson.*

Fields, at Cross Hill and in the neighbourhood of Shrewsbury.

Similar to the last, but of a larger size and stouter habit. *Root* tufted, woody, fibrous, scarcely creeping. *Culm* erect, round, jointed, striated, smooth, leafy. *Leaves* slender, rigid, acute, compressed, roughish at the edges and keel, upper ones broader and flatter, occasionally convolute. *Sheaths* close, smooth or slightly downy and furrowed, entirely divided. *Ligule* very short and cloven, auricled with a small tubercle. *Panicle* oblong, spreading in flower, branches angular, rough. *Glumes* much shorter than the paleæ, very unequal, lanceolate, acute, larger one 3-ribbed, smaller single-ribbed. *Paleæ* lanceolate, acute, *inferior valve* generally smooth, more or less pubescent (*β. Smith.*) along the margins, 5-ribbed, terminating in a short straight *awn*, *superior valve* resinoso-glandulose, pubescent near the apex, bifid, with 2 rough marginal ribs and incurved edges.

3. F. *pratensis*, Huds. *Meadow Fescue-grass.* Panicle patent, branched, spikelets linear, 5—10-flowered, florets cylindrical, awnless, sometimes awned; leaves linear; root scarcely creeping. *E. Bot. t.* 1592. *E. Fl. v.* 1. *p.* 148. *Hook. Br. Fl. p.* 50.

Moist meadows and pastures; not common. *Fl.* June, July. ♃.

Welbatch, near Shrewsbury; *Mr. T. Bodenham!* Hadnall, near Shrewsbury; Mr. *E. Elsmere, junr.!* Near Ludlow; *Plym. Agric. Surv.*

Near Lower Berwick and Shelton Rough, near Shrewsbury.

Root fibrous, tufted, scarcely creeping, fibres more or less downy. *Culms* several, erect, about 2 feet high, simple, leafy, round, striated, very smooth, bent at the lowest joint only. *Leaves* linear, acute, spreading, flat, striated, rough on the edges and upper surface. *Sheaths* striated, very smooth, entirely divided. *Ligule* short, obtuse, laciniate, decurrent, clasping the stem. *Panicle* nearly erect, simply or doubly branched, branches solitary or in pairs, unequal, subsecund, with compressed rough stalks, closed together after flowering. *Spikelets* linear, 5—10-flowered, with a rudimentary peduncle. *Glumes* unequal, lanceolate, acute, larger one keeled, smooth, edges membranous, 3 or 5 ribbed, smaller single ribbed. *Inferior palea* cylindrical, obtuse, keeled, smooth except the keel, 5-ribbed, the mid-rib occasionally elongated into a rough *awn*, in which case the membranous apex of the corolla is cloven and the awn apparently proceeds from a little below the cleft, *superior valve* bifid, with two rough marginal ribs and membranous incurved edges.

4. F. *arundinacea*, Schreb. *Tall Fescue-grass.* Panicle patent, very much branched, spikelets ovato-lanceolate, 4—5-flowered, florets cylindrical, subaristate; leaves linear-lanceolate; root creeping. *F. elatior, Linn. E. Bot. t.* 1593. *E. Fl. v. i. p.* 148. *Hook Br. Fl. p.* 50.

Moist meadows, banks of rivers, &c.; rare. *Fl.* June, July. ♃.

Near Ludlow; *Plym. Agric. Surv.*

Banks of River Severn, near Shrewsbury.

Root somewhat creeping, fibres downy. *Culms* erect, 4 feet high, reedy, striated, smooth, leafy. *Leaves* linear-lanceolate, smooth except the edges. *Sheaths* very long, striated, smooth, entirely divided. *Ligule* short, obtuse, laciniate. *Panicle* about 1 foot in length, repeatedly compound, spreading in every direction. *Spikelets* ovato-lanceolate, slightly compressed, more scabrous than in the last species, 4—5-flowered, with a rudimentary peduncle. *Glumes* lanceolate, *inferior* one very acute, 3-ribbed, rough, *superior* smaller, single-ribbed, rough. *Inferior palea* ovato-lanceolate, acute, scabrous, especially towards the apex, 5-ribbed, ribs rough, the mid-rib occasionally elongated into a rough *awn* proceeding from a little below the cloven membranous apex of the valve, *superior valve* bifid, with 2 rough marginal ribs.

After a careful comparison of the two preceding very closely allied species, I have been unable to discover any satisfactory characteristic distinctions, and am induced to consider them only as modifications of the same plant. The chief differences assuredly consist in the larger size of *F. arundinacea*, the ovato-lanceolate not linear form of the spikelets, and the number of the florets, yet all these may possibly depend on soil and situation producing a more vigorous growth and a suppression of developement in the florets. No dependance can be placed on the awns of the corolla since they are equally present or absent in both plants, nor on the roots, which are in both somewhat creeping, and the fibres downy. In deference, however, to the authority of our British Floras, they are here retained as separate species.

5. F. *gigantea*, Vill. *Tall Fescue-grass.* Panicle branched, drooping towards one side, spikelets ovato-lanceolate, compressed, 3—6-flowered, florets shorter than the awn; leaves linear-lanceolate, ribbed. *E. Bot. t.* 1820. *E. Fl. v.* 1. *p.* 144. *Bromus giganteus, Linn. Hook. Br. Fl. p.* 50.

Shady woods; not common. *Fl.* July, August. ♃.

Near Welbatch; *Mr. T. Bodenham!*
Shelton Wood, near Shrewsbury.

Root tufted, of many strong partly woolly fibres. *Culms* 3 or 4 feet high, erect, simple, leafy, round, striated, jointed, smooth. *Leaves* nearly upright, long, lanceolate, taper-pointed, broad, flat, with a strong pale mid-rib and numerous roughish lateral ones, interstices striated, edges rough. *Sheaths* striated, smooth, upper ones longer than their leaves, lower shorter, entirely divided. *Ligule* very short, brown or purplish, often jagged, with an acute clasping auricle at each side. *Panicle* a foot long, branches twice compound, angular and rough. *Spikelets* alternate, drooping, 3—6-flowered with the rudiments of another floret. *Glumes* very unequal, lanceolate, very acute, larger one 3-ribbed. *Inferior palea* lanceolate, scabrous, covered with minute resinoso-glandulose points in longitudinal rows, 5-ribbed, roughish, apex membranous, bifid, the mid-rib extended into a very long rough *awn* proceeding from a little below the bifid point, *superior valve* finely downy at the ribs, bifid.

39. VULPIA. *Gmel.* Vulpia.

1. V. *Myurus,* Gmel. *Wall Vulpia.* Panicle elongated, racemose, subsecund, with short erect branches, spikelets about 5-flowered, flowers shorter than the long awn, monandrous; culm leafy in its upper part, the sheath of the uppermost leaf enveloping the base of the panicle. *Festuca Myurus, Linn. E. Bot. t.* 1412. *E. Fl. v. i. p.* 143. *Hook. Br. Fl. p.* 49.

Walls, wastes and barren places; not common. *Fl.* June. ⊙.

Oakley Park, near Ludlow; *Mr. Spare.* Road-side leading from Blymhill to Shrewsbury; *Withering.*

Walls at Ludlow. Long Lane Quarry, near Cheney Longville.

Root fibrous. *Culm* about 1 foot high, erect, slender, round, striated, smooth, jointed, leafy. *Leaves* narrow, linear, setaceous, convolute, springing even from the upper part of the culm. *Sheaths* long and smooth. *Ligule* short, obtuse. *Panicle* often 4—5 inches long, branches short, angular and rough, spikelets 4—5-flowered, florets pedunculate, peduncle scabrous on alternate sides. *Glumes* very unequal, linear-lanceolate, keeled, larger one very acuminate, about as long as the paleæ 3-ribbed, smaller narrower and shorter, single-ribbed. *Inferior palea* somewhat scabrous, lanceolate, acuminate, tapering into a rough straight *awn,* nearly thrice the length of the valve, 5-ribbed, edges serrated near the apex, *superior valve* flat, resinoso-glandulose, apex bifid, marginal ribs ciliated, edges incurved.

2. V. *bromoides,* Gmel. *Barren Vulpia.* Panicle secund, spicæform, coarctate, nearly simple, erect, spikelets 5—6-flowered, flowers shorter than the awn, monandrous; culm above leafless. *Festuca bromoides, Linn. E. Bot. t.* 1411. *E. Fl. v. i. p.* 142. *Hook. Br. Fl. p.* 49.

Dry hilly pastures and wastes; not common. *Fl.* June. ⊙. (♂. Schrader.)
Fields behind House of Industry, Ironbridge; *Mr. F. Dickinson.*

Lyth Hill and Sharpstones Hill, near Shrewsbury. Volcanic sandstone rock at the base of Caer Caradoc, between that hill and the Lawley. Long Lane Quarry, near Cheney Longville.

Much resembling the last, but smaller. *Root* fibrous. *Culms* several, 6—8 inches high, bent at the lowest joints, leafy in the lower part, naked erect angular and smooth above. *Leaves* short, linear, setaceous, convolute, shorter than the angular, furrowed, lax, smooth *sheaths.* *Ligule* very minute. *Panicle* simple, scarcely branched, spikelets few, 5—6-flowered. *Glumes* very unequal, lanceolate,

acuminate, larger one about the size of the paleæ, 3-nerved, smaller single-nerved. *Inferior palea* linear-lanceolate, acuminate, scabrous, tapering into a rough strait *awn* about twice the length of the valve, 5-nerved, edges serrated near the apex, *superior* one flat, resinoso-glandulose, bifid, marginal ribs ciliated, edges incurved.

40. DANTHONIA. *DC.* Danthonia.

1. D. *decumbens,* DC. *Decumbent Danthonia.* Panicle of few, racemed, close, erect, about 4-flowered spikelets, glumes as long as the florets; ligule a tuft of hairs; awn short, straight and plane. *Poa decumbens. E. Bot. t.* 792. *Triodia decumbens, Beauv. E. Fl. v. i. p.* 131. *Hook. Br. Fl. p.* 47. *Festuca decumbens, Linn.*

Dry hilly pastures. *Fl.* July. ♃.
Summit of the Wrekin; *Mr. F. Dickinson.*

Woods beyond Haughmond Hill. Moist places near Bomere pool, near Shrewsbury. Shawbury Heath.

Root slightly creeping, with strong fibres. *Culm* decumbent, erect in flower, harsh and rigid, jointed, leafy, striated, very smooth. *Leaves* linear, acuminate, striated, rather glaucous, slightly hairy, ribs and edges near the apex rough. *Sheaths* long, striated, hairy especially near the top, entirely divided. *Ligule* a tuft of hairs. *Branches* of *panicle* angular, wavy and roughish. *Glumes* nearly equal, lanceolate, acute, smooth, about 4-ribbed, with thin margins, keels scabrous. *Inferior palea* broadly ovate with a small tuft of hairs and an intermediate depression on each side at the base, and a row of hairs on the margins near the exterior rib extending about half-way up the valve, edges membranous finely ciliated near the apex, 9-ribbed, 3 principal ones extending into the toothed summit, with 2 intermediate shorter ones on each side the central rib, and one on the exterior of the lateral ones, apex bifid, segments obtuse, ciliated, *awn* short, about equal to the segments, *superior valve* bifid with 2 ciliated marginal ribs and incurved edges.

41. BROMUS. *Linn.* Brome-grass.

1. B. *secalinus,* Linn. *Smooth Rye-Brome-grass.* Panicle spreading, peduncles slightly branched, spikelets oblongo-ovate, compressed, of about 10 or 12 subcylindrical glabrous rather remote florets, longer than the awn; sheaths nearly glabrous. *E. Bot. t.* 1171. *E. Fl. v. i. p.* 151. *Hook. Br. Fl. p.* 51.

Corn fields; not common. *Fl.* July, August. ⊙.

Hadnall; *Mr. E. Elsmere, junr.!* Oakley Park, near Ludlow; *Mr. Spare.* Fields near Stanwardine; *J. E. Bowman, Esq.! Dr. T. W. Wilson!*

Corn fields near Sharpstones hill.

Root fibrous, downy. *Culm* 2—3 feet high, round, smooth, striated, leafy, with downy joints. *Leaves* linear, pointed, flat, with many minutely hairy ribs, hairs longer on the upper side and edges. *Sheaths* striated, smooth, occasionally a little downy, close, very slightly divided at the summit only. *Ligule* very short, obtuse, torn. *Panicle* erect, lower branches whorled and slightly compound, upper one alternate and simple, all angular and rough. *Spikelets* pendulous in seed. *Florets* imbricated in flower, in seed spreading and exhibiting their distant mode of insertion. *Glumes* unequal, larger one broadly ovate, with 9 minutely scabrous ribs, smaller with 5 ribs. *Paleæ* unequal, scabrous, with minutely glandular points, *inferior valve* broadly ovato-oblong, obtuse, edges membranous, 9-ribbed, each rib with a double row of scabrous points, apex obtuse, bifid, *awn* shorter than the valve, rough and wavy, *superior valve* bifid, with 2 strong scabrous

marginal ribs, strongly toothed or fringed with distant bristles. *Styles* from opposite sides of the germen below the top. *Seed* elliptic-oblong, downy at the summit, convex and loose at the back, channelled in front, and attached to the inner valve of the corolla.

2. B. *racemosus,* Linn. *Smooth Brome-grass.* Panicle erect, peduncles simple, spikelets ovate, subcompressed, glabrous, florets imbricated, compressed; awn straight, about as long as the glume; leaves and sheaths slightly hairy. *E. Bot. t.* 1079. *E. Fl. v. i. p.* 154. *Hook. Br. Fl. p.* 52.

Meadows and pastures; not common. *Fl.* June, July. ⊙. or ♂.

Hadnall, near Shrewsbury; *Mr. E. Elsmere, junr.!* Pastures near Severn House, Coalbrookdale; *Mr. F. Dickinson.*

Root fibrous. *Culm* erect, round, striated, leafy, smooth, or occasionally downy with minute deflexed pubescence, most conspicuous at and near the joints. *Leaves* linear, pointed, ribbed, hairy, most so on the upper surface. *Sheaths* ribbed, slightly divided at the summit only, clothed with more or less copious deflexed pubescence, intermixed with minute glandular points. *Ligule* short, obtuse, torn. *Branches* of the *panicle* simple, rarely divided, angular and rough, with erect bristles. *Spikelets* 8—10-flowered. *Glumes* unequal, acute, larger 7-ribbed, smaller 3-ribbed, ribs of both rough. *Paleæ* covered, as are the glumes, with minutely elevated glandular dots, *inferior valve* broadly ovato-lanceolate, bifid, keeled, 7-ribbed, rough on the keel, ribs, and towards the margins, edges membranous, ciliated, *superior* one flat, the 2 marginal ribs fringed with distant bristles, bifid. *Awn* rough from below the bifid summit of the inferior palea. *Styles* distant, *stigmas* feathery. *Seed* elliptical.

Scarcely distinct, and only to be distinguished from *B. mollis* by its glabrous spikelets, and the less divided branches of the panicle.

3. B. *mollis,* Linn. *Soft Brome-grass.* Panicle erect, close, compound, spikelets ovate, subcompressed, pubescent, florets imbricated, compressed; awn straight, about as long as the glume; leaves and sheaths very soft, pubescent. *E. Bot. t.* 1078. *E. Fl. v. i. p.* 153. *Hook. Br. Fl. p.* 52.

Meadows, pastures, banks, road-sides; every where. *Fl.* June. ♂.

Root fibrous. *Culm* 1—2 feet high, erect, round, striated, leafy, jointed, clothed with dense short soft deflexed pubescence. *Leaves* linear, pointed, ribbed, hairy, most abundantly so on the upper surface. *Sheaths* ribbed, slightly divided at the summit only, with copious dense deflexed soft pubescence, interspersed with glandular points. *Ligule* short, obtuse. *Branches* of *panicle* simple or divided, angular, with dense erect pubescence. *Spikelets* rather tumid, about 8-flowered. *Glumes* unequal, broadly ovato-lanceolate, acute, covered with dense erect appressed pubescence, larger one 9-ribbed, smaller 7-ribbed, edges ciliated. *Paleæ* covered, as are the glumes, with glandulose dots, very unequal, *inferior valve* broadly elliptical, pubescent like the glumes, about 11-ribbed, bifid, edges ciliated, *superior* one flat, the 2 marginal ribs fringed with distant bristles, apex obtuse, notched.

4. B. *asper,* Linn. *Hairy Wood Brome-grass.* Panicle branched, drooping, spikelets linear-lanceolate, compressed, florets remote, subcylindrical, hairy; awns straight, shorter than the florets; leaves uniform, lower ones and sheaths hairy. *E. Bot. t.* 1172. *E. Fl. v. i. p.* 158. *Hook. Br. Fl. p.* 51.

Moist woods; not unfrequent. *Fl.* June, July. ⊙. or ♂.
Woods, Coalbrookdale; *Mr. F. Dickinson.*

Sutton Wood, Shelton Rough, Bicton, Haughmond Hill, near Shrewsbury. Benthal Edge.

Root woody, fibres strong, somewhat woolly. *Culm* 4—6 feet high, erect, stout, round, smooth in the lower part, striated above and covered with a short dense rigid deflexed pubescence, jointed, leafy. *Leaves* spreading, lanceolate, acute, long, broad, flat, ribbed, very rough and scabrous. *Sheaths* long, ribbed, copiously hairy, with long deflexed rigid hairs, slightly divided at the summit only. *Ligule* short, obtuse, truncate. *Panicle* large, elegant and drooping, branches compound, spreading, angular and scabrous with rigid erect teeth. *Spikelets* about 8-flowered. *Florets* pedunculate, peduncles with erect pubescence. *Glumes* very unequal, larger one lanceolate, subaristate, covered with glandular dots, 3-ribbed, ribs strong and rough, smaller almost subulate, single-ribbed. *Paleæ* glandular, as are the glumes, unequal, *inferior valve* lanceolate, hairy, the margins 3-ribbed, ribs rough, apex bifid, awned, *superior* flat, bifid, with 2 bristly marginal ribs. *Awn* straight, rough, about half the length of the paleæ, proceeding from below the bifid apex. *Styles* distant, from below the summit of the germen, *stigmas* feathery. *Seed* linear, blunt and downy at the summit, channelled above.

5. B. *sterilis,* Linn. *Barren Brome-grass.* Panicle mostly simple, drooping, spikelets linear-lanceolate, compressed, florets remote, subcylindrical, scabrous, shorter than the straight awn; leaves and sheaths pubescent. *E. Bot. t.* 1030. *E. Fl. v. i. p.* 159. *Hook. Br. Fl. p.* 51.

Waste grounds, fields, hedges; common. *Fl.* June, July. ⊙.

Root small, fibrous, woolly. *Culm* 2 feet high, erect, rather slender, round, smooth, jointed, leafy. *Leaves* spreading, linear, acute, flaccid, narrow, flat, ribbed, clothed on both sides with soft spreading pubescence. *Sheaths* long, strongly ribbed, those of the lower leaves clothed with spreading and deflexed pubescence, upper ones smooth, slightly divided at the summit only. *Ligule* oblong, laciniate. *Panicle* about 6 inches long, drooping, branches mostly simple, angular and wavy, scabrous with erect teeth. *Spikelets* pendulous, about 7-flowered, peduncles of florets scabrous on alternate sides. *Glumes* unequal, scabrous, subulato-lanceolate, very acuminate, edges membranous, larger 3-ribbed, smaller single-ribbed. *Inferior palea* scabrous, lanceolate, very acuminate, with 7 rough ribs, the mid-rib elongated into a straight rough slender *awn,* more than twice as long as the valve, and proceeding from below the bifid apex, *superior valve* flat, bifid, the 2 marginal ribs fringed with distant bristles. *Scales* 2, lanceolate, acute. *Styles* lateral, from below the apex of the seed, *stigmas* feathery. *Seed* lanceolate, channelled on the upper side, blunt and downy at the summit.

6. B. *erectus,* Huds. *Upright Brome-grass.* Panicle erect, spikelets linear-lanceolate, compressed, florets subcylindrical, remote, glabrous, awns straight shorter than the florets; root leaves very narrow, ciliated, sheaths hairy. *E. Bot. t.* 471. *E. Fl. v. i. p.* 157. *Hook. Br. Fl. p.* 50.

Road-sides, in the limestone districts; rare. *Fl.* July. ♃.
Farley and Much Wenlock Lime-quarries; *Mr. F. Dickinson.*
Road-side, near Lutwyche Hall, on Wenlock Edge.

Root strong, fibrous. *Culm* erect, 2—3 feet high, round, striated, smooth, leafy. *Leaves* numerous, radical ones very narrow, linear, long, ciliated with long hairs, stem leaves broader, pubescent. *Sheaths* long, striated, hairy, close, slightly divided near the summit only. *Ligule* short, mostly torn. *Panicle* erect, more or less branched, branches angular and rough with erect bristles. *Spikelets* 5—7-flowered, compressed. *Glumes* very unequal, *inferior valve* awl-shaped and keeled, *superior* lanceolate, acute, 3-ribbed, smooth, resinoso-glandulose. *Inferior palea* lanceolate, resinoso-glandulose, smooth, with 3 strong scabrous ribs and smaller intermediate ones, margins membranous, with a straight rough *awn* from a little below the bifid apex, *superior valve* with 2 stout lateral bristly ribs and membranous incurved margins.

Var. β. villosus. Inferior palea hairy. *Festuca montana β. villosa, Mert. and Koch.* 674.

Occurred in the same situation, and in equal abundance with the above.

42. HOLCUS. *Linn.* Soft-grass.

1. H. *lanatus,* Linn. *Meadow Soft-grass.* Glumes mucronate, imperfect floret with a curved twisted awn included within the glumes; joints of the culm without tufts of hairs; leaves and sheaths hairy; root fibrous. *E. Bot. t.* 1169. *E. Fl. v. i. p* 108. *Hook. Br. Fl. p.* 42.

Meadows and pastures; common. *Fl.* June, July. ♃.

Root tufted, fibrous. *Culm* erect, simple, round, jointed, leafy, clothed as is the whole herbage, except the upper portion of the stem, with soft deflexed hairs. *Leaves* numerous, long, linear-lanceolate, hairy on both sides, with long, furrowed, entirely divided *sheaths. Ligule* short and blunt. *Panicle* thrice compound, erect, spreading but rather dense, whitish or purplish, branches angular with erect pubescence. *Florets* pedunculate, with a few hairs at the base, peduncle smooth. *Glumes* unequal, hairy, longer than the florets, *outer* one narrow, lanceolate, acute, with a strong serrated keel and finely ciliated edges, *inner* one larger, ovato-lanceolate, acute, keel rough with a strong rough rib on each side, edges finely ciliated. *Paleæ* unequal, smooth, shining, *inferior valve* of *lower floret* ovate, obtuse, awnless, with a rough keel, *superior* one flat, with 2 rough ribs, the apex obtusely 3-lobed and ciliated, middle lobe longest, *inferior valve* of *upper floret* oblongo-ovate, somewhat acute, keeled, with a twisted curved *awn* proceeding from a little below the apex, scarcely protruding beyond the calyx, *superior* one erose, 3-ribbed.

2. H. *mollis,* Linn. *Creeping Soft-grass.* Glumes acuminate, imperfect floret with an exserted geniculated awn; joints of the culm with a tuft of hairs; leaves and sheaths pubescent; root creeping. *E. Bot. t.* 1170. *E. Fl. v. i. p.* 108. *Hook. Br. Fl. p.* 42.

Pastures and hedges; common. *Fl.* July. ♃.

Root widely creeping. *Culm* erect, round, smooth, jointed, leafy, sheathing. *Leaves* linear-lanceolate, slightly downy, with long, ribbed, downy, entirely divided *sheaths* arising from the tufted hairy joints. *Panicle* lax, branched, branches angular, covered with erect soft hairs. *Florets* pedunculate, hairy at the base, lower one smooth with a few scattered hairs on the back, upper one bristly with an awn from below the apex. *Glumes* longer than the florets, ovato-lanceolate, acuminate, bristly, resinoso-glandulose, *inferior* one the smaller, with a strong hairy keel, *superior* larger, with a hairy keel and a strong bristly rib on each side. *Paleæ* ovate, obtuse, ciliated at the summit. *Awn* proceeding from a little below the apex, rough, bent at right angles to and extending beyond the calyx.

43. AGROSTIS. *Linn.* Bent-grass.

1. A. *vulgaris,* With. *Fine Bent-grass.* Panicle loose, branches diverging, slightly rough, inferior palea 3-ribbed; ligule extremely short, truncate. *E. Bot. t.* 1671. *E. Fl. v. i. p.* 92. *Hook. Br. Fl. p.* 38.

Meadows, pastures, and dry borders of fields; common. *Fl.* June, July. ♃.

Root tufted, rooting from the lower joints. *Culms* numerous, 1—2 feet high, ascending, slender, finely striated, smooth, leafy. *Leaves* linear, narrow, taper-pointed, rough. *Sheaths* long, striated, smooth, entirely divided. *Ligule* extremely short, truncate, that of the uppermost leaves longer. *Panicle* purplish, delicate, slender, *rachis* smooth, *branches* divaricating, capillary, more or less rough with erect bristles. *Glumes* lanceolate, nearly equal, smooth, shining, rough on the keel, longer than the paleæ. *Paleæ* unequal, thin, delicate, membranous, *inferior valve* 3-ribbed, tridentate at the apex, awnless, (in some varieties awned) *superior valve* only half the size, 2-ribbed, bifid.

2. A. *alba,* Linn. *Marsh Bent-grass.* Panicle loose, branches patent, scabrous, inferior palea 5-ribbed; ligule oblong. *E. Bot. t.* 1189. *E. Fl. v. i. p.* 93. *Hook. Br. Fl. p.* 39.

Pastures, road-sides, and moist places; common. *Fl.* July, August. ♃.

A taller and stouter plant than the last, from which the shape of the ligule will distinguish it.

Root fibrous, sending out roots from the lower joints. *Culms* 2—3 feet high, decumbent below, afterwards ascending, smooth, striated, leafy. *Leaves* flat, broad, taper-pointed, very rough. *Sheaths* long, striated, smooth, occasionally rough, entirely divided. *Ligule* oblong, obtuse, torn. *Panicle* 2—6 inches long, rather contracted, *branches* patent, unequal, scabrous, pale green or purplish. *Glumes* nearly equal, lanceolate, smooth, except on the keel. *Paleæ* unequal, pale, thin and membranous, *inferior valve* larger, lanceolate, 5-ribbed, 5-toothed at the apex, *superior* smaller, 2-ribbed, bifid.

β. stolonifera. culm more extensively creeping, branches of the panicle densely tufted. *E. Fl. & Hook. Br. Fl. l. c. A. stolonifera, Linn. E. Bot. t.* 1532.

Wet clayey places.

Oakley Park, near Ludlow; *Mr. Spare.*

3. A. *Spica Venti,* Linn. *Silky Bent-grass.* Panicle spreading, glumes unequal, lanceolate, rough at the keel, inferior palea bifid, with a long straight awn below the apex, superior smaller with a small barren pedicel at its base. *E. Bot. t.* 951. *E. Fl. v. i. p.* 89. *Hook. Br. Fl. p.* 30.

Sandy fields occasionally flooded; rare. *Fl.* June, July. ⊙.

Oakley Park, near Ludlow; *Mr. Spare.*

Root fibrous. *Culm* 1—3 feet high, erect, jointed, frequently bent at the base, smooth, leafy. *Leaves* linear, striated, slightly downy above, rough beneath. *Sheaths* long, striated, smooth, entirely divided. *Ligule* long, lanceolate, frequently torn. *Panicle* large, silky in appearance, gracefully leaning to one side, and elegantly waving in the breeze, *branches* very slender, repeatedly subdivided, smooth. *Glumes* unequal, inferior one shorter, purplish. *Inferior palea* the largest, rough, 5-nerved, bifid, with a tuft of minute hairs at the base, *superior palea,* bifid at the apex. *Awn* straight, rough, proceeding from below the apex and more than thrice the length of the palea.

44. CALAMAGROSTIS. *Adans.* Small-reed.

1. C. *Epigejos,* Roth. *Wood Small-reed.* Glumes subulate, keel and edges rough, panicle erect, close, (open in flower) flowers crowded, unilateral, inferior palea with a dorsal awn nearly as long

as the glumes. *Arundo Epigejos, Linn. E. Bot. t.* 403. *E. Fl. v. i. p.* 169. *Hook. Br. Fl. p.* 36.

In shady moist places; rare. *Fl.* July. ♃.

Sides of a ditch on the borders of Aqualate Mere; *Turn. & Dillw. Bot. Guide.*

Near Battlefield Church; *Mr. E. Elsmere, junr.!*

Root creeping. *Culm* tall, erect, round, jointed, hollow, striated, furrows slightly pubescent. *Leaves* linear-lanceolate with a sharp taper point, roughish, a little glaucous underneath. *Sheaths* very large, clasping, striated, scarcely rough except the upper one. *Ligule* lanceolate, acute, thin, soon torn. *Panicle* 6—8 inches long, purple, branches rough. *Glumes* nearly equal, long and narrow, purplish, resinoso-glandulose, keel rough. *Inferior palea* about half as long as the glumes, membranous, flat, with 2 rough marginal ribs at each side, bifid and rough at the point, *superior valve* smaller, slightly bifid and rough. *Awn* from about the middle, inflexed, rough. *Hairs* as long as the glumes.

2. C. *lanceolata,* Roth. *Purple-flowered Small-reed.* Glumes lanceolate, keel smooth, panicle erect, loose, flowers scattered, spreading, inferior palea with a very short terminal awn below the bifid point. *Arundo Calamagrostis, Linn. E. Bot. t.* 2159. *E. Fl. v. i. p.* 170. *Hook. Br. Fl. p.* 37.

Moist margins of pools; rare. *Fl.* June. ♃.

North-margin of Croesmere mere, and Colemere mere; *J. E. Bowman, Esq.*

Borders of Aqualate mere; *Withering.*

Hancott Pool, near Shrewsbury.

Root fibrous, woolly. *Culm* 3—4 feet high, slenderer than the last, very smooth. *Leaves* narrow, pointed, bright green, roughish beneath, sometimes a little hairy on the upper side. *Sheath* smooth. *Ligule* oblong, obtuse, decurrent, mostly torn. *Panicle* much branched, loosely spreading every way as are the flowers also. *Glumes* lanceolate, acute, pale bronzed purple, resinoso-glandulose. *Paleæ* scarcely more than half as long, membranous, white, *inferior valve* notched or torn at the summit, keeled, the keel terminating in a very small rough *awn. Hairs* longer than the paleæ, shorter than the glume.

45. AVENA. *Linn.* Oat, or Oat-grass.

1. A. *fatua,* Linn. *Wild Oat.* Panicle erect, spreading, spikelets drooping, about 3-flowered, florets shorter than the glumes, rough and hairy at the base, awn long and stout. *E. Bot. t.* 2211. *E. Fl. v. i. p.* 163. *Hook Br. Fl. p.* 53.

Corn-fields; not frequent. *Fl.* June—August. ⊙.

Astley, near Shrewsbury; *Mr. E. Elsmere, junr.!*

Root fibrous. *Culm* 2—3 feet high, erect, round, striated, smooth, leafy. *Leaves* broad, linear-lanceolate, flat, acuminated, ribbed and striated, rough. *Sheaths* long, close, striated, smooth. *Ligule* short, obtuse, laciniate. *Panicle* large and spreading, *branches* simple or divided, slender, angular and rough, swollen beneath the spikelets. *Spikelets* large and drooping. *Glumes* large, membranous, ovato-lanceolate, shining at the margins, keeled, acuminate, numerously ribbed, each rib with a row of scabrous points on each side, resinoso-glandulose. *Paleæ* coriaceous, shorter than the glumes, *inferior valve* ovato-lanceolate, with long fulvous hairs at the base, somewhat scabrous, ribbed, bifid or laciniato-dentate at the apex, *superior valve* bifid, marginal ribs 2, finely ciliated. *Awn* from about the middle of the back of the inferior palea nearly 2 inches long, rough with minute points, geniculate, twisted below.

2. A. *pubescens,* Linn. *Downy Oat-grass.* Panicle erect, nearly simple, spikelets erect, about 3-flowered, florets rather longer than

the glumes, inferior palea unequally toothed at the apex; leaves plane, obtuse, and as well as the sheaths downy, ligule large, triangular, acute. *E. Bot. t.* 1640. *E. Fl. v. i. p.* 164. *Hook. Br. Fl. p.* 54.

Dry hilly places in limestone districts; not common. *Fl.* June, July. ♃.

Madeley pastures; *Mr. F. Dickinson.*

Lime quarries near the Five Chimnies, Wenlock Edge.

Root somewhat creeping. *Culm* 1—2 feet or more high, round, smooth, simple, leafy. *Leaves* flat, spreading, short, linear, obtuse, hairy. *Sheaths* long, close, striated, entirely divided, clothed with deflexed pubescence, lower ones hairy. *Ligule* large, triangular, acute. *Panicle* long, erect, slightly branched, *branches* angular and rough, swollen under the spikelets. *Florets* generally 3, each with a tuft of hairs at the base, upper ones on bifariously hairy, compressed pedicels, uppermost sterile. *Glumes* lanceolate, unequal, resinoso-glandulose, *inferior valve* smaller, single-ribbed, *superior* 3-ribbed, margins ciliated or rough. *Paleæ* lanceolate, unequal, *inferior valve* largest, 5-ribbed, covered with scabrous and glandular points, jagged or unequally toothed at the apex, *superior valve* membranous, folded, bifid. *Awn* arising from about the middle of the inferior palea, as long again, rough, geniculate, twisted below.

3. A. *flavescens,* Linn. *Yellow Oat-grass.* Panicle much branched, spreading, erect, spikelets about 3-flowered, florets longer than the very unequal glumes, inferior palea bifid with two terminal bristles; leaves acute, ligule short, jagged, fringed. *E. Bot. t.* 952. *E. Fl. v. i. p.* 166. *Hook. Br. Fl. p.* 54.

Dry meadows and pastures; frequent. *Fl.* July. ♃.

Root more or less creeping. *Culm* 1 foot high, slender, erect, bent at the lowest joint only, round, striated, smooth, leafy, occasionally with a tuft of long soft hairs immediately below the joints. *Leaves* short, linear, acute, flat, ribbed, hairy on both surfaces, mostly so on the upper side. *Sheaths* as long or longer than the leaves, ribbed, smooth. *Panicle* half-whorled, *branches* numerous, short, compound, angular and rough. *Spikelets* a little drooping, shining green on the back, brownish-yellow at the membranous edges, peduncle of the florets hairy and extended beyond the 3rd floret. *Glumes* very unequal, *larger valve* ovate, acute, 3-ribbed, keel rough, *smaller* single-ribbed. *Inferior palea* 5-ribbed, deeply cloven, edges ciliated, segments with 2 terminal bristles, *superior valve* notched, inflexed at the edges and downy on the marginal ribs. *Awn* from above the middle of, and twice as long as the inferior palea.

46. AIRA. *Linn.* Hair-grass.

* *Florets pedicellate, awn straight.*

1. A. *cæspitosa,* Linn. *Turfy Hair-grass.* Panicle spreading, branches scabrous, florets rather longer than the glumes, abrupt, hairy at the base, one of them on a hairy stalk which is elongated beyond the 2nd flower, awn straight, inserted near the base and about as long as the palea. *E. Bot. t.* 1432. *E. Fl. v. i. p.* 102. *Hook. Br. Fl. p.* 39.

Moist shady places; plentiful. *Fl.* June—August. ♃.

Root fibrous in dense tufts. *Culm* 3 feet high, erect, slightly compressed, with 2 joints, smooth, leafy. *Leaves* sheathing, narrow, rigid, furrowed and rough above, radical ones numerous. *Sheaths* very long, furrowed, smooth. *Ligule* oblong, acute, often cloven. *Panicle* large, spreading horizontally, *branches* numerous, capillary, elastic, angular, wavy and rough. *Flowers* solitary, small, purplish, shining, erect. *Glumes* glandulose, lanceolate, rather acute, pale and membranous at the edges, purple at the back, with a rough keel and a few serratures on the margin near the apex. *Florets* with a few longish hairs at the base, upper one on a hairy peduncle which is elongated beyond the 2nd flower. *Paleæ* ovate, obtuse, erose, *inferior valve* with 5 short teeth, *superior* deeply bifid with 2 marginal ciliated nerves, edges membranous, incurved. *Awn* rough, proceeding from the base as long or longer than the palea.

Var. β. major.

Shelton wood on the banks of the Severn, near Shrewsbury.

This variety differs from the preceding in being altogether a larger and stouter plant, with a taller more jointed culm and longer leaves. The panicle is 3 times as long, with more numerous branchlets and innumerable flowers which are smaller and more obtuse, the florets not exceeding the glumes in length, and the very short rough awn, scarcely equalling the length of the palea, proceeding from above the middle of the back of the inferior valve.

** *Florets pedicellate, awn geniculate.*

2. A. *flexuosa*, Linn.   *Waved Hair-grass.*   Panicle (when in flower) diffuse, triply forked, branches wavy, florets hairy at the base, about as long as the glumes, awn inserted near the base of, but much longer than the palea; leaves setaceous.   *E. Bot. t.* 1519.   *E. Fl. v. i. p.* 104.   *Hook. Br. Fl. p.* 40.

Heaths and hilly places; not uncommon.   *Fl.* July, August. ♃.

Hardwick and Grinshill; *Mr. E. Elsmere, junr.!* On the Wrekin abundant; *Mr. F. Dickinson.*

Middleton Hill. Stiperstones Hill. Caer Caradoc. Near Whitecliff Coppice, Ludlow.

Root of many long and strong fibres.   *Culm* erect, slender, round, jointed, smooth, leafy below. *Leaves* sheathing, slender, various in length according to the place of growth, upper ones with long furrowed *sheaths* somewhat glaucous and rough. *Panicle* erect, more or less regularly trichotomous, branches angular, wavy, rough. *Glumes* of a copper colour, membranous, ovate, with a long sharply acuminated point and strong rough keel, edges finely serrated near the apex. *Paleæ* bristly, *inferior valve* lanceolate, acute, with 4 rough nerves, edges incurved, serrated, point unequally 4-toothed, inner valve bifid, with 2 marginal rough ribs and incurved edges. *Awn* twisted, divaricating, rough, proceeding from a little above the base, much longer than the corolla.

*** *Florets sessile, inferior palea bifid.*

3. A. *caryophyllea*, Linn.   *Silvery Hair-grass.*   Panicle spreading, triple-forked, florets slightly villous at the base, sessile, shorter than the glumes, awn inserted a little below the middle of, and twice as long as the palea; leaves setaceous.   *E. Bot. t.* 812.   *E. Fl. v. i. p.* 106.   *Hook. B. Fl. p.* 40.

Gravelly hills and pastures; not unfrequent.   *Fl.* June, July. ⊙.

Lyth Hill. Sharpstones Hill, and Bicton, near Shrewsbury. Caer Caradoc. Near Whitecliff Coppice, Ludlow.

Root fibrous. *Culm* slender, erect, 2—6 inches or a foot high, jointed, round, smooth, leafy. *Leaves* short, slender, blunt, radical ones copious, occasionally a little rough, upper ones with close furrowed roughish *sheaths*. *Panicle* trichotomous, florets silvery-grey, branches slender, angular, compressed, channelled and smooth. *Glumes* nearly equal, ovato-lanceolate, irregularly notched, rough at the keel and near the pellucid point, edges finely serrated. *Paleæ* unequal, *inferior valve* lanceolate, acuminate, scabrous, bristly near the point and edges, bifid, *superior valve* bifid, with 2 smooth marginal ribs and incurved edges. *Awn* rough, twisted, slightly bent, proceeding from below the middle of the valve.

4. A. *præcox*, Linn.   *Early Hair-grass.*   Panicle close, erect, somewhat spiked; florets slightly villous at the base, sessile, about as long as the glumes, awn inserted below the middle of and twice as long as the corolla; leaves setaceous.   *E. Bot. t.* 1296.   *E. Fl. v. i. p.* 106.   *Hook. Br. Fl. p.* 41.

Hedge-banks and hilly pastures; not uncommon.   *Fl.* May, June. ⊙.

Oakley Park, near Ludlow; *Mr. Spare.*

Lyth Hill. Haughmond Hill, near Shrewsbury. Caer Caradoc and Lawley hills. Near Whitecliff Coppice, Ludlow.

Root fibrous, tufted. *Culm* 1—3 inches high, erect, or slightly reclining at the base, rigid, smooth, jointed, leafy. *Leaves* very short and narrow, blunt, channelled, sheathing. *Sheaths* angular, tumid, smooth. *Ligule* oblong, toothed, closely embracing the stem. *Flowers* pale silvery green. *Glumes* nearly equal, lanceolate, acuminate, irregularly notched, rough especially at the keel, edges finely serrated. *Paleæ* unequal, *inferior valve* lanceolate, narrow, very much acuminated, scabrous, bifid, *superior* one bifid with 2 smooth marginal ribs and incurved edges. *Awn* inserted below the middle nearer to the base than in the last species.

This species is very closely allied to *A. caryophyllea* from which it is chiefly distinguished by its smaller size, its close sub-spicate panicle, the tumid angular sheaths of its leaves, the narrower and more lanceolate acuminate form of its flowers and the insertion of the awn much nearer to the base.

47. ARRHENATHERUM. *Beauv.*   Oat-like-grass.

1. A. *avenaceum*, Beauv.   *Common Oat-like-grass.*   Florets sessile, hairy at the base, with the peduncle of a third floret between them; joints of the culm smooth; root knotty.   *Hook. Br. Fl. v. i. p.* 42.   *Holcus avenaceus, Scop.   E. Bot. t.* 813.   *E. Fl. v. i. p.* 109.   *Avena elatior, Linn.*

Hedges and pastures; frequent.   *Fl.* June, July. ♃.

Root fibrous, knotty, downy, from the swollen joints of the base of the culm. *Culm* tall, smooth, leafy, jointed, often bent at the lower joints. *Leaves* rough-edged, harsh, with long striated *sheaths*. *Ligule* short, abrupt. *Panicle* erect, a little drooping, *branches* half-whorled, directed to one side, angular, rough. *Glumes* unequal, *inferior valve* smaller, lanceolate, nearly smooth, with a rough keel, *superior* larger, about as long as the florets, acuminate, bifid, rough, 3-ribbed, keel and ribs rough. *Inferior palea* of the *lower floret* ovato-lanceolate, keeled about half way up, rough, resinoso-glandulose, 6-ribbed, bifid, with a twisted geniculated rough *awn* proceeding from above the base, twice as long as the valve, *superior* narrow with 2 rough ribs and edges incurved. *Inferior palea* of *upper floret* rough, 7-ribbed, keel and ribs rough, keel prolonged into a short awn proceeding from just below the bifid apex, *superior valve* as in the lower floret. *Seed* nearly cylindrical, coated with the hardened corolla. From between the two florets proceeds a small capillary peduncle of a third floret.

Var. β. A. bulbosum, Dumortier.   Lindl. Syn. p. 305.

In dry pastures; common.   *Fl.* June, July. ♃.

Fields, Meole Brace, near Shrewsbury.

Differs from the preceding in being a larger plant with a bulbous base to the culm, and occasionally hairy joints. The valves of the corolla are also smoother and more resinoso-glandulose.

CLASS IV.

TETRANDRIA.

" There are two books, from whence I collect my divinity; besides that written one of God, another of his servant Nature, that universal and public manuscript that lies exposed unto the eyes of all.   Those that never saw him in the one, have discovered him in the other.   This was the scripture and theology of the heathens.   And surely they knew better how to join and read these mystical letters, than we Christians who cast a more careless eye on these common hieroglyphicks, and disdain to suck divinity from the flowers of nature."

SIR THOMAS BROWNE.

CLASS IV.

TETRANDRIA.

(4 Stamens, equal in height.)

ORDER I. MONOGYNIA. 1 *Style.*

* *Perianth double. Corolla monopetalous, superior. Seed 1.*

1. DIPSACUS. *Flowers* aggregate. *Involucre* many-leaved. *Involucellum (outer calyx)* very minute, forming a thickened margin to the germen. *Calyx* cup-shaped, ciliated at the margin. *Receptacle* chaffy, spinous. *Fruit* tetragonous with 8 depressed pores.—*Nat. Ord.* DIPSACEÆ, *Juss.*—Name from δίψαω, *to be thirsty;* the connate leaves containing water in their concave bases.

2. KNAUTIA. *Flowers* aggregate. *Involucre* many-leaved. *Involucellum* a minute margin to the germen. *Calyx* cup-shaped, with radiate spines at the margin. *Receptacle* hairy. *Fruit* tetragonous with 4 depressed pores.—*Nat. Ord.* DIPSACEÆ, *Juss.*—Name in honour of *Christopher Knaut,* a Botanist of Saxony, who flourished in the latter half of the 17th century.

3. SCABIOSA. *Flowers* aggregate. *Involucrum* many-leaved. *Involucellum* membranaceous and plaited. *Calyx* with about 5 bristles at the margin. *Receptacle* chaffy. *Fruit* subcylindrical with 8 depressed pores.—*Nat. Ord.* DIPSACEÆ, *Juss.*—Name from *Scabies, the leprosy;* the infusion of the leaves of some species having been formerly employed in the cure of cutaneous diseases.

** *Perianth double. Corolla monopetalous, superior. Seeds 2. (Leaves whorled).*

4. GALIUM. *Corolla* rotate, 4-cleft. *Fruit* a dry 2-lobed, indehiscent *pericarp,* without any distinct margin to the *calyx.*—*Nat. Ord.* RUBIACEÆ, *Juss.*—Name from γαλα, *milk;* the plants having been formerly used for curdling milk.

5. ASPERULA. *Corolla* funnel-shaped. *Fruit* without any distinct margin to the *calyx.*—*Nat. Ord.* RUBIACEÆ, *Juss.*—Name from *asper, rough;* in allusion to the roughness of some species.

6. SHERARDIA. *Corolla* funnel-shaped. *Fruit* crowned with the *calyx.*—*Nat. Ord.* RUBIACEÆ, *Juss.*—Name in honour of *James Sherard,* an English Botanist and Patron of Botany, whose fine garden at Eltham in Kent gave rise to the famous "*Hortus Elthamensis*" of Dillenius.

*** *Perianth double. Corolla monopetalous, inferior. Seeds 2 or many.*

7. PLANTAGO. *Corolla* 4-cleft, the segments reflexed. *Stamens* very long. *Capsule* of 2 cells, 2 or many-seeded, bursting all round transversely.—*Nat. Ord.* PLANTAGINEÆ, *Juss.*—Name of doubtful origin.

**** *Perianth double. Corolla of 4 petals.*

8. CORNUS. *Calyx* of 4 teeth. *Petals* without a nectary, superior. *Nut* of the drupe 2-celled, each cell 1-seeded.—*Nat. Ord.* CORNEÆ, *DC.*—Name from *cornu,* a *horn;* in allusion to the hardness of the wood.

***** *Perianth single.*

9. PARIETARIA. *Perianth* 4-fid, inferior. *Filaments* of the *stamens* at first incurved, then expanding with elastic force. *Fruit* 1-seeded, enclosed by the enlarged perianth. (One or more of the central florets without stamens.)—*Nat. Ord.* URTICACEÆ, *DC.*—Name from *paries,* a *wall;* in allusion to the usual place of its growth.

10. ALCHEMILLA. *Perianth* inferior, 8-cleft, the 4 alternate and outer segments the smallest. *Fruit* 1 or 2-seeded, surrounded by the persistent perianth.—*Nat. Ord.* ROSACEÆ, *Juss.*—Name from *àlkémelyeh,* Arabic, *alchemy;* in allusion to its pretended alchemical virtues.

11. SANGUISORBA. *Perianth* 4-lobed, superior, coloured, with 4 scales or bracteas at the base. *Fruit* 1 or 2-seeded, surrounded by the persistent base only of the perianth.—*Nat. Ord.* ROSACEÆ, *Juss.*—Name from *sanguis, blood,* and *sorbeo,* to *take up* or *absorb;* in allusion to its supposed vulnerary properties.

ORDER II. TETRAGYNIA. 4 *Styles.*

12. ILEX. *Calyx* 4—5 toothed. *Corolla* rotate, 4—5 cleft. *Stigmas* 4, sessile. *Berry* sphærical, including 4, 1-seeded *nuts.* (Some flowers destitute of pistil).—*Nat. Ord.* ILICINEÆ, *Br.*—Name probably from the leaves resembling those of the *Quercus Ilex,* the true Ilex of Virgil.

13. POTAMOGETON. *Flowers* sessile upon a *spike* or *spadix,* which issues from a sheathing *bractea* or *spatha.* *Perianth* single, of 4 scales. *Anthers* sessile, opposite the scales of the perianth. *Pistils* 4, which become 4 small *nuts;*—*Embryo* curved.—*Nat. Ord.* POTAMEÆ, *DC.*—Name from ποταμος, a *river,* and γειτων, a *neighbour;* in allusion to the places of their growth.

---

14. SAGINA. *Calyx* of 4 leaves. *Petals* 4, shorter than the calyx. *Capsule* 1-celled, 4-valved.—*Nat. Ord.* CARYOPHYLLEÆ, *Juss.*—The name signifying *meat which fattens,* is little applicable to any of the minute plants of this genus.

15. MŒNCHIA. *Calyx* of 4 leaves. *Petals* 4, as long as the calyx. *Capsule* 1-celled, opening with eight teeth at the extremity.—*Nat. Ord.* CARYOPHYLLEÆ, *Juss.*—Name in honour of *Conrad Mœnch,* Professor of Botany at Hesse Cassel.

TETRANDRIA—MONOGYNIA.

1. DIPSACUS. *Linn.* Teasel.

1. D. *Fullonum,* Linn. *Fuller's Teasel.* Leaves connate, scales of the receptacle hooked at the extremity, involucres spreading, somewhat deflexed. *E. Bot. t.* 2080. *E. Fl. v.* i. *p.* 192. Hook. Br. *Fl. p.* 63.

Waste places; rare and scarcely wild. *Fl.* July, August. ♂.

Oakley Park, near Ludlow; Mr. H. Spare.

Root fleshy, tapering, branched. *Stem* 4—5 feet high, very angular, prickly, branched above. *Leaves* large, oblong or oblongo-lanceolate, obtusely and irregularly serrated, sessile, connate at the base, especially the upper ones. *Involucre* spreading, about as long as the heads of flowers. *Flowers* in oval heads, pale purple or whitish. *Scales* of the receptacle rigid, hooked at the extremity, edges bristly.

The heads are used by the manufacturers of woollen cloths in raising the nap to the desired length, which is effected by passing the cloth under rapidly revolving cylinders covered with the teasels, by which process the hooked extremities of the scales slightly catch the cloth, and produce the requisite effect.

The chief difference between *D. Fullonum* and *D. sylvestris* appears to consist in the hooked extremities of the scales, but it has been stated that by long cultivation on a poor soil these hooks become obsolete, and the scales assume the appearance of those of *D. sylvestris,* consequently there is every reason to believe that *D. Fullonum* is only a variety of *sylvestris.*

2. D. *sylvestris,* Linn. *Wild Teasel.* Leaves opposite connate, scales of the receptacle straight at the extremity, involucres curved upwards. *E. Bot. t.* 1032. *E. Fl. v.* i. *p.* 193. Hook. Br. Fl. p. 64.

Hedges, copses, sides of rivers and canals; frequent. *Fl.* July, August. ♂.

About Broseley and Buildwas; *Dr. G. Lloyd.* Eyton; *T. C. Eyton, Esq.*

Hedges, woods, and moist places generally about Shrewsbury.

Root fleshy, tapering. *Stem* 4 feet high, erect, strongly ribbed and furrowed, ribs with ascending prickles, leafy, branched at the top. *Leaves* sessile, connate, linear-lanceolate, occasionally wavy or dentate, smooth, mid-rib strong, with distant prickles on the under side, those of the lower half decurved, whilst those of the upper are ascending. *Involucre* curved upwards, as long as the heads of flowers, mid-rib prickly. *Flowers* very numerous, in dense conical obtuse heads. *Corolla* light purple or lilac, 4-cleft. *Scales* of the receptacle ovato-acuminate, rigid, edges bristly, points straight.

3. D. *pilosus,* Linn. *Small Teasel.* Leaves petiolate with a small leaflet at the base on each side, scales of the receptacle straight

at the extremity, involucres shortly deflexed. *E. Bot. t.* 877. *E. Fl. v. i. p.* 193. *Hook. Br. Fl. p.* 64.

Moist hedges and shady places; rare. *Fl.* August, September. ♂.

On the Baschurch road a little beyond Leaton Shelf; *J. E. Bowman, Esq.* Side of a rill between Albrighton and Shiffnal; *E. Lees, Esq.* Road side between Ludlow and Caynham Camp, near the Sheet; *Rev. W. Corbett.* Lilleshall Abbey; *Bot. Guide.* Neachley Brook; *Dr. G. Lloyd.* Poughmill and dingle, near Burford; *Miss Mc. Ghie.* Buildwas; *Mr. F. Dickinson.*

Root tapering. *Stem* 3—4 feet high, angular, branching, rough with short deflexed prickles, those on the peduncles immediately below the heads longer and bristly. *Leaves* petiolate, somewhat connate by the winged expansion of the bristly edges of the petiole, broadly ovato-acuminate, coarsely serrated, with a small ovato-acuminate, serrated leaflet on each side at the base, smooth, mid-rib prickly. *Involucre* ovato-lanceolate, pubescent, covered all over with long bristles, deflexed. *Flowers* numerous, in small dense globular heads. *Corolla* white, unequally 5-cleft. *Scales* of the receptacle ovato-acuminate, rigid, covered with short dense pubescence and long bristles, points straight.

2. KNAUTIA. *Coult.* Knautia.

1. K. *arvensis,* Coult. *Field Knautia.* Heads of many flowers, involucellum with very minute teeth, calyx with 8—16 awn-like bristles. *Hook. Br. Fl. p.* 64. *Scabiosa arvensis, Linn. E. Bot. t.* 659. *E. Fl. v. i. p.* 195.

Pastures and corn-fields; frequent. *Fl.* July—August. ♃.

Root tapering, with spreading fibres. *Stems* 2—3 feet high, erect, round, branched, hairy as is the whole plant. *Leaves* opposite, lower ones with winged petioles, radical leaves obovato-lanceolate, nearly entire, stem leaves deeply pinnatifid, upper-most sessile, lobes lanceolate or linear-lanceolate. *Involucre* of numerous lanceolate rather obtuse hairy leaves. *Flowers* in convex terminal heads. *Corolla* pale bluish lilac, *outer florets* large, segments unequal, lower ones large, somewhat radiate, segments of the *inner florets* equal. *Receptacle* convex, bristly. *Fruit* on a short glandular stalk, enveloped in the hairy tube of the involucellum, and crowned by the persistent cup-shaped bristly calyx.

3. SCABIOSA. *Linn.* Scabious.

1. S. *succisa,* Linn. *Devil's-bit Scabious.* Heads of flowers nearly globose, corolla 4-cleft, segments equal; radical leaves ovate, entire, cauline ones lanceolate, dentate. *E. Bot. t.* 878. *E. Fl. v. i. p.* 194. *Hook. Br. Fl. p.* 64.

Moist pastures and boggy ground; not unfrequent. *Fl.* July—Sept. ♃.

Oakley Park, near Ludlow; *Mr. H. Spare.* Shadwell, a part of the Standhill Coppice near Wenlock, also near Muckley; *W. P. Brookes, Esq.* Wyre Forest, near Bewdley; *Mr. G. Jorden.* Fields, near Coalbrookdale; *Mr. F. Dickinson.*

Berrington and Bomere pools, near Shrewsbury. Sutton Spa. Twyford Vownog, near Westfelton. Kingsland, near Shrewsbury.

Root premorse. *Stem* a foot or more high, erect, nearly simple, smoothish below, covered above with erect, dense, appressed pubescence. *Leaves* dark-green, harsh with bristly hairs, *radical* ones ovate, entire, on winged petioles, *cauline* ones oblongo-lanceolate, more or less dentate, united at the base, uppermost lanceolate, entire, and sessile. *Involucrum* of numerous lanceolate hairy leaves. *Flowers* beautiful, deep purplish-blue. *Receptacle* hairy. *Fruit* crowned by the persistent bristly calyx and enveloped in the hairy tube of the involucellum, whose margin is membranous and toothed.

A white variety has been found near Walford by T. C. Eyton, Esq.

2. S. *columbaria,* Linn. *Small Scabious.* Heads of flowers somewhat convex, corolla 5-cleft, segments unequal; radical leaves oblongo-ovate, crenate or lyrate, cauline ones pinnatifid, segments linear. *E. Bot. t.* 1311. *E. Fl. v. i. p.* 195. *Hook. Br. Fl. p.* 64.

Waste places; in limestone districts. *Fl.* July, August. ♃.

Oakley Park, near Ludlow; *Mr. H. Spare.*

Root tapering, woody, fibrous. *Stem* 1—2 feet high, erect, branched above, round, hairy or pubescent. *Lower leaves* on rather long footstalks, *cauline* ones deeply cut into narrow, linear, or setaceous pinnæ. *Involucrum* of narrow hairy leaves. *Flowers* purplish-blue, outer ones the larger, somewhat radiant. *Receptacle* scarcely hairy. *Fruit* crowned by the persistent calyx of about 5 long, dark roughish bristles, and enveloped in the hairy *involucellum,* whose limb is membranous, crenate, and plaited.

4. GALIUM. *Linn.* Bed-straw.

* *Fruit glabrous. Flowers yellow.*

1. G. *verum,* Linn. *Yellow Bed-straw.* Leaves about 8 in a whorl, linear, grooved above, deflexed; flowers in dense panicles. *E. Bot. t.* 660. *E. Fl. v. i. p.* 208. *Hook. Br. Fl. p.* 65.

Dry banks, and borders of fields; common. *Fl.* July, August. ♃.

Root creeping, tawny. *Stem* erect, 1—2 feet high, somewhat woody, obsoletely tetragonous, branched at the base, downy. *Leaves* narrow, linear, in numerous whorls, deep green, revolute at the margin, scabrous, mucronate. *Flowers* small, yellow, in a dense terminal branched panicle. *Fruit* blackish.

Useful for coagulating milk for cheese, and for dying woollen goods of a yellow colour. The roots also yield a fine red colouring matter, scarcely inferior to madder.

2. G. *cruciatum,* Linn. *Crosswort Bedstraw, Mug-wort.* Leaves 4 in a whorl, ovate, hairy; flowers polygamous, axillary, corymbose, peduncles with two leaves. *E. Bot. t.* 143. *E. Fl. v. i. p.* 199. *Hook. Br. Fl. p.* 65.

Hedge-banks; common. *Fl.* May, June. ♃.

Root slender, creeping. *Stem* branched at the base, simple above, weak, angular, grooved, covered with dense spreading soft hairs. *Leaves* with 3 principal nerves, bearing at the back immediately below the apex on each side of the mid-rib small pale yellow oblong glandular bodies. *Flowers* small, about 8, on slender hairy corymbose stalks from the axils of the leaves, accompanied by two smaller ovate leaves upon the peduncle; some wanting the pistil, some 3-cleft, a few only 5-cleft.

** *Fruit glabrous. Flowers white.*

3. G. *Mollugo,* Linn. *Great Hedge Bed-straw.* Leaves 8 in a whorl, elliptical, mucronate, rough at the margin; flowers in loose spreading panicles, segments of the corolla mucronate. *E. Bot. t.* 1673. *E. Fl. v. i. p.* 208. *Hook. Br. Fl. p.* 67.

Hedges and thickets; rare. *Fl.* July, August. ♃.

Hedges of the London road on each side the town of Shiffnal; *J. E. Bowman, Esq.!* Near Oswestry; *Rev. T. Salwey.* One mile from Shiffnal towards Coalbrookdale, in hedges; *Mr. F. Dickinson.*

Stems 3—4 feet high, straggling, square, pale and swelling above the whorls, smooth or slightly hairy. *Leaves* in rather distant whorls, green above, pale

beneath, margins with prickles pointing forwards. *Flowers* white, segments of corolla 3-ribbed, tapering into a bluntish, hair-like point. *Fruit* small, smooth, globular.

*** *Fruit minutely tuberculated or granulated. Flowers white.*

4. G. *uliginosum,* Linn. *Rough Marsh Bed-straw.* Leaves 6 in a whorl, linear-lanceolate, mucronate, margins and the stem rough with reflexed prickles; fruit minutely granulated. *E. Bot. t.* 1972. *E. Fl. v. i. p.* 201. *Hook. Br. Fl. p.* 66.

Wet meadows, marshes, &c.; not very common. *Fl.* July, August. ♃.

Near Oswestry; *Rev. T. Salwey.*

Boggy ground at the foot of Caradoc Hill.

Stems 1—2 feet high, weak, slender, and branched, angular, edges rough with reflexed bristles. *Leaves* linear-lanceolate, tapering at the base, shortly acuminated into a mucro, margins revolute with reflexed prickles. *Inflorescence* in small terminal branched erect panicles. *Flowers* white. *Fruit* small, minutely granulated.

5. G. *palustre,* Linn. *White Water Bed-straw.* Leaves 4—6 in a whorl, oblongo-lanceolate, obtuse, tapering at the base, and as well as the lax spreading branched stem, more or less rough; fruit minutely dotted or tuberculated.

a. stem and leaves smoothish. *Hook. Br. Fl. p.* 65. *G. palustre, Linn. E. Bot. t.* 1857. *E. Fl. v. i. p.* 200.

β. nerves at the back and margins of the leaves, and angles of the stem, distinctly rough with mostly reflexed bristles. *Hook. l. c. G. Witheringii. E. Bot. t.* 2206. *E. Fl. v. i. p.* 200.

Sides of ditches, pools and rivulets; *a.* common; *β.* less common. *Fl.* July ♃.

Stems very variable in size, tall, weak, slender, angular, much branched, branches widely spreading, angles roughish to the touch, but without any apparent bristles, except perhaps a few beneath the whorls. *Leaves* varying in size and number on different parts of the plant, smooth except a few scattered bristles on the edges and back. *Panicles* terminal, widely spreading. *Fruit* of two turgid lobes, minutely dotted.

The *var. β.* appears to be chiefly distinguished by the more numerous and apparent reflexed bristles of the stem and leaves.

6. G. *saxatile,* Linn. *Smooth Heath Bed-straw.* Leaves 6 in a whorl, obovate, mucronate; stem very much branched prostrate, smooth; fruit minutely granulated. *E. Bot. t.* 815. *E. Fl. v. i. p.* 201. *Hook. Br. Fl. p.* 66.

Heathy spots and hilly pastures; abundant. *Fl.* June—August. ♃.

Near Oswestry; *Rev. T. Salwey.* Titterstone Clee hill; *Mr. J. S. Baly!* Beaumont Hill, Quatford; *W. P. Brookes, Esq.* Wrekin Hill; *Mr. F. Dickinson.*

Near Whitcliffe Coppice, Ludlow; Haughmond Hill; Bayston Hill; Lawley Hill; Caer Caradoc; Sharpstones Hill; and on most of the dry hills throughout the County.

Plant variable in luxuriance, turning blackish in drying. *Leaves* thick and soft in texture, smooth, minutely pitted above, rough at the margins with bristly serratures directed towards the apex. *Panicles* lateral and terminal, forked, dense. *Fruit* reddish after the fall of the corolla, if fertile, minutely grandulated.

**** *Fruit hispid. Flowers white.*

7. G. *Aparine,* Linn. *Goose-grass or Cleavers.* Leaves 6—8 in a whorl, lanceolate, mucronate, hispid, their margins and midrib as well as the angles of the stem very rough with reflexed bristles; stem weak; fruit hispid with hooked bristles. *E. Bot. t.* 816. *E. Fl. v. i. p.* 210. *Hook. Br. Fl. p.* 68.

Hedges; every where. *Fl.* June, July. ⊙.

Plant straggling among bushes. Under surface of the leaves smooth except the midrib. *Flowers* few, pale, buff-coloured, 2 or 3 together on short simple footstalks arising from the axils of the leaves. *Fruit* a double globe.

The expressed juice of the herb is considered antiscorbutic, and the seeds are said to form an excellent substitute for coffee.

5. ASPERULA. *Linn.* Woodruff.

1. A. *odorata,* Linn. *Sweet Woodruff.* Leaves about 8 in a whorl, lanceolate, mucronate; flowers few, panicled on long stalks. *E. Bot. t.* 755. *E. Fl. v. i. p.* 197. *Hook. Br. Fl. p.* 69.

Woods and shady places; not very common. *Fl.* May, June. ♃.

Cox Wood, near Coalbrookdale; *Mr. F. Dickinson.* Moston, near Lee Bridge; *W. W. Watkins, Esq.* At the Leasowes, near Hales Owen; *Dr. Withering, in Bot. Guide.* Near Oswestry; *Rev. T. Salwey.* Near Great Ness, and Fitz Coppice; *T. C. Eyton, Esq.* Oakley Park, near Ludlow; *Mr. H. Spare.* Abundant near Wenlock; *W. P. Brookes, Esq.!* Welbatch Coppice, near Shrewsbury; *Mr. T. Bodenham.!*

Whitcliffe Coppice, near Ludlow; Woods near Wrekin hill; Almond Park, abundant; Benthal Edge; Lyth Hill.

Root creeping. *Stem* simple, angular, smooth. *Leaves* with a broad flat mucro, lower ones approaching to obovate, their margins rough with bristles pointing upwards. *Panicles* generally in threes, forked, slightly subdivided. *Flowers* pure white. *Fruit* globular, rough with hooked bristles. *Whole plant* fragrant in drying.

6. SHERARDIA. *Linn.* Sherardia or Field Madder.

1. S. *arvensis,* Linn. *Blue Sherardia.* Leaves about 6 in a whorl; flowers terminal, sessile, capitate. *E. Bot. t.* 891. *E. Fl. v. i. p.* 196. *Hook. Br. Fl. p.* 69.

Corn fields on a gravelly soil; frequent. *Fl.* June—August. ⊙.

Whole plant hispid. *Stems* numerous, slender, spreading, decumbent, branched, square. *Leaves* obovato-lanceolate acute, upper ones forming an involucre to a small sessile *umbel* of pale blue or white *flowers. Calyx* persistent of 4 segments, the two opposite ones bifid. These bifid segments correspond to the line where the fruit divides into two 1-seeded portions, each of which is crowned with 3 teeth, one being the single entire tooth or segment of the calyx, the other two, each half of a bifid one. *Fruit* of 2 globular closely combined roughish seeds.

7. PLANTAGO. *Linn.* Plantain.

1. P. *major,* Linn. *Greater Plantain.* Leaves broadly ovate, mostly on longish footstalks; scape rounded, spikes long, cylindrical; dissepiment of the capsule plane, each cell many-seeded. *E. Bot. t.* 1558. *E. Fl. v. i. p.* 213. *Hook. Br. Fl. p.* 70.

Pastures and road-sides; frequent. *Fl.* June, July. ♃.

Root of numerous, long, stout fibres. *Leaves* all radical, more or less spreading, with 7—9 prominent parallel ribs united at the base into a footstalk as long or longer than the leaf, entire or toothed, glabrous or pubescent. *Spike* dense of numerous closely imbricated small *florets*, each with a concave lanceolate *bractea* at its base. *Corolla* of one piece with 4 reflexed dry thin segments, united at the base. *Capsule* small, ovate, 2-celled, with 6—8 seeds in each cell.

2. P. *media*, Linn. *Hoary Plantain.* Leaves ovate, sessile or tapering into short and broad footstalks; scape rounded, spike cylindrical; dissepiment of the capsule plane, each cell 1-seeded. *E. Bot. t.* 1559. *E. Fl. v. i. p.* 214. *Hook. Br. Fl. p.* 70.

Meadows and pastures, chiefly in the limestone districts. *Fl.* June, July. ♃. Oakley Park, near Ludlow; *Mr. H. Spare.*

Ludlow Castle. Between Ludlow and Caynham Camp. Hill Top, Wenlock Edge. Benthal Edge.

Root large and woody. *Leaves* spreading, pressed close to the ground, about 6-ribbed, clothed with a soft pubescence, entire or slightly dentate. *Scape* very long and slender, with appressed hairs. *Spike* short, dense and broad, shining and silvery. *Florets* numerous, closely imbricated. *Bracteas* pale, thin, and shining, with a green keel, one at the base of each floret. *Calyx* pale and shining. Segments of the *corolla* 4, lanceolate, reflexed, membranous, pale pink and shining. *Stamens* long, with dark purple filaments. *Capsule* 2-celled, each cell 1-seeded.

3. P. *lanceolata.* Linn. *Ribwort Plantain.* Leaves lanceolate; scape angular, spikes ovate or ovato-lanceolate; dissepiment of the capsule plane, each cell 1-seeded. *E. Bot. t.* 175. *E. Fl. v. i. p.* 214. *Hook. Br. Fl. p.* 70.

Meadows and dry pastures; abundant. *Fl.* June, July. ♃.

Root somewhat woody, with spreading fibres. *Leaves* all radical, spreading or erect, lanceolate, tapering at the base into a footstalk varying in length, about 5-ribbed, smooth or downy. *Spike* compact. *Bracteas* hairy, pale at the base, point and keel dark brown. Segments of *corolla* 4, lanceolate, pale, spreading and single-ribbed. *Stamens* long, with cream-coloured anthers. *Capsule* 2-celled, each cell with a single seed.

4. P. *Coronopus.* Linn. *Buck's-horn Plantain.* Leaves linear pinnatifid; scape rounded; dissepiment of the capsule with four angles, (thus forming 4 cells) each cell 1-seeded. *E. Bot. t.* 892. *E. Fl. v. i. p.* 216. *Hook. Br. Fl. p.* 71.

Dry sandy or gravelly soils; not very common. *Fl.* June, July. ☉.

Near the Church, Ness; *T. C. Eyton, Esq.* Ruyton-of-the-Eleven-Towns; *Rev. T. Salwey.* Wrockwardine; *Mr. F. Dickinson.*

Lyth Hill; Hopton Cliff, near Nesscliffe; Gamester lane, near Westfelton; Bayston Hill.

Root tapering. *Leaves* numerous, spreading flat on the ground, pale, very variable in size and pubescence, pinnatifid, segments acute, often toothed or again divided. *Scapes* with erect appressed hairs. *Spikes* dense, cylindrical, varying greatly in length. *Capsule* 4-celled, each cell containing a single oblong seed.

8. CORNUS. *Linn.* Cornel.

1. C. *sanguinea.* Linn. *Wild Cornel or Dogwood.* Arborescent, branches strait, leaves opposite, ovate, green on both sides;

cymes destitute of involucre. *E. Bot. t.* 249. *E. Fl. v. i. p.* 221. *Hook. Br. Fl. p.* 72.

Woods, hedges and thickets; common, apparently in this county not confined to any particular soil. *Fl.* June, July. ♄.

Between Bridgnorth and Kinlet; *T. C. Eyton, Esq.* Near Middle; *Dr. Wilson.* Near Ludlow; *Miss Mc Ghie.* Lincoln's Hill, near Ironbridge; *Mr. F. Dickinson.* Astley; *Mr. E. Elsmere, junr.!*

Fitz. Shelton Rough. Woods beyond Haughmond Hill. Hedges around Shrewsbury.

Shrub 5—6 feet high, branches opposite, straight, round, smooth, bark of the older ones dark-red. *Leaves* petiolate, entire, acute, strongly nerved, slightly hairy beneath, turning red before they fall. *Cymes* terminal. *Flowers* numerous, greenish-white. *Petals* revolute at the sides, inserted with the stamens into a glandular ring crowning the germen. *Fruit* dark purple.

9. PARIETARIA. *Linn.* Wall-Pellitory.

1. P. *officinalis*, Linn. *Common Pellitory of the Wall.* Leaves ovato-lanceolate, 3-nerved above the base; involucre 2-leaved, 7-flowered, the central one fertile, leaves of the involucre with 7 ovate segments. *E. Bot. t.* 597. *E. Fl. v. i. p.* 222. *Hook. Br. Fl. p.* 72.

Ruins and walls; frequent. *Fl.* during the summer months. ♃.

Oswestry; *Rev. T. Salwey.* Bromfield Churchyard; *Mr. H. Spare.*

Ludlow Castle. Wenlock Abbey. Buildwas Abbey. Haughmond Abbey. Council House, Town Walls and St. Giles's Church, Shrewsbury.

Root woody, with downy fibres. *Stem* 1—2 feet high, erect, often procumbent, branched, reddish, pubescent. *Leaves* alternate, on slender footstalks, hairy. *Flowers* small, hairy, in axillary clusters. *Involucrum* of two leaves, each cut into about 7 ovate segments, containing 3 flowers, of which the lateral ones have stamens and pistil, the central one a pistil only. Between the 2 portions is placed a fertile flower. *Filaments* transversely plicate, at first inflexed, possessing an elastic property, which on a hot summer's day causes them to become reflexed, and to discharge copiously the pollen. *Fruit* black, shining. *Pericarp* closely investing the seed.

10. ALCHEMILLA. *Linn.* Lady's Mantle.

1. A. *vulgaris*, Linn. *Common Lady's Mantle.* Leaves plaited, many-lobed, serrated. *E. Bot. t.* 597. *E. Fl. v. i. p.* 224. *Hook. Br. Fl. p. -*73.

Meadows; not very common. *Fl.* May—July. ♃.

Near Park Brook, Wyre Forest; *Mr. W. G. Perry.* Titterstone; *Mr. J. S. Baly.* Walford; *T. C. Eyton, Esq.* Oakley Park, near Ludlow; *Mr. H. Spare.* Near Bishop's Castle; *Gough's Camden.* Cox Wood, Coalbrookdale; *Mr. F. Dickinson.*

Fields near Shrewsbury. Sutton. Berrington Pool.

Root woody with long fibres. *Stems* more or less procumbent and spreading, alternately branched, round, hairy. *Radical leaves* large, on long hairy footstalks, roundish-kidney-shaped, 6—9 lobed, with a few scattered soft hairs in the plaits and on the ribs. *Stem leaves* with connate toothed stipules, upper ones sessile and very small. *Flowers* yellowish-green, in numerous lax, corymbose, terminal clusters. *Calyx-tube* hairy. *Germens* 1—2. *Seeds* 1—2. *Style* lateral, occasionally 2.

2. A. *arvensis*, Sm. *Field Lady's Mantle or Parsley Piert.* Leaves flat, trifid, pubescent, lobes deeply cut; flowers sessile, axillary. *E. Bot. t.* 1011. *E. Fl. v. i. p.* 225. *Hook. Br. Fl. p.* 73. *A. Aphanes*, Willd. *Aphanes arvensis*, Linn.

Fields and gravelly soils, wall-tops and hedge-banks; very common. *Fl.* May—October. ☉.

Root small, fibrous. *Stems* numerous, branched, leafy, prostrate and spreading, pubescent. *Leaves* alternate, shortly petiolate. *Stipules* large, deeply cut, pubescent. *Flowers* green, in axillary hairy tufts, inconspicuous by reason of the investing stipules. *Stamens* varying in number, generally 4. *Germens* 1 or 2.

11. SANGUISORBA. *Linn.* Burnet.

1. S. *officinalis*, Linn. *Great Burnet.* Glabrous, spikes ovate, stamens about as long as the perianth. *E. Bot. t.* 1312. *E. Fl. v. i. p.* 218. *Hook. Br. Fl. p.* 73.

Meadows and pastures; very rare. *Fl.* June, July. ♃.

Lord's meadows, Albrighton, near Shiffnal; *Dr. G. Lloyd.!* Oakley Park, near Ludlow; *Mr. H. Spare.*

Root strong, rather woody. *Stem* 1—2 feet high, erect, furrowed, branching upwards. *Leaves* pinnate with a terminal leaflet, alternate, petiolate, *leaflets* in pairs opposite, ovate or oblongo-ovate, cordate at the base, glabrous, petioled, strongly serrated, the terminal tooth almost reduced to a mucro; with two small toothed appendages at the base of each pair of petioles in the larger leaves. *Stipules* large, rounded, cut, attached to the dilated membranous base of the common petioles, which nearly surrounds the stem. *Heads of flowers* dense, crowded, dull purple, on long flower-stalks. Limb of the *perianth* in 4 ovate segments, its tube hairy, enveloping the quadrangular germen, and having at its base 4 hairy ciliated scales or bracteas, (*calyx* of many authors). *Seed* 1, rarely 2, elliptical.

TETRANDRIA—TETRAGYNIA.

12. ILEX. *Linn.* Holly.

1. I. *aquifolium*, Linn. *Common Holly.* Leaves ovate, acute, spinous, wavy and thickened at the margins; peduncles axillary, short, many-flowered, flowers subumbellate. *E. Bot. t.* 495. *E. Fl. v. i. p.* 228. *Hook. Br. Fl. p.* 74.

In hedges, woods, and on dry hills. *Fl.* May, June. ♄.

Some very large and tall trees on the northern flank of the Wrekin Hill; *E. Lees, Esq.* Hatton, near Shiffnal; *T. C. Eyton, Esq.*

Pimhill. Stiperstones Hill.

A handsome evergreen tree of slow growth but great beauty, with smooth greyish bark. *Leaves* alternate, petioled, deep shining green, very thick and rigid, upper ones entire, with only a terminal spine, lower ones with strong sharp spines. *Flowers* copious, white tinged externally with purple. *Calyx* small, ciliated at the edges. *Berries* bright scarlet.

The tree bears clipping, and is excellent for fences. The wood is hard, close-grained, and white, and much employed for turnery work, for drawing upon, &c. Of the bark, which abounds in mucilage, bird-lime is made by maceration in water. The green shining leaves, and beautiful clustered scarlet berries, are used in the decoration of our houses and churches at Christmas, and diffuse an air of enlivening cheerfulness, the more pleasing from the contrast presented by the denuded aspect of all nature during that wintry period. The custom is said to

Pl. 4 p 75.

Potamogeton

pectinatus.　perfoliatus.　crispus.

heterophyllus.　pusillus.　lucens.

rufescens.　natans.　oblongus.

have originated with the Druids, and to have been intended to afford the sylvan spirits a congenial protection from the frosts and cold winds, until the return of that joyous season, when the trees once more bud forth into beauty and verdure.

Innumerable varieties are cultivated by gardeners, mainly depending on the variegation of the leaves, the number and position of the spines, and the colour of the berries.

A variegated variety occasionally occurs in the hedges in Shropshire.

Withering (*Bot. Arr.* ii. 211) states that on the north side of the Wrekin Hill are some large trees which bear *yellow* berries.

13. POTAMOGETON. *Linn.* Pond-weed.

I. *Leaves all submersed.*

* *Fruit compressed, rostrate, keeled at the back.*

1. P. *crispus*, Linn. *Curled Pond-weed.* Stem zig-zag, branched, compressed, furrowed on the sides, and articulated at the insertion of the leaves; leaves sessile, alternate, linear-lanceolate, obtuse, submucronate, waved and serrated, with 3 nerves and transverse veins, and a few chain-like reticulations near the middle nerve; spike short, flowers few and distant, peduncles axillary, longer than the leaves, compressed and channelled; fruit ovato-acuminate, rostrate. *E. Bot.* t. 1012. *E. Fl.* v. i. p. 233. *Hook. Br. Fl.* p. 76.

Ditches and pools; not common. *Fl.* June, July. ♃.

Near Dowle's Brook, Wyre Forest; *Mr. W. G. Perry.* Pools in Coalbrookdale; *Mr. F. Dickinson.*

Ditches under Cross-hill, near Shrewsbury, in the old course of the river Severn. Bomere Pool, south-east part.

Whole plant submersed, of a bright green colour tinged with red, waving elegantly in the water. *Root* creeping. *Stipules* free, very broad and membranous, clasping the stem, retuse, finally deeply laciniated. *Flowers* yellowish green.

** *Fruit compressed, lenticular, with a small oblique point.*

a. *obtuse at the back.*

2. P. *pectinatus*, Linn. *Fennel-leaved Pond-weed.* Stem slender, round, copiously branched; leaves distichous, setaceous or linear, single-nerved, elongated and sheathing at the base by means of their adnate stipules; spike interrupted, many-flowered, peduncle elongated; fruit ovato-rotund, compressed, point slightly recurved. *E. Bot.* t. 323. *E. Fl.* v. i. p. 237. *Hook. Br. Fl.* p. 75. *P. marinus*, Linn. *P. filiformis*, Cham. and Schlecht.

Rivers, ponds, and ditches; not common. *Fl.* July. ♃.

Shrewsbury Canal, near Shrewsbury, and Uffington, in great abundance.

Root tuberous, with creeping scions. *Stems* zig-zag, very much branched, variable in length according to the stillness or rapidity of the stream. *Leaves* alternate, slender, tapering, acute, composed of two hollow compressed tubes, their solitary nerve connected by transverse alternate veins with the margins. Summit of the adnate *stipules* cloven. *Spikes* few, solitary, on long peduncles from the uppermost forks of the branches, rising just above the surface. *Flowers* 2 or 3 together, dull green.

3. P. *perfoliatus*, Linn. *Perfoliate Pond-weed.* Stem round, branched; leaves cordato-ovate, obtuse, amplexicaul, 7-nerved, with numerous smaller intermediate ones, all connected by transverse veins, near the middle nerve are small chain-like reticulations; spikes short, dense, and crowded, peduncles solitary, stout, somewhat compressed, opposite to and longer than the leaves; fruit ovato-rotund, point nearly erect, very slightly recurved. *E. Bot.* t. 168. *E. Fl.* v. i. p. 229. *Hook. Br. Fl.* p. 76.

Rivers and ponds; common. *Fl.* July, August. ♃.

Oakley Park, near Ludlow; *Mr. H. Spare.* Severn near Coalbrookdale; *Mr. F. Dickinson.*

In the river Severn, near Shrewsbury, in great abundance.

Root creeping. *Stems* variable in length according to the rapidity of the stream, more or less branched, reddish. *Leaves* olive-coloured, pellucid, wavy, involute in the bud, nerves reddish, and all uniting with the two principal lateral nerves next the mid-rib, which also converge and join the mid-rib immediately below the apex. *Stipules* oblong, obtuse, tight and close. *Peduncles* axillary. *Spikes* oblongo-ovate. *Flowers* dense, brown, with copious white pollen.

b. *keeled at the back.*

4. P. *lucens*, Linn. *Shining Pond-weed.* Stem round, branched; leaves alternate, tapering at the base into short footstalks, elliptic-lanceolate, suddenly contracted towards the apex, mucronate, membranaceous, flat and straight, not minutely waved and serrated at the margins, with about 7 principal ribs springing at various distances from the mid-rib, all connected with each other and the mid-rib by uninterrupted transverse diagonal veins, near the mid-rib are small chain-like reticulations; peduncles solitary, axillary, longer than the leaves, stouter than the stem, and swelling upwards, spikes as long as the peduncles, cylindrical, dense, many-flowered; fruit orbicular, minutely tuberculated, with a stout oblique recurved point. *E. Bot.* t. 376. *E. Fl.* v. i. p. 232. *Hook. Br. Fl.* p. 76.

Lakes and pools; rare. *Fl.* June, July. ♃.

Oakley Park, near Ludlow; *Mr. Spare.*

Chamisso and Schlechtendal include this species in a division of the genus, which has sometimes floating and coriaceous leaves, changing its name to *P. Proteus*, and considering *P. lucens* and *P. heterophyllus* as varieties. The form of the straight (not recurved) and all submersed leaves, the peculiar mode in which the ribs are all connected with each other and the mid-rib by the transverse veins, which run in a diagonal line directly and with scarcely any interruption through the lateral ribs to the margin, the solitary not crowded peduncles, the larger spike, and the keeled not obtuse fruit appear to keep this species widely distinct from *P. heterophyllus.*

Stem thickened upwards, slightly branched. *Leaves* distantly inserted, alternate, those subtending the peduncles opposite, all tapering at the base into the footstalks, which in the stem leaves are very short, in the upper ones rather longer. *Coriaceous leaves*, according to Hooker, rare, ovato-lanceolate, moderately acute, less evidently stalked than in *P. heterophyllus. Stipules* large, plicate, half the length of the leaves, with 2 prominent wings at the back. *Spikes* 2 inches long. *Flowers* green.

5. P. *pusillus*, Linn. *Small Pond-weed.* Stem slender, compressed, much branched; leaves alternate, sessile, narrow, linear,

mucronate, 3—5 ribbed, the lateral ribs uniting with each other and with the middle rib at various distances below the apex, but not together; spikes short, few flowered, peduncles elongated; fruit ovato-rotund, with a prominent recurved point. *E. Bot.* t. 215. *E. Fl.* v. i. p. 235.

Ditches and still waters; rare. *Fl.* July. ♃.

Snowdon Pool, near Beckbury; *Dr. G. Lloyd.* Canal near Newport; *Mr. F. Dickinson.* Pond at Hord's Park, near Bridgnorth; *Rev. A. Bloxam.*

Leaves long and very narrow, abruptly pointed, mucronate, alternate except beneath the flower stalks, where they are opposite. *Stipules* broader than the leaves, cloven. *Peduncles* axillary, much longer than the small ovate few-flowered *spikes. Flowers* brownish.

β. *major.* Stem more compressed; leaves broader; spike somewhat interrupted. *Hook. Br. Fl.* p. 75. *P. compressus*, Linn. *E. Bot.* t. 418. *E. Fl.* v. i. p. 234.

Plentiful in the Canal between Shrewsbury and Uffington, though rarely flowering.

II. *Upper leaves floating, lower ones submersed.*

* *Fruit compressed, lenticular, with a small oblique point.*

a. *obtuse at the back.*

6. P. *heterophyllus*, Schreb. *Various-leaved Pond-weed.* Stem round, very much branched; leaves alternate, submersed ones sessile, lanceolate, tapering at the base, submucronate, membranaceous, waved, minutely serrated and recurved, with 7 ribs connected by transverse diagonal veins, near the middle rib are small chain-like reticulations, floating leaves petiolate, elliptical, subcoriaceous, many-ribbed; peduncles crowded at the extremities of the branches, about as long as the leaves, swelling upwards or clavate, spikes cylindrical, dense, many-flowered; fruit suborbicular, very minutely tuberculated, point short, stout, very slightly recurved. *E. Bot.* t. 1285. *E. Fl.* v. i. p. 229. *Hook. Br. Fl.* p. 77.

In pools; rare. *Fl.* July. ♃.

Wet ditch near Colemere; *Rev. A. Bloxam.*

Berrington Pool, near Shrewsbury.

Root a small tuber. *Stem* zig-zag, thickened upwards, copiously branched. *Leaves* distantly inserted, those subtending the crowded peduncles opposite, and with the attendant stipules widely spreading. *Stipules* large, plicate, half as long as the leaves, membranous, broad, ovate, and blunt, with 2 stout principal ribs, and numerous parallel lateral ones. *Spikes* dense, about an inch long, the peduncles suddenly contracted below them. *Flowers* green.

7. P. *oblongus*, Viv. *Blunt-fruited broad-leaved Pond-weed.* Stem generally short, round, branched; all the leaves on long petioles, many-ribbed, distinctly cellular, leafless petioles none, submersed leaves alternate, elliptico-lanceolate, subcoriaceous, tapering at the base into the footstalk, floating leaves opposite, broadly oval, obtuse, subcordate at the base, and decurrent with the footstalks; spikes cylindrical, dense, many-flowered, peduncles solitary, as long as the

leaves, stout, not thickening upwards; fruit nearly orbicular, with a small stout slightly recurved point. *Hook. Br. Fl. p.* 78. *Cham. and Schlecht. in Linnæa, v.* ii. *p.* 214, *t.* 6, *f.* 19. *Viviani Fragm. Fl. Ital. i. t.* 2.

Wet ditches, small rivulets, and moist margins of pools; not uncommon. *Fl.* June, July. ♃.

Shawbury Heath.   Stiperstones Hill.   Bomere Pool, west side.   Ditch on Shomere Moss.

Approaching in appearance to *P. natans,* but essentially different in the form and size of the fruit, and in the places of its growth. *Leaves* with about 7 principal and several intermediate ribs, connected by innumerable transverse veins, the interstices filled with minute and delicate reticulations. *Stipules* large, membranous, obtuse, many-nerved. *Flowers* greenish.

8. P. *natans,* Linn.   *Broad-leaved Pond-weed.*   Stem long, round, branched; leaves alternate, on long petioles, many-nerved, distinctly cellular, submersed leaves linear, submembranaceous, generally wanting, the long leafless petioles only remaining, floating leaves, elliptical, coriaceous; spikes cylindrical, dense, many-flowered, peduncles solitary, shorter than the leaves, stout, not thickening upwards; fruit oblong, with a small oblique point. *Eng. Bot. t.* 1822. *E. Fl. v. i. p.* 229. *Hook. Br. Fl. p.* 78.

Stagnant waters, and pools; frequent. *Fl.* June, July. ♃.

Root creeping extensively in the mud. *Floating leaves* very variable in size and shape, more or less elongated, sometimes linear-lanceolate, obtuse at the base, and decurrent with the footstalk, with about 7 principal nerves and several intermediate ones, connected by transverse veins, the interstices filled with minute reticulations, opposite under the flower stalk, involute in bud. *Stipules* very large, 2—3 inches long, lanceolate, acute, concave, pale and membranous, with 2 principal nerves and numerous close parallel lateral ones. *Peduncles* suddenly contracted below the spike. *Flowers* olive-green.

b.   keeled at the back.

9. P. *rufescens,* Schrad.   *Reddish Pond-weed.*   Stem round, simple; submersed leaves alternate, sessile, lanceolate, obtusely pointed, membranaceous, pellucid, many-ribbed, connected by transverse diagonal veins, with numerous linear reticulations near the middle rib, floating leaves partly opposite, petiolate, elliptic-oblong, obtuse, coriaceous, tapering and decurrent at the base with the footstalks, many-ribbed, distinctly cellular; peduncles axillary, solitary, as long as the leaves, stout, not thickening upwards, spikes cylindrical, dense, many-flowered; fruit ovate, minutely tuberculated. *Hook. Br. Fl. p.* 78. *Cham. et Schlecht. in Linnæa, v.* ii. *p.* 210. *P. fluitans, E. Bot. t.* 1286. *E. Fl. v. i. p.* 230. *(not of Roth.)*

Ditches, slow streams, and ponds; not common. *Fl.* July, August. ♃.

Lilleshall Mill Pool; *Rev. Edw. Williams in Eng. Fl.*   Near Whitchurch; *Dr. T. W. Wilson!*

Ditches at Eyton-on-the-Wildmoors.   Pit near the Sharpstones Hill.

Whole plant generally of a reddish-olive colour, nearly submersed, a few of the uppermost leaves only floating. *Root* creeping. *Stem* long. *Lateral ribs* 6—7 in number on each side the mid-rib, not separate to the base of the leaf, but arising at various distances from various parts of the central rib, 2 of them

more conspicuous and stronger than the rest, arising from near the base and alone continued to a little below the apex, where they converge and unite with the mid-rib. *Peduncles* suddenly contracted below the spike. *Flowers* green.

14. SAGINA.   *Linn.*   Pearl-wort.

1. S. *procumbens,* Linn.   *Procumbent Pearl-wort.*   Perennial, glabrous, stems procumbent; leaves shortly mucronate; petals much shorter than the calyx. *E. Bot. t.* 880. *E. Fl. v. i. p.* 239. *Hook. Br. Fl. p.* 79.

Waste places and dry pastures; common. *Fl.* May—August. ♃.

Stems spreading, 2—4 inches long, sending out roots from different parts of the stem at the insertion of the leaves, and these throwing up new plants. *Leaves* linear, subulate, connate, membranous at the margins at their base, tipped with a short pellucid point or mucro. *Flowers* solitary, axillary and terminal, about an inch long. *Flowers* white, at first drooping, petals roundish.

2. S. *apetala,* Linn.   *Annual small-flowered Pearl-wort.*   Annual, stems slightly hairy, erect or ascending; leaves aristate, fringed; petals much smaller than the calyx. *E. Bot. t.* 881. *E. Fl. v. i. p.* 240. *Hook. Br. Fl. p.* 79.

Dry gravelly places; frequent. *Fl.* May, June. ☉.

Stem partly reclining, but not taking root, slender, hairy. *Leaves* as in the last species, but narrower, bristle-pointed, glaucous and slightly hairy at the margins. *Petals* obcordate, or wedge-shaped and truncated. *Seeds* somewhat wedge-shaped, rough with minute points.

15. MŒNCHIA.   *Ehrh.*   Mœnchia.

1. M. *erecta,* Sm.   *Upright Mœnchia.*   E. *Fl. v. i. p.* 241. *Hook. Br. Fl. p.* 80.   M. *glauca,* Pers.   *Sagina erecta, Linn. E. Bot. t.* 609.

On dry hills; rare. *Fl.* April, May. ☉.

Near Oswestry; *Rev. T. Salwey.*   Hawkestone; *Mr. F. Dickinson.*

Stems several, 2—4 inches high, erect, or slightly reclining at the base, glabrous. *Leaves* opposite, linear-lanceolate, connate at the base, acute, rigid, glaucous. *Flowers* erect, solitary, on long terminal stalks. *Sepals* large, acuminate, white and membranous at the edges. *Petals* white, lanceolate, entire, as long as the calyx, withering. *Seeds* numerous, minutely tuberculated.

CLASS V.

PENTANDRIA.

Fairest, yet deadliest of Flora's train.

CLASS V.

PENTANDRIA. 5 *Stamens.*

ORDER I. MONOGYNIA. 1 *Style.*

* *Perianth double, inferior. Corolla monopetalous. Germen deeply 4-lobed. Fruit with 4 (or fewer by imperfection) apparently naked seeds.*—*Nat. Ord.* BORAGINEÆ, *DC.* (Asperifoliæ, *Linn.*)

In this Natural Order the leaves are covered with asperities, consisting of rigid hairs proceeding from an enlarged indurated base.

† *Throat of the corolla naked.*

1. ECHIUM. *Calyx* in 5 deep segments. *Corolla* irregular, its throat dilated, open and naked. *Stigma* deeply cloven. *Nuts* obliquely pointed, wrinkled or rough.—Name from εχις, *a viper;* in allusion to its supposed virtues as an effectual remedy against the bite of that reptile.

2. PULMONARIA. *Calyx* with 5 angles, 5-cleft. *Corolla* funnel-shaped, its mouth naked. *Nuts* almost globular, even and polished, hairy.—Name from *pulmo, the lungs;* in allusion to a fancied resemblance of the spots on the leaves to that part of the body.

3. LITHOSPERMUM. *Calyx* in 5 deep segments. *Corolla* funnel-shaped, its mouth naked. *Nuts* ovate, pointed, even or wrinkled.—Name from λιθος, a *stone,* and σπερμα, a *seed,* from the stone-like texture of the nuts. The English name *Gromwell* is derived from the Celtic *graun,* a *seed,* and *mil,* a *stone.*

†† *Throat of the corolla more or less clothed with scales.*

4. SYMPHYTUM. *Calyx* 5-cleft. *Corolla* swollen upwards, its throat closed with connivent subulate fringed scales. *Nuts* wrinkled, perforated at the base.—Name from συμφυω to *unite;* from its healing qualities.

5. BORAGO. *Calyx* 5-cleft. *Corolla* rotate, the mouth closed with 5 obtuse and emarginate teeth. *Nuts* rugged or tuberculated.—Name corrupted from *Cor, the heart,* and *ago, to bring or apply to;* whence the adage, " I *Borage* always bring *Courage.*"

6. LYCOPSIS. *Calyx* 5-cleft. *Corolla* funnel-shaped with a curved *tube,* the mouth closed with convex, connivent hairy scales. *Nuts* angular, concave at the base.—Name from λυκος, a *wolf,* and οψις, *a face;* from a fancied resemblance in the flower to a wolf's head.

7. ANCHUSA. *Calyx* 5-cleft or 5-partite. *Corolla* funnel-shaped, *tube* straight, its mouth closed with convex, connivent hairy scales. *Nuts* roundish, wrinkled, concave at the base.—Name from αγχουσα, *paint;* from the red dye afforded by the roots of *Anchusa tinctoria,* formerly employed to stain the face.

8. MYOSOTIS. *Calyx* 5-cleft. *Corolla* salver-shaped, the lobes obtuse, the mouth half closed with short rounded valves. *Nuts* perforated at the base.—Name from μυς, μυος, *a mouse,* and ους, ωτος, *an ear;* from the shape of its leaves.

9. ASPERUGO. *Calyx* 5-cleft, unequal, with alternate smaller teeth. *Corolla* (short) funnel-shaped, its mouth closed with convex, connivent scales. *Nuts* covered by the folded and compressed calyx.—Name from *asper, rough;* in allusion to the extreme roughness of the herbage.

10. CYNOGLOSSUM. *Calyx* 5-cleft. *Corolla* (short) funnel-shaped, its mouth closed with convex, connivent scales. *Nuts* depressed, fixed to the *style* or central column.—Name from κυων, *a dog,* and γλωσσα, *a tongue;* from the shape and texture of the leaves.

** *Perianth double, inferior. Corolla monopetalous. Seeds covered with a distinct capsule.*

11. ANAGALLIS. *Calyx* 5-partite, persistent. *Corolla* rotate, *tube* none, *limb* nearly flat, 5-lobed. *Stamens* hairy. *Capsule* globose, 1-celled, bursting all round transversely.—*Nat. Ord.* PRIMULACEÆ. *Vent.*—Name from αναγελαω, *to laugh;* probably in allusion to the delightful anticipation of fine weather experienced on beholding the brilliant appearance of the delicate petals, which expand only in dry states of the atmosphere.

12. LYSIMACHIA. *Calyx* 5-partite, persistent. *Corolla* rotate, *tube* none, *limb* 5-partite. *Stamens* not distinctly hairy. *Capsule* globose, 1-celled, 10-valved.—*Nat. Ord.* PRIMULACEÆ. *Vent.*—Named in honour of King *Lysimachus,* according to some; according to others from λυσις, a *dissolving,* and μαχη, a *battle.*

13. PRIMULA. *Calyx* tubular, 5-toothed, persistent. *Corolla* salver-shaped, *tube* cylindrical, *mouth* open, *limb* 5-lobed, *lobes* emarginate. *Capsule* cylindrical, 1-celled, opening with 10 teeth.—*Nat. Ord.* PRIMULACEÆ. *Vent.*—Name from *primus, first;* in allusion to the early appearance of its blossoms.

14. HOTTONIA. *Calyx* 5—6-partite, persistent. *Corolla* salver-shaped, *tube* short, *limb* plane, 5—6-fid. *Stamens* 5—6, inserted at the mouth of the tube. *Stigma* globose. *Capsule* globose, 1-celled, (valveless, *Spr.*,—with 5 valves, *Sm.*) tipped with the long style.—*Nat. Ord.* PRIMULACEÆ. *Vent.*—Named after *Pierre Hotton,* a Professor at Leyden during the latter half of the 17th century.

15. MENYANTHES. *Calyx* 5-partite, persistent. *Corolla* funnel-shaped, *tube* short, dilated upwards, *limb* 5-partite, *segments* recurved, their disk bearded. *Stigma* capitate, 2—3 lobed. *Capsule* ovato-globose, 1-celled, 2-valved. *Seeds* parietal.—*Nat. Ord.* GENTIANEÆ, *Juss.*—Name from μηνη, a *month,* and ανθος, a *flower.*

16. ERYTHRÆA. *Calyx* 5-cleft, persistent. *Corolla* funnel-shaped, withering, *tube* narrow, longer than the calyx, *limb* 5-partite, short. *Anthers* at length spirally twisted. *Style* erect. *Stigmas* 2. *Capsule* linear, 2-celled, of 2 valves with inflexed margins.—*Nat. Ord.* GENTIANEÆ, *Juss.*—Name from ερυθρος, *red;* in allusion to the usual colour of the blossoms.

17. DATURA. *Calyx* tubular, ventricose, 5-angular, 5-toothed, deciduous, separating horizontally near the base. *Corolla* funnel-shaped, *tube* cylindrical, *limb* moderately spreading, with 5 angles, 5 plaits, and 5 acuminate teeth. *Stigma* 2-lobed. *Capsule* of 2 partially bipartite cells, 4-valved.—*Nat. Ord.* SOLANEÆ, *Juss.*—Name from *Tâtôrah,* the Arabic name of the plant, which in some parts of the East Indies is called *Dâturo.*

18. HYOSCYAMUS. *Calyx* tubular, ventricose below, 5-cleft, persistent. *Corolla* funnel-shaped, *tube* cylindrical, short, *limb* moderately spreading in 5 oblique unequal obtuse *lobes,* one broader than the rest. *Stigma* capitate. *Capsule* ovate, compressed, furrowed on each side, 2-celled, opening transversely by a convex lid.—*Nat Ord.* SOLANEÆ, *Juss.*—Name from υς, υος, *a hog,* and κυαμος, *a bean;* in allusion to the partiality of hogs for its fruit.

19. ATROPA. *Calyx* campanulate, 5-partite, persistent. *Corolla* campanulate, twice as long as the calyx, *limb* spreading in 5 equal *lobes.* *Stamens* distant, arcuate. *Berry* globose, 2-celled.—*Nat. Ord.* SOLANEÆ, *Juss.*—Name from *Atropos,* one of the Fates; in allusion to its deadly qualities, whence also its English name *dwale,* (*devil, Fr.; dolor, Lat.*)

20. SOLANUM. *Calyx* 5—10 partite, persistent. *Corolla* rotate, *tube* very short, *limb* 5-cleft, reflexed, plaited. *Anthers* erect, connivent, opening at the apex with two pores. *Berry* roundish, 2 or more celled.—*Nat. Ord.* SOLANEÆ, *Juss.*—Name of doubtful origin. According to some from *solamen,* on account of the *comfort* or *solace* derived from some species as a medicine.

21. VERBASCUM. *Calyx* erect, 5-partite, persistent. *Corolla* rotate, *tube* cylindrical, very short, *limb* spreading in 5 unequal obtuse *lobes.* *Stamens* unequal, declined, hairy at the base. *Capsule* ovato-globose, 2-celled, 2-valved.—*Nat. Ord.*—SOLANEÆ, *Juss.*—Name altered from *Barbascum,* from *barba,* a *beard;* in allusion to the shaggy nature of the foliage.

22. CONVOLVULUS. *Calyx* 5-partite, persistent. *Corolla* campanulate, 5-plaited. *Stigmas* 2. *Capsule* 1—3-celled, with as many valves, cells 1—2-seeded.—*Nat. Ord.* CONVOLVULACEÆ, *Juss.*—Name from *convolvo*, to *entwine*; whence, too, the English name *Bindweed*.

23. POLEMONIUM. *Calyx* urceolate, 5-cleft, persistent. *Corolla* rotate, *tube* short, *limb* 5-lobed, dilated. *Stamens* inserted upon the 5-teeth or valves which close the mouth of the corolla. *Stigmas* 3. *Capsule* trigono-ovate, 3-celled, 3-valved.—*Nat. Ord.*—POLEMONIACEÆ, *Juss.*—Name from πολεμος, *war*, because, according to Pliny, the plant caused a war between two kings who laid claim to its discovery.

24. VINCA. *Calyx* deeply 5-cleft, persistent. *Corolla* salvershaped, *tube* cylindrical, dilated upwards, marked externally with 5 lines, and pentagonal at the mouth, which is closed with spreading hairs and the connivent stamens, *limb* spreading, 5-partite, *segments* oblique, spirally imbricated in the bud. *Filaments* horizontal at the base, then geniculate and ascending, dilated beneath the anthers. *Anthers* connivent, woolly on the outside. *Style* common to the 2 germens, thickened upwards, bearing the capitate stigma, and again elongated and crowned at the apex with hairs. *Follicles* 2, erect, dehiscing longitudinally. *Seeds* naked (destitute of seed-down).—*Nat. Ord.* APOCYNEÆ, *Juss.*—Name from *vinco*, to *bind*; in allusion to the twisting growth of its roots and stems.

(See *Gentiana* in ORD. II.)

*** *Perianth double, superior. Corolla monopetalous.*

25. SAMOLUS. *Calyx* 5-partite, persistent. *Corolla* salvershaped, *tube* very short, *limb* spreading, 5-partite, with 5 setaceous, intermediate converging scales (sterile stamens) alternating with the lobes. *Capsule* ovate, half inferior, 1-celled, 5-valved, many-seeded. *Seeds* fixed upon a globose, central pedicellate receptacle.—*Nat. Ord.* allied to PRIMULACEÆ, *Br.*—Name, according to some, from the island of *Samos*, where *Valerandus*, a botanist of the 16th century, is alleged to have gathered our *Samolus Valerandi*. According to others, from *san*, *salutary*, and *mos*, a *hog* in Celtic, on account of its being employed by the ancients to cure diseases in hogs.

26. JASIONE. *Calyx* 5-partite, persistent. *Corolla* rotate, *tube* very short, *limb* in 5 deep lanceolate segments. *Anthers* united at their base. *Stigma* club-shaped. *Capsule* 2-celled, opening at the top. (*Flowers* pedicellate, collected into a head within a many-leaved involucre).—*Nat. Ord.* CAMPANULACEÆ, *Juss.*—Name, supposed from ιον, a *violet*, from the blue colour of the flowers, applied by Pliny to some esculent plant.

27. LOBELIA. *Calyx* 5-cleft, persistent. *Corolla* irregular, cleft longitudinally on the upper side, *limb* 5-fid, unequal, 2-lipped, upper lip bipartite, lower one larger, 3-fid. *Anthers* united. *Stigma* capitate, hairy. *Capsule* 2—3-celled, the upper free part 2-valved.—*Nat. Ord.* CAMPANULACEÆ, *Juss.*—Named in honour of *Matthias Lobel* or *L'Obel*, a *Fleming*, naturalized in England, where he published several learned botanical works.

28. CAMPANULA. *Calyx* 5-cleft, persistent. *Corolla* campanulate, longer than the calyx, 5-fid, *lobes* broad, acute and spreading. *Filaments* arched, dilated and ciliated at the base. *Style* glanduloso-villose. *Stigma* 2—3-fid. *Capsule* 3—5-celled, bursting by lateral pores, exterior to the calycine segments.—*Nat. Ord.* CAMPANULACEÆ, *Juss.*—Name from *campana*, a *bell*; in allusion to the form of the corolla.

29. WAHLENBERGIA. *Calyx* 5-cleft, persistent. *Corolla* campanulate, longer than the calyx, 5-fid, *lobes* acute and spreading. *Filaments* erect, slightly dilated and ciliated at the base. *Style* glabrous. *Stigma* simple. *Capsule* 3-celled, dehiscing at the free extremity within the calycine segments.—*Nat. Ord.* CAMPANULACEÆ, *Juss.*

30. PRISMATOCARPUS. *Calyx* 5-partite, erect, persistent. *Corolla* rotate, *limb* flat, 5-cleft. *Filaments* erect, very slightly dilated (not ciliated) at the base. *Style* glabrous. *Stigma* clavate, cleft at the apex. *Capsule* elongated, prismatical, 2—3-celled, dehiscing by valves at the extremity on the outside, and between the calycine segments.—*Nat. Ord.* CAMPANULACEÆ, *Juss.*—Name from πρισμα, πρισματος, a *prism*, and καρπος, *fruit*; in allusion to the form of the capsule.

31. LONICERA. *Calyx* 5-cleft, persistent, small. *Corolla* tubular, *tube* oblong, gibbous, *limb* in 5 deep unequal revolute segments. *Berry* globose, 1—3-celled, many-seeded.—*Nat. Ord.* CAPRIFOLIACEÆ, *Juss.*—Named in honour of *Adam Lonicer*, a German botanist.

**** *Perianth double, inferior. Corolla of 4 or 5 petals.*

32. RHAMNUS. *Calyx* urceolate, 4—5-cleft, persistent. *Petals* 4—5, minute, converging, sometimes wanting. *Stamens* opposite the petals. *Berry* globose, 2—4-celled, 2—4-seeded.—*Nat. Ord.* RHAMNEÆ, *Juss.*—Name from ραμνος, a *branch*, from its numerous branches.

33. EUONYMUS. *Calyx* flat, 4—5-cleft, having a peltate disk within. *Petals* 4—5. *Stamens* alternating with the petals, inserted upon an annular disk. *Capsule* with 3—5 angles, and as many cells and valves. *Seeds* with a coloured fleshy arillus.—*Nat. Ord.*

CELASTRINEÆ, *DC.*—Name from *Euonyme*, mother of the *Furies*; in allusion to the injurious effects produced by the fruit of this plant.

34. VIOLA. *Calyx* 5-partite, extended at the base. *Corolla* of 5 unequal petals, the under one spurred at the base. *Anthers* connate, 2 of them spurred behind. *Capsule* of 1 cell and 3 valves, many-seeded.—*Nat. Ord.* VIOLARIEÆ, *DC.*—Name of uncertain origin.

***** *Perianth double, superior. Corolla of 5 petals.*

35. RIBES. *Calyx* 5-cleft, ventricose, bearing the *petals* and the *stamens.* *Style* divided. *Berry* globose, 1-celled, many-seeded.—*Nat. Ord.* GROSSULARIEÆ, *DC.*—Name, *Ribes*, was a word applied by the Arabic Physicians to a species of *Rhubarb*, *Rheum Ribes*. Our older Botanists believed that it was our *Gooseberry*; and hence Bauhin called that plant *Ribes acidum*.

36. HEDERA. *Calyx* of 5 teeth. *Petals* broadest at the base. *Style* simple. *Berry* globose, with 3—5 cells and seeds, crowned by the calyx.—*Nat. Ord.* ARALIACEÆ, *Juss.*—Name of uncertain origin.

ORDER II. DIGYNIA. 2 *Styles.*

* *Perianth single.*

37. CHENOPODIUM. *Perianth* single, inferior, 5-cleft, persistent, carinate, connivent and enveloping the *fruit*. *Germen* nearly round, depressed, 1-celled. *Seed* solitary, horizontal, *embryo* annuliform.—*Nat. Ord.* CHENOPODEÆ, *DC.*—Name from χην, χηνος, a *goose*, and πους, a *foot*; from the shape of the leaves.

38. ORTHOSPERMUM. *Perianth* single, inferior, 5-cleft, persistent, but not enveloping the *fruit*. *Germen* oval, compressed, 1-celled. *Seed* solitary, vertical, *embryo* annuliform.—*Nat. Ord.* CHENOPODEÆ, *DC.*—Name from ορθος, *erect*, and σπερμα, *seed*; in allusion to the vertical position of the seed.

39. ULMUS. *Perianth* single, superior, persistent, 4—5-cleft. *Capsule* compressed, winged all round, (hence a *Samara*) 1-seeded. —*Nat. Ord.* ULMACEÆ, *Mirb.*—Name, according to Theis, from the Anglo-Saxon *Elm*. *Ulm* is, however, still the German word for this tree.

(See *Scleranthus* in CL. X. *Polygonum* in CL. VIII.)

** *Perianth double, inferior. Corolla monopetalous.*

40. GENTIANA. *Calyx* 4—5-cleft, persistent. *Corolla* sub-campanulate, funnel or salver-shaped, tubular at the base, destitute

of nectariferous glands. *Stamens* inserted in the tube of the corolla. *Styles* often combined. *Capsule* oblong, acuminated, 1-celled, 2-valved, many-seeded.—*Nat. Ord.* GENTIANEÆ, *Juss.*—Name from *Gentius*, King of Illyria, who, according to Pliny, brought into use the species so valued in medicine, the *bitter Gentian, G. lutea.*

41. CUSCUTA. *Calyx* cyathiform, 4—5-fid, persistent, fleshy at the base. *Corolla* campanulate, 4—5-lobed. *Stamens* inserted in the throat, alternate with the segments, and having at their base several inflexed scales. *Capsule* bursting all round transversely at the base, 2-celled, cells 2-seeded. *Embryo* spiral. (Parasitical leafless plants, with long twining filiform stems.)—*Nat. Ord.* CONVOLVULACEÆ. *Juss.*—Name, the same as κασσυθα, probably from the Arabic, *Keshout.*

*** *Perianth double, superior. Petals 5. Seeds 2.*—(*Nat. Ord.* UMBELLIFERÆ, *Juss.*)

The Natural Order UMBELLIFERÆ derives its name from the circumstance of the *flowers* being arranged in *umbels*, which are generally compound, and with or without *involucres*. The *Germen* is inferior and 2-celled. The *Calyx* is visible only just below the petals, in the form of a thickened margin, or minute teeth or segments, the whole of its tube being coherent with the germen, which it completely envelopes. The *Petals* are 5, variable in size, entire or obcordate, with an inflexed point between the lobes. *Stamens* 5, spreading, and together with the petals inserted beneath the dilated base of the style. *Styles* 2, united at their base into a 2-lobed fleshy disk, which covers the top of the germen. *Stigmas* capitate.

The characteristic distinctions of the genera depend on the various forms and markings or ridges of the mature *fruit*. This is composed of two single-seeded indehiscent *pericarps*, each crowned with its style, which eventually separate from the base upwards, and remain attached at their summits to a central, filiform and generally bipartite *column* or *axis*, which lies between them. The entire fruit, thus, as it were, suspended within the tube of the calyx, is termed *Cremocarp*; each of the two pericarps or portions, *Mericarp*; and the central column to which they are attached, *Carpophore*. Each mericarp is marked with five, more or less apparent, sometimes obliterated, longitudinal *ridges*. These are the primary ridges, because they are always, however indistinctly, present. Between the primary ridges are sometimes four others (*secondary ridges*) one in each interstice. Within the coat of the mericarps, and occupying the interstices between the ridges, are longitudinal, filiform, or clavate ducts or vessels filled with an oily or resinous substance, generally coloured, in which resides the aromatic scent peculiar to the fruit of this order of plants. These are called *vittæ*, and their number seems to be tolerably constant in each genus, except on the inner face or *commissure*, as it is termed, of the mericarps. The *seed* or *albumen* itself is, on the inner face, either plane or deeply furrowed and incurved.

I. *Umbels compound or perfect.*

A. *Fruit with many ridges (primary and secondary); secondary ones armed with prickles.*

† *Seed or Albumen plane, not furrowed, in front. (orthospermous.)* —Tr. DAUCINEÆ.

42. DAUCUS. *Calyx* of 5 teeth. *Petals* obcordate, with an inflexed point, exterior ones often radiant and deeply bifid. *Cremocarp* dorsally compressed. *Mericarps* with 5 *primary* filiform setose *ridges*, of which the three intermediate ones are dorsal, the two lateral ones on the plane inner face ; *secondary ridges* 4, equal, more prominent, with one row of prickles slightly connected at the base. *Interstices* beneath the secondary ridges univittate. *Commissural vittæ* two. *Seed* plane in front. *Involucres* many-leaved, *universal* trifid or pinnatifid, *partial* trifid or entire. *Carpophore* bifid.—Name, the δαυκος of Dioscorides.

†† *Seed or Albumen furrowed in front, (campylospermous.)* —Tr. CAUCALINEÆ.

43. TORILIS. *Calyx* of 5 teeth. *Petals* obcordate, with an inflexed point, outer ones radiant, deeply bifid. *Cremocarp* slightly laterally compressed. *Mericarps* with 5 *primary* bristly *ridges*, of which the 3 intermediate ones are dorsal, the 2 lateral ones on the inner face ; the *secondary ridges* obliterated by the numerous prickles which entirely occupy the interstices. *Interstices* univittate. *Commissural vittæ* two. *Seed* inflexed at the margin. *Universal involucre* various, *partial* of many leaves.—Name from τορευω, to *carve* or *emboss* ; in allusion to the appearance of the fruit.

B. *Fruit with few ridges ; (primary only.)*

† *Seed or albumen plane, not furrowed, in front. (orthospermous.)*

a. *Fruit dorsally compressed, with a single wing on each side.* —Tr. SELINEÆ.

44. HERACLEUM. *Calyx* of 5 teeth. *Petals* obcordate, with an inflexed point, exterior ones often radiant. *Cremocarp* remarkably and dorsally compressed, with a plane dilated margin. *Mericarps* obovate, with very slender *ridges*, dorsal ones 3, equidistant, lateral ridges 2, remote, contiguous to the dilated margin. *Interstices* univittate, *vittæ* club-shaped. *Commissural vittæ* two. *Seed* plane in front. *Universal involucre* deciduous, *partial* of many leaves. *Carpophore* bipartite.—Name from Ηρακλης, *Hercules*, who is supposed to have introduced this or some similar plant into use.

45. PASTINACA. *Calyx* nearly obsolete. *Petals* roundish, entire, involute, with a sharp point. *Cremocarp* much compressed

dorsally, with a broad flat border. *Mericarps* with very slender *ridges*, the 3 intermediate ones equidistant, the two lateral ones remote, contiguous to the dilated margin. *Interstices* univittate, *vittæ* filiform. *Seed* plane in front. *Universal* and *partial involucres* of few leaves.—Name from *pastus, food* ; in allusion to its nutritious qualities.

b. *Fruit dorsally compressed, with a double wing on each side.* —Tr. ANGELICEÆ.

46. ANGELICA. *Calyx* obsolete. *Petals* elliptic-lanceolate, entire, straight or incurved at the point. *Cremocarp* much dorsally compressed, with two broad dilated margins on each side. *Mericarps* with 3 filiform, elevated dorsal *ridges*, lateral ones expanded into the dilated wings or margins of the fruit. *Interstices* univittate. *Commissural vittæ* two. *Seed* subsemiterete. *Universal involucre* of few leaves or wanting, *partial* of many leaves. *Carpophore* free, bipartite.—Named *Angelic*, from its cordial and medicinal properties.

c. *Fruit rounded or roundish.*—Tr. SESELINEÆ.

47. SILAUS. *Calyx* obsolete. *Petals* obovate, subemarginate, with an inflexed point, appendaged, or sessile, and truncated at the base. *Cremocarp* subterete. *Mericarps* with 5 sharp, somewhat winged, equal *ridges*, the lateral ones marginal. *Interstices* with many *vittæ*. *Commissural vittæ* four. *Universal involucre* of few leaves or none, *partial* of many leaves.—Name of doubtful origin.

48. ŒNANTHE. *Calyx* of 5 teeth. *Petals* obcordate, with an inflexed point. *Cremocarp* subterete, crowned with the straight styles. *Mericarps* with 5 obtuse somewhat convex *ridges*, lateral ones marginal and rather broader. *Interstices* univittate. *Commissural vittæ* two. *Seed* tereti-convex. *Universal involucre* various, *partial* of many leaves. *Carpophore* none. *Flowers of the ray on long pedicels, sterile ; those of the disk sessile or shortly pedicellate, fertile.*—Name from οινη, a *vine*, and ανθος, a *flower* ; in allusion to the vinous odour of the blossoms.

49. ÆTHUSA. *Calyx* obsolete. *Petals* obcordate, with an inflexed point. *Cremocarp* ovato-globose. *Mericarps* with 5 elevated, thick, acutely carinated *ridges*, 2 lateral ones marginal and rather broader, bordered with a somewhat winged keel. *Interstices* univittate. *Commissural vittæ* two. *Seed* semi-globose. *Universal involucre* none, *partial* of 3 unilateral pendulous leaves. *Carpophore* free.—Name from αιθω, to *burn* ; in allusion to the acridity of its juice.

50. FŒNICULUM. *Calyx* obsolete. *Petals* roundish, entire, involute, narrower apex obtuse. *Cremocarp* subterete. *Mericarps* with 5 prominent, obtusely keeled *ridges*, the 2 lateral ones marginal

and rather broader. *Interstices* univittate. *Seed* subsemiterete. *Universal* and *partial involucres* none. *Carpophore* free.—Name from *fœnum, hay* ; its smell being compared to that of hay.

d. *Fruit laterally compressed.*—Tr. AMMINEÆ.

51. SIUM. *Calyx* of 5 teeth, or obsolete. *Petals* obcordate, with an inflexed point. *Cremocarp* laterally compressed, or contracted and subdidymous, crowned with the depressed base of the reflexed *styles*. *Mericarps* with 5 equal, filiform, rather obtuse *ridges*. *Interstices* with many *vittæ*. *Universal involucre* various, *partial* of many leaves. *Carpophore* bipartite, or adnate.—Name from *siw*, Celtic, *water* ; from the situations in which the plants grow.

52. PIMPINELLA. *Calyx* obsolete. *Petals* obcordate, with an inflexed point. *Cremocarp* laterally contracted, ovate, crowned with the swollen base of the reflexed styles. *Mericarps* with 5 equal filiform *ridges*, the 2 lateral ones marginal. *Interstices* with many *vittæ*. *Seed* gibbo-convex, plane in front. *Universal* and *partial involucres* none. *Carpophore* free, bifid.—Name altered, as Linnæus informs us, from *bipennula, twice pinnated* ; in allusion to the divisions of the leaves.

53. BUNIUM. *Calyx* obsolete. *Petals* obcordate, with an inflexed point. *Cremocarp* linear-oblong, laterally contracted, crowned with the conical base of the straight *styles*. *Mericarps* with 5 equal, filiform, obtuse *ridges*, 2 lateral ones marginal. *Interstices* with many *vittæ*. *Commissural vittæ* 2—4. *Universal involucre* of few leaves or 0, *partial* of few leaves. *Carpophore* free, bipartite.—Name from βουνος, a *hill*, where the plant delights to grow.

54. ÆGOPODIUM. *Calyx* obsolete. *Petals* obcordate, with an inflexed point. *Cremocarp* oblong, laterally compressed. *Mericarps* with 5 filiform *ridges*, 2 lateral ones marginal. *Interstices* without *vittæ*. *Seed* tereti-convex, plane in front. *Universal* and *partial involucres* none. *Carpophore* free, forked at the apex.—Name from αιξ, αιγος, a *goat*, and πους, a *foot* ; the leaves being cleft something like a goat's foot.

55. SISON. *Calyx* obsolete. *Petals* broadly obcordate, deeply notched and curved with an inflexed point. *Cremocarp* laterally compressed, ovate. *Mericarps* with 5 equal filiform *ridges*, 2 lateral ones marginal. *Interstices* univittate, *vittæ* short, club-shaped. *Seed* convex, plane in front. *Universal* and *partial involucres* of few leaves. *Carpophore* bipartite.—Name from *sizun*, Celtic, a *running stream* ; some of the plants formerly placed in this genus delighting in such situations.

56. HELOSCIADIUM. *Calyx* of 5 teeth or obsolete. *Petals* ovate, acute, or obtuse and apiculated, with the point straight or inflexed.

Cremocarp laterally compressed, ovate or oblong. *Mericarps* with 5 filiform, slightly prominent, equal *ridges*, 2 lateral ones marginal. *Interstices* univittate. *Seed* gibbo or tereti-convex, plane in front. *Universal involucre* of 1—3 leaves, *partial involucre* of 5—8 leaves. *Carpophore* entire, free.—Name from ελος, a *marsh*, and σκιαδιον, *an umbel* ; in allusion to the place of their growth.

57. APIUM. *Calyx* obsolete. *Petals* roundish, entire, with a small involute point. *Cremocarp* roundish, laterally compressed, didymous. *Mericarps* with 5 filiform, equal *ridges*, the lateral ones marginal. *Interstices* univittate, exterior ones frequently with 2—3 *vittæ*. *Seed* gibbous, convex, plane in front. *Universal* and *partial involucres* 0. *Carpophore* entire.—Name *apon*, Celtic, *water* ; from the places where the plant grows.

58. CICUTA. *Calyx* of 5 teeth, leafy. *Petals* obcordate, with an inflexed point. *Cremocarp* roundish, contracted at the side, didymous. *Mericarps* with 5 nearly plane, equal *ridges*, 2 lateral ones marginal. *Interstices* univittate, *vittæ* prominent. *Universal involucre* of few leaves or 0 ; *partial* of many leaves. *Carpophore* bipartite.—Name : *Cicuta* was a term given by the Latins to those spaces between the joints of a reed of which their pipes were made ; and the stem of this plant is similarly marked by hollow articulations.

†† *Seed or albumen furrowed in front. (campylospermous.)*

a. *Fruit elongated.*—Tr. SCANDICINEÆ.

59. CHÆROPHYLLUM. *Calyx* obsolete. *Petals* obcordate, with an inflexed point. *Cremocarp* laterally compressed or contracted, elongated and narrow, but without a beak. *Mericarps* with 5 obtuse equal *ridges*, 2 lateral ones marginal. *Interstices* univittate. *Seed* tereti-convex, *commissure* deeply furrowed. *Universal involucre* none, or of few leaves, *partial* of many leaves. *Carpophore* bifid.—Name from χαιρω, to *rejoice*, and φυλλον, a *leaf* ; in allusion to the agreeable odour of the leaves of several of the species.

60. ANTHRISCUS. *Calyx* obsolete. *Petals* obcordate, with an inflexed, generally short point. *Cremocarp* contracted on the side, rostrate. *Mericarps* subterete, without *ridges*, the beak alone with 5 *ridges*. *Seed* tereti-convex, deeply furrowed in front. *Universal involucre* none ; *partial* of many leaves.—Name given by Pliny to a plant, allied probably to this genus, but whose derivation we are ignorant of.

61. SCANDIX. *Calyx* obsolete. *Petals* obovate, with an inflexed point. *Cremocarp* laterally compressed, with a very long beak. *Mericarps* with 5 equal obtuse *ridges*, 2 lateral ones marginal. *Interstices* univittate. *Seed* tereti-convex, deeply furrowed in front. *Universal involucre* none or of few leaves ; *partial* of 5—7 leaves. *Carpophore*

forked at the apex.—Name from σκεω, *to prick*; because of the sharp and long points to the seeds, which in Shropshire have acquired the name of *Beggars' Needles.*

62. MYRRHIS. *Calyx* obsolete. *Petals* obcordate, with an inflexed point. *Cremocarp* laterally compressed. *Mericarps* with a deep furrow between them, covered by a pericarp formed of 2 membranes, the exterior having 5 equal, acutely carinated, elevated *ridges*, hollow within, closely adnate to the interior. *Interstices* without *vittæ*. *Universal involucre* none; *partial* of many leaves. *Carpophore* divided at the apex.—Name from *Myrrha, Myrrh*; the foliage of one species at least possessing an agreeable scent.

b. *Fruit turgid.*—Tr. SMYRNIEÆ.

63. SMYRNIUM. *Calyx* obsolete. *Petals* equal, elliptical or lanceolate, entire, acuminate, with an inflexed point. *Cremocarp* solid, laterally contracted. *Mericarps* reniformi-globose, didymous, each with 3 sharp and prominent dorsal *ridges*, the two lateral ones nearly obsolete, placed near the narrow commissure. *Interstices* with many *vittæ*. *Seed* involute. *Involucre* none.—Name from σμυρνα, synonymous with μυρρα, *myrrh*; from the scent of the juice.

64. CONIUM. *Calyx* obsolete. *Petals* obcordate, with an inflexed point. *Cremocarp* laterally compressed, ovato-rotund. *Mericarps* with 5 prominent, waved and crenated, equal *ridges*, the 2 lateral ones marginal. *Interstices* with many *stræ*, but without *vittæ*. *Seed* with a deep narrow furrow in front. *Universal involucre* of few leaves; *partial* of 3 leaves on one side. *Carpophore* bifid at the apex.—Name formed from κωνος, *a cone* or *top*; in allusion to the giddiness produced after taking the poisonous juice.

II. *Umbels imperfect.*

† *Fruit turgid.*—Tr. SANICULEÆ.

65. SANICULA. *Calyx* of 5 teeth, leafy. *Petals* erect, obovate, with long inflected connivent points. *Cremocarp* subglobose. *Mericarps* densely clothed with hooked prickles, without *ridges*, but with many *vittæ*. *Seed* semiglobose. *Universal involucre* lobed, *partial* of many leaves. *Carpophore* none.—Name from *sano, to heal*; allusive of the sanatory virtues formerly attributed to it.

†† *Fruit laterally compressed.*—Tr. HYDROCOTYLINEÆ.

66. HYDROCOTYLE. *Calyx* obsolete. *Petals* ovate, entire, acute, plane at the apex, stellate. *Cremocarp* laterally compressed, of two flat nearly orbicular *lobes*. *Mericarps* with 5 filiform *ridges*, of which the central *dorsal* one and the two *lateral* ones are often obsolete, the two *intermediate* ones arched. *Vittæ* none. *Seed* compressed, carinated. *Involucre* none.—Name from υδωρ, *water*, and κοτυλη, *a cup*; allusion to the depression in the centre of the peltate leaves.

ORDER III. TRIGYNIA. 3 *Styles.*

* *Flowers superior.*

67. VIBURNUM. *Calyx* 5-cleft. *Corolla* monopetalous, campanulate, 5-lobed. *Berry* inferior, usually 1-seeded. *(Leaves simple).*—Nat. Ord. CAPRIFOLIACEÆ, *Juss.*—Name of doubtful origin.

68. SAMBUCUS. *Calyx* 5-cleft. *Corolla* monopetalous, rotate, 5-lobed. *Berry* inferior, 3—4-seeded. *(Leaves pinnated).*—Nat. Ord. CAPRIFOLIACEÆ, *Juss.*—Name from σαμβυκη, a musical instrument, in the construction of which this wood is said to have been employed.

(See *Chenopodium* in ORD. II. *Stellaria* in CL. X.)

ORDER IV. TETRAGYNIA. 4 *Styles.*

69. PARNASSIA. *Calyx* deeply 5-cleft, persistent. *Petals* 5. *Nectaries* 5, opposite to the claws of the petals, fringed with globular-headed filaments. *Capsule* 1-celled, 4-valved, each valve bearing a longitudinal, linear receptacle with numerous *seeds*.—Nat. Ord. DROSERACEÆ, *DC.*—Named from *Mount Parnassus*; to which place, indeed, the plant is by no means peculiar.

ORDER V. PENTAGYNIA. 5 *Styles.*

70. LINUM. *Calyx* of 5 leaves, persistent. *Petals* 5. *Capsule* globose, mucronate, with 10 valves and 10 cells. *Seeds* ovate, compressed.—Nat. Ord. LINEÆ, *DC.*—Name from *Lin*, Celtic, *thread.*

(See *Cerastium* and *Spergula* in CL. X.)

ORDER VI. HEXAGYNIA. 6 *Styles.*

71. DROSERA. *Calyx* 5-cleft. *Petals* 5. *Capsule* 1-celled, 3-valved, many-seeded. *(Plants with leaves clothed with beautiful glandular hairs).*—Nat. Ord. DROSERACEÆ, *DC.*—Name from δροσος, *dew*; in allusion to the sparkling dewy appearance of the leaves when covered with the pellucid fluid exuded by the glands.

PENTANDRIA—MONOGYNIA.

1. ECHIUM. *Linn.* Viper's Bugloss.

1. E. *vulgare*, Linn. *Common Viper's Bugloss.* Stem herbaceous, simple, hispid with tubercles; leaves linear-lanceolate, hispid, single-ribbed; flowers in lateral, short, secund, recurved spikes; stamens longer than the corolla. *E. Bot. t.* 181. *E. Fl. v. i. p.* 269. *Hook. Br. Fl. p.* 98.
Sandy and gravelly fields; not common. *Fl.* June, July. ♂.
Shelderton, near Ludlow; *Mr. H. Spare.* Near Bridgnorth; *E. Lees, Esq.* In dry sandy fields in the parishes of Stapleton and Condover; *Rev. W. Corbett.* Near Oswestry; *Rev. T. Salwey.* Ruckley Wood, in Tong parish; *Dr. G. Lloyd.*

Near Bridgnorth, about the High Rock; *Rev. W. R. Crotch.* Near Fennymere; *T. C. Eyton, Esq.* Wilmore Hill, near Wenlock; *W. P. Brookes, Esq.!* Near Shotton; *Dr. T. W. Wilson.!*
Whole plant very hispid with strong bristles, arising from callous tubercles, intermixed with shorter rigid hairs. *Stem* erect or spreading, 2—3 feet high. *Root-leaves* spreading, petioled, those of the stem merely tapering at the base. *Flowers* very beautiful, at first reddish-purple, then brilliant blue.

2. PULMONARIA. *Linn.* Lungwort.

1. P. *officinalis*, Linn. *Common Lungwort.* Leaves scabrous, radical ones ovato-cordate, petiolate, upper ones of the stem, ovate, sessile. *E. Bot. t.* 118. *(excluding the root-leaves which belong to P. angustifolia.)* *E. Fl. v. i. p.* 262. *Hook. Br. Fl. p.* 99.
Woods and thickets; rare. *Fl.* May. ♃.
Wood below Lumhole Pool, near Coalbrookdale; perfectly wild; *Mr. F. Dickinson!* Buildwas Park, and in a coppice near a cottage at Lawley's Cross, on the left-hand side of the road from Wenlock to Buildwas; *W. P. Brookes, Esq.!*
Whole plant more or less hispid and tuberculated. *Stem* 9—12 inches high, simple, erect, hispid. *Stem-leaves* all more or less ovate, lower ones as also the radical leaves with winged petioles, upper ones sessile, all entire and hispid, with short rigid hairs arising from tubercles, marked with pale spots, which are generally naked except a green tubercle and bristle in the centre. *Clusters* terminal, corymbose, forked, with a single flower in the fork. *Flowers* at first red, then blue. *Mouth* of the *tube* slightly hairy. *Seeds* brown or blackish, downy.
A variety with white flowers has been found near Buildwas by *Miss Moseley!*

3. LITHOSPERMUM. *Linn.* Gromwell.

1. L. *officinale*, Linn. *Common Gromwell, Grey Mill, Grey Millet.* Stem erect, very much branched; leaves broadly lanceolate, acute, ribbed, hispid with tubercles and bristles above, bristly beneath; calyx-segments closed in fruit, twice as long as the fruit tube of the corolla as long as the calyx; nuts smooth. *E. Bot. t.* 134. *E. Fl. v. i. p.* 265. *Hook. Br. Fl. p.* 99.
In dry, waste, and uncultivated places, on gravelly and limestone soils; not very common. *Fl.* June. ♃
Near Oswestry; *Rev. T. Salwey.* Benthal Edge; *Mr. F. Dickinson.* Welbatch, near Shrewsbury; *Mr. T. Bodenham !* Dowle; *Mr. G. Jorden.* Whitcliffe, near Ludlow; *Mr. H. Spare!*
Ludlow Castle. Benthall Edge. Red Hill, near Shrewsbury.
Whole herb rough, with erect, appressed, rigid and tuberculated bristles. *Stems* annual, 1—2 feet high. *Leaves* sessile, entire, greyish-green, with conspicuous lateral transverse ribs. *Clusters* axillary and terminal, dense, finally elongated. *Flowers* pale-yellow. *Nuts* grey, highly polished, and of a stony hardness, seldom more than 2 or 3 ripening in each calyx.
Hooker states from an analysis of these seeds produced the following result:—The stony parts of 60 seeds, weighing upwards of 7 grains, were submitted to a redheat and reduced to 3 grains, of which four-tenths of a grain were pure silica. There was also a considerable quantity of phosphate of lime and iron.

2. L. *arvense*, Linn. *Corn Gromwell or Bastard Alkanet.* Stem erect, branched; leaves lanceolate, acute, ribbed, hispid with tuberculated bristles on both sides; tube of the corolla a little longer

than the calyx; calyx-segments patent in fruit, thrice as long as the fruit; nuts wrinkled. *E. Bot. t.* 123. *E. Fl. v. i. p.* 255. *Hook. Br. Fl. p.* 99.
Corn-fields; not common. *Fl.* May, June. ⊙.
Plentiful near Ludlow; *Mr. H. Spare.* Astley Lodge, near Hadnall; *W. W. Watkins, Esq.* Near Oswestry; *Rev. T. Salwey.* Welbatch, near Shrewsbury; *Mr. T. Bodenham !* Eyton; *T. C. Eyton, Esq.*
Whole herb of a brighter green than the preceding, but like that, rough with erect, appressed, rigid, and tuberculated bristles. *Stems* 1 foot high, branched and spreading, often decumbent. *Leaves* sessile, entire, lower ones tapering into footstalks, the lateral ribs less conspicuous than in the last species. *Clusters* terminal, leafy, at length elongated. *Flowers* white. *Nuts* brown, polished, wrinkled and pitted, usually all perfected.
The bright red bark of the root communicates its colour to oily substances, paper, linen, &c.

4. SYMPHYTUM. *Linn.* Comfrey.

1. S. *officinalis*, Linn. *Common Comfrey.* Stem winged above; leaves ovato-lanceolate, attenuated at the base, and very decurrent. *E. Bot. t.* 817. *E. Fl. v. i. p.* 264. *Hook. Br. Fl. p.* 100.
Banks of rivers and watery places; frequent. *Fl.* May, June. ♃.
Whitecliff and Oakley Park, near Ludlow; *Mr. H. Spare.* About Madeley; *Dr. G. Lloyd.* Longnor; *Rev. W. Corbett.* Burway meadows, Ludlow; *Mr. J. S. Baly.* Banks of the Severn, near Emstry; *Mr. F. Dickinson.* Preston Boats, and near Meole Brace, both near Shrewsbury. Confluence of the Teme and Corve, Ludlow.
Whole herbage rough with hairs. *Stem* 2—3 feet high, stout, branched above. *Root-leaves* ovate, somewhat rounded at the base, with winged petioles, upper ones very narrow and lanceolate, running down into winged appendages to the stem. *Racemes* in pairs, secund, drooping, very hairy. *Corolla* large, yellowish white, often purple. *Calyx-segments* ovato-lanceolate, hairy, shorter than the tube.
The large fleshy roots abound in pure insipid mucilage, and are considered serviceable in coughs and internal irritations.

5. BORAGO. *Linn.* Borage.

1. B. *officinalis*, Linn. *Common Borage.* Lower leaves obovate, attenuated at the base; segments of the corolla ovate, acute, spreading. *E. Bot. t.* 36. *E. Fl. v. i. p.* 265. *Hook. Br. Fl. p.* 100.
Among rubbish, in waste and cultivated grounds; scarcely indigenous. *Fl.* June, July. ♂.
Oakley Park, near Ludlow; *Mr. H. Spare.* Between Wellington and Ketley; *E. Lees, Esq.* Bromfield, Whitbatch, and Halton, near Ludlow; *Miss Mc Ghie.*
Whole plant very hispid with strong tuberculated bristles. *Stem* 1—2 feet high, branched, spreading. *Leaves* alternate, wavy, more or less toothed, those of the stem petiolate and eared at the base, uppermost ones narrow and sessile. *Racemes* terminal, drooping. *Corolla* large of a beautiful brilliant blue. *Stamens* very prominent, mucronate, from the inner side of a row of awl-shaped valves. *Stigma* capitate. *Nuts* rugged or tuberculated, in longitudinal rows, with a bordered scar.

6. LYCOPSIS. *Linn.* Bugloss.

1. L. *arvensis*, Linn. *Small Bugloss.* Leaves lanceolate, repando-denticulate, very hispid; calyx erect while in flower. *E. Bot. t.* 938. *E. Fl. v. i. p.* 268. *Hook. Br. Fl. p.* 100.
Corn-fields and hedge-banks; not unfrequent. *Fl.* May, June. ⊙.
Common, near Ludlow; *Mr. H. Spare.* Kemberton; *Mr. F. Dickinson.* Middle; *Dr. T. W. Wilson.*

Near Berrington pool. Shawwell, near Pimhill. Berwick road. Cross Hill, near Shrewsbury.

Whole plant very hispid with, as in most of the plants of this Natural Order, rigid bristles seated on white callous tubercles. *Stem* 1—2 feet high, erect, branched. *Lower leaves* tapering into petioles, *upper ones* sessile, semiamplexicaul. *Racemes* leafy. *Corolla* small, bright blue, *limb* slightly irregular and inclining.

7. ANCHUSA. *Linn.* Alkanet.

1. A. ***officinalis***, Linn. *Common Alkanet.* Leaves oblongo-lanceolate; spikes crowded unilateral; bracteas ovato-lanceolate, as long as the calyx. *E. Bot. t.* 662. *E. Fl. v. i. p.* 259. *Hook. Br. Fl. p.* 101.

Waste ground; rare. *Fl.* June, July. ♃.

Oakley Park meadows, near Ludlow; *Mr. H. Spare.*

Herbage rough with bristly hairs arising from callous tubercles, those of the stem more or less deflexed. *Stem* 1—2 feet high, erect. *Lower leaves* tapering at the base into petioles, *upper ones* sessile, semiamplexicaul. *Corolla* deep purple, the segments of the limb rather narrow. *Calyx-segments* narrow, longer than the tube, single-ribbed, hairy on both sides.

2. A. ***sempervirens***, Linn. *Evergreen Alkanet.* Leaves ovate, lower ones upon long stalks; peduncles axillary, each bearing two dense spikes with an intermediate flower, accompanied by two leaves. *E. Bot. t.* 45. *E. Fl. v. i. p.* 260. *Hook. Br. Fl. p.* 101.

Waste ground, among ruins, and by road-sides; rare. *Fl.* May, June. ♃.

Lanes near Oswestry; *Rev. T. Salwey!* Broseley, by the road-side, doubtfully wild; *Mr. F. Dickinson.* Castle walk, Ludlow; *Miss Mc Ghie.* Near the Leopard on the road-side between Wenlock and Broseley; *W. P. Brookes, Esq.!*

Whole plant rough with rigid hairs, those of the stem and peduncles more or less deflexed. *Stems* 18 inches high, annual, somewhat angular. *Leaves* broadly ovate, wavy, strongly ribbed, upper ones sessile, radical ones lasting through the winter. *Corolla* of a beautiful blue, rather salver than funnel-shaped, tube white, scarcely half as long as the limb, tumid, tetragonous at the base. *Calyx-segments* longer than the tube, 3-ribbed, hairy on the outside only. *Nuts* ovate, compressed, wrinkled.

8. MYOSOTIS. *Linn.* Scorpion-grass.

1. M. ***palustris***, Kiphoff. *Great Water Scorpion-grass, or Forget-me-not.* Fruit smooth; calyx with straight appressed bristles, when in fruit campanulate, open, shorter than the divergent pedicels; limb of the corolla flat, longer than the tube; pubescence of the stem spreading. *E. Bot. t.* 1973. *E. Fl. v. i. p.* 249. *Hook. Br. Fl. p.* 101.

Ditches and sides of rivers; abundant. *Fl.* all summer. ♃.

Root creeping, long, blackish. *Stem* 12—18 inches high, branching, often decumbent below, then ascending, clothed with pubescence more or less spreading, often nearly erect, sometimes nearly wanting, that of the branches more copious and decidedly spreading. *Leaves* sessile, elliptic-oblong, bluntish, clothed on both sides with short, close-pressed bristles, 3-ribbed. *Racemes* many-flowered, 2 or 3 together, on terminal leafless peduncles, which, as well as the pedicels and calyces, are covered with short, dense, erect, appressed bristles. *Calyx* divided about one-third the way down, segments broad, triangular, single-ribbed. *Flowers* the largest of our species, of a beautiful enamelled blue, with a yellow eye and a small white ray at the base of each segment. *Seeds* ovate, compressed, brownish, somewhat bordered, polished, and beautifully dotted.

2. M. ***cæspitosa***, Schultz. *Tufted Water Scorpion-grass.* Fruit smooth; calyx with straight appressed bristles, when in fruit campanulate, open, shorter than the divergent pedicels, limb of the corolla concave, equalling the tube; pubescence of the stem appressed. *E. Bot. Suppl. t.* 2661. *E. Fl. v. i. p.* 251. *Hook. Br. Fl. p.* 102.

Watery places; common. *Fl.* May—Aug. ⊙. or ♂. Hook. ♃. or ♂. Sm.

Whole plant of a slenderer, more copiously branched and tufted appearance, with narrower leaves and less conspicuous flowers than in *M. palustris.* *Root* fibrous, not creeping. *Stems* weak, lax, branched from the base, throwing out fibres from the lower joints, about one foot or more high. *Stem-leaves* linear oblong, obtuse, sessile, the lower ones tapering into footstalks, clothed with short, close-pressed bristles, 3-ribbed, and veiny. *Racemes* elongated, 2 or 3 together on terminal peduncles, with a leaf or two at the base of each raceme, many-flowered. *Pedicels* more erect when in fruit than *M. palustris.* Peduncles, pedicels and calyces covered with short appressed bristles. *Calyx* cleft scarcely half-way down, though more deeply than in *M. palustris,* segments more acuminate, 3-ribbed. *Seeds* as in the preceding.

3. M. ***sylvatica***, Hoffm. *Field Scorpion-grass.* Fruit smooth; calyx with spreading uncinate bristles, deeply 5-cleft, when in fruit ovate, closed, shorter than the divergent pedicels; limb of the corolla flat, longer than the tube; root-leaves on dilated stalks. *E. Bot. Suppl. t.* 2630. *E. Fl. v. i. p.* 252. *Hook. Br. Fl. p.* 103.

Dry shady places; common. *Fl.* May—July. ♃.

Root fibrous. *Stems* one or more from the root, 12—18 inches high, stout, angular, erect, branching above, clothed as is the whole herbage with copious, long, spreading, somewhat bristly hairs. *Radical leaves* oblongo-obovate, tapering into dilated footstalks. *Stem-leaves* oblongo-lanceolate, with an obtuse point, sessile. *Racemes* terminal, each with a sessile, ovate, acute leaf at the base of its stalk. *Pedicels* moderately divergent, longer than the calyx, and as well as the peduncles, clothed with erect hairs. *Calyx-tube* with spreading uncinate bristles, segments longer than the tube, lanceolate, acute, with erect bristles.

4. M. ***arvensis***, Hoffm. *Field Scorpion-grass.* Fruit smooth; calyx with spreading uncinate bristles, half 5-cleft, when in fruit ovate, closed, shorter than the divergent pedicels; limb of the corolla concave, equalling the tube; root-leaves on dilated footstalks. *E. Bot. Suppl. t.* 2629. *Hook. Br. Fl. p.* 103.

Cultivated ground, corn-fields, hedge-banks, &c.; common. *Fl.* June—Aug. ⊙.

Very similar in general appearance to the last, and a very questionable species, the distinctive characters being very trifling and vague.

Root fibrous. *Stems* one or more 6—18 inches high, stout, angular, branching, clothed, as is the whole herbage, with copious, long, spreading, somewhat bristly hairs. *Radical leaves* oblongo-obovate, obtuse, tapering into dilated footstalks. *Stem-leaves* oblong, acute, sessile. *Racemes* terminal and axillary, with a leaf at the base; the terminal one forked, generally with a solitary flower in the axil. *Pedicels* moderately divergent, longer than the calyx, and as well as the peduncles, clothed with erect hairs. *Calyx-tube* with spreading uncinate bristles, segments about equal to the tube, lanceolate, acute, with erect bristles.

5. M. ***collina***, Hoffm. *Early Field Scorpion-grass.* Fruit smooth; calyx with spreading uncinate bristles, when in fruit ventricose, open, equalling the divergent pedicels; limb of the corolla concave, shorter than the tube; raceme usually with a distant flower at the base. *E. Bot. Suppl. sub. fol.* 2629. *Hook. Br. Fl. p.* 103. *M. arvensis.* *E. Bot. t.* 2558. *E. Fl. v. i. p.* 252.

Dry hilly and open places; rare. *Fl.* April, May. ⊙.

Near Oswestry; *Rev. T. Salwey!*

Haughmond Hill.

Root fibrous. *Stems* 3—6 inches high, erect, branched and spreading from the base, slender, densely covered with spreading hairs. *Lower leaves* obovate, petiolate, upper ones oblong, obtuse, sessile, covered on both sides with soft silky hairs. *Racemes* single, many-flowered, with one distant flower at the base, clothed as well as the pedicels with erect appressed hairs. *Tube* of the calyx with uncinate bristles, segments lanceolate, with spreading bristles, single-ribbed.

6. M. ***versicolor***, Lehm. *Yellow and blue Scorpion-grass.* Fruit smooth; calyx with spreading uncinate bristles, when in fruit oblong, closed, longer than the almost erect pedicels; limb of the corolla concave, shorter than the exserted tube. *E. Bot. t.* 2558. *(ad calc.) t.* 480. *(fig. sinist.)* *E. Fl. v. i. p.* 254. *Hook. Br. Fl. p.* 104.

Hills, hedge-banks, and other dry places; not uncommon. *Fl.* Ap.—June. ⊙.

Hedge-banks, near Bayston Hill. Pimhill. Hancott Park. Haughmond hill, and about Shrewsbury, and other parts of the County generally.

Root fibrous. *Stems* variable in luxuriance according to situation, 3—8 inches high, erect, branched and spreading from the base, slender, nearly round, obscurely 4-angled, covered with partly appressed, partly spreading hairs. *Leaves* linear-oblong, narrow, with bluntish points, lower ones petiolate, upper ones sessile, very hairy on both sides. *Racemes* many-flowered, elevated either singly or in pairs on long terminal leafless peduncles, which, as well as the pedicels, are clothed with densely appressed hairs, at first revolute, afterwards erect. *Pedicels* about half the length of the calyx. *Tube* of the calyx with slightly spreading, almost erect, uncinate bristles, segments with erect appressed ones. *Corolla* small, limb shorter than the tube, lower or fully expanded flowers blue, upper or younger ones pale yellow. *Stigma* capitate.

9. ASPERUGO. *Linn.* Madwort.

1. A. ***procumbens***, Linn. *German Madwort.* *E. Bot. t.* 661. *E. Fl. v. i. p.* 266. *Hook. Br. Fl. p.* 105.

Waste places; rare. *Fl.* June, July. ⊙.

In a field near the confluence of the rivers Corve and Teme, Ludlow; *Turn. & Dillw. Bot. Guide.*

Stems procumbent, angular, rough with short hooked prickles. *Leaves* oblongo-lanceolate, solitary or opposite or 3—4 from nearly the same point of the stem, lower ones petiolate, all rough and slightly hispid. *Flowers* small, blue, axillary, solitary, valves white or reddish. *Peduncles* short, at first erect, then curved downward. *Calyx* small, much enlarged in fruit. *Seeds* finely granulated.

10. CYNOGLOSSUM. *Linn.* Hound's-tongue.

1. C. ***officinale***, Linn. *Common Hound's-tongue.* Stem-leaves lanceolate, attenuate at the base, sessile, downy; stamens shorter than the corolla. *E. Bot. t.* 921. *E. Fl. v. i. p.* 261. *Hook. Br. Fl. p.* 105.

Waste-grounds and by road-sides; common. *Fl.* June, July. ♂.

Whole plant downy and very soft to the touch, dull green, with a fetid smell like that of mice. *Stem* 18 inches to 2 feet high. *Root-leaves* tapering at each end, on long footstalks. *Clusters* terminal, panicled. *Flowers* dull crimson. *Fruit* depressed, rough with hooked prickles.

2. C. ***sylvaticum***, Hænke. *Green-leaved Hound's-tongue.* Stem-leaves lanceolate, broad at the base, sessile, shining, slightly hairy and scabrous, especially beneath; stamens shorter than the corolla. *E. Bot. t.* 1642. *E. Fl. v. i. p.* 261. *Hook. Br. Fl. p.* 105.

Shady places, by road-sides, hedges, &c.; rare. *Fl.* June, July. ♂.

Neighbourhood of Ludlow; *Miss Mc Ghie.*

The more or less shining bright green leaves, their different shape, and want of downy pubescence and scent, afford a ready distinction from the last species. *Radical leaves* ovato-lanceolate, on very long footstalks. *Flowers* at first reddish, afterwards of a dull blue.

11. ANAGALLIS. *Linn.* Pimpernel.

1. A. ***arvensis***, Linn. *Common Pimpernel or Poor Man's Weather-glass.* Stem procumbent, or nearly so; leaves opposite, sessile, ovate, obtuse, dotted beneath; margins of the corolla crenate, piloso-glandulose. *E. Bot. t.* 529. *E. Fl. v. i. p.* 281. *Hook. Br. Fl. p.* 105.

Corn-fields and cultivated ground; frequent. *Fl.* June, July. ⊙.

Stems square, branched from the lower part. *Flower-stalks* longer than the leaves, recurved after flowering. *Flowers* bright scarlet, with a violet-coloured mouth, occasionally pale pink. *Sepals* lanceolate, membranous at the edges. *Seeds* numerous, minutely pitted.

β. cærulea. Stem more erect; margins of the corolla toothed, scarcely at all glandulose. *Hook. Br. Fl. p.* 105. *A. cærulea, Schreb.* *E. Bot. t.* 1823. *E. Fl. v. i. p.* 281.

β. In corn-fields; rare.

In a corn-stubble between Tugford and Munslow, in the latter parish; *Rev. W. Corbett.* In a field near Hopton Hill; *T. C. Eyton, Esq.* Shotton, near Shrewsbury; *Dr. T. W. Wilson!*

Corn-fields, Hill Top, Wenlock Edge.

Professor Henslow has proved by cultivation from seed, *A. arvensis* and *cærulea* to be varieties of the same species,—*see Loudon's Mag. Nat. Hist. iii.* 537.

The variety *β. cærulea* differs only in having blue flowers, a more erect stem and narrower leaves, though in our Shropshire specimens the leaves appear as broad as those of *A. arvensis, a.* There is no perceivable difference in the calyx.

A variety with rose-coloured flowers was found in 1832, at Shotton, near Shrewsbury, by *W. W. Watkins, Esq.* who has not since observed it.

2. A. ***tenella***, Linn. *Bog Pimpernel.* Stem creeping, filiform; leaves petiolate, roundish, somewhat pointed, dotted beneath; margins of the corolla entire. *E. Bot. t.* 530. *E. Fl. v. i. p.* 282. *Hook. Br. Fl. p.* 106.

An elegant little plant, covering the mossy sides of all our Shropshire pools, and in boggy ground generally. *Fl.* July, August. ♃.

Felton farm, near Ludlow; *Mr. H. Spare.* In a bog at the N.E. base of the Wrekin; *E. Lees, Esq.* Near Oswestry; *Rev. T. Salwey.* Near Walford; *T. C. Eyton, Esq.* Rowley, near Wenlock; *W. P. Brookes, Esq.!* Oreton, common; *Mr. G. Jorden.*

Boggy ground at the foot of the Caradoc hill. Haughmond hill. "Fairy land," near Westfelton, Bomere and Abbots Betton pools.

Flowers on long, slender, axillary peduncles. *Corolla* of a delicate pale rose-colour, streaked with numerous darker veins, somewhat campanulate. *Leaves* small. *Stigma* acute.

12. LYSIMACHIA. *Linn.* Loosestrife.

* *Peduncles many-flowered.*

1. L. ***vulgaris***, Linn. *Great yellow Loosestrife.* Stem erect; panicle terminal; leaves ovato-lanceolate, nearly sessile, opposite or ter-quaternate. *E. Bot. t.* 761. *E. Fl. v. i. p.* 278. *Hook. Br. Fl. p.* 106.

Sides of rivers and pools; not common. *Fl.* July, August. ♃.

Oakley Park, near Ludlow; *Mr. H. Spare.* Banks of the Severn, between Coalport bridge and Apley; *E. Lees, Esq.* Ellesmere; *Rev. A. Bloxam.* Shomere pool, near Condover; *Mr. F. Dickinson.* Dowle, not plentiful; *Mr. G. Jorden.* Severn, below Bridgnorth; *Dr. G. Lloyd.!*

Berrington and Almond pools, near Shrewsbury, but sparingly.

Root creeping. *Stems* simple, erect, 3—4 feet high, varying in the number of the angles as the leaves are two or more together, covered with jointed glandular hairs or down. *Leaves* glabrous or nearly so above, downy beneath, varying greatly in form, the lower ones being narrow and lanceolate, and gradually increasing in breadth until they become ovato-lanceolate, and in some instances nearly ovate in the upper portion of the stem. *Clusters* of the panicle corymbose, partly axillary, partly terminal, downy. *Corolla* large, yellow. *Filaments* connate at the base.

On both sides of the leaves of this species in specimens from Bath, Barmouth, and from two localities in Shropshire, distant at least 30 miles from each other, are numerous scattered roundish tubercles of a deep crimson colour and semipellucid. Similar tubercles are observable on Shropshire specimens of *Lysimachia Nummularia,* and *Lysimachia thyrsiflora* from Wiltshire.

**** *Peduncles single-flowered.*

2. L. *Nemorum,* Linn. *Yellow Pimpernel, or Wood Loosestrife.* Stem procumbent, creeping at the base; peduncles axillary, solitary, longer than the leaves; sepals linear, subulate; stamens smooth; leaves opposite, nearly sessile, ovate, acute. *E. Bot. t.* 527. *E. Fl. v. i. p.* 279. *Hook. Br. Fl. p.* 106.

Moist woods and shady places; frequent. *Fl.* during the summer months. ♃.

Stems branched, square, smooth, red and pellucid. *Flowers* golden yellow. *Corolla* fringed with minute glandular hairs.

3. L. *Nummularia,* Linn. *Creeping Loosestrife, Money-wort or Herb Twopence.* Stem prostrate, creeping; peduncles axillary, solitary, shorter than the leaves; sepals ovate, acute; stamens glandular; leaves opposite, shortly petiolate, subcordate, obtuse. *E. Bot. t.* 528. *E. Fl. v. i. p.* 280. *Hook. Br. Fl. p.* 107.

Moist shady woods and wet places; not uncommon. *Fl.* June, July. ♃.

Oakley Park, near Ludlow; *Mr. H. Spare.* Near Oswestry; *Rev. T. Salwey.* Near Lee Bridge; *Dr. T. W. Wilson!*

Stems nearly simple, compressed, with four prominent angles, smooth, pale green. *Flowers* larger than in the last, pale yellow. *Corolla* fringed with minute glandular hairs, and bearing on both sides numerous scattered, crimson, semipellucid tubercles, similar to those alluded to under L. *vulgaris.*

13. PRIMULA. *Linn.* Primrose.

1. P. *Veris,* Linn. *Primrose.* Leaves toothed, wrinkled; scapes single or many-flowered; limb of the corolla flat or concave. *Linn. Sp. Pl.* 204. *Henslow Cat. Br. Pl.* 2nd ed. p. 46.

a. officinalis, Linn. *Common Cowslip or Paigle.* Leaves contracted below the middle; scape many-flowered; flowers umbellate, nodding; sepals acute; limb of the corolla very short, concave, mouth inflated. *Henslow Cat. Br. Pl. l. c.* Primula Veris, Linn. *E. Bot. t.* 5. *E. Fl. v. i. p.* 272. *Hook. Br. Fl. p.* 107.

Meadows and pastures; abundant. *Fl.* April, May. ♃.

Root premorse. *Leaves* obovato-oblong, contracted below the middle into winged footstalks, hoary, downy, and soft, irregularly waved and toothed at the margin. *Scapes* taller than the leaves, downy, bearing at their summits an umbel of flowers with small partial bracteas. *Limb* of the corolla deep yellow, with 5 orange spots, "the freckles" of Shakspeare, fragrant.

β. elatior, Linn. *Oxlip Primrose.* Leaves contracted towards the middle; scape many-flowered; flowers umbellate, erect, the outer ones only nodding; sepals acute; limb of the corolla expanded, flat, mouth inflated; tube much exserted. *Henslow Cat. Br. Pl. l. c.* P. *vulgaris, β.* Huds. P. *elatior, With. E. Bot. t.* 513. *E. Fl. v. i. p.* 271. *Hook. Br. Fl. p.* 107.

Woods and pastures; not common. *Fl.* April, May. ♃.

High pastures near Little Wenlock; *Withering.* Oakley Park, near Ludlow; *Mr. H. Spare.* Near Oswestry; *Rev. T. Salwey.* Near Great Ness; *T. C. Eyton, Esq.* Pastures near Preston Gobbalds' Wood.

Leaves and *scapes* as in the preceding. *Limb* of the corolla pale sulphur-coloured, darker at the base, more deeply lobed, inodorous.

γ. acaulis, Linn. *Common Primrose.* Leaves obovato-oblong, not contracted in the middle, toothed and wrinkled; scapes single-flowered; flowers erect; sepals acuminate; limb of the corolla expanded, flat, mouth not inflated; tube exserted. *Henslow Cat. Br. Pl. l. c.* P. *vulgaris,* Huds. *E. Bot. t.* 4. *E. Fl. v. i. p.* 271. *Hook. Br. Fl. p.* 107.

Woods, hedge-banks, and pastures; abundant. *Fl.* April, May. ♃.

Root as in the preceding. *Leaves* obovato-oblong, tapering gradually into broad short footstalks, nearly glabrous above, downy beneath, unequally toothed and waved at the margins. *Flowers* numerous, sulphur-coloured, with a darker spot in the middle, all on solitary scapes, about as long as the leaves, and which, if traced to the base, are found to spring from a common point, and to constitute a sessile umbel with a small bractea at the base of each partial stalk.

A variety with dull reddish-coloured flowers, I once gathered on Haughmond hill in 1831, which character it has preserved during cultivation in the garden ever since.

A variety has occurred once in the woods near Bomere pool, having the leaves of the usual form of those of *γ. acaulis,* and the flowers similar, but collected into an umbel elevated on a common stalk as long or longer than the leaves, and with or without a few single-flowered scapes also arising from the root. This variety agrees with specimens gathered near Cambridge, which appear to identify it as P. *vulgaris, β.* of Engl. Flora, to which Smith quotes P. *elatior,* of Hooker's Flora Londin. t. 9.

Linnæus considered the Cowslip, Oxlip and Primrose, as varieties only of the same plant, though most modern botanists have usually separated and described them as two, and sometimes three distinct species. In the experiments of the Hon. and Rev. W. Herbert, (*Hort. Trans.* iv. 19) and of Professor Henslow, (*Loudon's Mag. Nat. Hist.* iii. 406) the latter instituted especially with a view to a decision of the question, the 3 varieties, as well as the Polyanthus, were raised by cultivation from the seed of the common Cowslip. These results clearly and satisfactorily demonstrate the opinion of Linnæus to have been correct. Indeed the mere external characters of the plants and their subvarieties appear to approximate to this conclusion, as will be evident from the following tabular arrangement, in which it will be perceived that *var. β. elatior* connects *var. a. officinalis* with *var. γ. acaulis;* and the subvariety of *acaulis* (P. *vulgaris, β.* of Smith) connects *var. γ. acaulis* with *var. β. elatior:*—

	Leaves	Scape	Limb of Corolla	Tube of Corolla	Mouth of Corolla	Sepals
a. officinalis.	contracted below the middle.	umbellate, drooping, longer than the leaves.	concave.	not exserted.	inflated.	obtuse.
β. elatior.	contracted towards the middle.	umbellate, erect, exterior flowers only drooping; longer than the leaves.	flat.	much exserted.	inflated.	acute.
γ. acaulis.	not contracted but tapering gradually into the footstalks.	solitary, erect, as long as the leaves.	flat.	exserted.	not inflated.	acuminated.
subvar. acaulis.	not contracted but tapering gradually.	umbellate and solitary, erect, as long and longer than the leaves.	flat.	not exserted.	not inflated.	acuminated.

14. HOTTONIA. *Linn.* Water Violet.

1. H. *palustris,* Linn. *Common Water Violet or Featherfoil.* Flowers whorled on a long solitary cylindrical stalk; corolla longer than the calyx; leaves pectinated. *E. Bot. t.* 364. *E. Fl. v. i. p.* 277. *Hook. Br. Fl. p.* 108,

Ditches and ponds; frequent. *Fl.* May, June. ♃.

Shotton and Marton; *W. W. Watkins, Esq.* Near Hodnet; *Rev. W. Corbett.* Ellesmere; *Rev. A. Bloxam.* Ditches near Aston and Whittington; *Rev. T. Salwey.* March pond, Wilcot; and near Walford; *T. C. Eyton, Esq.* Ditches near Bomere pool. Hancott pool. Near Albrighton.

Root creeping. *Leaves* crowded, smooth, all submerged. *Flowers* large, handsome, pale purple, deep yellow at the summit of the tube, rising above the water on round peduncles, which, as well as the partial stalks, bracteas and sepals, are covered with glandular viscid pubescence. *Tube* of the calyx tumid at the base. *Bracteas* linear at the base of each pedicel.

15. MENYANTHES. *Linn.* Buckbean.

1. M. *trifoliata,* Linn. *Common Buckbean or Marsh Trefoil.* Leaves ternate; scape axillary, bearing an erect bracteated thyrsus of numerous conspicuous flowers. *E. Bot. t.* 495. *E. Fl. v. i. p.* 275. *Hook. Br. Fl. p.* 108.

Marshy places and boggy margins of pools; frequent. *Fl.* May—July. ♃.

Near the Poles Farm, near Ludlow; *Mr. H. Spare.* Pools about the south-west end of the Wrekin; *E. Lees, Esq.* Near Church Stretton; *Rev. W. Corbett.* Ellesmere; *Rev. A. Bloxam.* Morda, near Oswestry; *Rev. T. Salwey.* Snowdon pool, near Beckbury; *Mr. F. Dickinson.* Pits near Walford; March pond, Wilcot; *T. C. Eyton, Esq.* Marsh pool, near Wenlock; *W. P. Brookes, Esq.!* Button Oak, near Bewdley; *Mr. G. Jorden.*

Bomere, Berrington, and Almond pools, near Shrewsbury.

Rhizoma densely matted, articulated, sending out long, stout, fibrous roots from the joints, ascending. *Lower leaves* reduced to membranous sheaths, *upper ones* ternate, on long petioles, dilated and sheathing at the base, leaflets obovate, obscurely toothed. *Flowers* very elegant, white tipped externally with red, and beautifully bearded with the white filaments of the disk. *Seeds* ovate, smooth, and shining.

16. ERYTHRÆA. *Renealm.* Centaury.

1. E. *Centaurium,* Pers. *Common Centaury.* Stem nearly simple; leaves ovato-oblong; flowers nearly sessile, fasciculato-paniculate; calyx half the length of the tube of the corolla. *Eng. Fl. v. i. p.* 321. *Hook. Br. Fl. p.* 109. *Chironia Centaurium, Curt. E. Bot. t.* 417.

Dry pastures; frequent. *Fl.* July, August. ☉.

Oakley Park, near Ludlow; *Mr. H. Spare.* Smethcott and near Hadnal; *W. W. Watkins, Esq.* Near Oswestry; *Rev. T. Salwey.* Benthal Edge; *Mr. F. Dickinson.* Smethcot and Hardwicke; *Dr. T. W. Wilson.* Rowley, near Harley; *W. P. Brookes, Esq.* Near Dowle's Brook, Wyre Forest; *Mr. W. G. Perry.* Meadows, foot of the Caradoc Hill. Near Uffington. Bomere. Benthall Edge. Wrekin.

Stem 8—12 inches high, erect, quadrangular. *Radical leaves* numerous, broader and more obtuse than those of the stem, which are in distant pairs, 5-nerved. *Panicles* of flowers fascicled near the top of the stem, forming a sort of corymb. *Corolla* handsome, rose-coloured, opening only in sun-shine, and closing as soon as gathered. *Capsule* closely invested with the permanent tube of the corolla.

17. DATURA. *Linn.* Thorn-apple.

1. D. *Stramonium*, Linn. *Common Thorn-apple.* Herbaceous; leaves ovate, angulato-sinuate, nearly cuneiform and unequal at the base, glabrous; fruit ovate, erect, densely spiny. *E. Bot. t.* 549. *E. Fl. v. i. p.* 315. *Hook. Br. Fl. p.* 110.

Waste ground near towns and gardens; not common. *Fl.* July. ⊙.

Near Ludlow; *Miss Mc Ghie.* Near Pontesford Hill, and Old Heath, Shrewsbury; *D. Crawford, Esq.*

Stem 2—3 feet high, spreading, much branched and forked. *Leaves* from the forks, large, shortly petiolate, dull green. *Flowers* axillary, erect, white, fragrant, especially at night. *Capsule* large, the base beset with the circular reflexed remains of the deciduous *calyx*, divided below into 4 cells by 4 dissepiments, of which two only reach the summit which is 2-celled. *Seeds* black.

18. HYOSCYAMUS. *Linn.* Henbane.

1. H. *niger*, Linn. *Common Henbane.* Leaves amplexicaul, oblong, sinuated; flowers axillary, subsessile, unilateral. *E. Bot. t.* 591. *E. Fl. v. i. p.* 316. *Hook. Br. Fl. p.* 110.

Waste places, often near towns, villages, and monastic ruins. *Fl.* July. ♂. (⊙. Sm., Hook, and Huds. ♂. Linn.)

Shelderton, near Ludlow; *Mr. H. Spare.* Leaton Shelf; *Mr. E. Elsmere, junr.* Middle; *W. W. Watkins, Esq.* Coreley and Hopton Castle; *Rev. W. Corbett.* Merrington and Prescot; *T. C. Eyton, Esq.* Ludlow Race-course; *Miss Mc Ghie.* Limeworks near Wenlock; *W. P. Brookes, Esq.!* Near Oreton, plentiful; *Mr. G. Jorden.* Knockin and Lilleshall; *Dr. Du Gard.*

Haughmond Abbey. Hill Top, Wenlock Edge.

Root fusiform. *Stem* much branched, round, covered as is the whole herbage with copious, viscid, glandular hairs or down, emitting an oppressive and fetid odour. *Leaves* soft and pliant, the upper ones nearly entire. *Corolla* of a peculiar lurid colour approaching to a dingy yellow, most beautifully and delicately pencilled with dark-purple veins. *Calyx* strongly veined. *Seeds* numerous, singularly impressed or reticulated with deep dots.

The flowers, when in blossom, are all somewhat drooping and directed towards the under side of the stem, but after impregnation, and as the seeds advance towards maturity, they gradually turn upwards until the capsules surrounded by the persistent calyx are placed in an erect position; a wise provision of Nature to prevent the premature dispersion of the seeds, which would otherwise take place, were the capsule to remain pendulous, by reason of its transverse dehiscence. Highly narcotic.

19. ATROPA. *Linn.* Dwale.

1. A. *Belladonna*, Linn. *Common Dwale, or deadly Nightshade.* Stem herbaceous; leaves broadly ovate, entire; flowers solitary, axillary, on short peduncles. *E. Bot. t.* 592. *E. Fl. v. i. p.* 316. *Hook. Br. Fl. p.* 111.

Waste places, especially among ancient ruins; not common. *Fl.* June. ♃.

Lilleshall Abbey; *Dr. G. Lloyd.*

Root fleshy, creeping. *Stem* 3—4 feet high, round, branched, and slightly downy. *Leaves* lateral, mostly two together, of unequal size, acute and smooth. *Flowers* drooping, lurid purple. *Berry* shining, violet black, poisonous: thus em-

phatically described by Bluff and Fingerhuth *Comp. Fl. Germ.* "Bacca rotunda, nigricans, splendens, succulenta, cerasi sylvestris forma et magnitudine, intus purpureo-rosea, venenata, lethalis."

The whole plant, but especially the berries, is poisonous, and the most effectual remedy against its deadly operations is to administer copious draughts of vinegar, and to prevent the patient from sleeping.

20. SOLANUM. *Linn.* Nightshade.

1. S. *Dulcamara*, Linn. *Woody Nightshade or Bittersweet.* Stem shrubby, zig-zag, climbing; lower leaves cordate, upper ones auriculato-hastate; corymbs drooping, inserted opposite the leaves; berries ovate, red. *E. Bot. t.* 365. *E. Fl. v. i. p.* 318. *Hook. Br. Fl. p.* 111.

Moist hedges and thickets; common. *Fl.* June, July. ♃.

Root woody. *Stem* climbing to the height of many feet. *Leaves* smooth. *Flowers* elegant, rich purple, with 2 green, occasionally white, tubercles at the base of each segment. *Anthers* large, yellow, united into a pyramidal or cone-shaped figure. *Berries* poisonous.

2. S. *nigrum*, Linn. *Common Garden Nightshade.* Stem herbaceous; leaves ovate, bluntly toothed, and waved; umbels lateral, drooping; berries globular, black. *E. Bot. t.* 566. *E. Fl. v. i. p.* 319. *Hook. Br. Fl. p.* 111.

Cultivated ground; not common. *Fl.* June—October. ⊙.

On the road-side near the church, Shineton; *W. P. Brookes, Esq.!*

Cultivated ground near Shrewsbury.

Root fibrous. *Stem* 8—12 inches high, branched, angular, and downy. *Leaves* attenuate at the base, with a few scattered hairs on each side. *Umbels* simple, solitary, from the intermediate spaces between the leaves. *Pedicels* thickened upwards. *Flowers* white.

21. VERBASCUM. *Linn.* Mullein.

* *Leaves decurrent.*

1. V. *Thapsus*, Linn. *Great Mullein.* Leaves ovate or oblong, crenulate, densely tomentose on both sides; stem simple; spike of flowers very dense; stamens unequal, 3 hairy, 2 longer glabrous, hairs white. *E. Bot. t.* 549. *E. Fl. v. i. p.* 309. *Hook. Br. Fl. p.* 111.

Hedge-banks and waste ground; not uncommon. *Fl.* July, August. ♂.

Stems 4—5 feet high, erect, angular, winged, tomentose, occasionally branched. *Leaves* alternate, thick, soft and tomentose, radical ones petiolate. *Spike* long, cylindrical. *Flowers* nearly sessile. *Corolla* golden yellow.

The tomentum or down on all the species will, on examination under a microscope, be found to be composed of innumerable stellate hairs.

** *Leaves not decurrent.*

2. V. *Lychnitis*, Linn. *White Mullein.* Leaves oblongo-ovate, crenate, slightly hairy above, pulverulento-tomentose beneath; stem angular, panicled above; stamens equal, all hairy, hairs white. *E. Bot. t.* 58. *E. Fl. v. i. p.* 310. *Hook. Br. Fl. p.* 113.

Road-sides, pastures and wastes; rare. *Fl.* July, August. ♂.

Near Downton Hall near Ludlow; *Mr. H. Spare.*

Stem 3 feet high, erect, woolly. *Leaves* alternate, reticulated with veins, dark-green, radical leaves petiolate, those of the stem sessile. *Panicles* branched, many-flowered. *Flowers* on short bracteated pedicels, collected into small, densely woolly tufts. *Corolla* small, cream-coloured.

3. V. *virgatum*, With. *Large-flowered Primrose-leaved Mullein.* Leaves oblongo-lanceolate, doubly toothed, sessile, hairy, radical ones somewhat lyrate; stem branched from the base; flowers aggregated in small axillary clusters, nearly sessile; stamens equal, hairy, hairs purple. *E. Bot. t.* 550. *E. Fl. v. i. p.* 312. *Hook. Br. Fl. p.* 112.

Fields, wastes, and by road-sides; rare. *Fl.* August. ♂.

Ten miles from Ludlow, on the Shrewsbury road; *Turn. & Dillw. Bot. Guide.* Shineton, near Wenlock; *W. P. Brookes, Esq.!* Haughmond Hill, near Shrewsbury; *Mr. F. Dickinson.*

Whole plant green, not hoary, more or less clothed with stellate hairs. *Stem* 5—6 feet high, stout, slightly angular, glandular in the upper part. *Uppermost leaves* cordato-acuminate, semiamplexicaul, and partially decurrent. *Flowers* in axillary clusters of 2 or 3 slightly-stalked flowers together, rather distantly placed on an elongated terminal raceme. *Calyx* glandular. *Corolla* large, bright yellow.

4. V. *Blattaria*, Linn. *Moth Mullein.* Leaves oblong, crenate, acute, sessile, glabrous, radical ones obovato-oblong, obtuse, sinuate; stem branched at the summit; flowers in an elongated raceme on solitary, remote, glandulose, bracteated pedicels; stamens unequal, 3 shorter entirely hairy, 2 longer hairy on the inner side only, hairs purple. *E. Bot. t.* 393. *E. Fl. v. i. p.* 313. *Hook. Br. Fl. p.* 112.

Banks in a gravelly soil; rare. *Fl.* July. ♂. (⊙. Hook. and Sm.)

Five miles from Ludlow, on the Shrewsbury road; *Turn. & Dillw. Bot. Guide.* Field near Church Stretton; *Mr. T. Bodenham.* Near Wem; *Dr. T. W. Wilson!*

Stem 3—4 feet high, smooth, stout, slightly angular, glandular in the upper part. *Leaves* dark-green, shining, smooth, veiny, uppermost ones cordato-acuminate, semiamplexicaul, very slightly decurrent. *Flowers* solitary, stalked, in elongated racemes terminating the branches. *Calyx* glandular. *Corolla* yellow, streaked with purple at the base.

22. CONVOLVULUS. *Linn.* Bindweed.

1. C. *arvensis*, Linn. *Small Bindweed.* Stem climbing or prostrate; leaves sagittate, their lobes acute; peduncles 4-sided, 1—2-flowered; bracteas minute, remote from the flower. *E. Bot. t.* 312. *E. Fl. v. i. p.* 285. *Hook. Br. Fl. p.* 113.

Hedges, fields, and gardens; common. *Fl.* June, July. ♃.

Root creeping to a great depth, and very difficult of extirpation, hence it has obtained the local name of "Devil's Guts." *Stems* numerous, angular, more or less downy and branched. *Leaves* alternate, various in breadth, obtuse, petiolate. *Flower-stalks* axillary, angular, swelling upwards, with two minute *bracteas* above the middle. *Corolla* small, rose-coloured, or white with pink stripes, closing before rain. *Calyx segments* short, obtuse.

2. C. *sepium*, Linn. *Great Bindweed.* Stem climbing; leaves sagittate, their lobes truncate; peduncles 4-sided, single-flowered; bracteas large, cordate, close to the flower. *E. Bot. t.* 313. *E. Fl. v. i. p.* 285. *Hook. Br. Fl. p.* 113. *Calystegia. Br.*

Moist hedges and thickets; not uncommon. *Fl.* July, August. ♃.

Near Tenbury and Coreley; *Rev. W. Corbett.* Abundant near Ludlow; *Mr. H. Spare.* Near Oswestry; *Rev. T. Salwey.* Banks of Severn near Bewdley; *Mr. G. Jorden.* Banks of Severn near Marn Wood, Coalbrookdale; *Mr. F. Dickinson.*

Banks of the Severn, and in hedges near Shrewsbury, plentiful.

Roots long, fleshy, and creeping extensively. *Stems* twining to the height of several feet, angular, slightly branched and downy. *Leaves* alternate, large, very acute, petiolate. *Flower-stalks* axillary, angular, swelling immediately beneath the flower. *Bracteas* large, acute, appressed, as long as and enclosing the *calyx*, whose segments are acute. *Corolla* very large and elegant, of a delicate white, forming a most striking and beautiful ornament of our hedges.

23. POLEMONIUM. *Linn.* Jacob's Ladder.

1. P. *cæruleum*, Linn. *Jacob's Ladder, Greek Valerian.* Leaves pinnate, nearly glabrous; leaflets oblongo-lanceolate; flowers erect; root fibrous. *E. Bot. t.* 14. *E. Fl. v. i. p.* 287. *Hook. Br. Fl. p.* 113.

Banks and bushy places; rare. *Fl.* June, July. ♃.

By the brook side below the new bridge at Pitchford, doubtful if wild; *Rev. W. Corbett.* On banks of the Woofe, near Bridgnorth; *Rev. W. R. Crotch.*

Stem 1½—2 feet high, angular, hollow, simple, panicled above. *Leaves* alternate. *Flowers* blue, in terminal panicles. *Flower-stalks* and *calyx* a little downy.

24. VINCA. *Linn.* Periwinkle.

1. V. *minor*, Linn. *Lesser Periwinkle.* Stems procumbent; leaves oblongo-lanceolate, their margins as well as the small lanceolate segments of the calyx glabrous. *E. Bot. t.* 917. *E. Fl. v. i. p.* 339. *Hook. Br. Fl. p.* 114.

Dry hedges and shady woods; rare. *Fl.* May, June. ♃.

Sandy lanes near Wheathall, in the parish of Condover; *Rev. W. Corbett.* Wood below Lumhole pool, near Coalbrookdale, perfectly wild; *Mr. F. Dickinson.* Hatton Dingle, doubtful if wild; *T. C. Eyton, Esq.* Smethcote lane; *Dr. T. W. Wilson.*

Root creeping. *Herbage* smooth and shining. *Flowering branches* simple, erect. *Leaves* shortly petiolate, opposite, dark-shining green. *Flowers* on solitary axillary peduncles, longer than the leaves. *Corolla* violet blue.

2. V. *major*, Linn. *Greater Periwinkle.* Stems declinate; leaves broadly ovate, subcordate at the base, their margins as well as those of the elongated subulate segments of the calyx ciliated. *E. Bot. t.* 514. *E. Fl. v. i. p.* 340. *Hook. Br. Fl. p.* 114.

Hedge-banks, thickets, and woods; rare. *Fl.* April, May. ♃.

Near Harlescote lower gate; *Dr. T. W. Wilson.* Chesterton Camp; *Dr. G. Lloyd.*

Hedges of the Bank Farm, near Shrewsbury.

Root stout, creeping. *Herbage* nearly smooth, dark-green and shining. *Flowering stems* branched, ascending. *Leaves* petiolate, opposite. *Flowers* on solitary axillary peduncles about as long as the leaves. *Corolla* pale, but beautiful blue.

25. SAMOLUS. *Linn.* Brook-weed.

1. S. *Valerandi*, Linn. *Common Brook-weed.* Leaves obovate; racemes terminal and axillary, many-flowered; pedicels with a soli-

tary bractea about the middle. *E. Bot. t.* 703. *E. Fl. v. i. p.* 324. *Hook. Br. Fl. p.* 114.

Marshy and watery places; rare. *Fl.* July. ♃.

Trefonen, Oswestry; *Rev. T. Salwey!*

Herbage glabrous, slightly succulent, pale green. *Stem* 8—10 inches high, rounded, simple. *Leaves* alternate, lower ones obovate, petiolate, upper ones more approaching to ovate and on very short petioles, all obtuse, occasionally pointed, and more or less covered with minute white granules. *Flowers* small, white. *Capsule* surmounted by the segments of the calyx. *Seeds* black, turbinate, angular, abrupt.

26. JASIONE. *Linn.* Sheep's-bit.

1. J. *montana*, Linn. *Annual Sheep's-bit or Sheep's Scabious.* Leaves linear-oblong, waved, hispid; root simple, many-stemmed. *E. Bot. t.* 882. *E. Fl. v. i. p.* 297. *Hook. Br. Fl. p.* 114.

In dry sandy soils and hilly places; not uncommon. *Fl.* June, July. ⊙. and ♂.

Whitecliff, near Ludlow; *Mr. H. Spare.* Near Oswestry; *Rev. T. Salwey.* Grinshill; *Mr. E. Elsmere, junr.!* Longnor; *Rev. W. Corbett.* Quatford; *Mr. F. Dickinson.*

Pimhill. Haughmond Hill. Nesscliffe. Caradoc Hill. Woods near Bomere pool.

Stems 6—10 inches high, simple or branched, leafy and hispid, except in the upper part. *Leaves* sessile, alternate, variable in size and shape, the lower ones often approaching to an oblongo-obovate form and irregularly serrated. *Flowers* bright blue, on short pedicels, collected into terminal, dense, hemisphærical heads, surrounded by an involucre of many ovate, dentate, hispid leaves.

The aggregato-capitate inflorescence of this plant would at first sight appear to rank it with the *Compositæ*, but a more careful inspection will show it to be essentially distinct from that tribe, in consequence of each flower being pedicellate within the involucre, and furnished with a true calyx of its own.

27. LOBELIA. *Linn.* Lobelia.

1. L. *Dortmanna*, Linn. *Water Lobelia.* Leaves radical, subcylindrical and obtuse, of 2 parallel tubes; scape simple, nearly leafless; flowers racemed. *E. Bot. t.* 140. *E. Fl. v. i. p.* 298. *Hook. Br. Fl. p.* 115.

Lakes and pools, covering the bottom with its densely matted and verdant foliage; rare. *Fl.* July, August. ♃.

Shallow pools on the Longmynd, in the parish of Church Stretton; *Rev. W. Corbett.* End of Blackmere, near Ellesmere; *Rev. A. Bloxam.*

Bomere and Berrington pools.

Root a small, thick, fleshy stock, from which descend many whitish fibres, and sending forth creeping filiform runners. *Leaves* numerous, 2—3 inches long, all radical and submersed, smooth and slightly recurved, formed of two parallel tubes or cells. *Scape* rising above the water, bearing a simple raceme of few, alternate, distant, drooping, pale blue flowers. *Capsule* 5-angular, erect, crowned below the summit with the segments of the calyx.

28. CAMPANULA. *Linn.* Bell-flower.

* *Flowers panicled; capsule erect, lateral pore immediately below the calycine segment.* (Campanula.)

1. C. *rotundifolia*, Linn. *Round-leaved Bell-flower.* Stem round, glabrous; leaves petiolate, radical ones subrotundo-cordate, crenate,

(very soon withering) those of the stem linear, entire; panicle lax; calycine segments subulate, entire. *E. Bot. t.* 866. *E. Fl. v. i. p.* 288. *Hook. Br. Fl. p.* 116.

Heaths, walls, hedge-banks, dry and hilly pastures, and borders of fields; common. *Fl.* July, August. ♃.

Root woody, slightly creeping. *Stems* numerous, 10—12 inches high. *Radical leaves* varying in shape from roundish heart-shaped or reniform to ovate and lanceolate. *Panicle* lax, of a few drooping blue flowers, on slender tremulous stalks, with a subulate *bractea* at the base of each. *Whole plant* smooth, dark-green, delicately slender and graceful.

A variety with *white* flowers occasionally occurs.

Birches Lane; *W. W. Watkins, Esq.*

2. C. *patula*, Linn. *Spreading Bell-flower.* Stem 5-angular, scabrous with deflexed hairs; leaves roughish, dentato-crenate, radical ones obovato-lanceolate, subpetiolate, cauline ones linear-lanceolate, sessile; panicle spreading; calycine segments subulate, toothed at the base. *E. Bot. t.* 42. *E. Fl. v. i. p.* 289. *Hook. Br. Fl. p.* 116.

Borders of fields and hedge-banks, chiefly on a gravelly soil; frequent. *Fl.* July, August. ⊙. or ♂.

About the High Rock, Bridgnorth; *Rev. W. R. Crotch.* Hedge-banks neer Brockton and Munslow; *Mr. F. Dickinson.* Oakley Park, near Ludlow; *Mr. H. Spare.* Side of the Watling Street road near the turnpike; Wellington; and in shady lanes diverging from the Shrewsbury road; *E. Lees, Esq.* Common in the parishes of Condover, Berrington, Froddesley, Pitchford, Acton Burnel, and occasionally at Acton Scott; *Rev. W. Corbett.* Shelton bank; Montford Bridge bank; Pitchford; Condover; Leaton Shelf; Caynham Camp; *Turn. & Dillw. Bot. Guide.* Between Cound and Condover; *Dr. G. Lloyd.* Road-side near Larden; *W. P. Brookes, Esq.!*

Bickley Coppice, on the banks of the Severn, near Shrewsbury. Road-side between Astley and Shrewsbury. Field and hedge-banks bordering Berrington pool.

Root tapering. *Stem* erect, simple, branched above, 2 feet high. *Radical leaves* withering when the flowers appear, *cauline ones* alternate, all more or less rough. *Panicle* branched, diffuse, peduncles rigid, smooth or nearly so, occasionally hispid, with a subulate *bractea* at the base, and one or more above the middle. *Flowers* very elegant, of a fine reddish-purple, much larger than those of the preceding, narrower at the base, and dilating gradually and gracefully upwards.

** *Flowers axillary; capsule reflexed, pore at the base.** (Trachelium.)

3. C. *latifolia*, Linn. *Giant Bell-flower.* Stem somewhat angular and hairy; leaves ovato-lanceolate, acuminate, crenato-serrate, hispid, shortly petiolate; peduncles erect, 1-flowered, drooping in fruit; calyx glabrous, its segments erect, ovato-lanceolate, entire. *E. Bot. t.* 302. *E. Fl. v. i. p.* 291. *Hook. Br. Fl. p.* 117.

* As the genus *Campanula* has been divided so greatly, according to the dehiscence of the capsule, it is suggested whether the genus as now constituted should not be still further divided, since in the species included in our first section the lateral pore is immediately below the calycine segments; whilst in the second or *Trachelium* section the pore is situated quite at the base of the reflexed capsule.

Moist shady woods; rare. *Fl.* July, August. ♃.

In a thicket between the Wrekin and Little Wenlock; *E. Lees, Esq.* Side of the Teme, below Ludlow; *Dr. G. Lloyd.* In hedges in fields a little beyond Aldenham Mill, near the road from Wenlock to Bridgnorth; also in a hedge by the road-side from Muckley to Round Acton, near the brook; lane near Atterley, Wenlock;! *W. P. Brookes, Esq.* Bradley, near Wenlock; *Mr. F. Dickinson.*

Root fleshy, fibrous. *Stem* 2—3 feet high, erect, quite simple. *Peduncles* with a pair of *bracteas* about the middle. *Corolla* very large, deep blue, smooth. The finest and most stately of our species.

4. C. *Trachelium*, Linn. *Nettle-leaved Bell-flower.* Stem 5-angular, hispid; leaves acuminate, inciso-serrate, hispid, lower ones cordate on long petioles, upper ones oblong sessile; peduncles few-flowered; calyx hispid at the base and margins, its segments erect, ovato-lanceolate, entire. *E. Bot. t.* 12. *E. Fl. v. i. p.* 293. *Hook. Br. Fl. p.* 117.

Woods, thickets, and hedges; not uncommon. *Fl.* July, August. ♃.

About the High Rock, Bridgnorth; *Rev. W. R. Crotch.* Oakley Park, near Ludlow; *Mr. H. Spare.* At the northern and western bases of the Wrekin; *E. Lees, Esq.* Near Bridgnorth; *Rev. A. Bloxam.* Near Dowle's Brook, Wyre Forest; *Mr. W. G. Perry.* Woods about Coalbrookdale; *Mr. F. Dickinson.* Near Ludlow and Little Wenlock; *Dr. G. Lloyd.* Leaton Shelf, near Shrewsbury; *T. C. Eyton, Esq.* Hedges near Harley;! also between Wenlock and Shineton; *W. P. Brookes, Esq.* Near Shelton and Nobold; *Mr. T. Bodenham.*

Benthall Edge, Emstry, and Monkmoor, near Shrewsbury.

Root thick, and rather woody. *Stem* 2—3 feet high, simple. *Leaves* resembling those of the Nettle, whence its English name. *Peduncles* short, with a pair of leafy *bracteas* above the middle, where they are divided and bear about 3 flowers. *Corolla* large, deep blue, occasionally white, hispid on the exterior, and hairy within.

5. C. *glomerata*, Linn. *Clustered Bell-flower.* Stem angular, simple, hoary with deflexed appressed hairs; leaves ovato-lanceolate, cordate at the base, hairy, crenate, radical ones petiolate, cauline ones semiamplexicaul; flowers sessile, in terminal and axillary clusters; calyx erect, hoary, its segments lanceolate, entire. *E. Bot. t.* 90. *E. Fl. v. i. p.* 293. *Hook. Br. Fl. p.* 117.

In dry pastures, chiefly in the limestone districts; rare. *Fl.* July, August. ♃.

On the banks of the Corve, Bromfield; *Miss Mc Ghie.* On the road-side between Easthope and Lutwyche, near the Hill Top; lane leading from Wenlock to the Townsmill;! *W. P. Brookes, Esq.* Road-sides near Lutwych Hall, near Longville; *Mr. F. Dickinson.*

Root woody. *Stem* 10 inches to 2 feet high. *Flowers* rather large, erect, clusters subtended by 2 broad, concave, pointed *bracteas*. *Corolla* deep and rich purplish blue, hairy.

29. WAHLENBERGIA. *Schrad.* Wahlenbergia.

1. W. *hederacea*, Reichenb. *Ivy-leaved Wahlenbergia.* Stem weak, filiform, prostrate; leaves subrotundo-cordate, angulato-dentate, petiolate, glabrous; flowers solitary, on long, slender, terminal peduncles; calyx glabrous, erect, its segments subulate, entire. *Campanula, hederacea, Linn.* *E. Bot. t.* 73. *E. Fl. v. i. p.* 294. *Hook. Br. Fl. p.* 117.

In moist shady places; rare. *Fl.* July, August. ♃.

Titterstone Clee Hill; *Rev. W. Corbett.*

A most graceful little plant, growing in lax tufts, and creeping to a great extent by its entangled branchy stems. *Flowers* half an inch or more in length, at first drooping, then erect, pale purplish blue. *Capsule* almost globose, adhering to the calyx, opening in the upper free part between the persistent segments of the calyx.

30. PRISMATOCARPUS. *L'Heritier.* Venus's Looking-glass.

1. P. *hybridus*, L'Her. *Corn Venus's Looking-glass.* Stem simple or branched from the base; leaves oblong, crenate, waved; flowers solitary; calycine segments shorter than the widely spreading corolla. *Lindl. Syn. p.* 135. *Campanula hybrida, Linn.* *E. Bot. t.* 375. *E. Fl. v. i. p.* 294. *Hook. Br. Fl. p.* 118.

Corn-fields; rare. *Fl.* August. ⊙.

Dry corn-fields near to Nash Court, in the parish of Burford; *Rev. W. Corbett.* In a corn-field near the Windmill, Wenlock, not common; *W. P. Brookes, Esq.*

Whole plant 6—8 inches high, of a pale greyish green, rough with prominent, rigid, minute hairs. *Leaves* alternate, sessile. *Flowers* few, terminal, solitary, sessile. *Corolla* purple, with a pale centre. *Calycine segments* lanceolate, spreading, rough.

31. LONICERA. *Linn.* Honey-suckle.

* *Climbing; flowers in whorled heads.* (Caprifolium, *Juss.*)

1. L. *Caprifolium*, Linn. *Pale perfoliate Honey-suckle.* Flowers ringent, whorled, terminal, sessile; upper leaves connato-perfoliate. *E. Bot. t.* 799. *E. Fl. v. i. p.* 326. *Hook. Br. Fl. p.* 117.

Woods and thickets; rare. *Fl.* June. ♄.

Whitecliff Coppice, near Ludlow; *Mr. H. Spare.*

Stem climbing, somewhat branched, turning from left to right. *Leaves* in pairs, opposite, obovate, entire and smooth, glaucous beneath, *lower ones* subpetiolate, *upper ones* connato-perfoliate. *Flowers* in axillary and terminal whorls, about 6 in each whorl, yellowish with a blush-coloured tube, fragrant. *Berries* smooth, of an orange colour.

2. L. *Periclymenum*, Linn. *Common Honey-suckle or Woodbine.* Flowers ringent, capitate, terminal; leaves all distinct. *E. Bot. t.* 800. *E. Fl. v. i. p.* 327. *Hook. Br. Fl. p* 119.

Woods, hedges, and thickets; frequent. *Fl.* June—October. ♄.

Stem twining and climbing as in the preceding, with opposite branches. *Leaves* oval, obtuse, attenuated at the base, entire, sometimes downy, glaucous beneath. *Flowers* in terminal heads, pale yellow, externally red, fragrant, especially in the evening. *Berries* red, often roughish.

** *Erect; peduncles 2-flowered.* (Xylosteum, *Juss.*)

3. L. *Xylosteum*, Linn. *Upright Fly Honey-suckle.* Peduncles 2-flowered; berries distinct; leaves ovate, acuminate, entire, downy. *E. Bot. t.* 916. *E. Fl. v. i. p.* 327. *Hook. Br. Fl. p.* 119.

Thickets; rare. *Fl.* July. ♄.

Coalbrookdale; *W. P. Brookes, Esq.!* Woods near Ludlow; *Miss McGhie.*

An erect bushy *shrub*, 4—5 feet high, with round branches, downy when young. *Leaves* deciduous, opposite, petiolate, of a dull green colour. *Flowers* small, scentless, of a pale yellow, on axillary simple peduncles. *Bracteas* hairy, double, the 2 *outermost* lanceolate, spreading, *inner* a small concave scale under each germen. *Calyx* in obtuse lobes. *Corolla* downy. *Berries* bright scarlet.

32. RHAMNUS. *Linn.* Buckthorn.

1. R. *catharticus*, Linn. *Common Buckthorn.* Spines terminal; flowers 4-cleft, diœcious; leaves ovate, sharply serrated. *E. Bot. t.* 1629. *E. Fl. v. i. p.* 328. *Hook. Br. Fl. p.* 119.

Woods, thickets, and hedges; not common. *Fl.* May, June. ♄.

In a wood near Smethcot; *W. W. Watkins, Esq.* Donnington, near Shiffnal; *Dr. G. Lloyd.* Oreton, not common; *Mr. G. Jorden.* In Shropshire, frequent; *Withering.* Banks of the Severn, near Buildwas; *Mr. F. Dickinson.*

Shrub with alternate or nearly opposite spreading branches. *Leaves* with 4 or 6 strong lateral nerves parallel with the margin or mid-rib, serratures glandular. *Flowers* in dense terminal fascicles. *Barren flower: calyx-tube* campanulate, segments ovate, 2-ribbed. *Petals* 4, oblongo-ovate, inserted below the mouth of the calyx, and alternate with its segments. *Stamens* inserted just below the petals. *Germen* abortive. *Fertile flower: petals* linear, incurved above. *Stamens* abortive. *Styles* 4, united half-way up, spreading. *Stigmas* small, slightly decurrent along the inner edge of the styles. *Germen* superior. *Berries* black, nauseous, powerfully cathartic. *Seeds* ovate, acute at the lower extremity, rounded at the back, with two flat sides, forming the internal angle. *Embryo* with kidney-shaped *cotyledons*, laterally bent, surrounded by the *albumen.*

The unripe berries afford a yellow dye; and the bark a green one.

2. R. *Frangula*, Linn. *Berry-bearing Alder or Alder Buckthorn.* Unarmed; flowers perfect; leaves elliptical, entire. *E. Bot. t.* 250. *E. Fl. v. i. p.* 329. *Hook. Br. Fl. p.* 119.

Woods and thickets; not very common. *Fl.* May. ♄.

Wyre Forest; *E. Lees, Esq.* Bog at the foot of the Wrekin; *Mr. F. Dickinson.* Cupid's Ramble, near Westfelton. Bomere Pool, and on Shomere Moss.

Shrub 4—10 feet high, with alternate spreading branches. *Leaves* alternate, glabrous, with many transverse parallel nerves, petioles downy. *Flowers* in axillary clusters, about 3 in each cluster, on short peduncles. *Corolla* whitish-green, 5-cleft. *Anthers* purple. *Style* very short. *Stigma* capitate, cloven. *Berries* dark purple, with 2 large seeds.

33. EUONYMUS. *Linn.* Spindle-tree.

1. E. *Europæus*, Linn. *Common Spindle-tree.* Peduncles compressed, many-flowered; flowers mostly tetrandrous; branches glabrous; leaves oblongo-lanceolate, minutely serrated. *E. Bot. t.* 362. *E. Fl. v. i. p.* 330. *Hook. Br. Fl. p.* 119.

Woods, hedges, and thickets; not very common. *Fl.* May. ♄.

Hedges between Church Stretton and Craven Arms; *E. Lees, Esq.* Near Oswestry; *Rev. T. Salwey.* Near Adcot on the Perry; *T. C. Eyton, Esq.* Hedges, Welbatch, near Shrewsbury; *Mr. T. Bodenham!* Stottesden, thinly dispersed in the hedges; *Mr. G. Jorden.*

Woods beyond Haughmond Hill. Hedges between Craven Arms and Ludlow, abundant.

Shrub 3—5 feet high, with a green smooth bark. *Leaves* petiolate, glabrous, serratures glandular. *Peduncles* axillary, forked. *Flowers* small, greenish-white. *Fruit* obtusely angular, very beautiful, rose-coloured. *Arillus* orange-coloured.

34. VIOLA. *Linn.* Violet.

* *Stemless, or nearly so.*

1. V. *hirta*, Linn. *Hairy Violet.* Leaves oblongo-cordate, rough, as well as the petioles and capsules, with prominent horizontal hairs; sepals broadly ovate, obtuse; lateral petals emarginate, with a hairy central line, lower petal emarginate; creeping scions none. *E. Bot. t.* 894. *E. Fl. v. i. p.* 302.

Thickets; rather uncommon. *Fl.* April. ♃.

On the Moelydd, in the parish of Oswestry; *Rev. T. Salwey.* Shadwell, Standhill Coppice; and Wenlock; *W. P. Brookes, Esq.* Lincoln's Hill, Coalbrookdale; *Mr. F. Dickinson.*

Very abundant near Wenlock. Near Queen Eleanor's bower on Haughmond Hill. Sharpstones Hill. Woods near Bomere Pool.

Root rather woody. *Stem* none, except a few short simple horizontal runners forming leafy tufts but not taking root. *Leaves* oblongo-cordate, obtuse, crenate, veiny, with 3 principal nerves at the base, clothed on both sides with short dense hairs, especially visible on the young unfolded leaves, light hoary green. *Petioles* considerably longer than the leaves, erect, furrowed above, slightly winged at the sides and keeled underneath, rough with copious horizontal pubescence. *Stipules* chiefly radical, lanceolate, acute, more or less toothed, pale green. *Peduncles* taller than the leaves, erect, smooth, except a few scattered hairs chiefly below the bracteas, obsoletely quadrangular, grooved above. *Bracteas* lanceolate, acute, serrated, smooth, placed below the middle of the peduncle. *Flowers* solitary, drooping, obliquely reversed, greyish blue streaked with black, scentless, lateral and lower petals emarginate, lower one most so, lateral petals with a hairy line just above the claw. *Spur* nearly as long as the calyx, hooked at the point. *Calyx* smooth, except a few hairs on the keel, sepals broadly ovate, obtuse, 3-nerved, slightly membranous at the edges. *Stamens* flat. *Anthers* each tipped with a flat orange-coloured membrane, converging but not united. *Style* slender at the base, thickening upwards, bent at the point. *Stigma* hooked. *Capsule* rounded, hairy. *Seeds* many, round, smooth.

Readily distinguished by the procumbent not creeping scions, the horizontal direction of the copious pubescence, the situation of the bracteas below the middle of the peduncle, and the obtuseness of the sepals.

2. V. *odorata*, Linn. *Sweet Violet.* Leaves cordate, pubescent; petioles with reflexed hairs; sepals ovate, obtuse; lateral petals entire, with a hairy central line, lower petal emarginate; scions creeping. *E. Bot. t.* 619. *E. Fl. v. i. p.* 302. *Hook. Br. Fl. p.* 120.

Woods and hedges; frequent. *Fl.* March, April. ♃.

Very plentiful all along the Wenlock Edge from Ludlow to Wenlock. Shelton, near Shrewsbury. Uffington. Cardiston.

Root rather woody. *Stem* none, but the long trailing leafy scions spreading very far, and throwing out abundance of fibres. *Leaves* cordate, obtuse, crenate, veiny, with 3 principal nerves at the base, clothed more or less on both sides with short dense hairs, especially visible on the young unfolded leaves, dark green. *Petioles* longer than the leaves, erect, furrowed above, slightly winged at the sides and keeled underneath, covered with copious deflexed pubescence. *Stipules* chiefly radical, lanceolate, acute, more or less toothed, pale green. *Peduncles* taller than the leaves, erect, smooth, except a few scattered deflexed hairs chiefly below the bracteas, obsoletely quadrangular, grooved above. *Bracteas* lanceolate, acute, serrated, smooth, placed above the middle of the peduncle. *Flowers* solitary, drooping, obliquely reversed, of a deep violet purple, streaked in the mouth, fragrant, lateral petals with a hairy line above the claw, petals all entire except the lower one which is emarginate. *Spur* as long as the calyx, rounded, obtuse,

straight. *Calyx* smooth, except a few hairs on the keel, sepals ovate, obtuse, 3-nerved, slightly membranous at the edges. *Stamens* flat. *Anthers* each tipped with a flat orange-coloured membrane, converging but not united. *Style* slender at the base, thickening upwards, bent at the point. *Stigma* hooked. *Capsule* rounded, hairy. *Seeds* several, round, smooth.

Var. β. alba. Flowers white, fragrant. *E. Fl. v. i. p.* 303. *V. martia alba, Raii Syn.* 364. 2.

Hedge-banks; not very frequent. *Fl.* March, April. ♃.

Pulley, near Shrewsbury. Eyton-on-the-Wildmoors.

Var. γ. imberbis. Flowers white; lateral petals destitute of the hairy line, fragrant. *V. imberbis, Leighton in Loudon's Mag. Nat. Hist. vol.* 8, *p.* 277. *V. suavis, Bieb. Lindl. Syn. 2nd edition, and Loudon's Mag. Nat. Hist. vol.* 8, *p.* 384. *V. odorata β. imberbis, Henslow Cat. Br. Pl. 2nd edition.*

Woods, thickets, hedge-banks and wastes. *Fl.* March, April. ♃.

Eyton-on-the-Wildmoors. *T. C. Eyton, Esq.!*

Sharpstones Hill. Sutton Wood. Nobold—all near Shrewsbury.

Root rather woody. *Stem* none, but with long trailing leafy scions which spread far and throw out fibrous radicles. *Leaves* dark green, rounded heart-shaped, obtuse, crenate, veiny, somewhat wrinkled, clothed on both sides with short pubescence. *Petioles* longer than the leaves, clothed with short deflexed hairs. *Stipules* lanceolate with taper-pointed teeth. *Peduncles* taller than the leaves, covered with short deflexed hairs, but not so copiously as the petioles. *Bracteas* lanceolate, acute, more or less toothed, serrated towards their summits, slightly hairy at their keels, placed above the middle of the peduncle. *Flowers* solitary, drooping, obliquely reversed, white with a light dull-purplish spur, fragrant, lower petal emarginate, lateral petals entire, without a hairy line. *Spur* as long as the sepals, round, inflated, very obtuse, straight. *Calyx* smooth, except a few hairs on the margins of the membranous edges and on the spurs. *Stamens* flat. *Anthers* each tipped with a flat orange-coloured membrane, converging but not united. *Stigma* a hooked point. *Capsule* roundish, hairy. *Seeds* numerous, round, somewhat oblong.

This was first noticed by me in 1833 growing on the Sharpstones Hill, and described as possibly a new species in *Loudon's Mag. Nat. Hist.*; but on specimens being submitted to Professor Lindley, that acute Botanist determined it to be *V. suavis* of *Bieberstein*, and expressed his opinion that that species was itself only a variety of *V. odorata.* I have cultivated my plant since 1833 without observing any change, though I have been unsuccessful in causing its seeds or those of *V. suavis, Bieb.* to germinate.

By the kindness of Mr. Borrer, I possess a cultivated plant of *V. suavis, Bieberst.* which certainly appears to me different from our plant, being in all respects larger and more luxuriant, with a hairy line on the lateral petals, the flowers of a pale sky-blue colour, and the leaves always cucullate or folded inwards at the base.

3. V. *palustris*, Linn. *Marsh Violet.* Leaves cordate or kidney-shaped, quite glabrous, veiny beneath; sepals obtuse; spur very short; lateral petals with a hairy central line; scions none; root creeping. *E. Bot. t.* 444. *E. Fl. v. i. p.* 303. *Hook. Br. Fl. p.* 121.

Bogs and marshy ground, on the margins of pools; frequent. *Fl.* April, May. ♃.

Hope Bowdler and Cardington Hills; *Rev. W. Corbett.* Longmynd; *Rev. S. P. Mansell.* Benthall Edge; *Mr. F. Dickinson.*

Vownog bog, near Westfelton. Bomere pool. Almond park.

Root creeping, with long white fibres. *Leaves* smooth, with shallow marginal crenatures and long smooth petioles grooved above. *Stipules* chiefly radical, ovate, acute, more or less toothed. *Peduncles* about as long as the leaves, erect, smooth, obsoletely quadrangular, grooved above. *Bracteas* small, lanceolate, erect, serrated, smooth, placed a little below the middle of the peduncle. *Flowers* solitary, drooping, scentless, pale lilac, with dark purple streaks chiefly on the lower petal, which is shorter than the others, and as well as the lateral ones emarginate. *Spur* rounded. *Calyx* smooth, sepals ovate, obtuse, 3-ribbed, slightly membranous at the edges. *Anthers* orange-coloured, converging, distinct. *Stigma* hollow, tumid at the upper side. *Capsule* smooth.

** *Furnished with an evident stem.*

4. V. *canina*, Linn. *Dog's Violet.* Stem at length ascending; leaves cordate, acute, crenate; sepals acuminate; stipules lanceolate, ciliato-dentate; petals entire, lateral ones with a hairy central line; spur long; bracteas subulate, ciliato-dentate. *E. Bot. t.* 620. *E. Fl. v. i. p.* 304. *Hook. Br. Fl. p.* 121.

Hedge-banks, woods, thickets; abundant. *Fl.* April—August. ♃.

Root woody. *Stem* ascending after flowering, angular and smooth. *Leaves* cordate, more or less elongated, glabrous or occasionally hairy, crenate. *Petioles* longer than the leaves, erect, furrowed above, slightly winged at the sides, and keeled underneath, smooth. *Stipules* lanceolate, more or less dilated, acute, ciliato-dentate, pale green. *Peduncles* axillary, about as long as the leaves, erect, smooth, obsoletely quadrangular. *Bracteas* subulate, ciliato-dentate, smooth, placed towards the upper part of the peduncle. *Flowers* solitary, drooping, obliquely reversed, blue, with darker lines in the mouth, scentless, lateral petals with a hairy line above the claw. *Sepals* membranous at the edges and keeled, 3-nerved. *Capsule* elliptical, mucronate, bluntly trigonous.

Var. β. alba. Flowers white. *Raii Syn.* 364. *4.

This variety appears uncommon, and has occurred to my notice only twice, near Radbrook, and at Shelton Rough, both in the neighbourhood of Shrewsbury. Near Radbrook I also found, in 1834, a plant having the *four* upper petals bearded, which character it retains under cultivation.

5. *Viola tricolor*, Linn. *Pansy Violet or Heart's Ease.* Stem branched at the base, ascending, zig-zag; leaves crenato-serrate, obtuse, upper ones ovate, ovato-oblong or lanceolate, lower ones ovato-cordate; stipules lyrato-pinnatifid, central lobe with a single crenature on each side; sepals linear-lanceolate, 2 lower ones equal, larger than the rest, divergent in flower; petals longer than the sepals, lower one broadly obovate, almost cordate, flat; spur equal to the calycine appendages; sepals of the fruit recurved; capsule oval. *V. tricolor, Linn. E. Bot. t.* 1287. *E. Fl. v. i. p.* 306. *Hook. Br. Fl. p.* 121.

Gardens and cultivated grounds; common. *Fl.* all summer. ⊙. and ♂.

Root fibrous. *Stems* straggling, branched at the base, decumbent, then ascending, zig-zag, obsoletely triquetrous, two of the angles formed by an acute winged process proceeding from the stipules rough with minute deflexed pubescence, the face between them convex, the 3rd angle obtuse continued from the petiole, the faces somewhat flattened. *Leaves* alternate, petiolate, ovate, ovato-oblong, or lanceolate, lower ones somewhat cordate at the base, crenato-serrate, serratures tipped with a pale callous dot, edges ciliated, slightly pubescent on the veins. *Petioles* winged and ciliated, deeply channelled above. *Stipules* oblong, deeply laciniated, on the side next the leaf with 2 linear, erect or slightly expanded lobes, the upper one broader somewhat lanceolate, shorter than the terminal or central portion of the stipule, lower one narrow and linear, always more or less distant from the base ; on the exterior side with 5 or 6 lobes, superior one linear, broader, inclining to lanceolate, erect or slightly expanded, shorter than the terminal portion, the others gradually a little shorter and narrower, the two upper ones of these nearly erect and straight slightly expanded, the rest greatly expanded, often curved, the inferior one subulate and reflexed, proceeding from the very base of the stipule ; terminal portion spathulate, more or less similar to the leaf, though never cordate at the base, generally with a single notch or crenature on each side, never perfectly entire, but in the most entire state always with some irregularity in the margin ; all the lobes and crenatures ciliated at the margins and tipped with a pale callous tubercle bearing a mucro. *Peduncles* axillary, solitary, considerably longer than the leaves, tetragonous, somewhat laterally compressed, deeply grooved on the upper side, glabrous. *Bracteas* minute, ovate, with about 3 teeth on each side at the base, lowermost expanded, placed near the summit of the peduncle. *Sepals* linear-lanceolate, acuminate, edges membranous, more or less toothed or serrated, 3-nerved ; the two lower sepals considerably larger and longer, recurved and divergent in flower. *Calycine appendages* dilated, 3 upper ones nearly equal, 2 lower ones broader and larger, all unequally crenated or notched, slightly ciliated. *Petals* much longer than the sepals, 2 uppermost oblongo-obovate, deep velvety purple, 2 lateral ones obovate, or oval with a curved claw about one-fourth the length of the petal, pale yellow with about 3 purple streaks, and a tuft of white hairs at the summit of the claw ; lower one broadly obovate, almost obcordate, flat, the margin rounded, generally entire, sometimes slightly emarginate, the throat marked with 2 parallel hairy lines, deep yellow with about 7 deep purple streaks. *Spur* equal in length to the calycine appendages, rounded, inflated, straight. *Appendages of the anthers* suddenly curved upwards, then straight, filiform, obtuse. *Style* short, suddenly curved and remarkably thickened upwards ; *stigma* large, capitate, obliquely perforated, pubescent below. *Capsule* oval, smooth. *Sepals* of the *fruit* recurved. *Seeds* smooth and shining, ovate.

6. *Viola arvensis*, Murr. *Corn Pansy.*

Stem branched at the base, erect ; leaves crenato-serrate, obtuse, upper ones ovate, ovato-oblong or lanceolate, lower ones ovato-cordate ; stipules lyrato-pinnatifid, central lobe with 2 crenatures on each side ; sepals broadly lanceolate, upper and 2 lower ones equal, larger than the rest, lower ones straight, not divergent in flower ; petals shorter than the sepals, lower one wedge-shaped, retuse, concave ; spur shorter than the calycine appendages ; sepals of the fruit straight ; capsule nearly globular. *E. Bot. Suppl. t. 2712. Viola tricolor β. E. Fl. v. i. p. 306. Hook. Br. Fl. p. 121.*

Corn-fields ; common. *Fl.* all summer. ☉. and ♂.

Similar to *V. tricolor* in all respects except in the following characters:—*Stems* decumbent, somewhat geniculate at the base, then erect, but not zig-zag. *Lobes* of the *stipules* narrower and more linear, the central or terminal one usually with 2 notches or serratures on each side, sometimes in the lower ones where the terminal lobes are broader with 3 notches. *Bracteas* with 2 teeth only. *Sepals* broadly lanceolate, acuminate, more or less serrated, often ciliated, 3-nerved, the upper and 2 lower ones equal, somewhat longer and broader than the 2 lateral ones, straight, not divergent in flower. *Petals* shorter than the sepals, cream-coloured, the two uppermost rotundo-obovate, two lateral ones spathulate, the curved claw nearly equal to the limb, with a single purple streak and a tuft of white hairs at the summit of the claw ; lower petal broader and larger, wedge-shaped or battledore-shaped, longitudinally indented in the centre, thus rendering it concave, margin truncate or retuse, deep yellow towards the claw, with 5 dark purple streaks, the throat marked with 2 parallel lines of hairs. *Spur* shorter than

the calycine appendages, compressed, nearly straight, very slightly curved. *Appendages of the anthers* suddenly curved upwards, then straight, filiform, obtuse. *Capsule* nearly globular, smooth. *Sepals* of the *fruit* straight. *Seeds* smooth and shining, ovate.

7. *Viola lutea.* *Yellow Mountain Violet or Yellow Pansy.*

Stem branched at the base, straggling ; leaves crenato-serrate, obtuse, upper ones ovate, ovato-oblong or lanceolate, lower ones ovato-cordate ; stipules subpalmato-pinnatifid, central lobe elongated, linear or linear-lanceolate, perfectly entire ; sepals lanceolate, acuminate, 2 lower ones equal, larger than the rest, straight not divergent in flower ; petals considerably longer than the sepals, lower one obcordato-cuneiform, flat ; spur equal to or twice as long as the calycine appendages ; sepals of the fruit patent ; capsule globose.

Hilly pastures ; rare. *Fl.* June. ♃.

Titterstone Clee Hill ; *Mr. H. Spare* ; *Mr. J. S. Baly!* Caer Caradoc and the Brown and Titterstone Clee Hills ; *Rev. W. Corbett.* Longmynd ; *Rev. S. P. Mansell.* Near Kinton ; *J. F. M. Dovaston, Esq.!* Pentregaer in the parish of Oswestry ; *Rev. T. Salwey!*

Gamester lane, near Westfelton. Dovaston Common. Stiperstones Hill.

Root creeping. *Stem* very slender, weak, branched and straggling, triquetrous, 2 of the angles winged with deflexed pubescence, 3rd obsolete. *Leaves* alternate, petiolate, ovate, ovato-oblong or lanceolate, lower ones cordate at the base, crenato-serrate, serratures tipped with a pale callous dot, more or less downy, edges ciliated. *Petioles* winged and ciliated, deeply channelled above. *Stipules* subpalmato-pinnatifid, deeply laciniated, on the side next the leaf with a single, linear, expanded lobe, shorter than the terminal or central portion of the stipule, (sometimes there is a 2nd lobe which is smaller, subulate, and close to the first) ; on the exterior side with 3 linear, very much expanded lobes, upper one the largest, shorter than the terminal portion, the others gradually shorter and narrower, the inferior ones very narrow, often subulate and curved upwards, proceeding from the very base of the stipule ; terminal portion elongated, linear or linear-lanceolate, perfectly entire ; all the lobes ciliated at the margins and tipped with a pale callous tubercle bearing a mucro. *Peduncles* axillary, solitary, very long, rising far above the top of the stem, tetragonous, deeply channelled above, glabrous. *Bracteas* minute, ovate, acute, with about 2 teeth on each side at the base, placed near the summit of the peduncle. *Sepals* lanceolate, very much acuminated, membranous at the edges, obscurely 3-nerved, the 2 lower ones larger and longer, straight, not divergent in flower. *Calycine appendages* very much dilated, uppermost one shorter than the rest, intermediate ones narrow equal, 2 lower ones very much broader and larger, all very unequally and deeply notched. *Petals* of a very pale yellow or sulphur colour, very considerably longer than the sepals, 2 uppermost erect, oblongo-obovate, 2 lateral ones irregularly obovate, curved upwards, clawed, of a deeper but still pale yellow colour, marked above the claw with 3 dark purple streaks and a tuft of white hairs, upper margin of the claw hairy, lower petal obcordato-cuneiform, flat, slightly longitudinally indented in the middle, margin very much truncated, sometimes emarginate, deep yellow particularly towards the base which is marked with 7 dark purple streaks, the throat longitudinally bearded with 2 dense parallel hairy lines. *Spur* equal to or twice as long as the calycine appendages, narrow, rounded, inflated, straight or hooked at the extremity. *Appendages of the anthers* suddenly curved upwards, then straight, filiform, obtuse. *Capsule* globose, smooth. *Sepals* of the *fruit* patent. *Seeds* smooth and shining, ovate, very slightly roughened.

I feel totally unable satisfactorily to affix any synonyms to this plant, but have retained for the present the specific name *lutea.* Sir J. E. Smith in Engl. Flora has distinguished *V. lutea,* Huds. from *V. grandiflora, Linn.* by the spur of

the latter being twice as long as, whilst that of the former only equals the calycine appendages. In all the specimens which I have ever seen, either from North Wales, Scotland, or Shropshire, the length and direction of the spur have been very variable, sometimes equalling the calycine appendages, in other instances a little longer, and in others twice as long, and either straight, or curved and hooked at the extremity. M. Gay in the *Annales des Sciences Naturelles,* vol. 26, quoted in *Hooker's Comp. to Bot. Mag. v. i. p.* 158, considers our British *V. lutea* as "identical with *V. sudetica, Willd.* which is the true *V. grandiflora* of Linnæus." The *V. grandiflora, Linn.* which from his observations appears to be a very variable plant in its general form and appearance, he especially characterises by the larger lobe of the stipule, which is "constantly quite entire and never assumes an elliptical or an oval form." A comparison with the Linnæan Herbarium can alone decide the question.

In describing *V. tricolor, arvensis* and *lutea,* I have applied the excellent suggestions of M. Gay relative to the number and direction of the lateral laciniations and the form and notches of the terminal lobe of the stipules, and have found the distinctions derived from these to be generally constant. The stipules selected have been those situated about the middle of the stem, the uppermost and lowest ones being sometimes liable to slight variations. A careful examination has also supplied other characters which seem constant, and which possibly may be applied with success to the determination of other species, viz : the relative size of the sepals particularly of the 2 lower ones, the straightness or divergence of these latter when in flower, their straight patent or recurved direction as well as that of the uppermost sepal when in fruit, and the general form and flatness or concavity of the lower petal. The form and direction of the appendages of the anthers, which have been mentioned as affording good specific characters, seem to be scarcely prominent or dissimilar enough in our British species to be made available to their distinctive determination, although in the foreign species assuredly presenting, when combined with others, very admirable and characteristic points of discrimination.

35. RIBES. *Linn.* Currant or Gooseberry.

* *Without prickles.*

1. R. *rubrum,* Linn. *Common or Red Currant.* Erect ; racemes glabrous, pendulous ; flowers nearly plane ; leaves on long ciliato-glandulose peduncles, obtusely 5 lobed, doubly serrated. *E. Bot. t.* 1289. *E. Fl. v. i. p.* 331. *Hook. Br. Fl. p.* 122.

Woods and hedges ; perhaps scarcely wild. *Fl.* May. ♄.

Woods near Ludlow ; *Miss Mc Ghie.*

Stem bushy, erect, smooth. *Leaves* alternate, glabrous or nearly so. *Bracteas* small, solitary under each pedicel. *Flowers* greenish. *Fruit* red, smooth, shining, crowned, as in all this genus, with the withered flower.

2. R. *nigrum,* Linn. *Black Currant.* Erect ; racemes lax, downy, pendulous, with a separate simple flower-stalk at their base ; flowers campanulate ; leaves on long pubescent petioles, acutely 5-lobed, doubly serrated, serratures glandular, dotted with glands beneath. *E. Bot. t.* 1291. *E. Fl. v. i. p.* 333. *Hook. Br. Fl. p.* 123.

Woods and swampy places ; not common. *Fl.* May. ♄.

Woods near Ludlow ; *Miss Mc Ghie.* Aqualate Meer ; *Mr. F. Dickinson.*

In a boggy part of Almond Park, near Shrewsbury, apparently wild.

Stem low and spreading, smooth, pubescent when young. *Leaves* alternate, downy on the veins beneath, margins ciliated, yielding a peculiar scent when bruised. *Flowers* green. *Sepals* reflexed. *Fruit* larger than the last, black, gratefully and fragrantly acid.

** *Prickly.*

3. R. *Grossularia,* Linn. *Common Gooseberry.* Prickles 1—3 at the base of the young branches ; peduncles hairy, single-flowered, with a pair of minute bracteas ; leaves rounded, 5-lobed and cut ; fruit more or less hairy and glandulose. *E. Bot. t.* 1295. *E. Fl. v. i. p.* 335. *Hook. Br. Fl. p.* 123.

Hedges and thickets, scarcely wild ; common. *Fl.* April, May. ♄.

Stem bushy, branches widely spreading. *Leaves* alternate, in fascicles, hairy, mostly so beneath, on short pubescent ciliato-glandulose petioles. *Flowers* green. *Sepals* reflexed. *Berry* green.

36. HEDERA. *Linn.* Ivy.

1. H. *Helix,* Linn. *Common Ivy.* Leaves cordate, with 3—5 angular lobes, those of the flowering branches ovate, acuminate ; umbels erect. *E. Bot. t.* 1267. *E. Fl. v. i. p.* 334. *Hook. Br. Fl. p.* 123.

An evergreen of excessive elegance and an universal favourite, ever enfolding our hedges, rocks, ruins, and trunks of trees, in a rich entangled mantle of verdure and beauty. *Fl.* October, November. ♄.

Stem branched, very long, trailing on the ground, or climbing by densely tufted supporting fibres. *Leaves* dark green, shining and veiny, in trailing plants or on the climbing stems broad cordate and angularly lobed, but changing to an ovato-acuminate form on the flowering branches which are only developed when the plant has nearly surmounted the object against which it is climbing. *Flowers* small, pale green, in many-flowered umbels, their pedicels clothed with delicate starry pubescence, and bracteated at the base. *Sepals* very minute. *Petals* reflexed. *Berries* smooth and black.

The greatest elevation at which I recollect to have seen this plant growing in Shropshire was near the summit of the Caradoc Hill, where it most magnificently overspread the perpendicular and exposed face of one of the rocks.

PENTANDRIA—DIGYNIA.

37. CHENOPODIUM. *Linn.* Goose-foot.

* *Leaves plane, undivided.*

1. C. *polyspermum,* Linn. *Many-seeded Goose-foot.* Stem erect or procumbent, simple or branched ; leaves ovato-elliptical, entire, acute, obtuse or emarginate, mucronate ; flowers in axillary compound racemes, cymose, branches dichotomous, divaricate, with a solitary flower in the forks, leafless ; seeds depressed, blackish-brown, shining, round, obsoletely punctulate, obtuse at the margins. *E. Bot. tt.* 1481 & 1480. *E. Fl. v. ii. p.* 15. *a.* & *β. Hook. Br. Fl. p.* 141. *C. acutifolium, Sm. E. Fl. v. ii. p.* 15.

Waste places, partially overflowed during winter ; not common. *Fl.* August, September. ☉.

Astley, near Shrewsbury ; *Mr. E. Elsmere, junr.* Near Ludlow ; *Mr. H. Spare.*

Pit near Sharpstones Hill.

A very variable plant in all its parts as well as in its size, and its different states appear to have given rise to the idea of two distinct species, *C. polyspermum* and *C. acutifolium* of English Flora. Sir W. J. Hooker in Br. Fl. p. 141, expresses a doubt of these species being permanently distinct, and the following results of **a** careful examination of very numerous specimens from the Sharpstones Hill locality tend to confirm this opinion:—

Stems 3—12 or 18 inches high, erect, angular, branched, branches opposite **or** nearly so, divaricate, lower ones elongated and spreading on the ground. In many instances the upper portion of the stem had become checked and stunted in growth near the ground, in which case the lower branches were occasionally elongated, prostrate and simple. *Leaves* ovato-elliptical, smooth, deep green, veiny, paler beneath, tapering at the base into petioles, obtusely or acutely pointed or emarginate, all tipped with a small mucro. All these various forms of the leaves repeatedly occured on the same individual plant. The veins of the leaves and the stem were often tinged with red. *Racemes* shorter than the leaves, compound, cymose, spreading and leafless, nearly sessile in the axils of the leaves, from whence their branches immediately spring and divide in a dichotomous manner, with a solitary sessile flower (sometimes however supported on a pedicel) in the forks of the subdivisions, thus presenting at first sight the appearance of two distinct racemes springing from the same point. In some cases the racemes, especially those in the lower portion of the plant, were very large compound and spreading in a nearly regular repeatedly dichotomous manner and perfectly leafless ; in others, the racemes were small and very contracted, the divisions very few but still dichotomous and divaricate ; whilst in a third state, one-half of the cymose racemes appeared to have been by some means suppressed and reduced to a few flowers only which remained nearly sessile in the axil, whilst the other half had been in consequence unusually developed into a branch with small leaves at intervals bearing in their axils small contracted but still dichotomous racemes or clusters of flowers, and thus exhibiting the appearance of an elongated interrupted leafy spike. These different states of inflorescence were like the leaves all to be observed sometimes on the same specimen, nor did they appear to be accompanied by any particular form of leaf. *Calyx* of the fruit lax, not converging or enveloping the fruit. The *seeds* were similar in all the forms, dark or blackish-brown, shining, depressed, round, notched, obsoletely punctulate, obtuse at the margins. In a very large and luxuriant specimen from Astley, the leaves had a large tooth on one side, as remarked also by *Withering Arr. Br. Pl. 3rd ed. ii. p.* 273.

These forms were also distinctly recognized in a series of specimens sent from Battersea, near London. In one of these all the leaves were emarginate with a mucro in the sinus, and the inflorescence remarkably large, cymose, and dichotomously branched, precisely resembling the figure of *C. polyspermum* in E. Bot. t. 1480. The accurate Mr. Purton, in *Midl. Flora, vol.* 3, *p.* 24, states that the characters of both species were united in his specimens, and doubts if they were really distinct, attributing the decumbent character to age or a greater state of luxuriance.

** *Leaves plane, toothed, angled or lobed.*

2. **C. intermedium**, Mert. & Koch. *Upright Goose-foot.* Stem erect, scarcely branched ; leaves triangular, elongated and toothed at the base, acute, sinuato-dentate, teeth unequal, sharply acuminate, with 3 principal ribs at the base ; flowers in axillary and terminal, erect, straight, subsimple, nearly leafless spikes ; seeds depressed, exsculpato-punctate, obtuse at the margins. *Hook. Br. Fl.* 4th ed. *p.* 124. *Mert. & Koch. v.* ii. *p.* 297. *Meyer Fl. Alt. v.* i. *p.* 403. *Host. Fl. Aust. v.* i. *p.* 322. *C. rhombifolium, Reich. Fl. Excurs.* 3749. *Mühlenb. ap. Willd. en Hort. Berol. v.* i. *p.* 288. *C. urbicum, Gaud. Fl. Helv. v.* ii. *p.* 247. *Bot. Gall. v.* i. *p.* 397. *C. urbicum (of Sm. E. Fl. not of Linn.)* E. Bot. t. 717. *E. Fl. v.* ii. *p.* 10. *Hook. Br. Fl.* 3rd ed. *p.* 142.

Chenopodium polyspermum.

Page 122

Waste places ; rare. *Fl.* August. ☉.

Hadnall, near Shrewsbury ; *Mr. E Elsmere, junr.!*

Stem erect, scarcely branched, square, furrowed, often red. *Lower leaves* in pairs, opposite, petiolate, large, elongated on each side at the base into a bidentate lobe, deeply and unequally sinuato-dentate, smooth, veiny, bright green, paler beneath and covered with mealy particles which are also partially scattered on the upper surface. *Upper leaves* alternate, petiolate, triangular, nearly entire, acute, elongated on each side into a large single tooth above the subcuneate base, with, as in all the other leaves, 3 principal ribs at the base. *Flowers* yellowish green, sessile and crowded, in small, dense, rather remote clusters on the straight and erect spikes.

Very similar in general appearance to *Orthospermum rubrum (Chenopodium rubrum, Linn.)* but well distinguished by all the leaves invariably having 3 principal ribs at the base and by the horizontal, not vertical seed. Our plant agrees with the description of *C. urbicum* of *Engl. Flora,* which is assuredly not the *C. urbicum* of Linnæus, since that is said to have the leaves sparingly toothed : "foliis triangularibus subdentatis," *Sp. Pl.* 318, "foliis triangularibus ...... parce dentatis," *Bluff and Fing. Comp. Fl. Germ.* 2nd ed. v. i. *p.* 449. It is presumed that *C. urbicum, Linn,* is also a native of England, since the following synonyms belong to it, and not to *intermedium : Sm. Fl. Br. v.* i. *p.* 273. *Huds. Fl. Angl.* 104.

3. **C. album**, Linn. *White Goose-foot.* Stem erect, branched ; leaves ovate, inclining to rhomboid, sinuato-dentate, upper ones oblong, nearly entire ; flowers in compound, branched, somewhat leafy racemes ; seeds depressed, smooth and shining, margins obtuse. E. Bot. t. 1723. *E. Fl. v.* ii. *p.* 13. *Hook. Br. Fl.* p. 143.

Waste places, cultivated ground, dunghills, &c.; common, *Fl.* July. Aug. ☉.

Herbage covered with a whitish mealy substance. *Stems* 1—2 feet or more high, erect, more or less branched, obtusely angular and furrowed, streaked with green and white, sometimes reddish. *Leaves* very variable in width and in the teeth of their upper half. *Spikes* compound, partly leafy, clusters alternate, sessile, dense, aggregate. *Seed* round, black, shining and smooth.

β. Leaves green, more entire ; spikes elongated, more branched. *Hook. Br. Fl.* p. 143. *C. album* γ. δ. & ε. *Engl. Fl. v.* ii. *p.* 13. *C. viride, Linn.*

In similar situations and equally common.

38. ORTHOSPORUM. *Meyer.* Upright-seed Mercury.

1. **O. Bonus Henricus**, Meyer. *Good King Henry Mercury.* Leaves triangular, arrow-shaped, entire ; spikes compound, terminal and axillary, erect, leafless. *Chenopodium Bonus Henricus, Linn.* E. Bot. t. 1033. *E. Fl. v.* ii. *p.* 9. *Hook. Br. Fl.* p. 142.

Waste ground and road-sides ; not common. *Fl.* August. ♃.

Bromfield, near Ludlow ; *Mr. H. Spare.* Grinshill church-yard ; *Mr. E. Elsmere, junr.!* Welbach near Shrewsbury ; *Mr. T. Bodenham.* Near Buildwas Abbey ; *Mr. F. Dickinson.*

Uffington. Sandford, near Westfelton. Near St. Mary's Water-lane gateway, Shrewsbury.

Stem 12—18 inches high, spreading and branched below, striated. *Leaves* large, dark green, alternate, on long petioles, glabrous and shining above, unctuous and mealy beneath, margins wavy. *Spikes* tapering, compound, clusters dense and crowded. *Flowers* green, mealy. *Calyx* bordered with an abrupt membrane. *Stigmas* elongated, papillose on the upper side, spreading. *Seed* vertical, kidney-shaped.

39. ULMUS. *Linn.* Elm.

1. U. *campestris*, Linn. *Common small-leaved Elm.* Leaves rhomboid-ovate, acuminate, wedge-shaped and oblique at the base, always scabrous above, doubly and irregularly serrated, downy beneath, serratures incurved; branches wiry, slightly corky, when young, bright brown, pubescent; fruit oblong, deeply cloven, naked. *Lindl. Syn. p.* 226. *E. Bot. t.* 1886. *E. Fl. v. ii. p.* 20. *Hook. Br. Fl. p.* 144.

In hedges. *Fl.* March, April. ♄.
A large tree with rugged bark. *Flowers* in dense heads, each flower pentandrous, mealy, sessile, and subtended by a small scale or bractea. *Segments* of the perianth minutely fringed.
This species yields the best wood of all the elms.

2. U. *suberosa*, Ehrh. *Common Cork-barked Elm.* Leaves nearly orbicular, acute, obliquely cordate at the base, sharply regularly and doubly serrated, always scabrous above, pubescent below, chiefly hairy in the axils; branches spreading, bright brown, winged with corky excrescences, when young very hairy; fruit nearly round, deeply cloven, naked. *Lindl. Syn. p.* 226. *E. Bot. t.* 2161. *E. Fl. v. ii. p.* 21. *Hook. Br. Fl. p.* 145.

In hedges. *Fl.* March. ♄.
Remarkable for the cork-like covering to the branches, which is full of deep fissures. *Flowers* tetrandrous, stalked, segments of the perianth smooth.
The wood is considered inferior to that of the former.

3. U. *glabra*, Mill. *Smooth-leaved Elm.* Leaves ovato-lanceolate, acuminate, doubly and evenly crenato-serrate, cuneate and oblique at the base, becoming quite smooth above, smooth or glandular beneath, with a few hairs in the axils; branches bright-brown, smooth, wiry, weeping; fruit obovate, naked, deeply cloven. *Lindl. Syn. p.* 226. *E. Bot. t.* 2248. *E. Fl. v. ii. p.* 23. *Hook. Br. Fl. p.* 145.

β. *glandulosa*; leaves very glandular beneath. *Lindl. l. c. Hook. l. c.*
Woods and hedges.—β. near Ludlow; *Professor Lindley. Fl.* March. ♄.
A tall elegant tree, branches quite smooth, scarcely downy in their earliest stage of growth. *Flowers* pentandrous, nearly sessile, segments of the perianth short, fringed.

4. U. *montana*, Bauh. *Broad-leaved or Wych Elm.* Leaves obovate, cuspidate, doubly and coarsely serrated, cuneate and nearly equal at the base, always exceedingly scabrous above, evenly downy beneath; branches not corky, cinereous, smooth; fruit rhomboid-oblong, scarcely cloven, naked. *Lindl. Syn. p.* 227. *E. Bot. t.* 1887. *E. Fl. v. ii. p.* 22. *Hook. Br. Fl. p.* 145.

In woods and hedges. *Fl.* March, April. ♄.
Plentiful about the base of the Wrekin; also, on the banks of the Teme, between Ludlow and Tenbury; *E. Lees, Esq.*
Distinguished at first sight by its large spreading branches and broad leaves, appearing just as the "hop-like fruit" comes to perfection. *Flowers* larger and paler, in looser tufts than most of the species, 5, 6, or 7-androus, segments of the perianth 5, 6, or 7, acute, smooth.

40. GENTIANA. *Linn.* Gentian.

1. G. *Amarella*, Linn. *Autumnal Gentian.* Stem very much branched, many-flowered; leaves oblongo-lanceolate, acute; calycine segments 5, nearly equal; corolla somewhat salver-shaped, 5-cleft, fringed at the throat. *E. Bot. t.* 236. *E. Fl. v. ii. p.* 30. *Hook. Br. Fl. p.* 125.

In high limestone pastures; not very common. *Fl.* April—September. ☉.
In abundance on the Tike's Nest, on the banks of the Severn, between Ironbridge and Broseley; *E. Lees, Esq.* Caynham Camp, near Ludlow; and the Knowle limeworks, in the parish of Burford, near Court of Hill; *Rev. W. Corbett.* Trefonen, parish of Oswestry; *Rev. T. Salwey!* Oreton limequarries, plentiful; *Mr. G. Jorden.* Benthall Edge; *Dr. G. Lloyd.*
Root tapering, twisted, yellowish. *Stem* erect, square, 3—12 inches high, branched from the base. *Flowers* numerous. *Corolla* pale dingy purple, tube whitish, the mouth fringed with purplish tapering filaments covered with prominent dots.
A variety with perfectly double flowers has been observed by the *Rev. T. Salwey!* at Trefonen, Oswestry.

2. G. *campestris*, Linn. *Field Gentian.* Stem very much branched, many-flowered; leaves ovato-lanceolate, acute; calycine segments 4, unequal, the two exterior ones the larger; corolla salver-shaped, 4-cleft, fringed at the throat. *E. Bot. t.* 237. *E. Fl. v. ii. p.* 31. *Hook. Br. Fl. p.* 125.

Hilly places on limestone soil. *Fl.* August—October. ☉.
Limekiln wood, near the Wrekin; *T. C. Eyton, Esq.* Trefonen, parish of Oswestry; *Rev. T. Salwey!* Top of Wenlock Edge, above the road from Wenlock to Harley; *W. P. Brookes, Esq.*
Benthall Edge.
Similar to the last in general habit, but of more humble growth, paler herbage, and larger and paler 4-cleft flowers, and essentially distinguished by having the two outer and opposite calycine-segments ovate and very broad, covering the two inner, which are narrow and lanceolate.

41. CUSCUTA. *Linn.* Dodder.

1. C. *Epilinum*, Weihe. *Flax Dodder.* Stem simple; clusters about 5-flowered, bracteated; flowers globose, very fleshy and succulent; calyx campanulate, acute, spreading, nearly as long as the corolla; tube of the corolla globose, longer than the limb, whose segments are very acute; filaments very short; scales erect, appressed, laciniate. *Reich Ic. Bot. t.* 500.

On flax; rare.
Discovered by *J. E. Bowman, Esq.!* 29th July, 1836, on flax in a field near Croesmere, Ellesmere.
Stem filiform, simple, turning upwards from west to east, and adhering at various distances to the stem of the flax by means of small tubercles and papillæ. *Leaves* none. *Flowers* about 5, large, pale, fleshy, succulent, and nearly sessile, in conglomerated nearly sessile *clusters*, subtended by a broad, membranous, obovate, reflexed *bractea.* *Calyx* 5-cleft, segments broadly ovate, acute, spreading, nearly as long as the corolla. *Corolla* 5-cleft, tube globose, longer than the limb, segments acute, nearly erect. *Filaments* very short and very acute, with a broad, membranous, erect, appressed, laciniate *scale* beneath each. *Stigmas* 2. *Capsule* globose, 2-celled, cells 2-seeded. *Seeds* subtriquetrous from compression, covered with chaffy granulations and deeply pitted.

42. DAUCUS. *Linn.* Carrot.

1. D. *Carota*, Linn. *Wild Carrot.* Stem hispid; leaves tripinnate, leaflets pinnatifid, segments linear-lanceolate, acute; prickles of the 3 primary dorsal ridges bifarious, those of the lateral ones in a single row, prickles of the secondary ridges forked at the apex; umbels (generally) with a solitary red abortive flower in the centre, when in seed concave; involucres nearly equal in length to the umbels. *E. Bot. t.* 1174. *E. Fl. v. ii. p.* 39. *Hook. Br. Fl. p.* 136.

Pastures, borders of fields, and way-sides; very frequent. *Fl.* July, Aug. ♂.
Stem 2—3 feet high, branched, erect. *Leaves* alternate, on broad concave footstalks, dilated at the base, hispid. *Involucres* membranous at the edges. *Flowers* white.
The origin of our garden Carrot.

43. TORILIS. *Adans.* Hedge-Parsley.

1. *Torilis Anthriscus*, Gærtn. *Upright Hedge-Parsley.* Stem erect, branched, rough with appressed deflexed bristles; leaves bipinnate, leaflets lanceolate, inciso-serrate, terminal one elongated; umbels terminal; involucres of many small subulate rough leaves; fruit densely clothed with subulate incurved prickles. *E. Fl. v. ii. p.* 43. *Hook. Br. Fl. p.* 137. *Caucalis Anthriscus, Huds. E. Bot. t.* 987.

Hedges, borders of fields, and waste places; very common. *Fl.* July, Aug. ☉.
Stems 2—3 feet high. *Common petiole* channelled, dilated at the base, rough as well as the leaves and the rays of the umbels with erect appressed bristles. *Flowers* white or flesh-coloured. *Anthers* violet-coloured. *Styles* divaricated.

2. T. *infesta*, Spr. *Spreading Hedge-Parsley.* Stem erect, branched, rough with appressed deflexed bristles; leaves bipinnate, leaflets ovate, inciso-pinnatifid, serrated, terminal one elongated; umbels terminal, general involucre of one, partial of few subulate leaves; fruit rough with spreading hooked bristles, and 3 rows of straight appressed ones. *E. Fl. v. ii. p.* 43. *Hook. Br. Fl. p.* 137. *Caucalis infesta, Curt. E. Bot. t.* 1314.

Fields and way-sides; common. *Fl.* July. ☉.
Smaller than the last and more spreading, 6—12 inches high. *Leaves* harsh with bristly hairs. *Umbels* with 3—5 rays. *Flowers* cream-coloured.
Essentially distinguished by the bristles of the fruit and the involucres.

3. T. *nodosa*, Gærtn. *Knotted Hedge-Parsley.* Stem prostrate, branched, rough with appressed deflexed bristles; leaves bipinnate, leaflets ovate, pinnatifid, segments linear, acute; umbels lateral glomerate subsessile, opposite to the leaves; outer mericarp with long straight hooked bristles, inner one tuberculate. *E. Fl. v. ii. p.* 44. *Hook. Br. Fl. p.* 137. *Caucalis nodosa, E. Bot. t.* 199.

Waste places in the limestone districts; not common. *Fl.* May, June. ☉.
Near Ludlow; *Miss Mc Ghie.* Cornfields over limerocks around Wenlock; *W. P. Brookes, Esq.* Pastures near Bomere pool; *Mr. F. Dickinson.*
Leaves of a deep glaucous green, rough with bristly hairs. *Umbels* simple, dense, with several linear bracteas. *Flowers* nearly sessile, reddish. *Interior fruits* often entirely tuberculate.

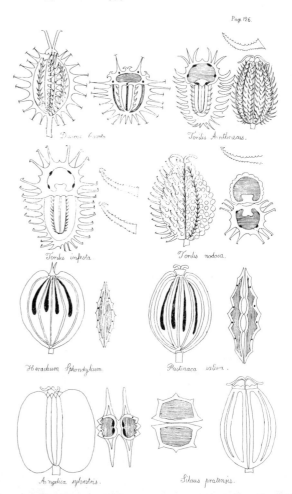

Page 126.

Daucus Carota.   Torilis Anthriscus.

Torilis infesta.   Torilis nodosa.

Heracleum Sphondylium.   Pastinaca sativa.

Angelica sylvestris.   Silaus pratensis.

44. HERACLEUM. *Linn.* Cow-Parsnep.

1. H. *Sphondylium*, Linn. *Common Cow-Parsnep or Hogweed.*
Stem furrowed, hispid with reflexed hairs ; leaves petiolate, with a
dilated clasping base, pinnated, rough, hairy, leaflets petiolate, pinnatifid, cut, sinuated, extreme one somewhat palmate, 3-lobed, sinuated and cut ; petals unequal ; fruit nearly glabrous. *E. Bot. t.*
939. *E. Fl. v. ii. p.* 102. *Hook. Br. Fl. p.* 135.

Hedges, moist meadows, and borders of fields ; very common. *Fl.* July. ♃.
Stem 4—5 feet high. *Leaves* coarsely serrated. *Umbels* flattish, many-rayed.
Flowers white. *Fruit* light-brown, with the 4 brownish vittæ very conspicuous on
the compressed back.

β. Leaves more deeply cut, lobes narrower. *E. Fl. v. ii. p.* 102.
Hook. Br. Fl p. 135.

In similar situations ; not common. *Fl.* July. ♂.
About Hayes, near Oswestry ; *Mr. Waring in Withering.* Westfelton ; *J. F.
M. Dovaston, Esq.*
Limerocks near Wrekin Hill. Hedge-banks between Craven Arms and
Cheney Longville.
Leaves deeply pinnatifid, the two lowest lobes elongated and spreading in a
radiated manner.

45. PASTINACA. *Linn.* Parsnep.

1. P. *sativa*, Linn. *Common Wild Parsnep.* Stem furrowed ;
leaves pinnate, shining, downy beneath ; leaflets ovate, cut and serrated, ultimate one 3-lobed ; universal and partial involucres wanting ; fruit oval. *E. Bot. t.* 556. *E. Fl. v. ii. p.* 101. *Hook. Br.
Fl. p.* 135.

Borders of fields and wastes ; not common. *Fl.* July. ♂.
Astley ; *Mr. E. Elsmere, junr.!*
Root fusiform. *Stems* 3-feet high, branched, roughish. *Petals* very convex,
involute, yellow.
The origin of the garden Parsnep.

46. ANGELICA. *Linn.* Angelica.

1. A. *sylvestris*, Linn. *Wild Angelica.* Stem striated, densely
pubescent above, as well as the umbels ; leaves bi-tripinnate, petioles
dilated and tumid at the base, leaflets equal, ovate, acutely serrated,
somewhat lobed, unequal at the base, pubescent. *E. Bot. t.* 1128.
E. Fl. v. ii. p. 81. *Hook. Br. Fl. p.* 134.

Moist woods, meadows, and margins of rivers ; frequent. *Fl.* July, Aug. ♃.
Captain's Coppice, Coalbrookdale ; *Mr. F. Dickinson.* Standhill Coppice,
Wenlock ; *W. P. Brookes, Esq.* Ludlow ; *Miss Mc Ghie.* Wyre Forest ; *Mr.
G. Jorden.*
Sutton Spa, near Shrewsbury ; Banks of Severn, near Shrewsbury ; and in
moist woods in the neighbourhood.
Stem 2-3 feet high, purplish, branched. *Umbels* large and convex. *Flowers*
white or pinkish.

47. SILAUS. *Besser.* Pepper-Saxifrage.

1. S. *pratensis*, Besser. *Meadow Pepper-Saxifrage.* Leaves
tripinnate, leaflets linear-lanceolate, opposite ; general involucre of

---

1 or 2 leaves, frequently wanting. *Hook. Br. Fl. p.* 133. *Peuce-
danum Silaus, Linn. E. Bot. t.* 2142. *Cnidium Silaus, Spreng.
E. Fl. v. ii. p.* 91.

Pastures and meadows ; not very common. *Fl.* July—September. ♃.
Oakley Park meadows, near Ludlow ; *Mr. H. Spare.* Shadwell, Standhill
Coppice, Wenlock ; *W. P. Brookes, Esq.* Astley, near Shrewsbury ; *Mr. E.
Elsmere, junr.!*
Burwood on Wenlock Edge. Pastures near New Inn, Shrewsbury. Sharp-
stones Hill.
Stem furrowed, 1—2 feet high, branched. *Partial umbels* small, distant.
Flowers pale-yellow. *Whole plant* of a smooth dark-green herbage, fetid when
bruised, and apparently rejected by cattle.

48. ŒNANTHE. *Linn.* Water-Dropwort.

1. Œ. *fistulosa*, Linn. *Common Water Dropwort.* Root tuberous, stoloniferous ; stem and petioles fistulose ; radical leaves bi-or
tripinnate, leaflets plane tri-or multifid, stem leaves pinnate, leaflets
linear, occasionally trifid ; universal involucre none ; fruit angulari-
turbinate. *E. Bot. t.* 363. *E. Fl. v. ii. p.* 68. *Hook. Br. Fl. p.*
131.

Ditches, margins of pools, and watery places ; not uncommon. *Fl.* July—
September. ♃.
Westminster pool, near Coalbrookdale ; *Mr. F. Dickinson.*
Bomere, Almond park, Oxon, Hancott and Mare pools, all near Shrews-
bury.
Stem 2-3 feet high, striated, glaucous, fistulose, contracted beneath the inser-
tion of each leaf. *Stem-leaves* distant, leaflets confined to the upper extremities.
Umbels small, in fruit globose, and bristly with the long rigid styles.

2. Œ. *crocata*, Linn. *Hemlock Water-Dropwort.* Root of large
fusiform tubers ; leaves tri-quadripinnate, leaflets cuneato-rhomboid,
nearly equal, multifid ; fruit cylindrical, oblong, striated, longer than
the pedicels ; universal involucre 1—2 leaved, partial of many leaves.
E. Bot. t. 2313. *E. Fl. v. ii. p.* 70. *Hook. Br. Fl. p.* 131.
Œ. apiifolia, Brot. Hook. Br. Fl. ed. ii. p. 129.

Watery, marshy places ; frequent. *Fl.* July. ♃.
Banks of Canal, near Ellesmere ; *Rev. A. Bloxam.* Banks of the Teme, about
half-a-mile below Ludlow ; *Dr. G. Lloyd.* Road-side between Buildwas and
Ironbridge, near to Marmwood ; *W. P. Brookes, Esq.* Moist shady places near
Welbach ; *Mr. T. Bodenham.* Banks of the Severn, near Coalbrookdale ; *Mr. F.
Dickinson.*
Sutton Spa, near Shrewsbury.
Readily distinguished from all the other species of this genus by the large
erect stem and the broad leaflets. *Stem* 3—5 feet high, stout, much branched.
Leaves dark shining green, glabrous. *Flowers* white.
The roots and stem are said to be full of a poisonous yellow juice, which,
however, I have never observed in our Shropshire plants. The colourless na-
ture of this juice gave rise to another species, *Œ. apiifolia*, which appears to have
differed in no other particular from this plant.

3. Œ. *Phellandrium*, Spreng. *Fine-leaved Water-Dropwort.*
Rhizoma jointed, with numerous whorled fibres at the joints ; stem
fistulose, much branched, branches divaricated ; leaves tripinnate,
leaflets ovate, pinnatifid and cut ; umbels lateral on peduncles oppo-

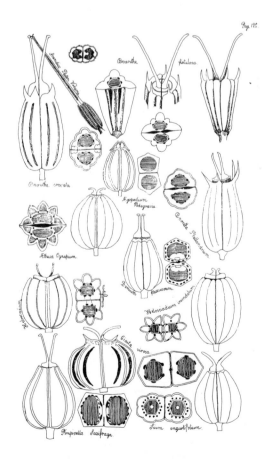

site the leaves; fruit ovato-oblong; general involucre none. *E. Fl. v. ii. p.* 71. *Hook. B. Fl. p.* 132. *Phellandrium aquaticum, Linn. E. Bot. t.* 684.

Ditches, stagnant waters, and marshy borders of pools; not very common. *Fl.* July—September. ♂. Sm. & Bluff. & Fing. ♃. Hook.

Astley, near Shrewsbury; *Mr. E. Elsmere, junr.!* Pits, near Whitley; *Mr. T. Bodenham!* Hancott pool, near Shrewsbury; *Mr. F. Dickinson.* Ellesmere; *H. Bidwell, Esq.* Pool by the side of the Canal between Lilleshall and Church Aston; *R. G. Higgins, Esq.*

Mare pool and Abbot's Betton pool, near Shrewsbury.

Stem 2-3 feet high, very thick below, furrowed, with numerous, widely spreading, divaricated branches. *Leaves* stalked, spreading, segments very fine and acute, dark-green and shining. *Flowers* white, outer ones largest and most irregular, and having the two exterior teeth of the calyx longer than the rest. *Fruit* minutely reticulated.

49. ÆTHUSA. *Linn.* Fool's Parsley.

1. *Æ. Cynapium,* Linn. *Common Fool's Parsley.* Stem striated; leaves bi-tripinnate, petiolate with a membranous sheathing base, leaflets pinnatifid, segments ovato-lanceolate, mucronate, cut; exterior pedicels twice as long as the fruit; partial involucres longer than their umbels. *E. Bot. t.* 1192. *E. Fl. v. ii. p.* 64. *Hook. Br. Fl. p.* 132.

Fields and gardens; frequent. *Fl.* July, August. ☉.

Stem 1-2 feet high, round, branched and very leafy. *Leaves* glabrous, darkgreen, when bruised emitting a nauseous odour and considered dangerous and unwholesome; margins of the segments thickened, and as well as the principal ribs, fringed on the upper side with a row of fine bristles, their points directed towards the apex. *Flowers* pure white, in stalked spreading umbels. *Fruit* pale-brown.

Similar in general appearance to Parsley, for which, when growing in gardens, it has been sometimes mistaken, and the mistake been productive of dangerous effects to the health of those who have partaken of it in their food. The partial involucres and the hatefully fetid smell of the bruised herbage will at all times afford a ready mode of discrimination to the most casual observer.

50. FŒNICULUM. *Hoffm.* Fennel.

1. *F. vulgare,* Gærtn. *Common Fennel.* Leaves biternate, leaflets lineari-filiform pinnatifid, segments awl-shaped; umbels of numerous rays, involucres none. *Hook. Br. Fl. p.* 132. *Anethum Fœniculum, Linn. E. Bot. t.* 1208. *Meum Fœniculum, Spreng. E. Fl. v. ii. p.* 85.

·Lime quarries; rare. *Fl.* July, August. ♃.

Lincoln's Hill, near Coalbrookdale; *Mr. F. Dickinson.* Madeley; *W. P. Brookes, Esq.*

Stem 3-4 feet high, erect, copiously branched, round, striated, smooth, solid. *Leaves* much divided, more or less drooping, of a deep glaucous green, on dilated, membranous petioles. *Umbels* very large, terminal. *Flowers* deep yellow. *Whole plant* aromatic.

51. SIUM. *Linn.* Water-Parsnep.

1. S. *latifolium,* Linn. *Broad-leaved Water-Parsnep.* Stem erect; leaves pinnate, leaflets oblongo-lanceolate mucronate, un-

equal at the base, equally and sharply serrated, serratures mucronate, umbels terminal. *E. Bot. t.* 204. *E. Fl. v. ii. p.* 56. *Hook. Br. Fl. p.* 130.

River-sides, ditches, and watery places; rare. *Fl.* July, August. ♃.

Pool's dam, Wenlock Abbey; *W. P. Brookes, Esq.*

Root fleshy, with numerous fibres, creeping. *Stem* 3-4 feet high, erect, angular, deeply furrowed, smooth, slightly branched. *Leaves* with 5—9 distant opposite leaflets, submersed ones often bipinnatifid with very narrow segments. *Umbels* large. *Flowers* white. *Universal involucre* of numerous lanceolate leaflets, edges membranous, often toothed, *partial one* similar but smaller. *Fruit* small.

2. S. *angustifolium,* Linn. *Narrow-leaved Water Parsnep.* Stem erect; leaves pinnate, leaflets ovate, unequally inciso-serrate; umbels pedunculate, opposite to the leaves. *E. Bot. t.* 139. *E. Fl. v. ii. p.* 56. *Hook. Br. Fl. p.* 130.

Ditches, rivulets, and canals; rare. *Fl.* July, August. ♃.

Madeley and Cox Wood, Coalbrookdale; *Mr. F. Dickinson.* Near Newport, *R. G. Higgins, Esq.*

Canal, and ditches between Shrewsbury and Uffington.

Root extensively creeping. *Stem* 2-3 feet high, erect, round, striated, much branched. *Lower leaves* with many pairs of opposite, unequally lobed, sessile leaflets, the lowermost pair distant; upper leaflets narrow, very unequal, laciniated. *Umbels* smaller than the last, of numerous rays. *Flowers* white. *Universal* and *partial involucres* of several, small lanceolate, reflexed leaflets.

52. PIMPINELLA. *Linn.* Burnet-Saxifrage.

1. P. *Saxifraga,* Linn. *Common Burnet-Saxifrage.* Stems slightly striated, pubescent; leaves pinnate, leaflets of the radical ones roundish-ovate, obtuse, inciso-serrate, lobed or laciniated; styles shorter than the germen; fruit ovate. *E. Bot. t.* 407. *E. Fl. v. ii. p.* 89. *Hook. Br. Fl. p.* 230.

Dry pastures and hills; frequent. *Fl.* July—September. ♃.

Whole plant pubescent, and very variable in the size, and also in the form of the foliage. *Root* woody, pleasantly aromatic. *Stem-leaves* few, lower and radical ones on long petioles. *Flowers* white.

α. *poteriifolia.* Radical leaves pinnate, leaflets ovate, roundish at the base, entire, inciso-serrate.

β. *intermedia.* Radical leaves pinnate, leaflets ovate, deeply and pinnatifidly cut, lobes ovate, inciso-serrate.

γ. *dissecta.* Radical leaves pinnate, leaflets ovate, variously laciniated, segments narrow.

All these varieties may occasionally be found growing together in the same locality, and possibly may be as Withering remarks, (Bot. Arr. 3rd ed. vol. 2, p. 312) referable only to different ages of the plant.

2. P. *magna,* Linn. *Greater Burnet-Saxifrage.* Stem furrowed, glabrous; leaves pinnate, leaflets oblongo-ovate, deeply serrated, subincised, the terminal ones 3-lobed; style longer than the germen; fruit oblong-ovate. *E. Bot. t.* 408. *E. Fl. v. ii. p.* 90. *Hook. Br. Fl. p.* 130.

Shady places, chiefly on limestone soil; not common. *Fl.* July, Aug. ♃.

Hadnall; *Mr. E. Elsmere, junr.!* Castle walk, Ludlow; *Miss Mc Ghie.*

Hill Top, Wenlock Edge.

Larger in all its parts than the foregoing, and the leaflets of the upper leaves much broader and less divided. *Flowers* white or cream-coloured.

53. BUNIUM. *Koch.* Earth-Nut.

1. B. *flexuosum,* With. *Common Earth-Nut.* Root a solitary globose tuber; leaves ternato-supra-decompound; radical ones petiolate, those of the stem subsessile; partial involucre unilateral; base of the style conical exserted; styles straight. *E. Bot. t.* 988. *E. Fl. v. ii. p.* 54. *Hook. Br. Fl. p.* 129.

Woods and pastures; frequent. *Fl.* May, June. ♃.

Oakley Park, near Ludlow: *Mr. H. Spare.* Standhill Coppice, near Wenlock; *W. P. Brookes, Esq.!* Cox Wood, Coalbrookdale; *Mr. F. Dickinson.* Badger Dingle, and Walford; *T. C. Eyton, Esq.*

Meadows generally about Shrewsbury.

Stem 12—18 inches high, solitary, erect, flexuose. *Segments of the leaves* very slender, linear or setaceous. *Flowers* pure white. *General involucre* of 1—3 leaves, often entirely wanting, *partial one* of several leaves. *Fruit* oblong, moderately ribbed, narrower upwards.

54. ÆGOPODIUM. *Linn.* Gout-weed.

1. *Æ. Podagraria,* Linn. *Gout-weed or Herb-Gerarde.* Root creeping; stem angular, furrowed, and branched; leaves bi-triternate, leaflets ovato-acuminate, unequal at the base, acutely and unequally serrated. *E. Bot. t.* 940. *E. Fl. v. ii. p.* 77. *Hook. Br. Fl. p.* 129.

Wet places and ditches; frequent. *Fl.* May—July. ♃.

Ludlow Castle; *Mr. H. Spare.* Apley, near Bridgnorth; *Rev. W. R. Crotch.* Coalbrookdale Woods; *Mr. F. Dickinson.*

Flash near Shrewsbury, banks of the river Severn, Shrewsbury Quarry, rough bank beyond White Horse fields, Shrewsbury. Westfelton.

Root thick, pungent, and aromatic. *Stem* 1-2 feet high, smooth, and hollow. *Leaves* with a dilated base, upper ones ternate. *Flowers* white.

55. SISON. *Linn.* Bastard Stone-Parsley.

1. S. *Amomum,* Linn. *Hedge Bastard Stone-Parsley.* Stem panicled, very much branched; leaves pinnate, leaflets of the lower ones ovate, lobed, inciso-serrate, of the upper ones lanceolate-linear; fruit roundish, ovate. *E. Bot. t.* 954. *E. Fl. v. ii. p.* 60. *Hook. Br. Fl. p.* 129.

Moist ground, chiefly calcareous; rare. *Fl.* August. ☉. or ♂.

Two miles from Ellesmere on the Wrexham road, and again near Overton; *J. E. Bowman, Esq.*

Stem 2-3 feet high, furrowed, pubescent. *Leaves* dark green, smooth. *Umbels* numerous, terminal, erect. *Flowers* cream-coloured. *Petals* broad.

56. HELOSCIADIUM. *Koch.* Marsh-wort.

1. H. *nodiflorum,* Koch. *Procumbent Marsh-wort.* Stem procumbent at the base, and rooting; leaves pinnate, leaflets ovato-lanceolate, equally obtusely serrated; umbels opposite to the leaves,

longer than the peduncles. *Hook. Br. Fl. p.* 128. *Sium nodiflorum.* *Linn. E. Bot. t.* 639. *E. Fl. v. ii. p.* 57.

Ditches and pools; common. *Fl.* July—September. ♃.

Stem 2 feet high, procumbent, rooting at the joints, stout, angular, furrowed, and hollow, with numerous widely spreading branches. *Leaves* distant, each of 2-3 pair of sessile ovato-lanceolate leaflets and an odd one, those of the upper leaves much more rounded, the terminal one often confluent with the next pair of leaflets. *Petioles* dilated and membranous at the base. *Umbels* opposite to the leaves, often nearly sessile, but generally on peduncles varying in length, but never equalling the umbels. *Involucres* dilated and membranous at the edges, *general* one of a single leaf, *partial* of 6 leaves. *Flowers* yellowish-white. *Petals* ovate, entire, with a slight incurved point. *Calyx* obsolete. *Styles* somewhat incurved.

2. H. *inundatum*, Koch. *Least Marsh-Wort.* Stem creeping; leaves pinnate, leaflets of the lower or submersed ones capillaceo-multipartite, of the upper ones cuneiform, inciso-trifid; umbels opposite to the leaves, generally of two rays. *Hook. Br. Fl. v. i. p.* 128. *Sium inundatum, Wiggers. E. Fl. v. ii. p.* 58. *Sison inundatum. E. Bot. t.* 277.

Ditches, margins of pools, and wet ground overflowed in winter and partially dried up in summer. *Fl.* May—July. ♃.

Shawbury heath. Hancott pool, near Shrewsbury. Almond pool.

Stem 4—8 inches long, slender, more or less branched. *Leaves* alternate, petioles dilated and clasping at the base. *Umbels* on peduncles about as long as the leaves, destitute of general involucre, *partial umbels* minute of 5 rays, scarcely longer than their *involucre* of 4 unequal leaflets. *Flowers* white. *Fruit* oblong, large in proportion to the size of the plant, striated.

57. APIUM. *Linn.* Celery.

1. A. *graveolens*, Linn. *Smallage or wild Celery.* Stem furrowed, branched; radical leaves pinnate, stem ones ternate, leaflets wedge-shaped, cut; umbels subsessile. *E. Bot. t.* 1210. *E. Fl. v. ii. p.* 76. *Hook. Br. Fl. p.* 127.

Marshy places; rare. *Fl.* August. ♂.

In Stank mead, near Bishop's Castle; *Gough's Camden.*

Stem 2 feet high, furrowed, smooth, widely spreading. *Leaves* pinnate or ternate, bright green, *leaflets* large, wedge-shaped, deeply lobed and cut at the extremity, entire in their lower part; *radical ones* on long stalks, their leaflets rounder and truncate at the base. *Umbels* terminal and lateral, subsessile, accompanied by 1 or 2 ternate leaves, *general* rays unequal, *partial* equal and more numerous. *Flowers* greenish-white.

The origin of our garden *Celery.*

58. CICUTA. *Linn.* Cowbane.

1. C. *virosa*, Linn. *Water Hemlock or Cowbane.* Leaves tripinnate, leaflets subternate, lanceolate, acute, serrated; umbels pedunculated; partial involucres linear-setaceous. *E. Bot. t.* 479. *E. Fl. v. ii. p.* 62. *Hook. Br. Fl. p.* 127.

Margins of ponds and lakes; not common. *Fl.* July, August. ♃.

Oxon pool, near Shrewsbury; *J. E. Bowman, Esq.* In Mr. Slaney's pool-dam at Hatton; *Withering.* Rivulet near Buildwas; *T. & D. Bot. Guide.* Ellesmere; *Rev. A. Bloxam.* Snowdon pool, near Beckbury; *Mr. F. Dickinson.* Lightmoor; *W. P. Brookes, Esq.!* Bomere pool; *Rev. E. Williams.*

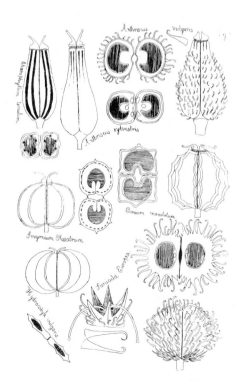

Hancott pool, near Shrewsbury.

Root tuberous, with many slender fibres, and as well as the lower part of the stem, hollow and divided by transverse partitions into large cells. *Stem* 3-4 feet high, branched, furrowed, smooth. *Leaves* on long footstalks dilated at the base, upper ones triternate. *Umbels* large, many-rayed, *partial* umbels of numerous slender rays. *Flowers* white.

A deadly poison to man, and according to Smith, to cattle also, though Hooker states that they seem to eat it with impunity.

59. CHÆROPHYLLUM. *Linn.* Chervil.

1. C. *temulentum*, Linn. *Rough Chervil.* Stem rough, spotted, swollen below the joints; leaves tripinnate, somewhat hairy on both sides, leaflets ovato-oblong, lobed and incised; partial involucres ovato-lanceolate, acuminate, ciliated, and reflexed; styles decurved, shorter than the convexo-conical base. *E. Bot. t.* 1521. *Hook. Br. Fl. p.* 138. *Myrrhis temulenta. E. Fl. v. ii. p.* 51. *M. temula, Spreng.*

Hedges and bushey places; common. *Fl.* June, July. ♃.

Stem 3 feet high, round, solid, striated, rough with deflexed hairs. *Leaves* dark-green, paler beneath, footstalks dilated and clasping at the base. *Rays* of the *general umbels* hairy, of the *partial* ones smooth. *General involucre* none or solitary, *partial* of several leaves. *Umbels* at first drooping. *Flowers* white. *Fruit* linear-oblong, striated.

60. ANTHRISCUS. *Pers.* Beaked-Parsley.

* Carpels smooth.

1. A. *sylvestris*, Koch. *Wild Beaked-Parsley.* Stem furrowed, striated, glabrous, slightly swollen below the joints; leaves triply pinnate, leaflets ovato-lanceolate, deeply cut; umbels terminal, stalked; partial involucres of several ovato-lanceolate cilated leaves. *Hook. Br. Fl. p.* 137. *Chærophyllum sylvestre, Linn. E. Bot. t.* 752. *E. Fl. v. ii. p.* 48.

Hedges and borders of fields; frequent. *Fl.* April—June. ♃.

Stem 3 feet high, branched, lower part downy, upper glabrous. *Umbels* at first slightly drooping, *rays* smooth. *Flowers* white. *Fruit* linear-oblong, with a short and inconspicuous few-ribbed beak.

2. A. *Cerefolium*, Koch. *Garden Beaked-Parsley.* Stem round, striated, slightly hairy at the joints only; leaves tripartite, decompound, leaflets ovate, pinnatifid, the segments obtuse; umbels lateral, sessile; partial involucres of 3 unilateral, linear leaves. *Hook. Br. Fl. p.* 138. *Scandix Cerefolium, Linn. E. Bot. t.* 1268. *Chærophyllum sativum. E. Fl. v. ii. p.* 48.

Waste ground; probably an escape. *Fl.* July. ☉.

On a dunghill near Ludlow Castle; *Miss Mc Ghie.* Broseley Hill; *Mr. F. Dickinson.* White Ladies, near Boscobel; *H. Bidwell, Esq.*

Stem 1½-2 feet high, slender. *Leaves* delicate, pale yellow-green, slightly hairy on both sides, on long channelled petiols dilated membranous and woolly at the base. *Umbels* axillary or opposite to the leaves, sessile, of few pubescent rays, *partial* umbels small. *Flowers* white. *Fruit* large, perfectly glabrous, linear, tapering upwards.

** *Carpels muricated.*

3. A. *vulgaris*, Pers.   *Common Beaked-Parsley.*   Stem round, smooth, swollen under each joint; leaves ternately decompound, segments obtuse; umbels opposite to the leaves; partial involucres of several small ciliated leaves; fruit ovately conical, hispid, about twice as long as the glabrous beak.   *E. Fl. v.* ii. *p.* 45.   *Hook. Br. Fl. p.* 138.   *Scandix Anthriscus.   E. Bot. t.* 818.

Waste places, by road-sides, hedge-banks, &c., especially near towns and villages; frequent.   *Fl.* May, June. ☉.

Stem 2 feet high, branched, swelling under each joint.   *Leaves* of a most beautiful light-green, slightly hairy.   *Petioles* membranous and woolly at the base.   *Umbels* on rather short smooth stalks, *partial* umbels small.   *Flowers* white.   *Fruit* rather large, slightly compressed, covered with hooked bristles, with a distinct furrow on each side extending to the beak.

61. SCANDIX. *Linn.* Shepherd's-Needle.

1. S. *Pecten Veneris*, Linn.   *Needle Chervil, Venus's Comb, or Shepherd's Needle.*   Fruit hispid, beak dorsally compressed, hispid at the edges; leaves tripinnate, leaflets deeply cut into many linear segments; umbels of few rays.   *E. Bot. t.* 1396.   *E. Fl. v.* ii. *p.* 46.   *Hook. Br. Fl. p.* 137.

Corn-fields; abundant.   *Fl.* June—September. ☉.

Stem 6—12 inches high, striated, roughish.   *Leaves* on long footstalks dilated clasping and ciliated at the base, of a light sh.ning green.   *Segments* bristly at the margins and mucronate.   *Flowers* white.   *Partial involucres* broad and laciniated.   Well distinguished by the very long and singular appearance of the fruit.

62. MYRRHIS. *Tourn.* Cicely.

1. M. *odorata*, Scop.   *Sweet Cicely.*   Stem round, striated, hairy; leaves ternately decompound, leaflets ovato-lanceolate, pinnatifid; segments inciso-serrate; umbels terminal; partial involucres lanceolate, ciliated; fruit large with very sharp ribs and deep furrows between them.   *E. Fl. v.* ii. *p.* 50.   *Hook. Br. Fl. p.* 139.   *Scandix odorata, Linn.   E. Bot. t.* 697.

Hilly pastures; rare.   *Fl.* May, June.

White Ladies, near Boscobel; *T. & D. Bot. Guide.*   Benthal Edge; *Mr. F. Dickinson.*   Near Furnace Mill, near Bewdley; *Mr. G. Jorden.*

Stem 2-3 feet high, branched, with deflexed hairs.   *Leaves* very large, of a bright green, slightly hairy, on hairy petioles dilated ribbed and membranous at the base.   *Umbels* of numerous downy rays, many of the *partial* ones abortive.   *Flowers* white.   *Fruit* dark-brown, very large, powerfully fragrant, and like the whole plant, highly aromatic.

63. SMYRNIUM. *Linn.* Alexanders.

1. S. *Olusatrum*, Linn.   *Common Alexanders.*   Stem round, furrowed; leaves bi or tri-ternate, petiolate with a very broad membranous base, leaflets broadly ovate, inciso-serrated and lobed.   *E. Bot. t.* 230.   *E. Fl. v.* ii. *p.* 76.   *Hook. Br. Fl. p.* 140.

Waste ground and among ruins; rare.   *Fl.* May, June. ♂.

Under the walls of Ludlow Castle.   Shrewsbury Castle Mount.

Stem 3-4 feet high, stout.   *Leaves* bright yellow-green.   *Flowers* yellow-green, in very dense, numerous, rounded umbels.   *Fruit* nearly black when ripe, aromatic and pungent.

64. CONIUM. *Linn.* Hemlock.

1. C. *maculatum*, Linn.   *Common Hemlock.*   Stem glabrous, spotted; leaves tripinnate, leaflets lanceolate, pinnatifid, with acute and often cut segments; partial involucres lanceolate, shorter than their umbels.   *E. Bot. t.* 1191.   *E. Fl. v.* ii. *p.* 65.   *Hook. Br. Fl. p.* 139.

Waste places, hedges, and road-sides; not unfrequent.   *Fl.* June, July. ♂.

Oakley Park, near Ludlow; *Mr. H. Spare.*   Ruckley Grange; *Dr. G. Lloyd.*   Callaughton, near Wenlock; Wenlock Abbey; Court House, Madeley; *W. P. Brookes, Esq.*   Stottesden and Prescott; *Mr. G. Jorden.*   Waste places, Coalbrookdale; *Mr. F. Dickinson.*   Higford, near Beckbury; *H. Bidwell, Esq.*   Near Newport; *R. G. Higgins, Esq.*

Ludlow Castle.   Red-barn, Meole, Bomere, Hancott, Preston Boats, and generally about Shrewsbury.

Root fusiform.   *Stem* 3—5 feet high, much branched upwards, striated and spotted with purple.   *Leaves* large and smooth, repeatedly divided, of a deep shining green, and on long footstalks dilated and sheathing at the base.   *Involucres* membranous at the margins.   *Flowers* white.

The leaves when bruised are extremely fetid and nauseous, and yield an extract which has been extensively employed in scrophulous and cancerous maladies, and for lowering the pulse in acute inflammatory disorders.

65. SANICULA. *Linn.* Sanicle.

1. S. *Europæa*, Linn.   *Wood Sanicle.*   Lower leaves palmate, lobes trifid, inciso-serrate; flowers all sessile.   *E. Bot. t.* 98.   *E. Fl. v.* ii. *p.* 36.   *Hook. Br. Fl. p.* 126.

Woods and thickets; frequent.   *Fl.* May, June. ♃.

Oakley Park, near Ludlow; *Mr. H. Spare.*   Standhill Coppice and Woods around Wenlock; *W. P. Brookes, Esq.!*   Wyre Forest; *Mr. G. Jorden.*   Near Oswestry; *Rev. T. Salwey.*   Chesterton; *Dr. G. Lloyd.*   Ball's Coppice, Walford; *T. C. Eyton, Esq.*

Haughmond Hill.   Benthall Edge.   Woods near Bomere pool.   Lyth Hill.

Root of numerous black tufted rather fleshy fibres.   *Stems* 12—18 inches high, ascending, furrowed.   *Leaves* chiefly radical, on long petioles, their serratures almost ciliate.   *Umbels* small, dense.   *Flowers* white.

66. HYDROCOTYLE. *Linn.* White-rot.

1. H. *vulgaris*, Linn.   *Common White-rot, Marsh Penny-wort.*   Stem creeping; leaves petiolate, peltate, orbicular, somewhat lobed and crenate; heads of about 5 flowers.   *E. Bot. t.* 751.   *E. Fl. v.* ii. *p.* 96.   *Hook. Br. Fl. p.* 126

Bogs, marshes, and banks of lakes; common.   *Fl.* May—July, ♃.

Root fibrous.   *Stem* prostrate, smooth and slender.   *Leaves* produced from the joints of the stem, on long erect petioles.   *Peduncles* axillary, shorter than the petioles, with a pair of bracteas at their base.   *Umbels* very small, their rays scarcely observable, with 3 or 4 minute lanceolate bracteas at their base.   *Flowers* white or reddish.

PENTANDRIA—TRIGYNIA.

67. VIBURNUM. *Linn.* Guelder-rose.

1. V. *Opulus*, Linn.   *Common Guelder-rose, or Water-Elder.*   Leaves glabrous, 3-lobed, acuminate and dentate; petioles glandulose.   *E. Bot. t.* 331.   *E. Fl. v.* ii. *p.* 107.   *Hook. Br. Fl. p.* 146.

Woods, hedges, and coppices; not unfrequent.   *Fl.* June, July. ♄.

Wenlock Edge; common; *W. P. Brookes, Esq.!*   Ludlow; Miss Mc Ghie.   Cox Wood, Coalbrookdale; *Mr. F. Dickinson.*   Astley; *Mr. E. Elsmere, junr.*   Near Newport; *R. G. Higgins, Esq.*

Woods beyond Haughmond hill.   Banks of river Severn, near Shrewsbury.   Coalbrookdale.

A small shrubby tree, glabrous in every part.   *Leaves* large, broad, subcordate at the base, bright green in summer, ch:nging to a rich crimson in autumn.   *Cymes* large.   *Flowers* white, central ones perfect, small; exterior ones abortive, dilated and plane.   *Berries* red.

68. SAMBUCUS. *Linn.* Elder.

1. S. *Ebulus*, Linn.   *Dwarf Elder or Dane-wort.*   Stem herbaceous; leaves pinnate, leaflets lanceolate, serrated; stipules foliaceous; cymes with 3 principal branches.   *E. Bot. t.* 475.   *E. Fl. v.* ii. *p.* 108.   *Hook. Br. Fl. p.* 146.

Pastures and waste places; not very common.   *Fl.* July. ♃.

At Tern Hill, near Whittington, and about Whittington Castle; *T. & D. Bot. Guide.*   Lincoln's Hill, near Coalbrookdale; *Mr. R. Dickinson.*   Hedge-bank near the Old Race-course, Bridgnorth; *Dr. G. Lloyd.*   Bitterley; Miss Mc Ghie.   Near Hughley Church, and on road-side, Burton, near Wenlock; *W. P. Brookes, Esq.!*   Exford's Green, near the Lyth Hill; *Mr. T. Bodenham.*   Albright Hussey, near Shrewsbury; *Dr. T. W. Wilson.!*   Near Llanymynech; *J. F. M. Dovaston, Esq.*

A field of several acres in extent entirely covered with it between Burwood and Middlehope, on Wenlock Edge.   Dingle in Shrewsbury Quarry.

Root fleshy, creeping.   *Stem* 2-3 feet high, simple, angular, and furrowed.   *Leaves* dark-green, leaflets very unequal at the base, veins downy beneath.   *Cymes* large, terminal, purplish.   *Anthers* large, purple.   *Berries* sphærical, black.

2. S. *nigra*, Linn.   *Common Elder.*   Stem arboreous; leaves pinnate, leaflets ovate, cuspidate, serrated; stipules none; cymes with 5 principal branches.   *E. Bot. t.* 476.   *E. Fl. v.* ii. *p.* 109.   *Hook. Br. Fl. p.* 147.

Woods, coppices, hedges, &c.; frequent.   *Fl.* June. ♄.

A small tree having the stem and branches full of pith.   *Branches* numerous, irregular, always opposite, bark smooth and grey.   *Leaves* deep-green, leaflets usually 2 pairs with an odd one, unequal at the base.   *Cymes* terminal, large, cream-coloured.   *Anthers* small, yellow.   *Berries* purplish black.

A variety with white or green berries, is said by Withering to occur in this county.

PENTANDRIA—TETRAGYNIA.

69. PARNASSIA. *Linn.* Grass of Parnassus.

1. P. *palustris*, Linn.   *Common Grass of Parnassus.*   Filaments of the nectary 9—13; radical leaves cordate, petiolate, cauline one

amplexicaul.   *E. Bot. t.* 82.   *E. Fl. v.* ii. *p.* 114.   *Hook. Br. Fl. p.* 148.

Bogs and wet places; not very common.   *Fl.* August—October. ♃.

Felton Farm, near Ludlow; *Mr. H. Spare.*   In the vicinity of Bitterley, at the base of the Clee Hill, in profusion; *E. Lees, Esq.*   Moist meadows near Ryton, in the parish of Condover; *Rev. W. Corbett.*   In a field between Great and Little Ness; *T. C. Eyton, Esq.*   Canton Rough, near Bridgnorth; *Dr. G. Lloyd.*   Near the Brickkilns, and at Maesbury, in the parish of Oswestry; *Rev. T. Salwey.*   North of Bridgnorth; *Rev. W. R. Crotch.*   Rowley, near Wenlock; *W. P. Brookes, Esq.!*   Croesmere; *H. Bidwell, Esq.*   Meadow near Ellerton Hall, through which the brook runs which divides the counties of Salop and Stafford; *R. G. Higgins, Esq.!*

Stem angular, 3—10 inches high.   *Leaves* chiefly radical, on long footstalks, cordate, entire, nerved; one on the stem below the middle, sessile.   *Flowers* solitary, terminal, large, yellowish-white, handsome.   *Petals* broadly obovate.   *Nectaries* each with an obovate scale, opposite the petals, fringed along the margin with white hairs terminated by yellow pellucid globular glands.

PENTANDRIA—PENTAGYNIA.

70. LINUM. *Linn.* Flax.

* *Leaves alternate.*

1. L. *usitatissimum*, Linn.   *Common Flax.*   Leaves lanceolate; calycine leaves ovate, acute, 3-nerved; petals crenate; stem subsolitary.   *E. Bot. t.* 1357.   *E. Fl. v.* ii. *p.* 118.   *Hook. Br. Fl. p.* 150.

Corn-fields; rare.   *Fl.* July. ☉.

In a wild lane near Coalbrookdale; *E. Lees, Esq.*

Stem 12—18 inches high, slender, branched above.   *Leaves* distant, 3-ribbed, with pellucid dots.   *Flowers* large, purplish blue.

2. L. *perenne*, Linn.   *Perennial Blue Flax.*   Leaves linear, acute; calycine leaves obovate, obtuse, obscurely 5-ribbed, glabrous; stems numerous from the same root.   *E. Bot. t.* 40.   *E. Fl. v.* ii. *p.* 118.   *Hook. Br. Fl. p.* 150.

Limestone districts; rare.   *Fl.* June, July. ♃.

In a wild tangled bushy lane between Little Wenlock and Buildwas; *E. Lees, Esq.*   Near Onibury, and on a hill above Totterton House, called Hucklement, looking towards Bishop's Castle; *Miss Mc Ghie.*   Once found (though probably not wild) near Welbatch, near Shrewsbury; *Mr. T. Bodenham.*

Root woody.   *Stems* numerous, ascending, 1 foot high.   *Flowers* similar to the last but smaller.

** *Leaves opposite.*

3. L. *catharticum*, Linn.   *Purging Flax.*   Leaves oblong; stem dichotomous above; petals acute.   *E. Bot. t.* 382.   *E. Fl. v.* ii. *p.* 119.   *Hook. Br. Fl. p.* 150.

Pastures; abundant.   *Fl.* June, July. ☉.

Stem slender, upright, 2—6 inches high.   *Flowers* gracefully pendulous before expansion, small, white.

PENTANDRIA—HEXAGYNIA.

71. DROSERA. *Linn.* Sun-dew.

1. D. *rotundifolia*, Linn. *Round-leaved Sun-dew.* Leaves radical, orbicular, spreading ; petioles hairy ; scapes much longer than the leaves ; seeds chaffy. *E. Bot. t.* 867. *E. Fl. v. ii. p.* 122. *Hook. Br. Fl. p.* 151.

Bogs and moist ground ; frequent. *Fl.* July. ♃.

Leaves in all our species covered with red pedunculated glands, exuding a viscid fluid which retains insects. *Scape* 2–5 inches high, glabrous. *Flowers* secund, small, in a simple raceme revolute when young and gradually becoming straight as the flowers expand ; the lowermost flower always expanding first, and the rest in succession. *Petals* white, expanded only in the sunshine, and soon closing, never again to re-open. *Styles* variable in number.

β. *ramosa.* A variety occurs not unfrequently at Bomere pool of stouter and taller habit, with more numerous leaves, and the raceme forked or branched. On some specimens a simple raceme occurs, as well as the forked one.

2. D. *longifolia*, Linn. *Spathulate-leaved Sun-dew.* Leaves radical, spathulate, very obtuse, erect, on long glabrous petioles ; scapes a little longer than the leaves ; seeds with a compact rough coat, not chaffy. *E. Bot. t.* 868. *E. Fl. v. ii. p.* 123. *Hook. Br. Fl. p.* 151.

Bogs and wet mossy places ; not common. *Fl.* July. ♃.

Whixall moss ; *J. E. Bowman, Esq.* Hatton, Hine Heath ; *T. & D. Bot. Guide.* Bomere pool ; *Rev. E. Williams.* Moss near Cold Hatton ; *T. C. Eyton, Esq.* Vownog bog, near Westfelton. Shomere moss, near Shrewsbury.

Well distinguished from the following by the rough and not loose coat to the seeds. *Styles* often 8. *Stigmas* deeply cloven.

3. D. *anglica*, Huds. *Great Sun-dew.* Leaves radical, linear-spathulate, erect, on very long glabrous petioles ; scape much longer than the leaves ; seeds with a loose chaffy coat. *E. Bot. t.* 369. *E. Fl. v. ii. p.* 123. *Hook. Br. Fl. p.* 152.

On bogs ; not common. *Fl.* July, August. ♃.

Whixall moss, between Ellesmere and Whitchurch ; *J. E. Bowman, Esq.* Vownog bog, near Westfelton.

Clearly distinguished from the preceding by the loose reticulated even chaffy coat of the seed.

CLASS VI.

HEXANDRIA.

" See, how the lily drinks
The latent rill, scarce oozing through the grass,
Of growth luxuriant ; or the humid bank,
In fair profusion, decks."
                                    THOMSON.

" Sometimes her head she fondly would aguise
With gaudy garlands, or fresh flow'rets dight
About her neck, or rings of rushes plight."
                                    SPENSER.

CLASS VI.

HEXANDRIA. 6 *Stamens.* *(equal in height.)*

ORDER I. MONOGYNIA. 1 *Style.*

* *Flowers complete, having a double perianth. (Calyx and Corolla.)*

1. BERBERIS. *Calyx* of 6 concave, coloured, inferior, deciduous leaves. *Petals* 6, each with 2 glands at the base. *Berry* 2-3-seeded.—*Nat. Ord.* BERBERIDEÆ, *Vent.*—Name, *Berbérys*, the Arabic name of the fruit.

2. PEPLIS. *Calyx* campanulate, persistent, with 6 large and 6 alternating smaller teeth. *Petals* 6, minute, inserted upon the calyx, often wanting. *Capsule* superior, 2-celled, many-seeded.—*Nat. Ord.* LYTHRARIEÆ, *Juss.*—Named from πεπλιον, anciently applied to the genus *Portulaca*, now to one somewhat similar in habit.

(See *Lythrum* in CL. XII.)

** *Perianth single, superior.*

3. GALANTHUS. *Perianth* petaloid, of 6 pieces, 3 outer ones spreading, 3 inner smaller, erect, emarginate. *Flowers* solitary from a spatha.—*Nat. Ord.* AMARYLLIDEÆ, *Br.*—Name from γαλα, milk, and ανθος, a *flower ;* in allusion to the pure whiteness of its elegant blossoms. The French name, *pierce-neige,* is very expressive.

4. NARCISSUS. *Perianth* superior, coloured, funnel-shaped, with a spreading 6-partite limb and a campanulate or cup-shaped *crown* or nectary enclosing the stamens. *Flowers* from a spatha, solitary or many-flowered.—*Nat. Ord.* AMARYLLIDEÆ, *Br.*—Name from ναρκη, stupor ; in allusion to the powerful and injurious smell of the flowers, or from the youth *Narcissus,* fabled to have been changed into this flower.

*** *Perianth single, inferior, petaloid, rarely herbaceous.*

5. CONVALLARIA. *Perianth* inferior, petaloid, deciduous, 6-cleft, globose or cylindrical. *Berry* 3-celled. *Seeds* 1-2 in each cell.—*Nat. Ord.* SMILACEÆ, *Br.*—Name, *convallis,* a *valley,* from the locality of the species.

6. ALLIUM. *Perianth* inferior, petaloid, of 6 ovato-lanceolate spreading pieces. *Capsule* triquetrous. *(Flowers* umbellate, arising from a 2-leaved spatha.)—*Nat. Ord.* ASPHODELINEÆ, *DC.*—Name from *all,* Celtic, *acrid or burning ;* in allusion to the pungency of its juices.

7. ORNITHOGALUM. *Perianth* inferior, petaloid, of 6 persistent pieces. *Stamens* alternately larger or dilated at the base. *Capsules* with 3 angles and 3 furrows. *(Flowers* racemose, or corymbose. *Bracteas* membranaceous.)—*Nat. Ord.* ASPHODELINEÆ, *DC.*—Name from ορνις, *a bird*, and γαλα, *milk.*

8. AGRAPHIS. *Perianth* divided to the base, 6-partite, segments campanulato-connivent reflexed at the extremities only. *Stamens* included, inserted in the segments below the middle. *Germen* ovate, obtusely triangular. *Style* filiform. *Stigma* obtuse, triangular. *Capsule* membranous, obtusely trigonous, crowned with the persistent style, 3-celled, cells few-seeded, valves 3. *Seeds* subglobose, marked with an unilateral raphe. *(Flowers* racemed, *bracteas* in pairs.)—*Nat. Ord.* ASPHODELINEÆ, *DC.*—Name from *a*, *without*, and γραφις, *a writing* ; from the foliage being destitute of any mark or figure.

9. NARTHECIUM. *Perianth* inferior, petaloid, of 6 linear-lanceolate spreading persistent pieces. *Stamens* opposite to the segments ; filaments bearded. *Germen* ovate, obsoletely triangular, attenuated into a conical style. *Stigma* simple, obtuse. *Capsule* 3-celled, 3-valved. *Seeds* elongated at each extremity into a filiform process.— *Nat. Ord.* JUNCEÆ, *Juss.*—Name from ναρθης, *a rod* ; probably from the elongated straight raceme of flowers.

10. FRITILLARIA. *Perianth* campanulate, drooping, 6-partite, segments straight excavated at the base with a nectariferous cavity. *Stigmas* 3. *Capsule* oblong, obtusely trigonous, 3-celled, 3-valved. *Seeds* flat.—*Nat. Ord.* LILIACEÆ, *Juss.*—Name from *fritillus*, a dice-board.

11. TULIPA. *Perianth* campanulate, erect, 6-partite, segments concave without nectaries. *Style* none. *Stigma* thickened, 3-lobed. *Capsule* oblong, trigonous, 3-celled, 3-valved. *Seeds* flat.—*Nat. Ord.* LILIACEÆ, *Juss.*—Name from *toliban*, Persian, a *turban*, whose gay colours are similar to those of a tulip.

12. ACORUS. *Flowers* densely arranged upon a lateral spadix. *Perianth* of 6 erect scales. *Stamens* opposite to the scales. *Germen* sessile, obtusely sexangular, mucronate, 3-celled. *Stigma* sessile. *Capsule* berried, triangular, obversely pyramidical, 1-celled, the dissepiments obliterated filled with gelatine, 2-3-seeded.—*Nat. Ord.* AROIDEÆ, *Juss.*—Name from *a*, *without*, and κοριον or κορη, *the pupil of the eye* ; the diseases of which it was supposed to remove.

**** *Perianth single, inferior, glumaceous.*

13. JUNCUS. *Perianth* persistent, 6-partite, glumaceous, the exterior segments a little longer and more obtuse than the interior.

Capsule 3-celled, 3-valved, valves with the dissepiments in their middle, many-seeded. *Seeds* very minute, affixed to the central margin of the dissepiments, subglobose. *(Leaves* rigid, mostly rounded, rarely plane, glabrous.)—*Nat. Ord.* JUNCEÆ, *Juss.*— Name from *jungo*, to *join* ; the leaves and stems of this genus having been employed as cordage.

14. LUZULA. *Perianth* persistent, 6-partite, glumaceous, segments nearly equal. *Capsule* 1-celled, 3-valved, valves without dissepiments, 3-seeded. *Seeds* erect, affixed to the base of the cells, and furnished with a large umbilical strophiole. *(Leaves* soft, plane, generally hairy.)—*Nat. Ord.* JUNCEÆ, *Juss.*—Name corrupted from *lucciola* or *luzziola*, Italian, a *glow-worm* ; in allusion to the sparkling brilliance by moonlight of the heads of flowers covered with dew.

(See *Peplis* in ORD. I.—*Polygonum* in CL. VIII.)

ORDER II. TRIGYNIA. 3 *Styles.*

15. RUMEX. *Calyx* of 3 leaves combined at the base. *Corolla* of 3 petals. *Stigmas* multifid, plumose. *Nut* triquetrous, covered by the enlarged petals which often bear tubercles.—*Nat. Ord.* POLYGONEÆ, *Juss.*—Name of unknown origin.

16. SCHEUCHZERIA. *Perianth* single, petaloid, of 6 permanent recurved leaves. *Anthers* elongated, recurved, on capillary filaments opposite to the leaves of the perianth. *Capsules* 3—6, inflated, 2-valved, 1-2-seeded.—*Nat. Ord.* ALISMACEÆ, *Juss.*—Named in honour of the 3 *Scheuchzers*, Swiss Botanists.

17. TRIGLOCHIN. *Perianth* of 6 concave deciduous leaves, the 3 outer more convex and gibbous at the base. *Stamens* inserted in the base of the leaves of the perianth, with their backs towards the pistil. *Capsules* 3—6, 1-seeded, united by a longitudinal angular receptacle from which they separate at the base.—*Nat. Ord.* ALISMACEÆ, *DC.*—Name from τρεις, *three*, and γλωχιν, *a point* ; in allusion to the appearance of the points of the capsules when separated from the receptacle.

18. COLCHICUM. *Perianth* single, tubular, very long, arising from a very short scape on the summit of a solid bulb, and enclosed in a tubular spatha ; *limb* campanulate, 6-partite, petaloid. *Stamens* inserted in the summit of the tube. *Styles* very long. *Capsule* 3-celled, cells united at the base, many-seeded.—*Nat. Ord.* MELANTHACEÆ, *Br.*—Name from *Colchis*, where it is said to grow abundantly.

(See *Elatine* in CL. VIII.)

ORDER III. HEXAGYNIA. 6 *Styles.*

19. ACTINOCARPUS. *Calyx* of 3 leaves, persistent. *Petals* 3. *Germens* 6—8. *Capsules* combined at the base, spreading in a radiated manner, 2-seeded. *Embryo* much curved.—*Nat. Ord.* ALISMACEÆ, *DC.*—Name from ακτιν, a *ray*, and καρπος, *fruit* ; in allusion to the curiously radiated fruit.

ORDER IV. POLYGYNIA. Many *Styles.*

20. ALISMA. *Calyx* of 3 leaves, persistent. *Petals* 3. *Capsules* many, clustered, distinct, indehiscent, 1-seeded. *Embryo* much curved.—*Nat. Ord.* ALISMACEÆ, *DC.*—Name from *alis*, Celtic, *water* ; in allusion to the situations in which the plants always grow.

HEXANDRIA—MONOGYNIA.

1. BERBERIS. *Linn.* Barberry.

1. B. *vulgaris*, Linn. *Common Barberry.* Racemes pendulous, spines 3-palmate ; leaves obovate ciliato-serrate. *E. Bot. t.* 49. *E. Fl. v. ii. p.* 184. *Hook. Br. Fl. v. i. p.* 157.

Hedges ; not unfrequent. *Fl.* June. ♄.

Near Baschurch ; *T. C. Eyton, Esq.* Buildwas Abbey ; *E. Lees, Esq.* Whitecliff, near Ludlow ; *Mr. J. S. Baly.* Hedges, near Madeley ; *Mr. F. Dickinson.* Between Lilleshall Church and Abbey ; *R. G. Higgins, Esq.*

Hedges near Shrewsbury, on the Bishop's Castle road. Hedges about Sutton, near Shrewsbury. Sharpstones Hill.

Root yellow, creeping extensively. *Shrub* 2-3 feet high, with upright twiggy angular stems filled with pith. *Leaves* tufted, petiolate, from lateral buds at the base of which are the spreading spines. *Racemes* from the centre of the leaf-buds. *Flowers* bright-yellow, with orange glands, smelling disagreeably. *Berries* oblong, slightly curved, red, tipped with the black style, gratefully acid. The *stamens* are highly curious in their formation, spreading from the pistil, reclining on and protected by the incurved petals. On the filaments being touched near the base on the inside, they immediately start forward and the anthers are brought into contact with the stigma. If the anther be fully matured it is burst by the violence of the motion and the pollen projected on the stigma. The stamens after a short time resume their original position and may be again stimulated.

This shrub has been charged as the cause of barrenness, blight, or mildew on corn growing in its neighbourhood, and is in consequence carefully eradicated by the agriculturist. Much has been written on the subject on both sides, but no certain conclusions have been arrived at. It is not improbable that the strong and far creeping yellow roots may, by their great extent, absorb a very considerable portion of nutriment, which would otherwise go to the support of the corn growing near, and which in consequence is rendered less productive. The leaves, fruit, and bark of the Barberry are also liable to be infested with parasitic fungi, which by their similarity in colour to those usually found on corn, may have given rise to the prejudice which the farmer entertains against it. The roots of some foreign species have been lately ascertained to be useful in the production of a yellow dye, and it is not unlikely that our native species may possess a similar property.

2. PEPLIS. *Linn.* Purslane.

1. P. *Portula*, Linn. *Water Purslane.* Flowers axillary, solitary ; leaves opposite, obovate. *E. Bot. t.* 1211. *E. Fl. v. ii. p.* 187. *Hook. Br. Fl. v. i. p.* 157.

Watery places ; not common. *Fl.* July, August. ☉.

Oreton, plentiful ; *Mr. G. Jorden.* Shawbury Heath ; *Mr. E. Elsmere, junr.!* Rudge heath ; *H. Bidwell, Esq.*

Ditches near Middleton Hill. Bomere pool, near Shrewsbury. Cold Hatton Heath.

Plant prostrate, 5-6 inches long, creeping, slightly branched. *Leaves* glabrous, tapering at the base. *Flowers* small, reddish.

3. GALANTHUS. *Linn.* Snowdrop.

1. G. *nivalis*, Linn. *Snowdrop.* Leaves two, carinate ; scape 1-flowered. *E. Bot. t.* 19. *E. Fl. v. ii. p.* 129. *Hook. Br. Fl. p.* 157.

Meadows, orchards, hedges, banks of rivers, &c. ; not unfrequent, but scarcely indigenous. *Fl.* February, March. ♃.

Near Harnage, above Cound ; *Turn. & Dilw. Bot. Guide.* Near Longdenupon-Tern ; *T. C. Eyton, Esq.!* Near to Little Millichope, in Munslow parish ; " I think really wild ;" *Rev. W. Corbett.* Farley Dingle, near Wenlock ; *W. P. Brookes, Esq.!* Cound Moor, Rushbury, Buildwas ; *Mr. F. Dickinson.* On the banks of a lane between Haygate and the Wrekin ; *E. Lees, Esq.*

Bulb ovate, tunicated. *Leaves* broadly linear, glaucous green. *Flowers* elegantly pendulous. *Petals* pure white, the 3 innermost blotched and streaked with green. " Veris nuntius."

4. NARCISSUS. *Linn.* Daffodil.

1. N. *poeticus*, Linn. *Narcissus of the Poets.* Spatha mostly single-flowered ; nectary very short, concave, membranous and crenate at the margin ; leaves with an obtuse keel. *E. Bot. t.* 275. *E. Fl. v. ii. p.* 131. *Hook. Br. Fl. p.* 158.

Orchards and meadows ; not common, and probably naturalized. *Fl.* May. ♃.

In an orchard on Felton farm, near Ludlow ; and also in a field to the right of the road leading to Wigmore over Whitecliff woods ; *Miss Mc Ghie.*

Bulb ovate. *Leaves* large, of a deep glaucous green, edges recurved. *Flowers* large, beautiful, white, powerfully fragrant. *Nectary* with a deep yellow border.

2. N. *biflorus*, Curt. *Pale Narcissus.* Spatha 2-flowered ; nectary very short, concave, membranous and crenate at the margin ; leaves acutely keeled. *E. Bot. t.* 276. *E. Fl. v. ii. p.* 132. *Hook. Br. Fl. p.* 158.

Orchards and fields ; not common, and probably naturalized. *Fl.* April, May. ♃.

In an orchard on Felton farm, near Ludlow, and also in a field to the right of the road leading to Wigmore over Whitecliff woods ; *Miss Mc Ghie.*

Bulb ovate, tunicated. *Leaves* narrower than those of the last, deep glaucous green, edges inflexed. *Flowers* similar in form but smaller, of a less pure white, without the coloured border of the nectary and with a less agreeable scent.

3. N. *Pseudo-narcissus*, Linn. *Common Daffodil.* Spatha single-flowered ; nectary campanulate, erect, crisped at the margin,

obsoletely 6-cleft, as long as the ovate spreading segments of the perianth. *E. Bot. t.* 17. *E. Fl. v. ii. p.* 132. *Hook. Br. Fl. p.* 158.

Moist woods and meadows; not unfrequent. *Fl.* March, April. ♃.

In pastures near Ellardine Moss; *Mr. F. Dickinson.* Close to Larden Hall; also at Madeley, abundantly; *W. P. Brookes, Esq.!* Near Ludlow; *Miss Mc Ghie.* Meadows at Dowle; *Mr. G. Jorden.* In a meadow between the Severn and the Uffington Coal-works; *W. W. Watkins, Esq.* Downton under Haughmond Hill; *Mr. J. S. Baly!* Covering almost a whole field near the turnpike at Shatterford, between Bridgnorth and Kidderminster; *E. Lees, Esq.* Banks of the Perry, above and below Milford; also at Hatton Dingle; *T. C. Eyton, Esq.*

Bulb nearly globular. *Leaves* rather glaucous, with an obtuse keel and flat edges. *Flowers* large, drooping, yellow, unpleasantly scented. *Nectary* deep yellow.

A variety with double flowers frequently occurs in orchards and about villages, in all probability escapes from gardens.

5. CONVALLARIA. *Linn.* Lily of the Valley.

1. C. *majalis*, Linn. *Lily of the Valley.* Scape semi-cylindrical; leaves 2, ovato-lanceolate, radical; flowers racemed, globoso-campanulate, drooping. *E. Bot. t.* 1035. *E. Fl. v. ii. p.* 154. *Hook. Br. Fl. p.* 158.

Woods and coppices; not unfrequent. *Fl.* May. ♃.

In Wyre Forest, but rather sparingly; *E. Lees, Esq.* "Sweeney mountain growing on a very steep face of sandstone rock at a considerable height and almost wholly concealed by an overgrowth of brush-wood and far from any cottage;" *Rev. T. Salwey!* Milson wood; *Mr. G. Jorden.* Burcott wood; *Rev. W. R. Crotch.* Coalbrookdale walks; *W. P. Brookes, Esq.* Old-brook-wood, Coalbrookdale; *Mr. F. Dickinson.* Abundantly in a meadow at Woodseaves, near Market Drayton; *R. G. Higgins, Esq.!* Pimhill.

Root fibrous, entangled, creeping. *Leaves* 2, occasionally 3, entire, smooth with a glaucous hue. *Flowers* secund, pure white, very elegant and fragrant. *Segments* recurved. *Berries* scarlet, globose.

6. ALLIUM. *Linn.* Onion.

* *Leaves rounded.*

1. A. *vineale*, Linn. *Crow Garlic.* Umbel bearing numerous bulbs; leaves fistulose; stamens deeply 3-cleft. *E. Bot. t.* 1974. *E. Fl. v. ii. p.* 137. *Hook. Br. Fl. p.* 160.

Dry pastures, waste places, &c.; not common. *Fl.* June. ♃.

Burway Lane, near Ludlow; *Miss Mc Ghie.* Near Buildwas; *Miss Moseley.* White-horse fields, near Shrewsbury.

Stem 1½-2 feet high. *Bulbs* numerous. *Spatha* of two rather small deciduous leaves. *Flowers* on longish peduncles which are thickened upwards, few, erect, reddish, green on the keels, shorter than the stamens whose filaments as well as the anthers are protruded.

** *Leaves plane.*

2. A. *ursinum*, Linn. *Broad-leaved Garlic or Ramsons.* Umbels without bulbs, level-topped; leaves ovato-lanceolate, petiolate; scape triangular. *E. Bot. t.* 122. *E. Fl. v. ii. p.* 137. *Hook. Br. Fl. p.* 160.

Moist woods; not uncommon. *Fl.* May, June. ♃.

The Hope, Shortwood, and other places near Ludlow; *Miss Mc Ghie.* Welbach Coppice, near Shrewsbury; *Mr. T. Bodenham!* Captain's Coppice, Coalbrookdale; *Mr. F. Dickinson.* Near Oswestry; *Rev. T. Salwey.* Profusely in the woods about the Arkell Hill, near Wellington; *E. Lees, Esq.* Below Milford, on the banks of the Perry; *T. C. Eyton, Esq.* Oakley Park, near Ludlow; *Mr. H. Spare.*

Bulb white, slender, oblong. *Leaves* 1 or 2, bright-green. *Scape* solitary. *Flowers* white, numerous, petals acute. *Spatha* of 2 ovato-lanceolate, membranous, ribbed leaves.

Odious from its strong and disagreeable odour.

7. ORNITHOGALUM. *Linn.* Star of Bethlehem.

1. O. *umbellatum*, Linn. *Common Star of Bethlehem.* Racemes corymbose,; peduncles longer than the bracteas; filaments subulate. *E. Bot. t.* 130. *E. Fl. v. ii. p.* 143. *Hook. Br. Fl. p.* 161.

Meadows and pastures; scarcely wild. *Fl.* April, May. ♃.

Whitecliff; Brook House Orchard near Ludlow; Ashford; *Miss Mc Ghie.* Field near Buildwas; *Mr. F. Dickinson.*

8—10 inches high. *Flowers* linear, acuminate, grooved. *Flowers* large, few, 6—9, lower pedicels very long so that their flowers reach to the same height with the upper ones thus forming a *corymb*, each having a membranous lanceolate *bractea.* *Segments* of the *perianth* green, with a white margin and white within.

8. AGRAPHIS. *Link.* Blue-bell.

1. A. *nutans*, Link. *Nodding Blue-bell.* Leaves linear, acuminate; flowers drooping; alternate stamens longer. *Scilla nutans*, Sm. *E. Bot. t.* 377. *E. Fl. v. ii. p.* 147. *Hyacinthus non-scriptus*, Linn. *Hook. Br. Fl. p.* 162.

Woods, copses, thickets, and hedge-rows; common. *Fl.* May. ♃.

Bulb globular, white. *Leaves* numerous, long, linear, channelled, acuminate, reflexed, of a pale shining green. *Scape* 1 foot high, taller than the foliage. *Raceme* drooping. *Flowers* pendulous, blue, with a pair of lanceolate bracteas at the base of and longer than each pedicel.

A variety with white flowers has been observed on Nesscliffe by *T. C. Eyton, Esq.*

9. NARTHECIUM. *Huds.* Bog-Asphodel.

1. N. *ossifragum*, Huds. *Lancashire Bog-Asphodel.* Leaves ensiform; pedicels with one bractea at the base and another above the middle; stamens shorter than the perianth. *E. Bot. t.* 535. *E. Fl. v. ii. p.* 151. *Hook. Br. Fl. p.* 163.

Wet and boggy grounds, margins of pools; frequent. *Fl.* July, August. ♃.

East side of Titterstone Clee Hill; *Mr. H. Spare.* Ozmere pool, near Whitchurch; *Dr. T. W. Wilson!* Western side of the Longmynd; *Rev. S. P. Mansel.* Catherton Marshes and Oreton; *Mr. G. Jorden.* Whixall Moss; *Rev. A. Bloxam.*

Vownog bog, near Westfelton. Bomere pool (west side.) Shomere moss. Stiperstones Hill.

Root creeping. *Stem* 6—12 inches high, decumbent at the base. *Leaves* all radical, uniform, equitant, striated. *Scape* with many scales or bracteas. *Flowers* in a lax spike, yellow.

10. FRITILLARIA. *Linn.* Fritillary.

1. F. *Meleagris*, Linn. *Common Fritillary.* Stem leafy in the upper part, single-flowered; leaves alternate, linear-lanceolate; points of the perianth inflexed; nectary linear. *E. Bot. t.* 622. *E. Fl. v. ii. p.* 139. *Hook. Br. Fl. p.* 163.

Meadows and pastures; naturalized. *Fl.* April. ♃.

Naturalized by J. F. M. Dovaston, Esq. in the fields near Westfelton.

Bulbs small, depressed, roundish, aggregate, lobed. *Stem* 1 foot high, simple. *Leaves* distant, erect, somewhat glaucous. *Flowers* terminal, pendulous, regularly chequered with pale and dark purple.

11. TULIPA. *Linn.* Tulip.

1. T. *sylvestris*, Linn. *Wild Tulip.* Stem 1-flowered, glabrous, somewhat drooping; leaves of the perianth ovato-acuminate, bearded at the extremity, margins of the 3 inner ones as well as the stamens hairy at the base; stigma obtuse. *E. Bot. t.* 63. *E. Fl. v. ii. p.* 140. *Hook. Br. Fl. p.* 163.

Near ruins; rare, probably naturalized. *Fl.* April. ♃.

Near Hopton Castle; *Miss Mc Ghie.*

Stem 12 inches high. *Leaves* linear-lanceolate, tapering at each extremity, amplexicaul, glaucous. *Flowers* large, yellow, fragrant. *Anthers* and *pollen* yellow.

12. ACORUS. *Linn.* Sweet Sedge.

1. A. *Calamus*, Linn. *Common Sweet Sedge.* Scape ancipitate, terminating above the spadix in a very long ensiform foliaceous point. *E. Bot. t.* 356. *E. Fl. v. ii. p.* 157. *Hook. Br. Fl. p.* 163.

Watery places, on the banks of rivers; very rare. *Fl.* June. ♃.

Tong Lodge Lake; *Dr. G. Lloyd.*

Whole plant aromatic. *Root* thick, rather spongy. *Leaves* bright-green, ensiform. *Spadix* 2-3 inches long, tapering. *Flowers* pale green.

13. JUNCUS. *Linn.* Rush.

* *Leaves none. Barren scapes resembling leaves. Panicle lateral. Flowers scattered.*

1. J. *conglomeratus*, Linn. *Common Rush.* Scapes faintly striated, soft; panicle much branched, densely conglomerate; leaflets of the perianth lanceolate, acute, nearly equal, about as long as the broadly ovate very obtuse capsule; stamens 3. *E. Bot. t.* 1835. *E. Fl. v. ii. p.* 160. *Hook. Br. Fl. p.* 164.

Marshy ground; common. *Fl.* July. ♃.

The stouter habit, darker colour, and densely globose panicle which usually bursts from the scapes considerably below the summit, distinguish this from all its allies. Intermediate states occur between this and *effusus*; but a careful attention to the size and form of the capsule, the less sharply acuminated and more spreading (in consequence of the peculiar shape of the fruit) leaflets of the perianth, will afford a constant and sure distinction.

2. J. *effusus*, Linn. *Soft Rush.* Scape very faintly striated, soft; panicle very much branched, divaricate, effuse; leaflets of the

perianth subulato-lanceolate, acuminate, nearly equal, rather longer than the obovate obtuse capsule; stamens 3 or 6. *E. Bot. t.* 836. *E. Fl. v. ii. p.* 162. *Hook. Br. Fl. p.* 164.

Marshy places; common. *Fl.* July. ♃.

Readily distinguished from the preceding by the loosely diffused panicle, the paler hue and slenderer habit of the whole plant, the more sharply acuminated and less spreading leaflets of the perianth which exceed the capsule in length, as well as by the different form and smaller size of the fruit. The absence of rigidity and glaucous hue in the scapes and the obtuse capsule separate it easily from *J. glaucus.*

3. J. *glaucus*, Sibth. *Hard Rush.* Scape deeply striated, glaucous, rigid; panicle much branched, erect; leaflets of the perianth subulato-lanceolate, acuminate, nearly equal, longer than the elliptical mucronate capsule; stamens 6. *E. Bot. t.* 665. *E. Fl. v. ii. p.* 160. *Hook. Br. Fl. p.* 163.

Wet places and by road sides; not unfrequent. *Fl.* July. ♃.

Astley; *Mr. E. Elsmere, junr.!* Welbach, near Shrewsbury; *Mr. T. Bodenham!* Coalbrookdale; *Mr. F. Dickinson.* Albrighton, near Shiffnal; *H. Bidwell, Esq.*

Canal side between Uffington and Shrewsbury.

This can never be confounded with *J. effusus* if attention be directed to the rigid habit and glaucous hue of the plant, the erect panicle, and the shape of the large black fruit.

** *Stems leafy. Leaves rounded or subcompressed and distinctly jointed internally. Panicle terminal. Flowers aggregated or fascicled.*

4. J. *acutiflorus*, Ehrh. *Sharp-flowered jointed Rush.* Leaves subcompressed; panicle very compound, pyramidal; clusters 5-6-flowered; leaflets of the perianth unequal, lanceolate, very acute, nearly as long as the narrow ovate subacuminate capsule. *E. Bot. t.* 2143. *E. Fl. v. ii. p.* 174. *Hook. Br. Fl. p.* 165. *J. articulatus*, Sm. *E. Bot. t.* 238.

Bogs and wet ground; common. *Fl.* June—August. ♃.

1-2 feet high. *Leaves* 3-4 on a stem, distinctly nodoso-articulate when dry. *Panicle* diffuse, in fruit spreading. *Flowers* several together, greenish-brown. *General bracteas* short, membranaceous, scarcely leafy. *Capsules* pale-coloured.

5. J. *lampocarpus*, Ehrh. *Shining-fruited jointed Rush.* Stem ascending and as well as the leaves compressed; panicle compound, spreading; clusters 4—6 or 8-flowered; leaflets of the perianth equal, rather obtuse, shorter than the acute triquetrous oblong-lanceolate capsule. *E. Bot. t.* 2143. *E. Fl. v. ii. p.* 175. *Hook. Br. Fl. p.* 166.

Boggy ground and watery places; frequent. *Fl.* July, August. ♃.

Near Coxwood pool, Coalbrookdale; *Mr. F. Dickinson.*

Canal between Shrewsbury and Uffington.

12—18 inches high. *Leaves* numerous. *Branches* of the panicle stiff, elongated, nearly simple. *Inner leaflets* of the perianth more obtuse and membranous at the edges than the outer ones. *Capsule* large, dark and shining, mucronate. This species often becomes foliaceous in the flower-scales.

6. J. *obtusiflorus*, Ehrh.   *Blunt-flowered jointed Rush.*   Stem and leaves rounded; panicle very compound, spreading and divaricated; clusters 3—6 flowered; leaflets of the perianth equal, rather obtuse, about equal in length with the oval trigonous capsule.   *E. Bot. t.* 2144.   *E. Fl. v. ii. p.* 176.   *Hook. Br. Fl. p.* 166.

Wet pastures; not common.   *Fl.* August. ♃.

Colemere Mere, near the limekiln; more or less abundant on the margins of most of the meres near Ellesmere; *J. E. Bowman, Esq.*

Stem 2-3 feet high, bearing usually 2 leaves.   *Panicle* repeatedly compound and entangled.   *Flowers* small.   *Leaflets* of the perianth all obtuse.   *Capsule* small, light-brown, shining, mucronate.

The number of the leaves on the stem and the obtuse leaflets of the perianth distinguish this from *acutiflorus*, whilst the very compound panicle and the relative length of the perianth and capsule keep it distinct from *lampocarpus*.

7. J. *uliginosus*, Sibth.   *Lesser Bog jointed Rush.*   Stem erect and often swollen at the base or decumbent and rooting; leaves bristle-shaped; panicle nearly simple, irregular; clusters few or many-flowered; leaflets of the perianth equal, oblong, subacute, nearly as long as the elliptical obtuse capsule.   *E. Bot. t.* 801.   *E. Fl. v. ii. p.* 169.   *Hook. Br. Fl. p.* 166.   *J. bulbosus, Linn.   J. subverticillatus, Wulf.   E. Fl. v. ii. p.* 170.

Boggy and swampy places; not unfrequent.   *Fl.* June—August. ♃.

Whixall moss; *Rev. A. Bloxam.*   Wrekin bogs; *Mr. F. Dickinson.*   Longmynd; *Mr. T. Bodenham!*

Shawbury heath.   Boggy ground at the base of Caradoc hill.   Stiperstones hill.

Extremely variable in size and appearance.   In dry places the *stem* is erect, 3-4 inches or more high, with a bulbous base and then constitutes the *J. bulbosus* of Linnæus.   In wet places, or places where water has lodged during the winter, the stems become decumbent and spreading, taking root at the joints, when it forms the *J. subverticillatus* of *Wulfen.*   The *flowers* vary in number from 2—8, are each supported on a very short stalk and are collected into clusters which owing to the obtuseness of the *capsule* have a depressed, spreading or whorled appearance   The *leaflets* of the perianth are of a dark chocolate colour with green keels and membranous edges.   In the decumbent variety the inflorescence is altogether of a paler green hue.   Both varieties often have some of the flowers foliaceous.   The almost simple panicle, the flattened clusters, and the obtuse form of the capsule afford certain distinguishing marks between this in all its varieties and its allies.

*** *Stems leafy.   Leaves plane or grooved above, not distinctly jointed.*

8. J. *compressus*, Jacq.   *Round-fruited Rush.*   Stem erect, compressed; leaves linear-setaceous grooved; panicle terminal, compound, subcymose, generally shorter than the bracteas; capsules roundish, ovate, obtuse, mucronate, longer than the obtuse incurved leaflets of the perianth.   *E. Fl. v. ii. p.* 165.   *Hook. Br. Fl. p.* 167.

Wet marshy places; not common.   *Fl.* August. ♃.

Banks of the river near Bridgnorth; *Rev. A. Bloxam.*

Root creeping horizontally, fibrous.   *Stem* simple, 6—12 inches high, cylindrical at the base, compressed upwards, leafy.   The first branch of the *panicle* longer than the others.   *Bracteas* foliaceous, the lowermost generally longer than the perianth.   *Leaflets* of the perianth obtuse, brown, scariose at the edges, the interior ones rather the broadest.

---

9. J. *bufonius*, Linn.   *Toad Rush.*   Stem erect or spreading, dichotomous above; leaves filiform, setaceous, grooved; panicle elongated; flowers solitary, unilateral, remote, mostly sessile; leaflets of the perianth unequal, lanceolate, very acuminate, exterior ones much longer than the oblong obtuse capsule.   *E. Bot. t.* 802.   *E. Fl. v. ii. p.* 168.   *Hook. Br. Fl. p.* 168.

Moist or watery, partially inundated places and by road-sides; frequent.   *Fl.* June—August. ☉.

Near Ludlow; *Miss Mc Ghie.*   Coalbrookdale; *Mr. F. Dickinson.*   Oakley Park, near Ludlow; *Mr. H. Spare.*

Monkmoor, Berrington, and various places near Shrewsbury.

Whole plant very pale-coloured and extemely variable in size according to situation.   *Root* fibrous.   *Stems* numerous and crowded, sometimes erect, often partially decumbent and spreading.   *Leaves* few, slender, only one on the stem generally near the middle.   *Bracteas* foliaceous.   *Flowers* green, pale and silvery from the white pellucid bracteas at their base, and the broad membranous and shining margins of the leaflets of the perianth.   *Seeds* very numerous.

**** *Leaves all radical.   Flowers terminal.*

10. J. *squarrosus*, Linn.   *Heath Rush.*   Stem simple, angular, striated; leaves setaceous, rigid, grooved; panicle terminal, elongated, compound; leaflets of the perianth ovato-lanceolate, acute, equal, as long as the elliptical ovate obtuse shortly mucronate capsule.   *E. Bot. t.* 933.   *E. Fl. v. ii. p.* 164.   *Hook. Br. Fl. p.* 168.

Sterile mountainous pastures and heathy ground; frequent.   *Fl.* June, July. ♃.

Oakley Park, near Ludlow; *Mr. H. Spare.*   Titterstone Clee Hill; *Mr. J. S. Baly.!*   Longmynd; *Mr. T. Bodenham.!*   White Sitch pool; *Mr. F. Dickinson.*   Shawbury Heath.   Stiperstones Hill.

Whole plant exceedingly rigid, 6—12 inches high.   *Roots* tufted, woody and fibrous.   *Leaves* subsecund and spreading, half as long as the scape, dilated and sheathing at the base.   *Bracteas* lanceolate, membranous, the lowermost terminating in a stiff leafy point.   *Leaflets* of the perianth glossy brown, membranous at the margins.   *Capsules* shining, of a pale brown.

14. LUZULA.   *De Cand.*   Wood-rush.

1. L. *sylvatica*, Bich.   *Great hairy Wood-rush.*   Leaves linear-lanceolate, hairy; panicle repeatedly compound, subcymose; peduncles elongated; fascicles 2-3-flowered; leaflets of the perianth lanceolate, aristate, as long as the ovate mucronate capsule; seeds elliptical, appendage obsolete.   *E. Fl. v. ii. p.* 180.   *Hook. Br. Fl. p.* 169.   *Juncus sylvaticus, Huds.   E. Bot. t.* 737.

Woods and hilly places; frequent.   *Fl.* March—June. ♃.

Root woody, fibrous.   *Stem* 12—18 inches high, striated, leafy.   *Leaves* broad, shining, many-nerved, the radical ones numerous and forming dense tufts.   *Flowers* small, variegated with brown and white.   *Floral bracteas* ciliated.

The clustered flowers and repeatedly compound panicle separate this from *L. pilosa* and *Forsteri*; whilst the absence of spiked heads affords a ready distinction from *campestris*.

2. L. *pilosa*, Willd.   *Broad-leaved hairy Wood-rush.*   Leaves linear-lanceolate, hairy; panicle slightly branched, subcymose; pe-

---

duncles reflexed, 1-flowered; leaflets of the perianth lanceolate, acuminate, rather shorter than the ovato-pyramidal obtuse mucronulate capsule; appendage of the seeds falcate.   *E. Fl. v. ii. p.* 178.   *Hook. Br. Fl. p.* 169.   *Juncus pilosus, Linn.   E. Bot t.* 736.

Woods; not very common.   *Fl.* April, May. ♃.

Cox Wood, Coalbrookdale; *Mr. F. Dickinson.*   Welbach Coppice, near Shrewsbury; *Mr. T. Bodenham.!*   Whitecliff, near Ludlow; *Mr. H. Spare.*

Root fibrous, stoloniferous.   *Stem* 6—12 inches high, slender, leafy.   *Leaves* of a bright shining green, many-nerved, hairy on the margin towards the base, radical ones numerous.   *Flowers* solitary, dark-brown, with two membranous acute *bracteas* under each.   *Falcate appendages* of the seeds meeting at the insertion of the pistil.

Well distinguished by its solitary flowers and falcate appendage of the seed. The reflexed peduncles afford a distinct mark of separation from *L. Forsteri.*

3. L. *campestris*, Br.   *Field Wood-rush.*   Leaves linear, hairy; spikes 3-4, ovate, dense, sessile and pedunculated; leaflets of the perianth lanceolate, aristate, longer than the subrotund obtuse mucronate capsule; seeds appendaged at the base.   *E. Fl. v. ii. p.* 181.   *Hook. Br. Fl. p.* 169.   *Juncus campestris, Linn.   E. Bot. t.* 672.

β. *congesta*; taller, spikes more numerous, shortly pedunculate, collected into an almost orbicular head.   *Hook. Br. Fl. p.* 170.   *L. congesta, Lej.   E. Fl. v. ii. p.* 181.   *E. Bot. Suppl. t.* 2718.

Dry pastures; very common.   β. wet marshy places; not unfrequent.   *Fl.* April, May. ♃.

Bomere and Hancott pools.   Shomere moss.

Root tough, creeping and tufted.   *Stems* 4—6 inches (in β. 12 inches or more) high, simple, erect, bearing 2 or 3 leaves.   *Radical leaves* numerous, flat, many-ribbed, dark-green, extremely hairy at the margins and tops of the sheaths.   *Spikes* of flowers erect, at length nodding, the lowermost sessile, the rest pedunculated.   *Bracteas* membranous, fringed, shining.   *Seeds* reniform, striated.

Intermediate states connecting the two varieties may often be observed.

HEXANDRIA—TRIGYNIA.

15. RUMEX.   *Linn.*   Dock and Sorrel.

A. *Flowers perfect.   Plants not acid.*

* *All the enlarged petals bearing tubercles.*

† *Margins entire, or nearly so.*

1. R. *Hydrolapathum*, Huds.   *Greater Water Dock.*   Enlarged petals ovato-deltoid, reticulated, nearly entire, each bearing a tubercle; whorls crowded, mostly leafless; leaves lanceolate, acute, tapering at the base, minutely crisped at the edges.   *E. Fl. v. ii. p.* 195.   *Hook. Br. Fl. p.* 171.   *R. aquaticus, Sm.   E. Bot. t.* 2104.

Margins of ponds; rare.   *Fl.* June, July. ♃.

Marbury Mere; *Mr. F. Dickinson.*

Sides of a pit near the New Park, Shrewsbury.

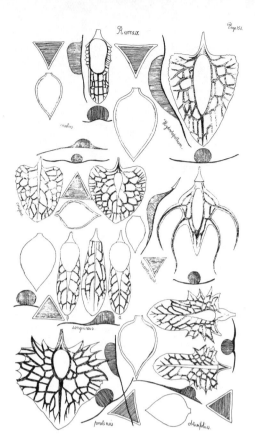

The largest of our Docks. *Stem* 3—5 feet high, erect, stout, hollow, deeply furrowed and angular. *Leaves* petiolate, the lowermost 1½ foot long, broad, smooth, slightly glaucous, coriaceous; *upper ones* smaller and narrower. *Panicles* branched, erect; *whorls* crowded, many-flowered, the lower ones accompanied by a leaf. *Enlarged petals* obtuse, reticulated with prominent veins, generally entire but often when the fruit is mature becoming wavy or notched, scarcely toothed at the base. *Tubercles* oblong, large. *Seed* elliptical, acute with thin sharp angles.

†† *Margins toothed.*

2. R. *acutus*, Linn. *Sharp Dock.* Enlarged petals narrow, oblong, reticulated, obscurely toothed, each bearing a tubercle; whorls distinct, leafy; leaves oblong, acute, cordate at the base, minutely crisped and wavy at the edges. *E. Bot. t.* 724. *E. Fl. v. ii. p.* 192. *Hook. Br. Fl. p.* 172.

Sides of rivers and watery places; not uncommon. *Fl.* June. ♃.
Hancott pool. Banks of river Severn.
Stem 2 feet high, deeply furrowed and angular, smooth, leafy, branched. *Leaves* petiolate, lower ones cordate at the base, upper ones narrow. *Branches* of the *panicle* lax, elongated, spreading; *whorls* distinct, all except the very uppermost accompanied by a small leaf. *Enlarged petals* obtuse, obscurely and sparingly toothed at the base. *Tubercles* very large in proportion to the size of the petals, red. *Seeds* ovate, acute, with 3 obtuse angles.

3. R. *maritimus*, Linn. *Golden Dock.* Enlarged petals subrhomboid, narrow, attenuated into an entire point, each bearing a tubercle, reticulated, the nerves prolonged on either side into straight setaceous teeth equalling and often exceeding the length of the petal; whorls many-flowered, approximate, leafy; leaves linear-lanceolate. *E. Bot. t.* 723. *E. Fl. v. ii. p.* 193. *Hook. Br. Fl. p.* 173. *R. aureus*, *With.*

Marshy sides of pools; very rare. *Fl.* July, August. ♃.
Hancott pool, near Shrewsbury; *Mr. A. Aikin.* Near the pool, Wilcot; *T. C. Eyton, Esq.*
Whole plant of a rich tawny golden hue. *Stem* 2 feet high, furrowed and angular, erect, smooth, leafy, branched. *Leaves* petiolate, long, narrow, acute, entire, flat, the margins very slightly if at all undulated. *Flowers* very numerous, in dense leafy *whorls*, the lower distinct, but the uppermost very near together and finally confluent. *Enlarged petals* fringed at each side with setaceous teeth (Smith says 4 in number; Bluff and Fingerhuth, *Comp. Fl. Germ.* 2; and in specimens from Warwickshire they are generally 2, sometimes 3) equalling and often exceeding them in length and spreading widely, giving the whole dense cluster a bristly appearance. *Tubercles* large, tawny, very prominent, oblong. *Seeds* small, elliptical, with 3 acute angles.

** *One enlarged petal only bearing a tubercle.*

† *Margins entire, or nearly so.*

4. R. *crispus*, Linn. *Curled Dock.* Enlarged petals broadly cordate, entire or crenulate, reticulated, one only bearing a perfect large coloured tubercle; whorls approximate, the upper ones leafless; leaves lanceolate, acute, waved and crisped at the margins. *E. Bot. t.* 1998. *E. Fl. v. ii. p.* 191. *Hook. Br. Fl. p.* 171.

Way-sides, rough waste ground and pastures; frequent. *Fl.* June, July. ♃.

Stem 2-3 feet high, erect, branched, leafy, furrowed and angular, smooth. *Leaves* petiolate, smooth, with a slight hoariness on the veins on the under surface, the lower ones broadest and subcordate at the base. *Panicle* branched, erect. *Whorls* numerous and crowded, a few of the lower ones accompanied by a leaf. *Enlarged petals* truly cordate, obtuse, reticulated with prominent veins, generally entire, but often on the same plant slightly crenulate at the base. *Tubercles* large, oblong, red or orange. *Seed* elliptical with 3 acute angles.

The Shropshire specimens which I have examined have uniformly only one of the enlarged petals with a perfect tubercle, the others having the mid-rib very slightly swollen at the base; occasionally indeed a single flower or so will be found with a second but much smaller tubercle, yet I have never detected any with all the petals tuberculated as described by Smith and most authors. Our plant appears to agree with *R. crispus* of Hook. Br. Fl. and with the specimens there mentioned as gathered in Lancashire and Switzerland.

5. R. *sanguineus*, Linn. *Blood-veined, and (β.) Green-veined Dock.* Enlarged petals small, linear, oblong, quite entire, reticulated, one only bearing a tubercle; whorls distinct on elongated leafless branches; leaves ovato-lanceolate, acuminate, subcordate at the base.
α. *sanguineus*; leaves with bright red veins. *R. sanguineus, Linn. E. Bot. t.* 1533. *E. Fl. v. ii. p.* 190. *Hook. Br. Fl. p.* 172.
β. *viridis*; leaves with green veins. *E. Fl. v. ii. p.* 190. *Hook. Br. Fl. p.* 172. *R. viridis, Sibth.* *R. Nemolapathum, Ehrh.*

α. shady places and road-sides; rare.
Underdale, near Shrewsbury; *T. & D. Bot. Guide.* Near the Race-course, Ludlow; *Miss Me Ghie.*
β. woods and dry shady places; very common. *Fl.* July. ♃.
Stems 2-3 feet high, erect, furrowed and angular, smooth, red, leafy and branched. *Branches* elongated, slender. *Leaves* all petiolate, smooth, veiny, slightly crisped and waved at the edges, lower ones largest cordate at the base, upper ones narrower attenuated at the base. *Panicle* elongated, *whorls* very numerous, the upper ones entirely leafless, 3 or 4 only of the lowermost being accompanied by a small leaf. *Enlarged petals* very obtuse, quite entire, only one bearing a large orange tubercle, the rest with the mid-rib thickened. *Seeds* elliptical, with 3 prominent blunt angles.

†† *Margins toothed.*

6. R. *pratensis*, Mert. and Koch. *Meadow Dock.* Enlarged petals unequal, cordato-ovate, reticulated, greatly dilated and toothed at the base, elongated into an entire triangular point, one principally tuberculated; whorls distinct, nearly leafless; leaves oblong, lanceolate, acuminate, subcordate at the base, wavy and crisped at the margins. *E. Bot. Suppl. t.* 2757. *Hook. Br. Fl. p.* 171.

Sides of pools; rare. *Fl.* June. ♃.
Hancott pool, near Shrewsbury.
Stem 3 feet or more high, erect, branched, leafy, deeply furrowed and angular, rough. *Leaves* all petiolate, smooth except on the under surface where the veins and mid-rib are rough. *Inflorescence* much branched, branches erect. *Whorls* many-flowered, distinct when in blossom but becoming apparently more approximate when in fruit, nearly leafless except two or three of the lowermost whorls and the bases of the branches which are accompanied by small leaves. *Enlarged petals* unequal and irregular, obtuse at the point; one, (the largest) bearing a large oblong orange-coloured tubercle, the rest with the mid-rib considerably thickened

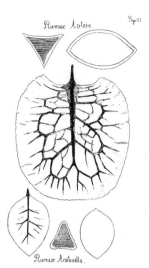

Rumex Acetosa

Rumex Acetosella

at the base, sometimes in one of them almost tuberculate. *Seeds* elliptical, with 3 acute angles.

This species appears to be nearly allied to *R. obtusiflorus* and *R. crispus*, from the former however the more elongated leaves, the unequal size and dilated form of the enlarged petals keep it distinct, whilst from *R. crispus* it is separated by the distinctly toothed and unequal enlarged petals, the less approximate whorls and the broader and less crisped leaves.

7. R. *obtusifolius*, Linn. *Broad-leaved Dock.* Enlarged petals oblongo-triangular, reticulated, toothed at the base, with an entire point, one principally tuberculated, whorls approximate, nearly leafless; radical leaves ovato-cordate, obtuse, wavy and crisped at the margins. *E. Bot. t.* 1999. *E. Fl. v. ii. p.* 192. *Hook. Br. Fl. p.* 173.

Waste ground, pastures, road-sides, &c.; very common. *Fl.* July, Aug. ♃.

Stem 3 feet high, erect, branched, round, deeply furrowed, leafy, rough. *Leaves* all petiolate, smooth, except on the under surface where the veins and also the petioles are rough; radical ones very large, broad, those of the stem narrower and acute, the uppermost lanceolate tapering at both ends. *Panicle* elongated. *Whorls* many-flowered, approximate except a few of the lower ones which are more distant and leafy. *Enlarged petals* obtuse, with about 3 teeth on each margin. *Tubercles* smaller than in most species, reddish. *Seed* elliptical, ovate, with 3 acute angles.

Distinguishable by its broad obtuse radical leaves and the oblongo-triangular form of the enlarged petals.

B. *Flowers diœcious. Plants acid.*

8. R. *Acetosa*, Linn. *Common Sorrel.* Enlarged petals orbiculari-cordate, membranous, reticulated, scarcely tuberculated; sepals reflexed; whorls approximate, leafless; leaves oblongo-sagittate. *E. Bot. t.* 127. *E. Fl. v. ii. p.* 173. *Hook. Br. Fl. p.* 173.

Meadows and pastures; frequent. *Fl.* May—July. ♃.

Root woody. *Stem* 1-2 feet high, erect, simple, leafy, striated. *Lower leaves* petiolate, upper ones narrow, smaller and sessile. *Stipule* tubular, membranous, laciniato-dentate. *Clusters* erect, compound, whorled, leafless. *Semiwhorls* with a clasping membranous *bractea* at their bases. *Barren flower* greenish with a reddish tinge. *Petals* ovate, obtuse, membranous at the edges, rather larger than the calyx. *Anthers* pale yellow. *Fertile flowers* on separate plants, red. *Enlarged petals* orbicular-cordate, membranous at the edges, red, entire, each bearing a very small almost obsolete tubercle. *Seed* elliptical, with 3 acute angles.

9. R. *Acetosella*, Linn. *Sheep's Sorrel.* Enlarged petals ovate not tuberculated; lower leaves lanceolato-hastate, lobes entire. *E. Bot. t.* 1674. *E. Fl. v. ii. p.* 197. *Hook. Br. Fl. p.* 173.

Hedge-banks and dry gravelly places; frequent. *Fl.* May—July. ♃.

Roots creeping. *Stems* variable in height from 2—8 inches, wavy, slender, branched. *Radical leaves* lanceolato-hastate, *cauline* ones often similar, generally lanceolate, all more or less petiolate. *Flowers* minute, in numerous leafless whorls, *petals* entire. Smaller in every part than the last, acid, and turning in the autumn of a rich red colour.

16. SCHEUCHZERIA. *Linn.* Scheuchzeria.

1. S. *palustris*, Linn. *Marsh Scheuchzeria.* *E. Bot. t.* 1801. *E. Fl. v. ii. p.* 199. *Hook. Br. Fl. p.* 174.

In the sphagnous swampy parts of mosses and marshes; very rare. *Fl.* May, June. ♃.

On the moss on the west side of Bomere pool, and also on the adjoining Shomere moss, both near Shrewsbury.

Root long, creeping, tough, with a lax white and shining cuticle. *Stem* 6—8 inches high, erect, wavy, simple, few, smooth, compressed. *Leaves* distichous, alternate, becoming considerably elongated after flowering, semicylindrical, obtuse, with a roundish depression or pore at the extremity on the upper side, spongy within, dilated at the base into a large membranous clasping obtuse stipule. *Raceme* terminal, of about 5 greenish flowers, each on a partial stalk with a leaf-like bractea at the base. *Perianth* and *stamens* reflexed. *Stigmas* lateral, oblong, downy, sessile at the outer margin of each germen. *Capsule* ovate, obtusely pointed, compressed, spreading. *Seeds* ovate, smooth.

This very rare plant was detected in an advanced state in August 1831, by my friend C. C. Babington, Esq. who in consequence has been stated in Loudon's Mag. Nat. Hist., and also in Watson's Botanist's Guide, as the first discoverer in this locality. Without wishing to depreciate in the slightest degree the merits of my excellent friend as a successful discoverer, I deem it but justice to state that the late John Jeudwine, Esq. M.A. Second Master of Shrewsbury School, told me in conversation that he had gathered this plant at Bomere *seven* years previously to Mr. Babington's discovery. Indeed it seems singular that the plant which is in some abundance should have escaped the observation of the accurate Rev. E. Williams, since he does not appear to have communicated the locality to Sir J. E. Smith, either for his English Botany or Flora.

17. TRIGLOCHIN. *Linn.* Arrow-grass.

1. T. *palustre*, Linn. *Marsh Arrow-grass.* Capsules 3 linear, angular, attenuated at the base. *E. Bot. t.* 366. *E. Fl. v. ii. p.* 200. *Hook. Br. Fl. p.* 174.

Wet marshy places, and on the sides of pools; not common. *Fl.* May, June. ♃.

Kinlet Woods, plentiful; *Mr. G. Jorden.* Canton Rough, near Bridgnorth; *Dr. G. Lloyd.* Caradoc Hill; *Mr. T. Bodenham.* Shadwell, Standhill Coppice, Wenlock; Rowley, near Harley; *W. P. Brookes, Esq.!* Near Newport; *R. G. Higgins, Esq.*

Hancott Pool, near Shrewsbury. Shawbury heath. Stiperstones Hill.

Root fibrous. *Leaves* all radical, distichous, linear, fleshy, slightly grooved on the upper side, sheathing and membranous at the base. *Scape* solitary, 6—14 inches high. *Raceme* elongated, lax, simple. *Flowers* numerous, small, greenish, erect, alternate, peduncled. *Capsules* separating at the base from the receptacle and becoming suspended by the extremity.

18. COLCHICUM. *Linn.* Meadow Saffron.

1. C. *autumnale*, Linn. *Common Meadow Saffron.* Bulb 1-3-flowered; segments of the perianth elliptical, styles thickened and curved at the summit, filaments unequal; leaves 3-4, plane, broadly lanceolate, erect. *E. Bot. t.* 133. *E. Fl. v. ii. p.* 202. *Hook. Br. Fl. p.* 174.

Meadows and pastures; frequent. *Fl.* August, September. *Fr.* April—June. ♃.

Between Oswestry and Llanymynech; *J. E. Bowman, Esq.* Albright Hussey; Brompton, near Atcham; *Mr. E. Elsmere, junr.!* Meadows near Uffington; *Dr. T. W. Wilson.* Castle field, 4 miles from Oswestry, on the Welshpool road; Hope Mead, near Bishop's Castle; Aston, near Oswestry; *Turn. & Dilw. Bot.*

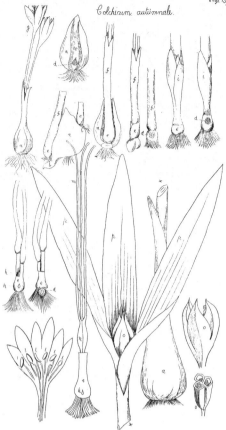

Colchicum autumnale.

Guide. Welbatch, near Shrewsbury; *Mr. T. Bodenham.* Meadows about Madeley, and near Rudge Heath; *Dr. G. Lloyd.* Stottesden and Oreton, very abundant; *Mr. G. Jorden.* Meadows at Sweeney and Llynclys, near Oswestry; *Rev. T. Salwey,* Hord's park, near Bridgnorth; *Rev. A. Bloxam.* Very abundant round Wenlock; *W. P. Brookes, Esq.* Fields near Buildwas; *Mr. F. Dickinson.* Near Ludlow; *Miss Mc Ghie.* Wellington and Hales Owen; *Withering.* Near Buildwas; *Miss Moseley.* Near Rudge heath; *H. Bidwell, Esq.*

Abundant on banks of the Teme, near Caynham Camp. Dorset Barn's fields, near Shrewsbury.

The vital economy of this plant is so exceedingly curious and beautiful, and the evidence of design is so visible as to merit a detailed description of its peculiar structure. The lowest portion of the stem is excessively and immoderately swollen into a cormus or solid bulb (*a*) which lies deeply buried in the earth, invested by the dried leaves and spathas of preceding years. The enveloping leaves are so firmly agglutinated as to be only distinguishable in a thickened scale or protusion. (*b*) In the axil of this scale, in a groove (*c*) on one side of the bulb, at a point a little above the life-knot or part from which the true roots depend, and connected with it by a bundle of horizontal fibres, (*d*) is attached a smaller bulb, (*e*) which, during the summer months, absorbs its chief nourishment from the parent bulb, gradually swelling and enlarging, and, so soon as the first chilly winds of autumn have breathed over the earth, elongating its summit and protruding through the soil a long tubular spatha or sheath, (*f*) which envelopes the entire bulb, stem and roots. In this sheath lie 2 or more perfectly formed flowers, (*g*) each consisting of an elongated tube crowned with a purple limb of 6 petals, and also the rudiments of one or two other blossoms. (*h*) On removing the membranous sheaths, we perceive at the base of the floral tube, immediately above the young roots, a few rudimentary leaves (*i*) closely encircling the slightly swollen bulb. On stripping off these leaves, a smaller bud or germ, (*k*) destined to become the bulb of a succeeding year, appears attached to that side of the young bulb which is farthest from the parent bulb of the present year. On entirely removing the leaves and opening the elongated tube of the flower, we discover that to the three inner divisions of its perianth are attached six stamens (*l*) surrounding three linear stigmas, whose filaments (*m*) are continued down the floral tube to its base, where they communicate with a 3-celled germen or ovary (*n*) containing the ovules. After fecundation the corolla withers, the young bulb becomes swollen, its roots burst through their membranous covering and protrude downwards. Throughout the winter it absorbs through the lateral attaching fibres the greater portion of the parent bulb, the surplus of which subsequently decomposes. The nutriment thus stored up remains dormant during the winter months, and until the first warming breezes of spring again stimulate into motion the vital juices, when the hitherto buried germen, protected from frost or accident by its several coats, is pushed upwards to the surface, the fully developed capsule (*o*) surrounded by shining green leaves, (*p*) displays itself, and on becoming fully matured opens its inflated cells, scattering the seeds over the earth.

HEXANDRIA—HEXAGYNIA.

19. ACTINOCARPUS. *Br.* Star-fruit.

1. A. *Damasonium, Br. Common Star-fruit.* Capsules 6, subulate, compressed, spreading, opening longitudinally; leaves subcordato-oblong, obtuse; 5-nerved. *Hook. Br. Fl.* 175. *Alisma Damasonium, Linn. E. Bot. t.* 1615. *E. Fl. v. ii. p.* 204.

Lakes and pools; rare. *Fl.* June, July. ♃.

Abundant in Ellesmere mere and adjoining ditches and canal; *J. E. Bowman, Esq.! Rev. A. Bloxam.*

---

158   HEXANDRIA—POLYGYNIA.   [*Alisma.*

Root of many long pale fibres. *Leaves* all radical, floating, on long broad tapering membranous ribbed petioles. *Scapes* 6—8 inches high, bearing one or two whorls of flowers. *Petals* white, very delicate, obcordate, each having a yellow spot at the base. *Capsules* with two *seeds* upon evident stalks one from the upper angle horizontal, the other from the lower angle of the axis erect, oblong, tubercled and transversely striated, compressed, with a deep furrow on each side occasioned by the form of the embryo within, which is cylindrical and bent double somewhat like a horse-shoe.

HEXANDRIA—POLYGYNIA.

20. ALISMA. *Linn.* Water-Plantain.

1. A. *Plantago,* Linn. *Great Water-Plantain.* Scape verticillato-paniculate; fruit depressed; capsules numerous, obtusely trigonal; leaves ovato-acuminate, subcordate, 7-nerved. *E. Bot. t.* 837. *E. Fl. v. ii. p.* 203. *Hook. Br. Fl. p.* 175.

Margins of rivers, pools and ditches; frequent. *Fl.* July. ♃.

Root fibrous. *Leaves* all radical, on long thick stalks broadly dilated at the base. *Scape* 2-3 feet high, panicled with innumerable whorled bracteated compound spreading branches. *Flowers* small, pale rose-colour.

2. A. *natans,* Linn. *Floating Water-Plantain.* Stem floating and rooting, leafy; peduncles simple, fruit globular; capsules 6—12 obliquely acuminated, striated; leaves elliptical, obtuse, 3-nerved. *E. Bot. t.* 775. *E. Fl. v. ii. p.* 204. *Hook. Br. Fl. p.* 175.

Lakes and pools; not very common. *Fl.* June, July. ♃.

Ellesmere, below the House of Industry; *Rev. A. Bloxam.* Astley; *Mr. E. Elsmere, junr.* Walford Pool; *T. C. Eyton, Esq.*

Hancott Pool, near Shrewsbury.

Root fibrous. *Stems* floating, various in length according to the depth of the water, throwing out roots leaves and flowers from each joint. *Leaves* small, on very long footstalks with membranous stipules at their base. *Peduncles* 1-2 from each joint, erect, simple, single-flowered. *Petals* large, orbicular, white with a yellow spot near the claw.

3. A. *ranunculoides,* Linn. *Lesser Water-Plantain.* Scape umbellate; fruit globose, squarrose; capsules numerous, acute with 5 angles; leaves linear-lanceolate, acute, 5-nerved. *E. Bot. t.* 326. *E. Fl. v. ii. p.* 205. *Hook. Br. Fl. p.* 176.

Sides of pools; not common. *Fl.* July, August. ♃.

Ellesmere and Colemere; *Rev. A. Bloxam.* Pools near Ellesmere, abundant; *Mr. F. Dickinson.* Shawbury heath; *Mr. E. Elsmere, junr.!* Bomere pool; *Rev. E. Williams.* Pool by the side of the Canal between Church Aston and Lilleshall, and in the ditch in the Duke of Sutherland's private road leading to Kinnersley; *R. G. Higgins, Esq.!*

Berrington and Hancott Pools. Moist ground beyond Haughmond Hill.

Root fibrous. *Leaves* all radical, on long semicylindrical petioles. *Scapes* 10-16 inches high, bearing one or two bracteated umbels. *Flowers* pale purple.

CLASS VIII.

OCTANDRIA.

—

" God of the changeful year!—amidst the glow
    Of strength and beauty, and transcendant grace,
Which, on the mountain heights, or deep below,
    In shelter'd vales, and each sequester'd place,
Thy forms of vegetable life assume,
   — Whether thy pines, with giant arms display'd,
Brave the cold north, or wrapt in eastern gloom,
    Thy trackless forests sweep, a world of shade;
Or whether, scenting ocean's heaving breast,
    Thy odoriferous isles innumerous rise;
Or, under various lighter forms imprest,
    Of fruits, and flowers, thy works delight our eyes;—
God of all life! whate'er those forms may be;
O! may they all unite in praising Thee!"

W. Roscoe.

CLASS VIII.

OCTANDRIA. 8 *Stamens.*

ORDER I. MONOGYNIA. 1 *Style.*

* *Flowers complete, (having Calyx and Corolla.)*

1. ACER. *Calyx* inferior, 5-cleft. *Petals* 5. *Germen* compressed, 2-lobed. *Capsules* 2, combined at the base, compressed, each terminating in a long spreading membranous wing (a *Samara*), 1-celled, 1-2-seeded.—*Nat. Ord.* ACERINEÆ, *Juss.*—Name from *acer, sharp* or *hard, (ac,* Celtic), on account of the hardness of the wood rendering it useful in fabricating spears, pikes, &c.

2. CHLORA. *Calyx* inferior, of 8 deep segments. *Corolla* of 1 petal, nearly rotate, 8-cleft. *Stigmas* 2, bifid. *Capsule* ovate, 1-celled, 2-valved, many-seeded.—*Nat. Ord.* GENTIANEÆ, *Juss.*—Name from χλωρος, *pale green;* in allusion to the pale very glaucous green hue of the herbage of the plant.

3. ERICA. *Calyx* inferior, of 4 leaves. *Corolla* of 1 petal, campanulate or ovate, often ventricose, 4-cleft. *Capsule* 4-celled, 4-valved; dissepiments from the middle of the valves. *(loculicidal dehiscence).*—*Nat. Ord.* ERICINEÆ, *DC.*—Name from ερικω, to *break;* in allusion to its supposed powers of destroying calculi in the bladder.

4. CALLUNA. *Calyx* inferior, of 4 coloured leaves, concealing the corolla, accompanied by 4 *bracteas* resembling an outer calyx. *Corolla* of 1 petal, campanulate, deeply 4-cleft, shorter than the calyx. *Capsule* 4-celled, 4-valved; dissepiments adhering to the axis of the fruit; valves opening at the dissepiments and separating from them. *(septifragal dehiscence).*—*Nat. Ord.* ERICINEÆ, *DC.*—Name from καλλυνω, to *cleanse* or *adorn;* in allusion to the use made of its twigs to form brooms, or to the extreme elegance of its beautiful flowers.

5. VACCINIUM. *Calyx* superior, 4-5-toothed. *Corolla* of 1 petal, 4-5-cleft, segments erect. *Anthers* elongated, opening by terminal pores. *Berry* globose, crowned with the persistent calyx, 4-celled, many-seeded.—*Nat. Ord.* ERICINEÆ, *DC.*—Name of uncertain derivation.

6. OXYCOCCUS. *Calyx* superior, 4-toothed. *Corolla* of 1 petal, deeply 4-cleft, segments reflexed. *Anthers* elongated, opening by terminal pores without dorsal appendages. *Berry* globose, crowned with the persistent calyx, 4-celled, many-seeded.—*Nat. Ord.* ERICINEÆ, *DC.*—Name from οξυς, *acid,* and κοκκος, a *berry;* in allusion to the grateful and highly agreeable acidity of the fruit.

7. ŒNOTHERA. *Calyx* superior, tubular, with a deeply 4-cleft limb; the segments reflexed, more or less combined. *Petals* 4. *Capsule* 4-valved, with many naked seeds.—*Nat. Ord.* ONAGRARIEÆ, *Juss.*—Name from οινος, *wine,* and θηρα, *searching* or *catching,* from the root having caught the perfume of wine.

8. EPILOBIUM. *Calyx* superior, of 1 leaf, 4-partite, deciduous. *Petals* 4. *Capsule* elongated, 4-sided, 4-celled, 4-valved, many-seeded. *Seeds* with a tuft of hairs at one extremity.—*Nat. Ord.* ONAGRARIEÆ, *Juss.*—Name from επι, *upon,* and λοβος, a *pod;* the flowers being placed upon the top of the elongated seed-vessel.

** *Flowers incomplete.*

9. DAPHNE. *Perianth* single, inferior, often coloured, tubular, limb 4-fid. *Berry* with 1 seed. (Plants intensely acrid).—*Nat. Ord.* THYMELEÆ, *Juss.*—Name in allusion to the Nymph *Daphne,* who was changed into a *Laurel,* some of the plants of this Genus having the habit of Laurels.

(See *Monotropa* in CL. X.)

DIGYNIA. 2 *Styles.*

(See *Polygonum* in ORD. II.—*Chrysosplenium* and *Scleranthus* in CL. X.)

ORDER II. TRIGYNIA. 3 *Styles.*

10. POLYGONUM. *Perianth* single, inferior, in 5 deep coloured persistent segments. *Stamens* 5—8. *Styles* 2-3. *Fruit* a one-seeded compressed or trigonous *nut. Embryo* lateral, incurved; cotyledons accumbent, not contorted.—*Nat. Ord.* POLYGONEÆ, *Juss.*—Name from πολυς, *many,* and γονυ, a *knee* or *joint,* from the numerous joints of the stem.

11. FAGOPYRUM. *Perianth* single, inferior, in 5 deep coloured persistent segments. *Stamens* 8. *Styles* 3. *Fruit* a one-seeded trigonous *nut. Embryo* axillary; cotyledons large, foliaceous, contorto-plicate.—*Nat. Ord.* POLYGONEÆ, *Juss.*—Name from φαγος, *edible,* and πυρος, *wheat;* in allusion to the economical uses to which the seeds are applied, as forming an excellent food for game, poultry, &c. and when ground, an excellent flour.

ORDER III. TETRAGYNIA. 4 *Styles.*

12. PARIS. *Calyx* of 4 leaves. *Petals* 4. *Cells* of the *anthers* fixed one on each side of the middle of a subulate *filament. Berry* 4-celled, each *cell* with several seeds in two rows.—*Nat. Ord.* SMILACEÆ, *Br.*—Name from *par, paris, equal;* in allusion to the regularity of its leaves and inflorescence.

13. ADOXA. *Calyx* half inferior, 2-4-cleft. *Corolla* superior, 4-5-cleft. *Stamens* 8-10, united in pairs; *anthers* terminal, 1-celled. *Berry* 4-5-celled. *(Flowers* 5, arranged in a head on a simple peduncle, one terminal and 4 lateral. In the terminal flowers the sepals are 2, petals 4, stamens 8, and carpels 4. In the lateral ones sepals 3, petals 5, stamens 10, and carpels 5*.)—*Nat. Ord.*—ARALIACEÆ, *Juss.*—Name from *a, without,* and δοξα, *glory;* on account of its insignificant appearance.

14. ELATINE. *Calyx* inferior, 3-4-partite, persistent. *Petals* 3-4. *Stamens* 6—8. *Styles* 3-4, very short. *Capsule* 3-4-valved, 3-4-celled, many-seeded. *Seeds* cylindrical, furrowed and transversely striated, attached to a central free receptacle.—*Nat. Ord.* CARYOPHYLLEÆ, *Juss.*—Name from ελατη, the *broad part of an oar,* possibly in allusion to the shape of the leaves.

(See *Sagina* in CL. IV.)

OCTANDRIA—MONOGYNIA.

1. ACER. *Linn.* Maple.

1. A. *Pseudo-platanus,* Linn. *Greater Maple* or *Sycamore.* Leaves 5-lobed, acute, unequally serrated; racemes pendulous, downy. *E. Bot. t.* 303. *E. Fl. v. ii. p.* 230. *Hook. Br. Fl. p.* 179.

Hedges, woods, plantations, &c.; common, but not truly wild. *Fl.* May, June. ♄.

A large *tree* with smooth, ash-coloured bark, spreading opposite branches, and ample leaves, handsome in appearance and quick in growth; wood white, soft, and valuable for turnery ware. *Leaves* on long petioles, broad and palmate, smooth, glaucous beneath, with a small hairy tuft at the bases of the veins. *Flowers* green, axillary, compound, many-flowered. *Flowers* green. *Fruit* with two long membranaceous wings. *Cotyledons* 1—4, circinately folded and incumbent on the radicle.

2. A. *campestre,* Linn. *Common Maple.* Leaves 5-lobed, obtuse, inciso-crenate; racemes erect, subtomentose. *E. Bot. t.* 304. *E. Fl. v. ii. p.* 231. *Hook. Br. Fl. p.* 179.

Hedges and thickets; not uncommon. *Fl.* May, June. ♄.

A small *tree* of humble growth with rough corky bark full of deep fissures, widely spreading branches and small leaves; wood compact, often beautifully veined. *Leaves* all petiolate and opposite, downy whilst young. *Racemes* at the extremities of the young branches, short, somewhat corymbose. *Flowers* green. *Fruit* downy, horizontally spreading with smooth oblong reddish wings.

2. CHLORA. *Linn.* Yellow-wort.

1. C. *perfoliata,* Linn. *Perfoliate Yellow-wort.* Leaves connato-perfoliate, ovate, glaucous; panicle forked, many-flowered. *E. Bot. t.* 60. *E. Fl. v. ii. p.* 218. *Hook. Br. Fl. p.* 179.

* Professor Henslow has skilfully investigated the anomalous structure of these flowers in Jardine's Mag. Zool. and Bot. vol. i. p. 259.

Limestone and gravelly soils; frequent. *Fl.* July—September. ☉.

Cox wood, Coalbrookdale; *Mr. F. Dickinson.* Road-side from Wenlock to Harley; also on Gleeton Hill, near Wenlock; *W. P. Brookes, Esq.!* Morfe, near Bridgnorth; *Rev. W. R. Crotch.* Totterton, and at Wheathill, Brown Clee Hill; *Miss Mc Ghie.* Canal bank near Burford, on the road leading from Tenbury to Cleobury Mortimer; *Rev. W. Crotch.* Meadows about the Clee Hill; *Rev. T. Salwey.* Oreton; *Mr. G. Jorden.* Between Welbatch and Pulley Common, near Shrewsbury; *Mr. A. Aikin.* Fields at Lythwood, near Shrewsbury; *Mr. T. Bodenham!* Astley; *Mr. E. Elsmere, junr.!* Ozmere, near Whitchurch; *Dr. T. W. Wilson!* Downton Hall, near Ludlow; *Mr. H. Spare.* Lilleshall lime works; *T. C. Eyton, Esq.* Near Wombridge; *Dr. Du Gard.* Buildwas; *Miss Moseley.* Between Lilleshall Abbey and Church Aston, near the Canal; *R. G. Higgins, Esq.!*

Lime rocks near the Wrekin. Benthal Edge. Uffington, near Shrewsbury. Wenlock Edge.

Whole plant very glaucous. Stem 12—18 inches high, simple, round. *Leaves* remote, acute. *Panicle* terminal, erect, leafy, repeatedly forked with an intermediate flower on a shorter peduncle in the axil. *Flowers* numerous, very elegant, bright yellow, with scarlet stigmas, open in sunshine only. *Seeds* numerous, angular, reticulated, the interstices sunk.

3. ERICA. *Linn. Heath.*

1. E. *Tetralix,* Linn. *Cross-leaved Heath.* Anthers with two acute serrulate appendages at the base, included; style about as long as the corolla; stigma obtuse; corolla ovate; leaves lanceolate, 4 in a whorl, tomentose above, margins revolute, ciliato-glandulose; flowers capitate. *E. Bot. t.* 1015. *E. Fl. v. ii. p.* 226. *Hook. Br. Fl. p.* 180.

Boggy ground; abundant. *Fl.* July, August. ♃.

Root creeping. *Stems* numerous, erect, 8—16 inches high, branched, leafy, hairy. *Leaves* crowded, spreading, shortly petiolate, covered on the upper surface and on the mid-rib beneath with dense white pubescence, margins revolute, nearly glabrous, fringed with glandular bristles. *Flowers* delicate, rose-coloured, on short tomentose peduncles. *Sepals* linear, tomentose, ciliato-glandulose. *Corolla* slightly downy at the mouth. *Style* in some flowers slightly exserted. *Seeds* numerous, oval, minutely pitted.

In June 1833 I found on the Vownog bog near Westfelton, a plant of this species having the corolla cleft into several divisions, and the place of the stamens occupied by petaloid segments bearing imperfectly developed lobes of anthers; a few perfect stamens were also present.

2. E. *cinerea,* Linn. *Fine-leaved Heath.* Anthers with two inciso-dentate appendages at the base, included; style exserted; stigma capitate; corolla ovate; leaves linear-lanceolate, 3 in a whorl, plane, glabrous, margins membranous, serrulate; flowers in whorled racemes. *E. Bot. t.* 1015. *E. Fl. v. ii. p.* 226. *Hook. Br. Fl. p.* 180.

Dry heaths; plentifully. *Fl.* July—October. ♃.

Stems 1 foot high, numerous, erect, branched chiefly at the base, leafy, pubescent. *Leaves* crowded, spreading, linear-lanceolate, glabrous, plane above, with a dorsal furrow beneath, very shortly petiolate, with young leafy shoots in the axils. *Flowers* 2 or 3 together on pubescent peduncles arising from the young axillary shoots, drooping, reddish-purple; the whole inflorescence forming elongated whorled leafy racemes. *Bracteas* 2 at the base of the calyx. *Sepals* glabrous, ovato-acuminate, serrulate and membranous at the margins. *Seeds* numerous, obovate, minutely pitted.

4. CALLUNA. *Salisb.* Ling.

1. C. *vulgaris,* Salisb. *Common Ling.* Anthers with two serrulate appendages at the base, included; style as long as the calyx; stigma capitate; corolla campanulate; leaves closely imbricated in 4 rows, with two small decurrent spurs at the base; flowers in elongated unilateral racemes. *E. Fl. v. ii. p.* 225. *Hook. Br. Fl. p.* 181. *Erica vulgaris, Linn. E. Bot. t.* 1013.

Heaths, moors, and hilly places; common. *Fl.* June—August. ♃.

A low, much branching, tufted *shrub* with pubescent stems. *Leaves* small, opposite, sessile, obtuse, glabrous, grooved above, with a dorsal furrow on the keel and serrulate margins. *Flowers* pale-lilac, on solitary short pubescent peduncles from the axils of the upper leaves; the *inflorescence* formed into elongated unilateral *racemes.* *Bracteas* dilated, membranous and ciliated at the base.

5. VACCINIUM. *Linn.* Whortle-berry.

1. V. *Myrtillus,* Linn. *Bilberry or Whin-berry.* Peduncles axillary, solitary, 1-flowered, without bracteas; calyx wavy; corolla urceolate, 5-cleft; stamens 8-10, filaments glabrous, anthers glabrous with two dorsal appendages; leaves ovate, thin, veined, smooth, margins serrated, deciduous; stem angular, glabrous. *E. Bot. t.* 456. *E. Fl. v. ii. p.* 219. *Hook. Br. Fl. p.* 181.

Woods and heathy places, in hilly situations; abundant. *Fl.* May. ♃.

Arkoll hill, near the Wrekin; *T. C. Eyton, Esq.* Limekiln woods, near Wellington; *E. Lees, Esq.* Wyre Forest; *Mr. W. G. Perry.* Lizard Wood; *Dr. G. Lloyd.* Woods near Willey Park; *W. P. Brookes, Esq.!* Hawkstone woods; *Mr. F. Dickinson.*

Whitecliffe woods, near Ludlow. Nesscliffe. Hills around Church Stretton. Stiperstones Hill.

A bushy *shrub* 1-2 feet high, with irregular smooth green leafy angular branches. *Leaves* on very short petioles, of a bright green, thin and veiny. *Flowers* on solitary axillary peduncles, drooping. *Calyx* wavy, scarcely toothed. *Corolla* urceolate, greenish with a red tinge, of a waxy transparent appearance, 5-cleft. *Anthers* included, their cells elongated and tubular, opening by pores at the extremity, and having 2 erect filiform appendages proceeding from the middle of their backs. *Style* exserted, stigma obtuse. *Berries* bluish black, glaucous, of a peculiar sweet rather sickly taste, much esteemed by some in tarts, &c. and great quantities are consequently annually brought to market.

2. V. *Vitis Idæa,* Linn. *Red Whortle-berry, Cow-berry.* Peduncles terminal, racemed, many-flowered, bracteated; calyx 4-5-cleft, ciliated; corolla campanulate, deeply 4-5-cleft; stamens 8-9; filaments hairy, anthers granulated, without dorsal appendages; leaves obovate, coriaceous, with glandular dots beneath, margins thickened minutely toothed and slightly revolute, evergreen; stems round, tomentose. *E. Bot. t.* 598. *E. Fl. v. ii. p.* 220. *Hook. Br. Fl. p.* 182.

Dry places on mountains; rare. *Fl.* May, June. ♃.

Stiperstones hill, abundant.

Roots creeping, *Stems* 3-4 inches high, erect, irregularly branched at the summit. *Leaves* resembling those of the Box, dark shining green, pale beneath, on very short downy petioles, rigid, obovate, notched at the extremity, with the obtuse

thickened apex of the mid-rib in the sinus.. *Flowers* drooping, flesh-coloured, each with a pale concave fringed deciduous *bractea* at the base of the downy pedicel. *Stamens* included. *Style* exserted, stigma obtuse. *Berries* globose, deep red, acid and bitter.

Gathered extensively for sale by the children of the cottagers and labourers in the neighbourhood of the above-named hill, and used for pies and jelly.

6. OXYCOCCUS. *Rich.* Cranberry.

1. O. *palustris,* Linn. *Cranberry.* Peduncles terminal, single-flowered, with two little bracteas below the middle; filaments ciliated at the margins, anthers granulated; leaves ovate, smooth, glaucous beneath, evergreen, margins entire and revolute; stem filiform, creeping. *E. Bot. t.* 319. *E. Fl. v. ii. p.* 221. *Hook. Br. Fl. p.* 182.

Wet turfy bogs, especially among *Sphagnum;* frequent. *Fl.* May, June. ♃.

Bog near Wenlock road, 1½ mile from Shrewsbury; *Mr. F. Dickinson.* Cranmere, near Bridgnorth; *W. P. Brookes, Esq.!* Whixall moss; *Rev. A. Bloxam.* Knockin heath; Cold Hatton heath; *T. C. Eyton, Esq.* Pilson Common in the parish of Chetwynd; *R. G. Higgins, Esq.* Bomere pool. Shomere moss. Hancott pool. Heathy ground near Wolf's head on Oswestry road. Stiperstones hill.

Stems straggling, slender, wiry, 8—10 inches long, branched. *Leaves* alternate, on very short petioles, small, ovate, subcordate at the base, convex and rigid. *Flowers* bright rose-colour, very elegant, drooping. *Berries* deep red, deliciously acid.

7. ŒNOTHERA. *Linn.* Evening-primrose.

1. Œ. *biennis,* Linn. *Common Evening-primrose.* Leaves ovato-lanceolate toothed; stem somewhat hairy; flowers sessile, subspicate, stamens about as long as the corolla; capsules nearly cylindrical, 4-toothed. *E. Bot. t.* 1534. *E. Fl. v. ii. p.* 211. *Hook. Br. Fl. p.* 182.

Waste places; doubtful if wild. *Fl.* July—September. ♂.

Coalbrookdale; *Mr. F. Dickinson.*

Root tapering. *Stem* 2-3 feet high, angular, rough with minute tubercles and hairy. *Leaves* alternate, acute, toothed, downy, the lower ones larger petiolate and wavy. *Flowers* yellow, fragrant, expanding in the evening. *Capsule* short, rough.

8. EPILOBIUM. *Linn.* Willow-herb.

* *Flowers irregular. Stamens bent down.*

1. E. *angustifolium,* Linn. *Rose-bay Willow.* Leaves scattered, linear-lanceolate, veined, glabrous; stem nearly simple, glabrous; root creeping; flowers irregular, subspicate; stamens declined; stigma 4-cleft, revolute. *E. Bot. t.* 1497. *E. Fl. v. ii. p.* 212. *Hook. Br. Fl. p.* 182.

Moist shady places; rare. *Fl.* July. ♃.

Near Ludlow paper mills; *Miss Mc Ghie.* Dowle; *Mr. G. Jorden.* Coppice at Buildwas; *Miss Moseley!* Oakley park, near Ludlow; *Mr. H. Spare.* Llanvorda, near Oswestry; *J. F. M. Dovaston, Esq.* Near Preston Boats, Shrewsbury; *Rev. W. R. Crotch!*

Road-side, Red-barn near Shrewsbury, doubtless an out-cast.

Stem 4—6 feet high, erect, roundish. *Leaves* nearly sessile, entire or slightly toothed, glaucous beneath. *Flowers* very handsome, crimson, numerous, in long terminal erect clusters with a small linear *bractea* at the base of each pedicel. *Capsule* hoary and purplish on the upper side. *Stigma* 4-cleft, revolute.

** *Flowers regular. Stamens erect. Stigma 4-cleft.*

2. E. *hirsutum,* Linn. *Great hairy Willow-herb.* Leaves semi-amplexicaul, ovato-lanceolate, deeply serrate, hairy; stem copiously branched, hairy; root creeping; stigma 4-cleft, erect. *E. Bot. t.* 838. *E. Fl. v. ii. p.* 213. *Hook. Br. Fl. p.* 183.

Watery places, sides of ditches and rivulets; frequent. *Fl.* July. ♃.

Whole herb, soft, downy and clammy. *Stem* 4 feet high, round, very much branched and bushy. *Lower leaves* opposite, *upper* ones sessile and alternate. *Flowers* corymbose, large, pink. *Petals* regular, cloven. *Capsule* pubescent.

3. E. *parviflorum,* Schreb. *Small-flowered Willow-herb.* Leaves sessile, lanceolate, slightly toothed, downy; stem nearly simple, very downy; root fibrous; stigma 4-cleft, erect; capsule glabrous or nearly so. *E. Bot. t.* 795. *E. Fl. v. ii. p.* 214. *Hook. Br. Fl. p.* 183.

Watery places, ditches, and sides of rivers; not very common. *Fl.* July, August. ♃.

Coalbrookdale, Cinder hill; *Mr. F. Dickinson.* Pool's dam, Wenlock Abbey; *W. P. Brookes, Esq.!* Near Ludlow; *Miss Mc Ghie.* Near Welbatch; *Mr. T. Bodenham!*

Meadow near Can Office, Shrewsbury. Meole Brook. Bickley Coppice. Canal between Shrewsbury and Uffington. Pit near Sharpstones Hill.

Stem 2 feet high, erect, simple, slightly branched above, round, clothed with a soft dense woolliness. *Lower leaves* opposite, *upper* ones alternate, soft and downy on both sides. *Flowers* small, pale purple, in long leafy clusters. *Capsule* glabrous or nearly so.

4. E. *montanum,* Linn. *Broad smooth-leaved Willow-herb.* Leaves shortly petiolate, ovate, acute, deeply toothed, glabrous; stem nearly simple, pubescent; root slightly creeping; stigma 4-cleft, erect; capsules pubescent. *E. Bot. t.* 777. *E. Fl. v. ii. p.* 214. *Hook. Br. Fl. p.* 183.

Dry shady banks; frequent. *Fl.* July. ♃.

Stem 1-2 feet high, erect, slightly branched above, reddish. *Lower leaves* opposite, broadly ovate, deeply toothed, thin, smooth except the veins which are slightly hoary; *upper* ones alternate and narrower. *Flowers* pale purple, few, in a terminal leafy somewhat unilateral corymb. *Capsule* pubescent. *Seeds* granulated.

A variety, though probably not a permanent one, with white flowers occurs occasionally in situations much exposed to the sun and light.

*** *Flowers regular. Stamens erect. Stigma undivided.*

5. E. *tetragonum,* Linn. *Square-stalked Willow-herb.* Leaves lanceolate, sessile, denticulate, glabrous; stem with 4 unequal angles, simple, nearly glabrous; root slightly creeping; stigma undivided; capsule downy. *E. Bot. t.* 1948. *E. Fl. v. ii. p.* 215. *Hook. Br. Fl. p.* 183.

Sides of ditches and watery places; not very common. *Fl.* July. ♃.
Hadnall; *Mr. E. Elsmere, junr.!* Whitecliff, near Ludlow; *Mr. J. S. Baly.!* Ditch in a field near Bayston hill. Hancott pool.

Stems 12—18 inches high, slightly downy in the upper part only. *Lower leaves* opposite, *upper* ones alternate, all glabrous and unequally toothed. *Flowers* few, pale purple. *Petals* cloven.

The clavate entire stigma, sessile leaves, and glabrous angular stem are the characters which distinguish this from *E. montanum.*

6. *E. palustre,* Linn. *Narrow-leaved Marsh Willow-herb.* Leaves narrow, lanceolate, sessile, nearly entire, and as well as the rounded erect stem subglabrous; root creeping; stigma undivided; capsule pubescent. *E. Bot. t.* 346. *E. Fl. v. ii. p.* 216. *Hook. Br. Fl. p.* 184.

Boggy places, sides of pools and ditches; frequent. *Fl.* July, August. ♃.
Old Canal between Horsehay and Coalbrookdale; *Mr. F. Dickinson.* Wet ground south side of Wrekin; *Dr. G. Lloyd.* Oakley park, near Ludlow; *Mr. H. Spare.*
Wildmoors, near Eyton. Bomere pool. Uffington. Canal between Shrewsbury and Uffington.

Stems 1 foot high, simple, sometimes branched. *Leaves* chiefly opposite, except a few of the uppermost which are alternate, obtuse, margins wavy or occasionally toothed, slightly pubescent on the upper surface, smooth beneath except the mid-rib. *Flowers* small, in leafy clusters. *Petals* pale purple with darker streaks. *Seeds* fusiform, crowned with a thickened margin from which the pappus arises.

9. DAPHNE. *Linn.* Mezereon and Spurge-Laurel.

1. D. *Mezereum,* Linn. *Common Mezereon.* Flowers subternate, lateral, sessile, appearing before the deciduous lanceolate leaves; tube of the perianth hairy. *E. Bot. t.* 1381. *E. Fl. v. ii. p.* 228. *Hook. Br. Fl. p.* 184.

Woods; very rare. *Fl.* March. ♄.
Clee Hill; *Miss Mc Ghie.*

A bushy *shrub* 3 feet high, with erect alternate smooth tough and pliant branches. *Leaves* appearing after the flowers, scattered, petiolate, smooth except the margins which are slightly downy. *Flowers* fragrant, pink, in tufts on the naked branches, bracteated. *Berries* scarlet, oval.

2. D. *Laureola,* Linn. *Spurge-Laurel.* Racemes axillary of about 5 flowers, drooping; leaves lanceolate, glabrous, evergreen. *E. Bot. t.* 119. *E. Fl. v. ii. p.* 229. *Hook. Br. Fl. p.* 184.

Woods and thickets, chiefly on a limestone soil. *Fl.* March. ♄.
Hincham coppice, and other woods near Ludlow; *Miss Mc Ghie.* Old Brook wood, Coalbrookdale; *Mr. F. Dickinson.* Wenlock Edge; *Rev. T. Salwey.* Woods around Wenlock; *W. P. Brookes, Esq.!*
Benthal Edge.

Shrub 1—3 feet high, with numerous stout erect smooth scarcely branched *stems,* naked below, bearing in the upper part only crowded tufts of dark evergreen shining elegantly drooping *leaves.* *Flowers* drooping, bracteated, powerfully fragrant, especially in the morning and evening. *Perianth* pale yellowish-green, funnel-shaped, segments of the *limb* obtuse. *Stamens* included, in two rows, alternately longer, *filaments* very short, *anthers* orange. *Berry* ovate, bluish-black.

OCTANDRIA—TRIGYNIA.

10. POLYGONUM. *Linn.* Persicària, Bistort, Knot-grass.

* *Stamens* 8. *Styles* 3. *Nut trigonous.*

1. P. *Bistorta,* Linn. *Bistort or Snake-weed.* Stem simple, bearing one cylindrical dense spike; leaves cordato-ovate, waved, radical ones decurrent into a winged petiole, cauline ones sessile on a tubular clasping sheath; stipules membranous, laciniate. *E. Bot. t.* 509. *E. Fl. v. ii. p.* 236. *Hook. Br. Fl. p.* 185.

Moist meadows; not uncommon. *Fl.* June. ♃.
Near Larden Hall; *W. P. Brookes, Esq.!* Meadows near Ludlow; *Mr. H. Spare.* In a field close to a pool at the end of Walker Street, Ludlow; *E. Lees, Esq.* 4 miles from Oswestry on the Welshpool road; *Turn. & Dilw. Bot. Guide.* Astley and Battlefield, near Shrewsbury; *Mr. E. Elsmere, junr.!* Coppice near Hanwood; *Mr. T. Bodenham!* Dowle and Stottesden, plentiful; *Mr. G. Jorden.* Meadow adjoining Whittington Castle; *Rev. A. Bloxam.* Meadows at Sweeney, near Oswestry; *Rev. T. Salwey.* Field near Toll Bar, Horshay; *Mr. F. Dickinson.*
In a field opposite the east end of Meole Brace church, on the other side of the brook. Sutton, near Shrewsbury.

Root large, tortuose, fleshy and woody. *Stem* 1-1½ feet high, erect, leafy, striated, smooth. *Leaves* smooth, glaucous and hoary on the veins beneath. *Flowers* flesh-coloured, on stout peduncles with small brown membranous *bracteas* at their base.

2. P. *aviculare,* Linn. *Knot-grass.* Stem very much branched, procumbent; leaves lanceolate or elliptical, petiolate, plane; ochreæ two-lobed, at length torn, with distant nerves; flowers nearly solitary, axillary; fruit acutely triquetrous, opake, insculpto-striated with minutely elevated points. *E. Bot. t.* 1252. *E. Fl. v. ii. p.* 238. *Hook. Br. Fl. p.* 185.

Waste places and way-sides; common. *Fl.* May—September. ⊙.

Stem very variable in size, generally procumbent and much branched, sometimes when growing among corn, drawn up and erect, round, striated, smooth. *Leaves* alternate, proceeding from the joints of the stem, variable in size, entire, obtuse, smooth, somewhat glaucous. *Ochreæ* membranous, acute, often red, with brownish ribs. *Flowers* axillary, solitary or 2 or 3 together, small, beautifully variegated with crimson white and green.

3. P. *Convolvulus,* Linn. *Climbing Buck-wheat.* Stem twining, angular; leaves cordato-sagittate; the three exterior segments of the perianth obtusely keeled; fruit triquetrous, opake, sculpto-rugose with minutely elevated points. *E. Bot. t.* 941. *E. Fl. v. ii. p.* 239. *Hook. Br. Fl. p.* 186.

Corn-fields; frequent. *Fl.* July, August. ⊙.

Stem very long, climbing, branched, angular and rough. *Leaves* alternate, petiolate, smooth. *Racemes* axillary and terminal, lax and elongated. *Flowers* about 4, greenish, subverticillate, upon long stalks jointed near to the flower, reflexed in fruit. *Fruit* triquetrous with concave faces, elliptical, dark purplish-brown, opake, covered with the persistent enlarged perianth which is longer than its stalk.

** *Stamens 5 or 6. Styles mostly 2. Nuts compressed.*

4. P. *amphibium,* Linn. *Amphibious Persicaria.* Spikes dense, ovato-cylindrical; flowers pentandrous; styles forked; leaves petiolate, oblongo-lanceolate, rough at the margins; ochreæ tubular, clasping; fruit ovate, subcompressed, shining. *E. Bot. t.* 436. *E. Fl. v. ii. p.* 232. *Hook. Br. Fl. p.* 186.

Var. α. natans. Stem floating; leaves broad, oblongo-lanceolate, submucronate, tapering at the base into the petioles, glabrous and shining, margins roughish with appressed bristles; petioles four or more times longer than the smooth ochreæ

Shallow margins of rivers and pools; frequent. *Fl.* July, August. ♃.

Var. β. aquaticum. Stem rooting, somewhat erect; leaves narrow, lanceolate, obtuse, cordate at the base, glabrous, margins rough with appressed bristles; petioles about twice as long as the smooth ochreæ.

Muddy and broken margins of rivers, canals, &c.; not so common as the preceding variety. *Fl.* July, August. ♃.

Var. γ. terrestre. Stem erect; leaves very narrow, lanceolate, acuminate, obtuse, subcordate at the base, covered on both sides and on the margins with appressed bristles; petioles shorter than the scabrous ochreæ.

On partially overflowed ground, frequent, but rarely flowering.

The above states of this plant can scarcely claim to be considered as distinct varieties, the transition into each other being so very gradual, and evidently depending entirely on their places of growth being more or less watery. Indeed I have more than once gathered specimens with some of the branches spreading into the water and partaking of the characters of *β.* whilst other branches spreading on the bank have answered to those of *γ.* In all, the *stems* are various in length, more or less branched, trailing and throwing out whorls of fibrous *roots* from the joints. *Leaves* alternate, bright green. *Ochreæ* long, tubular, clasping the stem, in *α.* and *β.* extended beyond the point from which the petiole springs into a truncated membrane, but in *γ.* gradually tapering into the dilated base of the petiole. In *γ.* the bristles which cover the entire upper surface of the leaf are often found arranged in parallel lines from the base to the apex presenting a denser appearance than those on the other portions. *Spikes* solitary, raised above the water on *peduncles* proceeding from the extremities of the stem and branches. *Flowers* crimson, very elegant, nearly sessile, with a small ovato-acuminate coloured *bractea* at the base of each.

5. P. *Persicaria,* Linn. *Spotted Persicaria.* Spikes oblongo-cylindrical, compact, erect or somewhat nodding; flowers hexandrous; leaves broadly lanceolate, with appressed hairs on the margins and ribs; ochreæ with appressed hairs on the ribs, margins denticulate with long ciliæ; peduncles and perianths smooth; fruit generally compressed, sometimes triquetrous. *E. Bot.t.* 756. *E. Fl. v. ii. p.* 233. *Hook. Br. Fl. p.* 186.

Moist ground and waste places; common. *Fl.* August. ⊙.

Root fibrous. *Stem* erect, 1-2 feet high, branched, smooth, round, slightly swollen above the joints. *Leaves* alternate, shortly petiolate, entire, often marked

with a dark crescent-shaped spot above the middle, the ribs and margins clothed with appressed hairs, both surfaces covered with tubercles from whence appressed hairs sometimes arise and doubtless when, owing to accidental circumstances, these hairs are very numerous or are confined more to one surface than to the other the plant assumes the appearances indicated in the varieties *β.* and *γ.* of *Smith's English Flora.* *Ochreæ* membranous, with strong ribs covered with erect appressed hairs, margins denticulate, the teeth terminated by long ciliæ. *Spikes* terminal and lateral, dense. *Flowers* greenish with rose-coloured tips, on short smooth pedicels arranged in whorls encompassed with membranous clasping hairy ciliato-denticulate *bracteæ.* *Fruit* broadly ovate, dark-brown, shining, compressed, sometimes triquetrous, and covered, when seen under a powerful lens, with minute impressed dots.

6. P. *lapathifolium,* Linn. *Pale-flowered Persicaria.* Spikes oblongo-cylindrical, compact, erect or somewhat nodding; flowers hexandrous; leaves oblongo-lanceolate, wavy, with appressed hairs on the margins and ribs, resinoso-glandulose beneath; ochreæ nearly glabrous, obtuse, ribbed, lower ones not fringed, upper ones with very short ciliæ; peduncles and perianths glanduloso-scabrous; fruit compressed, concave at each side. *E. Bot. t.* 1382. *E. Fl. v. ii. p.* 234. *Hook. Br. Fl. p.* 186.

Waste and moist places; common. *Fl.* July, August. ⊙.

Root fibrous. *Stem* 1-2 feet high, erect or often decumbent, somewhat zig-zag, very much branched, round, smooth, often spotted, swelled above the joints. *Leaves* alternate, shortly petiolate, entire, wavy, marked with a black spot in the middle, the ribs margin and petioles rough with appressed hairs, profusely covered on the under side with resinoso-glandulose dots. *Ochreæ* membranous with strong ribs, nearly glabrous, margins of the lower ones very obtuse, of the upper or floral ones somewhat denticulate, the teeth terminated by short ciliæ. *Spikes* terminal and lateral, dense. *Flowers* greenish-white, on short smooth *pedicels,* semiverticillate, encompassed with membranous, glandular, lobed *bracteæ.* *Fruit* broadly ovate, dark-brown, shining, compressed, the sides concave, covered with minute appressed dots.

7. P. *Hydropiper,* Linn. *Biting Persicaria.* Spikes filiform, interrupted, nodding; flowers hexandrous; leaves lanceolate, undulate, glabrous, resinoso-glandulose beneath; ochreæ glabrous, ciliated, glandular, upper ones scarcely ciliated; perianths with glandular dots; fruit compressed, (sometimes triquetrous) gibbous on one side. *E. Bot. t.* 989. *E. Fl. v. ii. p.* 235. *Hook. Br. Fl. p.* 186.

Ditches and wet places; common. *Fl.* August, September. ⊙.

Root fibrous. *Stem* 1-2 feet high, erect, rooting from the lower joints, much branched, slightly swelled above the joints. *Leaves* lanceolate, attenuated below, wavy, glabrous, margins with hairs pointing forward, under surface with resinous glands. *Ochreæ* lax, glabrous, ciliated and slightly glandular, obscurely ribbed, floral ones scarcely at all ciliated. *Spikes* terminal and lateral, lax, leafy below. *Flowers* variegated with red white and green, on smooth pedicels accompanied by tubular sheathing abrupt coloured *bracteæ.* *Perianths* rough with large resinoso-glandulose spots. *Fruit* ovate, acute, dark purplish-brown, opake, punctato-rugose.

10. FAGOPYRUM. *Gærtn.* Buckwheat.

1. F. *vulgaris,* Meisn. *Common Buckwheat.* Stem erect, without prickles; leaves cordato-sagittate, acute; flowers in cymose pani-

cles; fruit triquetrous, acuminate, angles acute, entire. *Polygonum Fagopyrum, Linn. E. Bot. t.* 1044. *E. Fl. v. ii. p.* 239. *Hook. Br. Fl. p.* 186.

Introduced by cultivation into fields and waste places. *Fl.* July—September. ☉.

Root tapering. *Stem* 12—18 inches high, erect, waved, branched, smooth except a downy line along one side. *Leaves* petiolate, alternate, upper ones sessile, entire, wavy, glabrous. *Ochreæ* membranous, short, abrupt, glabrous or sometimes downy. *Spikes* terminal and lateral, many-flowered, spreading. *Peduncles* with a downy line on one side. *Flowers* variegated with red green and white, on smooth pedicels. *Fruit* brown, smooth and shining.

OCTANDRIA—TETRAGYNIA.

11. PARIS. *Linn.* Herb Paris.

1. P. *quadrifolia*, Linn. *Common Herb Paris.* Leaves broadly ovate, 4 in a whorl. *E. Bot. t.* 7. *E. Fl. v. ii. p.* 241. *Hook. Br. Fl. p.* 187.

Moist shady woods; not common. *Fl.* May, June. ♃.

At the foot of Tick wood, near Buildwas; *Mr. F. Dickinson.* Treflach, in the parish of Oswestry; wood below the Blodwell rocks; *Rev. T. Salwey!* In an alder bed near to the parsonage at Coreley, but in the parish of Burford; *Rev. W. Corbett.* Cliff wood, near Bridgnorth; *Purt. Mid. Fl.* The Pell wood, Totterton; *T. C. Eyton, Esq.* Wood 10 miles from Shrewsbury on the Ludlow road; *Turn. & Dillw. Bot. Guide.* Raglit wood; *Rev. S. P. Mansel.* In the Limekiln woods, near Wellington; *E. Lees, Esq.* On Wenlock Edge and woods around Wenlock; Coppice between Presthope and Hughley; *W. P. Brookes, Esq.!* Arkoll wood, near the Wrekin; *Dr. Du Gard.* Near Buildwas; *Miss Moseley.*

Whole herb smooth and of a dull green. *Root* fleshy, creeping. *Stem* 1 foot high, naked except 4 or 5, (rarely 3, 6, or 8) whorled large ovate acute petiolate *leaves* at its summit spreading horizontally with three principal ribs and many veins. *Flower* single, terminal, on a peduncle about 2 inches in length. *Sepals* 4, (rarely 3 or 5) green, linear-lanceolate, 3 ribbed, spreading. *Petals* 4, (rarely 3 or 5,) subulate, yellowish. *Stamens* 8, occasionally 7, 9 or 10, rarely 6 or 11. *Stigmas* 4, sometimes 3 or 5, rarely 6, downy. *Berry* globose, depressed, bluntly tetragonous, purplish-black.

The roots are said to be purgative, and the berry, which is esteemed poisonous, has been employed in the cure of inflammation in the eyes.

Professor Henslow (Loudon's Mag. Nat. Hist. vol. 5, p. 429) has shewn, after an examination of 1500 specimens, that this plant, so anomalous in the division of its parts to the law which prevails in monocotyledons, ranges in limits within which the number of parts developed in the whorls of leaves, sepals, petals and stigmas, are nearly 3 and 6, and in the whorl of stamens nearly 6 and 12; the common variety with 4 leaves, 4 sepals, 4 petals, 8 stamens, and 4 stigmas, and that with 5 leaves and the other parts regular, exceeding nine-tenths of the above number of specimens.

12. ADOXA. *Linn.* Moschatell.

1. A. *moschatellina*, Linn. *Tuberous Moschatell. E. Bot. t.* 463. *E. Fl. v. ii. p.* 242. *Hook. Br. Fl. p.* 187.

Woods and shady places; not very common. *Fl.* March—May. ♃.

Oakley Park near Ludlow, rare; *Mr. H. Spare.* In the greatest profusion on the sides of the road between Wellington and Newport; also close to the spring at the east end of St. Kenelm's chapel among the Clent Hills; *E. Lees, Esq.* Dingles near Hord's Park; *Purton's Midl. Fl.* Donnington, near Shiffnal; *Dr. G. Lloyd.* Near the ascent of Tinker's Hill from the paper mills, Ludlow; *Miss Mc Ghie.* Lanes near Oswestry; *Rev. T. Salwey.* Cox-wood, Coalbrookdale; *Mr. F. Dickinson.* Walford and near Yeaton; *T. C. Eyton, Esq.*

Abbot's Well, Haughmond Hill. Shelton Rough. Lyth hill. Almond Park, all near Shrewsbury.

Root composed of white fleshy imbricated tooth-like scales, producing fibres and runners from the interstices. *Stem* solitary, erect, simple, about a span high. *Radical leaves* 2-3 on very long petioles, biternate, broadly and unequally lobed and cut, lobes obtuse, mucronate; *cauline ones* 2, opposite, smaller and on shorter petioles, ternate. *Flowers* pale green, giving out a musky odour in the morning and evening when moist with dew. *Stamens* arising from a fleshy ring surrounding the upper portion of the germen, in pairs, one placed on either side of the sinus of the contiguous petals, sometimes so combined as to constitute 4 or 5 forked *stamens* bearing a single-celled *anther* on the apex of each ramification. The lower part of the *berry* invested with the persistent calyx whose segments surround the middle portion.

13. ELATINE. *Linn.* Water-wort.

1. E. *hexandra*, DC. *Small hexandrous Water-wort.* Leaves opposite, spathulate; flowers alternate, pedicellate, erect, hexandrous, tripetalous; calyx longer than the petals, segments roundish; capsule turbinate, concave at the summit, 3-celled; seeds about 12 in each cell, straight, ascending. *Hook. Br. Fl. p.* 187. *E. Hydropiper. E. Bot. t.* 955. *E. tripetala. E. Fl. v. ii. p.* 243.

Margins of pools and lakes; rare. *Fl.* July, August. ☉.

Ellesmere mere between the House of Industry and Otley Park; *J. E. Bowman, Esq.!*

Bomere pool (east side) near Shrewsbury.

A minute plant 2-3 inches long, generally growing under the water near the margins and almost imbedded in the sand and mud. *Roots* of numerous long white fibres proceeding from the base and lower joints of the procumbent alternately branched angular smooth pale and pellucid *stem. Leaves* entire, their upper surface covered with minute prominent points. *Peduncles* solitary, axillary, 1-flowered, about as long as the leaves. *Flowers* small. *Calyx segments* fleshy, green, unequal. *Petals* roundish, inflexed, rose-coloured, smaller than the *calyx. Seeds* most beautifully ribbed and transversely striated, very slightly curved.

<hr>

CLASS IX.

ENNEANDRIA.

—

" The more we extend our researches into the vegetable kingdom, the more will every susceptible mind be excited to proceed. We shall find the most delicate and elaborate processes in ceaseless progression on the mountains and in the valleys—the meadows and the recesses of our woods, all subject to immutable laws. We shall find colours unrivalled, odours inimitable, and forms exhaustless in variety and grace, daily developed in the grand laboratory of Nature, demanding only to be seen to extort our unqualified admiration, and leading us irresistibly to contemplate the glory of that Almighty Being from whom so many wonders emanate."

GREVILLE.

CLASS IX.

ENNEANDRIA. 9 *Stamens.*

ORDER I.　HEXAGYNIA.　6 *Styles.*

1. BUTOMUS. *Perianth* single, coloured, 6-partite, inferior. *Capsules* 6, many-seeded. *Seeds* fixed to the inner lining of the capsule.—*Nat. Ord.* ALISMACEÆ, *DC.*—Name from βους, *an ox,* and τεμνω, *to cut;* on account of the sharp leaves injuring the mouths of cattle which browse upon them.

ENNEANDRIA—HEXAGYNIA.

1.　BUTOMUS.　*Linn.*　Flowering-rush.

1. B. *umbellatus,* Linn. *Common Flowering-rush.* Leaves linear, subulate, trigonous; spatha of 3 leaves; flowers umbellate. *E. Bot. t.* 651. *E. Fl. v. ii. p.* 245. *Hook. Br. Fl. p.* 188.

Ditches and margins of rivers and canals; not common. *Fl.* June, July. ♃.

In the Severn near Buildwas; *Mr. F. Dickinson.* Town's mills, Bridgnorth; *Purton's Midl. Fl.* In the drains on the Wild-moors, near Eyton, tolerably common; *T. C. Eyton, Esq.* Banks of the Severn, near Cressage Bridge; *W. P. Brookes, Esq.*

Canal between Shrewsbury and Uffington. River Severn, near Shrewsbury.

Root white, tuberous, spreading horizontally. *Leaves* all radical, 2-3 feet long, linear, acuminate, acutely trigonous. *Scape* solitary, longer than the leaves, round and smooth. *Umbel* of many handsome rose-coloured *flowers,* on pedicels 3-4 inches long, interspersed at the base with many scarious sheathing *bracteas,* and beneath these having a membranous *spatha* of 3 leaves. *Germen* ovate, compressed. *Style* about as long as the germen, stigma recurved, cleft. *Seeds* small, very numerous, parietal or fixed to the inner surface of the pericarp.

CLASS X.

DECANDRIA.

" Nor is the mead unworthy of thy foot,
Full of fresh verdure, and unnumber'd flowers,
The negligence of Nature, wide, and wild;
Where, undisguised by mimic art, she spreads
Unbounded beauty to the roving eye."

THOMSON.

CLASS X.

DECANDRIA. 10 *Stamens.*

ORDER I. MONOGYNIA. 1 *Style.*

1. MONOTROPA. *Perianth* single, of 4-5 leaves, cucullate at the base. *Anthers* 1-celled, 2-lipped. *Capsule* superior, 4-5-celled. *Seeds* numerous, invested with a long arillus.—*Nat. Ord.* ERICINEÆ, *DC.*—Name from μονος, *one,* and τρεπω, *to turn;* the flowers all pointing one way.

2. PYROLA. *Calyx* 5-cleft, persistent. *Petals* 5, often connected at the base. *Anthers* opening with 2 pores at the summit. *Capsule* superior, 5-celled. *Seeds* numerous, invested with a long arillus.— *Nat. Ord.* ERICINEÆ, *DC.*—Name from *pyrus, a pear;* from a fancied resemblance in its leaves to those of a Pear-tree.

3. ANDROMEDA. *Calyx* deeply 5-cleft. *Corolla* 1-petaled, ovate or campanulate. *Anthers* with awns. *Capsule* superior, 4-5-celled, the dissepiments from the middle of the valves.—*Nat. Ord.*—ERICINEÆ, *DC.*—Named in allusion to the fable of *Andromeda,* whose condition was considered analogous to that of this beautiful tribe of plants growing in dreary wastes and situations, the imagined abode of preternatural beings.

4. ARCTOSTAPHYLOS. *Calyx* small, 5-parted. *Corolla* of 1-petal, ovate, with a small 5-cleft revolute limb. *Stamens* smooth, *anthers* without pores. *Berry* smooth, *seeds* solitary.—*Nat. Ord.* ERICINEÆ, *DC.*—Name from αρκτος, *a bear,* and σταφυλις, *a berry or grape;* in allusion to the rugged localities in which the plants occur, inhabited by bears and wild beasts.

(See *Vaccinium* in Cl. VIII.)

ORDER II. DIGYNIA. 2 *Styles.*

5. SCLERANTHUS. *Calyx* of one piece, 5-cleft, persistent. *Corolla* 0. *Stamens* inserted upon the calyx, 5 frequently abortive or wanting. *Capsule* 1-seeded, covered by the indurated calyx.—*Nat. Ord.* PARONYCHIEÆ, *St. Hil.*—Name from σκληρος, *hard,* and ανθος, *a flower;* in allusion to the indurated nature of the floral envelope.

6. CHRYSOSPLENIUM. *Calyx* superior, 4-5-cleft, somewhat coloured, persistent. *Corolla* 0. *Stamens* 8—10, inserted upon the calyx. *Capsule* with 2 beaks, many-seeded.—*Nat. Ord.* SAXIFRAGEÆ, *DC.*—Name from χρυσος, *gold,* and σπληυ, the *spleen;* in allusion to the supposition of the plants being a *sovereign* remedy for that disease.

7. SAXIFRAGA. *Calyx* superior, or inferior, or ½ inferior, in 5 persistent segments. *Corolla* of 5 petals. *Capsule* with 2 beaks, 2-celled, many-seeded, opening between the beaks. *Seeds* upon a receptacle attached to the dissepiment.—*Nat. Ord.* SAXIFRAGEÆ, *Juss.*—Name from *saxum, a stone,* and *frango to break;* in allusion to the roots penetrating the crevices of the rocks and stones among which the plants grow.

8. SAPONARIA. *Calyx* monophyllous, tubular, 5-toothed, without *bracteas* at the base. *Petals* 5-6, clawed. *Capsule* oblong, 1-celled. —*Nat. Ord.* CARYOPHYLLEÆ, *Juss.*—Name from *sapo, soap;* in allusion to the mucilaginous juice of the plant, which has similar properties to soap.

9. DIANTHUS. *Calyx* monophyllous, tubular, 5-toothed, with about 4 imbricated opposite *scales* or *bracteas* at the base. *Petals* 5, clawed. *Capsule* cylindrical, 1-celled.—*Nat. Ord.* CARYOPHYLLEÆ, *Juss.*—Name from Ζευς, Διος, Jupiter, and ανθος, *a flower;* expressive of the high estimation in which these beautiful flowers were held, being such as to warrant their dedication to Deity itself.

ORDER III. TRIGYNIA. 3 *Styles.*

10. SILENE. *Calyx* monophyllous, tubular, often ventricose, 5-toothed. *Petals* 5, clawed, mostly crowned at the mouth, and the *limb* generally notched or bifid. *Capsule* 3-celled, 6-toothed, many-seeded.—*Nat. Ord.* CARYOPHYLLEÆ, *Juss.*—Name supposed to arise from σιαλον, *saliva;* in allusion to the viscid moisture on the stalks of many species; hence too, the English name of *Catchfly.*

11. STELLARIA. *Calyx* of 5 sepals. *Petals* 5, deeply cloven. *Capsule* 1-celled, opening with 6 teeth, many-seeded.—*Nat. Ord.* CARYOPHYLLEÆ, *Juss.*—Name from *stella,* a *star;* because the corolla is spread in a star-shaped manner.

12. ARENARIA. *Calyx* of 5 sepals. *Petals* 5, entire or slightly emarginate. *Capsule* 1-celled, opening with 6 valves, many-seeded. *Seeds* with a naked hilum.—*Nat. Ord.* CARYOPHYLLEÆ, *Juss.*— Name from *arena, sand;* the greater number of species growing in sandy soils.

13. MOEHRINGIA. *Calyx* of 4-5 sepals. *Petals* 4-5, entire or slightly emarginate. *Capsule* 4—6-valved, many-seeded. *Seeds* with an appendage at the hilum.—*Nat. Ord.* CARYOPHYLLEÆ, *Juss.*

14. ALSINE. *Calyx* of 5 rarely 4 sepals. *Petals* 5 rarely 4, entire or slightly emarginate. *Stamens* glandular at the base. *Capsule* 3-valved, many-seeded. *Seeds* naked at the hilum.—*Nat. Ord.* CARYOPHYLLEÆ, *Juss.*—Name from αλσος, *a grove.*

15. LEPIGONUM. *Calyx* 5-partite. *Petals* 5, entire. *Capsule* 3-valved, many-seeded. *Leaves* with *stipules.*—*Nat. Ord.* CARYOPHYLLEÆ, *Juss.*—Name from λεπις, *a scale,* and γονυ, *a knee or joint;* from the appearance of the stipulate leaves which proceed from the joints of the stem.

(See *Polygonum* in CL. VIII.)

ORDER III. PENTAGYNIA. 5 *Styles.*

16. COTYLEDON. *Calyx* 5-partite. *Corolla* monopetalous, tubular, 5-cleft. *Capsules* 5, each with a *gland* or nectariferous scale at its base.—*Nat. Ord.* CRASSULACEÆ, *DC.*—Name from κοτυλη, *a cup,* to which the leaves bear a distant resemblance.

17. SEDUM. *Calyx* in 5 (sometimes 4—8) deep segments often resembling the leaves. *Petals* 5, patent. *Germens* 5, each with a nectariferous scale at its base.—*Nat. Ord.* CRASSULACEÆ, *DC.*— Name from *sedo,* to *sit;* from the humble growth of these plants on their native rocks.

18. OXALIS. *Calyx* of 5 sepals. *Petals* 5, often united by the bases of their claws. *Filaments* often combined below, 5 outer ones shorter. *Capsule* angular, 5-celled : *cells* 2 or many-seeded. *Seeds* with an elastic *arillus.*—*Nat. Ord.* OXALIDEÆ, *DC.*—Name from οξυς, *sharp* or *acid.* The leaves of O. acetosella produce oxalic acid in the state of binoxalate of Potash.

19. LYCHNIS. *Calyx* monophyllous, tubular, 5-toothed. *Petals* 5, clawed. *Capsule* half 5 celled or 1-celled, opening at the apex with 5 or 10 teeth.—*Nat. Ord.* CARYOPHYLLEÆ, *Juss.*—Name from λυχνος, *a lamp;* the thick cottony substance of the leaves of some species, or some similar plant, having been employed as wicks to lamps.

20. CERASTIUM. *Calyx* of 5 sepals. *Petals* 5, bifid. *Capsule* 1-celled, many-seeded, cylindrical, the apex opening by 10 or 8 erect teeth.—*Nat. Ord.* CARYOPHYLLEÆ, *Juss.*—Name from κερας, *a horn,* from the long and curved capsules of some species.

21. MALACHIUM. *Calyx* of 5 sepals. *Petals* 5, bifid. *Capsule* 1-celled, many-seeded, 5-valved, valves bifid at the apex.—*Nat. Ord.* CARYOPHYLLEÆ, *Juss.*—Name from μαλακος, *soft;* in allusion to the soft texture of the foliage.

22. SPERGULA. *Calyx* of 5 sepals. *Petals* 5, entire. *Capsule* ovate, 5-celled, 5-valved, many-seeded.—*Nat. Ord.* CARYOPHYLLEÆ, *Juss.*—Name from *spargo,* to *scatter;* from the wide dispersion of the seeds.

(See *Silene* and *Stellaria* in ORD. III.—*Adoxa* in CL. VIII.)

DECANDRIA—MONOGYNIA.

1. MONOTROPA. *Linn.* Bird's-Nest.

1. M. *Hypopitys,* Linn. *Yellow Bird's-nest.* Lateral flowers with 8, terminal one with 10 stamens. *E. Bot. t.* 713. *E. Fl. v. ii. p* 249. *Hook. Br. Fl. p.* 191.

Woods ; very rare. *Fl.* June, July. ♃.
"Woods between Coalbrookdale and Coalport, but very rare. I have only found two specimens;" *A. Aikin, Esq.*

Root fibrous, said to be parasitical, but not as yet accurately ascertained. *Stem* stout, erect, 6—9 inches high, simple or slightly branched, instead of *leaves* having numerous ovate scattered *scales* of the same dingy yellow colour as the stem. *Raceme* terminal, a continuation of the stem, at first drooping, then erect. *Flowers* on short scaly or bracteated *pedicels,* large, of the same colour as the rest of the plant. *Stamens* alternately smaller. *Germen* 4-5-lobed, ovate. *Stigma* large, peltate. *Seeds* very minute, rarely perfect, enveloped in a reticulated *arillus.*

2. PYROLA. *Linn.* Winter-green.

1. P. *rotundifolia,* Linn. *Round-leaved Winter-green.* Flowers drooping, racemed ; leaves obovato-rotundate, slightly crenate ; stamens ascending, shorter than the style ; style extruded, bent down, curved upwards at the extremity ; stigma large, with 5 erect points. *E. Bot. t.* 213. *E. Fl. v. ii. p.* 255. *Hook. Br. Fl. p.* 192.

Moist woods ; rare. *Fl.* July. ♃.
Whitecliff wood, near Ludlow ; *Mr. H. Spare.*

Stems 6—8 inches high, leafy chiefly in the lower part. *Leaves* obovato-rotundate, bluntly and very obscurely crenate, slightly decurrent at the base, smooth, shining and reticulated above, paler beneath. *Flowers* in a short terminal raceme, pure white, fragrant, with a small *bractea* at the base of each pedicel. *Stigma* large, annular, with a central protuberance having 5 notches.

Distinguishable by the shape and crenulation of its leaves, the direction and relative length of the stamens and style, and the peculiar form of the stigma.

2. P. *minor,* Linn. *Lesser Winter-green.* Flowers drooping, racemed ; leaves ovato-rotundate, crenate ; stamens erect, as long as the style ; style very short, straight, included ; stigma large, with 5 divergent rays. *E. Fl. v. ii. p.* 257. *Hook. Br. Fl. p.* 192.

Woods ; rare. *Fl.* July. ♃.
Whitecliff wood, near Ludlow, and coppice below Delbury common ; *Rev. T. Salwey.* Woods near Ludlow ; *Rev. A. Bloxam.*

Attention to the leaves stamens and style appears to be the only means for accurately discriminating the three too nearly allied species, *P. rotundifolia, media,* and *minor.*

3. ANDROMEDA. *Linn.* Andromeda.

1. A. *polifolia,* Linn. *Marsh Andromeda.* Leaves alternate, lanceolate, their margins revolute, glaucous beneath ; flowers in short terminal racemes. *E. Bot. t.* 713. *E. Fl. v. ii. p.* 251. *Hook. Br. Fl. p.* 193.

Mosses and bogs ; rare. *Fl.* July. ♄.

Moss near Cold Hatton ; *T. C. Eyton, Esq.!* Whixall moss ; *J. E. Bowman, Esq.!* Bog near Ellesmere, and Whixall moss ; *Rev. A. Bloxam.* Ellardine moss, near Hawkstone ; *Mr. F. Dickinson.* Birch bog, near Ellesmere ; *Turn. & Dilw. Bot. Guide.*

A small smooth evergreen *shrub,* 8—10 inches high, with alternate rigid leafy branches. *Leaves* irregularly scattered, shortly petiolate, dark bluish-green above, very glaucous with a prominent rib beneath. *Flowers* drooping, oval or urceolate, the limb in 5 small reflexed segments, rose-coloured, on reddish peduncles proceeding from the extremities of the branches.

4. ARCTOSTAPHYLOS. *Kunth.* Bear-berry.

1. A. *Uva-Ursi,* Linn. *Red Bear-berry.* Stems procumbent ; leaves obovate, entire, evergreen ; racemes terminal. *E. Bot. t.* 714. *E. Fl. v. ii. p.* 253. *Hook. Br. Fl. p.* 193.

Dry stony mountainous places ; extremely rare. *Fl.* May. ♂
Devil's Arm-chair, Stiperstones hill ; *A. Aikin, Esq.*

Stems very long and trailing, round, branched, smooth. *Leaves* alternate, petiolate, obovate, obtuse, entire, stiff, rigid, evergreen, shining and wrinkled on the upper surface, veiny and paler beneath, glabrous except the slightly revolute margins which are minutely downy. *Flowers* in small crowded terminal racemes, of a beautiful rose-colour. *Berry* small, red, austere, mealy.

DECANDRIA—DIGYNIA.

5. SCLERANTHUS. *Linn.* Knawel.

1. S. *annuus,* Linn. *Annual Knawel.* Calyx of the fruit with erecto-patent rather acute segments ; stems spreading ; root annual. *E. Bot.* 351. *E. Fl. v. ii. p.* 282. *Hook. Br. Fl. p.* 194.

Dry hilly stony places ; not uncommon. *Fl.* April—July. ☉?
Bromfield, near Ludlow ; *Mr. H. Spare.* Beaumont hill, Quatford ; *W. P. Brookes, Esq.!* Near Newport ; *R. G. Higgins, Esq.*
Bayston hill ; Sharpstones hill ; Haughmond hill.

Root small, tapering. *Stems* numerous, straggling, slender, dichotomously branched, pubescent with decurved hairs chiefly on opposite sides. *Leaves* linear-subulate, keeled, entire, opposite and combined at the base by a membranous ciliated margin. *Flowers* green, inconspicuous, in axillary leafy clusters. *Calyx* urceolate, ribbed, with 5 ovato lanceolate acute segments, white and membranous at the edges, spreading in flower, erecto-patent in fruit.

The membranous edges of the calyx-segments are narrow and diminish upwards, leaving the acute apex free ; whilst in *S. perennis* the membrane is very broad and does not diminish upwards, but entirely surrounds the incurved apex, rendering it obtuse, and presenting the appearance of a white membranous calyx with a thickened green mid-rib.

6. CHRYSOSPLENIUM. *Linn.* Golden-saxifrage.

1. C. *alternifolium,* Linn. *Alternate-leaved Golden-saxifrage.* Leaves alternate, lower ones subreniform upon very long petioles, crenate, hairy. *E. Bot. t.* 54. *E. Fl. v. ii. p.* 260. *Hook. Br. Fl. p.* 194.

Banks of rivers ; not common. *Fl.* March, April. ♃.
Banks of Severn, between Preston Boats and Uffington, near Shrewsbury ; *Mr. Robert Edwards.* Pasford brook, near Rudge heath ; *H. Bidwell, Esq.* By side of a rill at the eastern end of St. Kenelm's chapel, among the Clent hills,

near Hales Owen ; *E. Lees, Esq.* Longnor, near Attingham ; *Turn. & Dilw. Bot. Guide.* Near Oswestry ; *Rev. T. Salwey.* Near Bridgnorth ; Faintree ; *Purton's Midl. Fl.* Detton ; *Mr. G. Jorden.* Tong ; *Dr. G. Lloyd.* Rocks by the side of the Teme, above Ludlow ; *Rev. A. Bloxam.* By the side of a rivulet near the Hope, Ludlow ; *Miss Mc Ghie.* In the wood near the Lodge, Ludlow ; *Rev. T. Salwey.* Foot of Caer Caradoc ; *Mr. F. Dickinson.* Rivulet near the Marsh Hall ; *J. F. M. Dovaston, Esq.*
Banks of the Severn at Shelton rough, near Shrewsbury.

Whole plant of a full deep cheerful green, and of a stiff erect habit ; the lower portion of the stem, lower leaves and their long petioles covered with glandular hairs, the upper leaves and stem glabrous. *Root* fibrous, creeping. *Stems* 3-4 inches high, erect, branched near the summit only. *Leaves* rough on both sides with hairs, paler on the under surface ; those of the stem few and distant, the uppermost glabrous, smaller and crowded under the umbellate nearly sessile deep yellow *flowers.*

In all the Shropshire specimens which I have examined of both species of *Chrysosplenium,* the calyx has been uniformly found to be 4-cleft and the stamens 8 ; 4 of them alternating with, and 4 opposite to the sepals.

2. C. *oppositifolium,* Linn. *Common Golden-saxifrage.* Leaves opposite, lower ones subcordato-rotundate upon short petioles, wavy, glabrous. *E. Bot. t.* 490. *E. Fl. v. ii. p.* 260. *Hook. Br. Fl. p.* 194.

Sides of rivulets and wet shady places ; not uncommon. *Fl.* May—July. ♃.
Cox wood, Coalbrookdale ; *Mr. F. Dickinson.* Wood near the Lodge, Ludlow ; *Rev. T. Salwey.* Shortwood, and near most rivulets in the neighbourhood of Ludlow ; *Miss Mc Ghie.* Tong, and near Rudge heath ; *Dr. G. Lloyd.* Detton ; *Mr. G. Jorden.* Hurst coppice, Ness ; Leaton Shelf and Grafton ; *T. C. Eyton, Esq.* Oakley park, near Ludlow ; *Mr. H. Spare.* Badger Dingle ; *H. Bidwell, Esq.*
Sutton Spa ; Welbatch ; Shelton rough, all near Shrewsbury.

Whole plant of a pale glaucous green, and of a straggling, decumbent, (especially in the lower part of the stem) habit, entirely glabrous. *Root* creeping with very long fibres. *Stems* 4—6 inches high, weak and straggling, more or less decumbent in the lower portion and throwing out roots from the joints, very much branched at the base, slightly so at the summit. *Leaves* all opposite and glabrous, in position and number similar to those of the preceding. *Flowers* pale lemon-colour.

7. SAXIFRAGA. *Linn.* Saxifrage.

1. S. *granulata,* Linn. *White Meadow Saxifrage.* Flowering stem erect, panicled, leafy ; calyx spreading ; radical leaves reniform on long petioles, obtusely lobed, those of the upper part of the stem nearly sessile, acutely lobed ; root granulated ; capsule ½ inferior. *E. Bot. t.* 500. *E. Fl. v. ii. p.* 269. *Hook. Br. Fl. p.* 197.

Hedge-banks, meadows, and pastures ; not uncommon. *Fl.* May, June. ♃.
Oakley park, near Ludlow ; *Mr. H. Spare.* The Lodge, near Ludlow, and by side of old turnpike road from Ludlow to the Lodge ; *Rev. T. Salwey.* Donnington ; *Dr. G. Lloyd.* Foot of the High Rock, Bridgnorth ; *Rev. W. R. Crotch.* Near Ludlow abundant, and occasionally by the road-side near Leebotwood ; *Rev. W. Corbett.* Stockton and Brockton ; *Mr. F. Dickinson.* Road-side near Welbatch and about Welbatch, tolerably common ; *Mr. T. Bodenham.* Quatford ; *W. P. Brookes, Esq.!* White horse fields, Shrewsbury ; *J. F. M. Dovaston, Esq.*
Hedge-bank between Craven Arms and Cheney Longville. Meadow near Harwood's boat-house and Severn hill, Shrewsbury.

Root of many long dark fibres proceeding from a slender *rhizoma* covered at intervals with clusters of numerous small pale scaly *bulbs,* which are invested with a tough hairy membrane. This membrane may, with attention, be traced to be the base of an undeveloped petiole, and a transverse section of the bulbs proves them to be transformed leaf-buds in the axil. *Stems* 10—12 inches high, simple, panicled above, clothed as is the entire plant with pale loosely-spreading glandular hairs. *Leaves* on long petioles, mostly radical, glanduloso-pilose. *Flowers* several, in a corymbose panicle, nearly erect, large, each on a short peduncle with lanceolate *bracteas* at the base. *Calyx* obtuse, moderately spreading. *Petals* spathulate, white with 3 green ribs, of which the lateral ones are branched. *Stigmas* large, downy.

2. S. *tridactylites,* Linn. *Rue-leaved Saxifrage.* Stem erect, panicled, leafy ; pedicels single-flowered ; calyx spreading ; leaves cuneate, 3—5-fid, the uppermost reduced to undivided bracteas ; capsule inferior. *E. Bot.* 501. *E. Fl. v. ii. p.* 271. *Hook. Br. Fl. p.* 198.

Walls, roofs, and dry barren ground ; not uncommon. *Fl.* April—June. ☉.
Oakley Park, near Ludlow ; *Mr. H. Spare.* Shotton ; *W. W. Watkins, Esq.* Near Oswestry ; *Rev. T. Salwey.* Wenlock Abbey ; *W. P. Brookes, Esq.!* Llanymynech church-yard wall ; Shamber Wên ; *J. F. M. Dovaston, Esq.*
Hill top, Wenlock edge ; Haughmond Abbey ; Walls and house tops, Shrewsbury.

Whole plant covered with viscid glandular pubescence. *Root* small, tapering. *Stems* 3-4 inches high, alternately branched. *Leaves* fleshy, the base of the lower ones elongated and tapering, upper ones sessile and less divided. *Flowers* small on simple bracteated pedicels. *Calyx* short and obtuse. *Petals* pure white, scarcely exceeding the calyx-segments in length. *Stigmas* downy. *Capsule* almost wholly inferior.

3. S. *hypnoides,* Linn. *Mossy Saxifrage.* Stem erect, panicled, leafy ; radical leaves 3 or 5-cleft, those of the procumbent shoots undivided or 3-cleft, all bristle-pointed and more or less fringed ; calyx spreading, segments ovate pointed ; petals roundish obovate or oblong, 3-ribbed with or without lateral veins. *E. Fl. v. ii. p.* 277. *Hook. Br. Fl. p.* 198.

In rocky, mountainous situations ; rare. *Fl.* May—July. ♃.
Titterstone Clee hill ; *Mr. H. Spare.* On Titterstone Clee hill among the rocky masses below the Giant's Chair ; *E. Lees, Esq.*

Shoots numerous, long, entangled, procumbent, leafy, proceeding from the crown of the *root. Stem* erect, 3—5 inches high, slightly leafy. *Flowers* 3—7 in a corymbose panicle, white, each on a viscid and glandular *pedicel,* with an awl-shaped *bractea* at the base.

8. SAPONARIA. *Linn.* Soapwort.

1. S. *officinalis,* Linn. *Common Soapwort.* Flowers faciculato-corymbose ; calyx cylindrical, glabrous ; petals retuse, crowned ; leaves ovato-lanceolate, ribbed ; stem erect. *E. Bot. t.* 1060. *E. Fl. v. ii. p.* 284. *Hook. Br. Fl. p.* 203.

Road-sides, meadows, river sides and waste places ; not common. *Fl.* July, August. ♃.
Lane between Prado and Knockin ; *Rev. T. Salwey.* By road-side between Llanymynech and the new bridge ; Nescliffe ; *Turn. & Dilw. Bot. Guide.* Banks of the Severn above and below Bridgnorth ; *Purton's Midl. Fl.* Dowle ; *Mr. G.*

Jorden. Near Cockshutt and Marbury; *Mr. F. Dickinson.* Near Ludlow; *Miss Mc Ghie.* Longnor, and near Leebotwood; *Rev. W. Corbett.* Dovaston; Nesscliffe; Devolog, near river Vyrnwy; *J. F. M. Dovaston, Esq.*
Moist meadow near Pimley, near Shrewsbury. Haughmond Abbey.
Root fleshy, extensively creeping. *Stem* 1-1½ foot high, stout, cylindrical, smooth. *Leaves* ovato-lanceolate, acute, strongly ribbed, connate at the base. *Flowers* large, handsome, pale pink.
A variety with double flowers appears not uncommon, occurring at Longnor; near Leebotwood; and Haughmond Abbey.

9. DIANTHUS. *Linn.* Pink.

* *Flowers clustered.*

1. D. *Armeria*, Linn. *Deptford Pink.* Flowers clustered, fascicled; scales of the calyx lanceolate, downy, as long as the tube. *E. Bot. t.* 317. *E. Fl. v. ii. p.* 286. *Hook. Br. Fl. p.* 203.
Waste places; rare. *Fl.* July, August. ♃.
Canal bank, in the parish of Burford, once only; *Rev. W. Corbett.* On the hill towards Cleobury; *Miss Mc Ghie.* Near Ketley; *Withering.*
Root tapering. *Stem* 1-1½ foot high, leafy, with straight stiff branches upwards. *Leaves* linear, opposite and connate, pubescent, upper ones acute. *Flowers* small, scentless, only one expanded at a time in each fascicle. *Limb* of the *petals* rose-coloured, speckled with white dots, crenate at the margin.

** *Flowers solitary; one or more on the same stem.*

2. D. *plumarius*, Linn. Stem 2—5-flowered, flowers solitary; scales of the calyx subrotundo-ovate, shortly mucronate, four times shorter than the tube; leaves lineari-subulate, glaucous, minutely serrulate; petals digitato-multifid as far as the middle, with an entire obovate intermediate space; flowerless stems procumbent, rooting, very much branched, densely cæspitose. *Koch. Syn. p.* 98. *Reich. Excurs.* 5040. *Bluff. & Fing. Comp. Fl. Germ. 2nd. ed. t. i. pt.* 2, *p.* 88.
Ruins and old walls; rare. *Fl.* July. ♃.
Old walls at Ludlow; *Rev. T. Salwey.* Walls of Ludlow Castle; *Turn. & Dilw. Bot. Guide.* Ludlow and Ludford; *Mr. H. Spare.* Ludford church-yard wall; *Mr. J. S. Baly!*
Haughmond Abbey.
Whole herbage smooth and glaucous. *Root* rather woody, branching at the crown. *Stem* 1-1½ foot high, branched above. *Leaves* lineari-subulate, acute, grooved, opposite and connate, finely serrulate at the margins. *Flowers* fragrant, *petals* pale-pink, sometimes white, the outer edge deeply and unequally laciniated. *Calyx-teeth* ovate, submucronate, membranous and ciliated at the margins. *Styles* very much exserted, curved.
A variety with white flowers grows on Ludford church-yard wall! *Mr. J. S. Baly.*
The only true stations known for *D. Caryophyllus*, Linn. are the Kentish Castles.

3. D. *deltoides*, Linn. *Maiden Pink.* Flowers solitary; scales of the calyx 2, ovato-cuspidate, half the length of the tube; leaves bluntish, serrulate at the margins and keel, slightly pubescent; petals glabrous, unequally notched. *E. Bot. t.* 61. *E. Fl. v. ii. p.* 288. *Hook. Br. Fl. p.* 204.

Hilly pastures; rare. *Fl.* June—August. ♃.
Haughmond hill; Woodbatch, near Bishop's Castle; *Turn. & Dilw. Bot. Guide.* Roman walls at Chesterton; *H. Bidwell, Esq.* Quatford ferry, near Bridgnorth; *W. P. Brookes, Esq.!* Ludlow and Ludford; *Mr. H. Spare.* Plantation, World's End, Church Stretton; *Rev. S. P. Mansel.* Close to Quatford church; on the Morfe, near Bridgnorth; *Purton's Midl. Fl.* Pulley Common; *A. Aikin, Esq.*
Sharpstones hill, near Shrewsbury.
Root woody, much branched and tufted at the crown. *Stems* numerous, branched, decumbent, slender, straggling, leafy, ascending in flower, 4—6 or 8 inches high, scabrous. *Leaves* linear-lanceolate, short, of a deep-green colour, grooved, opposite and connate. *Flowers* scentless, solitary, very beautiful. *Petals* rose-coloured, with darker and white spots intermixed in a circle around the mouth of the flower. *Calyx-teeth* lineari-lanceolate, serrulate at the points. *Styles* exserted.

DECANDRIA—TRIGYNIA.

10. SILENE. *Linn.* Catchfly.

1. S. *inflata*, Sm. *Bladder Campion.* Stem erect; flowers numerous, panicled; petals deeply cloven with narrow segments scarcely crowned; calyx inflated, bladdery, reticulated; leaves ovato-lanceolate. *E. Fl. v. ii. p.* 292. *Hook. Br. Fl. p.* 204. *Cucubalus Behen, Linn. E. Bot. t.* 164.
Fields and road-sides; common. *Fl.* June—August. ♃.
Whole plant very glaucous and smooth. *Stem* 2-3 feet high, erect, branched. *Leaves* sessile, single-ribbed, variable in size and shape. *Panicle* terminal, repeatedly forked, many-flowered, with a pair of membranous *bracteas* at each division. *Flowers* slightly drooping. *Petals* pure white. *Capsule* ovate, rigid, enveloped by the pale inflated reticulated calyx.

Var. β. hirsuta. Lower part of the stem rough with hairs; leaves covered with minute tubercles and hairs, margins fringed with bristles; flowers more densely and irregularly panicled; peduncles smooth. *S. inflata β, Sm. E. Fl. v. ii. p.* 293. *(not of Hook. Br. Fl.)*
In similar situations, and flowering at the same time with the last; not common.
Westfelton. Road-sides near Meole Brace.

2. S. *nutans*, Linn. *Nottingham Catchfly.* Stem erect; flowers panicled, secund, cernuous, branches opposite; calyx cylindrical, ventricose but not bladdery; petals deeply cloven, their segments linear crowned; cauline leaves lanceolate, pubescent. *E. Bot. t.* 465. *E. Fl. v. ii. p.* 296. *Hook. Br. Fl. p.* 206.
Rocks; rare. *Fl.* June, July. ♃.
Hawkstone; *Turn. & Dilw. Bot. Guide.*
Root fleshy and tapering with one or more upright flowering stems and several recumbent shoots. *Stem* 1-1½ foot high, simple, rough with dense reflexed pubescence. *Leaves* rough with pubescence, the lower ones spathulate, acute. *Branches* of the *panicle* bracteated, glanduloso-pilose. *Flowers* sweet-scented in an evening. *Petals* rather large, white. *Calyx* swelling upwards, membranous, with 10 glanduloso-pilose ribs. *Capsule* obovate, 6-toothed, covered by the persistent calyx. *Seeds* subreniform, concentrically rugose.

11. STELLARIA. *Linn.* Stitchwort.

1. S. *nemorum*, Linn. *Wood Stitchwort.* Stem ascending, hairy upwards; leaves petiolate, cordate, acuminate, upper ones sessile; panicle dichotomous; sepals lanceolate; petals deeply bifid, twice as long as the calyx; capsule oblong, longer than the calyx. *E. Bot. t.* 92. *E. Fl. v. ii. p.* 300. *Hook. Br. Fl. p.* 207.
Moist woods; rare. *Fl.* May, June. ♃.
Near Oswestry; *Rev. T. Salwey.* Lilleshall Abbey; *R. G. Higgins, Esq.* In a hedge or thicket rather near to the Boiling well, beyond Lynney, towards Burway; and in a field near the Paper Mills, Ludlow; *Miss Mc Ghie.*
Root slender, creeping. *Stem* 1-2 feet high, weak, spreading, round, hairy in the upper part. *Leaves* pale-green, rough with very minute elevated dots, having jointed hairs more or less scattered over both surfaces, on long hairy *petioles.* Close to the ciliated margins of the leaf a strong rib runs into which all the other ribs merge. *Flowers* numerous, on pubescent glandular peduncles. *Sepals* obscurely 5-ribbed, slightly hairy at the base, white at the edges. *Petals* narrow, deeply bifid, pure white. *Capsule* straight, opening with 6 valves. *Seeds* brown, singularly tuberculated, especially on the margins.

2. S. *media*, With. *Common Chickweed or Stitchwort.* Stem procumbent, with an alternate line of hairs on one side; leaves ovate, petiolate, upper ones sessile; flowers axillary and terminal; stamens 5—10; petals deeply bifid, shorter than the calyx; capsule ovato-oblong, as long as the calyx. *E. Bot. t.* 537. *E. Fl. v. ii. p.* 301. *Hook. Br. Fl. p.* 207.
Cultivated ground and waste places; common. *Fl.* almost the whole year. ⊙.
Root tapering, capillary. *Stem* copiously branched and spreading from the base, brittle, smooth, with alternate lines of decurved hairs between each pair of leaves. *Leaves* opposite, ovate, acute, single-ribbed, glabrous, and except the uppermost ones on dilated footstalks fringed with hairs. *Flowers* on solitary terminal and axillary stalks, glanduloso-pilose on one side, deflexed after flowering. *Sepals* glandulso-pilose, single-ribbed, slightly membranous at the margins. *Petals* small, white, shorter than the calyx, deeply bifid. *Stamens* 3—5 or 10, each arising from a gland. *Nectaries* 5, connected by a thickened glandular border, gibbous at the base of the *filaments.* *Capsule* ovato-oblong, as long as the calyx, opening with 6 valves. *Seeds* numerous, orbicular, compressed, notched, concentrically rugose.

3. S. *holostea*, Linn. *Greater Stitchwort.* Stem square, nearly erect, angles rough; panicle dichotomous; petals inversely heart-shaped, bifid, twice as long as the calyx; calyx single-ribbed; leaves sessile, lanceolate, much acuminated, margins recurved, and together with the mid-rib finely serrulated; capsule globular. *E. Bot. t.* 211. *E. Fl. v. ii. p.* 301. *Hook. Br. Fl. p.* 207.
Woods and hedges; frequent. *Fl.* May, June. ♃.
Root creeping. *Stem* 1-2 feet high, simple, weak and recumbent in the lower part, erect above, rigid and brittle, square, angles rough. *Leaves* opposite, sessile, spreading almost horizontally, somewhat glaucous, single-ribbed. *Panicle* dichotomous with an intermediate flower, of few flowers. *Peduncles* long with rough pubescence, bracteated. *Flowers* large, handsome, white. *Petals* broad, deeply cloven, *anthers* yellow. *Calyx* smooth, edges membranous, with a pellucid mid-rib extending rather beyond the apex. *Capsule* globular, smooth. *Seeds* 5 or 6, large, orbiculari-reniform, or subreniform, covered with minutely raised processes, concentrically arranged.

4. S. *graminea*, Linn. *Lesser Stitchwort.* Stem diffuse, quadrangular, glabrous; leaves sessile, lanceolate, acute, glabrous, ciliated at the base; corymb dichotomous; bracteas scarious, ciliated at the margin; sepals 3-ribbed; petals bipartite, equalling (sometimes exceeding) the calyx; capsule oblong, longer than the calyx. *E. Bot. t.* 803. *E. Fl. v. ii. p.* 302. *Hook. Br. Fl. p.* 207.
Dry pastures, fields and sides of pathways; common. *Fl.* May, June. ♃.
Stem 1-1½ foot high, of a grass-green colour, not glaucous. *Corymb* large and much branched. *Flowers* small, white, *anthers* reddish. *Capsule* smooth. *Seeds* nearly orbicular, about 12 in number.
Similar to the last, but of weaker and slenderer habit, smaller in all its parts and perfectly smooth. Essentially distinguished by the 3-ribbed sepals, the smaller petals, form of the capsule and greater number of the seeds.
A variety having the petals twice the length of the calyx has occurred to observation on the Stiperstones Hill, and in one or two other places, but seems to be different in no other respect.

5. S. *glauca*, With. *Glaucous Marsh Stitchwort.* Stem erect, weak, quadrangular; leaves sessile, linear-lanceolate, acute, glabrous; corymb subdichotomous; bracteas scarious, glabrous at the margin; sepals 3-ribbed; petals bipartite, longer than the calyx; capsule oblongo-ovate, equalling the calyx. *E. Bot. t.* 825. *E. Fl. v. ii. p.* 303. *Hook. Br. Fl. p.* 207.
Margins of lakes, &c.; rare. *Fl.* June, July. ♃.
Blackmere, near Ellesmere; *Rev. A. Bloxam.*
Stem about a foot high, quite smooth. *Leaves* of a glaucous hue. *Flowers* often axillary and solitary, *petals* twice as long as the calyx.
This species appears intermediate between *S. holostea* and *S. graminea.* From the first it is distinguished by its smoothness, the 3-ribbed sepals and the form of the capsule; from the latter by its glaucous hue, the glabrous margins of the bracteas, and the larger petals.

6. S. *uliginosa*, Murr. *Bog Stitchwort.* Stem diffuse, quadrangular, glabrous; leaves sessile, oblongo-lanceolate, glabrous, slightly ciliated at the base; panicle dichotomous; bracteas scarious, glabrous at the margins; sepals 3-ribbed; petals bipartite, shorter than the calyx; capsule ovate, nearly equalling the calyx. *E. Bot. t.* 1074. *E. Fl. v. ii. p.* 303. *Hook. Br. Fl. p.* 208. *Larbræa aquatica, St. Hil. DC. Lindl. S. graminea, β. Linn.*
Springs, ditches, and watery spots; frequent. *Fl.* June. ⊙.
Root fibrous. *Stems* weak, much branched, smooth. *Leaves* pale and glaucous, with a strong rib running round the margin, into which the lateral veins merge. *Flowers* small, white, on smooth *peduncles*, swelling under the flowers. *Bracteas* opposite, small, membranous.

12. ARENARIA. *Linn.* Sandwort.

1. A. *serpyllifolia*, Linn. *Thyme-leaved Sandwort.* Stems erect or procumbent; leaves ovate, acute, subscabrous, sessile, 5-nerved, ciliated; flowers solitary; calyx hairy, segments 3-ribbed, 3 outer ones obscurely 5-ribbed and ciliated; capsule obpyriform, longer than the calyx; seeds reniform, rugose. *E. Bot. t.* 923. *E. Fl. v. ii. p.* 307. *Hook. Br. Fl. p.* 209.

Walls; frequent. *Fl.* June. ⊙.

Root small and tapering. *Stems* 2—6 inches in length, erect or procumbent, copiously branched, covered with short dense deflexed pubescence. *Leaves* small, opposite and connate, rather rigid. *Flowers* small, white, on short pubescent peduncles from the forkings of the upper part of the stem or the axils of the leaves. *Petals* rather shorter than the calyx. *Sepals* lanceolate, acute, membranous at the edges, the 3 outer ones obscurely 5-ribbed and ciliated, the 2 inner ones 3-ribbed, plane at the edges. *Capsule* with 6 teeth. *Seeds* reddish-brown.

13. MOEHRINGIA. *Fenzl.* Möhringia.

1. M. *trinervis*, Clairville. *Three-nerved Mohringia.* Stems procumbent; leaves ovate, petiolate, 3 (rarely 5) nerved, ciliated; flowers solitary; calyx obscurely 3-ribbed, rough on the keel, margins of the 3 outer sepals ciliated; capsule ovate, shorter than the calyx; seeds reniform, smooth. *Arenaria trinervis, Linn. E. Bot. t.* 1483. *E. Fl. v. ii. p.* 307. *Hook. Br. Fl. p.* 209.

Shady woods and moist places; not very common. *Fl.* May. ⊙.

Near Oswestry; *Rev. T. Salwey.* Near Newport; *R. G. Higgins, Esq.* Oakley park, near Ludlow; *Mr. H. Spare.* Lincoln's Hill, Coalbrookdale; *Mr. F. Dickinson.* Ludlow; *Miss Mc Ghie.*

Shelton rough, near Shrewsbury. Whitecliff coppice, near Ludlow. Haughmond hill.

Root small, tapering. *Stems* 10—12 inches high, weak, much branched, covered with decurved pubescence. *Leaves* opposite and connate, upper ones sessile, lower ones smaller on long winged petioles. *Peduncles* longer than the leaves, pubescent, from the forkings and extremities of the stems, in fruit spreading, the upper part deflexed. *Petals* oblongo-obovate, white, scarcely longer than the calyx. *Sepals* lanceolate, acute, membranous at the edges, 3-ribbed, the 3 outermost ciliated, the other 2 smooth, all rough on the keel. *Capsule* with 6 revolute teeth. *Seeds* black.

14. ALSINE. *Mert. & Koch.* Alsine.

1. A. *tenuifolia*, Wahlenb. *Slender-leaved Alsine.* Stems dichotomous; flowers fasciculate; leaves subulate, 3-ribbed; pedicels many times longer than the calyx; sepals lanceolato-subulate, 3-ribbed, membranous at the margins; petals oval, narrow at the base, shorter than the calyx. *Arenaria tenuifolia, Linn. E. Bot. t.* 219. *E. Fl.v. ii. p.* 208. *Hook. Br. Fl. p.* 219.

Corn-fields; rare. *Fl.* June. ⊙.

Fields near Sharpstones hill; *Mr. F. Dickinson.*

Root tapering. *Stem* 4—6 inches high, much branched, slender and glabrous. *Leaves* combined at the base. *Flowers* small, white, on capillary, erect peduncles. *Capsule* equal to the calyx in length, 3-valved. *Seeds* roundish, compressed, minutely tuberculated, especially on the margins.

15. LEPIGONUM. *Fries.* Scaly-jointed Sandwort.

1. L. *rubrum*, Fries. *Red Scaly-jointed Sandwort.* Stems prostrate; leaves narrow, acute, plane, somewhat fleshy, tipped with a very minute bristle; stipules ovate, cloven; flowers solitary, axillary; calyx glanduloso-pilose, membranous at the edges, nerveless; capsule ovate, as long as the calyx; seeds angular, compressed, rough with minutely elevated points. *Arenaria rubra, Linn. E. Bot. t.* 852. *E. Fl. v. ii. p.* 311. *Hook. Br. Fl. p.* 210.

Gravelly or sandy soils; frequent. *Fl.* June, July. ⊙.

On Lilleshall hill; *R. G. Higgins. Esq.* Bromfield, near Ludlow; *Mr. H. Spare.* Harmer Hill; *Dr. T. W. Wilson.* Red lake, Ketley; *Dr. Du Gard.*

Haughmond hill; Bayston hill; Pimhill; Sharpstones hill.

Root tapering. *Stems* numerous, prostrate, much branched and spreading, smooth. *Leaves* mostly opposite, often in pairs with clusters of others in their axils. *Stipules*, a pair of ovate acute white membranous connate scales. *Flowers* numerous, from the axils of the upper leaves, on short solitary glanduloso-pilose peduncles slightly reflexed after flowering. *Petals* ovate, red, about as long as the calyx. *Capsule* of 3-valves. *Seeds* numerous.

DECANDRIA—PENTAGYNIA.

16. COTYLEDON. *Linn.* Pennywort.

1. C. *Umbilicus*, Huds. *Wall Pennywort.* Leaves peltate, crenate, depressed in the centre; stem with a simple raceme of pendulous flowers; upper bracteas minute, entire; root tuberous. *E. Bot. t.* 325. *E. Fl. v. ii. p.* 314. *Hook. Br. Fl. p.* 211.

Shady rocks and old walls; frequent. *Fl.* June—August. ♃.

Near Oswestry; *Rev. T. Salwey.* Harmer Hill; *T. C. Eyton, Esq.* Dorrington rectory wall; *H. Bidwell, Esq.* Near Tibberton Church; *R. G. Higgins, Esq.* Church Preen and Kenley; *W. P. Brookes, Esq.!* Quatford and Rowton, on the walls by the turnpike road; *Purton's Midl. Fl.* Aston Botterell; *Mr. G. Jorden.* Upon old walls and rocky hedgerows, in several parts of the Country, especially Cardington and Stretton; *Rev. W. Corbett.* Roadside between Ellesmere and Shrewsbury; rocks near Bridgnorth; *Rev. A. Bloxam.* Hawkstone and Grinshill; *Mr. J. S. Baly.!* Quatford, near Bridgnorth; Hill called the Helion, near Rushbury; *Mr. F. Dickinson.* Sandstone rocks overhanging the Severn, between Apley Terrace and Bridgnorth; also about Ludlow; *E. Lees, Esq.* About Middle and Shotton; *W. W. Watkins, Esq.*

Haughmond hill; Lythe hill; Nesscliffe.

Whole plant succulent and smooth. *Stem* 6—12 inches high, rounded, leafy chiefly in the lower part. *Leaves* on rather long petioles, succulent and brittle, irregularly crenate. *Flowers* very numerous, cylindrical, drooping, pale yellowish-green, on short pedicels with minute solitary bracteas at their base.

17. SEDUM. *Linn.* Orpine or Stonecrop.

* *Leaves plane.*

1. S. *Telephium*, Linn. *Orpine or Live-long.* Leaves oval-oblong, plane, coarsely serrated; corymbs leafy; stems erect. *E. Bot. t.* 1319. *E. Fl. v. ii. p.* 316. *Hook. Br. Fl. p.* 212.

Rocky pastures; not common. *Fl.* July, August. ♃.

Near Oswestry; *Rev. T. Salwey.* Near the summit of the Wrekin; *W. P. Brookes, Esq.!* Titterstone Clee hill; *Mr. H. Spare.* Near All Stretton; *Rev. W. Corbett.*

Haughmond Hill and Abbey. Lyth hill.

Stems 1-1½ feet high, erect, simple, spotted with red. *Leaves* scattered, sessile, spreading, fleshy, with a stout mid-rib, somewhat glaucous especially on the under surface, which is marked with minute darker green strokes. *Flowers* crimson, in dense corymbose terminal or partly axillary tufts, interspersed with leaflike bracteas.

** *Leaves terete. Flowers white or reddish.*

2. S. *dasyphyllum*, Linn. *Thick-leaved Stonecrop.* Leaves opposite, (except on the flowering stems) ovato-globose, fleshy; panicles glutinous. *E. Bot. t.* 656. *E. Fl. v. ii. p.* 316. *Hook. Br. Fl. p.* 212.

Walls and rocks; rare. *Fl.* June. ♃.

Ludlow; *Mr. H. Spare.*

Stems slender, procumbent below, slightly viscid. *Leaves* short, singularly thick and fleshy, glaucous, with a reddish tinge and dotted. *Flowering-stems* 2-3 inches high. *Panicle* simple, terminal. *Flowers* tinged with rose-colour. *Petals* and pistils 5—8.

3. S. *Anglicum*, Huds. *White English Stonecrop.* Leaves alternate, ovate, gibbous, fleshy, produced at the base; cymes few-flowered; petals very sharp at the point. *E. Bot. t.* 171. *E. Fl. v. ii. p.* 317. *Hook. Br. Fl. p.* 212.

Rocks; rare. *Fl.* June, July. ⊙.

Harmer hill; *T. C. Eyton, Esq.* Stiperstones hill; *Mr. F. Dickinson.*

Stems 2-3 inches high, much branched, procumbent below. *Leaves* crowded, glaucous green often tinged with red. *Flowers* white, starlike. *Anthers* purple.

4. S. *album*, Linn. *White Stonecrop.* Leaves scattered, oblongo-cylindrical, produced at the base, obtuse, spreading; cyme much branched. *E. Bot. t.* 1578. *E. Fl. v. ii. p.* 320. *Hook. Br. Fl. p.* 212.

Rocks, walls and roofs; rare. *Fl.* July. ♃.

Walls at Ludlow; *Rev. T. Salwey.* Old walls about Ludford; *Miss Mc Ghie.*

Ludlow Castle and town walls (south side.)

Stems prostrate below, flowering-stem only erect, 3—6 inches high. *Leaves* very succulent, pale glaucous green, occasionally tinged with red. *Flowers* crowded, white or slightly tinged with rose-colour.

*** *Leaves terete. Flowers yellow.*

5. S. *acre*, Linn. *Biting Stonecrop or Wall Pepper.* Leaves erect, alternate, ovate, gibbous, fleshy, produced at the base; cymes trifid, glabrous, leafy. *E. Bot. t.* 839. *E. Fl. v. ii. p.* 318. *Hook. Br. Fl. p.* 213.

Walls, rocks, and roofs; frequent. *Fl.* June. ♃.

Near Oswestry; *Rev. T. Salwey.* Near Newport; *R. G. Higgins, Esq.* Clee hill; *Mr. G. Jorden.* Acton Burnell; *Mr. F. Dickinson.* About Middle; *W. W. Watkins, Esq.*

Walls about Ludlow; Haughmond Abbey; Shrewsbury.

Stems 2-3 inches high, with many erect leafy branches, forming wide spreading tufts. *Leaves* small, obtuse, smooth and succulent, imbricated. *Flowers* of a golden yellow. *Whole herb* hot and pungent, when chewed.

6. S. *reflexum*, Linn. *Crooked Yellow Stonecrop.* Stems pendulous; leaves scattered, awl-shaped, mucronate, fleshy, nearly cylindrical, slightly flattened, produced at the base, lowermost recurved; flowers cymose; sepals ovate. *E. Bot. t.* 695. *E. Fl. v. ii. p.* 320. *Hook. Br. Fl. p.* 213.

Walls and roofs of houses; not common. *Fl.* July. ♃.

Near Oswestry; *Rev. T. Salwey.* Wenlock Abbey; *W. P. Brookes, Esq.!* Oreton; *Mr. G. Jorden.* Buildwas Abbey; *Dr. G. Lloyd.*

Ludlow Castle and town walls (south side.) Alberbury priory walls. Shrewsbury Abbey wall, and roof of house in the Abbey Foregate.

Stems long, straggling, pendulous. *Leaves* thick and fleshy, scarcely glaucous, upper ones erect, lower ones recurved, all mucronate, with a depressed pore or mark at the back immediately below the apex. *Flowers* bright yellow, in large dense terminal cymes, outermost branches frequently recurved.

7. S. *Forsterianum*, Sm. *Welsh Rock Stonecrop.* Stems recumbent; leaves spreading in many rows, lanceolate, mucronate, flattened, produced at the base, those of the branches in rosaceous tufts; flowers cymose; sepals elliptical. *E. Bot. t.* 1802. *E. Fl. v. ii. p.* 322. *Hook. Br. Fl. p.* 214.

Rocks; frequent. *Fl.* June, July. ♃.

Haughmond hill; Caer Caradoc; volcanic sand-stone rock at the base of the Caradoc, between that hill and the Lawley.

Stems recumbent with short erect branches densely covered with numerous leaves forming glaucous rosaceous tufts. *Flowering-stems* erect, 4—6 inches high, leafy. *Leaves* of the sterile branches minutely tuberculated, densely imbricated, those of the fertile ones rather more distant, all lanceolate mucronate and flattened, bearing a depressed pore or mark at the back of the apex. *Flowers* bright yellow, in small compact terminal cymes.

The following localities have been communicated for *S. rupestre*, Linn. but as the plant which I have seen growing in some of them, as well as the dried specimens from others, agree with the plant which occurs on most of the loftier hills around the plain of Shrewsbury, and on Craig Breidden, in Montgomeryshire, which corresponds in character to *S. Forsterianum*, Sm. and are very different from the plant which I possess as the *S. rupestre* from the Cheddar rocks, Somersetshire, I conclude that they should all be referred to *S. Forsterianum.*

Callaughton, near Wenlock; *W. P. Brookes, Esq.!* Summit of Bury hill, between Ludlow and Bishop's Castle; *E. Lees, Esq.* Titterstone Clee hill; *Turn. & Dilw. Bot. Guide.* Longmynd; *Rev. T. Salwey.* Caer Caradoc; *Mr. F. Dickinson.*

18. OXALIS. *Linn.* Wood-sorrel.

1. O. *Acetosella*, Linn. *Common Wood-sorrel.* Rhizoma scaly; scape single-flowered, longer than the leaves; leaves all radical, ternate, leaflets inversely heart-shaped, hairy; capsule ovate with 5 prominent angles, cells 2-seeded, seeds ovate, smooth. *E. Bot. t.* 762. *E. Fl. v. ii. p.* 323. *Hook. Br. Fl. p.* 214.

Woods and shady places; frequent. *Fl.* May. ♃.

Root of many long branched fibres proceeding from the joints of the rhizoma. *Rhizoma* slender, fleshy, covered with thick fleshy ovate truncated reddish scales, most numerous and imbricated in the upper portion. These scales are filled with a minutely granular substance and are the swollen bases of the petioles of the leaves of former seasons which were once articulated with them on the now truncated apex. *Leaves* on long petioles, leaflets of a delicate bright-green, paler and often purplish on the under side, hairy, obcordate, with a reflexed obtuse point in the sinus, drooping at night. *Flowers* singularly elegant and delicate, bell-shaped, white, streaked with slender purplish veins, on scapes taller than the leaves bearing a pair of connate *bracteas* rather above the middle. *Seeds* black and shining.

Var. β. purpurea. Flowers purple. *Dill. in Raii Syn.* *281-*2. *E. Fl. v. ii. p.* 323.

Woods; rare. *Fl.* May. ♃.

This variety presents no difference that I am aware of, except in the colour of its flowers. It occurs in great plenty on the wooded parts of the Hopton end of Nescliffe hill, growing of equal size, intermixed with and blossoming at the same time with the common white variety. It retains its colour under cultivation.

2. O. *corniculata,* Linn. *Yellow procumbent Wood-sorrel.* Stem branched, branches procumbent, peduncles axillary, mostly 2-flowered, shorter than the leaves; leaves alternate, ternate, leaflets inversely heart-shaped, hairy, stipules united to the base of the footstalks; capsule linear-oblong, cells many-seeded, seeds elliptical, transversely rugose. *E. Bot. t.* 1726. *E. Fl. v. ii. p.* 324. *Hook. Br. Fl. p.* 214.

Shady waste ground; very rare. *Fl.* May—October. ⊙.

"When I first knew Shrewsbury it was abundant near the Welsh Bridge; but after 3 or 4 years disappeared, probably having escaped from some garden;" *A. Aikin, Esq.* Oakley Park, near Ludlow; *Mr. H. Spare.*

Root fibrous. *Stems* spreading widely on the ground, downy. *Leaves* on long spreading petioles. *Stipules* lanceolate. *Flowers* small, yellow.

19. LYCHNIS. *Linn.* Catchfly.

1. L. *Flos Cuculi,* Linn. *Meadow Lychnis or Ragged Robin.* Petals deeply 4-fid, segments linear palmato-divergent; crown bipartite, segments subulate erect with an acute central tooth; calyx 10-ribbed, segments shorter than the corolla; capsule 1-celled. *E. Bot. t.* 573. *E. Fl. v. ii. p.* 326. *Hook. Br. Fl. p.* 214.

Moist meadows and pastures; frequent. *Fl.* June. ♃.

Stem 1-2 feet high, erect, angular, rough with deflexed bristles, viscid in the upper part. *Leaves* lanceolate, nearly smooth, lower ones petiolate. *Panicle* terminal, forked. *Calyx* and *flower-stalks* reddish-purple, the former smooth, 10-ribbed. *Petals* rose-coloured. *Capsule* with 5 teeth.

A variety with white flowers has been found in Chetwynd moss by *R. G. Higgins, Esq.!*

2. L. *dioica,* Linn. *Red or White Campion.* Flowers diœcious; segments of the calyx shorter than the corolla; capsule 1-celled. *E. Fl. v. ii. p.* 328. *Hook. Br. Fl. p.* 215. *a.* flowers red. *L. dioica. E. Bot. t.* 1579. *L. diurna, Sibth. Ox. L. sylvestris, Hoppe. De Cand.* β. flowers white. *E. Bot. t.* 1580. *L. vespertina, Sibth. Ox.*

Hedges and fields; common. *Fl.* May—September. ♃.

Root tapering. *Stem* 1-2 feet high, erect, panicled above, clothed with spreading hairs, swollen and slightly viscid at the joints. *Leaves* opposite and connate, ovate or ovato-lanceolate, soft and hairy, lower ones petiolate. *Panicle* terminal, many-flowered. *Calyx* in the anther-bearing flowers subcylindrical, in the fruit-bearing ones larger and ovate, hairy, ribbed and veined. *Petals* with broad dilated claws crowned with two obtuse or notched teeth at the margin and two others near the centre.

3. L. *Githago,* Larn. *Corn-cockle Campion.* Flowers solitary; petals retuse, destitute of a crown; calyx 10-ribbed; segments foliaceous, much longer than the corolla; capsule 1-celled. *Agrostemma Githago, Linn. E. Bot. t.* 741. *E. Fl. v. ii. p.* 325. *Hook. Br. Fl. p.* 214.

Corn-fields; frequent. *Fl.* June, July. ⊙.

Root tapering. *Stem* 1-2 feet high, erect, branched, clothed with erect appressed long hairs. *Leaves* opposite and connate, sessile or semiamplexicaul, linear-lanceolate, rough with hairs. *Flowers* large, purple, solitary at the extremities of the branches. *Calyx* covered with very long hairs, *segments* long slender and acute, twice as long as the *petals,* becoming indurated when in fruit. *Capsule* with 5 teeth. *Seeds* numerous, reniform, granulated.

20. CERASTIUM. *Linn.* Mouse-ear Chickweed.

1. C. *vulgatum,* Linn. *Broad-leaved Mouse-ear Chickweed.* Stem nearly erect, glanduloso-pilose; leaves ovate; flowers in a dichotomous panicle with an intermediate flower, those of the branches conglomerate; petals about as long as the calyx, cloven, segments acute; sepals lanceolate, acute, membranous at the margins, 3-ribbed lateral ribs unbranched, and together with the bractea herbaceous, glanduloso-pilose throughout; capsule cylindrical, curved, about twice as long as the calyx, fruit-stalks about as long as the calyx; seeds small, tuberculated. *E. Bot. t.* 789. *E. Fl. v. ii. p.* 330. *Hook. Br. Fl. p.* 215. *Babington in Jardine's Mag. Zool. & Bot. ii. p.* 198. *C. viscosum, Huds. C. glomeratum, Thuill. Koch. Syn. Fl. Germ.* 121.

Fields, dry banks, &c.; common. *Fl.* April—September. ⊙.

Root fibrous. *Whole plant* stiff and erect, of a pale yellowish hue covered with long spreading hairs tipped with glands. *Stems* 6—8 inches high, erect, branched at the base. *Leaves* ovate, often broad, usually obtuse, but somewhat pointed with a stout mucro, 3-ribbed, lower ones tapering into petioles. *Flowers* in terminal dichotomous panicles with an intermediate flower in the fork, those of the branches aggregated, each on a short pedicel never exceeding the calyx in length. *Petals* white, scarcely longer than the calyx. *Stamens* 10. *Seeds* roundish-ovate, compressed, notched and gibbous, concentrically tuberculated.

2. C. *viscosum,* Linn. *Narrow-leaved Mouse-ear Chickweed.* Stem spreading, hairy; leaves oblongo-lanceolate; flowers in a dichotomous panicle with an intermediate flower, those of the branches lax; petals about as long as the calyx, cloven, segments rounded; sepals oblong, obtuse, 3-ribbed, lateral ribs branched, and together with the bractea hairy, membranous at their margins and glabrous apices; capsule cylindrical, curved, about twice as long as the calyx, fruit-stalks longer than the calyx; seeds large, tuberculated. *E. Bot. t.* 790. *E. Fl. v. ii. p.* 330. *Hook. Br. Fl. p.* 215. *Babington in Jardine's Mag. Zool. & Bot. v. ii. p.* 198. *C. vulgatum, Huds. C. triviale, Link. Koch. Syn. Fl. Germ.* 122.

Fields, banks, walls, &c.; common. *Fl.* May—September. ♃.

Root fibrous, strong. *Whole plant* much larger in every part than the last, lax and spreading, of a dark-green colour, covered with short dense hairs a few only of which are sometimes glandular. *Stems* 12—14 inches long, diffuse, prostrate, the extremities ascending. *Leaves* oblongo-lanceolate, frequently acute, lower ones tapering into petioles. *Flowers* in terminal dichotomous panicles with an intermediate flower in the fork, those of the branches lax, each on a pedicel longer than the calyx. *Petals* white, scarcely longer than the calyx. *Stamens* 10.

3. C. *semidecandrum,* Linn. *Little Mouse-ear Chickweed.* Stem nearly erect, glanduloso-pilose; leaves oblongo-ovate; flowers in a subdichotomous panicle with an intermediate reflexed or spreading flower, those of the branches subconglomerate; petals shorter than the calyx, slightly cloven, segments obtuse; sepals lanceolate, acute, single-ribbed, and together with the bractea glanduloso-pilose, broadly membranous at the margins and apices; capsule cylindrical, straight, the extremity only slightly curved, longer than the calyx, fruit-stalks longer than the calyx, at first reflexed, but ultimately erect; seeds small, tuberculated. *E. Bot. t.* 1630. *E. Fl. v. ii. p.* 331. *Hook. Br. Fl. p.* 216. *Babington in Jardine's Mag. of Zool. & Bot. v. ii. p.* 198.

Dry waste places; not unfrequent. *Fl.* March, April. ⊙.

Root fibrous. *Whole plant* very small, 2-3 inches high, much branched and spreading, at length erect, hairy and glandular, of a pale green colour. *Leaves* with an obtuse point, single-ribbed, the lower ones tapering into petioles. *Flowers* on pedicels two or three times as long as the calyx. *Petals* very minute, white. *Stamens* 5. *Seeds* roundish-ovate, compressed, notched and gibbous, concentrically tuberculated.

4. C. *arvense,* Linn. *Field Chickweed.* Stems ascending, prostrate below; leaves linear-lanceolate, bluntish; flowers in terminal panicles; sepals and bractea lanceolate, slightly acute and broadly membranous at their margins and apex; capsule at length longer than the calyx. *E. Bot. t.* 93. *E. Fl. v. ii. p.* 333. *Hook. Br. Fl. p.* 216. *Babington in Jardine's Mag. Zool. & Bot. v. ii. p.* 198.

Dry gravelly places; rare. *Fl.* April—August. ♃.

Oakley Park, near Ludlow; *Mr. H. Spare.*

Root strong, creeping. *Stems* decumbent below, the flowering part ascending, covered with fine deflexed hairs. *Leaves* narrowly lanceolate, often nearly linear, the edges fringed below, placed closely upon the lower parts of the stem, but much more distant above. *Flowers* in a di-or trichotomous panicle, 7—15, each on long pedicels, which, as well as the general peduncle, are clothed with minute spreading glandular hairs. *Petals* white, twice as long as the calyx. *Capsule* oblong.

21. MALACHIUM. *Fries.* Water Chickweed.

1. M. *aquaticum,* Fries. *Water Chickweed.* Stems decumbent and climbing, rooting at the base; leaves cordato-ovate, acuminate, sessile and semiamplexicaul; panicle dichotomous, spreading, glanduloso-pilose; bracteas herbaceous; petals bipartite, rather longer than the calyx. *Cerastium aquaticum, Linn. E. Bot. t.* 538. *E. Fl. v. ii. p.* 335. *Hook. Br. Fl. p.* 217.

Sides of rivers and ditches; not common. *Fl.* July, August. ♃.

Snowdon Pool; *H. Bidwell, Esq.* Near the Pool, Wilcot; *T. C. Eyton, Esq.*

Hedges between Canal and river Severn, near New Inn, between Shrewsbury and Uffington. Shelton. Crow-meole, both near Shrewsbury.

Stems 2-3 feet long, branched and straggling, angular, glanduloso-pilose. *Leaves* in pairs, opposite, sessile except the lower ones which are shortly petiolate and have short scattered hairs on their surfaces and margins, the rest generally glabrous with minutely elevated dots, and having a rib running along the margins into which the other ribs merge. *Sepals* ovate, acuminate, herbaceous, very slightly membranous at the edges, 3-ribbed, glanduloso-pilose. *Flowers* in dichotomous *panicles* with an intermediate flower in the forks, white, each on a glanduloso-pilose *pedicel* twice as long as the calyx, having a pair of leaf-like glanduloso-pilose *bracteas* at its base. *Capsule* scarcely longer than the calyx, opening with 5 bifid valves. *Seeds* numerous, roundish, singularly rugoso-tuberculated.

22. SPERGULA. *Linn.* Spurrey.

1. S. *arvensis,* Linn. *Field Corn Spurrey.* Leaves whorled with minute membranaceous stipules at their base; flowers panicled; seeds compressed, covered with minute elevated points, and surrounded with a white border. *E. Bot. t.* 1535. *E. Fl. v. ii. p.* 336. *Hook. Br. Fl. p.* 217.

Corn-fields; frequent. *Fl.* June—August. ⊙.

Root tapering. *Stems* 6—12 inches high, round, branched from the base, more or less glandular, swollen at the joints. *Leaves* long, narrow, linear, furrowed, glandular, arranged in two fascicles from each joint, spreading in a whorled manner with a pair of membranacous stipules between them. *Panicle* subdichotomous with an intermediate flower in the fork, *flowers* on long glandular stalks, reflexed in fruit, with a pair of membranaceous *bracteas* at their base. *Sepals* glandular, membranous at the margin. *Petals* white, rather longer than the calyx. *Seed* round, dark-brown, compressed, with minute dots, margin surrounded with a white border.

2. S. *vulgaris,* Bönningh. *Common Corn Spurrey.* Leaves whorled with minute membranaceous stipules at their base; flowers panicled; seeds compressed, sharply keeled, covered with white papillæ. *Reich. Fl. Excurs.* 3665. *Reich. Pl. Crit. vi. ic.* 705. *Bluff. & Fing. Comp. Fl. Germ. 2nd. ed. t. i. pt. ii. p.* 129.

Corn-fields with the preceding. *Fl.* June—August. ⊙.

Astley, near Shrewsbury; *Mr. E. Elsmere, junr.!*

In all respects similar to the last, but distinguished by the seeds.

3. S. *nodosa,* Linn. *Knotted Spurrey.* Leaves subulate, opposite, connate, glabrous, lower ones sheathing, upper ones bearing clusters of young leaves; petals longer than the calyx. *E. Bot. t.* 694. *E. Fl. v. ii. p.* 338. *Hook. Br. Fl. p.* 218.

Wet places; rare. *Fl.* July, August. ♃.

Near Oswestry; *Rev. T. Salwey.* West end of Croesmere mere; *J. E. Bowman, Esq.*

Root fibrous. *Stems* 3-4 inches high, branched and decumbent at the base, round, glabrous, swollen at the joints. *Lower leaves* about ⅝ of an inch long, *upper ones* gradually smaller, glabrous, sub-mucronate. *Flowers* large, white, 2-3 on the terminal branches, on long smooth peduncles.

CLASS XI.

DODECANDRIA.

—

"Nature is not with us a thing incidentally alluded to,—a thing to be voluptuously enjoyed when we find ourselves in the flowery lap of May; ours is a living, permeating, perpetual affection. We seek after communion with her as one of the highest enjoyments of our existence; we seek it to soothe the ruffling of our spirits; to calm our world-vexed hearts; to fill us with the Divine Presence and over-shadowing of beauty. The love of her is with us as a daily attraction; the knowledge of her a daily pursuit; we have advanced her cognizance and admiration into a science."

W. Howitt.

CLASS XI.

DODECANDRIA. 12 (—19) *Stamens.*

ORDER I. MONOGYNIA. 1 *Style.*

1. Lythrum. *Calyx* inferior, tubular, with 12 teeth alternately smaller. *Petals* 6, inserted upon the calyx. *Capsule* oblong, 2-celled, many-seeded.—*Nat. Ord.* Lythrarieæ, *DC.*—Name from λυθρον, *blood;* probably from the red colour of the flowers.

ORDER II. DIGYNIA. 2 *Styles.*

2. Agrimonia. *Calyx* inferior, turbinate, 5-cleft, clothed beneath the limb with hooked bristles. *Petals* 5, inserted upon the calyx. *Stamens* 7—20. *Fruit* of 2 small indehiscent capsules, invested by the hardened calyx.—*Nat. Ord.* Rosaceæ, *Juss.*—Name corrupted from *Argemone,* given by the Greeks to a plant supposed to cure the cataract in the eye called αργημα.

ORDER III. TRIGYNIA. 3 *Styles.*

3. Reseda. *Calyx* of 1 piece, 4—6-partite. *Petals* more or less divided and unequal, equal in number to the calyx-segments. *Capsule* of 1 cell opening at the top.—*Nat. Ord.* Resedaceæ, *DC.*—Name from *resedo,* to calm; from its supposed sedative qualities.

(See *Euphorbia* in CL. XXI.)

Tetragynia. 4 *Styles.*
(See *Tormentilla* in CL. XII.)

ORDER IV. DODECAGYNIA. 12 *Styles.*

4. Sempervivum. *Calyx* 12-cleft. *Petals* 12. *Capsules* 12.—*Nat. Ord.* Crassulaceæ, *DC.*—Name from *semper,* always and *vivo,* to live; from its continual verdure and tenacity of life.

DODECANDRIA—MONOGYNIA.

1. LYTHRUM. *Linn.* Purple-Loosestrife.

1. L. *Salicaria,* Linn. *Spiked Purple-Loosestrife.* Leaves opposite and verticillate, lanceolate, cordate at the base; flowers in whorled leafy spikes; stamens 12; calyx without bracteas at the base. *E. Bot. t.* 1061. *E. Fl. v. ii. p.* 343. *Hook. Br. Fl. p.* 219.

Ditches, watery places, and margins of rivers; frequent. *Fl.* July. ♃.

Stems erect, 2-3 feet high, acutely quadrangular, smooth or pubescent, the pubescence more copious in the upper part and deflexed. *Leaves* sessile, decurrent by their principal ribs, lanceolate, entire, smooth above, pubescent beneath and on their thickened edges. *Spikes* long, leafy, of very elegant and conspicuous purple *flowers*, in axillary, 6-flowered semiwhorls. *Calyx* striated, with 6 long and reddish teeth and 6 minute alternate ones. *Petals* oblong, cuneiform. *Stamens* within the tube of the calyx, 6 long and 6 short ones.

DODECANDRIA—DIGYNIA.

2. AGRIMONIA. *Linn.* Agrimony.

1. A. *Eupatoria*, Linn. *Common Agrimony.* Calyx of the fruit obconical, deeply furrowed to the base, exterior bristles spreading; leaves interruptedly pinnate, leaflets hoary beneath, terminal leaflet pedicellate. *E. Bot. t.* 1335. *E. Fl. v. ii. p.* 346. *Hook. Br. Fl. p.* 220.

Borders of fields, wastes and road-sides; common. *Fl.* June, July. ♃.
Root strong, tapering. *Whole herb* of a dark-green colour, clothed with soft silky hairs interspersed with glands and emitting, when bruised, an aromatic scent. *Stems* 2 feet high, erect, round, nearly simple. *Leaves* alternate, of several pairs of deeply and coarsely serrated leaflets, with smaller intermediate ones. *Stipules* rounded, palmate. *Flowers* numerous, yellow, in long simple dense *spikes*, each on a short pedicel with a 3-cleft *bractea* at its base, erect, reflexed in fruit. *Calyx* persistent, with erect soft hairs and glandular in the lower part, and a dense whorl of hooked bristles around the mouth.

DODECANDRIA—TRIGYNIA.

3. RESEDA. *Linn.* Rocket.

* *Sepals, petals, and styles,* 4.

1. R. *Luteola*, Linn. *Dyer's Rocket, Yellow Weed, or Weld.* Leaves lanceolate, undivided; calyx 4-partite. *E. Bot. t.* 320. *E. Fl. v. ii. p.* 347. *Hook. Br. Fl. p.* 220.

Waste places; frequent. *Fl.* July. ⊙.
Root tapering. *Stem* 2 feet or more high, branched, round, smooth. *Leaves* sessile, lanceolate, obtuse, somewhat undulate, and with a small tooth on each side at the base. *Racemes* elongated. *Flowers* yellowish, with prominent *stamens*, on short pedicels with *bracteas* at the base. *Nectary* large, green, crenate, on the upper side of the flowers; 3 of the *petals* 3-cleft, segments linear, 2 lower petals entire. *Capsules* broad, depressed. *Seeds* roundish, compressed, black and shining.

** *Sepals and petals* 5-6. *Styles* 4.

2. R. *lutea*, Linn. *Base Rocket, Wild Mignonette.* Leaves 3-cleft or pinnatifid, lower ones pinnated; calyx 6-partite, petals 6, very unequal. *E. Bot. t.* 321. *E. Fl. v. ii. p.* 348. *Hook. Br. Fl. p.* 220.

Waste places in limestone districts; not common. *Fl.* July, August. ⊙. or ♃.
Road-side from Broseley to Ironbridge, opposite the Foundry; *W. P. Brookes, Esq.!* Between Broseley and the Ironbridge; *E. Lees, Esq.*
Root woody. *Stem* 2 feet high, branched, smooth. *Leaves* very variable, some bipinnatifid. *Flowers* deep yellow, each on a pedicel equal in length to the calyx, with a small *bractea* at the base. Two upper *petals* with 2 wing-like lobes, lateral ones unequally bifid, lower ones entire. *Capsule* oblong, wrinkled.

3. R. *alba*, Linn. *White Base Rocket.* Leaves pinnatifid, wavy; calyx 6-partite; petals 6, equal, trifid. *Bluff. & Fingerh. Comp. Fl. Germ.* 2nd ed. v. i. s. ii. p. 159. *Reich. Excurs.* 4448.

Wastes; rare, possibly naturalized. *Fl.* June. ⊙. or ♂.
Broseley hill; *Mr. F. Dickinson!*
Stem 18 inches—2 feet high, branched, round, striated, smooth. *Leaves* alternate, smooth, pinnatifid; segments linear, mucronate, margins serrulate, the terminal segment elongated. *Racemes* dense. *Flowers* white, each on a pedicel as long or longer than the calyx with a narrow *bractea* at the base, equal in length to the pedicel. *Petals* deeply trifid, segments linear. *Seeds* reniform, very minutely dotted.

Very similar to *R. fruticulosa, Linn.*, but differing chiefly in the number of its parts. In the Herbarium of the Shropshire and North Wales Natural History and Antiquarian Society is a specimen with a label inscribed " Reseda alba, Portmarnock, Dublin, *Mackay*," which has 6 sepals and 6 peta's, and with which our plant appears to agree in all respects, nevertheless I perceive Mr. Mackay in his Flora Hibernica has inserted the plant found in the above locality as *R. fruticulosa,* although in his Catalogue of the Plants of Ireland he assigns it to *R. alba.*

DODECANDRIA—DODECAGYNIA.

4. SEMPERVIVUM. *Linn.* Houseleek.

1. S. *tectorum*, Linn. *Common Houseleek.* Leaves ciliated; offsets spreading; petals entire and hairy at the margins. *E. Bot. t.* 1320. *E. Fl. v. ii. p.* 350. *Hook. Br. Fl. p.* 221.

House tops and walls; frequent, but most probably in no case wild. *Fl.* July. ♃.
Root fibrous. *Leaves* in rosaceous tufts, oblong, acute, keeled, ciliated, very succulent. *Stem* from the centre of a tuft, erect, round, downy, with many alternate, sessile leaves. *Cyme* of many large, rose-coloured, very beautiful flowers. " The number of *stamens* is in reality 24; of which 12, inserted 1 at the base of each *petal*, are perfect; the rest alternating with the *petals*, small and abortive; some, bearing *anthers*, open longitudinally and laterally, producing, instead of pollen, *abortive ovules*; others resemble a cuneate pointed scale, in the inside of which, upon a longitudinal receptacle, are likewise ranged *abortive ovules*, in the same manner as in the real germen;—thus exhibiting the most complete transition from stamens to germens, in the same individual flower." *Hooker.*

CLASS   XII.

I C O S A N D R I A.

"Let no flower of the spring pass by us: let us crown ourselves with rosebuds before they be withered."
                                        WISDOM OF SOLOMON ii. 7-8.

" *   *   *   a circling row
Of goodliest trees loaden with fairest fruit,
Blossoms and fruits at once of golden hue
Appear'd, with gay enamel'd colours mixt."
                                        MILTON.

"See the country, far diffused around,
One boundless blush, one white-empurpled shower
Of mingled blossoms; where the raptured eye
Hurries from joy to joy, and, hid beneath
The fair profusion, yellow Autumn spies."
                                        THOMSON.

CLASS XII.

ICOSANDRIA.   *20 or more stamens, placed on the calyx.*

ORDER I.   MONOGYNIA.   *1 Style.*

1. PRUNUS. *Calyx* inferior, 5-cleft. *Petals* 5. *Drupe* covered with bloom, *nut* smooth, deeply furrowed at its inner edge. *Vernation* convolute.—*Nat. Ord.* ROSACEÆ, *Juss.*—Name προυνη, the Greek name for the tree.

2. CERASUS. *Calyx* inferior, 5-cleft. *Petals* 5. *Drupe* not covered with bloom, *nut* smooth, not furrowed at its inner edge. *Vernation* conduplicate.—*Nat. Ord.* ROSACEÆ, *Juss.*—Name κερασος, the Greek name of the tree.

(See *Cratægus* in ORD. PENTAGYNIA.)

ORDER II.   PENTAGYNIA.   *5 Styles, (variable in most of the Genera.)*

3. CRATÆGUS. *Calyx-segments* superior, acute. *Petals* roundish, *Styles* 1—5. *Fruit* oval or round, concealing the upper end of the cells which are bony.—*Nat. Ord.* ROSACEÆ, *Juss.*—Name from κρατος, *strength;* in allusion to the extreme hardness of the wood.

4. PYRUS. *Calyx* superior, 5-fid, withering. *Petals* 5. *Styles* 2—5. *Fruit* fleshy, (a *Pome* or *Apple*) with 5 cartilaginous 2-seeded cells.—*Nat. Ord.* ROSACEÆ, *Juss.*—Name from *peren,* Celtic, a *pear.*

5. SPIRÆA. *Calyx* inferior, 5-cleft, persistent. *Petals* 5. *Capsules* 3—12, 1-celled, 2-valved, with few *seeds.*—*Nat. Ord.* ROSACEÆ, *Juss.*—Name supposed to be the σπειρεια of Theophrastus.

ORDER III.   POLYGYNIA.   *Many Styles.*

6. ROSA. *Calyx* urn-shaped, fleshy, contracted at the orifice, terminating in 5 segments. *Petals* 5. *Pericarps* (or *Carpels*) numerous, bristly, fixed to the inside of the calyx.—*Nat. Ord.* ROSACEÆ, *Juss.*—Name from the Celtic *Rhos,* (from *rhodd, red*); whence also the Greek name for a *rose,* ροδον, was probably derived.

7. RUBUS. *Calyx* 5-cleft. *Petals* 5. *Fruit* superior, of several single-seeded juicy *drupes,* placed upon a protuberant spongy *receptacle.*—*Nat. Ord.* ROSACEÆ, *Linn.*—Name of uncertain origin; perhaps from the Latin *ruber* or the Celtic, *rub, red.*

8. FRAGARIA. *Calyx* 10-cleft, segments alternately smaller. *Petals* 5. *Fruit* consisting of many minute naked *nuts* placed upon a large fleshy deciduous *receptacle.*—*Nat. Ord.* ROSACEÆ, *Juss.*—Name from *fragrans, odorous;* on account of the fragrant smell of the fruit.

---

9. COMARUM. *Calyx* 10 or more cleft, segments alternately smaller. *Petals* 5 or more, shorter than the calyx. *Pericarps* inserted upon a large, spongy, permanent *receptacle.*—*Nat. Ord.* ROSACEÆ, *Juss.*—Name from κομαρος, a term applied by Theophrastus to some plants of the *Arbutus* tribe.

10. POTENTILLA. *Calyx* 10-cleft, segments alternately smaller. *Petals* 5. *Fruit* consisting of numerous minute naked *nuts* placed upon a small dry *receptacle.*—*Nat. Ord.* ROSACEÆ, *Juss.*—Name from *potens, powerful;* from the medicinal properties attributed to some of the species.

11. TORMENTILLA. *Calyx* 8-cleft, segments alternately smaller. *Petals* 4. *Fruit* consisting of numerous minute *nuts,* placed upon a dry *receptacle.*—*Nat. Ord.* ROSACEÆ, *Juss.*—Name from *tormina,* the dysentery, in the cure of which it was employed on account of its astringent qualities.

12. GEUM. *Calyx* 10-cleft, segments alternately smaller. *Petals* 5. *Pericarps* with long geniculated *awns.*—*Nat. Ord.* ROSACEÆ, *Juss.*—Name from γευω, *to yield an agreeable flavour;* the roots of *G. urbanum* being aromatic.

ICOSANDRIA—MONOGYNIA.

1. PRUNUS. *Linn.* Plum.

1. P. *spinosa,* Linn.   *Black-thorn or Sloe.*   Flowering buds 1 2 or 3 together, single-flowered; branches spinous, younger ones pubescent; leaves elliptic. *E. Bot. t.* 842. *E. Fl. v. ii. p.* 357. *Hook. Br. Fl. p.* 223.

Hedges; frequent. *Fl.* April, May. ♄.

A rigid bushy *shrub,* the *branches* with numerous shorter ones spreading generally at right angles and terminating in *spines* which bear flower and leaf-buds, *bark* smooth, blackish, *young branches* pubescent. *Leaves* mostly glabrous, serrated, slightly downy near the ribs and base when young, generally appearing after the blossoms, though some bushes may be observed on which the young leaves are contemporaneous with the blossoms, and which correspond nevertheless in other characters with this species. *Peduncles* mostly solitary, glabrous. *Calyx-segments* ovate, obtuse, irregularly toothed, with a single pellucid nerve. *Petals* round or oval, with scarcely any claw. *Fruit* globose, black, austere, and astringent.

2. P. *insititia,* Linn.   *Wild Bullace-tree.*   Flowering buds 1-2-flowered; peduncles pubescent; branches ending in a spine, younger ones velvety; leaves elliptical or ovato-lanceolate. *E. Bot. t.* 481. *E. Fl. v. ii. p.* 356. *Hook. Br. Fl. p.* 223.

Woods and hedges. *Fl.* May. ♄.
Buildwas; *Miss Moseley.*   Eudon Burnell, near Bridgnorth; *Purton's Midl. Fl.*
Almond park; Shomere wood; Shelton wood, near Shrewsbury.

A small bushy *tree* with greyish brown *bark,* and spreading nearly erect *branches* terminating in a spine more or less blunt, the *younger branches* velvety. *Leaves* serrated, downy beneath, appearing at the same time with the blossoms. *Peduncles* 1 or 2 together, slightly pubescent. *Calyx-segments* oblong, obtuse, irregularly toothed, with three branched pellucid nerves. *Petals* somewhat obovate, clawed. *Fruit* globose, black.

3. P. *domestica,* Linn.   *Wild Plum-tree.*   Flowering buds 1-2-flowered; peduncles pubescent; branches without spines, glabrous; leaves ovato-lanceolate. *E. Bot. t.* 1783.   *E. Fl. v. ii. p.* 355. *Hook. Br. Fl. p.* 223.

Woods and hedges, scarcely wild. *Fl.* May. ♄.
Munslow and Sutton; *Mr. F. Dickinson.*   Wyre Forest; *Mr. W. G. Perry.* Hedges about Shrewsbury.

A small *tree* with smooth *bark* and spreading erect *branches* without spines. *Leaves* very downy beneath, appearing with the blossoms, serrated. *Calyx-segments* oval, irregularly serrated. *Petals* oblong-ovate with scarcely any claw.

2. CERASUS. *Tournef.* Cherry.

1. C. *Padus,* DC.   *Bird Cherry.*   *Hag-berry.*   Flowers in pendulous racemes; leaves deciduous, obovate or oval, glabrous, with 2 glands at the summit of the footstalk. *Lindl. Syn.* 90. *Henslow Cat. Prunus Padus,* Linn.   *E. Bot. t.* 1383.   *E. Fl. v. ii. p.* 354. *Hook. Br. Fl. p.* 223.

Woods and hedges; rare. *Fl.* May. ♄.
Mile house, near Oswestry; *J. F. M. Dovaston, Esq.*   Near Ness and Valeswood; *T. C. Eyton, Esq.*
Westfelton.

A small *tree* with a hard close-grained wood and smooth even *branches. Leaves* acute, doubly serrated, when bruised smelling like rue. *Stipules* linear, serrated, deciduous. *Flowers* pure white, *petals* rotundo-obovate, jagged or toothed at the edges. *Calyx-segments* small, reflexed, serrated. *Fruit* small, black, austere and bitter.

2. C. *Avium,* Mœnch.   *Wild Cherry.*   Flowers in nearly sessile umbels; leaves ovato-lanceolate, somewhat downy beneath. *Lindl. Syn. p.* 90. *Henslow Cat.   Prunus Cerasus,* Linn.   *E. Bot. t.* 706. *E. Fl. v. ii. p.* 354. *Hook. Br. Fl. p.* 224.

Woods; frequent. *Fl.* May. ♄.
Ludlow; *Miss Mc Ghie.*   Jiggers' bank, Coalbrookdale; *E. Lees, Esq.* Walford; *T. C. Eyton, Esq.* Westbury; *Mr. F. Dickinson.*
Bomere woods; coppices about Church Stretton; Haughmond hill; Almond park.

The origin of the Garden Cherry. *Leaves* veiny, with copious glandular irregular serratures, and two unequal glands at the base or on the footstalks. *Flowers* white, about 3 in each *umbel,* on long simple stalks. *Stipules* and *bracteas* pale with glandular teeth or fringes, deciduous. *Fruit* globular, red or black. *Nut* hard, very smooth.

ICOSANDRIA—PENTAGYNIA.

3. CRATÆGUS. *Linn.* Hawthorn.

1. C. *Oxyacantha,* Linn. *Hawthorn, White-thorn or May.* Spiny; leaves glabrous, cut in 3 or 5 deeply serrated segments, cuneate at the base; flowers corymbose; style 1 or 2. *Hook. Br. Fl. p.* 224. *Mespilus Oxyacantha, Gært. E. Bot. t.* 2504. *E. Fl. v. ii. p.* 359.

Woods and hedges. *Fl.* May, June. ♄.

Variable in the form of the *leaves,* in the downiness of the *calyx,* and in the colour of the *flower* and *fruit,* the latter usually red.

4. PYRUS. *Linn.* Pear, Apple, and Service.

1. P. *communis,* Linn. *Wild Pear-tree.* Leaves simple, ovate, serrated; peduncles corymbose; fruit turbinate. *E. Bot. t.* 1784. *E. Fl. v. ii. p.* 361. *Hook. Br. Fl. p.* 220.

Hedges. *Fl.* April, May. ♄.

The origin of our *Pear.* A tall handsome *tree. Branches* at first erect, then curved downwards and pendulous, thorny in a truly wild state. *Flowers* snow-white.

2. P. *Malus,* Linn. *Crab Apple.* Leaves ovate, acute, serrated; flowers in a sessile umbel; styles combined below; fruit globose. *E. Bot. t.* 179. *E. Fl. v. ii. p.* 362. *Hook. Br. Fl. p.* 225.

Woods and hedges. *Fl.* May. ♄.

The origin of our *Apple. Branches* spreading, irregular, horizontal, destitute of thorns. *Flowers* beautiful, variegated with white and rose-colour.

3. P. *torminalis,* Sm. *Wild Service-tree.* Leaves ovate or cordate, lobed and serrated, lower lobes spreading; peduncles corymbose. *E. Fl. v. ii. p.* 362. *Hook. Br. Fl. p.* 225. *Cratægus torminalis, Linn. E. Bot. t.* 298.

Woods and hedges; rare. *Fl.* April, May. ♄.

In a field near Marm wood, Coalbrookdale; *Mr. F. Dickinson.* Wyre Forest; *Mr. G. Jorden,* and *E. Lees, Esq.* Woolston lane, Westfelton; Fitz; *J. F. M. Dovaston, Esq.*

A *tree* of slow growth, *wood* hard, *bark* smooth. *Leaves* deciduous, petiolate, 5—7-lobed, the lower pair broadest and most distant, dark-green, downy beneath. *Flowers* white, in terminal corymbose downy panicles. *Fruit* small, greenish-brown, spotted, agreeably acid.

4. P. *aucuparia,* Gærtn. *Quicken-tree, Mountain-ash, or Rowan-tree.* Leaves pinnated, glabrous, leaflets serrated; flowers corymbose; fruit small, globose. *E. Fl. v. ii. p.* 364. *Hook. Br. Fl. p.* 225. *Sorbus aucuparia. E. Bot. t.* 387.

Hilly woods; not common. *Fl.* May, June. ♄.

Tentree hill, Coalbrookdale; *Mr. F. Dickinson.* "Plentiful in Wyre Forest, but mostly young trees or offsets from old roots. A curious contorted specimen grows on a rock of Caer Caradoc;" *E. Lees, Esq.* Ludlow; *Miss Mc Ghie.* Faintree, near Bridgnorth; *Purton's Midl. Fl.*

Nesscliffe; Pimhill; Haughmond hill; Stiperstones hill.

A *tree* of slow growth, *wood* compact, *nut* very hard, *branches* smooth round greyish. *Leaflets* oblong, smooth above, downy beneath. *Flowers* corymbose, cream-coloured, fragrant. *Berries* globose, scarlet, sour and bitter.

The branches are often superstituously hung up in houses on a particular day to ensure to the inmates good luck and protection against evil spirits.

5. P. *Aria,* Sm. *White Beam-tree.* Leaves ovate, cut and serrated, white and downy beneath; flowers corymbose; fruit globose. *E. Bot. t.* 1858. *E. Fl. v. ii. p.* 367. *Hook. Br. Fl. p.* 225. *Cratægus Aria, Linn.*

Hilly places; rare. *Fl.* June. ♄.

Craig y Rhu, in the parish of Oswestry; Blodwell rocks; *Rev. T. Salwey!*

Tree of moderate size, *branches* white and downy when young, reddish brown and smooth when old, *wood* very hard. *Leaves* on downy stalks, often minutely lobed at the margin. *Flowers* white, in large downy corymbs. *Fruit* globular, scarlet, dotted, mealy, acid and astringent.

5. SPIRÆA. *Linn.* Spiræa, Dropwort, or Meadow Sweet.

1. S. *salicifolia,* Linn. *Willow-leaved Spiræa.* Shrubby; leaves elliptico-lanceolate, unequally serrated, glabrous, without stipules; racemes terminal, compound. *E. Bot. t.* 1468. *E. Fl. v. ii. p.* 367. *Hook. Br. Fl. p.* 226.

Moist situations; rare. *Fl.* July. ♄.

Margin of Colemere mere, near Ellesmere; a doubtful station; *J. E. Bowman, Esq. in Watson's Bot. Guide.*

A small *shrub* 3-4 feet high. *Stems* branched, round, smooth, wandlike. *Leaves* alternate, smooth on both sides, paler beneath, on short bordered petioles. *Flowers* flesh-coloured, in erect crowded racemes interspersed with hairy *bracteas.*

2. S. *Filipendula,* Linn. *Common Dropwort.* Herbaceous; root tuberous; leaves interruptedly pinnate, leaflets oblong, pinnatifido-incised, serrated; stipules adnate to the petioles; flowers paniculato-cymose; capsules pubescent, parallelly appressed. *E. Bot. t.* 284. *E. Fl. v. ii. p.* 369. *Hook. Br. Fl. p.* 226.

Dry pastures in the limestone districts; rare. *Fl.* July. ♃.

Gleeton hill, near Homer, Wenlock; *W. P. Brookes, Esq.!*

Stem a foot or more high, round, smooth, panicled at the summit. *Leaves* chiefly radical, depressed or spreading, dark green, elegantly pinnate, leaflets opposite or alternate with smaller intermediate ones. *Stipules* of the root leaves linear, acute, entire, those of the stem leaves rounded and cut. *Flowers* yellowish or cream-coloured, tinged with red.

3. S. *Ulmaria,* Linn. *Meadow Sweet, Queen of the Meadows.* Herbaceous; root fibrous; leaves interruptedly pinnate, leaflets ovate undivided, serrated, terminal one larger palmate 3—5-fid; stipules adnate to the petioles; flowers paniculato-cymose; capsules glabrous, spirally contorted. *E. Bot. t.* 960. *E. Fl. v. ii. p.* 369. *Hook. Br. Fl. p.* 226.

Meadows, banks of rivers and ditches; frequent. *Fl.* July. ♃.

Stem 3-4 feet high, branched, angular, smooth. *Leaves* of few large, opposite *leaflets* with very small intermediate ones, dark green, smooth above, more or less downy beneath. *Stipules* rounded, deeply toothed. *Flowers* yellowish or cream-coloured, sweet-scented.

ICOSANDRIA—POLYGYNIA.

6. ROSA. *Linn.* Rose.

"All the British species are prickly *shrubs,* with pinnated leaves. *Inflorescence* ternate: primordial *peduncle* continuous; lateral ones with a joint near the base, accompanied by two bracteas, and capable of producing there another pair of flowers, and so on; but rarely, beyond a third series; the larger bunches being composed of independent fascicles, which terminate alternate, often leafless, ramifications. Such compound bunches are produced on strong shoots only; on the feebler ramuli the flowers grow three together; on the weakest solitary. The primordial fruit has the shortest stalk, is the largest, and is very generally more produced at the base and less at the apex, than the subordinate ones."—*Borrer in Hook. Br. Fl.*

Mr. Woods in his elaborate *Synopsis of the British Species of Rosa* in *Linn. Trans. xii.* 159, has furnished the student with many useful hints in the investigation of this genus, from which the following remarks are abridged:—The stems petioles and nerves of the leaves are furnished with *prickles* and *setæ.* The *prickles* are either straight or hooked, equal or unequal; their form being taken from those which grow on the strong parts of the plant, and from those which are largest and with the most extended base. In *R. spinosissima* they are straight; in *R. villosa* straightish with a very slight curve downwards; in *R. gracilis* and *tomentosa,* falcate or bent as a scythe; in *R. canina* hooked or uncinate like a claw or sickle; and in *R. arvensis,* a sort of obtuse elliptical cone, with a straight or curved mucro, a peculiarity of form not found in *R. systyla.* The *setæ* are distinguishable from the prickles by their smaller size, and by being always straight and tipped with a *gland;* the stalk always obviously exceeding in length the diameter of the gland it supports. *Glands* rarely occur on the stems, but chiefly on the stipules, petioles, nerves, under surface and serratures of the leaves, fruitstalk, receptacle and calyx. They are rarely quite sessile, the footstalk however seldom exceeding in length the diameter of the gland. They are generally most abundant on the early and imperfectly formed leaflets, and sometimes fall off or dry up towards autumn. The presence or absence of *hairs* on the petioles and upper and lower surfaces of the leaf is of considerable importance, and it is observable that these always accompany each other. The shape of the *leaflet* is taken principally from the terminal one, which is the most perfect; all those of the earlier leaves are uncertain in shape, always rounder than the others, sometimes retuse: these are to be rejected, and the shape of the leaflet deduced from those expanded later in the season. The simple or compound serratures of the leaflets must be especially attended to. The *stipules* are linear-decurrent on the petiole, and generally edged with glands; in some species these continue unchanged, or nearly so in those leaves which accompany the inflorescence, and no stipules are found unaccompanied by leaves; whilst in others the leaflets gradually diminish in number, till at last they are entirely deficient, and the two stipules unite (having often previously increased remarkably in breath) and form a *bractea.* The *bractea* is described from the usual form of the first which is found entirely devoid of leaflets. The *calyx* is either *simple,* the segments being undivided or without pinnæ; or *compound,* two of the segments having pinnæ on each side, one on one side only, and the remaining two uniformly entire. The pinnæ are always very narrow at the base, expanded upwards, and go off at a considerable angle; always occurring before the contraction of the calyx-segment at the point of the flower. The calyx-segments are liable to become monstrous, either by growing out into leaves, or by a suppression of the pinnæ. In the first case the divisions mostly take place after the contraction of the segment, which in the bud marks the termination of the petals; or if it occur in the lower and broader part, it carries the appearance of a division, not of a pinna, being wide at the base, contracted upwards, and lying parallel to the line

of the segment when the flower is expanded. In the second case the receptacle is generally small, and the segments are expanded towards the apex. In both the receptacle is but little contracted at the summit, and assumes somewhat of a turbinate form. The *simple* calyx must be determined with caution; as even in some species whose calyx is generally simple, a small pinna may be observed.

The following descriptions of the Shropshire species are derived from Hooker's British Flora and Wood's Synopsis.

* *Shoots setigerous, prickles scarcely curved.*

1. R. *spinosissima,* Linn. *Burnet-leaved Rose.* Prickles crowded, unequal, mostly straight, intermixed with setæ; leaflets small, simply serrated, their disk eglandulose; calyx simple; fruit nearly globular. *E. Bot. t.* 187. *Woods Linn. Trans. xii. p.* 178. *Lindl. Ros. p.* 50. *E. Fl. v. ii. p.* 375. *Hook. Br. Fl. 4th ed. p.* 198. *R. pimpinellifolia, Linn. Sabine. Ser. in De Cand. Prod.*

Heaths, &c.; rare. *Fl.* May. ♄

Near Hanmer Hall; *Mr. F. Dickinson.* Near the Race Course, Ludlow; *Miss Mc Ghie.* Shotton; *W. W. Watkins, Esq.*

Increasing fast by suckers. *Shrub* upright 1—3 feet high, occasionally taller in the shade or when drawn up in hedges; branches spreading, tortuous, much divided; lowermost often lengthened and trailing. *Prickles* horizontal or deflexed, very numerous on every part, although old bushes are sometimes denudated; extremely unequal in size; the larger not rarely compressed and somewhat falcate, the smaller and often all, straight, gradually dwindling down to setæ. *Leaflets* rigid, most frequently 7, but varying in number from 5 to 11, sometimes even to 15, and in figure from orbicular and subretuse to ovate and acute, the more numerous usually the longer, and the more finely serrated; they are mostly flat, their hue full green or somewhat glaucous, paler beneath; serratures simple, generally broad, very rarely irregular; petioles usually glandulose, and with a few straight prickles, sometimes naked, sometimes with a few chaffy scales, rarely downy or hairy; *stipules* fringed with glands, narrow at the base, dilated, leafy and divaricated at the points. *Flowers* numerous, solitary. *Peduncle* gradually thickened upwards, becoming fleshy and coloured with the fruit, naked or glandulose, sometimes setose, more rarely prickly. *Calyx-segments* shorter than the petals, acuminate, entire, or with a few gland-tipped teeth and occasionally a minnte *pinna* or two. *Petals* cream-coloured, with or without crimson blotches on the outside, sometimes suffused with pink, rarely full pink or deeper red. *Stigmas* somewhat depressed, pale or red, mixed with hairs. *Fruit* varying from the size of a large cherry to that of a large pea, globose, or more often spheroidal, in some varieties obovate, in others ovate and urceolate; dark purple or blood-red or full black; firm, not pulpy when ripe, of a sweet taste and with a purple juice; the spreading or erect persistent segments of the calyx are affixed to a prominent ring, and often somewhat fleshy at the base.

** *Shoots mostly without setæ.*

1. *Leaves glandulose.*

a. *Prickles uniform or nearly so; setæ none or very few.*

2. R. *villosa,* Linn. *Villous Rose.* Prickles uniform, nearly straight; leaflets doubly serrated, downy, glandulose; calyx slightly pinnate; root-shoots straight. *Woods l. c. p.* 189. *E. Fl. v. ii. p.* 382. *Hook. Br. Fl. 4th ed. p.* 199. *Linn. Herb. R. mollis. E. Bot. t.* 2459. *Lindl. Syn. p.* 100. *R. mollissima, Willd. R. heterophyllus, Woods l. c. p.* 195. *R. pulchella, Woods l. c. p.* 196.

Hedges, &c.; not uncommon. *Fl.* June, July. ♄.

Near Oswestry; *Rev. T. Salwey.* Between Aston and Bishop's Castle; and by the road-side near Walcott; *E. Lees, Esq.* Near Millechope, near Rushbury; *Mr. F. Dickinson.* Wyre Forest, and Stottesden, abundant; *Mr. G. Jorden.*

Sharpstones hill; Kingsland; and hedges about Shrewsbury.

Root stoloniferous. *Shoots* upright or ascending, not arched; *bush* sometimes 6—8 feet high, but usually of more humble growth; branches irregular, ascending, variously tinged with purple, and cæsious in various degrees whilst young. *Prickles* not numerous, subulate from a dilated base, sometimes very slightly curved, often in pairs at the base of the petioles. *Leaflets* 5 or 7, ovate or elliptical, not acuminate, sometimes subretuse, hoary with down and glandulose, most plentifully so beneath; primary serratures often rather distant, especially towards the base of the leaflet, with their points frequently somewhat divaricate, and the secondary ones sometimes scarcely more than a fringe of glands; *petioles* and *stipules* downy and glandulose, the former mostly with feeble straight prickles; the latter linear, scarcely dilated towards the points; those nearest the flowers coalescing into broadly ovate, elliptical, or sometimes lanceolate pointed *bracteas.* *Flowers* 1—3 together on the ramuli, often in large bunches on strong shoots. *Peduncle* and *calyx-tube* cæsious, setose, more rarely naked. *Calyx-segments* downy, setose and glandulose, simple or more generally sparingly pinnate, very often leafy at the point, about as long as the *petals,* persistent, connivent, erect or somewhat spreading on the fruit. *Petals* generally of a vivid full pink or deeper red, often fringed with distant glands, sometimes white with crimson blotches on the outside. *Styles* included, *stigmas* prominent. *Fruit* mostly pendulous; broadly elliptical or nearly globose, lateral ones often urceolate; when ripe, purplish-red approaching to crimson, with a cæsious bloom. The turpentine scent perceptible in most of the glandulose-leaved roses is powerful in this species.

3. R. *tomentosa,* Sm. *Downy-leaved Rose.* Prickles mostly uniform, straight or curved; leaflets doubly serrated, downy, glandulose; calyx copiously pinnate. *E. Bot. t.* 990. *Woods l. c. p.* 197. *E. Fl. v. ii. p.* 384. *Hook. Br. Fl.* 4*th ed. p.* 200. *Lindl. Syn. p.* 100. *Hook. in Fl. Lond. N. S. t.* 124. *Pers. De Cand. Ser.*

Hedges and thickets; not unfrequent. *Fl.* June, July. ♄.

Wyre Forest, near Bewdley; *E. Lees, Esq.* Buildwas; *Miss H. Moseley.* "Frequent in this county;" *Dr. Smith in Turn. & Dilw. Bot. Guide.* Sandford heath; Westfelton; Whip lane, near Osbaston; *J. F. M. Dovaston, Esq.*

A most variable species, best distinguished from *R. villosa* by the copiously pinnate *calyx-segments,* which generally, but not invariably, spread widely on the fruit. The *fruit* too is mostly of a more slender figure; and the *leaflets* are usually more narrowly elliptical and more pointed. These vary much in the quantity of *glands* and denseness of pubescence; their upper surface being often very hoary, and sometimes, although rarely, quite naked. The peduncle and calyx-tube are seldom without numerous *setæ.* Some of the varieties throw up suckers freely; others sparingly; others not at all.

b. *Prickles various, intermixed with setæ.*

4. R. *rubiginosa,* Linn. *True Sweet-briar.* Prickles numerous, larger uncinate, smaller subulate; leaflets doubly serrated, hairy, glandulose beneath, mostly rounded at the base; calyx-segments and pinnæ elongated, persistent; primordial fruit pear-shaped. *E.*

Bot. t. 991. *E. Fl. v. ii. p.* 385. *Hook. Br. Fl.* 4*th ed. p.* 200. *R. rubiginosa, a. Lindl. Ros. p.* 86. *Hook. Scot. i. p.* 157. *De Cand. Wapl. Fries. R. Eglanteria, Woods l. c. p.* 206. *Huds. R. suavifolia, Lightf.*

Hedges, &c.; rare. *Fl.* June, July. ♄.

Bromfield road, and many other places about Ludlow; *Miss Mc Ghie.* Buildwas; *Miss H. Moseley.*

Stoloniferous; 4—6 feet high, compact and densely branched in general, and the shoots seldom arched. *Prickles* numerous; the large uncinate ones on the stem and branches mixed irregularly with abundance of smaller, some slightly curved, and some straight, subulate and setaceous, and some real *setæ,* which last, however, are not always present; the flowering-twigs are occasionally unarmed, but have more usually binate uncinate *prickles* near the base of the *leaves,* and others scattered, varying in size and curvature. *Leaflets* flat, or often concave, pale bright-green, more or less hairy, ovate, or broadly elliptical, or often almost round, occasionally narrower and more pointed, but scarcely tapering to the base; sprinkled copiously beneath, on the edges and on the petioles with fragrant viscid *glands,* which are found also on the backs and edges of the *stipules.* *Peduncles* and often the *calyx-tube* beset with *setæ,* of which those at the base of the latter are usually larger. *Calyx-segments* setose and glandulose, with a lengthened leafy point and narrowly lanceolate *pinnæ,* spreading almost at right angles with gland-pointed teeth. *Petals* deep pink, equal to the calyx or rather shorter. *Styles* included, slightly hairy; *stigma* scarcely protuberant. *Fruit* changing first to yellow, then to orange-red or scarlet, its substance thin, scarcely pulpy, and almost insipid when ripe, when in bunches the primordial is pear-shaped, the secondary obovate, but less tapering at the base; the others elliptical.

2. *Leaves eglandulose.*

a. *Styles distinct, included or nearly so.*

5. R. *canina,* Linn. *Common Dog-Rose.* Prickles uniform, hooked; leaves naked or slightly hairy, their disk eglandulose; calyx-segments fully pinnate, deciduous; styles not united; shoots assurgent. *a. δ. ε. Lindl. Ros. p.* 98 *(excl. some syns.) Hook. Scot. i. p.* 157. *Fries. Hook. Br. Fl.* 4*th ed. p.* 201.

Thickets, hedges, &c.; very common. *Fl.* June, July. ♄.

The Shropshire Roses answering to the above character may be subdivided as follows:—

Var. α. Leaflets naked, carinate, serratures simple. *R. canina, Woods l. c. p.* 223. *E. Fl. v. ii. p.* 394.

a. green. *α. Woods. R. canina, E. Bot. t.* 992.

b. grey. *β. Woods.*

A straggling *shrub* 6—8 feet high, with spreading olive-coloured branches armed with uncinate *prickles.* *Petioles* pubescent, protected by falcate prickles and a few minute scattered glands. *Stipules* linear, serrated, glabrous, those adjacent to the flowers broader, and at length, on the suppression of the leaflets, changing into elliptical acuminate *bracteas.* *Leaflets* 7, narrowly elliptical, carinate, with a small twisted point, the younger ones somewhat shining as if varnished, glabrous, with simple (never double) unequal, acuminate serratures. *Peduncles* glabrous, solitary or two, rarely three, on the ramuli, generally four on the young shoots. *Calyx-tube* elliptical, tawny, glabrous. *Calyx-segments* triangulari-ovate, glabrous; *pinnæ* linear-lanceolate, with here and there a glandular tooth. *Flowers*

generally reddish, rarely white, a little redder in the bud than in the expanded flower. *Styles* included, *stigmas* depressed. *Fruit* elliptical, glabrous, shining, red.

In the *subvariety b.* the young leaves are covered with a waxy substance and till rubbed are of a glaucous green entirely without gloss. *Root-shoots* are more freely produced in this than in *a,* and there are sometimes as many as 8 flowers in a cyme. The plant is 8—10 or even 15 feet high, leaflets broader and the little point at the end always a little twisted, a character observable in a slight degree in *a,* but more conspicuous here.

Var. β. sarmentacea. Leaflets naked, carinate, serratures compound. *R. sarmentacea, Woods. l. c. p.* 213. *E. Bot. Suppl. t.* 2595. *R. canina, Fl. Lond.*

a. green. *β. Woods. R. sarmentacea, Swartz?*

b. grey. *α. Woods. R. glaucophylla, Winch.*

Shrub 8—10 feet high; branches diffuse, olive-coloured, *prickles* hooked, sometimes few and scattered, sometimes solitary or in substipular pairs. *Petioles* without pubescence, slightly glandulose, and armed with falcate prickles expanded at the base. *Stipules* spathulate, glabrous, serrated, the serratures sometimes glandular; those adjacent to the flowers much broader and at length changing into ovate acuminate *bracteas,* which are more numerous on the cymes but narrower at the base. *Leaflets* 5 or 7, elliptical, subacuminate, glabrous, shining above, the nerve sometimes prickly beneath, serratures unequal, generally divaricate, irregularly serrulate. *Peduncles* 1—8, glabrous or frequently with a few setæ, shorter than the bracteas. *Calyx-tube* narrowly elliptical, tawny, glabrous. *Calyx-segments* glabrous, triangulari-elliptical, acuminate; *pinnæ* lanceolate-linear, inciso-serrate. *Flowers* of a delicate pink. *Styles* included; *stigmas* depressed. *Fruit* elliptical, red, shining.

In the *subvariety b.* the leaves have a grey waxy appearance.

Var. γ. surculosa. Leaflets naked, flat, serratures simple. *R. surculosa, Woods. l. c. p.* 228. *R. venosa, Swartz? R. canina, β. E. Fl.*

a. green. *β. Woods.*

b. grey. *α. Woods.*

Shrub 8 feet high, straggling; branches diffuse, dark-purple or very tawny, the younger ones glaucescent, sometimes copiously prickly, sometimes nearly unarmed. *Prickles* very strong, uncinate, scattered, sometimes solitary, sometimes in substipular pairs. *Petioles* glabrous except a few scattered hairs on the upper side, armed with strong uncinate prickles. *Stipules* spathulate or linear, sometimes serrated, sometimes glanduloso-ciliate at the base, sometimes entire except at the apex, glabrous, sometimes hairy at the margin, those adjacent to the flowers broader and at length passing into elliptical acuminate *bracteas.* *Leaflets* 7, with a few hairs on the upper side of the nerve, elliptical or subrotund, acuminate, unequal and cordate or ovate at the base, serrated, glabrous beneath, dark, the younger ones somewhat purple. *Peduncles* 1—24, with scattered slender setæ or hairs. *Receptacle* ovate, tawny, glabrous, the disk convex. *Calyx-segments* triangulari-elliptical, acute, divided almost to the base; *pinnæ* lanceolate or linear-lanceolate, nerved, entire. *Flowers* pale pink. *Styles* somewhat protruded, hairy, with a round head of *stigmas.* *Fruit* broadly elliptical, red.

The only British species which can be mistaken for *R. surculosa,* are *R. canina, R. systyla,* and *R. arvensis,* and from each of them it may perhaps be difficult accurately to distinguish it, though in habit it is very different from either. From *canina* it may always be known by the porrect styles, the entire pinnæ of the calyx-segments, the peduncles almost always furnished with hairs or setæ, the

shape and flatness of the leaflets, the strong and hooked prickles of the footstalk, and the glands of the peduncle never extending themselves on the receptacle or calyx. These marks are however all more or less uncertain, and a better distinction in the living plant is found in the enormous surculi, 8—10 or even 20 in number, covered with beautiful blue wax and bearing great cymes of flowers; whilst in *canina* the number of flowers is never more than four. From *R. arvensis* it may be known by the styles, which are here hairy and but just protruded, not smooth and collected into a long cylinder, as in that plant. It is also a much more upright plant, the surculi being rather erect than decumbent. From *R. systyla* also a due attention to the styles will distinguish it, and the shape and flatness of the leaflet give a decidedly different appearance to the present plant.

Var. ε. Forsteri. Leaflets more or less hairy, not flat. *R. collina, Woods. l. c. p.* 219. *R. Forsteri, E. Fl. v. ii. p.* 392. *Borr. in E. Bot. Suppl. t.* 2611.

a. concave, green, *γ. Woods. R. campestris, Swartz.*

b. carinate, grey.

1. hairy beneath only. *β. Woods. R. Forsteri, E. Bot. Suppl. t.* 2611.
2. hairy on both sides.

Shrub 6—8 feet high, branches somewhat straggling, olive-coloured. *Prickles* uncinate, pale, often weak, nearly equal, solitary or in substipular pairs. *Petioles* downy, hoary, armed with falcate prickles, sometimes with a few scattered glands. *Stipules* linear, serrated or glanduloso-serrated towards the apex, glabrous above, those adjacent to the flowers larger and at length passing into ovato-lanceolate *bracteas.* *Leaflets* 5 or 7, elliptical, with very large (but never double) serratures, pale and cartilaginous at the apex, hairy beneath, glabrous above, glaucous and not shining. *Peduncles* 1—5, glabrous, shorter than the bracteas. *Receptacle* elliptical, glabrous, olive-coloured. *Calyx-segments* ovato-lanceolate, hairy towards the apex, pinnate; *pinnæ* linear-lanceolate, incised, generally entire at the margin. *Flowers* of a pale pink. *Styles* included, *stigmas* convex. *Fruit* elliptical, bright scarlet.

The forms above mentioned are not so defined, but that connecting varieties may be found. In all of them, the *ramification* varies in denseness, and the *shoots* are more or less arched or erect according to the vigour of the plant; the *prickles* are not very numerous, hooked in various degrees and compressed, and their base considerably dilated; the *leaflets* vary in width, their serratures, although scarcely compound, except in *β.* are mostly irregular in size; the *bracteas* vary in size; the *peduncle* and *calyx-tube* are most commonly naked, their *setæ,* when present, feeble and not numerous; the *calycine-segments* are free from *glands,* or more or less copiously fringed with them; the *styles* are hairy; the *fruit* is coral-red, or more scarlet, soft and pulpy when ripe, with a pleasant somewhat acid taste.

b. *Styles united in a column, mostly exserted.*

6. R. *arvensis,* Huds. *Trailing Dog-rose.* Prickles uncinate, those of the ramuli feeble; leaves simply serrated, deciduous, (glaucescent beneath), their disk eglandulose; calyx-segments sparingly pinnate, deciduous; styles united, hairless; shoots trailing. *E. Bot. t.* 188. *Woods. l. c. p.* 232. *Lindl. Ros. p.* 112. *E. Fl. v. ii. p.* 397. *Hook. Fl. Lond. N. S. t.* 123. *Linn. Hook. Br. Fl.* 4*th ed. p.* 202.

Woods, hedges, thickets, &c.; common. *Fl.* June, July. ♄.

Bush scarcely a yard high when unsupported, with trailing shoots often many feet in length, and much divided entangled feeble *ramuli*, which occasionally produce rugged excrescences and take root. *Prickles* numerous, not much dilated at the base, uncinate, those on strong shoots often compressedly conical with a straight or curved point; those on the ramuli few and scattered, small, more or less curved. *Leaflets* thin, nearly flat, coarsely serrated, dull green, paler and somewhat glacous beneath, naked on both sides, or slightly hairy beneath, chiefly on the mid-rib; on some plants they are elliptical, ovate, or almost round, on others much elongated; *petioles* hairy or glandulose, or both, with falcate *prickles*. *Flowers* copiously produced, often in large bunches, with lanceolate *bracteas*, white, large and handsome, opening flat, with a slight fragrance at first, but soon becoming unpleasant. *Peduncle* long, sprinkled with almost sessile glands. *Calyx-segments* reflexed by the time the petals fall, broad and short, with an acute point shorter than the petals, and a few small, entire, lanceolate *pinnæ*. Column of *styles* often overtopping the stamens; persistent *stigmas* in a round head. *Fruit* small, spherical, ovate or elliptical, sometimes long and slender, its length varying almost in accordance with that of the leaflets; blood-red when ripe, with an orange-red pulp of a pleasant peculiar flavour.

R. arvensis is distinguished from all the other British species by its trailing habit.

7. RUBUS. *Linn.* Bramble.

Of the fruticose or Bramble division of this difficult genus Mr. Borrer distinguishes and describes 10 species in *Hooker's British Flora*, and accompanies them with the following observations, to direct the attention of the student to those parts upon which their characteristic distinctions are founded:—"*Shrub-like* plants, or *herbs* with perennial *roots*. The latter offer nothing very peculiar. In some species of the former, the *stem* is upright or merely curved at the top; but in the greater number it is either prostrate, or, as is more generally the case, assurgent, arched, and decurved, and the ends of the shoot and of the side-branches, if it produce any, unless prevented by circumstances from reaching the ground, take root in the latter part of the year. In the winter the shoot is partially destroyed, the part next to the original root surviving to produce flowering-branches during the ensuing summer, and usually dying after the fruit has been perfected; young shoots meanwhile springing up by its side. The rooted ends also become distinct plants at various distances, often many yards from the parent root. This mode of growth adds much to the difficulties in the discrimination of the species; since an acquaintance with both the leafy shoot and the floriferous stem, formed in the second year from its remains, is necessary. The best characters are found in the figure, the arms, and the leaves of the former. The *leaves* in all the British species of this division are, occasionally at least, quinate, and, with one exception, digitate or somewhat pedate from a partial junction of the stalks of the two lateral pairs of leaflets; the margins serrated, for the most part unequally and irregularly; the prickles on the leaf-stalks more curved than those on the stem. In some species the *inflorescence* is remarkable; but in general the panicle varies so much as to afford no good distinction. Nor can the arms of the calyx, nor the form of its segments be depended on. The *petals* in all are delicate and crumpled, and in several species vary very considerably in size and width. There are some differences in the *fruit*, but they are rarely discriminative. In examining the figure of the leaves, the central leaflet is to be regarded; the lateral ones are always smaller and of a narrower proportion. In several species the leaves occasionally survive a mild winter, and are found the next season subtending flowering branches. The leaves of these branches are of less determinate figure; the number of their *leaflets* is reduced as they approach the inflorescence, and their place is supplied in the upper part of the panicle by first trifid, and then, simple *bracteas*, formed by the coalescence of the *stipules*. These last are usually long and narrow, entire or

sometimes toothed or jagged, and issue from the petiole for the most part a little above its base. They afford no distinguishing characters. No less than 48 supposed species of the genus are described and figured in the elaborate *Rubi Germanici* of *Weihe* and *Nees von Esenbeck*, nearly all of which are probably found in Britain."

Professor Lindley enumerates in the first edition of his *Synopsis of the British Flora* 24 species in this division, which in the 2nd edition he reduces to 18, arranging them under 5 sections, characterized by the smoothness or hairiness of the stems, the green or white colour of the under surface of the stem-leaves, and the direction of the calyx of the fruit. In each section the species or forms are distinguished from the typical state by the longer or shorter inflorescence, and the jaggedness of the leaves. His reasons for this distribution are given in his remarks on the dubious character of these plants, which, as Sir W. J. Hooker justly observes, "deserve to be quoted, as they are the words of one who has made this Genus, and the whole family to which it belongs, the object of his peculiar study." Dr. Lindley says "I am bound to declare, that I can come to no other conclusion than that with which I first started, namely, that we have to choose between considering *R. suberectus*, *fruticosus*, *corylifolius*, and *cæsius*, as the only genuine British species, or adopting in a great measure the characters of the learned German Botanists above mentioned, who have so much distinguished themselves in the elaboration of this Genus. So clear is my opinion upon this point, that, if it had been possible to prove the four species to which I have alluded to be themselves physiologically distinct, I should at once have reduced all the others to their original places; but as it is in the highest degree uncertain whether *R. fruticosus*, *corylifolius*, and *cæsius*, are not as much varieties of each other, as those it would be necessary to reject, I have thought it better to steer a middle course, until some proof shall have been obtained, either one way or the other."

To the examination of the Shropshire Brambles I devoted two successive summers, and collected specimens of every form which came under my notice, in which any conspicuous differences were observable. But, notwithstanding some care and attention, the strange and inconstant manner in which similar forms, and even what were apparently distinct ones, in innumerable instances varied or ran into each other, precluded me from arriving at any definite conclusions as to the limits and true characters of the estimated species, and in fact left me in a complete maze of doubt and ignorance whether they should not all be considered as modifications endlessly varied of one and the same species. In this state of utter incertitude I resolved to distribute my specimens among, and solicit the opinions of Nees von Esenbeck, Professor Lindley, and Mr. Borrer. These eminent Botanists, with the utmost liberality, most cheerfully and speedily responded to my wishes. In consequence, however, of their severally referring the same forms, in many instances, to different and opposite species, I found that no other alternative was left to me than to describe from the specimens submitted to them the various forms of the genus, and append to each description their opinions and remarks. By this plan, indeed, the genus would be apparently extended by the addition of many forms, which in all probability could have no real claims to rank as true species, still upon the whole it appeared to be the best, that under the circumstances could be adopted, inasmuch as it would excite the attention of Botanists to the subject, and thereby in some degree tend towards the final resolution of the great and important question of the limits of the species. In carrying out my plan I have adopted the nomenclature of Esenbeck, as well by reason of his priority of authorship, as on account of his references to the plates of his Monograph, and of his present views with respect to the species contained in that work. In those cases in which specimens were not forwarded to Esenbeck, but only to Lindley and Borrer, I have assumed the names respectively assigned by them. The following excellent remarks contained in Professor Esenbeck's letter must not be omitted:—"J'ai soigneusement comparé vos échantillons avec ceux de mon herbier qui servirent de modèles aux tables de notre Monographie, et j'ai trouvé avec quelque plaisir, que les formes

proposées dans cet ouvrage se reproduisent partoût et dans des payes très-eloignées l'une de l'autre. Je ne suis pas d'avis que toutes les formes proposées par feu mon ami Mr. Weihe comme espèces, doivent être considérées comme telles, mais, à mon opinion, c'est absolument necessaire de rechercher la plupart des formes différentes que nous présente un genre quelconque, avant que de s'engager à juger des espèces, et d'en oser fixer les limites. Je ne sais pas qui a fait plus de tort à la science, si c'est feu Mr. Weihe en distinguant comme espèces la plupart des formes de Ronce qui venaient se lui présenter, ou Mr. Koch qui vient de les reduir toutes à une seule. Enfin, il s'agit dans ce cas des objêts purement d'observation et je ne doute point, qu'en observant fidèlement la nature nous trouverons enfin le juste milieu de ces extrèmes. En vous faisant part de la nomenclature de vos Rubi suivant la Monographie de Weihe et Nees d'Esenbeck jè saisirai en plusieurs lieux l'occasion pour vous donner mon avis sur celles qui, proposées comme espèces dans cet ouvrage, ne sont plutôt que des variétés d'une seule espèce."

A. *Stems mostly biennial, woody.*

I. *Stems erect or nearly so, not rooting.*

* *Leaves pinnate.*

1. R. *idæus*, Linn. *Raspberry.* Stem nearly erect, round, downy, with innumerable straight and decurved prickles; leaves pinnate, with 5 or 3 rotundo-ovate, acuminate leaflets, cordate at the base, rigid, plicate and downy above, white and densely cottony beneath; flowers in pendulous clusters; petals as short as the calyx; calyx downy, spreading in flower, slightly deflexed in fruit; fruit roundish. *E. Bot. t.* 2443. *E. Fl. v. ii. p.* 408. *Hook. Br. Fl. p.* 246.

Moist margins of woods and moist heaths; frequent. *Fl.* May, June. ♄.

Linley, near Bishop's Castle; *W. P. Brookes, Esq.* At Ingerton, by the sides of rivulets, plentiful; *Mr. G. Jorden.* In great abundance in the Arkoll wood, near the Wrekin hill; *E. Lees Esq.* In a wood near Shawbury; *W. W. Watkins, Esq.*

Shawbury heath. Shomere wood. Banks of Canal near the Queen's head, near Oswestry. Sutton Spa. Almond park. Cupid's Ramble, near Westfelton.

Stem erect, round, downy, densely covered with innumerable prickles, dilated at the base, straight and decurved, dark-purple. *Leaves* pinnate, of 5 or 3 leaflets, rotundo-ovate, sharply acuminate, cordate at the base, deeply inciso serrate, rigid, plicate and downy above, white and densely cottony beneath, with numerous small prickles scattered on the mid-rib and channelled petioles. *Flowering-stems* erect, slightly curved at the top, not rooting, with numerous prickles. *Flowers* from lateral branches which are pubescent, with numerous crimson, straight, deflexed prickles, dilated at the base. *Leaves* ternate, plicate, inciso-serrate. *Lower flowers* solitary, on simple drooping or deflexed, downy, and prickly peduncles from the axil of the leaves; *upper ones* terminal, about 3, pendulous, peduncles of all with numerous, curved, deflexed, crimson prickles. *Calyx* downy, plentifully covered with prickles, segments ovate, acute, spreading in flower, slightly deflexed in fruit. *Fruit* roundish, depressed, crimson, of a fragrant and delicious flavour.

A variety with amber-coloured fruit occurs with the preceding, and apparently equally wild at Almond park wood. It differs principally in having the prickles always pale, the leaflets obovate, acute, somewhat cuneiform at the base, soft and flexible, and the fruit roundish, conical, and of an amber-colour.

** *Leaves digitate or pedate.*

2. R. *suberectus*, And. *Upright Bramble.* Stem nearly erect, not rooting, obsoletely angular; prickles uniform, few, small; leaves digitate, 5-nate, leaflets flexible, lower pair sessile or nearly so; panicle nearly simple. *And. in Linn. Trans. xi. p.* 218. *t.* 16. *E. Bot. t.* 2575. *E. Fl. v. ii. p.* 406. *Hook. Br. Fl. 4th ed. p.* 203.

Somewhat boggy heaths, sides of streams, &c.; rare. *Fl.* June—August. ♄.

Whixall Moss; *J. E. Bowman, Esq.* "I think I have met with this about the base of the Wrekin, but am not quite certain;" *E. Lees, Esq.*

"*Stems* nearly upright, about 3 or 4 feet high, bluntly angular, brittle, reddish, leafy, destitute of hairs, but armed with scattered, deflexed, scarcely hooked, small prickles, generally not above a line or two long; when larger they are dilated or elongated at the base. The *flowering-stems* bear several lateral, alternate, spreading, simple or compound branches, whose leaves are ternate, the uppermost simple, each branch hairy, and more or less prickly, terminating in a long, simple, bracteated, hairy cluster of 10 or 12 large, white, upright *flowers*. *Bracteas* lanceolate, hairy, not hoary, sometimes cut or lobed. *Flower-stalks* hairy and sparingly glandular; the lower ones only sometimes bearing a prickle or two. *Calyx* densely woolly within, externally hairy, totally destitute of prickles; its segments moderately spreading while in flower, afterwards reflexed, but not closely. *Petals* crumpled. *Berry* of a rather small number of dark red, or blood-coloured, not purple *grains*, said to be agreeably acid, with some flavour of the Raspberry, ripening later than that fruit. The *foliage* of this species is in one respect peculiar. Some of the *leaves* on the barren stems, though generally of 5 *leaflets*, the 2 lowermost of which are quite sessile, are often furnished with a pair of similar small leaflets on the central stalk below the terminal one, so that the whole *leaf* is partly digitate, partly pinnate, combining the foliage of the Bramble and the Raspberry. The *leaflets* are all of a deep green, ovate, or heart shaped, pointed, sharply serrated, quite smooth above, paler with hairy ribs and veins beneath. *Stipules* linear-lanceolate. *Footstalks* sparingly and minutely prickly."—*Smith.*

I have never seen specimens of this species which I insert on the authority of Mr. Bowman in *Watson's New Botanist's Guide.*

3. R. *plicatus*, W. and N. *Plaited-leaved Bramble.* Stem not rooting, barren one erect, angular, smooth, flowering one slightly curved, obsoletely angular; prickles confined to the angles, small, distant, slightly curved and deflexed; leaves digitate, 5-nate, often pinnate, leaflets cordato-ovate, cuspidate, plicate, flexible, decurved, shining above, paler and pubescent beneath, lower pair on very short stalks; panicle nearly simple, corymbose, hairy, with few weak prickles; floral leaves ternate, obovato-cuneate, or simple and ovato-acuminate; calyx spreading in flower, reflexed in fruit, nearly smooth and shining; berry small, nearly globular, red. *R. plicatus, Weihe & Nees ab Esenbeck Rubi Germanici* 1. *t.* 1. *E. Bot. Suppl. t.* 2724. *Hook. Br. Fl. 4th ed. p.* 203. *R. suberectus* β. *Hook. Br. Fl. 2nd & 3rd eds.* *R. suberectus (not of And.) Lindl. Syn. 2nd ed. p.* 92.

Moist boggy places; rare. *Fl.* July—September. ♄.

Boggy part of Almond Park wood, near Shrewsbury.

Barren-stem 3-4 feet high, erect, pale succulent green, bluntly angular, perfectly smooth and destitute of hairs or bristles, but with a few scattered reddish sessile glands. *Prickles* confined to the angles of the stem, small, dark red, distant, slightly curved and deflexed, dilated at the base. *Leaves* digitate, of 3 or 5 petiolate leaflets, the lowermost pair of leaflets on very short but still distinct stalks, often pinnate by the intervention of another pair of nearly sessile leaflets below the central leaflet, whose stalk is very long and hairy; *common petiole* flattened and channelled above, rounded beneath, with a row of reddish recurved distant prickles on each side, and a similar one on the under side, continued also along the mid-ribs of the leaflets, with a few scattered hairs and sessile glands: *leaflets* very large, broadly cordato-ovate, cuspidate, sharply and deeply unequally serrated, thin and flexible, plicate, of a deep shining green and slightly hairy above, paler beneath, the ribs and nervures clothed with dense soft shining pubescence, elegantly decurved, somewhat fragrant. *Stipules* linear-lanceolate, ciliated and glandular. *Flowering-stems* slightly curved at the summit, obsoletely angular, almost round, reddish, smooth, except a few sessile reddish glands. *Prickles* few, scattered, recurved; *branches* simple or compound, lateral, alternate, spreading, each terminated by a simple, bracteated raceme of erect flowers, the uppermost lateral ones generally overtopping the terminal flower; *panicle* hairy or downy, with a few small scattered prickles and sessile reddish glands. *Lower leaves* ternate, obovate, somewhat cuneiform at the base, inciso-serrate, more hairy on both sides, lateral leaflets nearly sessile, central one stalked; *uppermost leaves* single, ovate, acuminate, petioles downy. *Calyx* spreading in flower, reflexed in fruit, segments generally broad and short, with an acute, short point, often however very greatly elongated, densely hoary and white on the interior and edges, with a few scattered hairs and sessile glands on the exterior, which is of a shining green and destitute of prickles. *Petals* white, obovate, crumpled. *Fruit* small, nearly globular, bright-red, grains large, of a gratefully acid flavour, somewhat resembling that of the raspberry.

Nees ab Esenbeck identifies our plant as *R. plicatus, W. & N.* and remarks "Votre echantillons se rapproche beaucoup du *R. fastigiatus,* qui n'est qu' une varietié du *R. plicatus.*" Professor Lindley determines it to be the *R. suberectus* of his *Synopsis 2nd ed.* which he there considers identical with *R. plicatus, W. & N.* The *R. plicatus of E. Bot. Suppl.* he states to be nothing but a small form of *R. affinis, W. & N.,* the *R. nitidus* of *E. Fl.* The figure of *R. plicatus* in *E. Bot. Suppl.* appears to me a tolerably correct general representation of our plant; certainly the figure of *R. suberectus* in *E. Bot.* 2572 does not belong to our plant, nor indeed to specimens sent to me by Mr. Wm. Wilson from Woolston Moor, Lancashire, as the *true suberectus,* which to me seems much more similar to Shropshire specimens marked by Professor Lindley as the *R. fissus* of *2nd. ed.* of his *Synopsis.* Mr. Wilson to whom Shropshire specimens of the above and of *R. fissus,* with Lindley's names attached, had been sent writes in reply: "I am a little surprised to observe how much confusion exists in respect to *R. suberectus* which (the true plant of Anderson I mean) is surely a very distinct species, almost as much so as *R. idæus* from any other British Rubus. Mr. Borrer has lately arrived at correct notions of it, but your specimen marked by Lindley does not argue that he is correct in the tribe."

Mr. Borrer at first refered our plant to *R. suberectus, And.* but after a subsequent re-examination writes "I doubt whether your *R. suberectus* can be *R. plicatus, W. & N.* since the latter is figured with recurved and somewhat hooked prickles, few, but yet more numerous than those of *R. suberectus,* and stalked lower leaflets. In all these characters the figure agrees with *R. nitidus, E. Fl. (R. plicatus, E. Bot. Suppl. R. suberectus β. Hook. Br. Fl. ed. 2nd & 3rd.)* Koehler's plant mentioned by Weihe & Nees in their "Adnot" to *R. plicatus* is probably *R. suberectus,* which I suspect the authors themselves never saw growing. Upon re-examination I am not at all certain that your plant which both Lindley and I called *R. suberectus* is not really *R. plicatus, E. Bot. Suppl.* as Nees says it is of *Rubi Germanici.* I still think the *true R. suberectus* not referable to any "species" in the German work."

My friend Mr. C. C. Babington, who examined the Smithian Herbarium at the Linnæan Society for this purpose, informs me that it does not contain any Shropshire specimen of *R. plicatus,* consequently it will be impossible to ascertain

what plant was meant in *English Flora.* The *R. nitidus* from Shropshire appeared to be a variety of *R. fruticosus.*

The habit of this plant is totally different from that of any other Bramble which I have ever seen, being erect, 3-4 feet high, of a bright succulent green, with dark-red short and thick prickles; the leaves very large, shining, and gracefully decurved, very much resembling in general appearance those of the Mulberry.

4. R. *fissus,* Lindl. *Cleft Bramble.* Stem reclining, angular, hairy; prickles scattered, not confined to the angles, numerous, slender, straight and somewhat curved and deflexed; leaves digitate, 5-nate, often pinnate, leaflets cordato-ovate or ovate-subcordate at the base, cuspidate, plicate, coriaceous, hairy above, paler and hoary (but not white) with dense pubescence beneath; lower pair sessile or nearly so; panicle nearly simple, corymbose, hairy, with few weak prickles; floral leaves ternate and oblongo-ovate, or simple and ovate, acute; calyx spreading in flower, erect and loosely clasping the fruit, nearly smooth and shining; berry small, nearly globular. *R. fissus, Lindl. Syn. 2nd ed.*

Moist boggy places; rare. *Fl.* July—September. ♄.

Boggy part of Almond park wood, near Shrewsbury, growing near *R. plicatus* and *idæus.*

Barren-stem reclining, angular, hairy and somewhat glaucous, with innumerable minute red sessile glands. *Prickles* numerous, slender, straight and somewhat curved, deflexed, unequal, very slightly dilated at the base, not confined to the angles of the stem. *Leaves* digitate, of 5 leaflets, the lowermost pair sessile or nearly so, overlapping the intermediate pair which is on short hairy petioles, often pinnate by the intervention of another pair of sessile leaflets below the central leaflet whose stalk is very long hairy and prickly; *common petiole* flattened channelled and hairy above, rounded beneath, with scattered recurved slender prickles which occur also on the mid-rib of the leaflets. *Central leaflet* broadly cordato-ovate cuspidate, when destitute of the additional pair of leaflets, but when these are present, ovate subcordate at the base cuspidate, sharply and deeply unequally serrated, coriaceous, plicate, of deep green but not shining, hairy on the upper surface, especially on the veins and at the base where the leaflet is united to the petiole, paler and somewhat hoary (but not white) beneath and covered with soft dense shining pubesence, somewhat fragrant. *Stipules* linear-lanceolate, ciliated and glandular. *Flowering stems* arched, obsoletely angular, reddish, smooth, except a few sessile reddish glands. *Prickles* numerous, slender, scattered, recurved. *Branches* simple or compound, lateral, alternate, spreading, each terminated by a simple bracteated raceme of erect *flowers,* the uppermost lateral ones generally overtopping the terminal flower; *panicle* hairy or downy, with a few scattered recurved prickles and sessile reddish glands. *Lower leaves* ternate, oblongo-ovate, acute, inciso-serrate, plicate, hairy chiefly on the veins above, paler and more hairy beneath, central one stalked, lateral leaflets sessile, *uppermost leaves* single, petiolate, ovate, acute, petioles downy. *Calyx* spreading in flower, erecto-patent and loosely clasping the fruit, segments generally broad and short with an acute short though often elongated point, densely hoary and white on the interior and edges, exterior of a shining green, smooth except a few scattered hairs and sessile glands with occasionally one or two weak prickles. *Petals* white, obovate, crumpled. *Fruit* small, nearly globular, bright red, of few grains, gratefully acid.

This, though found growing in the same locality with *R. plicatus* and partaking of many of its characters, is essentially different in its mode of growth, being of a weaker, straggling, and reclining habit, seldom rising more than 18

inches or 2 feet from the soil; whilst the former is erect and attains the height of 3 or 4 feet. Judging from a comparison with dried specimens of the true *suberectus,* kindly communicated by Mr. W. Wilson, it seems to approach that species very closely, particularly in its general appearance and form, and in the character of the lower pair of leaflets being sessile, though a closer examination will show it to be distinct, inasmuch as the prickles are not as in *suberectus* confined to the angles, but scattered on all sides of the stem, and the calyx of the fruit is erecto-patent, loosely enveloping the fruit, and not reflexed. These latter characters independently of its different growth separate it also from *R. plicatus.*

Of this Lindley says, "if an erect bush, it is a mere form of *suberectus;* possibly a hybrid between it and *vulgaris;* if its mode of growth be arched it is *R. fissus.* The two species, although different enough when growing, are much alike when dried, and in that state difficult to be distinguished."

Mr. Borrer writes, "The sessile lower leaflets seem to distinguish this from *R. plicatus, R. fastigiatus,* &c.; and if this difference hold good, and if it be not a more hairy variety of *R. suberectus,* it may bear Lindley's new name *fissus.*"

II. *Stem arched or prostrate, rooting, angular, smooth or nearly so; prickles chiefly confined to the angles.*

5. R. *affinis,* W. & N. *Related Bramble.* Stem arched, angular, prickly, smooth; leaves 5-nate, cordato-ovate, cuspidate, flat at the base, somewhat wavy towards the points, between pubescent and downy beneath; panicle compound with cymose branches; calyx naked, reflexed. *W. & N. t.* 3. *A. & B.*

Waste places and hedges; not uncommon. *Fl.* July—September. ♄.

The Shropshire specimens submitted to Nees ab Esenbeck were referred by him to his varieties β. and γ. of *R. affinis.* He writes, "Je n'ose pas de rapporter cette espèce au *R. plicatus,* puisque je n'en ai jamais trouvé des formes intermédiaires. Le *R. nitidus* en est une varietée." Of *var.* γ. Mr. Borrer says, "This I should call *R. corylifolius,* pretty confidently." This last certainly approaches very closely to the specimens of *R. rhamnifolius,* mentioned under that species as referred by Mr. Borrer to *corylifolius, Sm.,* but like them is totally different from the specimen of *R. dumetorum,* also referred to the same species by that Botanist.

Var. β. *W. & N.* Side of the road leading along the foot of Haughmond hill to the Abbey.

Stem arched, angular, furrowed, hairy, with a few minute reddish sessile glands. *Prickles* chiefly confined to the angles, slender, slightly curved, deflexed, dilated at the base. *Leaves* digitate, 5-nate, lowermost leaflets on short but distinct stalks, petioles hairy, with a few sessile glands, and numerous hooked prickles, continued also along the mid-ribs of the leaflets. *Central leaflet* on a longish stalk, ovate, narrower towards the base, cuspidate, coarsely serrated, dark-green above, nearly smooth or with a few scattered hairs, paler beneath, somewhat hoary, covered with a dense soft shining pubescence. *Flowering-stems* obsoletely angular, hairy, prickles small, slightly recurved, deflexed. *Panicle* long, straggling, compound, lower branches distant, upper ones somewhat cymose, hairy, somewhat hoary, with numerous sessile glands, prickles numerous, weak and slender. *Lower leaves* 5-nate, upper ones ternate, uppermost simple, large, ovate, subcordate at the base, petiolate. *Calyx* white, hairy and hoary on both sides, with a few minute glands on the exterior, usually unarmed, but occasionally with a solitary weak prickle, reflexed in fruit. *Fruit* of many large black grains.

Var. γ. *W. & N.* Codsall wood, near Albrighton, Shiffnal; *H. Bidwell, Esq.!*

Hedges on the road-side leading from Atcham towards Preston Boats. Hedges about the Flash, near Shrewsbury.

Stem rising to 3 or 4 feet high, then arching, green and shining, obsoletely angular, furrowed, smooth. *Prickles* scattered, short, weak, purple, decurved, dilated at the base. *Leaves* digitate, 5-nate, coriaceous, lowermost leaflets sessile, somewhat overlapping the intermediate pair. *Petiole* pubescent, with not very numerous, short, weak, hooked prickles, one or two of which appear on the mid-ribs of the leaflets. *Central leaflet* on a long stalk, broadly cordate, cuspidate, coarsely and unequally serrated, nearly smooth and shining above, white and hoary beneath with short dense shining velvety pubescence. *Flowering stems* obsoletely angular, pubescent, prickles few and scattered. *Panicle* long and straggling, compound, branches cymose, densely pubescent, somewhat hoary, prickles few, weak, scattered. *Lower floral leaves* 5-nate, upper ones 3-nate, uppermost simple, very large, rotundo-cordate, often 3-lobed, petiolate. *Calyx* reflexed both in flower and fruit, white and hoary on both sides with a few weak prickles at the base. *Fruit* of many large black grains.

6. R. *rhamnifolius,* W. & N. *Buckthorn-leaved Bramble.* Stem arched, angular and furrowed, nearly naked; prickles chiefly confined to the angles, straight, horizontal or deflexed, uniform, dilated at the base; leaves digitate, 5-nate, leaflets ovate or suborbiculari-cordate, cuspidate, coriaceous, slightly hairy above, hoary white and downy beneath, lowermost pair distinct or overlapping, nearly sessile; panicle long, compound, divaricate, hoary and hairy, prickly; floral leaves ternate and simple, often reduced to bracteas; calyx spreading in flower, reflexed in fruit, white and hoary, unarmed; berry large, of many grains, black. *R. rhamnifolius, W. & N. t.* 6. *E. Bot. Suppl. t.* 2604. *E. Fl. v. ii. p.* 401. *Lindl. Syn. 2nd ed. p.* 94.

Hedges, thickets, &c.; common. *Fl.* July—September. ♄.

Judging from a comparison of the specimens referred by Esenbeck, Lindley, and Borrer to this species, two very different forms appear to be included in it, of which I insert descriptions, premising that all these Botanists concur in naming the first *rhamnifolius,* but that Esenbeck and Lindley only refer the second to *rhamnifolius,* whilst Borrer calls it *R. corylifolius, Smith,* though totally different from the specimen mentioned under *R. dumetorum* which he likewise referred to *corylifolius, Sm.*

First form. *Barren-stem* arched, angular, furrowed, smooth, a few scattered hairs occasionally, sometimes very hairy with numerous minute glands interspersed, reddish. *Prickles* chiefly confined to the angles but not entirely, large, nearly uniform, straight, sometimes slightly deflexed, very much dilated at the base. *Leaves* digitate of 5 petiolate leaflets, thick and coriaceous, flat, lowermost pair distinct not overlapping on very short stalks; common petiole flattened and channelled above, rounded beneath, hairy, hairs tufted and spreading, armed with numerous very strong decurved prickles which are continued along the mid-ribs of the leaflets. *Central leaflet* on long petiole, ovate or suborbiculari-cordate, cuspidate, coarsely irregularly and acutely doubly serrated, with a few scattered hairs chiefly on the ribs on the upper surface, deep green, paler hoary or densely tomentose but not white beneath and with longer soft shining pubescence. *Stipules* linear-lanceolate, hairy and ciliated. *Flowering stems* obsoletely angular, somewhat furrowed, covered with clustered hairs, with a few glands interspersed. *Leaves* ternate, uppermost simple, ovate, gradually diminishing in size but appearing at the foot of each branch of the panicle up to its summit. *Panicle* long, compound, the lower branches distant, the upper ones closer together and as well as their subdivisions all divaricate, with a petiolate floral leaf at the base of each branch, the subdivisions bracteated, prickly and densely hairy, prickles numerous, deflexed, hairs clustered and spreading, the subdivisions almost hoary. *Calyx* densely hoary, hairy, shining and whitish on both sides without prickles or glands, spreading in flower, deflexed in fruit. *Fruit* of numerous small grains, black.

High Holborn, near Albrighton, Shiffnal; *H. Bidwell, Esq.!*

Sharpstones Hill.

Second form. *Barren-stem* arched, angular, furrowed, smooth, often purplish and glaucous. *Prickles* strong, straight or slightly deflexed, almost entirely on the angles with one or two on the faces, dilated at the base. *Leaves* digitate, 5-nate, the lowermost pair nearly sessile and overlapping the intermediate ones. *Central leaflet* broadly cordate, cuspidate, acutely coarsely and doubly serrated, nearly smooth above, white tomentose and with soft velvety pubescence beneath. *Stipules* linear-lanceolate, hoary and hairy. *Flowering-stem* obsoletely angular, white and hoary, prickly. *Leaves* ternate, uppermost simple, frequently reduced entirely to linear bracteas. *Panicle* as in the last, though often contracted and corymbose, white and hoary with minute starry clustered hairs and fewer prickles. *Calyx* white and hoary, unarmed, spreading in flower, deflexed in fruit. *Fruit* as in the preceding.

High Holborn, near Albrighton, Shiffnal; *H. Bidwell, Esq.!*

Hedges about the Flash, near Shrewsbury. Generally about Shrewsbury. Crowmeole, near Shrewsbury. Sharpstones Hill.

7. R. *discolor*, W. & N. *Discoloured Bramble.* Stem arched, angular, deeply furrowed, silky; flowering-stem hairy not white; prickles confined to the angles, large, stout, straight and horizontal, sometimes slightly curved and deflexed; leaves digitate, 5-nate, coriaceous, almost plane, lower leaflets on short stalks, central one rotundo-obovate, abrupt, with a long tapering point, (suborbicular acuminate *W. & N.*) coarsely doubly serrated, margins undulate, smooth and shining above, white and tomentose beneath, with reticulated veins; panicle decompound, dense at the summit, hairy, not white; floral leaves ternate, upper ones simple; calyx reflexed in flower and fruit, hoary, unarmed; fruit nearly globular, of many grains. *W. & N. t.* 20. *Lindl. Syn.* 1*st ed. p.* 93. *R. fruticosus, Lindl. Syn.* 2*nd ed. p.* 95. *Hook. Br. Fl.* 4*th ed. p.* 204.

Hedges, thickets, &c.; frequent. *Fl.* August, September. ♃.

Hedges at Crowmeole, and generally about Shrewsbury. Hedges between Atcham, and Preston Boats.

Barren-stem arched, angular, deeply furrowed, covered with stellate clusters of minute silky hairs. *Prickles* confined to the angles, large, very much dilated at the base, generally straight and horizontal, sometimes slightly curved and deflexed, silky. *Leaves* digitate, 5-nate, coriaceous, almost plane, lower pair on short footstalks, hairy as are all the others, and armed with numerous small hooked prickles, which are continued along the mid-ribs of the leaflets. *Common petiole* channelled above, with scattered minute stellate clusters of silky hairs, armed with hooked prickles. *Central leaflet* on a long stalk, rotundo-obovate, abrupt with a long tapering point, coarsely doubly serrated, margins somewhat undulate, smooth, shining and deep green above, white tomentose and veiny beneath. *Flowering stem* angular, hairy, with few scattered hooked prickles. *Panicle* decompound, dense and clustered at the summit, hairy but not white. *Calyx* reflexed in flower and fruit, segments broad with a very short point, hoary, without glands setæ or prickles. *Flowers* rose-coloured, often white, handsome. *Fruit* large, globular, of many grains, black.

Esenbeck remarks on this plant:—"*R. discolor, W. & N.* Bona species. In

tuo specimine folia solito minora sunt et medium rami sterilis foliolum obovatum. Ad sepes regionis Rhenanæ inferioris frequens occurrit hæc species, in Silesia eandem haud reperi."

8. R. *fruticosus*, Linn. *Shrubby Bramble.* Stem arched, angular, deeply furrowed, silky or smooth and glaucous; flowering stem hoary and white; prickles confined to the angles, large, stout, straight and horizontal, often curved and deflexed; leaves digitate, 5-nate, lower pair on short stalks, leaflets coriaceous, decurved at the margins and points, oblongo-ovate, narrower at the base, acuminate not abrupt, doubly serrated, smooth above, white hoary and veiny beneath; panicle decompound, narrow, elongated at the summit, white and hoary, floral leaves very large, truly oblongo-ovate; calyx spreading in flower and early fruit, finally deflexed, hoary, unarmed. *Hook. Br. Fl.* 4*th ed. p.* 204. *Lindl. Syn.* 1*st ed. p.* 92.

Hedges and thickets; frequent. *Fl.* July—September. ♄.

Hedges near Shrewsbury.

Barren-stem arched, angular, deeply furrowed, covered with stellate clusters of minute silky hairs, often smooth and glaucous. *Prickles* confined to the angles, large, very much dilated at the base, stout, straight and horizontal, often curved and deflexed, silky. *Leaves* digitate, 5-nate, coriaceous, decurved at the margins and points, lower pair on short downy footstalks, petiole downy armed with sharp hooked prickles. *Central leaflet* on longish stalk, oblongo-ovate, narrower at the base, thus somewhat inclining to obovate, acuminate, not abrupt as in the last, doubly serrated, margins and point decurved, smooth above, white hoary and veiny beneath. *Flowering-stem* angular, downy, white and hoary, with numerous very strong somewhat curved deflexed prickles. *Panicle* decompound, long, narrow, elongated and naked at the summit, branches erect and closer to the stem than in the last, white and hoary. *Calyx* spreading in flower and in the early stage of the fruit, but finally reflexed, segments broad with a very short point, white and hoary, without glands setæ or prickles. *Flowers* rose-coloured, handsome. *Floral leaves* very large, oblongo-ovate, acute, lower ones ternate, upper simple. The mature fruit I have not observed.

Probably this and the last are forms only of the same plant.

III. *Stem arched or prostrate, rooting, angular, hairy, but without glands or setæ.*

* *Prickles chiefly confined to the angles.*

9. R. *carpinifolius*, W. & N. *Hornbeam-leaved Bramble.* Stems decumbent or arched, angular and furrowed, hairy; prickles confined to the angles, uniform, deflexed, curved; leaves digitate, 5-nate, coriaceous, lower leaflets on very short stalks, central leaflet ovate, acuminate, plicate, slightly hairy above, pale green and velvety beneath; panicle compact, hairy, branches ascending, corymbose, hoary, hairy, inconspicuously glandular and prickly; calyx spreading in flower and fruit, segments with short points, hoary, hairy and inconspicuously glandular, unarmed; fruit small, of few loosely set grains, black. *W. & N. t.* 13. *E. Bot. Suppl. t.* 2664. *Hook. Br. Fl.* 4*th ed. p.* 203.

Hedges, thickets, &c. *Fl.* July—September. ♄.

Sharpstones Hill.

Barren stem decumbent or arched, angular and furrowed, usually much stained with purple, and with abundance of spreading somewhat fascicled soft hairs, remains of which are generally left in the flowering state of the plant, and a few minute glands. In our Shropshire plant the stem is nearly smooth, with a few scattered fascicles of hairs and a gland or two. *Prickles* numerous, confined to the angles, remarkably tipped with yellow, of moderate size, uniform, mostly deflexed and curved. *Petioles* hairy with a few glands interspersed, armed with numerous strongly hooked prickles. *Leaves* digitate 5-nate, leaflets on rather short stalks, the central one the longest, ovate, acuminate, plicate towards the lateral nerves, edges often somewhat deflexed, inciso-serrate, coriaceous, of a dark dull green and sprinkled with hairs above, beneath paler green and clothed with soft velvety shining hairs, those towards the ends of the shoots and on the flowering stems usually hoary or almost white. *Flowering-stems* angular, hairy and inconspicuously glandular, with not very numerous nearly straight deflexed prickles. *Leaves* ternate, upper ones simple. *Panicle* variously branched, mostly compact and narrow, occasionally almost simple and raceme-like, the principal stalk very hairy, the smaller branches and flower-stalks covered with dense whitish woolly pubescence intermixed with longer hairs, inconspicuous glands, small straight deflexed prickles, and often a few setæ. *Flowers* small, white. *Calyx* spreading, scarcely reflexed in any stage, points short rarely elongated, hoary and hairy, with a few inconspicuous glands, small straight prickles, and often a few setæ. The calyx of our Shropshire plant is without prickles or setæ. *Fruit* small, grains not very numerous or large, loosely set, of a full black, shining.

The above description was drawn up from that in *E. Bot. Suppl.*, compared with a Shropshire specimen of which Mr. Borrer says, "I think this *R. carpinifolius, E. Bot. Suppl.*, and a specimen of his from Weihe seems the same. My Sussex specimens have the barren shoots more hairy and the calyx prickly."

10. R. *leucostachys*, Sm. *White-spiked Bramble.* Stem arched, angular and furrowed, hairy; prickles confined to the angles, uniform, straight or curved, horizontal or deflexed; leaves digitate, 5-nate, leaflets flat and coriaceous, all stalked, central one roundish, ovate, cordate at the base, abruptly pointed, jagged and unequally serrated, hoary and with shining tawny hairs beneath; panicle elongated, narrow, densely shaggy, with inconspicuous nearly sessile imbedded glands, and armed with setæ and prickles; calyx densely hoary, shaggy and glandular, unarmed, reflexed in flower. *E. Fl. v. ii. p.* 403. *Lindl. Syn.* 2*nd ed. p.* 95. *Hook. Br. Fl.* 4*th ed. p.* 204. *E. Bot. Suppl. t.* 2631.

Hedges, thickets, &c.; frequent. *Fl.* July, September. ♃.

Albrighton, near Shiffnal; *H. Bidwell, Esq.!*

Hedges near Shrewsbury. Sharpstones Hill.

Barren-stem arched, angular and furrowed, more or less hairy, often nearly naked. *Prickles* confined to the angles, large, tolerably numerous, uniform, straight or curved, horizontal or deflexed. *Leaves* digitate, 5-nate, on hairy petioles, armed with numerous strong decurved prickles, leaflets flat and coriaceous, all stalked, roundish ovate, mostly cordate at the base, with an abrupt point, jagged and unequally serrated, upper surface dark green and even with a few scattered hairs, hoary beneath with shining tawny hairs. *Flowering stem* angular, somewhat furrowed, hairy, with numerous small nearly sessile glands which never exceed the length of the hairs. *Leaves* chiefly ternate. *Panicle* elongated and narrow, the branches short, contracted, more or less spreading, the lowermost often more elongated and very compound, densely shaggy and shining, with numerous inconspicuous small nearly sessile glands imbedded among the hairs and setæ, pric-

kles more or less numerous slender and deflexed. *Calyx* densely hoary and shaggy, with inconspicuous imbedded glands and occasionally a seta or two, reflexed. *Petals* white.

Determined by Professor Lindley.

11. R. *vulgaris*, W. & N. *Common Bramble.* Stem arched, angular, furrowed, more or less hairy; prickles strong, straight, deflexed, confined to the angles; leaves digitate, 5-nate, lower leaflets on very short stalks, leaflets elliptical or suborbicular cuspidate, obtusely wedge-shaped or subcordate at the base, slightly hairy above, green and pubescent beneath; panicle short, branches simple or compound, spreading, naked or without leaves at the summit, pubescent or hairy, glandular and prickly; floral leaves ternate or simple, confined to the lower branches; calyx reflexed in fruit, tomentose, hoary on the edges and interior, glandular and prickly. *R. vulgaris, W. & N. t.* 14. *Lindl. Syn.* 2*nd ed. p.* 93.

Wastes, thickets, &c.; *Fl.* July—September. ♄.

Sides of the lane between Atcham and Preston Boats, near Shrewsbury. Almond Park, near Shrewsbury. Sharpstones Hill. Shawbury Heath.

Barren-stem arched, angular, furrowed, more or less hairy and glandular, glands often inconspicuous or wanting, with or without setæ. *Prickles* strong, straight, deflexed, confined to the angles. *Leaves* digitate, 5-nate, lower pair of leaflets on very short stalks. *Petioles* hairy, with a few glandular bristles interspersed, and more or less numerous hooked and deflexed prickles which are continued along the mid-ribs of the leaflets. *Central leaflet* elliptical or suborbicular, cuspidate, obtusely wedge-shaped or subcordate at the base, coarsely but simply and uniformly serrated, with a few scattered hairs on the upper surface, green with short soft pubescence and hairy veins beneath. *Flowering-stem* more or less hairy, glandular, bristly and prickly. *Panicle* short and few-flowered, branches simple or compound, spreading, one or two of the lower ones accompanied by a ternate or simple leaf, the rest only bracteated, densely pubescent or hairy, glandular and prickly. *Calyx* of the fruit reflexed, greenish, tomentose, hoary on the edges and inside, glandular, with scattered weak bristles or prickles, especially at the base. The mature fruit I have not seen.

Our specimens were determined by Lindley and Esenbeck. Mr. Borrer writes of this plant: "I do not know *R. vulgaris.* If suberect not rooting, *R. plicatus, E. Bot. Suppl.* (and as I still believe *W. & N.*) If arched and rooting, the doubtful plant mentioned in *E. Bot. Suppl.* under *R. plicatus*, and probably *R. nitidus, W. & N.*" Lindley in *Syn.* 2*nd ed. p.* 93. quotes to this, *R. corylifolius, Sm.*, and observes that it is extremely variable in the roundness of the leaflets, and their degree of hairiness.

The suborbicular, cuspidate leaves, and the short panicle, naked at the summit, appear to be the prevailing characteristics of this species.

** *Prickles not confined to the angles.*

12. R. *villicaulis*, Koehler. *Hairy-stemmed Bramble.* Stem prostrate or arched, angular, hairy and shaggy; prickles not confined to the angles, unequal, long, slender, straight and horizontal, sometimes curved and deflexed, often hairy; leaves digitate, 5-nate, lower leaflets on short stalks, central leaflet orbicular, subcordate at the base, shortly cuspidate, hairy above, hoary and whitish beneath with

soft shining hairs; panicle compound, narrow, naked at the summit, branches divaricate, hoary, hairy, glandular and very prickly; floral leaves ternate, upper ones large, orbicular, entire or 3-lobed; calyx spreading in flower, reflexed in fruit, hoary, hairy, glandular, setose and prickly; fruit large, grains numerous, black.

Hedges, woods, &c.; not very common. *Fl.* July—September. ♃.

Crowmeole, near Shrewsbury.

Stem prostrate unless supported, then arched and declining, angular, hairy and shaggy. *Prickles* numerous, not confined to the angles, unequal, long, slender, dilated at the base, generally straight and horizontal, sometimes slightly curved and deflexed, often hairy. *Leaves* digitate, 5-nate, thick and coriaceous, petiolate, lower leaflets on very short stalks, central one on a longer one. *Petiole* hairy, with numerous strong straight and decurved prickles. *Central leaflet* orbicular, subcordate at the base, shortly cuspidate, coarsely serrated, hairy above, hoary and whitish beneath with soft shining hairs. *Flowering-stem* obsoletely angular, hairy, with numerous unequal straight or curved and deflexed long slender prickles. *Panicle* compound, narrow, straggling, naked at the summit, branches divaricate, clothed with shining hoary hairs, and numerous very long slender stout straight deflexed prickles with glandular setæ interspersed. *Lower floral leaves* ternate, upper ones large and orbicular, entire or 3-lobed. *Calyx* hairy, hoary, glandular, setose and prickly on the exterior, hoary within, segments spreading in flower, reflexed in fruit. *Flowers* large and white. *Fruit* large, of numerous small grains, black.

Esenbeck writes thus of our Shropshire plant: "*R. villicaulis*, Koehler. *var. rotundifolia*, foliolis brevicuspidulatis ad quam spectat *R. vulgaris*, ε. *mollis*, *W. & N.* Folia in tuis minora sunt et acumine admodum brevi prædita. Ad *R. villicaulem*, pertinent *R. macroacanthus*, *W. & N*, *R. pubescens*, *W. & N*, et probabiliter *R. sylvaticus*, *W. & N.*"

IV. *Stem arched or prostrate, rooting, angular, hairy glandular and setose.*

* *Prickles chiefly confined to the angles.*

13. R. *Radula*, W. & N. *Hairy Bramble.* Stem arched, angular, furrowed, very hairy, with innumerable glands and setæ; prickles chiefly but not entirely confined to the angles, unequal, stout, recurved; leaves digitate, 5-nate, thick opaque and coriaceous, lower leaflets on short stalks, central one ovato-acuminate, jagged, coarsely serrated, smooth above, white and softly pubescent beneath; panicle more or less elongated, naked at the summit, decompound, branches divaricate, densely hairy, shaggy, glandular, with scarcely any setæ, but with numerous long straight deflexed prickles; calyx spreading in flower, reflexed in fruit, segments with very long points, white, hoary, hairy, glandular and setose; fruit large, of many large black grains. *W. & N. t.* 39.

Hedges, thickets, &c.; frequent. *Fl.* July—September. ♃.

Harriett's Hayes, near Albrighton, Shiffnal; *H. Bidwell, Esq.!*

Sharpstones Hill. Preston Boats, near Shrewsbury. Haughmond Abbey.

Barren-stem arched, angular, furrowed, very hairy, and with innumerable glandular bristles and setæ. *Prickles* chiefly but not entirely confined to the angles, unequal, stout, recurved. *Petioles* hairy glandular and setose, with many strong recurved prickles. *Leaves* digitate, 5-nate, thick opaque and coriaceous, lower

leaflets on short stalks, central one on longer stalk, ovato-acuminate, jagged, strongly serrated, smooth above, white beneath with soft shining velvety pubescence. *Flowering-stem* obsoletely angular, densely hairy, glandular and setose, with many strong recurved scattered prickles. *Panicle* more or less elongated, naked at the summit, decompound, branches divaricate, densely hairy shaggy and glandular, with or without very few setæ, but with many long slender straight or slightly curved deflexed prickles. *Calyx* reflexed in fruit, segments with very long points, white, hoary, hairy and glandular, with a few scattered setæ. *Flowers* white. *Fruit* large, globular, of many large grains, black.

"*Rubus Radula*, *W. & N.* Bona species, complectens *R. rudem*, *Hystricem* and *pygmæum*, *W. & N.*" *Esenbeck.*

"Very different from the specimens which Lindley called states of *R. Radula.* I should think it a state of *R. Kohleri*, or perhaps of *R. echinatus*, *Lindl.* (*Syn. ed.* 1) if that is to be distinguished." *Borrer.*

The specimens marked *R. Radula* by Professor Lindley approach in the form and clothing of the leaves that named by him *R. diversifolius*, but referred by Esenbeck to *R. dumetorum*, and described as such in this work, but differ from that in the prickles being more uniform and not passing insensibly into the setæ, which however are present and very numerous. Specimens from hedges near Upper Berwick named *R. rudis*, *W. & N.* by Lindley correspond with those named *R. Radula* by Esenbeck, except in their having more numerous and crowded prickles on the barren stems. These, however, are likewise referred by Mr. Borrer to *R. Kohleri*, with a remark that "a specimen from Weihe as *R. rudis* is not the same as this or that referred by Lindley to his *R. echinatus*, the numerous setæ on it being remarkably regular in length and in dispersion."

14. R. *Leightoni*, Lees' MSS. *Leighton's Bramble.* Stem arched, angular and furrowed, hairy, copiously glandular and setose; prickles chiefly confined to the angles, nearly uniform, straight, horizontal; leaves digitate, 5-nate, thin and flexible, lower leaflets petiolate, central leaflet rotundo-obovate, somewhat cuneate, abrupt and sinuato-serrated at the anterior margin, with a short cuspidate point, smooth above, grey with soft close pubescence beneath; panicle moderately long, straggling, naked at the summit, compound, branches divaricate, densely hairy, glandular, setose and prickly; calyx spreading in flower, reflexed in fruit, hoary, hairy, glandular, setose and prickly; flowers pink.

In the hedge near the upper part of the field in front of the Can Office, Shrewsbury, on the left hand side of the footpath leading down the field to the ferry over the river Severn. *Fl.* July. ♃.

Barren-stem arched, angular, furrowed, green, clothed with scattered hairs, copious pedicellate glands and numerous short setæ. *Prickles* nearly uniform, chiefly but not entirely confined to the angles, straight and horizontal, dilated at the base. *Petioles* with copious glands and setæ, a few hairs and hooked prickles, the partial stalks with more numerous hairs glands setæ and prickles, the latter being also continued along the mid-ribs of the leaflets. *Leaves* digitate, 5-nate, thin and flexible, lower leaflets on short stalks, central one on a very long one, rotundo-obovate somewhat cuneate at the base, the anterior margin very abrupt and truncate, sinuato-serrated, terminated by a short narrow cuspidate point, very sharply and irregularly serrated, smooth and pale-green above, grey with soft close pubescence beneath. *Flowering-stem* nearly round, very hairy and setose, with scattered glands and straight deflexed prickles. *Leaves* ternate, upper ones single. *Panicle* moderately long, rather straggling, naked at the summit, compound,

branches divaricate, densely hairy and shaggy, with innumerable glands and a few scattered setæ interspersed, prickles very numerous, long, slender, straight or slightly curved, deflexed. *Calyx* spreading in flower, reflexed in fruit, very hoary hairy glandular and setose, armed with many scattered short weak prickles, segments broadish with an attenuated point. *Flowers* pink. The mature fruit I have never seen.

This was gathered by me in July 1836 and specimens forwarded to Professor Lindley, Esenbeck and Mr. Borrer whose opinions I subjoin. "I think a hybrid between *R. fruticosus* and *leucostachys* or *R. fruticosus* and *rudis* perhaps rather the latter:—if really a plant of common occurrence, like the others, I should say it is a new form of the genus, combining the characters of my sections III** and V, *Synopsis 2nd ed.*" Lindley.

"*R. Kohleri var* :?, although resembling *R. fruticosus* in shape and hoariness of leaf." Borrer.

"*R. Lejeunei*, *W. & N.* affinis, nescio an species an forma tantummodo, distincta folioli medii rami sterilis figura ad *R. Linguæ* ejusdem hujus folioli figuram accedente. Inflorescentia est *R. Lejeunei.*" Esenbeck.

The bush from which the specimens were gathered has through changes consequent on the sale of the property been unfortunately cut down, and I have in vain repeatedly and anxiously searched the spot during the last two summers for its reappearance. A description has been here introduced with a view of directing the attention of Botanists to its rediscovery in other portions of the County. The present name was assigned to it by my friend Mr. Edwin Lees, of Worcester, whose enthusiastic and zealous labours in this difficult tribe have been successfully rewarded by an extensive and accurate knowledge of our British forms. Weihe and Esenbeck's descriptions of the leaves of *R. Lingua* is "foliis quinatis, foliolis cuneatis cuspidatis inæqualiter attenuato-serratis glabriusculis;" and of the inflorescence of *R. Lejeunei*, "panicula divaricata, floribus magnis roseis."

** *Prickles not confined to the angles, passing insensibly into setæ.*

15. R. *Kohleri*, W. & N. *Kohler's Bramble.* Stem decurved, somewhat angular and furrowed, very hairy, glandular and setose; prickles numerous, unequal, passing insensibly into setæ, decurved or straight; leaves digitate 5-nate, lower leaflets on short stalks, central one obovate with a long acuminate point, coarsely and unequally serrated, with dense soft shining pubescence beneath; panicle moderately long, decompound, naked at the summit, branches divaricate, corymbose, densely hairy, glandular, setose and very prickly, bracteas foliaceous; calyx reflexed in flower, spreading in fruit, hoary, hairy, glandular, setose and prickly; fruit of numerous small grains, black. *W. & N. t.* 25. *E. Bot. Suppl. t.* 2605. *Lindl. Syn. p.* 94. *R. Kohleri, a, Hook. Br. Fl. 4th ed. p.* 204. *R. glandulosus, E. Fl. v. ii. p.* 403. *(excl. syn.)*

Wastes, thickets, &c.; not common. *Fl.* July—September. ♃.

Coalbrookdale; *Mr. F. Dickinson!* Wyre Forest, *E. Lees, Esq.*

Base of the Wrekin.

Barren-stem decurved or prostrate, somewhat angular and furrowed, very hairy, glandular and setose. *Prickles* copiously scattered on every part of the stem, unequal, passing insensibly into setæ which are very numerous decurved or straight. *Petioles* hairy, glandular, and setose, with numerous hooked prickles, which are also continued along the mid-ribs of the leaflets. *Leaves* digitate, 5-nate, lower leaflets on short footstalks, central one on a longer one, obovate with a

long acuminate point, coarsely and irregularly serrated, dark green and with scattered hairs above, paler and with dense soft shining pubescence beneath. *Flowering-stem* obsoletely angular, almost round, densely hairy and shaggy, with innumerable setæ glands and decurved prickles. *Leaves* ternate, the central leaflet much more obovate than those on the barren stem with the anterior margin jagged, uppermost or floral ones often simple. *Panicle* moderately long, naked at the summit, decompound, branches divaricate, corymbose, densely hairy, glandular, setose, and very prickly, with numerous trifid foliaceous bracteas interspersed. *Calyx* reflexed in flower, spreading in fruit, hoary, hairy, glandular, setose and prickly, segments often with very elongated points. *Petals* white or pale pink. *Fruit* black and shining, of numerous small grains.

Determined by Professor Lindley.

16. R. *echinatus*, Lindl. *Echinate Bramble.* Stem arched, angular, densely covered with glands and setæ mixed with numerous short nearly straight prickles; leaflets 5 roundish cordate, coarsely and unequally serrated, taper-pointed, green and velvety beneath; panicle spreading, prickly and glandular, leafy at the base, the branches corymbose, bracteæ 3-toothed and entire; prickles of the peduncle scattered, with very few setæ. *Lindl. Syn. ed. i. p.* 94.

Thickets, &c.; not common. *Fl.* July—September. ♃.

Almond Park, near Shrewsbury.

Barren-stem arched, angular, somewhat furrowed, without hairs, but densely covered with copious glands, and innumerable setæ interspersed, with scattered, short, straight, deflexed prickles, which are very numerous and unequal and pass insensibly into the setæ. *Petioles* hairy, glandular, and setose, with many strong unequal decurved prickles. *Leaves* digitate, 5-nate, thin and flexible, lower leaflets on very short stalks arising from those of the intermediate pair; central leaflet on a long stalk, rotundo-ovate, somewhat narrower at the base and cordate, with a long acuminate point, coarsely and unequally serrated, with a few scattered hairs above, paler beneath, green, with soft shining pubescence. *Flowering-stem* obsoletely angular, slightly hairy, with many glands and setæ, and numerous straight deflexed prickles. *Panicle* spreading, hairy, glandular and setose, crowded with long slender straight prickles. *Calyx* hoary, densely armed with glands setæ and prickles, segments with long leafy points. *Bracteas* 3-fid.

Our specimens were determined by Professor Lindley and confirmed by Mr. Borrer, who says, "I have taken this for *R. echinatus*, *Lindl.* and have doubted whether it is a mere variety of *R. Kohleri.*"

17. R. *fusco-ater*, W. and N. *Brownish-black Bramble.* Stem prostrate, obsoletely angular, hairy, glandular and setose; prickles scattered, unequal, recurved and deflexed; leaves pedate 5-nate, lower leaflets arising from petioles of intermediate pair, central leaflet elliptical, subcordate at the base, with a long acuminate point, slightly hairy above, pale green and hairy on the veins beneath; panicle short, naked at the summit, densely hairy, glandular and setose, lower branches racemose; bracteas lanceolate, simple or 2-3-fid; calyx reflexed in flower, spreading in fruit, hoary, hairy, glandular and setose. *W. & N. t.* 26. *Lindl. Syn. 2nd ed. p.* 93.

Woods, &c.; *Fl.* July—September. ♄.

Buildwas; *Miss H. Moseley!* Albrighton, near Shiffnal; *H. Bidwell, Esq!*

Woods at the base of the Wrekin. Haughmond Abbey.

Barren stem decurved or prostrate, obsoletely angular, furrowed, very hairy glandular and setose. *Prickles* numerous, scattered, unequal, somewhat recurved and deflexed. *Leaves* pedate 5-nate, lower leaflets on very short stalks arising from those of the intermediate pair which are frequently 2-lobed in consequence of the adhesion of the lower leaflets. Common and partial *petioles* hairy, glandular and setose, with scattered recurved prickles which are also continued along the midribs of the leaflets. *Central leaflet* elliptical, slightly narrower and subcordate at the base, with a very long acuminate point, coarsely serrated, pale opaque green and with scattered soft hairs above, paler green beneath, the veins very pale and conspicuous, hairy. *Flowering stem* obsoletely angular, very hairy, glandular setose and prickly, the prickles small, scattered, straight or curved and deflexed. *Panicle* short, loose and straggling, naked at the summit, densely hairy glandular and setose, the lower branches racemose, interspersed with lanceolate simple or 2-3-fid *bracteas*. *Floral leaves* ternate, petiolate. *Calyx* reflexed in flower, spreading in fruit, hoary, hairy, glandular and setose, segments with elongated points. *Flowers* white.

Professor Lindley refers this to his *R. fusco-ater*. Mr. Borrer writes "I should say *R. pallidus*. I refer both *R. pallidus* and *R. fusco-ater* to *R. Kohleri*."

18. R. *pallidus*, W. & N. *Pale Bramble*. Stem prostrate, obsoletely angular, hairy, glandular, and setose; prickles scattered, distant, nearly uniform, straight, horizontal; leaves pedate 5-nate, lower leaflets arising from petioles of intermediate ones; leaflets very elongated and narrow, central one oblongo-elliptical with a long acuminate curved point, hairy above, pale green and densely pubescent beneath; panicle short, naked at the summit, hairy, glandular, prickly and with few setæ, lower branches in corymbose racemes, bracteas lanceolate, simple or 2-3-fid; calyx reflexed in flower, somewhat spreading in fruit, very hoary, hairy, glandular and setose; fruit of few large grains, black. *W. & N. t.* 29. *Lindl. Syn. 1st ed. p.* 94. *R. Kohleri γ. pallidus. Hook. Br. Fl. 4th ed. p.* 204. *(excl. syn. of E. Fl.)*

Woods, thickets &c.; frequent. *Fl.* July—September. ♄.

Foot of Haughmond Hill.

Barren stem decurved or prostrate, obsoletely angular and furrowed, very hairy, glandular and setose. *Prickles* not so numerous as in the last, more scattered, distant and less unequal, straight and horizontal, dilated at the base. *Leaves* pedate 5-nate, lower leaflets on very short stalks arising from those of the intermediate pair which are frequently 2-lobed from the adhesion of the lower ones. Common and partial *petioles* hairy glandular and setose, with scattered straight and curved deflexed prickles which are also continued along the midribs of the leaflets. *Leaflets* narrow and very much elongated, central one on a long stalk, oblongo-elliptical, with a long acuminate curved point, coarsely and doubly serrated, upper surface with scattered hairs, paler beneath and densely pubescent, veins pale conspicuous and hairy. *Flowering stems* obsoletely angular, hairy, glandular, very setose and prickly, the prickles larger, more scattered, straight and deflexed. *Panicle* short, straggling, naked at the summit, hairy, glandular and prickly, with a few setæ, lower branches in corymbose racemes, interspersed with lanceolate simple 2-3-fid *bracteas*. *Floral leaves* ternate and simple, petiolate. *Calyx* reflexed in flower, somewhat spreading in fruit, very hoary, hairy, glandular and setose, segments broader and with shorter points than the last. *Flowers* white. *Fruit* of a few large black grains.

Very nearly allied in habit and appearance to the last, and with that in all

probability forms of the same plant varying from situation. Esenbeck says of it "*R. pallidus W. & N.* Dubia species, sed nostris speciminibus bene congrua."

19. R. *Schleicheri*, W. & N. *Schleicher's Bramble*. Stem arched, obsoletely angular, nearly round, with scattered hairs and glandular bristles; prickles scattered, very unequal, diminishing insensibly into setæ, straight and horizontal or slightly recurved; leaves quinato-pedate or ternate, lateral leaflets often 2-lobed, obovato-acuminate, central leaflet roundish, narrower and subcordate at the base, acuminate, hairy above, green soft and pubescent beneath; panicle compound, upper branches single-flowered, very hairy, glandular, setose and prickly; floral leaves ternate, central leaflet obovato-acuminate, upper ones simple; calyx reflexed in flower, erect and clasping the fruit, hoary, hairy, glandular and setose; fruit nearly globular, grains large, black. *W. & N. t.* 23.

Hedges, &c.; *Fl.* July—September. ♄.

Hedges between Atcham and Preston Boats, near Shrewsbury. Sharpstones Hill.

Barren stem arched, obsoletely angular, nearly round, clothed with scattered hairs and glandular bristles. *Prickles* scattered, very unequal, diminishing insensibly into innumerable setæ, straight and horizontal or slightly recurved. *Leaves* quinato-pedate or ternate, the lateral leaflets somewhat 2-lobed, obovato-acuminate, on very short hairy stalks, central leaflet on a longer stalk which as well as the common petiole is hairy, with a few glandular bristles interspersed, prickles sharp and recurved continued along the midribs. *Central leaflet* roundish, narrower and subcordate at the base, sometimes elliptical, nearly obovate, acuminate, irregularly serrated, hairy above, green soft and pubescent beneath. *Flowering stem* obsoletely angular, hairy, very glandular and setose, with numerous unequal straight or slightly curved deflexed prickles. *Panicle* compound, branches erect, upper ones single-flowered, very hairy, glandular setose and prickly. *Floral leaves* ternate, central leaflet obovate-acuminate, upper ones simple 3-lobed or entire. *Calyx* reflexed in flower, erect and clasping the fruit, segments with rather long points, hoary, hairy, glandular and setose, the setæ most numerous about the base. *Flowers* large, white. *Fruit* nearly globular, grains large, black.

"*R. Schleicheri W. & N.* Videtur bona species." *Esenbeck.* "I cannot distinguish it from *R. dumetorum, W. & N.*" *Borrer.*

20. R. *dumetorum*, W. & N. *Bramble of the thickets*. Stem arched, angular, prickly, hairy and glandular; prickles scattered, straight, diminishing insensibly into setæ; leaves 5-nate, leaflets orbicular, narrower and cordate at the base, acuminate, sharply and irregularly serrated, rugose and hairy above, with dense, velvety, pale (but not white) pubescence beneath; panicle long and straggling, hairy, glandular, setose and prickly, with short contracted branches; floral leaves ternate or single; sepals hoary, hairy and glandular, with a few setæ about the base, reflexed in flower, erect in fruit; fruit of few large and black grains. *R. dumetorum W. & N. t.* 45. *A. & B. R. diversifolius, Lindl. Syn. 2nd ed. p.* 94.

Hedges, &c.; common. *Fl.* June—September. ♄.

Pattingham; Albrighton near Shiffnal; *H. Bidwell, Esq.!*

Haughmond hill. Sharpstones hill.

Stem several feet in length, arched, angular, prickly, hairy and glandular, the *prickles* straight, dilated at the base, not confined to the angles, very variable in size, diminishing insensibly into setæ. *Petioles* hairy, with a few glands, channelled above, armed with 3 rows of uniform decurved prickles. *Stipules* linear-lanceolate, acuminate, hairy and glandular. *Leaflets* 5, central one on a hairy slightly glandular and prickly stalk, intermediate pair unequal at the base on very short stalks, lowermost pair smaller equal at the base sessile on the stalks of the intermediate pair; central leaflet very large, orbicular, narrower and heartshaped at the base, acuminated, sharply and irregularly serrated, thick, rugose and slightly hairy above, under surface paler, covered with dense soft velvety (but not white) pubescence, the mid-rib armed with small hooked prickles. *Panicle* long and straggling, hairy, glandular setose and prickly, branches contracted, shorter than the ternate or single leaf from the axil of which they arise, hairy, sometimes nearly tomentose, glandular setose and prickly. *Flowers* few, on short hairy glandular and prickly bracteated pedicels. *Sepals* hoary hairy and glandular, with a few setæ about the base, acuminate, very white and hoary on the margins and interior, reflexed in flower and during the early state of the fruit, but becoming erecto-patent and often clasping the fruit as it advances towards maturity. *Petals* white, delicate, veiny and wrinkled. *Fruit* of few grains, large and black.

Specimens of this plant submitted to Prof. Lindley were pronounced by him to be his *R. diversifolius*. Of similar specimens Mr. Borrer says "*R dumetorum W. & N.* I incline to refer it to *R. corylifolius, Smith.* although in some respects it is more like *R. cæsius*. I have a very different thing from the garden of the Horticultural Society as from the authentic bush of *R. diversifolius*."

Specimens in every respect similiar but from a different locality were sent to Prof. Nees ab Esenbeck who says of them: "*R. dumetorum, W. & N. var β. nemorosus* et *var. β. nemorosus* ad *var. α ferocem* accedens; si calyces fructûs sint erecti."

From the latter locality specimens were forwarded to Mr. Borrer with Esenbeck's remarks; he also concurs in naming them *R. dumetorum W. & N.* and identifies them with the specimens marked by Lindley *R. diversifolius, Lindl.* In Bluff & Fingerh. Comp. Fl. Germ. 2nd ed. vol. 1. pt. 2. p. 190. *R. corylifolius, Sm.* is quoted as a synonyme to *R. dumetorum, W. & N. α vulgaris.* Lindley in 1st ed. of his Syn. says that the figure of *R. corylifolius, Sm.* in E. Bot. t. 827, represents *R. vulgaris, W. & N.* Our plant certainly does not correspond with that of *E. Bot.*

V. *Stem prostrate, rooting, round, glaucous.*

21. R. *cæsius* Linn. *Dew-berry*. Stem prostrate, round or nearly so, glaucous; prickles scattered, slender, straight, deflexed, unequal, passing insensibly into setæ; leaves digitate 3-nate, more rarely 5-nate, with the outermost pair sessile, leaflets broadly ovate, cuspidate, doubly serrated, thin and flexible, slightly hairy above, paler and downy beneath; panicle corymbose, prickly, downy, setose and glandular; floral leaves ternate; calyx spreading in flower, erect and clasping the fruit, hoary and glandular; berry of few grains, large, black and glaucous. *E. Bot. t.* 826. *E. Fl. v. ii. p.* 409. *Hook. Br. Fl. 4th ed. p.* 205. *W. & N. t.* 46. *A. B. & C.*

Ditches, borders of fields, hedges, &c.; frequent. *Fl.* July—September. ♄.

Frequent about Albrighton (Shiffnal) *H. Bidwell, Esq!* Coalbrookdale; *Mr. F. Dickinson.*

Roadside near the Flash; Crowmeole, and generally about Shrewsbury.

Barren stem prostrate, trailing, weak, round, glaucous. *Prickles* scattered, slender, straight, deflexed, unequal, diminishing insensibly into setæ. *Leaves* ternate, lowermost leaflets on very short hairy footstalks, central one on long hairy somewhat prickly stalk; sometimes there is an additional pair of sessile leaflets. Common *petiole* flattened, channelled and hairy above, rounded beneath, with scattered deflexed prickles. *Central leaflet* broadly ovate, cuspidate, doubly serrated, thin and flexible, deep green, with a few scattered hairs, paler and downy beneath. *Stipules* linear-lanceolate, ciliated, hairy and glandular. *Flowering stem* erect, round or nearly so, glabrous, sometimes pubescent, with setæ and pedicellate glands especially in the upper part; prickles weak, deflexed, slightly curved. *Leaves* ternate, the uppermost single 3-lobed. *Flowers* in terminal and axillary corymbs, peduncles pubescent, prickly setose and glandular, interspersed with small scattered bracteas. *Calyx* spreading in flower, erect and clasping the fruit, segments large, ovate, with a long point, hoary glandular and green on the exterior, except the edges which are white, with a few setæ near the base, very white and hoary on the inside. *Petals* white, obovate, crumpled. *Fruit* of few large grains, black, glaucous.

R. cæsius, Linn. on the authority of Nees ab Esenbeck.

B. *Stem herbaceous or nearly so.*

22. R. *saxatilis*, Linn. Stone Bramble. Barren stem flagelliform, prostrate, unarmed, hairy; flowering stem erect, simple, herbaceous, angular, hairy or downy; prickles few, weak, straight, and horizontal; leaves ternate, leaflets ovate, acuminate, coarsely inciso-serrate, slightly hairy above, paler and hairy on the veins beneath; panicle terminal, corymbose, of few flowers; calyx spreading in flower, inflexed in fruit, slightly hairy, unarmed; fruit of few large crimson grains. *W. & N. t.* 9. *E. Bot.* 2233. *E. Fl. v. ii. p.* 411. *Hook. Br. Fl. 4th ed. p.* 205.

Stony mountainous places; rare. *Fl.* June. ♄.

Stiperstones hill; *A. Aikin, Esq.*

"*Root* rather woody. *Herb* of a light green, slightly downy or hairy, not hoary, throwing out a few very long round trailing *runners*, either naked or leafy, taking root at the extremity, where they, in the following spring, send up one or two simple herbaceous *flowering stems* 3—6 inches high, which are angular, slightly hairy, and often armed with a few small, weak, spreading *prickles*. These stems bear 2 or 3 alternate, long-stalked, ternate, serrated or notched *leaves*, not unlike those of a strawberry, the mode of growth of the two plants being also very similar. *Stipules* elliptic-oblong. *Panicle* terminal, corymbose, simple, downy, seldom a little prickly, of 3 or 4 upright small, greenish-white flowers. *Calyx* angular at the base; its segments oblong, downy within, externally somewhat hairy, but not at all prickly or glandular; spreading in flower, inflexed but not closely in fruit. *Fruit* of 1, 2 or 3 large crimson grains, agreeably acid, but not perfumed. *Seeds* large, tumid, pitted, and elegantly wrinkled." *Smith.*

8. FRAGARIA. *Linn.* Strawberry.

1. F. *vesca*, Linn. *Wood Strawberry*. Calyx of the fruit reflexed; hairs of the peduncles widely spreading, those of the pedicels close-pressed, silky. *E. Bot. t.* 1524. *E. Fl. v. ii. p.* 414. *Hook. Br. Fl. p.* 253.

Woods, thickets, and hedge banks; frequent. *Fl.* May, June. ♃.

Root rather woody, sending forth trailing runners which root at intervals. *Stems* 4-8 inches high, erect, clothed with soft spreading hairs. *Leaves* ternate, chiefly radical, on long channelled petioles covered with spreading hairs, leaflets oblong, wedge-shaped, coarsely toothed, lateral leaflets unequal at the base, nearly sessile, all more or less hairy on both sides. *Flowers* erect, white, the pedicel of the lowest or first expanded flower always with spreading hairs whilst those of the subsequent flowers are appressed. *Fruit* drooping, deep scarlet, gratefully acid.

9. COMARUM. *Linn.* Marsh Cinquefoil.

1. C. *palustre*, Linn. *Purple Marsh Cinquefoil.* Root creeping; stem ascending; leaves pinnate, leaflets 5-7 lanceolate, deeply and acutely serrated, more or less hairy above, very hairy and hoary beneath; petals lanceolate, acuminate, much shorter than the calyx. *E. Bot. t.* 172. *E. Fl. v. ii. p.* 434. *Hook. Br. Fl. p.* 253.

Spongy marshes, bogs and sides of pools; frequent. *Fl.* July. ♃.

Pits about Albrighton, near Shiffnal; *H. Bidwell, Esq.* Ellesmere; *Rev. A. Bloxam.* Near Longnor; *Rev. W. Corbett.* Walford; *T. C. Eyton, Esq.* Pond near the Canal, Church Aston; *R. G. Higgins, Esq.* Hopton village and neighbourhood; *Miss Mc Ghie.* Near Oswestry; and in a field between Bedstone and the Heath, Ludlow; *Rev. T. Salwey.* Marton near Baschurch; Moston near Lee bridge; *W. W. Watkins, Esq.*

Hancott, Bomere, Berrington, and Almond park pools.

Stem a foot or more high, round, densely downy and silky in the upper part. *Lower leaves* on channelled petioles more or less downy; *upper ones* ternate, nearly sessile, with a pair of ovate acute *stipules*, which in the lower leaves are converted into a dilated membranous appendage at the base of the *petiole*; all silky on both sides with soft hairs, which on the under surface are longer and accompanied by a white hoary nap. *Flowers* on panicled downy stalks, of a deep dingy purple.

Our plant appears to be the var. β. of Eng. Flora.

10. POTENTILLA. *Linn.* Cinquefoil.

† *Leaves pinnate.*

1. P. *anserina*, Linn. *Silver-weed.* Stem creeping; leaves interruptedly pinnate, leaflets lanceolate, sharply serrated, silky especially beneath; peduncles axillary, solitary; stipules of the stem leaves sheathing, multifid. *E. Bot. t.* 861. *E. Fl. v. ii. p.* 418. *Hook. Br. Fl. p.* 253.

Moist meadows and road sides; frequent. *Fl.* June, July. ♃.

Leaves varying much in degree of silkiness, sometimes silky and white on both sides. *Flowers* large, yellow, on an erect simple stalk. *Calyx* hairy, outer segments notched at each side.

†† *Leaves digitate.*

2. P. *argentea*, Linn. *Hoary Cinquefoil.* Stems ascending, tomentose, corymbose at the top; leaves quinate, leaflets cuneiform, deeply inciso-serrated, white and tomentose beneath, margins revolute; seeds smooth. *E. Bot. t.* 89. *E. Fl. v. ii. p.* 418. *Hook. Br. Fl. p.* 254.

Hilly pastures; not common. *Fl.* June. ♃.

Beaumont hill, Quatford; *W. P. Brookes, Esq.* Ruckley wood; *Dr. G. Lloyd.*

Haughmond Abbey.

Root woody. *Stems* several, 6—12 inches high, nearly simple. *Leaves* on downy channelled petioles, with a pair of lanceolate very acuminate *stipules* at the base, leaflets dark green above, densely tomentose beneath. *Flowers* in terminal downy corymbs. *Calyx* covered with long silky hairs. *Corolla* small, yellow. *Seeds* numerous, ovate, smooth, often slightly rugulose. *Receptacle* hairy.

3. P. *verna*, Linn. *Spring Cinquefoil.* Stems procumbent, hairy; radical leaves 5—7-nate, leaflets obovato-cuneiform, deeply serrated in the upper part, green on both sides, hairy beneath and on the edges; petals obcordate, longer than the calyx; seeds smooth. *E. Bot. t.* 37. *E. Fl. v. ii. p.* 421. *Hook. Br. Fl. p.* 254.

Dry pastures; rare. *Fl.* May, June. ♃.

Near Downton Castle; *Turn. & Dilw. Bot. Guide.* Clee Hill on the Bridgnorth road, on the bank going up to Steventon Cottage, and to Ashford towards Caynham house, on the bank going up Whitecliff woods; *Miss Mc Ghie.*

Root woody, creeping. *Stems* spreading in circular patches, somewhat branched, clothed with erect and spreading silky hairs. *Radical leaves* on long hairy petioles, minutely granulated on the upper surface, paler beneath, serratures of the leaflets obtuse, about four on each side of the terminal tooth which is smaller. *Stem leaves* of 3 narrower leaflets, uppermost nearly sessile, entire and simple. *Stipules* hairy, combined with the petioles, lower ones narrow, upper ones broader. *Flowers* 2 or 3 at the end of weak leafy branches on solitary simple peduncles. *Calyx* very hairy, outer segments 3-ribbed. *Petals* bright yellow. *Seeds* smooth. *Receptacle* hairy.

4. P. *reptans*, Linn. *Common creeping Cinquefoil.* Stems filiform, creeping; leaves quinate, leaflets obovato-cuneiform, deeply serrated, glabrous above, clothed with appressed hairs beneath; peduncles axillary, single-flowered, longer than the leaf; seeds granulated. *E. Bot. t.* 862. *E. Fl. v. ii. p.* 424. *Hook. Br. Fl. p.* 255.

Meadows, pastures and waysides; common. *Fl.* June—August. ♃.

Root woody, tapering. *Stem* various in length, prostrate, round, clothed with appressed hairs, rooting at intervals by radicles proceeding from the joints. *Leaves* opposite, in pairs from each joint, on long channelled hairy petioles, quinate often ternate, leaflets obovato-cuneiform, coarsely serrated, serratures ciliated, with a callous crimson tip, hairy on both sides, generally bearing in the axils a tuft of leaves. *Stipules* of the exterior leaves ovate, more or less acuminated, entire, hairy; of the interior or axillary leaves, linear. The exterior leaf is often suppressed and only developed in the form of a small lanceolate petiolate entire leaf scarcely larger than the stipules, which are present and retain their usual size and ovate shape: this is perhaps the appearance intended by *Smith* in *Engl. Fl.* when he speaks of the stipules being "accompanied by two opposite 3-lobed or undivided entire leafy *bracteas.*" *Flowers* on simple round axillary peduncles longer than the leaves, clothed with erect appressed silky hairs. *Sepals* 10, ovate, acute, hairy, outer ones rather the larger, inner ones paler, all spreading in flower, in fruit the inner ones erect. *Petals* 5, yellow, round, deeply notched, with a very minute claw. *Receptacle* hairy. *Seeds* when viewed under a powerful lens minutely granulato-scabrous.

††† *Leaves ternate.*

5. P. *Fragariastrum*, Ehrh. *Strawberry-leaved Cinquefoil.* Stems procumbent; leaves ternate, leaflets obovato-rotund cuneate at the base, deeply serrated, silky on both sides especially beneath,; petals obcordate as long as the calyx; seeds transversely wrinkled, hairy at the scar and inner margin. *E. Fl. v. ii. p.* 425. *Hook. Br. Fl. p.* 255. *Fragaria sterilis Linn. E. Bot. t.* 1785.

Woods, banks and dry pastures; common. *Fl.* March, April. ♃.

Root woody, running deep into the ground. *Stems* procumbent, spreading to a considerable distance, but not creeping or taking root, clothed with long spreading silky hairs. *Leaves* ternate, on petioles with long spreading silky hairs; central leaflet petiolate, equal at the base, lateral ones nearly sessile, unequal at the base, deeply and broadly serrated, central and terminal tooth considerably smaller than the others. *Stipules* 2 united to the base of the petioles, broadly linear, acute. *Flowers* white, calyx silky. *Receptacle* hairy.

Whole plant with the habit and appearance of a Strawberry, but sufficiently distinguished by the procumbent not creeping shoots, the minute central tooth of the leaflets and the dry hairy receptacle.

TORMENTILLA. *Linn.* Tormentil.

1. T. *officinalis*, Sm. *Common Tormentil.* Stem ascending, dichotomous; leaves ternate those of the stem sessile, leaflets lanceolate inciso-serrate; stipules deeply incised; seeds smooth, wrinkled in the upper part. *E. Bot. t.* 863. *E. Fl. v. ii. p.* 428. *Hook. Br. Fl. p.* 255. *Potentilla Tormentilla, Sibth, Nestl.*

Moors and borders of pools; frequent. *Fl.* June—August. ♃.

Root large and woody. *Stems* weak, slender, with appressed hairs. *Leaves* dark green above, paler beneath, hairy on the ribs and margins on both surfaces, all sessile except one or two radical ones, which are on long slender stalks and with broader and roundish inciso-serrated leaflets. *Stipules* smaller than the leaflets, very variable in shape and number of incisions. *Flowers* on long slender axillary and terminal hairy peduncles longer than the leaves, small, bright yellow. *Calyx* hairy, ribbed. *Receptacle* densely hairy. *Seeds* smooth, longitudinally wrinkled chiefly on the upper part

2. T. *reptans*, Linn. *Trailing Tomentil.* Stem prostrate; leaves ternate and quinate on footstalks, obovato-cuneiform, inciso-dentate. *E.Bot. t.* 864. *E. Fl. v. ii. p.* 428. *Hook. Br. Fl. p.* 255. *Potentilla nemoralis, Nestl.*

Hedge banks, borders of fields and waste places. *Fl.* June, July. ♃.

Side of Stafford road, one mile from Newport; *Mr. F. Dickinson.*

"*Stems* 2 feet long, prostrate, but not creeping. *Stipules* lanceolate, entire. *Leaves* on long hairy footstalks; leaflets 3 or 5, obovate, hairy, light green, more or less deeply cut or serrated. *Flowers* of a full yellow, twice the size of the foregoing, on long slender stalks; the lowermost with 5 petals and 10 sepals." *Smith.*

This plant I am totally ignorant of. It is frequently confounded with *Potentilla reptans* from which Reichenbach in 3814, *Excurs Fl.* states it may be distinguished by the lanceolate entire or bifid stipules, and the sharply serrated leaflets. Bluff & Fingerhuth *Comp. Fl. Germ.* 2*nd ed. v. i. pt.* 2. *p.* 207. mention the seeds as being striato-rugose. Smith in *Engl. Fl.* considers it "totally different from *P. reptans.*"

12. GEUM. *Linn.* Avens.

1. G. *urbanum*, Linn. *Common Avens, Herb Bennet.* Flowers erect, calyx and corolla patent in flower, petals small, obovate; awns glabrous; stipules large, rounded, lobed and cut; cauline leaves ternate, radical ones interruptedly pinnate and lyrate, carpophore none. *E. Bot. t.* 1400. *E. Fl. v. ii. p.* 429. *Hook. Br. Fl. p.* 255.

Woods and hedges; common. *Fl.* June. ♃.

Root of many strong fibres. *Stem* 2 feet high, erect, tetragonous, roughish with deflexed hairs, branched above. *Radical leaves* on long hairy stalks, interruptedly pinnate, somewhat lyrate, terminal deeply 3-lobed. *Cauline leaves* 3-nate, petiolate, all notched and serrated, scabrous and hairy on both surfaces. *Flowers* on long terminal solitary peduncles, small, bright yellow, erect. *Calyx* reflexed in fruit. *Head of fruit* sessile. *Germen* hairy especially at the back and front, style glabrous ending in a curved articulation, terminated by a glabrous awn, which falling off leaves a hook behind.

2. G. *rivale*, Linn. *Water Avens.* Flowers drooping; calyx and corolla erect in flower, petals broadly obovate, emarginate, with a long claw; awns feathery; stipules small, ovate, entire or toothed; cauline leaves ternate, radical ones interruptedly pinnate and lyrate; carpophore elongated. *E. Bot. t.* 106. *E. Fl. v. ii. p.* 431. *Hook. Br. Fl. p.* 256.

Marshy meadows and moist woods; not unfrequent. *Fl.* May—July. ♃.

Caynton wood; *H. Bidwell, Esq.* Standhill Coppice, near Wenlock; *W. P. Brookes, Esq.!* Wern ddû, near Llanymynech, and Bagley brook Shrewsbury; *T. & D. Bot. Guide.* Near Ludlow; *Rev. A. Bloxam.* Sides of river Perry, near Halston; *J. F. M. Dovaston, Esq.* Stanton, near Moreton Corbett; *Mr. E. Elsemere, junr.!* Moors, near Lubstree park; *R. G. Higgins, Esq.* Oreton; *Mr. G. Jorden.* By the side of the road between Wellington and the Lime-quarries, below the Limekiln woods; *E. Lees, Esq.* Corfton, Corvedale; *Miss Mc Ghie.* Wrekin bog; *Mr. F. Dickinson.* Maulbrook, near Walton; Faintree, near Bridgnorth; *Purton's Midl. Fl.* Meadows in Corve dale; *Rev. T. Salwey.*

Old course of the river Severn under Cross Hill, near Shrewsbury.

A shorter but stouter plant than the last with larger flowers and a more purplish hue. *Herbage* hairy, the pubescence of the stem and of the peduncles of the radical leaves deflexed, that of the cauline ones erect. *Terminal leaflets* of the radical leaves more rounded and less deeply and acutely cut. *Calyx* purplish brown, erect in flower and fruit. *Petals* broadly obcordate, clawed, dull purplish orange, elegantly marked with darker veins. *Head of fruit* pedicellate. *Germen* and lowest part of style covered with long hairs and a few pedicillate glands, above that a smooth portion, then a curved articulation terminated by a hairy awn, which falling off with the stigma leaves a hook behind.

A variety with semi-double flowers often occurs.

CLASS XIII.

POLYANDRIA.

"Can it be believed, that Nature bestowed beauty on the foliage of a flower but with a view to please? The fruit might be produced, in the same process, without any richness and diversity of colour. No other animals are sensible of their grace but the human; and yet the austere man of business, or the vain man of pleasure, will arraign another with a face of importance for his admiration of a flower. He calls the taste trifling and useless. But is not a refusal to be pleased with such appearances, like the malignant unthankfulness of a sullen guest, who refuses to taste the most delicious dainties prepared for his entertainment?"

DR. V. KNOX.

CLASS XIII.

POLYANDRIA.   *Many Stamens, inserted upon the receptacle.*

ORDER I.   MONOGYNIA.   1. *Style.*

* Petals 4.

1. PAPAVER.   *Calyx* of 2 deciduous leaves.   *Petals* 4.   *Stigma* sessile, radiated.   *Capsule* superior, 1-celled, *seeds* on parietal *receptacles* projecting towards the centre and escaping by pores beneath the permanent *stigma.—Nat. Ord.* PAPAVERACEÆ, *Juss.*—Named because it was administered with *pap (papa,* Celtic.) to induce sleep.

2. CHELIDONIUM.   *Calyx* of 2 deciduous leaves.   *Petals* 4. *Stigma* 2-lobed.   *Pod* superior, linear, knotted, 1-celled, 2-valved. *Seeds* crested.—*Nat. Ord.* PAPAVERACEÆ, *Juss.*—Name from χελιδων, a *swallow;* probably from the plant flowering about the time of the arrival of those birds.

** Petals 5.

3. HELIANTHEMUM.   *Calyx* of 3 equal leaves, or 5, of which the 2 outer ones are smaller.   *Petals* 5.   *Stigma* capitate.   *Capsule* 3-valved.—*Nat. Ord.* CISTINEÆ, *Juss.*—Name from ἥλιος, *the sun,* and ανθος, *a flower;* from the flowers expanding in the clear sunshine.

4. TILIA.   *Calyx* 5-partite, deciduous.   *Petals* 5, with or without a *nectary* at the base.   *Fruit* coriaceous, 5-celled, without valves; cells 1—5, 2-seeded.—*Nat. Ord.* TILIACEÆ, *Juss.*—Name of obscure origin,

*** Petals numerous.

5. NYMPHÆA.   *Calyx* of 4-5 leaves.   *Petals* numerous, in several rows, inserted as well as the *stamens* upon a fleshy *disk* or covering to the germen, (so as apparently to arise from it.)   *Berry* many-celled, many-seeded, deliquescent.—*Nat. Ord.* NYMPHÆACEÆ, *DC.*— Name, the Νυμφαια of the Greeks, so called from its inhabiting the waters, as the *Nymphs* or Naiads were wont to do.

6. NUPHAR.   *Calyx* of 5-6 leaves.   *Petals* numerous, in one row, inserted as well as the *stamens,* upon the *receptacle.   Berry* superior, many-celled, many-seeded.—*Nat. Ord.* NYMPHÆACEÆ, *DC.*—Name, the Νουφαρ of Dioscorides, applied to this plant. The *Arabic* name is *Naúfar,* according to Förskal.

ORDER II.   PENTAGYNIA.   *Styles variable,* 2—6.

7. HELLEBORUS.   *Calyx* of 5 persistent leaves.   *Petals* 8—10, small, tubular, 2-lipped, nectariferous.   *Pericarps* or *follicles* nearly erect, mány-seeded.—*Nat. Ord.* RANUNCULACEÆ, *Juss.*—Name

from ελειν, *to injure,* and βορα, *food ;* from the poisonous nature of the plant.

8. DELPHINIUM. *Calyx* of 5 petaloid leaves, coloured, deciduous, irregular, upper leaflet produced at the base into a *spur.* *Petals* 4, 2 upper ones with appendages included within the spur, or all the petals coalescing into a single spurred one. *Pericarps* or *follicles* beaked, 1—3, many seeded.—*Nat. Ord.* RANUNCULACEÆ, *Juss.*—Name from *Delphinus* or Δελφιν, *a Dolphin ;* from the shape of the upper calycine leaf.

9. ACONITUM. *Calyx* of 5 petaloid sepals, irregular, upper leaflet helmet-shaped. *Petals* 5 ; 2 upper petals cuculliform, nectariferous, on long stalks, concealed within the helmet-shaped leaflet ; remaining 3 small, linear, often deficient. *Pericarps* or *follicles* beaked, 3—5, many-seeded.—*Nat. Ord.* RANUNCULACEÆ, *Juss.*—Name, derived it is said, from *Acone* in Bithynia, or from ακονη, a *rock* or *stone ;* from the place of its growth.

10. AQUILEGIA. *Calyx* of 5 petaloid, deciduous, coloured leaves. *Petals* 5, funnel-shaped, terminating below in a horn-shaped spur or nectary. *Follicles* 5, separate, many-seeded.—*Nat. Ord.* RANUNCULACEÆ, *Juss.*—Name from *Aquila,* an *eagle ;* whose claws the nectaries resemble.

11. STRATIOTES. *Spatha* of 2 leaves. *Calyx* 3-cleft. *Corolla* of 3 petals. *Berry* inferior, angular, with 6 cells, many-seeded. *Nat. Ord.* HYDROCHARIDEÆ, *Rich.*—Name from στρατος, an *army ;* on account of the numerous sword-like leaves.

(See *Reseda* in CL. XI. and *Trollius* and *Caltha* in ORD. POLYGYNIA.)

ORDER III. POLYGYNIA. *Many Styles.*

* *Germens small, roundish,* 1-*seeded.*

12. THALICTRUM. *Calyx* subpetaloid, of 4-5 leaves. *Corolla* 0. *Pericarps* without awns.—*Nat. Ord.* RANUNCULACEÆ, *Juss.*—Name from θαλλω, *to be green* or *flourishing.*

13. CLEMATIS. *Calyx* petaloid, of 4—6 leaves. *Petals* 0. *Pericarps* terminated by a long, mostly feathery awn.—*Nat. Ord.* RANUNCULACEÆ, *Juss.*—Name from κλημα, *the shoot of a vine ;* which its long branches somewhat resemble.

14. ANEMONE. *Involucre* of 3 divided leaves, more or less remote from the flower. *Calyx* petaloid, of 5—9 leaves. *Corolla* 0. —*Nat. Ord.* RANUNCULACEÆ, *Juss.*—Name from ανεμος, *the wind ;* from the exposed situations in which many of the species grow.

15. ADONIS. *Calyx* of 5 leaves. *Petals* 5—10, without a nectary. *Pericarps* without awns.—*Nat. Ord.* RANUNCULACEÆ, *Juss.*—Name from the bright red colour of its petals having suggested the idea of its being stained by the blood of the beautiful youth *Adonis.*

16. RANUNCULUS. *Calyx* of 5, (rarely 3) leaves. *Petals* 5 (rarely many), with a nectary at the base. *Pericarps* without awns. —*Nat. Ord.* RANUNCULACEÆ, *Juss.*—Name from *rana,* a *frog ;* from the plants delighting in moist situations.

** *Germens elongated, many-seeded.*

17. TROLLIUS. *Calyx* of 5 or many petaloid leaves. *Petals* 5 or many, small, ligulate, with a tubular depression above the contracted base. *Follicles* numerous, many-seeded.—*Nat. Ord.* RANUNCULACEÆ, *Juss.*—Name from *troll* or *trolen,* German, a *ball* or *globe ;* in allusion to the globular form of the flowers.

18. CALTHA. *Calyx* of 5 or more petaloid leaves. *Petals* 0. *Follicles* several, compressed, spreading, many-seeded.—*Nat. Ord.* RANUNCULACEÆ, *Juss.*—Name from καλαθος, a *cup,* which the flowers resemble.

(See *Helleborus* in ORD. II.)

POLYANDRIA—MONOGYNIA.

1. PAPAVER. *Linn.* Poppy.

* *Capsules hispid.*

1. P. *Argemone,* Linn. *Long prickly-headed Poppy.* Filaments dilated upwards ; capsules elongated, clavate, ribbed, hispid with scattered erect setæ ; stigma 4-5-rayed ; stem leafy, many-flowered ; leaves bipinnatifid. *E. Bot. t.* 643. *E. Fl. v. iii. p.* 10. Hook. Br. Fl. p. 259.

Corn-fields ; not common. Fl. May, June. ☉.
Astley, near Shrewsbury ; *Mr. E. Elsmere, junr.!* Near Ludlow ; *Mr. H. Spare.* Weston, near Hawkstone ; *W. W. Watkins, Esq.* Whiston, near Albrighton, Shiffnal ; *H. Bidwell, Esq.* Near Newport ; *R. G. Higgins, Esq.* Pulley ; Red-barn ; Hancott pool ; near Shrewsbury. Westfelton.
Stem 8—12 inches high, erect, round, clothed with erect appressed bristles. *Lower leaves* on long petioles with spreading bristles, bipinnatifid, segments lanceolate, mucronate, ribs on both sides hispid with long bristles, *stem leaves* similar but narrower. *Petals* pale scarlet, black at the base, narrow. *Ribs of the capsule* answering in number to the rays of the *stigma ;* interstices even, not furrowed.

** *Capsules glabrous.*

2. P. *dubium,* Linn. *Long smooth-headed Poppy.* Filaments subulate ; capsule oblong, angular, ribbed, attenuated at the base,

---

glabrous ; stigma 7-rayed ; stem many flowered, hispid with spreading bristles, those of the flower stalks appressed ; leaves bipinnatifid. *E. Bot. t.* 644. *E. Fl. v. iii. p.* 11. Hook. Br. Fl. p. 259.

Corn-fields, hedge banks, &c. ; not unfrequent. Fl. June. ☉.
Quatford, near Bridgnorth ; *Rev. W. R. Crotch.* Corn-fields, Bridgnorth ; *Mr. F. Dickinson.* Near Newport ; *R. G. Higgins, Esq.* Near Ludlow ; *Miss Mc Ghie.* Cressage ; *W. P. Brookes, Esq.!*
Generally about Shrewsbury.
Stem 1-2 feet high, erect, round, leafy, hispid with spreading bristles in the lower part. *Lower leaves* on broad dilated hairy petioles, bipinnatifid, segments broad, ovate, hairy on both sides, mucronate, *upper leaves* sessile, with narrower and more elongated segments. *Petals* unequal, two of them broader than long, and two obovate, marked at the base with 2 dark purple streaks, palish scarlet. *Capsule* with as many smooth ribs as the stigma has rays, perfectly smooth and glaucous.

3. P. *Rhœas,* Linn. *Common red Poppy.* Filaments subulate ; capsule nearly globose, glabrous ; stigma many-rayed ; stem many-flowered, bristly, its bristles and those of the flowerstalks spreading ; leaves pinnatifid. *E. Bot. t.* 645. *E. Fl. v. iii. p.* 11. Hook. Br. Fl. p. 259.

Corn-fields ; not unfrequent. Fl. June, July. ☉.
Near Ludlow ; *Mr. H. Spare.* Eyton and Walford ; *T. C. Eyton, Esq.* Belswardine ; *W. P. Brookes, Esq.!*
Generally about Shrewsbury.
Stems several, 1-2 feet high, branched, erect, round, clothed with spreading horizontal bristles. *Lower leaves* large, spreading on the ground, upper ones smaller, all pinnatifid, bristly on both sides, segments mucronate, of a glaucous green. *Calyx* rough with erect bristles. *Petals* large, deep scarlet. *Stigma* of 10—12 rays.

4. P. *somniferum,* Linn. *White Poppy.* Filaments dilated upwards ; capsule globose, glabrous ; stigma many-rayed ; stem many-flowered, smooth and glaucous ; leaves amplexicaul. *E. Bot. t.* 2145. *E. Fl. v. iii. p.* 11. Hook. Br. Fl. p. 259.

Waste places ; probably not indigenous. Fl. July. ☉.
Occasionally about Bromfield, near Ludlow ; *Mr. H. Spare,*
Stem 3-4 feet high, erect, branched, leafy. *Leaves* oblong, sinuated and unequally toothed, clasping the stem with their cordate base, smooth and glaucous as is the whole herbage. *Peduncle* clothed with spreading horizontal bristles. *Petals* large, usually white with a bluish cast, and a purple eye, but varying much in colour. *Calyx* smooth.

2. CHELIDONIUM. *Linn.* Celandine.

1. C. *majus,* Linn. *Common Celandine.* Peduncles umbellate ; leaves deeply pinnatifid, segments broadly ovate, obtuse, dentato-lobate ; petals entire. *E. Bot. t.* 1581. *E. Fl. v. iii. p.* 4. Hook. Br. Fl. p. 260.

Waste places, hedge banks and crevices of walls near towns and villages ; frequent. Fl. May, June. ♃.
Walford ; *T. C. Eyton, Esq.* Buildwas and Acton Burnell ; *Mr. F. Dickinson.* Near Newport ; *R. G. Higgins, Esq.* Priors Halton ; *Mr. H. Spare.* Near

---

Ludlow ; *Miss Mc Ghie.* Near Oswestry ; *Rev. T. Salwey.* Harley ; *W. P. Brookes, Esq.!*
Shrewsbury ; Meole brace ; Westfelton ; &c.
Root spindle-shaped. *Stem* 2 feet high, erect, round, branched, swollen and clothed with long white hairs above the joints. *Leaves* alternate, on hairy petioles dilated at the base, very deeply pinnatifid, smooth except on the ribs beneath, segments 2 or 3 pair, with a larger terminal tripartite one, all deeply dentato-lobate, obtuse. *Flowers* on slightly hairy long peduncles, opposite the leaves, bearing an umbel of 3—6 rays with solitary blossoms, each with a roundish obtuse ciliated *bractea* at its base. *Petals* yellow. *Calyx* hairy. *Filaments* dilated upwards. *Pod* long, linear, knotted from the contained *seeds,* which are numerous, disposed in 2 rows, black and shining, each with a white glandular crest.

3. HELIANTHEMUM. *Tourn.* Rock-rose.

1. H. *vulgare,* Gærtn. *Common Rock-rose.* Shrubby, procumbent, stipulate ; leaves opposite, oval or linear oblong, green above, hoary beneath, margins ciliated subrevolute ; racemes terminal, bracteated ; calyx-leaves 5, inner furrowed and scariose at the edge ; style bent at the base, somewhat clavate at the apex ; seeds black. *Benth.—Lindl. Syn. p.* 37.—*Cistus Helianthemum,* Linn. *E. Bot. t.* 1321. *E. Fl. v. iii. p.* 26.

Dry hilly gravelly or rocky places ; not very common. Fl. July, August. ♃.
Plealey banks ; *A. Aikin, Esq.* Oreton ; *Mr. G. Jorden.* Whitecliff, near Ludlow ; *Mr. H. Spare.* Near Oswestry ; *Rev. T. Salwey.*
Haughmond Hill.
Root woody. *Stems* numerous, weak, straggling, procumbent then slightly ascending, downy. *Leaves* on short petioles, varying in shape and breadth from oval to linear oblong, glabrous and green or with scattered bristly hairs, margins ciliated subrevolute, hoary with stellate down beneath. *Stipules* lanceolate, ciliated, green on both sides. *Flowers* in terminal racemes, with a small lanceolate ciliated *bractea* at the base of each partial stalk. *Petals* bright yellow, expanding only in clear sunshine. *Inner calyx-segments* large, membranous, glabrous or stellately downy, with 3 strong bristly ribs, 2 *outer* ones minute, green, lanceolate, ciliated.

4. TILIA. *Linn.* Lime.

1. T. *Europæa,* Linn. *Common Lime or Linden-tree.* Nectaries none ; leaves obliquely subrotundo-cordate, acuminate, quite glabrous, except a woolly tuft at the origin of each vein beneath ; young branches smooth ; cymes many flowered ; fruit coriaceous, downy. *E. Bot. t.* 610. *E. Fl. v. iii. p.* 17. Hook. Br. Fl. p. 262. *T. intermedia,* DC. *Lindl.*

Woods and hedge rows. Fl. July. ♄.
Wyre Forest ; *Mr. G. Jorden.* Cox wood, Coalbrookdale ; *Mr. F. Dickinson.*
A large and handsome *tree ;* its *flowers* "at dewy eve distilling odours," yellowish-green, on a stalked *cyme* springing from a large lanceolate foliaceous *bractea,* which falls off with the fructified cymes. *Leaves* large and broad, unequally cordate at the base, margins acutely and unequally serrated, point elongated and acute. *Flowers* 6—10. *Capsule* turbinate, densely downy, generally 1-celled, 1-seeded.

2. T. *parvifolia,* Ehrh. *Small-leaved Lime-tree.* Nectaries

none; leaves rotundo-cordate, acuminate, smooth above, glaucous beneath, with scattered as well as axillary hairy blotches, young branches hairy; flowers in compound many-flowered umbels, fruit roundish, brittle, nearly glabrous. *E. Bot. t.* 1705. *E. Fl. v. iii. p.* 20. *T. microphylla, Vent.*

Woods and hedges. *Fl.* August. ♄.

Ludlow Castle; *Mr. F. Dickinson.* "There is a magnificent tree of this species, (perhaps planted) equalled neither in size nor beauty by few oaks, at the north end of Blackmere near Ellesmere." *J. E. Bowman, Esq.* in *Watson's New Bot. Guide.* Near Oswestry; *Rev. T. Salwey.*

Distinguishable from the preceding by its much smaller *leaves*, which are more regularly cordate though still unequal at the base, sharply serrated, dark green and smooth above, though sometimes with scattered hairs on the veins. *Flowers* 5—7, in double umbels. *Germen* densely woolly. *Capsules* sparingly perfected, turbinate, 1-seeded.

5. NYMPHÆA. *Linn.* White Water-Lily.

1. N. *alba*, Linn. *Great White Water-Lily.* Leaves oval, cordate, entire; stigma of 16 ascending rays. *E. Bot. t.* 160. *E. Fl. v. iii. p.* 14. *Hook. Br. Fl. p.* 263.

Lakes and pools; frequent. *Fl.* July. ♃.

Snowden Pool; *H. Bidwell, Esq.* Pool on Hughley Common Farm; *W. P. Brookes, Esq.* Ellesmere; *Rev. A. Bloxam.* Walford and Fenny-mere; *T. C. Eyton, Esq.* Hopton; *Mr. H. Spare.* Hayes, near Oswestry; Llynklys pool, parish of Llanyblodwell; *Rev. T. Salwey.* Marton pool, near Baschurch; *W. W. Watkins, Esq.*

Bomere, Almond park, Hancott and Berrington pools.

Root tuberous. *Leaves* floating on foot-stalks proportioned to the depth of the water, (Mr. Dovaston having observed them in Llynklys Pool of the extraordinary length of *fourteen feet*,) oval, cordate, the lobes nearly parallel and approximate at the base, entire, smooth. *Flowers* very large and beautiful, pure white, closing towards evening and reclining or sinking beneath the surface of the water; *petals* elliptic-oblong, passing insensibly into the dilated *filaments.*

6. NUPHAR. *Sm.* Yellow Water-Lily.

1. N. *lutea*, Sm. *Common Yellow Water-Lily.* Leaves oblongo-cordate, entire; sepals 5, concave, subrotund, twice the length of the petals; petals spathulato-ovate, obtuse, fleshy; stigmatic disk entire. *E. Fl. v. iii. p.* 15. *Hook. Br. Fl. p.* 263. *Nymphæa lutea, Linn. E. Bot. t.* 159.

Lakes and pools; frequent. *Fl.* July. ♃.

Tong pool; *H. Bidwell, Esq.* Marbury mere, near Whitchurch; *Mr. F. Dickinson.* Snowden pool, near Beckbury; *Dr. G. Lloyd.* Marton pool, near Baschurch; *W. W. Watkins, Esq.*

Bomere, Almond park, Hancott and Berrington pools; Ellesmere Canal; Westfelton.

Leaves very large, floating, oblongo-cordate, the lobes approximate; *petioles* 2-edged, flattened on the upper surface. *Peduncles* cylindrical. *Flowers* cupped, golden yellow, strongly scented.

POLYANDRIA—PENTAGYNIA.

7. HELLEBORUS. *Linn.* Hellebore.

1. H. *viridis*, Linn. *Green Hellebore.* Stem few-flowered, leafy; leaves digitate; calyx spreading. *E. Bot. t.* 200. *E. Fl. v. iii. p.* 58. *Hook. Br. Fl. p.* 263.

Meadows and pastures; rare. *Fl.* April, May. ♃.

Clee Hill, near Bitterley; *Miss Mc Ghie.* Stottesden; *Mr. G. Jorden.* Near the Wall in a field adjoining Lilleshall Abbey; *R. G. Higgins, Esq.!*

Herbage annual, bright green. *Stem* 1 foot high, branched above. *Radical leaves* large, on a broad stalk, truly digitate; *upper ones* sessile at the ramifications, the lobes frequently combined, *segments* linear-lanceolate, acute, deeply and irregularly incised and serrated. *Flowers* few, terminal and axillary, stalked, mostly solitary, large, greenish-yellow.

2. H. *fœtidus*, Linn. *Stinking Hellebore.* Stem many-flowered, leafy; leaves pedate; calyx converging. *E. Bot. t.* 613. *E. Fl. v. iii. p.* 58. *Hook. Br. Fl. p.* 264.

Pastures and thickets; rare. *Fl.* April. ♃.

Clee Hill, near Bitterley; *Miss Mc Ghie.* Eyton; *T. C. Eyton, Esq.* In a coppice close to the left side of road from Wenlock to Buildwas, just below Mr. Aston's of Fayerley; *W. P. Brookes, Esq.!* One of the outer ditches of Whittington Castle; *Rev. A. Bloxam.*

Herbage evergreen, dark green, fetid. *Stem* 2 feet high, much branched. *Lower leaves* large, petiolate, spreading, pedate, *segments* 7—9, lanceolate, acute, serrated; *upper ones* paler, the lobes combined, gradually passing into entire *bracteas.* *Flowers* numerous, in drooping panicles, globose, pale-green tinged with purple.

8. DELPHINIUM. *Linn.* Larkspur.

1. D. *Consolida*, Linn. *Field Larkspur.* Stem erect, branched; racemes lax, few-flowered; petals combined into one spurred piece; pedicels longer than the bracteas; capsule glabrous. *E. Bot. t.* 1839. *E. Fl. v. iii. p.* 30. *Hook. Br. Fl. p.* 264.

Corn-fields; probably naturalized. *Fl.* June, July. ⊙.

Davenport woods; *Rev. W. R. Crotch.*

Preston Boats hamlet, near Shrewsbury; probably an outcast.

Herbage pubescent. *Stem* 18 inches to 2 feet high, erect, with alternate spreading *branches.* *Leaves* sessile or nearly so, deeply multifid, divisions 3-cleft, subdivided into narrow linear acute segments. *Flowers* in terminal, few-flowered, bracteated racemes, of a vivid and permanent blue.

9. ACONITUM. *Linn.* Wolf's-bane.

1. A. *Napellus*, Linn. *Common Wolf's-bane or Monk's-hood.* Upper leaflet of the calyx arched at the back; spur of the nectary nearly conical, bent down; lobes of the leaves cuneate, pinnatifid; germens 3—5, glabrous. *E. Bot. Suppl. t.* 2730. *E. Fl. v. iii. p.* 31. *Hook. Br. Fl. p.* 264.

Banks of rivers, woods, &c.; doubtfully wild. *Fl.* June, July. ♃.

"Wood bordering Colemere mere, but it may have been planted there." *J. E. Bowman* in *Watson's New Bot. Guide.*

Abundantly scattered and with every present appearance of being wild, but doubtful if originally so, for above a mile along the banks of the Ludwych river between Ludlow and Caynham Camp, where it was first discovered in 1819 by Rev. Edward Whitehead, Fellow of Corpus Christi College, Oxford, and recorded in the Appendix to Purton's Midland Flora.

Stem 2-3 feet high, erect, simple, downy. *Leaves* alternate, on short petioles, 5-lobed, lobes cut into numerous linear acute smooth segments, paler beneath. *Flowers* in a solitary, terminal, erect cluster, each on a simple bracteated pedicel, dark blue, veiny, hairy externally and internally. *Follicles* smooth, veined.

10. AQUILEGIA. *Linn.* Columbine.

1. A. *vulgaris*, Linn. *Common Columbine.* Spur of the petals incurved; capsules hairy; stem leafy, many-flowered; leaves nearly glabrous; styles as long as the stamens. *E. Bot. t.* 97. *E. Fl. v. iii. p.* 33. *Hook. Br. Fl. p.* 264.

Woods, coppices, &c. *Fl.* June. ♃.

Hedge opposite the Cold bath, near the Lodge, Ludlow; near Oswestry; *Rev. T. Salwey.* Oakley Park meadows; *Mr. H. Spare.* Near Dowles Brook and Park Brook, Wyre Forest; *Mr. W. G. Perry.* Clee Hill, near Bitterley; *Miss Mc Ghie.* Abundant in Wyre Forest; *E. Lees, Esq.* Chirk banks; *J. F. M. Dovaston, Esq.* Standhill Coppice, Wenlock; *W. P. Brookes, Esq.!* In a small coppice, near the Lawley Hill; *Rev. W. Corbett.* Footpath between Old Factory and Cross Hill, near Shrewsbury; *Dr. Du Gard.*

Side of Canal, near Queen's head between Westfelton and Oswestry.

Stem erect, 1-2 feet high, branched above, nearly glabrous. *Lower leaves* on long hairy petioles, biternate, leaflets roundish, wedge-shaped, 3-lobed, crenate, glaucous and hairy beneath. *Flowers* pendulous, on downy bracteated pedicels, blue or dull red purple.

11. STRATIOTES. *Linn.* Water-Soldier.

1. S. *aloides*, Linn. *Aloe-like Water-Soldier.* Leaves sword-shaped, triangular, aculeato-serrate. *E. Bot. t.* 379. *E. Fl. v. iii. p.* 34. *Hook. Br. Fl. p.* 265.

Lakes and ditches; rare. *Fl.* July. ⊙.

Lilleshall Pond; *Miss Mc Ghie.*

A singular plant having the general appearance of an Aloe, stemless; increasing rapidly by long simple *runners*, each bearing a terminal gemma, which rooting in the mud, throws up numerous *radical leaves* and a solitary central *scape* 4—6 inches long, compressed, 2-edged. In the summer the plant rises to the surface to blossom, and afterwards becomes submerged, when the whole herbage is destroyed, the crown of the root only remaining, from which runners are sent forth, and fresh plants produced. *Flowers* white, large and handsome, from a compressed 2-leaved *spatha.* *Pollen* globose, rough. Sometimes the *flowers* are diœcious, and sometimes the *stamens* are on the same flower, with 5-6-cleft *styles.*

POLYANDRIA—POLYGYNIA.

12. THALICTRUM. *Linn.* Meadow-Rue.

1. T. *minus*, Linn. *Lesser Meadow-Rue.* Leaves 3-4-pinnate, leaflets roundish, glabrous, trifid and toothed, glaucous beneath;

panicle diffuse, its branches alternate or whorled, flowers mostly drooping. *E. Bot. t.* 11. *E. Fl. v. iii. p.* 41. *Hook. Br. Fl. 4th ed. p.* 217.

Stony pastures; rare. *Fl.* June, July. ♃.

Near Button Oak in Wyre Forest; *E. Lees, Esq.*

Stem zigzag, 1 foot high, mostly glaucous. *Leaflets* small. *Fruit* narrow, ovate, sulcate.

2. T. *flavum*, Linn. *Common Meadow-Rue.* Stem erect, branched, furrowed, leaves bipinnate, leaflets broadly obovate or wedge-shaped, trifid; panicle compact, subcorymbose, flowers erect. *E. Bot. t.* 367. *E. Fl. v. iii. p.* 42. *Hook. Br. Fl. p.* 266.

Banks of rivers, ditches and moist meadows; rare. *Fl.* June, July. ♃.

Bank of Severn opposite Buildwas; *W. P. Brookes, Esq.!* Oakley Park meadows; *Mr. H. Spare.*

Stem 2-3 feet high. *Flowers* very numerous, yellow. *Lobes of leaves* varying in breadth.

13. CLEMATIS. *Linn.* Traveller's Joy.

1. C. *Vitalba*, Linn. *Common Traveller's Joy.* Stem climbing; leaves pinnate, leaflets cordato-ovate, inciso-lobate, petioles twining; peduncles rather shorter than the leaves. *E. Bot. t.* 612. *E. Fl. v. iii. p.* 39. *Hook. Br. Fl. p.* 266.

Hedges; rare. *Fl.* May, June. ♄.

Whitecliff, near Ludlow; *Mr. H. Spare.* Harley Churchyard; *W. P. Brookes, Esq.!* Lincoln Hill; *Mr. F. Dickinson.* Brown Hill, near Ruyton; *T. C. Eyton, Esq.* Between Wellington and Dothill park; *E. Lees, Esq.*

Shrub with woody, angular, branched, entangled *stems*, climbing and supporting themselves by the twining *petioles*, which act as tendrils. *Flowers* greenish-white, in terminal and axillary, forked, downy panicles, fragrant. *Fruit* in conspicuous and beautiful tufts, with long wavy feathery silky awns.

14. ANEMONE. *Linn.* Anemone.

1. A. *nemorosa*, Linn. *Wood Anemone.* Leaves ternate, leaflets lanceolate, lobed and cut; involucre similar to them, petiolate; stem single-flowered, calycine-leaflets 6, elliptical, obtuse; pericarps downy, awnless. *E. Bot. t.* 355. *E. Fl. v. iii. p.* 37. *Hook. Br. Fl. p.* 266.

Moist woods and pastures; abundant. *Fl.* April, May. ♃.

Root tuberous. *Leaves* on long petioles, often wanting. *Flowers* white, tinged with purple on the outside, sometimes purplish on both sides. *Pericarps* beaked with the style of nearly their own length.

2. A. *apennina*, Linn. *Blue Mountain Anemone.* Leaves triternate, leaflets lanceolate, cut and toothed; involucre ternate and cut, petiolate; stem single-flowered, calycine leaflets 12—14, linear-lanceolate, obtuse; pericarps awnless. *E. Bot. t.* 355. *E. Fl. v. iii. p.* 37. *Hook. Br. Fl. p.* 266.

Woods; rare, probably not indigenous. *Fl.* April. ♃.

Shortwood, near Ludlow; *E. Lees, Esq.*

Similar to the last, but with broader richer and more hairy *foliage.* *Flowers* of a splendid light bright blue and very elegant, the calycine leaflets widely and stellately spreading.

3. A. *ranunculoides*, Linn. *Yellow Wood Anemone.* Leaves ter-or quinate, leaflets subtrifid, cut and toothed; involucres 3, shortly petiolate, ternate, leaflets subtrifid, incised; stem 1-2-flowered; calycine segments 5, elliptical; pericarps without awns. *E. Bot. t.* 1486. *E. Fl. v. iii.p.* 38. *Hook. Br. Fl. p.* 266.

Woods; rare. *Fl.* April. ♃.

Badger Dingle; *H. Bidwell, Esq.!*

Root tuberous. *Leaflets* ovato-lanceolate, bright green, glabrous except on the margins. *Peduncle* and *calycine segments* silky on the under surface. *Flowers* bright yellow. *Pericarps* pubescent, nearly equal to the style.

15. ADONIS. *Linn.* Pheasant's Eye.

1. A. *autumnalis*, Linn. *Corn Adonis* or *Pheasant's Eye.* Petals concave, connivent, scarcely longer than the glabrous patent calyx; pericarps reticulated, collected into an ovate head; stem branched. *E. Bot. t.* 308. *E. Fl. v. iii. p.* 43. *Hook. Br. Fl. p.* 267.

Corn-fields; rare. *Fl.* September, October. ⊙.

Near Ludlow; *Mr. H. Spare.* Banks of Severn, a short distance below Coalport; *E. Lees, Esq.*

Stem erect, round, striated, generally smooth. *Leaves* alternate, sessile, thrice compound, with linear acute segments. *Petals* 6—10, obcordate, bright scarlet, with a dark spot at the base; *anthers* violet.

16. RANUNCULUS. *Linn.* Crowfoot.

* *Pericarps transversely wrinkled. Petals white.*

1. R. *fluitans*, Lam. *River Crowfoot.* Leaves all submersed, trichotomous, segments setaceous, elongated, parallel; petals 5—10, rotundo-obovate, more than twice the length of the calyx; carpels transversely wrinkled, glabrous, obovate, with a short obtuse straight lateral point. *Babington, Ann. Nat. Hist. iii.* 229. *Reich. Fl. Exc.* 4577. *Bluff & Fing. Comp. Fl. Germ. 2nd. ed. t. i. pt. ii. p.* 286. *R. aquatilis, δ. E. Fl. v. iii. p.* 55. *R. aquatilis, var.* 5. *fluriatilis, With Arr. ed. iii. v. ii. p.* 507. *R. aquatilis, var. γ. Pollich. Pal. Elect. v. ii. p.* 120. *R. fluviatilis, Sibth. Fl. Oxon.* 176.

Shallows of the river Severn; abundant. *Fl.* June. ♃.

Root of many long fibres, proceeding from the lower joints of the stem. *Stem* 3—5 feet or more in length, round, smooth, hollow, leafy, branching towards the extremities, floating. *Leaves* alternate, submersed, repeatedly trichotomous, the extreme divisions often only bifid, segments setaceous, elongated, parallel, compressed, obtuse, with 2 or 3 minute bristles at the apex; lower leaves on long semicylindrical petioles dilated at the base, membranous and surrounding the stem; upper nearly sessile, amplexicaul by the broad dilated membranous somewhat hispid appendages of the very short petiole. Not uncommonly at the extremity of the stem is observed, on long petioles dilated membranous and clasping at the base, a tripartite leaf or two, with divergent elliptical lobes, the central one shorter and entire, lateral ones bifid, smooth, dark green and shining above, paler beneath. *Flowers* on round, smooth, solitary peduncles, as long as and opposite the upper

leaves, appearing only for a few days and falling on the first rise of the water. *Sepals* 5, concave, ovate, obtuse, reflexed, smooth, 3-ribbed, paler and membranous at the edges. *Petals* 5—10, rotundo-obovate, more than twice the length of the sepals, of a delicate shining white with a yellow claw, beautifully pencilled with pellucid rays, crenulate or wavy at the margin. *Nectary* thickened and raised, ovate, tubular. *Seeds* obovate, smooth, transversely wrinkled, crowned with a short obtuse straight point.

This plant forms a transient but very elegant and conspicuous ornament of the river Severn during the month of June, copiously expanding its large pure white blossoms, and gracefully undulating its bright-green elongated stems and hair-like leaves, in the rapid and shallow currents, strikingly reminding us of the "tresses fair" of Sabrina, alluded to in Milton's Comus:—

" Sabrina fair,
Listen where thou art sitting
Under the glassy, cool, translucent wave,
In twisted braids of lilies knitting
The loose train of thy amber-dropping hair."

When the plant grows on the moist sandy margins of the river, entirely out of the water, it assumes a dense cæspitose appearance, the stems branching, straggling and creeping in all directions, by means of the long and stout white fibrous roots, which are copiously developed from almost every joint. The leaves with their petioles become reduced in length to about 1½ or 2 inches, their divisions more or less divaricated, but always in the same plane, and their ultimate segments linear, flattened, obtuse and dilated at their extremities, thus putting on a linear-spathulate form. The flowers are also smaller.

2. R. *aquatilis*, Linn. *Water Crowfoot.* Submersed leaves trichotomous, petiolate, segments setaceous, spreading on all sides, floating leaves reniform, 3—5-lobed, lobes crenate; petals 5, rotundo-obovate, many times longer than the calyx; carpels transversely rugose, hispid or glabrous, unequally ovate, with an obtuse terminal point. *Babington, Ann. Nat. Hist. iii.* 227. *Reich. Fl. Excurs. n.* 4576. *Bluff & Fing. Comp. Fl. Germ. ed. ii. t. i. pt. ii. p.* 285.

Pools, ditches, and shallow stagnant waters; common. *Fl.* May, June. ♃.

Var. α. heterophyllus, Wallr. Floating leaves reniform. *Bab. l. c. R. aquatilis, E. Bot. t.* 101. *R. aq. α. E. Fl. v. iii. p.* 54. *Hook. Br. Fl. 4th ed. p.* 218. *R. heterophyllus, Sibth. Fl. Oxon. p.* 175.

Root of many long fibres from the lower joints of the stem. *Stems* many, very variable in length, branched, especially at the extremities. *Leaves* on rather short petioles, dilated and clasping at the base, *submersed* ones repeatedly trichotomous, segments setaceous, variable in length, spreading in all directions, forming a more or less spherical mass; *upper* or floating leaves varying greatly in form, sometimes reniform, sometimes cordato-subrotund, and sometimes truncated and not cordate at the base, 3—5 lobed, lobes very variable in form and number of crenatures. *Flowers* on round, smooth, solitary *peduncles*, opposite to and longer than the upper leaves. *Sepals* 5, concave, ovate, obtuse, reflexed. *Petals* 5, white with a yellow claw, pencilled with pellucid rays, margins entire. *Seeds* unequally ovate, compressed, smooth or hispid, transversely wrinkled, crowned with the short straight blunt style. *Receptacle* hairy.

Var. *β. pantothrix.* Floating leaves none, all the leaves trichotomous, segments setaceous, spreading on all sides. *Bab. l. c. R.*

aquatilis, β. E. Fl. v. iii. p. 54. *Hook. Br. Fl. 4th ed. p.* 218. *R. aquatilis, Sibth. Fl. Oxon, p.* 175.

Equally common with the preceding.

Var. γ. tripartitus. Upper leaves trichotomous, petiolate, segments setaceous, spreading on all sides; lower leaves on very long petioles, tripartite, lobes wedgeshaped, deeply cut into numerous linear acute segments; flowers white, small; fruit hispid, transversely wrinkled.

Hancott pool; near Shrewsbury.

Probably only an accidental variety.

3. R. *hederaceus*, Linn. *Ivy Crowfoot.* Stem creeping; leaves rotundo-reniform, with 3—5 rounded entire lobes; petals small, scarcely longer than the calyx; stamens 5—10; pericarps glabrous. *E. Bot. t.* 2003. *E. Fl. v. iii. p.* 54. *Hook. Br. Fl. p.* 267.

Wet places, and where water has stood; not very common. *Fl.* throughout the summer. ♃.

Lane between Plealey and the Oaks, near Shrewsbury; *Mr. T. Bodenham!* Walford; *T. C. Eyton, Esq.* Between the Wrekin and Coalbrookdale; *E. Lees, Esq.* Near Oswestry; *Rev. T. Salwey.*

Pulley Common; Bicton; Shelton rough, near Shrewsbury; Seeches, Westfelton; Shrawardine pool.

Stem prostrate, branched, spreading in all directions, and rooting at each joint, round, smooth and hollow. *Leaves* opposite or alternate, petiolate, dark green, smooth and frequently marked with a dark spot. *Petioles* amplexicaul at the base, with a pair of large pale membranous *stipules. Flowers* on solitary peduncles, as long as and opposite to the leaves. *Sepals* 5, ovate, concave, obtuse, reflexed. *Petals* 5, obovate, minute, white, somewhat longer than the sepals. *Seeds* ovate, compressed, keeled, glabrous, transversely wrinkled.

** *Pericarps not transversely wrinkled. Nectary with a small scale. Petals yellow.*

† *Leaves undivided.*

4. R. *Lingua*, Linn. *Great Spear-wort.* Leaves lanceolate, acuminate, sinuato-denticulate, sessile, amplexicaul; stem erect, glabrous. *E. Bot. t.* 100. *E. Fl. v. iii. p.* 46. *Hook. Br. Fl. p.* 267.

Boggy margins of pools; not very common. *Fl.* July. ♃.

Snowden pool, near Beckbury; *H. Bidwell, Esq.* Colemere, near Ellesmere; *Rev. A. Bloxam.* Near Longnor; *Rev. W. Corbett.* Marbury Mere and near Hampton Bank, Coalbrookdale; *Mr. F. Dickinson.* Littlehales, near Newport; *R. G. Higgins, Esq.* Tong Lodge lake; *Dr. G. Lloyd.* Wooton, near Oswestry; *Rev. T. Salwey.* In a pool adjoining Cound Hall. Pits, near Albrightlee. Marton pool. Whitemere, near Ellesmere; *Rev. E. Williams's, MSS.*

Bomere, Almond park, Mare and Hancott pools, near Shrewsbury.

Root of many whorled fibres proceeding from the lower joints of the stem. *Stem* erect, 2-3 feet high, striated, smooth, jointed, branched upwards. *Leaves* alternate, dilated and clasping the stem with their membranous base, hairy on both sides, somewhat scabrous beneath. *Peduncles* round, striated, with long silky erect appressed hairs, single-flowered. *Sepals* 5, coloured, hairy externally, spreading. *Petals* 5, broad and round, narrower at the base, margin minutely crenate or wavy, bright yellow. *Seeds* oval, compressed, smooth, minutely pitted,

surrounded with a margin, and crowned by a broad beak; dorsal margin clothed with erect bristles.

5. R. *Flammula*, Linn. *Lesser Spear-wort.* Leaves linear-lanceolate, nearly entire or subserrated, petiolate, lower ones ovato-lanceolate; stem declined at the base and rooting. *E. Bot. t.* 387. *E. Fl. v. iii. p.* 45. *Hook. Br. Fl. p.* 267.

Sides of pools, ditches, and watery places; abundant. *Fl.* July, August. ♃.

Root fibrous, from the lower joints of the stem. *Stem* 6—18 inches long, more or less reclining and decumbent, branched, smooth or slightly hairy. *Leaves* particularly the lower ones, on rather long footstalks dilated and clasping the stem with their membranous base, varying in form from ovate or ovato-lanceolate to linear-lanceolate, entire or coarsely serrated, smooth or with scattered hairs. *Flowering-stems* dichotomously branched; *peduncles* round, with erect appressed hairs, single-flowered. *Sepals* 5, spreading, hairy. *Petals* 5, golden yellow, with a minute *nectary* at the base. *Seeds* small, rotundo-obovate, compressed, margined and keeled, minutely pitted, crowned with a short stout beak.

Var. β. reptans. Leaves linear, quite entire, smooth, of the same breadth at either end, their sheaths and the carpels very smooth; stem creeping, filiform, rooting at almost every joint. *Hook, Br. Fl. 4th ed. p.* 219. *R. Flammula δ. E. Fl. v. iii. p.* 45. *R. reptans, Linn. Lindl. Syn. Suppl. 2nd ed. p.* 319.

By the side of Eaton Marscot pool; *Rev. E. Williams's MSS.*

6. R. *Ficaria*, Linn. *Pilewort Crowfoot, Lesser Celandine.* Root with fasciculated tubers; leaves cordate, petiolate, angular or crenate; stem 1-flowered; sepals generally 3. *E. Bot. t.* 584. *E. Fl. v. iii. p.* 47. *Hook. Br. Fl. p.* 268. *Ficaria verna, Huds. F. ranunculoides, DC.*

Pastures, woods, shady places, &c.; abundant. *Fl.* April, May. ♃.

Root fibrous, with many fleshy oblong fasciculated *tubers. Stem* erect or recumbent, 3—8 inches long. *Leaves* petiolate, bright shining green, somewhat succulent, frequently marked with a dark spot. *Sepals* 3, rarely 4 or 5, ovate, obtuse. *Petals* generally 8, frequently 9, sometimes 10 or 11, more rarely 6 or 7, lanceolate, obtuse, expanding in a stellate manner, bright glossy yellow, with a watery spot at the base. *Nectary* a small hollow, closed by an obtuse scale.

†† *Leaves divided. Pericarps smooth. Perennial.*

7. R. *auricomus*, Linn. *Wood Crowfoot.* Leaves pubescent, radical ones petiolate, reniform, tripartite and cut, stem leaves sessile, divided to the base into linear subdentate segments; calyx pubescent, shorter than the petals; head of fruit globose. *E. Bot. t.* 624. *E. Fl. v. iii. p.* 48. *Hook. Br. Fl. p.* 268.

Woods, coppices, bushy and shady places; not unfrequent. *Fl.* April, May. ♃.

Hedges near Welbatch; *Mr. T. Bodenham.!* Longnor and Coreley; *Rev. W. Corbett.* Twyford, near Westfelton; *J. F. M. Dovaston, Esq.* Captain's Coppice, Coalbrookdale; *Mr. F. Dickinson.* Near Lubstree park; *R. G. Higgins, Esq.* Near Oswestry; *Rev. T. Salwey.*

Foot of Haughmond Hill. Radbrook, near Shrewsbury. Buckley farm, near Oswestry.

Root fibrous. *Stem* about 1 foot high, erect, branched, pubescent above. *Flowers* solitary, terminal, of a bright golden yellow, the *petals* seldom all perfect, their place being supplied by the erect dilated coloured *sepals*. *Fruit* pubescent. This species is not acrid.

A slight variety of shorter and stouter habit of growth, and with the lobes of the leaves more acute, more deeply and acutely cut and toothed, occurs in abundance in the marshy meadows below Cross Hill, near Shrewsbury, the ancient bed of the river Severn.

8. R. *sceleratus*, Linn. *Celery-leaved Crowfoot*. Leaves glabrous, lower ones petiolate, 3-partite, lobes cut very obtuse, upper ones nearly sessile, in 3 linear obtuse cut segments; calyx hairy, reflexed, as long as the petals; head of fruit oblong. *E. Bot. t.* 681. *E. Fl. v. iii. p.* 48. *Hook. Br. Fl. p.* 268.

Sides of pools, ditches and watery places; not uncommon. *Fl.* May—August. ♃. *Hooker.* (⊙. *Smith.*)

Whiston Marshes; *H. Bidwell, Esq.* Marbury Meer, and near Ellardine Moss; *Mr. F. Dickinson.* Near Oswestry; *Rev. T. Salwey.*

Bomere pool; Shelton; Abbots Betton; Uffington; Battlefield; all near Shrewsbury. Whittington.

Root fibrous, proceeding in whorls from the lowest joints. *Stem* 1-2 feet high, stout, succulent, hollow, repeatedly branched, smooth below, hairy above. *Lower leaves* glabrous and shining, upper ones hairy. *Flowers* numerous, small, on solitary hairy stalks. *Fruit* minute, wrinkled.

9. R. *acris*, Linn. *Upright Meadow Crowfoot*. Stem erect, many-flowered, with deflexed hairs below; radical leaves subrotund, palmato-tripartite, segments 3-fid, deeply inciso-dentate, upper ones 3-partite, segments linear; peduncles round; sepals erecto-patent; receptacle glabrous. *E. Bot. t.* 652. *E. Fl. v. iii. p.* 52. *Hook. Br. Fl. p.* 268.

Meadows, pastures, &c.; common. *Fl.* June, July. ♃.

Root of many long simple fibres. *Stem* somewhat swollen at the base, 2-3 feet high, erect, dichotomously branched, round, hollow, leafy, with deflexed hairs in the lower part, glabrous or with scattered erect appressed hairs above. *Lower leaves* on long rounded erect petioles, channelled above, clothed with soft spreading hairs, dilated at the base with a membranous hairy amplexicaul appendage, subrotund, palmato-tripartite, lateral lobes deeply tripartite, segments 3-fid, deeply inciso-dentate, teeth acuminate, each tipped with a white callous point, dark green and hairy above on the margin, paler beneath with numerous scattered imbedded roundish green glandular dots, hairy only on the ribs: *upper leaves* on shorter petioles, with similar but narrower, more linear, elongated and acuminated divisions, *uppermost* or floral leaves sessile, with 3 linear-lanceolate segments. *Peduncles* round, slightly compressed, with erect appressed pubescence, simple, single-flowered, opposite to the leaves and thus rendering the stem dichotomous. *Sepals* 5, erecto-patent, elliptical, concave, 3-ribbed, lateral ribs branched, membranous and revolute at the margins, pale yellow tinged with green, with silky hairs on the exterior. *Petals* 5, subrotundo-cordate, slightly emarginate, deep yellow, shining except a paler duller radiated spot at the base, veiny. *Nectary* an oblong emarginate or retuse yellow scale, covering a slight depression in the thickened fleshy claw of the petal. *Seeds* lenticular, ovate, compressed, glabrous, with a margined small recurved beak, about one fourth the length of the whole seed. *Receptacle* glabrous.

Var. β. minor. Segments of the radical leaves broader; petals pale yellowish white, with a darker yellow radiated spot at the base.

A single specimen of this was gathered in the fields between Shrewsbury and Preston Boats, 20th May, 1833, (the flowers and a leaf of which were dried for the herbarium) and transplanted into my garden, where it has retained its characters ever since. It differs from *acris* in being altogether a smaller plant, not more than 18 inches high, the segments of the radical *leaves* broader and not so acuminate; the *flowers* not half the size, the *petals* more regularly cordate; the *petals* and *sepals* pale yellow on the exterior, very pale yellowish white and shining on the interior surface, with a darker yellow radiated spot at the base occupying about one third of the petal. In all other respects it agrees with *acris*.

10. R. *repens*, Linn. *Creeping Crowfoot*. Stem erect, many-flowered, with erecto-patent hairs, scyons creeping; radical leaves ternate, almost biternate, leaflets petiolate, 3-partite, segments 3-fid, inciso-dentate; peduncles angular and furrowed; sepals erecto-patent; receptacle hairy. *E. Bot. t.* 516. *E. Fl. v. iii. p.* 51. *Hook. Br. Fl. p.* 268.

Pastures and cultivated ground; frequent. *Fl.* June—August. ♃.

Root of many long branched fibres, thickened or swollen at the crown. *Primary stem* 10—12 inches high, erect, obsoletely angular, hollow, clothed with erecto-patent roughish hairs which become more appressed above, leafy, dichotomously branched in the upper part. From the axils of the radical leaves proceed stout *secondary stems* or runners, prostrate, depressed, covered with long somewhat hispid hairs, rooting at the joints and elevating flowering stems. *Radical leaves* on long depressed hairy *petioles* channelled above, with a pair of broad membranous amplexicaul silky adnate *stipules*, ternate, almost biternate, central lobe on a long channelled stalk, lateral ones on shorter stalks, all unequally trifid below the middle, their segments unequally 3-fid, inciso-dentate, tipped with a pale callous tubercle, hairy and soft on both surfaces especially the under one. *Upper leaves* sessile by the amplexicaul adnate stipules, all the lobes on very long channelled stalks, segments narrower and more elongated; *uppermost* or floral leaves of 3 lanceolate entire segments. *Peduncles* angular and furrowed, with scattered appressed hairs, simple, single-flowered, opposite the leaves, thus rendering the stem dichotomous. *Sepals* 5, erecto-patent, broadly elliptical, concave, obtuse, membranous and reflexed at the margins, 5-ribbed, lateral ribs branched, pale yellow tinged with green and dark streaks, externally silky. *Petals* 5, rounded, cuneate at the base, entire or sometimes slightly emarginate, bright shining yellow with a duller watery radiated spot at the base. *Nectary* a slight depression in the thickened fleshy claw, covered by a convex irregularly obcordate scale contracted and sinuated at the base and with very divergent lobes. *Seeds* lenticular, nearly round, compressed, glabrous, margined, with a scarcely recurved beak about one third the length of the whole seed. *Receptacle* hairy.

11. R. *bulbosus*, Linn. *Bulbous Crowfoot*. Stem erect, many-flowered, with erect appressed hairs, bulbous at the base; radical leaves ternate, almost biternate, leaflets petiolate, 3-partite, segments 3-fid, inciso-dentate; peduncles pentangular and deeply furrowed; sepals reflexed; receptacle hairy. *E. Bot. t.* 515. *E. Fl. v. iii. p.* 50. *Hook. Br. Fl. p.* 269.

Meadows and pastures; frequent. *Fl.* May. ♃.

Root of long stout white fibres. *Stem* swollen at the base into a solid bulb, erect, 8—18 inches high, obsoletely angular and furrowed, hollow, branched above, clothed with long soft erect appressed silky hairs. *Radical leaves* ternate, on long depressed petioles, channelled above, covered with long spreading soft silky hairs, and clasping the bulbous stem with their dilated membranous bases,

lateral leaflets nearly sessile, central one petiolate, 3-partite, lobes trifid, segments ovato-lanceolate, inciso-dentate, tipped with pale callous tubercles, margins ciliated, softly hairy on both surfaces. *Stem-leaves* alternate, on shorter petioles dilated and clasping at the base, with a linear-lanceolate leaf on each side, all the lobes on more equal stalks, segments narrower and more acuminate; *uppermost* sessile, of 3 linear-lanceolate leaflets. *Peduncles* opposite to the leaves, pentangular, deeply furrowed, clothed with erect appressed hairs, single-flowered. *Sepals* 5, ovato-lanceolate, obtuse, 3-ribbed, central rib branched, with long silky hairs on the exterior, concave, membranous at the edges, reflexed from about the middle. *Petals* rounded, cuneate at the base, margin entire, bright shining yellow, with a radiated watery spot above the base. *Nectary* a slight depression, covered by an obcordate convex fleshy scale, the lobes broad and round, contracted and sinuated at the base. *Seeds* lenticular, round, compressed, punctato-striate, margined, with a short stout straight beak. *Receptacle* hairy.

††† *Leaves divided. Pericarps tuberculated or muricated. Annual.*

12. R. *hirsutus*, Curt. *Pale hairy Crowfoot*. Root fibrous; stem erect, many-flowered, with spreading hairs; radical leaves ternate, leaflets 3-fid, inciso-dentate, upper ones 3-lobed or 3-partite, lobes cut; peduncles furrowed; sepals reflexed; pericarps orbicular, compressed, margined, with a series of tubercles near the margin. *E. Bot. t.* 1501. *E. Fl. v. iii. p.* 50. *Hook. Br. Fl. p.* 269.—*R. Philonotis, Ehrh.*

Corn-fields; rare. *Fl.* June—October. ⊙.

Welbatch; *Mr. T. Bodenham.* Near Albrighton, (Shrewsbury); *Mr. E. Elsmere, junr.!* In a field going from Ashford Village to Saltmoor, on the right hand side of the lane; *Miss Mc Ghie.*

Stem 4—18 inches high, dichotomously branched, round, furrowed, clothed with long silky spreading hairs. *Lower leaves* on long petioles clasping by their dilated membranous base, ternate, leaflets subrotund, 3-fid, inciso-dentate, hairy on both sides, margins ciliated; *upper ones* subsessile, 3-fid, segments narrow. *Peduncles* with erect appressed hairs, single-flowered. *Sepals* ovato-acuminate, externally hairy, deflexed. *Petals* obovate, bright shining yellow. *Seeds* orbicular, compressed, with a triply keeled margin, surmounted by a short stout beak, minutely tuberculated, and with an exterior irregular row or two of larger tubercles near the margin.

13. R. *arvensis*, Linn. *Corn Crowfoot*. Root fibrous; stem erect, many-flowered, smooth; radical leaves 3-fid, dentate, cauline ones 3-nate, leaflets subpetiolate, 3-multi-fid, segments cuneiform, toothed at the summit, upper ones linear, entire; sepals spreading; pericarps plane, compressed, margined, muricated. *E. Bot. t.* 135. *E. Fl. v. iii. p.* 53. *Hook. Br. Fl. p.* 269.

Corn-fields; not very common. *Fl.* June. ⊙.

Windmill Hill, near Wenlock; *W. P. Brookes, Esq.!* Welbatch; *Mr. T. Bodenham.* Near Little Wenlock; *Mr. F. Dickinson.* Astley; *Mr. E. Elsmere, junr.!* Walford; *T. C. Eyton, Esq.* Oakley Park, near Ludlow; *Mr. H. Spare.* Cupid's Ramble, and Lady Hills, Westfelton; *J. F. M. Dovaston, Esq.*

Uffington; Bomere; Berrington; Shrewsbury; Westfelton moor. Buckley Farm, near Oswestry; Corn-field between Battlefield and Albright Hussey.

Stem 1 foot high, erect, more or less branched, round, nearly smooth and pale green, as is the whole plant. *Leaves* alternate, upper ones partly opposite, divided in a tripartite mode. *Flowers* small, on solitary peduncles opposite to the leaves. *Sepals* lanceolate, spreading, hairy. *Petals* obovate, pale lemon-colour.

Pericarps large, plane, compressed, densely covered with large prominent prickles, crowned by a long spinous beak and surrounded by a prominent margin.

14. R. *parviflorus*, Linn. *Small-flowered Crowfoot*. Root fibrous; stem prostrate, with horizontal hairs; leaves cordato-rotund, lower ones 5-lobed, crenato-dentate, upper ones 3-lobed or entire; peduncles opposite the leaves; sepals erect in flower, finally reflexed; petals minute, seldom all perfected; pericarps orbicular, compressed, margined, muricated. *E. Bot. t.* 120. *E. Fl. v. iii. p.* 53. *Hook. Br. Fl. p.* 269.

Dry hedge-banks; not uncommon. *Fl.* May, June. ⊙.

Wenlock; *W. P. Brookes, Esq.!* Ellesmere; *Rev. A. Bloxam.* Welbatch; *Mr. T. Bodenham.* Near Diddlebury, in Corvedale; *Rev. W. Corbett.* Albright Hussey, near Shrewsbury; *Mr. E. Elsmere. junr.!* Near Oswestry; *Rev. T. Salwey.* Twyford, near Westfelton; *J. F. M. Dovaston, Esq.* Between Shrewsbury and the Weeping Cross turnpike. Ditch banks about Habberley. Alberbury. Plentifully in a pasture between Pitchford and Cound Stank, 1800. New Inn, near Ludlow. Uffington. Whittington. Oswestry. Norton, near Atcham. Plentifully in a clover field near Cronkhill, 1807; *Rev. E. Williams's MSS.*

Shrewsbury; Pulley; near Sharpstones hill; Westfelton; Queen Eleanor's bower, Haughmond hill; near the castle, on the summit of Haughmond hill immediately above the Douglas Crag; Montford Village; Shrawardine; Tre Vawr Clawdd, near Oswestry; School house, near Lea-Cross, near Hanwood; and generally on the hedge-banks in the immediate vicinity of villages, hamlets, or houses, throughout the county.

Root fibrous. *Stems* prostrate, numerous, very variable in length, spreading, from 3 or 4 to 18 inches, round, hollow, leafy, clothed with soft horizontal hairs. *Leaves* alternate, cordato-rotund, hairy on both surfaces, on hairy petioles, channelled above, dilated and clasping the stem with a hairy membranous appendage on each side, *lower ones* 5-lobed, lobes unequally crenato-dentate, crenatures obtuse and tipped with a pale callous tubercle, lobes and crenatures of the *upper* leaves narrower and more acuminate; *uppermost* leaves nearly sessile, simple, lanceolate, entire. *Peduncles* solitary, opposite and about equal to the leaf, hairy or glabrous, bearing a simple minute flower. *Sepals* ovate, obtuse, membranous and recurved at the edges, 3-ribbed, hairy externally, erect on the first expansion of the flower, but soon becoming reflexed as the carpels increase in size. *Petals* seldom all perfected, elliptical, obtuse, with a long claw, thick, deep shining yellow with a pale watery mark above the claw, 5-ribbed, equal to the sepals. *Nectary* an obtuse truncated thick scale. *Seed* roundish, lenticular, compressed, very minutely tuberculated, armed with incurved prickles arising from larger scattered tubercles, surrounded with a smooth sharp margin and crowned with a broad recurved smooth beak.

17. TROLLIUS. *Linn.* Globe-flower.

1. T. *Europæus*, Linn. *Mountain Globe-flower*. Leaves palmato-fid, lobes 5, rhomboid, tripartite, inciso-serrate; peduncles angular and furrowed; sepals about 12, petaloid, concave, converging into a globe; petals 10, ligulate, half the length of the sepals; fruit aggregated into flattened heads. *E. Bot. t.* 28. *E. Fl. v. iii. p.* 56. *Hook. Br. Fl. p.* 269.

Moist pastures; rare. *Fl.* June, July. ♃.

In the upper part of the parish of Oswestry; *Rev. T. Salwey.!* Meadow, near Walford; *T. C. Eyton, Esq.* Meadows at the Hayes, near Oswestry; *T. &*

D. Bot. Guide. Aston; Maesbury; very common; *J. F. M. Dovaston, Esq.* Meadows between Halston and Whittington. In a small meadow on the left hand side of the road adjoining Maestermin bridge, between Halston and Hardwick; *Rev. E. Williams's MSS.*

Fairy-land, Westfelton. Meadows adjoining turnpike road opposite the lodge of Aston park, near Oswestry.

Root fibrous. *Stems* several, 1½-2½ feet high, erect, branched above, round, somewhat compressed, hollow, smooth, leafy. *Radical leaves* on very long round smooth channelled footstalks, dilated membranous and clasping at the base, deeply palmato-fid, lobes 5, rhomboid, deeply 3-partite, segments irregularly but acutely inciso-serrate, smooth veiny and deep-green above, pale beneath. *Stem leaves* alternate, similar, the lobes narrower and more of an ovato-lanceolate form, nearly sessile by their dilated membranous amplexicaul base; *uppermost* of 3 sessile ovato-lanceolate serrated acute leaflets. *Peduncles* smooth, angular and furrowed, the angles corresponding with the number of sepals, single-flowered. *Sepals* about 12, large, handsome, bright yellow, oblongo-obovate, petaloid, concave, converging into a globe, the exterior ones more rounded, margins irregularly crenated or notched, arranged in whorls of 3, the innermost whorl of 5. *Petals* 10, half as long as the sepals, deep yellow, alternate with the innermost whorls of sepals, ligulate, somewhat dilated and rounded at the apex, thick and fleshy. *Nectary* a deep groved depression with a fleshy elevated margin, above the base of the claw. *Fruit* of numerous, cylindrical, compressed, keeled *follicles* with a bifid slightly recurved beak, dehiscing by a suture on the edge next the axis, seated on a depressed spongy *receptacle*, forming flattened heads. *Seeds* black and shining.

18. CALTHA. *Linn.* Marsh-marigold.

1. C. *palustris*, Linn. *Common Marsh-marigold.* Leaves orbiculari-cordate or reniform, minutely crenate; sepals 5, petaloid, roundish oval. *E. Bot. t.* 506. *E. Fl. v. iii. p.* 59. *Hook. Br. Fl. p.* 269.

Marshy places, ditches, &c.; common. *Fl.* March—June. ♃.

Root thick, somewhat tuberous, with many simple fibres. *Stem* 12—18 inches high, erect, round or somewhat compressed, hollow, leafy, branched, furrowed. *Leaves* orbiculari-cordate or reniform, crenate, lowermost on long obsoletely triangular footstalks, upper ones nearly sessile, alternate. *Stipules* membranous, withering. *Flowers* 3—5, large, bright yellow, on alternate solitary peduncles. *Sepals* 5, roundish oval.

CLASS XIV.

DIDYNAMIA.

———

"Here's flowers for you;
Hot lavender, mints, savory, marjoram;
* * * these are flowers
Of middle summer."

SHAKSPEARE.

CLASS XIV.

DIDYNAMIA. 4 *Stamens; 2 longer than the other* 2.

ORDER I. GYMNOSPERMIA. *(γυμνος, naked, and σπερμα, seed.) Germen a pericarp deeply 4-lobed, closely investing the* 4 *seeds, and thus presenting the appearance of* 4 *apparently naked seeds.* (All belonging to the *Nat. Ord.* LABIATÆ, *Juss.*)

A. *Calyx 5-toothed, nearly equal, (not 2-lipped.)*

* *Stamens equal in length.*

1. MENTHA. *Calyx* equal, 5-toothed, its *mouth* naked or rarely villous. *Corolla* nearly regular, 4-cleft, its *tube* very short. *Stamens* distant, exserted or included. *Filaments* naked. *Anthers* with 2 parallel cells.—Name,—μινθα or μινθη, an ancient Greek term.

** *Inferior stamens shorter.*

2. NEPETA. *Calyx* tubular, 15-ribbed, its mouth a little oblique, 5-toothed. *Corolla* with the tube exserted; upper *lip* emarginate or bifid; lower 3-fid, lateral lobes reflexed, middle one broad, concave, sinuate.—Name, according to some from *Nepi*, a town in Italy; according to others from *Nepa*, a *scorpion*, for whose bite the plant was considered a cure.

3. GLECHOMA. *Calyx* tubular, 15-ribbed, its mouth somewhat oblique, 5-toothed. *Corolla* with the tube exserted, upper *lip* obcordate, margins recurved; lower 3-lobed, lateral lobes spreading, recurved, middle one broad, plane, emarginate. *Anthers* before bursting, approaching in pairs and forming a cross.—Name, γληχων, —applied by the Greeks to *Pennyroyal.*

*** *Inferior stamens longer.*

† *Mouth of the calyx with a ring of hairs.*

4. ORIGANUM. *Calyx* ovato-tubular, 10—13 ribbed, 5-toothed. *Corolla* with the upper *lip* erect, nearly plane, emarginate; lower 3-lobed, patent, lobes nearly equal. *Flowers* in cylindrical or oblong crowded bracteated spikes.—Name from *opos*, a *hill*, and *γανος, joy*; from the hilly situations in which it grows.

5. TEUCRIUM. *Calyx* tubular, 5-toothed, nearly equal or 2-lipped. *Corolla* with the upper *lip* bipartite; lower one patent 3-fid. *Stamens* much exserted. *Cells* of the *Anthers* confluent, spreading. —Name from *Teucer*, Prince of Troy, who first employed the plants medicinally.

6. MARRUBIUM. *Calyx* tubular, 10-ribbed, teeth 5—10, spreading. *Corolla* scarcely longer than the calyx; upper *lip* straight, linear, cloven; lower one 3-lobed, middle lobe the largest, emarginate. *Stamens* included within the tube.—Name of doubtful origin.

†† *Mouth of the calyx naked.*

7. AJUGA. *Calyx* ovate, 5-ribbed, 5-cleft, teeth nearly equal. *Corolla* with the tube exserted: upper *lip* short, erect, entire or emarginate; lower one larger, patent, trifid. *Stamens* ascending, protruded above the upper lip.—Name altered from the *Abiga (abigo, to drive away)* of the Latins, a medicinal plant allied to this.

8. BALLOTA. *Calyx* salver-shaped, equal, with 10 ribs and 5 broad mucronated teeth. *Corolla* with the upper *lip* erect, concave; lower one trifid, middle lobe the largest, emarginate. *Cells* of the *Anthers* spreading.—Name βαλλωτη, from βαλλω, *to reject*; on account of its disagreeable smell.

9. LEONURUS. *Calyx* tubular, with 5 or 10 ribs, equal, with 5 subulate teeth. *Corolla* with the upper *lip* very hairy above, entire; lower one patent, trifid. *Anthers* sprinkled with shining dots.—Name from λεων, a *lion*, and ουρα, a *tail*; from a fancied resemblance in the plant to a Lion's tail.

10. GALEOBDOLON. *Calyx* campanulate, 5-ribbed, nearly equal, 5-toothed. *Corolla* with the upper *lip* incurved, arched, entire; lower one smaller, in 3 nearly equal acute lobes.—Name from γαλεη, a *weasel*, and βδολος, a *fetid scent*; formerly considered synonymous with *Galeopsis*, from which genus it is now removed.

11. GALEOPSIS. *Calyx* campanulate, equal, 5-toothed, teeth mucronate. *Corolla* with the *tube* exserted, the throat inflated: upper *lip* arched; lower one with 3 unequal lobes, having two teeth on its upper side.—Name from γαλεη, a *weasel*, and οψις, *aspect* or *appearance*; from a resemblance in the lips of the flower to the snout of an animal.

12. LAMIUM. *Calyx* campanulate, 10-ribbed, 5-toothed, nearly equal. *Corolla* with the throat inflated: upper *lip* entire, arched; lower one patent, 2-lobed, with one or two teeth on each side at the base.—Name from λαιμος, the *throat*; on account of the shape of the flower.

13. BETONICA. *Calyx* ovate, 10-ribbed, teeth equal, awned. *Corolla* with the *tube* exserted, cylindrical: upper *lip* ascending; lower one patent, trifid, its middle lobe entire, or nearly so.—Name altered from *Bentonic*, Celtic : (*ben, head,* and *ton, good or tonic,*) from its cephalic properties.

14. STACHYS. *Calyx* subcampanulate, 10-ribbed, teeth 5, nearly equal, acuminate. *Corolla* with the *tube* as long as the calyx : upper *lip* mostly arched, entire; lower one 3-lobed, with the two lateral lobes reflexed.—This genus scarcely differs from *Betonica* but in the shorter tube of its corolla. Name from σταχυς, a *spike*; in allusion to the nature of the inflorescence.

B. *Calyx 2-lipped. Inferior stamens longer.*

* *Calyx-teeth acute.*

15. THYMUS. *Flowers* whorled or capitate. *Calyx* 10-ribbed, tubular, 2-lipped : upper *lip* 3-toothed; lower one bifid, throat hairy. *Corolla* with the upper *lip* erect, nearly plane, notched; lower one patent and trifid. Name from θυμος, *strength*; from its balsamic odour, strengthening the animal spirits.

16. ACINOS. *Whorls* few-flowered. *Calyx* tubular, 13-ribbed, gibbous at the base below; upper *lip* 3-, lower 2-fid, throat hairy. *Corolla* with the upper *lip* nearly plane; lower one trifid, middle lobe nearly entire.—Name applied by the Greeks to some aromatic plant.

17. CALAMINTHA. *Flowers* axillary, somewhat solitary, or often in loose bracteated *cymes*. *Calyx* tubular, 13-nerved, nearly equal at the base : upper *lip* 3-toothed; lower one bifid, the throat mostly hairy. *Corolla* with the upper *lip* nearly plane, emarginate; lower one 3-fid, middle lobe emarginate.—Name from καλος, *good*, and μινθα, *mint*; from the supposition of its scent driving away serpents.

18. CLINOPODIUM. *Whorls* many-flowered, with numerous linear *bracteas*, forming a sort of *involucre*. *Calyx* tubular, 13-nerved, nearly equal at the base, often curved; upper *lip* 3-toothed; lower one bifid. *Corolla* with the upper *lip* nearly plane, emarginate; lower one 3-lobed, middle lobe emarginate.—Name from κλινη, a *bed*, and ποδιον, a *support*; from the compact and stalked head of flowers.

** *Calyx-teeth nearly obsolete.*

19. PRUNELLA. *Calyx* ovate : upper *lip* plane, more or less distinctly 3-toothed; lower one bifid. *Corolla* with the upper *lip* nearly entire, arched; lower one 3-lobed. *Filaments* with two teeth at the extremity, one bearing the *anther*.—Name from the German *bräune*, the *quinsy*, whence comes *Brunella* of Ray, softened into *Prunella*.

20. SCUTELLARIA. *Calyx* broadly ovate, having a conspicuous concave tooth or scale on the upper side; its 2 nearly equal entire *lips* closed after flowering. *Corolla* with the tube much exserted :

upper *lip* straight, arched; lower one trifid.—Name from *scutella*, a *little dish* or *cup*, which the calyx with its appendage or ear somewhat resembles.

(See *Salvia* in CL. II.)

ORDER II. ANGIOSPERMIA. (αγγειον, a *vessel* or *capsule*, and σπερμα, *seed*.) *Seeds enclosed in a distinct capsule.*

* *Calyx 4-cleft.*

21. EUPHRASIA. *Calyx* tubular, 4-fid, persistent. *Corolla* ringent, upper *lip* truncate or emarginate, lower 3-fid. *Anthers* spurred at the base. *Capsule* oblong, compressed, 2-celled, many-seeded. *Nat. Ord.* SCROPHULARINEÆ, *Juss.*—Name from ευφρασια, *joy*; in allusion to its properties.

22. RHINANTHUS. *Calyx* inflated. Upper *lip* of the *Corolla* compressed laterally; lower one plane, 3-lobed. *Capsule* of 2-cells, obtuse, compressed, with many imbricated, flat and margined *seeds*. —*Nat. Ord.* SCROPHULARINEÆ, *Juss.*—Name from ριν, a *nose*, and ανθος, a *flower*; in allusion to the beaked upper lip of the corolla which is very remarkable in the *R. Elephas*.

23. MELAMPYRUM. *Calyx* tubular, 4-fid. *Corolla* ringent, upper *lip* laterally compressed, emarginate, margin reflexed; lower lip tridentate. *Capsule* ovate, oblique, opening at one side, 2-celled, cells 2-seeded. *Seeds* gibbous at the base, smooth.—*Nat. Ord.* SCROPHULARINEÆ, *Juss.*—Name from μελας, *black*, and πυρος, *wheat*; from the black seeds bearing a resemblance in size and shape to grains of wheat.

24. LATHRÆA. *Calyx* campanulate. *Corolla* tubular, 2-lipped, the upper *lip* concave. A depressed *gland* is at the base of the *germen*. *Capsule* 2-valved, 1-celled, having two spongy *receptacles* in the middle of each valve.—*Nat. Ord.* OROBANCHEÆ, *Rich.*—Name from λαθραιος, *hid* or *concealed*; from the downward direction taken by the stem in its growth and its remaining subterraneously concealed.

** *Calyx 5-cleft, (in Pedicularis irregular.)*

25. PEDICULARIS. *Calyx* inflated, 5-cleft, or unequally 2-3-lobed, jagged, somewhat leafy. Upper *lip* of the *Corolla* laterally compressed, arched; lower one plane, 3-lobed. *Capsule* oblique, compressed, 2-celled. *Seeds* angular.—*Nat. Ord.* SCROPHULARINEÆ, *Juss.*—Name from *pediculus*, a *louse*; in allusion to its supposed property of producing the lousy disease in sheep which feed upon it.

26. ANTIRRHINUM. *Calyx* 5-partite. *Corolla* personate, gibbous at the base, (no distinct spur,) its *mouth* closed by a projecting palate. *Capsule* 2-celled, oblique, opening by 3 pores at the extremity. —*Nat. Ord.* SCROPHULARINEÆ, *Juss.*—Name from αντι, *in the place of*, and ριν, a *nose*; in allusion to the apex representing the nose in the masklike capsule.

27. LINARIA. *Calyx* 5-partite. *Corolla* personate, spurred at the base; its mouth closed by a projecting palate. *Capsule* ventricose, 2-celled, opening by valves or teeth.—*Nat. Ord.* SCROPHULARINEÆ, *Juss.*—Name from *linum*, *flax*, which the leaves of some species resemble.

28. SCROPHULARIA. *Calyx* 5-lobed (or in *S. vernalis* deeply 5-cleft.) *Corolla* subglobose; its *limb* contracted with 2 short *lips*; the upper with 2 lobes and frequently a small scale or abortive stamen *(staminodium)* within it; the lower 3-lobed. *Capsule* 2-celled, 2-valved, the margins of the valves turned inwards.—*Nat. Ord.* SCROPHULARINEÆ, *Juss.*—Name from the *Scrophula*, a disease which this plant was supposed to cure.

29. DIGITALIS. *Calyx* in 5, deep, unequal segments. *Corolla* campanulate, inflated beneath; *limb* obliquely 4—5-lobed, unequal. *Capsule* ovate, 2-celled, many seeded.—*Nat. Ord.* SCROPHULARINEÆ, *Juss.*—Name from *digitale*, the *finger of a glove*, which its flowers resemble.

30. LIMOSELLA. *Calyx* 5-cleft, equal. *Corolla* shortly 5-cleft, campanulate, equal. *Stamens* nearly equal. *Stigma* capitate. *Capsule* globose, 2-valved.—*Nat. Ord.* SCROPHULARINEÆ, *Br.*—Name from *limus, mud*; the plant growing in muddy places.

31. VERBENA. *Calyx* tubular, with 5 teeth, one of them generally shorter than the rest. *Corolla* tubular, with the *limb* rather unequal, 5-cleft. *Stamens* included, (sometimes only 2.) *Seeds* 2 or 4, enclosed in a thin evanescent pericarp.—*Nat. Ord.* VERBENACEÆ, *Juss.*—Name *ferfaen*, Celtic, derived from *fer*, to *drive away*, and *faen*, a *stone*; from its supposed virtue in curing the complaint so called.

*** *Calyx lateral in 2, generally combined, often bifid segments.*

32. OROBANCHE. *Calyx* of 2 lateral, often combined and bifid segments, bracteated. *Corolla* ringent, 4—5-cleft. A *gland* at the base of the *germen* beneath. *Stigma* capitate. *Capsule* 2-valved, bearing numerous minute *seeds*, on parietal longitudinal *receptacles*. —*Leafless, brown or purplish, herbaceous, scaly plants, often attached to the roots of other plants.*—*Nat. Ord.* OROBANCHEÆ, *Vent.*—Name from οροβος, a *leguminose* or *pea-like plant*, and αγχειν, to

strangle; in allusion to the injury effected by it on plants of that description, to which it is often attached.

DIDYNAMIA—GYMNOSPERMIA.

1. MENTHA. *Linn.* Mint.

* *Stems with reflexed pubescence.*

† *Calyx all over hairy.*

1. M. *sylvestris,* Linn. *Horse-mint.* Leaves subsessile, ovate or ovato-lanceolate, acute, coarsely and unequally serrated, downy, shaggy and hoary beneath; spikes lineari-cylindrical, whorls dense, lowermost slightly remote; bracteas subulate, ciliated on the margins and keel with erect hairs; calyx 10-ribbed, with long silky erect hairs, teeth lineari-setaceous; pedicels with reflexed hairs. *E. Bot. t.* 686. *E. Fl. v. iii. p.* 74. *Hook. Br. Fl. p.* 276.

Moist waste ground; rare. *Fl.* August—September. ♃.

Oakley park meadows, near Ludlow; *Mr. H. Spare.* By the side of the road in the village of Monk's Hopton; on the left hand side of the road between Uckington and Overley Hill, about one or two hundred yards below the turning to Walcot; *Rev. E. Williams's MSS.*

Stems 2-3 feet high, erect, bluntly quadrangular, covered with soft silky white deflexed hairs, most densely so below each joint and on the peduncles, with opposite, erecto-patent branches. *Leaves* opposite, subsessile, spreading, varying from ovate to ovato-lanceolate, acute, sharply coarsely and strongly serrated, serratures with a callous tip, downy on the upper surface especially on the veins, under surface shaggy with dense soft white hairs and whitish glandular mealiness. *Spikes* several, terminal, erect, paniculate, lineari-cylindrical, generally acute. In the ovate-leaved varieties the spikes are thicker and obtuse. *Whorls* numerous, many-flowered, dense, bracteated, lower one somewhat distant. *Bracteas* subulate, ciliated on the margins and keel with erect silky hairs; outer pair largest. *Pedicels* with reflexed hairs. *Calyx* 10-ribbed, with erect hairs, teeth lineari-setaceous. *Corolla* pale purple, externally downy.

2. *M. hirsuta,* Linn. *Hairy Mint.* Leaves petiolate, ovate, acute, coarsely and unequally serrated, hairy on both surfaces, especially on the ribs beneath; inflorescence capitate or whorled; bracteas lanceolate; calyx about 13-ribbed, with erect hairs, teeth subulate from a triangular base; pedicels with reflexed hairs. *E. Bot. t.* 447. *Hook. Br. Fl. p.* 277. α. and ζ. *E. Fl. v. iii. p.* 79. *M. sativa, L. E. Bot. t.* 448.

Banks of rivers, ditches and watery places; common. *Fl.* Aug. Sept. ♃.

Stem 2-3 feet high, erect, quadrangular and furrowed, clothed with rather short soft white deflexed hairs, most densely so in the upper part, with opposite spreading branches, similarly clothed. *Leaves* opposite, petiolate, hairs of petiole erect, ovate, subcordate at the base, acutely coarsely and unequally serrated, serratures with callous tips, margins thickened, upper surface with scattered hairs, under surface paler and with imbedded glandular dots, hairy especially on the ribs. *Inflorescence* sometimes in a single terminal subglobose head, sometimes capitate with one or more pair of stalked axillary whorls below the head, and sometimes (*var.* ζ. *Sm.*) the head is absent and all the whorls are axillary and sessile.

Whorls many-flowered, dense, crowded, bracteated. *Bracteas* lanceolate, acute, with erect hairs. *Pedicels* densely covered with white reflexed hairs. *Calyx* tubular, furrowed, about 13-ribbed, resinoso-glandulose, with erect hairs, teeth short, subulate from a triangular base. *Corolla* bluish purple, externally downy. *Stamens* generally, but not invariably exserted.

Withering (*Bot. Arr. v. iii. p.* 522.) mentions that the variety with the heads and leaves smaller rounder and blunter, and the flowers smaller, not so much branched, of which Dillenius in *Raii Syn.* 233. *t.* 10. *f.* 1. gives a figure, is found "near the Mill at Lilleshall, Shropshire." This is the var: β. of *English Flora,* of which Sir J. E. Smith thus speaks in *Tr. Linn. Soc. v.* 5. *p.* 193. "What I have marked β. is so distinguished only in deference to Dillenius, and more with respect to the sweet smell he attributes to it, than to any thing in its form. I have not been so fortunate as to find this variety in any old herbarium." Dillenius thus describes its odour: "Odor valde gratus et aromaticus, mala aurantia plane æmulans." I have never gathered any specimen which in the least agrees with this description, and have introduced it in the hope that some Botanist in the neighbourhood of Lilleshall will direct his attention to this point and enable us to determine what plant Dillenius meant.

3. M. *arvensis,* Linn. *Corn-Mint.* Leaves shortly petiolate, ovate or elliptical rather obtuse, tapering at the base, with shallow serratures, hairy on both surfaces, especially on the midrib above; inflorescence in sessile axillary whorls; bracteas 4, lanceolate; calyx campanulate, with horizontally spreading hairs, teeth triangular; pedicels smooth or with reflexed hairs. *E. Bot. t.* 2119. *E. Fl. v. iii. p.* 86. α. *and* β. *Hook. Br. Fl. p.* 277.

α. borders and furrows of corn-fields. β. banks of rivers and wet places; common. *Fl.* August, September. ♃.

Stem 6—18 inches high, erect, bluntly quadrangular and furrowed, with rather short, soft, white deflexed hairs, principally so on alternate sides between each pair of leaves, with opposite spreading branches. *Leaves* opposite, shortly petiolate, petioles with spreading hairs, ovate or elliptical, rather obtusely pointed, tapering at the base into the winged petiole, with shallow serratures more or less obtuse and callously tipped, margins ciliated, hairy on both surfaces, especially on the midrib above, paler beneath, resinoso-glandulose. *Inflorescence* in small, many-flowered, sessile, axillary whorls. *Bracteas* 2 to each semiwhorl, small, lanceolate, hairy beneath. *Pedicels* smooth or with a few reflexed hairs. *Calyx* campanulate, slightly furrowed, resinoso-glandulose, with horizontally spreading hairs, teeth triangular. *Corolla* pale bluish-purple, externally downy. *Stamens* in α. generally included, in β. exserted. In the var. β. the plant is larger, more branched, less hairy, the leaves nearly smooth and shining.

†† *Calyx smooth except the teeth.*

4. M. *piperita,* Sm. *Pepper-Mint.* Leaves petiolate, ovato-lanceolate, acute, strongly serrated, glabrous above, resinoso-glandulose and hairy on the ribs beneath; spikes bluntish, interrupted, whorls stalked, lax; bracteas lanceolate, ciliated; calyx slender, 10-ribbed, glabrous, resinoso-glandulose, teeth lanceolate-subulate, ciliated with erect hairs; pedicels glabrous. *E. Bot. t.* 687. α. *and* β. *E. Fl. v. iii. p.* 77. *Hook. Br. Fl. p.* 276.

Watery places; rare. *Fl.* August, September. ♃.

Canal bank at Weston Lullingfields Wharf; *J. E. Bowman, Esq!* Buildwas brooks; *Mr. F. Dickinson.* Ditches at Hadnall, near Shrewsbury; *Mr. E.*

Elsmere, junr. (*Var.* α. *E. Fl. l. c.*) On some waste ground in the village of Haughton, near Sundorn. (*Var.* β. *E. Fl. l. c.*) Buildwas mill-stream below the mill and Lawley's Cross mill-stream above Buildwas; *Rev. E. Williams's MSS.*

Wet ditch by turnpike road-side at Nobold, near Shrewsbury.

Stem 18 inches to 2 feet high, erect, quadrangular and furrowed, roughish with recurved hairs, branched. *Leaves* opposite, petiolate, petioles ciliated, ovato-lanceolate, acute, sharply and strongly serrated, serratures with callous tips, margins thickened, dark green and glabrous above, paler beneath with innumerable scattered shining glandular resinous dots, hairy on the ribs. *Spikes* rather short and blunt, interrupted in the lower part with one or more distant whorls; *whorls* stalked, rather lax. *Bracteas* lanceolate, ciliated. *Pedicels* purple, round, glabrous. *Calyx* slender, furrowed, 10-ribbed, copiously covered with prominent resinoso-glandular dots, smooth, except the lanceolate-subulate *teeth* which are ciliated with erect hairs. *Corolla* pale purple, externally glabrous. *Stamens* included.

** *Stems with spreading pubescence.*

5. M. *rotundifolia,* Linn. *Round-leaved Mint.* Leaves sessile, elliptical, obtuse, crenato-serrate, wrinkled and downy above, densely shaggy or woolly beneath; spikes lineari-cylindrical, interrupted, whorls dense except the lowermost; bracteas lanceolate, hairy beneath; calyx campanulate, with spreading hairs, teeth lanceolate-subulate, with erect hairs; pedicels with reflexed hairs. *E. Bot. t.* 446. *E. Fl. v. iii. p.* 74. *Hook. Br. Fl. p.* 276. *Sole's Menth.* 7. *t.* 3.

Moist places, in waste ground; not common. *Fl.* August, September. ♃.

On a common at Elsmoor (Ellesmere?)—Wenlock Abbey; *Mr. Sole.* Bishop's Castle and Clunton; *T. & D. Bot. Guide.* Botterells Aston and Norton; *Mr. G. Jorden.* Hopton Castle churchyard; *Miss Mc Ghie.* By the side of the road in Longnor and Presthope. Under the walls of Haughmond Abbey garden. Under the wall of Diddlebury churchyard. In Munslow and Stokesay churchyards. By the side of the road between Culmington and Diddlebury. Under the walls of Woodhouse Friars, near Cleobury Mortimer. By the side of the roads in Haughton, near Sundorne, between Mr. Dickson's and Mr. Ore's house. In Brocton, by the turning to Easthope. On the left hand side of the road between Stretton and the Marsh. Roadside in Astley, near Shrewsbury; *Rev. E. Williams's MSS.*

Preston Gobalds churchyard near Shrewsbury.

Stem 2-3 feet high, erect, quadrangular, clothed with dense, spreading, soft, almost woolly hairs slightly deflexed in the upper part, branched. *Leaves* opposite, sessile, lower ones elliptical obtuse, somewhat cordate at the base, crenato-serrate; upper ones rounder, more sharply serrated, wrinkled and downy above, densely shaggy or woolly beneath, the veins naked and prominent. *Spikes* paniculate, lineari-cylindrical, interrupted, more or less obtuse, whorls dense and crowded except the lower ones. *Bracteas* lanceolate, hairy beneath. *Calyx* small, campanulate, with spreading hairs, teeth lanceolato-subulate, as long as the tube, with erect hairs. *Pedicels* with minute deflexed hairs. *Corolla* pale purple. *Herbage* of a peculiar acrid odour.

*** *Stems smooth or with slight recurved pubescence.*

6. M. *viridis,* Linn. *Spear-mint.* Leaves sessile, lanceolate, acute, serrated, glabrous, paler and resinoso-glandulose beneath; spikes interrupted; bracteas setaceous, somewhat hairy; calyx narrow, campanulate, resinoso-glandulose, smooth, teeth subulate,

smooth or slightly hairy; pedicels smooth. *E. Bot. t.* 2424. *E. Fl. v. iii. p.* 76. *Hook. Br. Fl. p.* 277. *Sole's Menth.* 11. *t.* 5.

Marshy places; rare. *Fl.* August. ♃.

By the side of Buildwas mill-stream below the mill. Eaton brook under Newman Hall; *Rev. E. Williams's MSS.*

Stem 2-3 feet high, erect, branched, acutely angular, smooth, often purplish. *Leaves* generally smooth, sometimes very slightly hairy on the ribs beneath, margins thickened, serratures with callous tips. *Spikes* elongated, acute, the whorls all slightly distant. *Pedicels* perfectly smooth, round and polished. *Calyx* furrowed, 10-ribbed, resinoso-glandulose, always quite smooth except the teeth which though generally smooth are sometimes slightly hairy. *Corolla* light purple, smooth.

7. M. *rubra,* Sm. *Tall Red Mint.* Leaves petiolate, broadly ovate, acute, strongly serrated, with scattered hairs on both surfaces, paler and resinoso-glandulose beneath; inflorescence in shortly stalked axillary dense whorls; bracteas linear, ciliated and hairy beneath; calyx tubular, resinoso-glandulose, smooth in the lower part, teeth ovato-acuminate, with erect hairs on the margins and ribs; pedicels smooth. *E. Bot. t.* 1413. *E. Fl. v. iii. p.* 83. *Hook. Br. Fl. p.* 277. *Sole's Menth.* 47. *t.* 21.

Banks of rivers; rare. *Fl.* September. ♃.

By the side of Pitchford Ford mill-stream near the cascade. By the side of the road in Monk's Hopton; *Rev. E. Williams's MSS.* Meole brook; *J. E. Bowman, Esq.* In various waters in Colebrookdale, as well as in a wet place between Kidderminster and Bridgnorth. *Mr. Sole.*

Stem 2-3 or more feet high, erect, somewhat wavy or zigzag, quadrangular, furrowed, nearly smooth, with here and there a little recurved pubescence, branched. *Leaves* opposite, petiolate, petioles channelled, ciliated with spreading hairs, broadly ovate, acute, strongly serrated, of a deep shining green, serratures with callous tips, margins thickened, with scattered minute hairs on both surfaces, paler beneath, resinoso-glandulose. *Inflorescence* in shortly stalked axillary dense whorls. *Bracteas* linear, ciliated and hairy beneath, outermost larger. *Pedicels* round, perfectly smooth and shining. *Calyx* tubular, perfectly smooth in the lower part, about 10-ribbed, resinoso-glandulose, upper part occasionally hairy, teeth ovato-acuminate, ciliated on the margins and on the ribs with erect hairs. *Corolla* purplish red, externally resinoso-glandulose.

8. M. *gentilis,* Linn. *Bushy Red Mint.* Leaves petiolate, ovato-elliptical, acute, strongly serrated, with scattered hairs on both surfaces, paler, resinoso-glandulose and with stronger hairs on the ribs beneath; inflorescence in shortly stalked axillary small whorls; bracteas lanceolate, hairy beneath; calyx short, campanulate, resinoso-glandulose, smooth in the lower part, teeth lanceolato-acuminate, with erect hairs on the margins and ribs; pedicels smooth. *E. Bot. t.* 2118. *E. Fl. v. iii. p.* 84. *Hook. Br. Fl. p.* 277. *Sole's Menth.* 41. *t.* 18.

Watery places; rare. *Fl.* August. ♃.

In an orchard adjoining Tugford church. By the side of the road in Shipton, Hungerford, and Shineton by the bridge Below Buildwas mill and by the mill stream. By the side of the road in Lilleshall; *Rev. E. Williams's MSS.*

Stem 18 inches to 2 feet high, erect, quadrangular, furrowed, with recurved pubescence, copiously branched. *Leaves* opposite, petiolate, petioles winged, channelled and hairy, ovato-elliptical, acute, tapering at the base into the petiole, strongly coarsely and acutely serrated, serratures with callous tips, margins thickened, with scattered hairs on both surfaces, and stronger ones on the ribs beneath, light pale green, rather paler and resinoso-glandulose on the under surface. *Inflorescence* in distant, axillary, small, dense, shortly stalked whorls. *Bracteas* lanceolate, hairy beneath. *Pedicels* round, purplish, smooth. *Calyx* short, campanulate, 10-ribbed, resinoso-glandulose, perfectly smooth in the lower part, upper part occasionally hairy, teeth lanceolato-acuminate, ciliated on the margins and ribs with erect hairs. *Corolla* pale purple.

**** *Stems prostrate, pubescent.*

9. M. *Pulegium*, Linn. *Penny-royal.* Leaves petiolate, ovate, obtuse, with shallow serratures, nearly glabrous, resinoso-glandulose; inflorescence in sessile, axillary, large, many-flowered whorls; bracteas none; calyx slender, tubular, strongly 10-ribbed and furrowed, resinoso-glandulose, with horizontal hairs, teeth lanceolato-subulate, with erect hairs; pedicels with minute horizontal pubescence. *E. Bot. t.* 1026. *E. Fl. v. iii. p.* 88. *Hook. Br. Fl. p.* 278.

Wet commons and margins of brooks; rare. *Fl.* August, September. ♃.

Wet fields, near Llanymynech; *T. & D. Bot. Guide.*

On Rodington heath 1797, and on Evenwood common, (both since inclosed.) *Rev. E. Williams's MSS.*

Stem 6—8 inches long, procumbent or prostrate, bluntly quadrangular, furrowed, minutely pubescent, copiously branched. *Leaves* petiolate, ovate, obtuse, recurved, with a few shallow unequal serratures, margins thickened, minutely bristly, resinoso-glandulose. *Whorls* very large, globose, distant, axillary, sessile, many-flowered. *Bracteas* none. *Pedicels* with dense, short, horizontal pubescence. *Calyx* tubular, slender, nearly cylindrical, furrowed, prominently 10-ribbed, resinoso-glandulose, covered with horizontal spreading hairs, teeth lanceolato-subulate, ciliated with erect hairs; mouth closed with an internal ring of converging white hairs. *Corolla* light purple, twice as long as the calyx, externally hairy. *Stamens* exserted. *Herbage* with a strong peculiar acrid odour.

2 NEPETA. *Linn.* Cat-mint.

1. N. *Cataria*, Linn. *Cat-mint.* Stem erect, hoary and pubescent; leaves petiolate, cordate, acute, inciso-serrate, greenish above, hoary, pubescent and mealy beneath; flowers in dense, spiked, subpedunculate whorls; calyx pubescent, 15-ribbed, teeth lanceolate, 3 upper ones longer; corolla scarcely longer than the calyx, upper lip small, bifid, lower one 3-lobed, central lobe transversely dilated, sinuate; seeds oval, smooth. *E. Bot. t.* 137. *E. Fl. v. iii. p.* 71. *Hook. Br. Fl. p.* 283.

Hedges, and waste places; not uncommon. *Fl.* July, August. ♃.

Beckbury; *H. Bidwell, Esq.* Wenlock Abbey; *W. P. Brookes, Esq.* Munslow; *Mr. F. Dickinson.* Baschurch; *T. C. Eyton, Esq.* Lilleshall Abbey and road side near Lubstree park; *R. G. Higgins, Esq.* Oreton; *Mr. G. Jorden.* Ludlow; *Miss M'Ghie.* Alderton near Middle; *W. W. Watkins, Esq.* By the side of the road between the Cross Houses and Cound: about Morvill: Cardiston: Ensdon, &c. Ludlow: in Condover: between Shrewsbury and Nobold; *Rev. E. Williams's MSS.*

Sides of the road between Ensdon and Nesscliffe. Near Cross Gates on turnpike road from Shrewsbury to Welchpool. Benthall Edge. Near confluence of rivers Teme and Corve, Ludlow.

Root woody. *Herbage* hoary and pubescent. *Stem* 1½—2 feet high, erect, tetragonous, with opposite branches. *Leaves* opposite, on long channelled hairy petioles, cordate, acute, deeply inciso-serrate, greenish above, softly and densely downy and hoary on both sides, under surface with a glandular mealiness which, on being bruised, emits a peculiar aromatic scent. *Flowers* in dense terminal spikes of subpedunculated, many-flowered, bracteated whorls. *Calyx* tubular, hoary and pubescent, 15-ribbed, teeth subulate, about half the length of the tube, three upper ones longest. *Corolla* yellowish white, externally pubescent, tube short, glabrous; *upper lip* small, deeply bifid into 2 rounded entire lobes; *lower lip* 3-lobed, central lobe transversely dilated, margin sinuate, palate hairy, lateral lobes rounded entire. *Seeds* oval, triquetrous, brown, smooth.

"The seeds of this plant will remain in the earth without vegetating for many years: I have proved them for more than twenty years;" *Mr. G. Jorden.*

3. GLECHOMA. *Linn.* Ground-Ivy.

1. G. *hederacea*, Linn. *Ground-Ivy.* Stems procumbent; leaves petiolate, cordato-reniform, crenate; whorls axillary, pedicellate, secund, 3-flowered; bracteas 2, scarcely equalling the pedicels; calyx hairy and glandular, 15-ribbed, teeth ovate, apiculate, 3 upper ones larger; corolla twice as long as the calyx, upper lip obcordate, margins recurved, lower ones 3-lobed, central lobe broad, dilated, emarginate, wavy; seeds oblong, with impressed dots. *E. Bot. t.* 853. *E. Fl. v. iii. p.* 89. *Hook. Br. Fl. p.* 283.

Hedges and waste places; common. *Fl.* April, May. ♃.

Herbage with an aromatic fragrance. *Root* fibrous, from the lower side of the joints of the long runners, which send up from the opposite sides flowering *stems* 6—12 inches high, erect or ascending, unbranched, quadrangular, rough with deflexed hairs. *Leaves* in pairs, opposite, on long hairy petioles connate at the base by a ciliated process, cordato-reniform, crenate, hairy above, paler (often purplish) beneath, with imbedded glandular dots, ribs only hairy. *Flowers* in axillary, pedicellate, secund, few-flowered whorls. From the summit of a short compressed peduncle, clothed with deflexed hairs, bearing 2 opposite ciliated subulate bracteas, connected by two intermediate obtuse teeth or projections ciliated with long hairs, arise two similarly bracteated peduncles, each bearing a flower on a plain pedicel. Between the 2 secondary peduncles is a 3rd flower on a simple pedicel without bracteas, clothed with deflexed hairs. The intermediate flower blossoms first, and the lateral ones, which are often entirely suppressed, open after the other has withered away. *Calyx* tubular, campanulate, somewhat oblique, the tube twice as long as the teeth, externally hairy and glandular, 15-ribbed, teeth 5, ovate, apiculate, spreading, nearly equal, the 3 upper ones the larger. *Corolla* large, pale blue, hairy externally, 2-lipped; *upper lip* smaller, straight, obcordate, margins recurved, *lateral lobes* spreading, oval, retuse, recurved; central lobe broad, dilated, narrower at the base, emarginate, wavy and irregular at the margin, marked with darker irregular streaks; *tube* laterally compressed, twice the length of the calyx, naked within; *throat* inflated, streaked with dark purple, the *palate* beautifully clothed with white and purple capillary projections. *Filaments* naked, inferior ones shortest, elongated beyond the anthers. *Anthers* approximating in pairs, their cells divergent, so as to appear cruciform: *pollen* white, grains broadly oval. *Seeds* pale, oblong, obsoletely triquetrous, rounded at the apex, covered with impressed dots.

4. ORIGANUM. *Linn.* Marjoram.

1. O. *vulgare*, Linn. *Common Marjoram.* Heads of flowers roundish, panicled, crowded, trichotomous; bracteas ovate, ciliated,

eglandulose; calyx-teeth nearly equal; leaves ovate, obtuse, subserrated, ciliated. *E. Bot. t.* 1143. *E. Fl. v. iii. p.* 107. *Hook. Br. Fl. p.* 278.

Dry hilly pastures; rare. *Fl.* July, August. ♃.

Dowle wood; *Mr. G. Jorden.* Wyre Forest, near Dowles brook; also in profusion by the roadside at Burford, near Tenbury; *E. Lees, Esq.* Ludlow; *Miss Mc Ghie.* On the left hand side of the road between Willey and the Bridgnorth turnpike road. By the side of the road between Frodesley and Frodesley Green; *Rev. E. Williams's MSS.*

Benthall Edge.

Root creeping. *Stem* 18 inches to 2 feet high, erect, obsoletely quadrangular, with recurved hairs, branched above. *Leaves* opposite, ovate, subcordate at the base, obtuse, subserrated or wavy and ciliated at the margins, with scattered hairs above, paler beneath, punctato-glandulose, hairy on the ribs; petioles with erect hairs. *Inflorescence* paniculate in dense, roundish, crowded, trichotomous heads of sessile flowers, each with an ovate, entire, slightly hairy, minutely ciliated, eglandulose, coloured *bractea* at the base. *Calyx* tubular, with appressed hairs, resinoso-glandulose, teeth smooth, ovate, ciliated, nearly equal; mouth closed with an internal wavy ring of prominent white hairs as long as the calyx-teeth. *Corolla* purple, 3 times as long as the calyx, externally hairy. Fragrant and aromatic.

4. TEUCRIUM. *Linn.* Germander.

1. T. *Scorodonia*, Linn. *Wood Germander* or *Sage.* Stem erect; leaves oblongo-ovate, cordate, crenato-serrate, wrinkled; flowers in lateral and terminal one-sided racemes; calyx 10-ribbed, pubescent, subbilabiate, upper tooth the largest; tube of the corolla exserted. *E. Bot. t.* 1543. *E. Fl. v. iii. p.* 69. *Hook. Br. Fl. p.* 278.

Woods and dry hedge banks; frequent. *Fl.* July, August. ♃.

Root creeping. *Stem* 18 inches to 2 feet high, erect, branched above, acutely quadrangular, with spreading hairs. *Leaves* opposite, on short hairy petioles, oblongo-ovate, cordate, crenato-serrate, dark green, wrinkled and pubescent above, veiny hairy and paler with a resinoso-glandulose mealiness on the under surface. *Flowers* in terminal and lateral one-sided erect racemes. *Calyx* tubular, somewhat swollen at the base in front, pubescent, resinoso-glandulose, 10-ribbed, 2-lipped, upper lip with a large single ovate acuminate tooth, lower lip unequally 4-toothed, the two lower ones longer than the lateral ones, all apiculate and ciliato-glandulose; mouth with an internal annular single row of hairs shorter than the calyx teeth. *Pedicels* short, with minute deflexed pubescence, and a minute ovate acute *bractea* longer than and at the base of each pedicel. *Corolla* yellowish-white, 2-lipped, upper lip truncate, emarginate, with a rounded auricle or tooth on each side, lateral lobes small, oblong; central lobe externally hairy and resinoso-glandulose, very large and dilated in front, concave, wavy; tube longer than the calyx, swollen in front, externally pubescent. *Filaments* purplish-red, hairy. Whole plant glutinous and of an aromatic bitter odour, similar to that of hops.

2. T. *Chamædrys*, Linn. *Wall Germander.* Stem ascending; leaves ovate, cuneate and entire at the base, tapering into the petiole, inciso-crenate, ciliated; flowers axillary, semiwhorls 2-3-flowered; calyx 10-ribbed, hairy and tuberculate, teeth nearly equal; tube of the corolla shorter than the calyx. *E. Bot. t.* 680. *E. Fl. v. iii. p.* 70. *Hook. Br. Fl. p.* 279.

Ruined walls; rare. *Fl.* July. ♃.

Wenlock Abbey; *T. and D. Bot. Guide.* Oreton, nearly destroyed by quarrying for limestone; *Mr. G. Jorden.*

Walls of the Keep of Whittington Castle near Oswestry.

Root creeping. *Stem* decumbent, then ascending, with opposite branches, 8—10 inches high, obsoletely quadrangular, with soft spreading horizontal hairs. *Leaves* opposite, ovate, cuneate and entire at the base and tapering into a short petiole, inciso-crenate, crenatures ciliated with a callous point, hairy on both sides, deep shining green above, paler beneath. *Flowers* axillary, pale pink with darker streaks; semiwhorls of 2-3 flowers, each on a short hairy pedicel without bracteas. *Calyx* tubular, gibbous on the lower side at the base, pentangular, 10-ribbed, copiously covered with minute globular opaque white tubercles, which also appear on the exterior of the corolla and exterior base of the anthers, interspersed with horizontal spreading hairs, teeth shorter than the tube, nearly equal, lanceolate, mucronate, erect, upper one the larger, reflexed, 2 lowest ones smallest, parallel and straight, mouth with an annular appendage of white hairs. *Corolla* 2-lipped, upper lip deeply bifid reduced to 2 erect lanceolate acuminate ciliated teeth, lateral lobes ovate, acute, concave, erecto-patent, central lobe very large, wedgeshaped, contracted at the base, concave, anterior margin wavy, hairy on the exterior, the palate with two longitudinal short rows of white hairs; tube shorter than the calyx, glabrous without, pubescent within; throat scarcely inflated. *Filaments* pubescent, very much exserted, inferior ones longest; *anthers* red brown.

4. MARRUBIUM. *Linn.* White Horehound.

1. M. *vulgare*, Linn. *White Horehound.* Stem erect, hoary and tomentose; leaves petiolate, roundish ovate, obtuse, rugose, undulate and irregularly crenate; whorls distant, dense, globose, many-flowered; calyx 10-ribbed, stellately hairy, segments of 10 subulate, spreading, revolutely hooked spines; corolla scarcely longer than the calyx; upper lip erect, deeply bifid into 2 linear obtuse segments, lower 3-fid, central lobe transversely dilated, reflexed and serrulate at the margins. *E. Bot. t.* 410. *E. Fl. v. iii. p.* 104. *Hook. Br. Fl. p.* 283.

Waste places and way-sides; not common. *Fl.* August. ♃.

Near the Lea Farm House on the side of the road from Wenlock to Church Stretton; *W. P. Brookes, Esq.!* Morfe near Bridgnorth; *Rev. W. R. Crotch.* Llanymynech Hill; *J. F. M. Dovaston, Esq.* Oreton; *Mr. G. Jorden.* Ludlow; *Miss M'Ghie.* About Bridgnorth. Between Cross Houses and Cound; *Rev. E. Williams's MSS.*

Shawbury Heath. Hopton Cliff. Side of Turnpike Road near Ensdon House. Benthall Edge.

Root woody. *Herbage* white, hoary and densely tomentose. *Stems* 1—1½ or 2 feet high, numerous, erect, branched, bluntly tetragonous. *Leaves* opposite on dilated channelled petioles, roundish-ovate, obtuse, undulate and irregularly crenate or obtusely toothed, deeply wrinkled and dark green above, strongly veined and tomentose beneath; upper leaves nearly sessile. *Flowers* in distant, dense, globose, sessile, many-flowered whorls, having numerous subulate *bracteas* with rigid hooked apices at their base. *Calyx* tubular, 10-ribbed, densely hoary with matted stellate hairs, teeth half the length of the tube, of 10 spreading, revolutely hooked, rigid, subulate and alternately shorter spines; mouth internally closed with an annular appendage of long white hairs. *Corolla* scarcely longer than the calyx, greenish-white, externally pubescent, tube glabrous, upper lip erect, deeply bifid into 2 linear obtuse segments, pubescent on the inside; lower lip 3-fid, lateral lobes narrow, acute, entire; central one transversely dilated, serrulate and reflexed at the margins.

Bitter and aromatic and used extensively for coughs and asthmatical affections.

6. AJUGA. *Linn.* Bugle.

1. A. *reptans,* Linn. *Common Bugle.* Stem solitary, with creeping scyons; leaves obovate, broadly crenate, diminishing upwards into coloured bracteas. *E. Bot. t.* 489. *E. Fl. v. iii. p.* 66. *Hook. Br. Fl. p.* 279.

Moist pastures and woods; common. *Fl.* May, June. ♃.

Root woody, with long fibres. *Stem* erect, solitary, 6—8 inches high, furrowed, smooth, throwing out from the base, long leafy prostrate *scyons.* *Radical leaves* obovate, tapering into a dilated petiole, broadly crenate, smooth or with a few scattered hairs; upper leaves sessile or nearly so, wavy or crenate, diminishing into coloured *bracteas* bearing axillary whorls of sessile *flowers.* *Calyx* 5-ribbed, smooth, teeth 5, nearly equal, ovate or ovato-lanceolate, ciliated on the margin, hairy on the interior. *Corolla* blue, upper lip very short, obtuse, emarginate, lower lip 3-fid, lateral lobes small, central one glabrous, 2-lobed, tube incurved, pubescent, with an internal transverse annular appendage of erect hairs near the base.

A variety with *white flowers* has been found at Hawkstone by *Mr. J. S. Baly!*

7. BALLOTA. *Linn.* Horehound.

1. B. *ruderalis,* Fries. *Rubbish Horehound.* Calyx-tube narrow and elongated, gracefully dilated upwards, calyx-teeth ovate, gradually acuminate, aristate, erecto-patent. *Fries Nov. 2nd ed.* 194. *Drej. Fl. Hafn.* 205. "B. *nigra, Linn. Fl. Suec. n.* 529. *and B. alba, n.* 530. B. *nigra, Reich. Fl. Excurs.* 2210. *Reich. Iconog.* VIII. 1039. *Reich! Fl. Exsic. n.* 526. *Peterm. Fl. Lips.* 448. *Babington Prim. Fl. Sarn. p.* 75. *(according to the Author.)* B. *nigra* α. *vulgaris, Bluff and Fing. Comp. 2nd ed. t. i. pt.* 2. *p.* 375. B. *nigra* β, *ruderalis, Koch Syn.* 572. B. *vulgaris,* "*Link.*" *Host. Fl. Austr.* 2. 171. B. *nigra* α. *Huds. Fl. Angl.* 260. *With. 3rd ed. p.* 533.

Waste places and hedges. *Fl.* July, August. ♃.

Similar to the succeeding species in general appearance and character but distinguishable by the above marks. Of this addition to our British Flora, I have only a single specimen with white flowers, gathered in 1835, at Bomere, near Shrewsbury. Great confusion prevails in accurately determining the *true B. nigra* of *Linnæus,* as he appears to have described two different species as *nigra,* in his *Spec. Plant.* and *Fl. Suec.* Many authors have consequently rejected the name *nigra* altogether, and distinguished the species of these two works under new names. Fries *(Nov. 2d. ed.)* gives to the *Fl. Suec.* plant the name *B. ruderalis,* and to that of the *Sp. Plant. B. fœtida.* On examining the works of British Botanists we perceive some indications which would lead us to conclude that *both* these species are indigenous to Britain, and that their *B. nigra* is the *B. nigra, Fl. Suec.* and *B. ruderalis, Fries,* being universally described "calycibus acuminatis:" whilst their *var.* β. *flore albo* would seem to belong to *B. nigra, Sp. Pl.* and *B. fœtida* of *Lamarck* and *Fries,* as being characterized "calycibus subtruncatis." Our Shropshire plants appear to favour this view, since specimens from Shrewsbury (described below under *B. fœtida)* have the calyx-teeth short, broadly ovate, suddenly acuminate, mucronate and carinato-reflexed, and consequently are the *B. fœtida Lam.* and *Fries,* and *B. nigra, Sp. Pl.* A variety with white flowers

from Newport coincides in these characters, and therefore belongs to the same species. But in the plant with white flowers from Bomere (described above as *B. ruderalis,*) the calyx-tube is longer, narrower, and more gracefully dilated upwards, the teeth elongated, ovato-lanceolate, gradually acuminate and aristate, *erecto-patent* not *carinato-reflexed:* this then is necessarily a white variety of *B. ruderalis, Fries,* and *B. nigra, Fl. Suec.*

2. B. *fœtida,* "Lam." *Stinking Horehound.* Calyx-tube shorter and stouter, calyx-teeth broadly ovate, short, suddenly acuminate, mucronate, carinato-reflexed. *Fries Nov. 2d ed.* 195. *Reich. Fl. Excurs.* 2212. *Reich. Iconog. VIII.* 1041. *Reich! Fl. Exsicc. n.* 527. *Drej. Fl. Hafn.* 205. *Peterm. Fl. Lips.* 448. B. *nigra* α. *fœtida. Koch. Syn.* 572. B. *nigra,* "*Linn. Sp. Pl. ed.* I. 1582." *E. Bot. t.* 46. *E. Fl. v. iii. p.* 102. *Fl. Brit.* 635. B. *nigra* β. *fœtida, Bluff and Fing. Comp. 2d ed. t.* 1. *pt. ii. p.* 375.

Waste places and hedges; frequent. *Fl.* July, August. ♃.

Root fibrous, somewhat creeping. *Stem* about 2 feet high, erect, with opposite branches, tetragonous, striated, clothed with soft decurved hairs. *Leaves* opposite on short channelled hairy petioles, ovate, acute, coarsely and unequally serrated, serratures ciliated, rugose, covered on both sides with minute glandular mealy particles, and short dense soft pubescence. *Flowers* axillary. *Semiwhorls* on very short hairy pedicels bearing numerous sessile flowers, each of the exterior ones with a pair of rigid, subulate, hairy *bracteas* at the base, about half as long as the calyx. *Calyx* funnelshaped, pentangular, 10-ribbed, clothed with erect hairs interspersed with innumerable horizontal pedicellate glands, segments 5, broadly ovate, short, suddenly acuminate, mucronate, carinato-reflexed. *Corolla* longer than the calyx, 2-lipped, upper lip erect, oblong, concave, the anterior margin recurved, hairy within and without, the anterior and about half of the lateral margins minutely notched, lower lip 3-fid, lateral lobes oval, emarginate, margins recurved, central lobe elongated, larger and broader, contracted at the base, emarginate, margins crenulate, concave, with a central longitudinal elevation, pinkish purple with white veins; tube shorter than the calyx, with a transverse oblique annular appendage above the base on the interior; throat scarcely dilated, with deflexed hairs on the exterior. *Seeds* obovate, triquetrous, rounded at the apex, smooth and shining, resting on a fleshy 4-toothed receptacle.

Var. β. *alba.* Flowers white. "B. *alba, Linn. Sp. Pl. ed.* 2. 814." and according to Smith, *Fl. Suec.* 206. B. *nigra* β. *Huds. Fl. Angl.* 260. *E. Fl. l. c.*

Newport; *R. G. Higgins, Esq.!* By the side of the road near Buildwas (old) bridge. In the pound at Diddlebury. Near the blacksmith's shop in Berrington. In Bearstone near Drayton; *Rev. E. Williams's MSS.* Buildwas; *Miss H. Moseley,* Weston near Hawkstone; *W. W. Watkins, Esq.*

Possibly some of these localities (except the Newport one) may on examination prove referable to B. *ruderalis.*

8. LEONURUS. *Linn.* Motherwort.

1. L. *Cardiaca,* Linn. *Motherwort.* Leaves petiolate, cordato-ovate, acuminate, 3-lobed, inciso-dentate, upper ones ovate or lanceolate, 3-lobed, entire. *E. Bot. t.* 286. *E. Fl. v. iii. p.* 105. *Hook. Br. Fl. p.* 280.

Waste places; rare. *Fl.* August. ♃.

Morfe, near Bridgnorth; *Rev. W. R. Crotch.* On the left-hand side of the road about fifty yards beyond Linley turnpike towards Bridgnorth, (since destroyed); *Rev. E. Williams's MSS.*

Root fibrous. *Stem* 2-3 feet high, erect, acutely quadrangular, deeply channelled, with minute deflexed pubescence on the angles. *Leaves* opposite, spreading horizontally, on dilated channelled petioles with erect appressed hairs, lower ones cordato-ovate, acuminate, divided to the middle in 3 inciso-dentate lobes; upper ones ovate or ovato-lanceolate, 3-lobed, toothed or entire, wrinkled and downy above, paler and hoary beneath especially on the prominent nerves. *Whorls* numerous, axillary, sessile, many-flowered. *Bracteas* subulate, rigid, hairy. *Calyx* tubular, pentangular, 10-ribbed, nearly glabrous or slightly hairy on the ribs, resinoso-glandulose, teeth large, subulate, spiny, rigid, at length spreading, the two anterior ones reflexed. *Corolla* white with a reddish tinge, upper lip helmet-shaped, ovate, obtuse, entire, densely shaggy with long white hairs on the exterior; lower lip 3-lobed, lobes oblong, central one longest entire acute, lateral ones reflexed, tube with an internal oblique annular appendage of erect hairs above the base. *Seeds* triquetrous, angles acute, truncated at the apex and surmounted by a dense tuft of white hairs, smooth, with minutely impressed dots.

9. GALEOBDOLON. *Huds.* Weasel-snout.

1. G. *luteum,* Huds. *Yellow Weasel-snout or Archangel.* Leaves ovate, acuminate, subcordate, deeply serrated, petiolate; calyx 5-ribbed; tube of corolla about as long as the calyx, subrecurved, with an internal oblique annular appendage, upper lip elongated, entire, lateral lobes somewhat reflexed, entire. *E. Bot. t.* 787. *E. Fl. v. iii. p.* 97. *Hook. Br. Fl. p.* 280.

Woods and shady places; frequent. *Fl.* May, June. ♃.

Root creeping, tuberous. *Stems* numerous, 18 inches high, erect, simple, quadrangular and channelled, angles rough with deflexed hairs. *Leaves* opposite, ovate, acuminate, subcordate or truncated at the base, deeply serrated, serratures coarse and ciliated, bright green, slightly hairy on both surfaces, under one paler. *Petioles* winged, dilated, ciliated and connate at the base. *Whorls* axillary, of 10 or more sessile flowers, extreme lateral ones of each semiwhorl with 2-3 subulate ciliated *bracteas.* *Calyx* tubular, campanulate, 5-ribbed, teeth 5 nearly equal, ovate, very acute and spinulose, hairy and glandular, about as long as the tube of the corolla. *Corolla* large, yellow, subrecurved, 2-lipped, upper lip oblong, elongated, arched or helmet-shaped, very narrow at the base, entire and ciliated with long white silky hairs, externally pubescent, lower lip 3-fid, recurved and spreading, of a deeper yellow, marked as is also the throat with brown spots and stripes, lateral lobes ovate, acute, entire, central lobe longer and narrower, ovate, acute, entire, the point elongated; tube scarcely inflated, rough on the interior with minute protuberances and an oblique annular appendage of white hairs above the base.

10. GALEOPSIS. *Linn.* Hemp-Nettle.

1. G. *Ladanum,* Linn. *Red Hemp-nettle.* Stem not swollen below the joints; leaves lanceolate or ovato-lanceolate, serrated; upper lip of corolla ovate, crenulate. *E. Bot. t.* 884. *E. Fl. v. iii. p.* 94. *Hook. Br. Fl. p.* 280.

Limestone districts. *Fl.* September, October. ⊙.

Gleeton Hill near Wenlock; *W. P. Brookes, Esq.!* Field near Welbach coppice; *Mr. T. Bodenham.* Roadside 2 miles beyond Wenlock on Stretton road;

Mr. F. Dickinson. Buildwas; *Miss H. Moseley.* Bromfield near Ludlow; *Miss Mc Ghie.* Ditch-banks, road-sides and corn-fields about Much Wenlock, Buildwas, Ledwich, &c.; *Rev. E. Williams's MSS.*

Wenlock Edge at the Hill Top.

Root fibrous. *Stem* 1 foot or more high, erect, quadrangular, furrowed, red, roughish with deflexed hairs, internodes of an equal thickness, branches in opposite pairs crossing each other. *Leaves* opposite, shortly petiolate, spreading, lanceolate or ovato-lanceolate, somewhat acute, distinctly serrated, serratures callously tipped, glabrous and furrowed along the veins on the upper surface, hairy and prominently veiny beneath. *Flowers* in dense sessile distant axillary whorls, the uppermost or terminal one largest. *Bracteas* linear-lanceolate, spinous, clothed with short appressed erect hairs. *Calyx* tubular, campanulate, with about 20 ribs, pubescent, with erect appressed hairs, 5-toothed, tube and teeth nearly equal, teeth subulate, spinous, rigid, nearly equal, three uppermost larger. *Corolla* very much exserted, twice as long as the calyx, rose-coloured, variegated with crimson and white; upper lip ovate, galeate, obtuse, crenulate, externally hairy, lower lip 3-fid, lateral lobes roundish, obtuse, crenulate, reflexed, central lobe broader, obtuse, crenulate, with a small obtuse tooth on each side at the base. *Seeds* triquetrous, truncated and rounded at the apex.

Var: β. "Calyx very hairy; Stem thickened upwards." *E. Fl. l. c.*

In the same localities with the preceding variety. *Rev. E. Williams's MSS.*

2. G. *versicolor,* Curt. *Large-flowered Hemp-nettle.* Stem hispid, swollen below the joints; leaves oblongo-ovate, acuminate, coarsely serrated; calyx campanulate, teeth shorter than the tube; corolla very much exserted, upper lip roundish oval, galeate, crenulato-denticulate. *E. Bot. t.* 667. *E. Fl. v. iii. p.* 96. *Hook. Br. Fl. p.* 281.

Corn-fields; frequent. *Fl.* July August. ⊙.

Albrighton, near Shiffnal; *H. Bidwell, Esq.* Corn-fields, near Prior's Lee; *W. P. Brookes, Esq!* Near Baschurch and Llanymynech, abundantly; Caynham Court, near Ludlow; *T. & D. Bot. Guide.* Corn-fields, near Westbury; *Mr. F. Dickinson.* Walford; *T. C. Eyton, Esq.* Lee-Gomery, near Wellington; *E. Lees, Esq.* Buildwas; *Miss H. Moseley.* Ludlow; *Miss Mc Ghie.*

Westfelton Moors. Bomere. Base of Middleton hill. Little Calcott.

Root fibrous. *Stem* 1-2 feet high, erect, oppositely branched, tetragonous, swollen beneath the joints, hispid with long deflexed jointed hairs. *Leaves* opposite on short hairy chanelled petioles, oblongo-ovate, acuminate, hairy on both sides, coarsely serrated, serratures callously tipped. *Flowers* in dense terminal whorled spikes, large, very handsome and showy. *Semiwhorls* axillary, of about 7 sessile flowers in two rows, three in the upper and 4 in the lower one, with 10 unequal lanceolate and subulate *bracteas* at their base; the flowers in the upper row always expanding before the others, and of these the central one first. *Calyx* tubular-campanulate, with 12-13 strong ribs and finer intermediate ones, hairy, tube longer than the segments which are subulate, spinous, 3 uppermost longest, the sinuses with long jointed hairs, margins ciliato-glandulose. *Corolla* large, very much exserted, twice as long as the calyx, 2 lipped, tube straight, as long as the calyx, throat inflated, both pubescent on the exterior, naked within: *upper lip* roundish-oval, galeate, anterior margin crenulato-denticulate, with long jointed hairs interspersed with shorter pedicellate glands on the back, pale yellow: *lower lip* 3-fid, lateral lobes oval, reflexed, margins wavy, very pale yellow, with a deeper yellow tinge at the base, central lobe larger and broader, subrotundo-quadrate, crenulate, more or less emarginate, with a longitudinal central indentation, margins deflexed, contracted and with a small obtuse tooth on each side at the base, deep purple, with darker veins and a white narrow border on the margin, palate orange, streaked with purple and crimson and having on each side just behind the sinuses of the lobes an obtuse hollow protuberance. *Filaments* slightly ciliated, lower ones longest; *anthers* having the interior valves ciliated with short white hairs. *Seeds* imbedded in an obliquely truncated, irregularly toothed, fleshy receptacle.

I sowed seeds of this plant in my garden in 1834, where it has now completely naturalized itself, but I cannot perceive, on a comparison with the dried specimens from which the seeds were taken, any change in its characters, or any tendency to run into *G. Tetrahit*, (of which it has been considered a variety,) of which species I have never observed a single specimen in my garden. The above description was drawn up from one of these naturalized plants, (1838).

3. G. *Tetrahit*, Linn. *Common Hemp-nettle.* Stem hispid, swollen below the joints; leaves oblongo-ovate, acuminate, coarsely serrated; calyx tubular, teeth and tube nearly equal; corolla exserted, "upper lip erect ovate entire." *E. Bot. t.* 207. *E. Fl. v. iii. p.* 95. *Hook. Br. Fl. p.* 280.

Corn-fields; frequent. *Fl.* August. ⊙.

Of a more rigid habit and smaller growth than the preceding with which it agrees in many particulars, but from which it seems very different though the distinguishing marks be difficult of description. The more prominent features are the smaller and tubular not campanulate calyx of which the tube and teeth are about equal, the smaller, slenderer, and less exserted corolla of a purplish colour, often also white.

Var: with white flowers. Hancott, near Shrewsbury.

11. LAMIUM. *Linn.* Dead-nettle.

1. L. *album*, Linn. *White Dead-nettle or Archangel.* Leaves cordato-ovate, acuminate, deeply serrated, petiolate; calyx 10-ribbed; tube of corolla as long as the calyx, inflated, recurved, with an oblique annular appendage of hairs within above the base, throat somewhat dilated, upper lip oblong, arched, narrower at the base, margin irregularly crenate, lateral lobes recurved with 3 teeth; anthers hairy; whorls of 16—18 (white) flowers; bracteas 3-4; seeds truncated at the apex with acute angles. *E. Bot. t.* 768. *E. Fl. v. iii. p.* 90. *Hook. Br. Fl. p.* 281.

Waste ground, borders of fields, &c.; common. *Fl.* May, July. ♃.

Root creeping. *Stems* numerous, 12—18 inches high, erect, with a few opposite branches near the base spreading at right angles, quadrangular, roughish with deflexed hairs. *Leaves* in pairs, opposite, cordato-ovate, acuminate, deeply and doubly serrated, serratures coarse and ciliated, deep green, upper surface when viewed under a microscope minutely granulated, hairy, under surface paler, hairy on the ribs. *Petioles* broadly winged, very much dilated, ciliated and connate at the base. *Whorls* axillary, of about 16—18 sessile *flowers*, the exterior lateral ones of each semi-whorl with 3-4 subulate hairy *bracteas*. *Calyx* tubular, campanulate, 10-ribbed, teeth 5 equal, subulate, hairy and glandular, about as long as the tube of the corolla. *Corolla* large, white, recurved, 2-lipped, *upper lip* oblong, arched or helmet-shaped, narrower at the base, margin irregularly crenate in front, covered externally with long white silky hairs; *lateral lobes* with a reflexed triangular tooth, a central subulate appendage, and another smaller tooth near the base of the lower lip; *lower lip* broadly dilated, very deeply emarginate, contracted at the base, margin irregularly crenate, cream-coloured, spotted with

green and with green streaks at the base almost within the throat; *tube* inflated in front, recurved, interior rough with minute protuberances, and an oblique annular appendage of white hairs above the base. *Filaments* with glandular hairs, inferior ones longest. *Anthers* black, clothed on the exterior side with long white hairs and one or two small tuberculate bodies at their base; *pollen* greenish-yellow, grains oval. *Seeds* smooth and shining, oblong, triquetrous, truncated at the apex with acute angles.

2. L. *purpureum*, Linn. *Red Dead Nettle, or Archangel.* Leaves cordate, obtuse, crenato-serrate, petiolate, uppermost crowded; calyx 10-ribbed; tube of corolla as long as the calyx, not inflated, straight, with a transverse annular appendage of hairs within above the base, throat dilated, upper lip oval, scarcely arched, not contracted at the base, margin entire, lateral lobes with two teeth, very slightly recurved; anthers hairy; whorls of about 14—16 (purple) flowers; bracteas 2; seeds truncated at the apex with acute angles. *E. Bot. t.* 1933. *E. Fl. v. iii. p.* 92. *Hook. Br. Fl. p.* 281.

Borders of fields, waste and cultivated ground; common. *Fl.* May—September. ⊙.

Root fibrous. *Stems* numerous, 8—10 inches high, weak, curved and declining at the base, with a few short leafy branches spreading at right angles, then erect, quadrangular, rough-edged, naked in the middle, densely leafy at the top. *Leaves* in pairs, opposite, deflexed, broadly cordate, obtuse, crenato-serrate, serratures coarse and obtuse, deep green, floral ones tinged with purple, unpleasantly scented, upper surface minutely granulated, hairy, under surface paler, hairy on the ribs. *Petioles* broadly winged, dilated but not connate at the base. *Whorls* axillary, of about 14—16 sessile *flowers*, the exterior lateral ones of each semi-whorl with two minute subulate *bracteas*, hairy at their summits. *Calyx* tubular-campanulate, 10-ribbed, teeth 5, equal, subulate, slightly hairy and glandular, about as long as the tube of the corolla. *Corolla* small, purplish-red, straight, 2-lipped, *upper lip* oval, scarcely arched, not contracted at the base, margin entire, covered externally as is also the throat with dense white pubescence; *lateral lobes* very slightly recurved, entire, with the exception of two unequal teeth near the base of the lower lip; *lower lip* broadly dilated, very deeply emarginate, contracted at the base, marked with darker purple spots, which also appear in the inflated throat; *tube* not inflated in front, interior very slightly rough with protuberances, and a transverse annular appendage of white hairs above the base. *Filaments* smooth, inferior ones longest. *Anthers* dark purple, clothed on the exterior side with tufts of white hairs and 6—8 small oval white tuberculate bodies at their base. *Pollen* of a red-lead colour, grains larger than in L. *album*, elliptical. *Seeds* smooth and shining, oblong, triquetrous, truncated at the apex with acute angles.

A variety with white blossoms occurred on the ditch bank on the right hand side of the road between Shrewsbury and Underdale; *Rev. E. Williams's MSS.*

3. L. *amplexicaule*, Linn. *Henbit Nettle.* Leaves orbicular or reniform, deeply lobed, lower ones petiolate, floral ones amplexicaul; calyx 5-ribbed; tube of corolla 3 times the length of the calyx, not inflated, straight, naked within, throat dilated, upper lip oval, arched, not contracted at the base, margin entire, lateral lobes reflexed, nearly entire or with 3 minute shallow teeth; anthers hairy; whorls of about 20—22 (purple) flowers; bracteas 2-3; seeds rounded at the apex. *E. Bot. t.* 770. *E. Fl. v. iii. p.* 92. *Hook. Br. Fl. p.* 282.

Waste places, sandy and gravelly fields and gardens; not common. *Fl.* March—July. ⊙.

Welbach, near Shrewsbury; *Mr. T. Bodenham.* Ness; *T. C. Eyton, Esq.* Turnip-field above Lydbury, opposite Walcot; *Miss Mc Ghie.* Dodington, near Whitchurch; *W. W. Watkins, Esq.*

Side of the turnpike road close to Nesscliffe village on the Oswestry side. Fields opposite the Bank Farm, by the bridle road to Meole Brace, near Shrewsbury. Monkmoor, near Shrewsbury. Between Ludlow and Bromfield. Gardens in Coleham, near Shrewsbury. Lyth Hill.

Root fibrous. *Stems* numerous, straggling, 6-12 inches long, very much branched at the base, branches in pairs, opposite, spreading at right angles, quadrangular, roughish with deflexed bristles. *Leaves* opposite, in distant pairs, *lower* ones on short somewhat dilated petioles, channelled above, rounded beneath, hairy, *upper* or *floral* ones amplexicaul, orbicular or reniform, deeply lobed, lobes generally but not invariably 9, very irregular, obtuse, the central or terminal one trifid, edges ciliated, upper surface minutely glandular, hairy, lower one paler, hairy. *Whorls* axillary, of about 20—22 sessile flowers, the exterior lateral ones of each semiwhorl with 2-3 hairy *bracteas*, the outermost lanceolate acute, occasionally broader and cut, the others subulate. *Calyx* tubular, campanulate, 5-ribbed, teeth 5, equal, lanceolate acuminate, as long as the tube, clothed externally with erect soft silky hairs. *Corolla* large, purplish-red, straight, hairy, 2-lipped; *upper lip* oval, arched, not contracted at the base, margin entire, covered externally with long dense dark purple hairs; *lateral lobes* reflexed, nearly entire or with three minute shallow teeth, marked with a round dark purple spot; *lower lip* very broadly dilated, very much contracted and reflexed at the base, deeply emarginate, lobes divergent, concave, margins irregularly crenate, each marked with one or two dark purple spots with a larger one on the contracted base; *throat* inflated; *tube* not inflated, slender and elongated, about three times the length of the calyx, naked within. *Filaments* very slightly hairy at the base, inferior ones longest. *Anthers* black, with tufts of long white hairs on their exterior sides; *pollen* of a red-lead colour, grains elliptical. *Seeds* obovato-oblong, triquetrous, rounded at the apex, smooth shining and minutely dotted.

The herbage is slightly but pleasantly aromatic when bruised. The earlier flowers never expand their corollas, though their seeds are fertile.

12. BETONICA. *Linn.* Betony.

1. B. *officinalis*, Linn. *Wood Betony,* Spike interrupted, short; leaves oblongo-cordate, crenate; middle lobe of lower lip of corolla, transversely dilated, crenato-undulate; calyx glabrous, teeth about two-thirds as long as the tube. *Linn. Sp. Pl.* 810. *Fl. Brit.* 632. *Reich. Fl. Excurs* 2179. *Sibth. Fl. Oxon.* 185. *With. 3rd ed.* 350. *Stachys Betonica, Benth. β. glabrata, Bluff and Fing. Comp. Fl. Germ. 2nd ed. t. i. pt.* 2. *p.* 362.

Woods; frequent. *Fl.* July, August. ♃.

Coppices around Wenlock; *W. P. Brookes, Esq.* Cox wood, Coalbrookdale; *Mr. F. Dickenson.* Ball's coppice near Walford; *T. C. Eyton, Esq.* Near Newport; *R. G. Higgins, Esq.* Wyre forest; *Mr. G. Jorden.*

Bomere wood; Bickley coppice; and woods generally about Shrewsbury.

Root woody, fibrous. *Stem* erect, 12—18 inches high, simple, slender, bluntly quadrangular, with soft deflexed bristly hairs. *Lower leaves* on very long channelled petioles with spreading hairs, oblongo-cordate, obtuse, deeply and obtusely crenate, hairy on both sides, rugose; *upper ones* remote, narrower, nearly sessile, the crenatures becoming more acute and dentate. *Inflorescence* in a dense whorled interrupted truncate terminal spike, the lowermost whorls frequently dis-

tant. *Bracteas* lanceolate, slightly spiny, hairy externally, as long as the calyx. *Calyx* tubular, 5-ribbed with finer intermediate ones, glabrous, teeth equal, subulate, spiny, very slightly hairy, about two-thirds the length of the tube, the sinuses rounded and ciliated with long silky hairs. *Corolla* crimson, twice as large as the calyx, externally hairy, tube curved, naked within; *upper lip* ovate, obtuse, entire, erect; *lower lip* trifid, lateral lobes small, obtuse, central lobe very much larger, transversely dilated, crenate, undulate.

A variety with white flowers occurred by the side of the road between Ruckley and Causeway wood; *Rev. E. Williams's MSS.*

I presume our Shropshire plant is the true B. *officinalis, Linn*, to which Reichenbach (*l. c.*) assigns a glabrous calyx with teeth nearly equalling the tube, the middle lobe of the inferior lip transversely dilated, emarginate, and the lateral ones abbreviated. He considers the *fig.* 1142 of *E. Bot.* as referable to B. *hirta, Leyss.* which has the calyx hairy in the upper part, the teeth twice as short as the tube, the middle lobe of the inferior lip rounded and crenate. A nearly allied species B. *stricta, Ait.* has the calyx hairy, the teeth half as long again as the tube; the middle lobe of the inferior lip crenulate, subemarginate, and the lateral ones oblong, deflexed. Many authors consider the above as modifications only of the same species and possibly all may be found in Britain. Mr. Babington thus writes to me on this subject:—"Your B. *officinalis* appears to me to belong to the true plant; it agrees with *Reich. Icon.* viii. 952. and the specimen in his *Fl. Exsicc. No.* 990. with the exception of the few hairs which exist on the outer part of the upper portion of the calyx. In B. *hirta*, I learn from his figure and specimen that the calyx is *much* more hairy and the teeth *much* shorter. *E. Bot. t.* 1142, appears to belong to the true B. *officinalis. Linn.* I have never seen either of the other species in England. *Koch. (Syn.)* combines *No.* 2179 B. *officinalis, L.*; 2180 B. *hirta, Leyss*; 2181 B. *stricta, Ait.*; and 2182 B. *incana, Ait*; of *Reich. Fl. Excurs.* as varieties of B. *officinalis.* *Wimm.* and *Grab.* (*Fl. Siles.*) keep them distinct, at least, they describe B. *stricta* and refer to B. *officinalis* as different."

13. STACHYS. *Linn.* Woundwort.

1. S. *sylvatica*, Linn. *Hedge Woundwort.* Stem erect; leaves petiolate, ovato-acuminate, cordate, coarsely serrated; semiwhorls 3-4-flowered; calyx glanduloso-pilose; upper lip of corolla entire or emarginate; seeds glabrous. *E. Bot. t.* 416. *E. Fl. v. iii. p.* 99. *Hook. Br. Fl. p.* 282.

Woods and shady places; frequent. *Fl.* July, August. ♃.

Root creeping. *Stem* 2-3 feet high, erect, branched above, tetragonous, hispid with decurved hairs, glanduloso-pilose in the upper part. *Leaves* opposite, petiolate, ovato-acuminate, cordate at the base, coarsely serrated, serratures somewhat rounded, obtusely mucronate, softly hairy on both sides, the upper floral ones gradually diminishing into a linear entire shape; petioles about as long as the leaf, channelled above, with spreading hairs, connate at the ciliated base. *Flowers* in a terminal whorled spike, deep dull red purple, the palate variegated with darker streaks and spots and white intermixed. *Semiwhorls* axillary, of 3-4 flowers on very short glanduloso-pilose peduncles, with a pair of minute subulate glanduloso-pilose *bracteas* at the base of each. *Calyx* tubular campanulate, 10-ribbed, glanduloso-pilose externally ½5-fid, teeth equal or nearly so, the upper one slightly longer, lanceolate, very acute, spiny, widely spreading, much shorter than the tube. *Corolla* 2-lipped; *upper lip* oval, erect, somewhat galeate, externally glanduloso-pubescent, anterior margin entire or emarginate, *lower lip* trifid, lateral lobes oval, entire, deflexed, central lobe larger, produced, subrotund, contracted at the base, emarginate or irregularly notched, slightly glanduloso-pilose on the exterior, tube very long, with a gibbous protuberance just above the base, where on the interior is an oblique transverse annular appendage of white hairs, glabrous in front, the back with deflexed pubescence, throat and palate rough with hairs. *Filaments* ciliato-tomentose about the middle, inferior ones longest, laterally decurved and protruded from the corolla after shedding the pollen, the superior ones continuing erect. *Anthers* dusky, glabrous, *pollen* white, oblong. *Seeds* oval, shining, glabrous, triquetrous, rounded at the apex, resting on a pale fleshy concave receptacle with a raised oblique irregularly notched margin. *Herbage* with a fetid disagreeable smell.

2. S. *palustris*, Linn. *Marsh Woundwort.* Stem erect; leaves nearly sessile, linear-lanceolate, sinuato-subcordate at the base, coarsely serrated; semiwhorls 4-6-flowered; calyx with erect hairs; upper lip of corolla entire; seeds minutely dotted. *E. Bot. t.* 1075. *E. Fl. v. iii. p.* 100. *Hook. Br. Fl. p.* 282.

River banks and moist watery places; frequent. *Fl.* August. ♃.

By the side of the bridge near the Water Mill, Harley; *W. P. Brookes, Esq.* Near Haughmond Abbey; *Mr. F. Dickinson.* Near Newport; *R. G. Higgins, Esq.*

Canal between Shrewsbury and Uffington. Meole Brook, near Brace Meole.

Root creeping. *Stem* 2 feet high, erect, slightly branched above, tetragonous, hispid with decurved bristles. *Leaves* opposite, nearly sessile, on very short petioles connate at the base, linear-lanceolate, sinuato-subcordate at the base, coarsely serrated, serratures with an obtuse glandular mucro, softly hairy on both sides; upper leaves semiamplexicaul. *Flowers* in a terminal whorled spike, dull purple, palate variegated with white spots and darker streaks. *Semiwhorls* axillary, of 4—6 flowers, on very short hairy peduncles, with a pair of minute subulate hairy *bracteas* at the base of each. *Calyx* tubular-campanulate, 10-ribbed, clothed externally with erect hairs, ⅓5-fid, teeth equal or nearly so, lanceolate, spreading, spiny, glanduloso-pilose on the margins, slightly shorter than the tube. *Corolla* 2-lipped, *upper lip* ovate, erect, externally hairy, anterior margin entire; *lower lip* 3-fid, lateral lobes deflexed, irregularly crenate, central lobe very large and dilated, subrotund, emarginate, tube rather longer than the calyx, slightly hairy on the throat, with an oblique transverse annular appendage of white hairs above the base on the interior. *Filaments* hairy, inferior longest. *Seeds* oval, triquetrous, shining, minutely dotted.

Var. β. intermedia. Leaves on longer but still short petioles, ovato-lanceolate, crenato-serrate, sinuato-cordate at the base.

Sides of the Shrewsbury canal, and Meole brook near Shrewsbury.

This variety is apparently intermediate between *palustris* and *ambigua* and is similar in all respects to the first except in the above characters, and in having the upper lip of the corolla more obtuse, and the lateral lobes of the lower lip larger and more rounded.

3. S. *arvensis*, Linn. *Corn Woundwort.* Stem decumbent or ascending, hispid; leaves ovate, obtuse, cordate at the base, crenate, 3-ribbed, lower ones petiolate; whorls 4—6-flowered, distant; calyx hispid, segments lanceolate, spinulose; corolla scarcely longer than the calyx, lips nearly equal, upper lip erect, galeate, entire; lower one 3-fid, central lobe subrotund, entire; seeds rotundo-obovate, dull, tuberculate. *E. Bot. t.* 1154. *E. Fl. v. iii. p.* 101. *Hook. Br. Fl. p.* 283.

Corn fields; frequent. *Fl.* July, August. ♃.

Root fibrous. *Stems* 6—8 inches high, several, weak, with opposite branches chiefly from the base, decumbent or ascending, tetragonous, rough with long horizontal spreading hairs. *Leaves* opposite, on short dilated channelled hairy petioles, ovate, obtuse, crenate, cordate at the base, hairy on both sides, with 3 principal ribs at the base; upper or floral leaves nearly sessile, narrower, cuneate at the base, mucronate at the apex, the crenatures gradually passing into dentate serratures. *Flowers* in distant axillary whorls, but crowded towards the extremities of the branches, 4—6-flowered, on very short hairy pedicels. *Calyx* tubuloso-campanulate, ⅓5-fid, 10-ribbed, ribs hispid with long spreading horizontal hairs, teeth as long as the tube, lanceolate, nearly equal, spinulose, 3-ribbed, margins ciliated with long horizontal purplish hairs. *Corolla* pale purplish, scarcely longer than the calyx, 2-lipped, lips nearly equal, upper one erect, subgaleate, subrotund, entire, hairy externally; lower lip 3-lobed, lateral lobes ovate, obtuse, entire, central one largest, subrotund, entire, tube externally glabrous, with an oblique internal annular appendage of hairs above the base. *Seeds* rotundo-obovate, triquetrous, dull brownish black, covered with minute dots and larger black scattered tubercles.

14. THYMUS. *Linn.* Thyme.

1. Th. *Serpyllum*, Linn. *Wild Thyme.* Stems branched, decumbent; leaves lineari-elliptical, obtuse, petiolate, minutely serrulate, ciliated in the lower part; flowers capitate. *E. Bot. t.* 1514. *E. Fl. v. iii. p.* 108. *Hook. Br. Fl. p.* 278.

Dry hilly pastures; frequent. *Fl.* July, August. ♃.

Root creeping. *Stem* somewhat woody and shrubby, recumbent, copiously branched and entangled, young branches reddish, clothed on alternate sides between each joint with minute decurved pubescence. *Leaves* opposite, lineari-elliptical, obtuse, attenuated at the base into a very short dilated petiole, glabrous, punctato-glandulose, margins thickened and slightly recurved, minutely serrulate, ciliated in the lower part and on the petioles with long spreading distant bristly hairs. *Inflorescence* in dense, whorled, leafy heads terminating the wavy and ascending branches. *Semiwhorls* axillary, sessile, 3—5-flowered. *Bracteas* subulate, ciliated. *Pedicels* with reflexed pubescence. *Calyx* tubular, swollen in front, externally hairy and resinoso-glandulose, 10-ribbed, 2-lipped, upper lip with 3 triangular teeth, ciliated, lower lip with 2 larger lanceolate-subulate ciliated teeth, mouth closed with an internal ring of converging white hairs. *Corolla* purple, externally hairy; *upper* lip obtuse, emarginate, twice as long as broad. *Seed* globose.

15. ACINOS. *Mœnch.* Basil Thyme.

1. A. *vulgaris*, Pers. *Common Basil Thyme.* Stems spreading, ascending, hairy on two opposite sides; leaves petiolate, ovate, subserrated, revolute and ciliated at the margins; whorls distant, 6-flowered; calyx pedicellate, 13-ribbed, hispid and glandular, ventricose at the base, 2-lipped, 3 upper segments triangular, recurved, 2 lower subulate, straight; corolla scarcely longer than the calyx, upper lip subrotund, emarginate, lower 3-fid, central lobe broad, emarginate; seeds obovate, smooth. *Hook. Br. Fl. p.* 284. *Thymus Acinos, Linn. E. Bot. t.* 411. *E. Fl. v. iii. p.* 109.

Dry gravelly or limestone soils; not very common. *Fl.* August. ⊙.

Buckley; *H. Bidwell, Esq.* Near Wenlock; *Mr. F. Dickinson.* Rushley wood; *Dr. G. Lloyd.* Banky pasture above the mill stream near Cound paper mill; *Rev. E. Williams's MSS.*

Whitecliff, near Ludlow.

Root somewhat woody. *Stems* 6—8 inches long, spreading, ascending, oppositely branched from the base, tetragonous, clothed with spreading white hairs on two opposite sides, which alternate between the joints. *Leaves* on short winged petioles, ovate, acute, coarsely and shallowly serrated, recurved and ciliated at the margins, pubescent above, hairy on the veins beneath and glandularly mealy; upper leaves narrower. *Flowers* on short pedicels, in distant, axillary, 6-flowered whorls, with very minute *bracteas* at the base. *Calyx* tubular, ventricose at the base on the under side, 13-ribbed, ribs armed with a single row of short incurved bristles, interstices with glandular mealy dots, 2-lipped, segments unequal, about one-third as long as the tube, 3 upper ones shorter, triangular, recurved, 2 lower ones subulate, straighter or incurved, all ciliated on the margins; mouth closed with an appendage of long white hairs, which also appear on the interior surface of the 3 upper segments, whilst that of the 2 lower ones is glabrous. *Corolla* bluish-purple, upper lip subrotund, emarginate, lower lip 3-partite, lateral lobes obtuse, entire, central broader, emarginate, throat white, variegated with darker purple streaks, slightly pubescent. *Seeds* obovate, triquetrous, smooth.

16. CALAMINTHA. *Mœnch.* Calamint.

1. C. *officinalis*, Mœnch. *Common Calamint.* Leaves on longish petioles, broadly ovate, acute, crenato-serrated; flowers in subdichotomous axillary cymes, 5-6-flowered; calyx tubular, ventricose in front, distinctly 2-lipped, segments of upper lip 3, triangular, partially recurved, of lower lip 2, twice as long, lanceolate-subulate, straight or incurved; hairs of mouth not prominent; central lobe of lower lip of corolla obcordate; seeds dark brown, covered with impressed dots. *Hook. Br. Fl. p.* 284. *Melissa Calamintha, Linn.—Thymus Calamintha, Scop. E. Bot. t.* 1676. *E. Fl. v. iii. p.* 110.

Waysides and borders of fields; not common. *Fl.* July—October. ♃.

Lincoln's Hill, Coalbrookdale; *Mr. F. Dickinson.* In Shineton churchyard. Sides of roads about Coalbrookdale. Harnage. Between Uffington and Sundorne. Minsterley. Haughmond Hill; *Rev. E. Williams's MSS.*

Red Hill, near Welbach. Near the Old Hall at Westfelton. Fields near Bomere pool. Sharpstones Hill. Near Ludlow.

Root woody. *Stems* 18 inches high, numerous, erect, oppositely branched, tetragonous, rough with spreading and recurved pubescence. *Leaves* opposite, on channelled hairy petioles of about their own length, broadly ovate, acute, crenato-serrate, pubescent on both surfaces, the nerves on the under surface hairy, interstices with resinous dots. *Flowers* in terminal lax secund racemes, on short axillary pubescent peduncles, subdichotomously divided into 5-6 longer hairy pedicels, with subulate *bracteas* at their base. *Calyx* tubular, ventricose at the base on the under side, 13-ribbed, ribs with erect bristles, interstices with glandular mealiness and large resinous dots interspersed, 2-lipped, segments of upper lip 3 triangular partially recurved, of lower lip 2 twice as long straight or incurved, lanceolato-subulate, all ciliated on the margins, mouth closed with white hairs not prominent. *Corolla* twice as long as the calyx, light purple with violet dots, externally pubescent and with scattered resinous dots, upper lip erect, oval, emarginate, lower lip trifid, lateral lobes rounded, entire, central one broader, obcordate. *Seeds* nearly round, triquetrous, dark brown, covered with impressed dots.

Judging from a comparison of the above with a dried specimen of *C. Nepeta* from Hastings, Sussex, communicated by Mr. Borrer, the following appear to be the distinctive marks of the latter:—Leaves smaller on very short petioles, ovate or ovato-deltoid, obtuse, serrated; flowers in dichotomous, axillary, many-flowered cymes; calyx sub-campanulate, not swelling in front, indistinctly 2-lipped, segments all subulate and straight, the 2 lower ones slightly longer than the 3 upper ones; hairs of mouth prominent; central lobe of lower lip of corolla rounder and entire; seeds paler brown, less conspicuously dotted.

17. CLINOPODIUM. *Linn.* Wild Basil.

1. C. *vulgare*, Linn. *Wild Basil.* Leaves shortly petiolate, ovate, crenate, subrugose; flowers in branched or forked whorls; bracteas setaceous; *E. Bot. t.* 1041. *E. Fl. v. iii. p.* 106. *Hook. Br. Fl. p.* 284.

Hills and dry shady places; not uncommon. *Fl.* August. ♃.

Near Rudge; *H. Bidwell, Esq.* Lincoln's Hill, Coalbrookdale; *Mr. F. Dickinson.* Milford, near Baschurch; *T. C. Eyton, Esq.* Lane between Buildwas and Little Wenlock; *Dr. G. Lloyd.* Ludlow; *Miss Mc Ghie.*

Gamester Lane, near Westfelton. About Shrewsbury generally.

Root creeping, fibrous. *Stem* about a foot high, erect, wavy, ascending, tetragonous, hairy, with opposite branches. *Leaves* opposite, ovate, obtuse, deflexed, hairy on both sides, subrugose, crenate, on short hairy petioles. *Flowers* in crowded axillary and terminal many-flowered whorls, on branched or forked pedicels subtended by setaceous hairy *bracteas.* *Calyx* tubular, hairy, 2-lipped, segments of upper lip 3, acute, of lower one 2, longer, incurved; throat closed with converging hairs. *Corolla* twice the length of the calyx, large, purple, upper lip ovate, obtuse, bifid, slightly reflexed at the margins, lower lip 3-fid, throat pubescent, lateral lobes ovate, obtuse, entire, central one longer, subrotund, emarginate. *Seeds* roundish, black, glabrous.

18. PRUNELLA. *Linn.* Self-heal.

1. P. *vulgaris.* Linn. *Self-heal.* Leaves petiolate, oblongo-ovate, obtuse, crenate; upper lip of calyx truncate, its teeth almost obsolete. *E. Bot. t.* 961. *E. Fl. v. iii. p.* 115. *Hook. Br. Fl. p.* 285.

Moist pastures; frequent. *Fl.* July, August. ♃.

Root creeping. *Stem* 8—10 inches high, ascending, branched at the base, rounded, channelled on opposite sides, with erect appressed pubescence, especially on the edges of the furrows. *Leaves* opposite, oblongo-ovate, obtuse, crenate or wavy, hairy on both sides, on ciliated channelled petioles. *Flowers* in a terminal, dense, oblong, imbricated, whorled spike, sessile in the upper pair of leaves, bright blue with darker veins. *Semiwhorls* of 3 flowers, on short pubescent peduncles, in the axil of a broad, cordato-acuminate, hairy *bractea.* *Calyx* reddish green or purple, unequally 2-lipped, 10-ribbed, ribs ciliated, upper lip wedge-shaped, truncate, lateral margins rounded, minutely ciliated, sharply incurved, the teeth reduced to 3 minute points or mucros terminating the 3 principal ribs, with an intermediate notch in the margin on each side the midrib; lower lip ovato-lanceolate, deeply bifid, mucronate. *Corolla* somewhat longer than the calyx, nearly equally 2-lipped, upper lip erect, roundish, galeate, margins emarginate, externally slightly hairy; lower lip 3-fid, lateral lobes, oblong, entire, reflexed, central lobe roundish, contracted at the base, anterior margin minutely toothed; throat glabrous internally and externally, dilated; tube with a transverse horizontal appendage of white hairs above the base. *Filaments* glabrous, inferior ones longest, bifid at the apex, the inferior or innermost segments bearing the anther; *pollen* white. *Seeds* oblong, triquetrous, rounded at the apex, shining, glabrous, imbedded in a fleshy 4-toothed receptacle.

19. SCUTELLARIA. *Linn.* Skull-cap.

1. S. *galericulata*, Linn. *Common Skull-cap.* Stem divaricately

branched; leaves shortly petiolate, lanceolate, obtuse, cordate at the base, obtusely denticulate and revolute; flowers axillary, secund, shortly pedicellate; corolla large, pubescent. *E. Bot. t.* 523. *E. Fl. v. iii. p.* 113. *Hook. Br. Fl. p.* 285.

Banks of rivers, pools, &c.; frequent. *Fl.* July, August. ♃.

Root creeping, with fibrous joints. *Stem* 10—18 inches high, erect, tetragonous, glabrous, furrowed, with opposite divaricated branches. *Leaves* shortly petiolate, lanceolate, obtuse, cordate at the base, obtusely denticulate and slightly revolute at the margin, somewhat wrinkled and glabrous above, pubescent on the veins beneath. *Whorls* 2-flowered, each on an axillary solitary pubescent peduncle, with two setaceous *bracteas* at the base, both drooping to one side. *Calyx* short, tubular, truncate, equally 2-lipped, entire, with deflexed pubescence, sometimes glabrous, and a concave tooth near the base of the upper lip. *Corolla* many times longer than the calyx, light blue variegated with white, externally pubescent, 2-lipped, mouth closed. *Seeds* round, obtusely triquetrous, granulated, yellow.

2. S. *minor*, Linn. *Lesser Skull-cap.* Stem with erect branches; leaves shortly petiolate, radical ones broadly ovate, obtuse, upper ones ovato-lanceolate, obtuse, cordate and subhastate at the base, margins entire and revolute; flowers axillary, secund; corolla small, glabrous. *E. Bot. t.* 524. *E. Fl. v. iii. p.* 114. *Hook. Br. Fl. p.* 285.

Moist heathy places and sides of pools, &c.; not common. *Fl.* July, August. ♃.

Boggy places on the Titterstone Clee Hill; *Rev. W. Corbett.* Wyre Forest; *Mr. G. Jorden.* Toterell wood, Uncless, Wyre Forest; *Mr. W. G. Perry.* On the boggy spots of ground at the bottom of Haughmond Hill; *Rev. E. Williams's MSS.*

Boggy ground on the summit of Haughmond Hill. Margins of Oxon pool, near Shrewsbury.

Root creeping, fibrous. *Stems* 6—8 inches high, erect, very much branched from the base, tetragonous, slightly pubescent. *Leaves* opposite, on very short petioles, lower ones broadly ovate, obtuse, cordate at the base, upper ones lanceolate, obtuse, cordate and often with a notch at the base on each side, thus rendering them subhastate, entire and revolute at the margins, pubescent above, paler and dotted beneath, the veins only hairy. *Whorls* 2-flowered, each on a short axillary solitary pubescent peduncle, drooping towards one side, with two setaceous *bracteas* at the base. *Calyx* 2-lipped, truncate, entire, pubescent. *Corolla* many times longer than the calyx, pale reddish almost white. *Seeds* round, obsoletely subtriquetrous, granulate.

DIDYNAMIA—ANGIOSPERMIA.

21. EUPHRASIA. *Linn.* Eyebright.

1. E. *officinalis*, Linn. *Common Eyebright.* Leaves ovate, with 5 deep teeth on each side; lobes of the lower lip emarginate. *E. Bot. t.* 1416. *E. Fl. v. iii. p.* 123. *Hook. Br. Fl. p.* 286.

Pastures both on plains and hills; common. *Fl.* July. ⊙.

Root fibrous. *Stem* variable in height from 1—8 inches, nearly simple or oppositely and decussately branched, obsoletely tetragonous, with decurved pubescence. *Leaves* opposite, sessile, ovate, dentate, sometimes glabrous, sometimes slightly hairy or hispid with short bristles, teeth generally 5 on each side, those of

the upper leaves sharply mucronate. *Flowers* axillary, solitary, sessile, crowded at the extremities of the branches. *Calyx* tubular, 4-ribbed, glabrous, ¼4-fid, segments subulate, equal, acute, the ribs and segments either smooth or with erect bristles. *Corolla* bluish-white with dark violet streaks and a yellow throat, externally pubescent, tube as long as the calyx, ringent, 2-lipped; *upper lip* erect, slightly arched, 2-lobed, lobes sinuato-dentate; *lower lip* 3-lobed, lobes deeply emarginate, nearly equal, the central one rather the longest. *Anthers* mucronulate at the base, hairy. *Capsule* as long as the calyx, of 2 oblong compressed lobes, with a short obtuse point (the base of the style) in the sinus, hairy at the apex. *Seeds* oblong, attenuated at both ends, with a loose white coat, longitudinally wrinkled and transversely striated.

2. E. *Odontites*, Linn. *Red Eyebright.* Leaves linear-lanceolate, distantly serrated; lobes of the lower lip entire. *Bartsia Odontites, Huds. E. Bot. t.* 1415. *E. Fl. v. iii. p.* 119. *Hook Br. Fl. p.* 286.

Corn-fields; frequent. *Fl.* July, August. ⊙.

Root fibrous. *Stem* erect, 8—18 inches high, obsoletely tetragonous, pubescent, with opposite branches. *Leaves* in opposite pairs, linear-lanceolate, distantly serrated, scabrous. *Flowers* axillary, solitary, secund, on very short peduncles. *Calyx* tubular, scabrous with erect appressed bristles, ¼4-fid, segments equal, 2 broader than the rest, lanceolate, acute. *Corolla* dull pink, externally pubescent, tube as long as the calyx, ringent, unequally 2-lipped, *upper lip* erect, truncate, very slightly emarginate, *lower lip* smaller, 3-lobed, lobes nearly equal, central one longest, rounded, entire, reflexed. *Anthers* mucronulate at the base, hairy. *Capsule* oblong, hairy. *Seeds* oval, longitudinally striated and rugulose.

22. RHINANTHUS. *Linn.* Yellow Rattle.

1. R. *Crista-Galli*, Linn. *Common Yellow Rattle.* Leaves oblongo-lanceolate, serrated; flowers in lax spikes; segments of the upper lip of the corolla small, roundish; style included; nectary ovate, curved upwards; seeds with a broad membranous radiated border. *E. Bot. t.* 657. *E. Fl. v. iii. p.* 120. *Hook. Br. Fl. p.* 286.

Meadows and pastures; common. *Fl.* June. ⊙.

Root fibrous. *Stem* very variable in size, 6-18 inches high, erect, tetragonous, smooth, streaked with purple, more or less branched, branches opposite. *Leaves* opposite, sessile, oblongo-lanceolate, deeply serrated, margins of the serratures thickened, recurved and armed with minute bristles, dilated and somewhat cordate at the base, scabrous on both sides, veiny, curiously variegated with darker green veins on the under surface. *Bracteas* opposite, lowermost ovate, dilated and inciso-serrated at the base, the apex more or less elongated or acuminated, uppermost nearly rhomboid, acute, inciso-serrated, all very shortly petiolate. *Flowers* in a long terminal bracteated spike, in pairs, opposite, each pair crossing the next, on short peduncles arising each from the axil of the bractea. *Calyx* bladdery, veiny, roundish-ovate, compressed, 4-fid, the keels and margins of the segments pubescent. *Corolla* 2-lipped, compressed; *upper lip* galeate, pubescent on the exterior, yellow, with 2 small bluish roundish lateral teeth at the apex; *lower lip* 3-fid, lateral lobes oval, erect, intermediate one broader and complicate; *tube* straight, shorter than the calyx. *Nectary* an ovate green thick scale, curved upwards, at the base of the lower edge of the compressed germen. *Filaments* glabrous, lower ones longest. *Anthers* externally with long white curiously flexuose hairs. *Capsule* included in the persistent calyx and surrounded at the base with the remains of the tube of the corolla, obliquely orbiculari-ovate, compressed, the lower cell the

largest, glabrous, notched and tipped with the persistent base of the style. *Seeds* obliquely oval, large, laterally compressed, minutely dotted, with a broad membranous radiated border.

23. MELAMPYRUM. *Linn.* Cow-wheat.

1. M. *cristatum*, Linn. *Crested Cow-wheat.* Spikes densely imbricated, 4-sided; bracteas rotundo-cordate, acuminate, deeply incised, ciliated, complicate; calycine segments straight, subulate, two upper ones longest; corolla 3 times as long as the calyx, lower lip obtuse, with 3 equal obtuse converging teeth; capsule semi-orbicular, with an oblique point. *E. Bot. t.* 41. *E. Fl. v. iii. p.* 124. *Hook. Br. Fl. p.* 287.

Woods and thickets; very rare. *Fl.* July. ⊙.

" Woods in Shropshire, very rare." *Purton's Midl. Flora.*

Stem 12—18 inches high, erect, tetragonous, pubescent, with many opposite erecto-patent branches. *Leaves* in opposite pairs, sessile, linear-lanceolate, acuminate, entire, rough with minute hairs above, glabrous and veiny beneath, except the margins and nerves. *Flowers* in dense 4-sided terminal spikes, solitary, sessile, variegated with cream colour and light purple, palate yellow. *Bracteas* rotundo-cordate, acute, lower ones acuminated with a long recurved leafy point, complicated and keeled, margins deeply incised, teeth very acute, often again cut, ciliated. *Calyx* glabrous, except a row of hairs along the ribs of each segment, which is straight, subulate, ciliated, 2 upper ones longest, about one-third the length of the corolla. *Upper lip* of *Corolla* obtuse, emarginate, reflexed at the margin; *lower lip* obtuse, with 3 equal obtuse converging teeth. *Capsule* compressed, concealed by the bractea, semi-orbicular, one side nearly straight, the other convex and pubescent, with a minute oblique recurved point, veiny, glabrous. *Seeds* 4, oblong, glabrous.

2. M. *pratense*, Linn. *Common Yellow Cow-wheat.* Flowers axillary, secund; bracteas lanceolate, with 1-2 spreading acute teeth on each side at the base; calycine segments nearly equal, lanceolate, 2 upper ones ascending, 2 lower ones straight; corolla 3 times as long as the calyx, closed, lower lip protruded, with 3 straight obtuse teeth; capsule obliquely ovate, somewhat acuminate. *E. Bot. t.* 113. *E. Fl. v. iii. p.* 125. *Hook. Br. Fl. p.* 287.

Woods and thickets; common. *Fl.* July, August. ⊙.

Stem 8—12 inches high, erect, tetragonous, glabrous, with opposite spreading branches. *Leaves* opposite, in distant pairs, on very short petioles, lanceolate, acuminate, obtuse, glabrous. *Flowers* solitary, axillary, secund, on very short peduncles, large, pale yellow. *Bracteas* with 1-2 spreading acute teeth on each side at the base. *Calyx* glabrous, segments nearly equal, lanceolate, acuminate, bristly at the edges, 2 upper ones ascending, not so deeply divided, 2 lower ones straight, narrower, 3 times shorter than the corolla. *Upper lip* of *corolla* obtuse, slightly emarginate, reflexed and pubescent at the margin; *lower lip* protruded, with 3 straight obtuse teeth. *Capsule* obliquely ovate, somewhat acuminate, compressed, glabrous, veiny. *Seeds* 4, oblong, large.

" In the Forest of Wyre are varieties with orange and yellow flowers, occupying distinct and extensive habitats;" *Mr. G. Jorden.*

β. *montanum.* Smaller, bracteas quite entire. *Hook. Br. Fl. p.* 287. *M. montanum, Johnst. Fl. Berwick, v. i. p.* 136.

Chiefly in mountainous pastures. *Fl.* July, August. ⊙.

3. M. *sylvaticum*, Linn. *Lesser-flowered Yellow Cow-wheat.* Flowers axillary, secund; bracteas linear-lanceolate, acuminate, entire; calycine segments equal, lanceolate, straight; corolla twice as long as the calyx, lips equal, slightly open, lower lip with three large and deep obtuse straight teeth; capsule obliquely ovate, shortly pointed. *E. Bot. t.* 804. *E. Fl. v. iii. p.* 126. *Hook. Br. Fl. p.* 287.

Woods; very rare. *Fl.* July. ⊙.

Standhill coppice near Wenlock; *W. P. Brookes, Esq.* Woods near Hanwood; near Bedston, Ludlow; *T. and D. Bot. Guide.* Whitcliff near Ludlow; *Mr. H. Spare.*

Stem 8—10 inches high, erect, tetragonous, slightly pubescent, oppositely branched. *Leaves* narrow, linear-lanceolate, very much acuminated, acute, roughish above and on the midrib and slightly recurved margins beneath. *Flowers* axillary, solitary, secund, on short peduncles. *Bracteas* similar to the leaves, entire. *Calyx* glabrous, segments lanceolate, rough with erect bristles, rather more than half the length of the corolla, all equal. *Corolla* deep yellow, very small, lips nearly equal in length, a little open, *upper lip* elevated, *lower* one pointing downwards, with three large and deep teeth.

24. LATHRÆA. *Linn.* Toothwort.

1. L. *squamaria*, Linn. *Greater Toothwort.* Stem simple; flowers pendulous in one-sided racemes; lower lip of the corolla 3-cleft. *E. Bot. t.* 50. *E. Fl. v. iii. p.* 127. *Hook. Br. Fl. p.* 288. *Bowman in Linn. Trans. v. xvi. p.* 399.

Woods and coppices, parasitic on the roots of Hazels, Ash, and other trees; rare. *Fl.* April, May. ♃.

Benthall Edge, Coalbrookdale; Shrubbery at Bitterley Court, Ludlow; *T. and D. Bot. Guide.* Near Ludlow; *Rev. A. Bloxam.* In the plantations at the Lodge near Ludlow; *Rev. W. Corbett.* Naturalized to excess on hazel-roots at the Nursery, Westfelton; *J. F. M. Dovaston, Esq.* Coalbrookdale woods; *Mr. F. Dickinson.* Caynham Camp; Oakley Park near Ludlow; *Miss Mc Ghie.* In a thicket and dingle by Hord's Park, Bridgnorth; *Purton's Midl. Flora.*

Root red-brown, smooth, solid and woody, spindle-shaped, terminating in many forked fibres, each furnished at and near its extremity with very minute brown semiglobular and succulent tubercles, either solitary or in groups of 2 or 3, which fixing themselves upon the roots of the trees whereon the plant grows, send down from their under surface or point of attachment a straight or curved funnel-shaped process, which penetrates through the bark to various depths into the alburnum, but never into the solid wood. A single moniliform duct thickening and branching upwards traverses the cellular tissue of this penetrating process and tubercle, and conveys the nutriment of the parasite along the fibres of the root into the subterranean stem. Similar fibres bearing tubercles are thrown out from between the scales of the subterranean stem. The embryo stem at first takes a downward direction in common with the root, until coming into contact with the roots and fibres of the tree it spreads horizontally. *Subterranean stem* densely imbricated with broad, rotundo-ovate, transversely dilated, obtusely pointed and irregularly wavy *scales (true leaves)* in opposite pairs, thick, fleshy, succulent and convex, slightly marked externally with longitudinal parallel striæ, perpendicularly traversed internally with many hollow irregularly corrugated cells, which are lined with innumerable, minute, oval, transparent, pedicellate and sessile glandular papillæ, *(cuticular absorbents,)* marked externally by 4 longitudinal depressions indicating as many septa or valves within with hollow intermediate spaces. Between the incurved lower edge of the leaf and the underside of the leafstalk is a very narrow interstitial passage, which leads into an inclosed wider inner space running along the whole underside of the leaf, and communicating by oblong narrow orifices with the base of each of the leaf-cells. The solid portion of the leaves consists of cellular substance formed into hollow dodecahedral compartments, containing a watery juice, in which are several oval or pearshaped transparent detached bladders filled with a glutinous or mucilaginous fluid, probably a kind of liquid sugar. *Flowering-stems* erect, rising 10, 12 or 14 inches above the surface, bearing 30—40 flowers disposed in 3 distinct yet perfectly unilateral rows, smooth, shining, tinged with bluish-purple, and having a few flowerless bracteas below, but gradually becoming hairy upwards, thickest about the middle. *Bracteas* broader than their length, rotundo-obovate, pointed, sessile, smooth, shining and rather fleshy, one at the base of each pedicel, destitute of the internal cells and bladders of the subterranean leaves. *Pedicels* and *calyx* copiously covered with jointed hairs tipped with a globular or oval gland. *Calyx* 4-cleft, segments acute, 2 lower ones narrower, soft and thin, not at all succulent. *Corolla* longer than the calyx, of a dullish pink purple; tube very wide, glabrous; lips about equal, *upper* one truncate, entire or very slightly notched, externally pubescent, collapsed and enveloping the stamens and greater part of the style; *lower lip* glabrous, 3-lobed, lobes about equal, rounded or wavy, incurved or collapsed, palate hairy. *Filaments* of longer stamens very much thickened upwards. *Anthers* hairy. *Pollen* white, globose. *Germen* oval, pointed, triquetrous, with a dilated obtusely pointed yellow scale or *nectary* on the under side at the base. *Stigma* exserted, large, capitate, 4-lobed, papillose. *Capsule* ovate, acuminate, triquetrous, succulent, glabrous.

25. PEDICULARIS. *Linn.* Lousewort.

1. P. *palustris*, Linn. *Marsh Louse-wort.* Stem solitary, branched upwards, erect; calyx broadly ovate, pubescent, in 2 nearly equal inciso-dentate crisped lobes. *E. Bot. t.* 399. *E. Fl. v. iii. p.* 129. *Hook. Br. Fl. p.* 288.

Marshy pastures and margins of pools; not very common. *Fl.* July. ♃ ?

Snowden pool, near Beckbury; *H. Bidwell, Esq.* Near Ellesmere; *Rev. A. Bloxam.* Littlehales; *R. G. Higgins. Esq.* Ludlow; *Miss Mc Ghie.* Wyre Forest; *Mr. W. G. Perry.* Oakley Park, near Ludlow; *Mr. H. Spare.* Fennemere pool, near Baschurch; *W. W. Watkins, Esq.*

Bomere and Oxon pools, near Shrewsbury.

Root small, by some said to be annual. *Stem* solitary, erect, 12—18 inches high, angular, with alternate spreading branches, glabrous or nearly so. *Leaves* of the stem alternate, of the branches opposite, on dilated flattened petioles, subbipinnatifid, segments obtusely notched, margins revolute, glabrous. *Flowers* axillary, solitary, on very short peduncles, of a deep rose-colour, large and handsome. *Calyx* broadly ovate, ribbed and veiny, in 2 nearly equal lobes, the lobes in the upper part variously and unequally inciso-dentate, margins crisped and revolute, somewhat ciliated, ventricose in fruit. *Capsule* obliquely ovate, pointed, compressed, glabrous. Seeds numerous, obovate, reticulato-rugoso-punctate.

2. P. *sylvatica*, Linn. *Pasture Lousewort or Dwarf Red Rattle.* Stem branched from the base, spreading; calyx oblong, angular, glabrous, in 5 unequal foliaceous dentate segments. *E. Bot. t.* 400. *E. Fl. v. iii. p.* 130. *Hook. Br. Fl. p.* 288.

Moist pastures and heaths; common. *Fl.* July. ♃.

Root fusiform. *Stems* numerous, spreading flat on the ground, the extremities ascending in flower, generally simple, sometimes branched, round, glabrous and succulent. *Leaves* alternate, on dilated flattened succulent petioles, deeply pin-

natifid, lobes oval, deeply variously and acutely toothed and incised, margins revolute, glabrous; radical ones ovate, undivided, crenated, recurved. *Flowers* axillary, solitary, on very short peduncles, of a delicate pink or pale rose-colour, very elegant. *Calyx* 5-angular, glabrous and veiny, with 5 unequal teeth, the upper one linear acute, the others dilated upwards into trifid foliaceous lobes, margins densely ciliated. *Corolla* very much exserted, more than twice the length of the calyx, *upper lip* compressed laterally, erect, the apex somewhat falcate and elongated into a short truncated beak, with a triangular tooth on each side; *lower lip* 3-lobed, lobes spreading, nearly equal, rotundo-obovate, margins irregularly crenate or notched, with a few scattered ciliæ, central lobe rather the smaller, marked in the throat with two longitudinal stripes of white edged with darker pink, communicating with two longitudinally hairy folds continued down the tube in front to the point of attachment of the *filaments* which are hairy at the base. *Calyx* of the *fruit* ventricose. *Capsule* obliquely ovate, pointed, compressed, glabrous.

A variety with a salvershaped 5-cleft regular corolla and 5 stamens found by Miss Bage of Bangor, near Hanwood, near Shrewsbury is in the Herbarium of the Shropshire and North Wales Natural History and Antiquarian Society.

26. ANTIRRHINUM. *Linn.* Snapdragon.

1. A. *majus*, Linn. *Great Snapdragon.* Leaves lanceolate, alternate, those of the branches opposite; flowers racemose; sepals ovate obtuse, much shorter than the corolla. *E. Bot. t.* 129. *E. Fl. v. iii. p.* 136. *Hook. Br. Fl. p.* 288.

Old walls and ruins; not uncommon. *Fl.* July, August. ♃.

Walls of Ludlow castle; *Rev. W. Corbett.* Oswestry town-walls; *J. F. M. Dovaston, Esq.* Walls of Sundorn garden near Shrewsbury. Hardwick House near Ellesmere; *Rev. E. Williams's MSS.*

Shrewsbury town-walls. Ludlow town-walls.

Stem 1-2 feet high, branched, shrubby. *Leaves* lanceolate, acute, recurved, entire, glabrous. *Flowers* large, purplish-red, often white, with a downy yellow palate, white in front. *Capsule* ovate, pubescent, protuberant at the base in front. *Seeds* foveolato-rugose.

2. A. *Orontium*, Linn. *Lesser Snapdragon.* Leaves linear-lanceolate, mostly alternate; spikes few-flowered, lax; sepals linear, longer than the corolla. *E. Bot. t.* 1155. *E. Fl. v. iii. p.* 137. *Hook. Br. Fl. p.* 289.

Corn-fields; rare. *Fl. July, August.* ⨀.

About Ludlow; *Miss Mc Ghie.*

Stem 1 foot high, hairy, more or less branched. *Leaves* linear-lanceolate, entire, shortly petiolate. *Flowers* axillary, nearly sessile, purple. *Calyx* deeply 5-fid, segments linear, pubescent, longer than the corolla. *Capsule* ovate, hairy, protuberant in front. *Seeds* covered on one side with a convex minutely dotted shield projecting at the margins, naked on the other side, minutely tuberculate, and with deep central longitudinal depression.

27. LINARIA. *Juss.* Toadflax.

1. L. *Cymbalaria*, Mill. *Ivy-leaved Toadflax.* Stems trailing; leaves reniform-orbicular, 5-lobed, alternate, globose. *Hook. Br. Fl. p.* 289. *Antirrhinum Cymbalaria, Linn. E. Bot. t.* 502. *E. Fl. v. iii. p.* 131.

Naturalized on old walls; frequent. *Fl.* all summer. ♃.

On a wall at Oldport near Oswestry; *T. and D. Bot. Guide.* Old walls at Ludlow; *Mr. F. Dickinson.* Astley; *Mr. E. Elsmere, jun.* Wall in Lilleshall; *R. G. Higgins Esq.* Ludlow Castle; *Mr. H. Spare.* South side of Cound churchyard wall. Walls round Hall Hussey House, near Shrewsbury; *Rev. E. Williams's MSS.*

Council-house garden wall in the Water-lane, Shrewsbury. Albrighton near Shrewsbury.

Root fibrous. *Stem* long, filiform, round and smooth. *Leaves* petiolate, deep shining green above, frequently purple beneath. *Flowers* on axillary stalks, as long as the leaves, violet and blue, palate yellow, spur short and pointed. *Capsule* roundish. *Seeds* black, wrinkled.

Mr. Dovaston informs me that this plant was first introduced into Shropshire by the late Richard Hill Waring, Esq., a very elegant botanist, at the Hayes near Oswestry.

2. L. *Elatine*, Desf. *Sharp-pointed Fluellen* or *Toadflax.* Stems procumbent, hairy; leaves ovato-hastate, inferior ones ovate, opposite; spur of the corolla straight; peduncles glabrous. *Hook. Br. Fl. p.* 289. *Antirrhinum Elatine, Linn. E. Bot. t.* 692. *E. Fl. v. iii. p.* 132.

Corn-fields; not common. *Fl.* July—September. ⨀.

Corn-stubbles, common; *Rev. W. Corbett.* Milloon and Hailstones, West-felton near Oswestry; *J. F. M. Dovaston, Esq.* Astley; *Mr. E. Elsmere, jun.* Hopton Hill; *T. C. Eyton, Esq.* Corn-field, at Bonninghall Park; *R. G. Higgins, Esq.* Over Wood; *Mr. G. Jorden.* Buildwas; *Miss H. Moseley.* Corn-fields about Eaton Mascott, Berrington, &c. in a clay soil; *Rev. E. Williams's MSS.*

Fields near the south-east base of Sharpstones Hill.

Root fibrous. *Stem* slender, round, hairy, branched, procumbent. *Leaves* hairy, on short hairy petioles; lower ones opposite, ovate, toothed on the hinder margin; upper ones hastate with a single auricle on each side at the base, becoming more sagittate above, whilst the uppermost are rounded and entire at the base. *Peduncles* solitary, axillary, very slender, longer than the leaves, glabrous except on the slightly swollen part below the calyx, spreading and deflexed. *Sepals* lanceolate, acute, hairy, margins membranous at the base. *Corolla* small, yellow, upper lip purple; *spur* subulate, straight, as long as the corolla. *Capsule* globose, minutely pitted. *Seeds* oval, pale, very singularly wrinkled.

3. L. *vulgaris*, Mœnch. *Yellow Toadflax.* Stem erect; leaves linear-lanceolate, scattered, crowded; spikes terminal, flowers imbricated; calyx glabrous; spur shorter than the corolla. *Hook. Br. Fl. p.* 289. *Antirrhinum Linaria, Linn. E. Bot. t.* 658. *E. Fl. v. iii. p.* 134.

Hedges and corn-fields; frequent. *Fl.* July, August. ♃.

Root creeping. *Stem* 1-2 feet high, erect, round, simple, glaucous, densely clothed with scattered, sessile, linear-lanceolate, acute, 3-ribbed *leaves. Spikes* terminal, erect, imbricated. Flowers on short bracteated pedicels, yellow, palate downy and orange; *spur* pointing perpendicularly downwards. *Sepals* ovate, acuminate, glabrous. *Capsule* ovate. *Seeds* black, orbicular, compressed, tuberculoso-scabrous, surrounded by a membranous margin.

A variety with two spurs once occurred to my notice near Shrewsbury.

4. L. *minor*, Desf. *Least Toadflax.* Stem erect with spreading branches; leaves linear-lanceolate, obtuse, mostly alternate, glandu-

loso-pubescent; flowers solitary, axillary; calyx longer than the spur. *Hook. Br. Fl. p.* 290. *Antirrhinum minus, Linn. E. Bot. t.* 2014. *E. Fl. v. iii. p.* 135.

Corn-fields; rare. *Fl.* June—August. ⨀.

Coreley and Burford parishes; *Rev. W. Corbett.* Near Ludlow; *Miss Mc Ghie.* Corn-fields, Oswestry; *Rev. T. Salwey.*

Root fibrous. *Stem* erect, 4—12 inches high, round, somewhat zigzag, glanduloso-pubescent, branched. *Leaves* generally alternate, except a few of the lower ones, linear-lanceolate, obtuse, tapering into the petiole. *Flowers* on solitary, axillary, erect, glanduloso-pubescent peduncles as long as the leaves. *Corolla* with the tube upper lip and very short spur purplish, lower lip white, palate yellow. *Sepals* linear-lanceolate, obtuse, glanduloso-pubescent. *Capsule* ovate, oblique, shorter than the persistent sepals. *Seeds* oblong, tuberculoso-scabrous, with prominent longitudinal wrinkles.

28. SCROPHULARIA. *Linn.* Figwort.

* *Staminodium 2-lobed.*

1. S. *nodosa*, Linn. *Knotted Figwort.* Stem acutely quadrangular; leaves ovato-acuminate, triangular, acute, subcordate, doubly and acutely serrated, basal serratures the longest; sepals roundish ovate, obtuse, margins with a very narrow nearly entire membrane; staminodium transversely oblong, emarginate. *E. Bot. t.* 1544. *E. Fl. v. iii. p.* 137. *Hook. Br. Fl. p.* 290.

Moist shady ground, banks of rivulets, &c. *Fl.* July. ♃.

Root tuberous, thick and knotty, with many long fibres. *Stem* 2-3 feet high, erect, acutely quadrangular, nearly simple, glabrous. *Leaves* opposite, on short winged petioles, ovato-acuminate, triangular, acute, subcordate at the base which is sinuated or cut away to the two lateral ribs and tapers into the petiole, doubly acutely and unequally serrated, the basal serratures the longest and often with a tendency to become auriculate, glabrous, veiny. *Flowers* in dichotomous axillary and terminal bracteated panicles; peduncles and pedicels with stipitate glands. *Bracteas* lanceolate, acute, entire, glabrous. *Sepals* roundish ovate, obtuse, with a very narrow membranous nearly entire or very slightly notched or laciniated margin, glabrous. *Tube* of *corolla* green, limb dark-red, with a transversely oblong emarginate scale (*staminodium*) adnate to the upper lip. *Capsule* ovate. *Seeds* elliptical, longitudinally furrowed, transversely rugose.

** *Staminodiun entire.*

2. S. *aquatica*, Linn. *Water Figwort, Water Betony.* Stem quadrangular, broadly and acutely winged; leaves oblong or oblongo-ovate, rotundato-obtuse, cordate at the base, doubly and obtusely crenato-serrate, crenatures obtusely mucronate, basal ones the smallest, glabrous, the lower leaves sometimes auricled at the base; leaves sepals and corolla pellucido-punctate; cymes many-flowered; sepals subrotund, very obtuse, with a broad laciniate membranous margin; staminodium subrotundo-reniform, entire; capsule ovate, somewhat acute. *Stevens in Ann. Nat. Hist. vol.* 5. *p.* 1. *Linn. Herb. Sp. Pl.* 864. *Curt. Fl. Lond.* 5. *t.* 44. *E. Bot. t.* 854. *Sm. Fl. Br.* 663. *E. Fl. v. iii.* 139. *(diagn. only.) Sm. Herb. n.* 2. *Hook. Scot.* 189. *Br. Fl. ed.* 3. 554. *ed.* 4. 239. *With.* 3*d ed.* 3. 554. *ed.* 7. 3. 738. *Grev. Fl. Edin.* 137. *Purton's*

Midl. Flora 1. 293. *Kroch. Fl. Siles.* 2. 393. *Duby Bot. Gall.* 1. 346. *Sebast et Mauri Fl. Roman.* 205. *Loisel. Fl. Gall.* 2. 35. *Pollinus Fl. Veron.* 325. *S. Scorodonia (aquatica? Sm. not.) Linn. Herb. S. Balbisii, "Hornem. Fl. Hafn.* 2. 577." *Bluff. and Fing. Comp. Fl. Germ. ed.* 2. *t.* 1. *pt.* 2. *p.* 389. *Koch. Syn.* 515. *Guss. Fl. Sic. Prod.* 2. 172.

Sides of streams, wet ditch banks, &c. *Fl.* July. ♃.

Coalbrookdale pools; *Mr. F. Dickinson.* Eyton; *T. C. Eyton, Esq.*

Sutton Spa. Longden. Welbach. Red Hill. Canal between Shrewsbury and Uffington. Nobold; all near Shrewsbury. Weston Cotton. Maesbury; near Oswestry.

Root somewhat creeping, strong, jointed, with long strong fibres from the joints. *Stem* 2-3 feet high, erect, quadrangular, angles broadly and acutely winged, glabrous. *Leaves* opposite, on dilated winged petioles, oblong or oblongo-ovate, obtuse, cordate at the base which is slightly sinuated but not quite down to the lateral ribs and tapers into the petioles, doubly and obtusely crenato-serrate, crenatures obtusely mucronate, the basal ones smaller, glabrous, veiny, with copious pellucid dots. The leaves are sometimes auricled on both sides at the base with a small leaflet. *Flowers* in dichotomous, short, distant, axillary and terminal bracteated panicles; peduncles and pedicels with stipitate glands. *Bracteas* lanceolate, acute, entire, glabrous or with a few glands. *Sepals* subrotund, very obtuse, with a broad membranous jagged or laciniate margin, glabrous, pellucido-punctate. *Corolla* dark-red, the lobes and upper portion pellucido-punctate. *Staminodium* large, subrotundo-reniform, entire, adnate to the upper lip. *Capsule* globose, acute.

In March 1839, I perceived that a specimen in my herbarium gathered in 1835, at Sutton near Shrewsbury as *S. aquatica*, differed very materially from that species as described in the English Flora, and on a comparison with the works of some continental Botanists detected it to be *S. Balbisii, Hornemann.* About the same time Mr. C. A. Stevens of Rochester recognized the same species in a specimen from Scotland, and in order to elucidate our doubts examined the Linnæan and Smithian Herbaria. In the Linnæan Herbarium is a specimen (probably from a garden) marked by Linnæus himself *S. aquatica*, with the usual reference to his *Sp. Plant*, and which is identical with our Shropshire and Scotch plants. There is also a similar specimen marked by Linnæus *S. Scorodonia*, without however the usual numerical reference to the *Sp. Plant*; to which is added by Smith in pencil "*aquatica?*" In the Smithian Herbarium is a specimen (No. 1.) marked *S. aquatica, Linn.*, but which in reality is a totally different plant from the specimen so named in the Linnæan Herbarium. From this specimen Smith seems to have drawn up his *description* of the species in English Flora, though it is manifest that his *diagnosis* belongs to the true *S. aquatica, L.* as it is merely a translation of that in the *Sp. Plant.* On the same sheet is another specimen (No. 2.) which proves to be *S. Balbisii*, and identical with our Shropshire and Scotch plants; and hence the confusion in the English Flora. It is evident therefore, until some further information be elicited from cultivation or otherwise, that for the present *S. Balbisii, Hornem.* and *S. aquatica, Linn Herb*, and of *Sm. Herb. No.* 2. must be regarded as identical and as constituting the *true S. aquatica* of *Linnæus*; and the *S. aquatica* of *Sm. Herb. No.* 1. and *E. Fl.* as a distinct and unnamed species. The only specimen of the latter in the *Sm. Herb.* being amongst the "*Plantæ Officinales*" of Ehrhart, Mr. Stevens very properly proposes to name it, *S. Ehrharti.* Many of the continental botanists have described Ehrhart's plant as *S. aquatica, L.* and the *true S. aquatica, L.* as *S. Balbisii, Hornem. S. aquatica, L.* seems to be very generally distributed through Shropshire, but I have never observed any traces of *S. Ehrharti*, of which species the following are the essential characters as given by Mr. Stevens in the Annals:—

S. Ehrharti, Stev. Leaves ovate or ovato-lanceolate, subcordate at the base, acute, serrated; stem and petioles winged; panicle terminal, lateral cymes lax, few (5-6) flowered; sepals subrotund with a broad scarious margin; staminodium bifid, segments divaricating; capsule globose very obtuse. *S. aquatica, Ehrh. Pl. Offic. n.* 156. *Sm. Herb. No.* 1. *Fl. Dan. t.* 507. *Kunth Fl. Berol.* 260. *Bluff and Fing. Comp.* 389. *Reich. Fl. Excurs.* 2562. *Koch Syn.* 515. *Peterm. Fl. Lips.* 459. *Host. Austr.* 2. 203. *Wimm and Grab. Siles.* 2. 226.

Root fibrous. *Stem* erect, 2 feet or more high, simple, square, winged at the angles. *Leaves* ovate, ovato-oblong or lanceolate, slightly cordate at the base, acute, simply and finely serrated. *Panicle* of many, mostly alternate, dichotomous, few-flowered cymes. *Peduncles* and *pedicels* divaricating, slightly glandulose. *Bracteas* foliaceous, lanceolate, acute, simple or 3-partite, in which latter case the segments are lanceolate. *Staminodium* obreniform, bifid, lobes divaricating. *Sepals* with a broad torn membranous margin. *Capsule* globose, very obtuse.

*** *Staminodium wanting.*

4. S. *vernalis*, Linn. *Yellow Figwort.* Stem bluntly quadrangular, hairy; leaves broadly cordate, acute, doubly inciso-serrate, hairy on both sides; peduncles axillary, solitary, dichotomous; flowers cymose, aggregate; sepals oblong, somewhat acute; staminodium none. *E. Bot. t.* 567. *E. Fl. v. iii. p.* 140. *Hook. Br. Fl. p.* 290.

Waste places; very rare. *Fl.* April, May. ♃.

On a stone wall in a lane south of the house at Llanforda near Oswestry.

Whole plant of a light pleasant green, covered with soft spreading hairs each tipped with a globular gland. *Root* tuberous, scaly. *Stem* 2 feet high, quadrangular, hollow. *Leaves* opposite or 3 together, on long winged hairy petioles, veiny. *Flowers* pale yellow. *Corolla* ovate, inflated, contracted at the mouth, 5-lobed. *Calyx* with 5 deep acute segments, hairy. *Stamens* from the base of the corolla and with the *styles* very much exserted. *Capsule* ovate, acute. *Seeds* numerous, minute, elliptical, longitudinally furrowed and transversely rugose.

29. DIGITALIS. *Linn.* Foxglove.

1. D. *purpurea*, Linn. *Purple Foxglove.* Leaves ovato-lanceolate, crenate, tomentose beneath, inferior ones attenuated into a petiole; sepals ovato-oblong, shortly acuminate, 3-ribbed, pubescent; corolla campanulate, externally glabrous, upper lip very obtuse, truncate or slightly emarginate, segments of lower lip rounded. *E. Bot. t.* 1297. *E. Fl. v. iii. p.* 141. *Hook. Br. Fl. p.* 291.

Hilly and woody places, dry banks, walls, &c.; common. *Fl.* June, July. ♂.

Root fibrous. *Stem* erect, 3—5 feet high, somewhat angular, pubescent, mostly simple. *Leaves* alternate, on long winged hairy petioles, ovato-lanceolate, obtuse, crenate, crenatures obtusely mucronate, pubescent and rugose above, tomentose and strongly veined beneath. *Spikes* terminal, erect, very long; flowers numerous, on pubescent pedicels, with a linear-lanceolate *bractea* at the base, secund, drooping. *Corolla* large, purple, externally pale and glabrous, internally hairy, marked with darker purple spots surrounded with a white stained or watery nimbus. *Upper sepals* narrower.

A variety with white flowers occurs near the base of the Stiperstones Hill, and on the Wrekin; and in 1834, I gathered at Aber Waterfall, Caernarvonshire, a specimen bearing two distinct spikes, one of purple flowers, the other of purple and white flowers intermixed.

30. LIMOSELLA. *Linn.* Mudwort.

1. L. *aquatica*, Linn. *Common Mudwort.* Leaves lanceolate spathulate, on long petioles; peduncles axillary, crowded, shorter than the petioles, single-flowered. *E. Bot. t.* 357. *E. Fl. v. iii. p.* 145. *Hook. Br. Fl. p.* 240.

Muddy places, where water has stagnated; rare. *Fl.* July. ⊙.

Sandy muddy margins of river Severn at Shelton Rough, near Shrewsbury.

A small, stemless plant with long fibrous *roots* and sending out naked runners which produce new plants. *Leaves* all radical, on dilated petioles 1—3 inches long, erect, sheathing at the base, lanceolate or spathulate, 3—5 nerved. *Peduncles* axillary, recurved after flowering. *Corolla* white or pale-rose colour. *Anthers* purplish-blue, 1-celled. *Capsule* ovate. *Seeds* numerous, oblong, longitudinally furrowed, transversely rugose.

31. VERBENA. *Linn.* Vervain.

1. V. *officinalis*, Linn. *Common Vervain.* Stamens 4; stem erect, somewhat hispid; leaves rough, lanceolate, inciso-serrate or trifid, with the segments cut; spikes filiform, somewhat panicled, flowers rather remote. *E. Bot. t.* 767. *E. Fl. v. iii. p.* 71. *Hook. Br. Fl. p.* 291.

Road-sides and waste places; not very common. *Fl.* July. ♃.

Lincoln's Hill, Coalbrookdale; *Mr. F. Dickinson.* Baschurch; *T. C. Eyton, Esq.* Under Lilleshall church-yard wall; *R. G. Higgins, Esq.* Near Ludlow; *Dr. G. Lloyd.*

Maesbury and Weston Cotton, near Oswestry. Preston Gobald's church-yard. Wenlock Edge, near Middlehope. Ensdon. Kingsland. Lane leading from Bishop's Castle turnpike-road into Hanwood turnpike-road at Port Hill, near Shrewsbury. Old Heath. Wenlock road, near Lord Hill's column; and road-sides generally about Shrewsbury.

Root woody, somewhat creeping. *Stem* 1—2 feet high, ascending, angular and furrowed, hispid, with minute erect bristles. *Leaves* opposite, lanceolate, inciso-serrate or 3-fid, segments cut, tapering at the base into short broad petioles, hispid, especially on the ribs beneath. *Spikes* several, opposite and terminal, slender. *Flowers* small, blue.

32. OROBANCHE. *Linn.* Broom-rape.

1. O. *major*, Linn. *Greater Broom-rape.* Calyx 2-nerved, equally bipartite, segments bifid, nearly equalling in length the tube of the corolla; corolla campanulate, ventricose at the base in front, arcuate on the back, lips undulate obsoletely denticulate, upper lip galeate undivided, lower lip 3-lobed, lateral lobes ovate subrhomboid acute, central lobe larger and elongated, rotundo-ovate, obtusely pointed; stamens inserted into the base of the corolla, glabrous below, glandulo3o-pubescent above; style glanduloso-pubescent, decurved above, stigma distantly 2-lobed. *E. Fl. v. iii. p.* 147. *Hook. Br. Fl.* 4*th ed. p.* 241. *Babington. Prim. Fl. Sarn. p.* 65. *O. Rapum, Thuill. Reich. Fl. Excurs.* 2426. *Koch. Syn. p.* 533.

On the roots of Broom and Furze; not very common. *Fl.* June, July. ♃.

In a field opposite Quatford church; *H. Bidwell, Esq.* Willey Park; *W. P. Brookes, Esq.* Plantations at Neach Hill, near Shiffnall; *Dr. G. Lloyd.*

Near Shortwood, Ludlow; *Miss Mc Ghie.* Downton-hill Woods; *Mr. H. Spare.* Naturalized on the canal-side near the Queen's Head turnpike between Westfelton and Oswestry; *J. F. M. Dovaston, Esq.*

Sharpstones Hill near Shrewsbury.

Whole plant of a dark brown colour. *Stem* 2½ feet high, erect, thick, fleshy and rigid, furrowed, more or less hairy. *Lower scales* fleshy, broadly ovate, imbricated; those of the stem distant, lanceolato-acuminate, patent, glanduloso-pubescent, margins irregularly serrated. *Spikes* cylindrical, 18 inches long, glanduloso-pubescent, crowded with 70—80 sessile bracteated *flowers*. *Bracteas* lanceolate, very much acuminated, piloso-glandulose on the exterior, glabrous within, one at the base of and longer than the flower. *Upper lip* of *corolla* large, entire, rounded, irregularly notched or wavy, margin minutely denticulate; lower lip 3-lobed spreading, lateral ones ovate subrhomboid acute irregularly notched and wavy, minutely denticulate, central one elongated, rotundo-ovate, obtusely pointed, contracted at the base, denticulate, all with a longitudinal depression in the centre giving them a somewhat collapsed and wavy appearance. *Calyx* about equal in length to the tube of the corolla, piloso-glandulose, equally and deeply 2-lobed, each lobe deeply divided into two lanceolate acuminate segments, the upper one the longest, each with a strong prominent mid-rib and a smaller inconspicuous one on each side, serrated. *Stamens* attached to the base of the tube, broad, dilated at the base, channelled, glabrous on both sides below, glanduloso-pubescent above. *Style* very long, glanduloso-pubescent, decurved above and bearing two large distant globular glanduloso-pubescent lobes or *stigmas*. *Capsule* elliptical, glanduloso-pubescent. *Seeds* minute, very numerous, oval, dark brown, curiously and deeply reticulated and pitted.

2. O. *elatior*, Sutton. *Tall Broom-rape.* Sepals many-nerved, equally bifid, united in front, nearly equalling the tube of the corolla; corolla curved, tubuloso-campanulate, lips undulate, upper one 2-lobed, lobes patent?, lobes of the lower lip ovate, nearly equal, patent; stamens inserted above the base of the corolla, densely hairy on the inner side from the base to the middle; style slightly glanduloso-pilose. *E. Bot. t.* 568. *E. Fl. v. iii. p.* 148. *Hook. Br. Fl. p.* 241.

Clover fields; very rare. *Fl.* July, August. ♃.

Hill sides near Hope Bowdler; *A. Aikin, Esq.* "20 or 30 plants nearly a yard high each, observed in 1824, in a clover field immediately at the northern or rather north-west base of the Wrekin, not far from a round hill crowned with fir-trees, which terminates the undulation of the Wrekin to the west." *E. Lees, Esq.*

"Taller and of a more yellowish hue than the former, with *flowers* of a lighter purple, more wavy in their margins; their upper lip lobed, commonly three times more numerous in the *spike*, and of a smaller size. *Stamens* downy in the lower half withinside, and smooth at the top." *Smith.*

Of this very rare British plant, I have never seen Shropshire specimens. Smith and Sutton (*Linn. Trans. iv.* 173.) both describe the style as *glabrous*; though the foreign botanists state it to *slightly glanduloso-pubescent* and this is believed to be correct.

CLASS XV.

TETRADYNAMIA.

—

> " But now the sounds of population fail,
> No cheerful murmurs fluctuate in the gale,
> No busy steps the grass-grown footway tread,
> But all the bloomy blush of life is fled.
> All but yon widow'd solitary thing,
> That feebly bends beside the plashy spring ;
> She, wretched matron, forc'd, in age, for bread,
> To strip the brook with mantling cresses spread."
>
> GOLDSMITH.

CLASS XV.

TETRADYNAMIA. 6 *Stamens,* 4 *long and* 2 *short.*—*(Nat. Ord.* CRUCIFERÆ, *Juss.)*

ORDER I. SILICULOSA. *Fruit a short pod or pouch (Silicula.)*

* *Cotyledons incumbent* (o‖)

1. CORONOPUS. *Silicula* 2-lobed, without valves or wings. *Seeds* solitary in each cell.—Name from κορωνη, a *crow,* and πους, a *foot;* from the resemblance of the cut leaves.

2. CAPSELLA. *Silicula* laterally compressed, obcordato-cuneate ; the *valves* sharply keeled, without wings, many-seeded. Name the diminutive of *capsula,* a *little box.*

3. LEPIDIUM. *Silicula* with the cells one-seeded ; the *valves* keeled. *Petals* equal. Name from λεπις, a *scale;* from the appearance of the fruit.

4. SUBULARIA. *Silicula* oval, pointless, many-seeded ; *valves* turgid. *Cotyledons* linear, curved. Name from *subula,* an *awl ;* the leaves being subulate or awl-shaped.

5. CAMELINA. *Silicula* subovate, many-seeded ; *valves* inflated. *Filaments* simple. Name from χαμαι, *dwarf* or *humble,* and λινον, *flax.*

** *Cotyledons accumbent.* (o=)

6. THLASPI. *Silicula* laterally compressed, emarginate ; *valves* winged at the back, many-seeded.—Name from θλαω, to *flatten;* on account of its compressed seed-vessels.

7. TEESDALIA. *Silicula* emarginate ; the *valves* keeled ; the *cells* 2-seeded. *Filaments* having a little scale within at the base.— Name in honour of *Mr. Robert Teesdale,* a Yorkshire botanist.

8. COCHLEARIA. *Silicula* oval or globose, many-seeded ; the *valves* turgid. *Filaments* simple. *Seeds* not margined. *Calyx* patent.—Name from *cochlear,* a *spoon,* from the shape of its leaves.

9. DRABA. *Silicula* entire, oval or oblong ; *valves* plane or slightly convex ; *cells* many-seeded. *Seeds* not margined. *Filaments* simple.—Name from δραβη, *acrid,* as are the leaves of many of this tribe.

ORDER II. SILIQUOSA. *Fruit a long narrow pod (Siliqua.)*

* *Cotyledons incumbent.* (o‖)

10. SISYMBRIUM. *Siliqua* rounded or angular. *Calyx* patent, sometimes erect.—Name σισυμβριον, given by the ancients to some plant, perhaps allied to this.

11. ERYSIMUM. *Siliqua* 4-sided. *Seeds* not margined. *Stigma* capitate, sometimes emarginate, with the lobes patent. *Calyx* erect.

—Name from ερυω, *to cure;* on account of the supposed virtues of the plant.

12. HESPERIS. *Siliqua* 4-sided or 2-edged. *Stigma* nearly sessile, the lobes connivent. *Calyx* erect.—Name from εσπερος, *the evening;* at which time the flowers yield a powerful fragrance.

** *Cotyledons accumbent.* (o=)

13. CARDAMINE. *Siliqua* linear; the *valves* flat, generally seperating elastically, nerveless. *Seed-stalks* slender.—Name from καρδια, *the heart,* and δαμαω, *to fortify;* from its supposed strengthening qualities.

14. ARABIS. *Siliqua* linear, crowned with the nearly sessile *stigma; valves* veiny or nerved. *Seeds* in one row. *Calyx* erect.— So named, because originally an Arabian genus.

15. TURRITIS. *Siliqua* elongated, 2-edged; *valves* nerved or keeled. *Seeds* in a double row.—Name from *turris,* a *tower;* from the pyramidal growth of the plant.

16. BARBAREA. *Siliqua* 4-angled and somewhat 2-edged. *Seeds* in a single row. *Calyx* erect. *Glands* between the shorter *filaments.* —Name from the plant being formerly dedicated to St. Barbara.

17. NASTURTIUM. *Siliqua* nearly cylindrical (sometimes short); *valves* concave, neither nerved nor keeled. *Calyx* patent.—Name from *Nasus tortus,* a *convulsed nose,* an effect supposed to be produced by the acrid and pungent quality of the plant.

18. CHEIRANTHUS. *Siliqua* compressed or 2-edged. *Calyx* erect, opposite sepals saccate at the base. *Stigma* placed on a *style,* 2-lobed, lobes patent or capitate.—Name from the Arabic *Kheyry,* not however originally applied to this genus.

*** *Cotyledons conduplicate* (o >>)

19. BRASSICA. *Siliqua* linear or oblong, *valves* convex with a straight dorsal nerve. *Seeds* in a single row in each *cell.*—Name from *bresic,* Celtic, a *cabbage.*

20. SINAPIS. *Siliqua* linear or oblong, *valves* convex with 3 or 5 straight nerves. *Seeds* in a single row in each *cell.*—Name from σιναπι, *mustard-seed.*

21. DIPLOTAXIS. *Siliqua* linear, *valves* plane with a single central nerve. *Seeds* in a double row in each *cell.*—Name from διπλοος, *double,* and ταξις, *arrangement.*

22. RAPHANUS. *Siliqua* with a 2-valved, minute, pedicelliform, sterile, basal joint; beak very large, fertile, moniliform, transversely seperating. *Seeds* in a single row.—Name from ρα, *quickly,* and φαινομαι, *to appear,* from its rapid vegetation.

TETRADYNAMIA—SILICULOSA.

1. CORONOPUS. *Gærtn.* Wart-cress.

1. C. *Ruellii,* Sm. *Common Wart-cress, Swine's Cress.* Silicula entire, cristato-muricated; style prominent; corymbs dense, few-flowered, shorter than the leaves; leaves pinnatifid, segments obtuse, incised. *E. Bot. t.* 1660. *Hook. Br. Fl. p.* 297. *Senebiera Coronopus, DC. E. Fl. v. iii. p.* 179. *Cochlearia Coronopus, Linn.*

Waste ground and road-sides; not very common. *Fl.* June. ⊙.
About Berrington, Pitchford, West Coppice, Battlefield, Bewford, Ensdon, Atcham, all near Shrewsbury; *Rev.* E. Williams's MSS. Ludlow; *Mr.* H. Spare. Meole Brace. Holyhead Road between Shrewsbury and Shelton. Waste ground at top of the middle walk of Shrewsbury Quarry.
Root tapering. *Stems* spreading on the ground, dichotomously branched, smooth. *Leaves* glaucous green, smooth, deeply pinnatifid, segments linear-lanceolate, obtuse, incised. *Corymbs* opposite to and shorter than the leaves. *Pedicels* short, erect or erecto-patent, smooth. *Flowers* small, white. *Silicula* reniform, laterally compressed, the margin cristato-muricated, the sides reticulato-rugose. *Style* short and thick, but prominent. *Seeds* oblong, curved.

2. CAPSELLA. *DC.* Shepherd's Purse.

1. C. *Bursa-Pastoris, DC. Common Shepherd's Purse. DC. Syst. Veg. v.* ii. *p.* 283. *Hook. Br. Fl. p.* 298. *Thlaspi Bursa-pastoris, Linn. E. Bot. t.* 1435. *E. Fl. v. iii. p.* 173.

Cultivated ground, waste places, waysides, &c.; most abundant. *Fl.* all the summer. ⊙.
Root tapering. *Stem* one or several, variable in height, erect, angular, roughish, glaucous above, alternately branched, leafy. *Radical leaves* spreading in a circle on the ground, lanceolate, pinnatifid, often entire, tapering into a petiole, hairy on both sides, margins ciliated and dentate; *stem leaves* lanceolato-sagittate, amplexicaul, hairy, ciliated and dentate. *Flowers* white in terminal corymbs, minute, on erect pedicels, in fruit afterwards elongated into a raceme with patent pedicels. *Sepals* erect, ovate, obtuse, membranous at the margins, with a few scattered hairs. *Petals* twice the length of the sepals, obovate, obtuse or truncate, entire, spreading, white. *Silicula* obcordato-cuneate, laterally compressed, keeled, smooth. *Seeds* 10 or 12 in each cell, oblong, minutely tuberculated.

3. LEPIDIUM. *Linn.* Pepper-wort.

1. L. *campestre,* Br. *Common Mithridate Pepper-wort.* Silicula ovate, emarginate, rough with minute scales; style scarcely longer than the notch; cauline leaves sagittate, toothed. *E. Fl. v. iii. p.* 166. *Hook. Br. Fl. p.* 300. *Thlaspi campestre, Linn. E. Bot. t.* 1385.

Corn-fields, dry gravelly places; common. *Fl.* July. ⊙.
Stems 10—18 inches high, erect, simple, branched above, pubescent. *Radical leaves* obovate, obtuse, entire or slightly wavy at the margins, petiolate; *cauline leaves* amplexicaul, sagittate, obtusely pointed, wavy and toothed, pubescent. *Flowers* white, on short hairy pedicels, divaricate in fruit. *Silicula* ovate, winged and keeled, the wings very narrow at the sides, much dilated at the apex which is

notched, with the short and stout *style* in the sinus; *stigma* large, rather broader than the style. The scales on the valves when viewed under a lens are in reality minute hollow globular blisters of the cuticle perfectly resembling small pearls, and which becoming depressed assume a scalelike appearance. *Seeds* obovate, smooth.

2. L. *Smithii,* Hook. *Smooth Field Pepper-wort.* Silicula ovate, emarginate, winged, glabrous, quite smooth or occasionally very minutely scaly on the back; style much exserted beyond the notch; cauline leaves sagittate, toothed. *Hook. Br. Fl. p.* 300. *L. hirtum, Sm. E. Fl. v. iii. p.* 167. *Thlaspi hirtum, E. Bot. t.* 1803.

Rocky and waste places; not common. *Fl.* June, July. ♃.
Astley; *Mr.* E. Elsmere, *jun.* Donnington, near Shiffnal; *Dr.* G. Lloyd. Long lane quarries near Cheney Longville. Hedge banks, Meole and Sutton near Shrewsbury. Ebury encampment on Bayston Hill near Shrewsbury.
Root woody. *Stems* several, 8—10 inches high, simple or branched from above the base, as well as in the upper part, spreading, hairy. *Radical leaves* obovate, petiolate; *cauline ones* sagittate, toothed, the teeth larger than in the last species, especially at the base, thus rendering them sinuato-dentate, hairy on both sides. *Flowers* white, on short hairy pedicels, divaricated and reflexed in fruit. *Silicula* similar to the last but the *style* twice the length of the border, tapering upwards, the *stigma* not exceeding its breath, often without the blisters or scales, though a few imperfect ones may occasionally be perceived. *Seeds* obovate, smooth.

4. SUBULARIA. *Linn.* Awl-wort.

1. S. *aquatica,* Linn. *Awl-wort. E. Bot. t.* 732. *E. Fl. v. iii. p.* 157. *Hook. Br. Fl. p.* 301.

Shallow margins of lakes and pools; very rare. *Fl.* July. ♃.
Hancott pool near Shrewsbury; *T. and D. Bot. Guide.*
Root of numerous, long, white fibres. *Leaves* radical, awlshaped, spreading, 1—3 inches long. *Scape* 2—4 inches high. *Flowers* few, white, small. *Silicula* erect, ovali-elliptical, smooth; valves convex, turgid. *Seeds* oval, smooth; *embryo* curved above the base of its linear long *cotyledons.*

5. CAMELINA. *Crantz.* Gold of Pleasure.

1. C. *sativa,* Crantz. *Common Gold of Pleasure.* Silicula obovate, margined; stigma simple; leaves lanceolate, sagittate. *E. Fl. v. iii. p.* 164. *Hook. Br. Fl. p.* 303. *Alyssum sativum. E. Bot. t.* 1254. *Myagrum, Linn.*

Fields; rare. *Fl.* June, July. ⊙.
Corn-fields at Hord's Park, near Bridgnorth; *Rev.* A. Bloxam. "Among flax, common;" *Rev.* E. Williams's MSS.
Stem 2-3 feet high, erect, simple, panicled above, somewhat angular, scabrous with stellate hairs. *Leaves* alternate, amplexicaul, sagittate, scabrous, lanceolate, obtusely mucronate, margins somewhat recurved, with distant minute denticulations. *Flowers* small, yellow. *Silicula* obovate, margined, 4-ribbed, smooth and veiny. *Style* rather long. *Seeds* 6 or 8 in a cell.

6. THLASPI. *Linn.* Penny-cress.

1. T. *arvense,* Linn. *Mithridate Mustard or Penny-cress.* Silicula orbicular with a broad longitudinal wing; seeds concentri-

cally rugose and striated; leaves oblong, sagittate, toothed, glabrous. *E. Bot. t.* 1659. *E. Fl. v. iii. p.* 171. *Hook. Br. Fl. p.* 298.

Corn-fields and roadsides; not common. *Fl.* June, July. ⊙.
On a bank by the road-side near the bridge, Shineton; *W. P.* Brookes, *Esq.* Fields, Welbach near Shrewsbury; *Mr.* T. Bodenham. About Ludlow; *Miss* Mc Ghie. Near Shotton; *W. W.* Watkins, *Esq.*
Near Ascott. Fields of Bank Farm. Pulley. Bayston Hill. Red Hill; all near Shrewsbury.
Stem 6—18 inches high, erect, decurrently winged, smooth, branched above. *Leaves* alternate, obtuse, smooth; *radical ones* obovate, entire; *stem-leaves* oblong, sagittate and amplexicaul at the base, dentate and somewhat sinuated, uppermost narrower. *Flowers* extremely small, white. *Pedicels* of the *fruit* erecto-patent, generally longer than the silicula. *Silicula* very large, erect, orbicular, smooth and veiny, the wings equalling the cells in breadth and rising high above the minute *style* placed in a deep sinus or notch at the apex. *Seeds* broadly ovate, black, concentrically rugose and striated, minutely pitted when viewed under a powerful lens.

7. TEESDALIA. *Br.* Teesdalia.

1. T. *nudicaulis,* Br. *Naked-stalked Teesdalia.* Petals unequal; radical leaves lyrato-pinnatifid, segments rounded. *E. Fl. v. iii. p.* 170. *Hook. Br. Fl. p.* 299. *Iberis nudicaulis, Linn. E. Bot. t.* 327.

Rocky places; frequent. *Fl.* May, June. ⊙.
Shotton lane near Shrewsbury; *Dr.* T. W. Wilson. Pentregaer near Oswestry; *Rev.* T. Salwey. Harmer Hill; *T. C.* Eyton, *Esq.* Hawkstone; *Mr.* F. Dickinson. On Lilleshall Hill. High ground at Woodcote; *R. G.* Higgins, *Esq.* At the junction of 3 roads between Cound paper-mill and Venus Bank. Rock between Ruyton near Baschurch and the Platt Mill. By side of road between Quatford and Dudmaston; *Rev.* E. Williams's MSS.
Plentifully in the lane leading to Pimhill. Sandy heath near Wolf's Head turnpike gate near Nesscliff. Pontesford Hill, south side. Haughmond Hill. Sharpstones Hill. Grinshall. Sandford near Oswestry.
Leaves numerous, spreading on the ground, almost entirely radical, lyrato-pinnatifid, segments rounded, smooth. *Stems* several, 2—4 inches high, erect or spreading, bearing sometimes a small entire or cut leaf or two. *Flowers* corymbose, white. *Stamens* always 6. *Silicula* rotundo-obovate, laterally compressed, concave on one side, convex on the other, winged and keeled, apex notched with the minute *stigma* in the sinus; *valves* smooth, veiny. *Seeds* round, compressed, very minutely dotted.

8. COCHLEARIA. *Linn.* Scurvy-grass.

1. C. *Danica,* Linn. *Danish Scurvy-grass.* Silicula broadly ovate, obscurely veiny; seeds 6 in each cell, tuberculated; leaves all petiolate, nearly deltoid. *E. Bot. t.* 697. *E. Fl. v. iii. p.* 177. *Hook. Br. Fl. p.* 301.

Walls, &c. *Fl.* May, June. ⊙.
Town Walls Oswestry, near the Coney Green; *J. F. M.* Dovaston, *Esq.* Walls of the Dominican Friary, and on the walls and roofs in Chester-street and on Coton-hill, Shrewsbury.
Stems 3—6 inches long, simple or branched in the lower part, spreading or prostrate, angular, leafy, purplish. *Lower leaves* on rather long slender petioles,

angulato-cordate; *upper* ones on shorter and more dilated petioles, 3-lobed, entire, angulato-cordate at the base, resembling the leaves of Ivy in miniature. *Clusters* lax and short. *Flowers* small, pure white. *Sepals* elliptical, obtuse, membranous at the edges, smooth. *Petals* twice the length of the sepals, erect, obovate. *Anthers* yellow. *Silicula* ovato-elliptical when viewed from the back of the cells, broadly ovate when transversely to the partition, tipped with the short *style*, obscurely veiny. *Seeds* 6 in each cell, ovate, minutely tuberculated.

2. C. *Armoracia*, Linn. *Horse-radish.* Silicula oblong; radical leaves oblong, on long petioles, crenate, cauline ones lanceolate, serrated or entire. *E. Bot. t.* 2323. *E. Fl. v. iii. p.* 177. *Hook. Br. Fl. p.* 301.

Banks of rivers; doubtfully wild. *Fl.* May. ♃.

Banks of the river Severn, Buildwas; *W. P. Brookes, Esq.!* Steventon Farm near Ludlow, and near Newport; *Miss Mc Ghie.* Banks of Severn between Underdale and Monkmoor near Shrewsbury; *Rev. E. Williams's MSS.*

Banks of river Severn beneath Shrewsbury Castle Mount.

Root very long, cylindrical, white and pungent. *Stem* 2 or more feet high, erect, branched, furrowed, glabrous. *Radical leaves* very large, on long erect channelled petioles, oblong, obtuse, crenate, veiny; *upper leaves* lanceolate, obtuse, entire or serrated, sometimes deeply and finely pinnatifid in consequence of the parenchymatous matter being suppressed and the ribs only developed. *Flowers* numerous, white, in corymbs which become elongated into long clusters. *Sepals* oblong, obtuse, erecto-patent. *Petals* 3 times as long as the sepals, oblongo-obovate, the limb spreading. *Silicula* minute, narrow oblong, somewhat compressed, nodulose, seldom perfected, on long erect appressed smooth round *pedicels*. *Seeds* very minute, ovate.

9. DRABA. *Linn.* Whitlow-grass.

1. D. *verna*, Linn. *Common Whitlow-grass.* Scapes naked; petals deeply cloven; leaves lanceolate, more or less toothed, hairy. *E. Bot. t.* 586. *E. Fl. v. iii. p.* 158. *Hook. Br. Fl. p.* 302. *Erophila vulgaris, DC.*

Walls, rocks, dry banks, &c.; common. *Fl.* March—May. ⊙.

Leaves several, spreading in a star-like form close to the ground, oblong, lanceolate, entire or toothed, clothed with simple as well as stellate hairs. *Scapes* hairy below. *Silicula* ovali-elliptical, smooth, veiny. *Seeds* numerous, ovate, tuberculated.

TETRADYNAMIA—SILIQUOSA.

10. SISYMBRIUM. *Linn.* Hedge-Mustard.

1. S. *officinale*, Linn. *Common Hedge-mustard.* Leaves lyrato-pinnatifid, runcinate, segments subrhomboid, acuminate; sepals linear-oblong, erect; petals spreading, twice as long as the sepals; siliqua subulate, obsoletely tetragonous, pubescent, erect, closely appressed. *E. Fl. v. iii. p.* 196. *Hook. Br. Fl. p.* 307. *Erysimum officinale, Linn. E. Bot. t.* 735.

Waste places, way-sides, &c.; very common. *Fl.* June, July. ⊙.

Root tapering. *Stem* 1-2 feet high, erect, with horizontally spreading branches, round, rough with minute deflexed bristles. *Leaves* lyrato-pinnatifid,

somewhat runcinate, segments subrhomboid, the apex more or less acuminated, sinuato-dentate, rough with hairs. *Flowers* minute, yellow, corymbose, afterwards elongated into long straight close clusters. *Sepals* erect, linear-oblong, obtuse, hairy. *Petals* twice the length of the sepals, spathulate, emarginate, the limb spreading, claw erect as long as the sepals. *Siliqua* on a very short pedicel, erect, closely appressed, subulate, densely pubescent, subtetragonous. *Seeds* ovate, glabrous, shining.

2. S. *Sophia*, Linn. *Fine-leaved Hedge-Mustard or Flaxweed.* Leaves deeply bipinnatifid, pubescent, segments very narrow, linear; sepals linear, erecto-patent; petals erect, shorter than the sepals; siliqua linear, obsoletely tetragonous, nodulose, erecto-patent. *E. Bot. t.* 963. *E. Fl. v. iii. p.* 197. *Hook. Br. Fl. p.* 307.

Waste places, cultivated ground, road-sides; not unfrequent. *Fl.* August. ⊙.

Stanley; *Mr. G. Jorden.* Ness; *T. C. Eyton, Esq.* By the roadside between Wellington and Wrockwardine; *E. Lees, Esq.* About Ludlow; *Miss Mc Ghie.* Lane between Bayston Hill and the Sharpstones Hill; *Mr. T. Bodenham.* Cressage; *W. P. Brookes, Esq.!*

Cultivated ground, roadsides, Pulley, and generally about Shrewsbury.

Root tapering, branched. *Herbage* clothed with minute spreading pubescence. *Stem* erect, 1-2 feet or more high, round, branched. *Leaves* sessile, deeply bipinnatifid, segments very narrow, linear, entire or incised, acute, the midrib deeply channelled above. *Flowers* minute, corymbose, elongated in fruit. *Sepals* erecto-patent, linear, obtuse, convex. *Petals* yellow, erect, shorter than the sepals, broader and obtuse above, entire or notched. *Siliqua* linear, obsoletely tetragonous, nodulose, erect, somewhat curved, on erecto-patent pedicels. *Seeds* oblong, minutely tuberculated. *Nectaries* none.

3. S. *Thalianum*, Gaud. *Common Thale-cress.* Leaves toothed, hispid with forked hairs, radical ones oblongo-spathulate, cauline ones lanceolate or linear-lanceolate; sepals lanceolate, obtuse, erect; petals erect, twice as long as the sepals; siliqua linear, obsoletely tetragonous, smooth, curved upwards. *Hook. Br. Fl. p.* 307. *Arabis thaliana, Linn. E. Bot. t.* 901. *E. Fl. v. iii. p.* 209.

Dry banks and gravelly soils; not very common. *Fl.* April, May. ⊙.

On old walls, Kenley Common; *W. P. Brookes, Esq.* Nobold near Shrewsbury *Mr. T. Bodenham.* Wrockwardine; *Mr. F. Dickinson.* Ludlow; *Miss Mc Ghie.* Bromfield, near Ludlow; *Mr. H. Spare.*

Pulley; Bank Farm; Bishop's Castle road; Nobold; all near Shrewsbury.

Root fibrous. *Stem* erect, 6—12 inches high, branched upwards, round, hispid with spreading deflexed hairs in the lower part, glabrous above, glaucous. *Radical leaves* spreading in a circle on the ground, oblongo-spathulate or lanceolate, obtuse, tapering into a long petiole, hispid on both sides with forked hairs, coarsely toothed; *cauline* ones alternate, one at the base of each branch, lanceolate, obtuse, sessile, toothed or nearly entire, ciliated and hispid. *Sepals* erect, lanceolate, obtuse, convex. *Petals* white, erect, spathulate, entire, twice as long as the sepals. *Filaments* with a minute gland or nectary at their base externally, those of the shorter stamens twice the size of the others, spreading, scarcely recurved. *Pedicels* of the fruit round, horizontally spreading, glabrous, about half as long as the siliqua. *Siliqua* narrow, linear, obsoletely tetragonous, smooth, glaucous, curved upwards. *Seeds* numerous, oval, smooth.

11. ERYSIMUM. *Linn.* Treacle-mustard.

1. E. *cheiranthoides*, Linn. *Wormseed Treacle-mustard.* Leaves

sessile, lanceolate, entire or obsoletely denticulate, with stellato-tripartite hairs; siliqua nearly erect, pedicel spreading; stigma undivided, nearly sessile. *E. Bot. t.* 942. *E. Fl. v. iii. p.* 200. *Hook. Br. Fl. p.* 307.

Fields and waste places; rare. *Fl.* July, August. ⊙.

Shelderton near Ludlow; *Rev. T. Salwey.* Roadside near the stables at Condover Hall near Shrewsbury; *Rev. E. Williams's MSS.*

Stem erect, branched, angular, rough with closely deflexed simple bristles. *Leaves* sessile, alternate, lanceolate, with shallow and distant denticulations, clothed with minute crowded stellate bristles. *Flowers* numerous, small, yellow. *Siliqua* tetragonous, *valves* internally downy.

2. E. *Alliaria*, Linn. *Garlic Treacle-mustard, Jack by the Hedge, or Sauce alone.* Leaves petiolate, cordate, sinuato-dentate, glabrous; siliqua nearly erect, pedicel horizontally spreading; stigma undivided. *E. Bot. t.* 79. *E. Fl. v. iii. p.* 201. *Hook. Br. Fl. p.* 308.

Hedge banks and waste places; common. *Fl.* May, June. ♂.

Root somewhat woody. *Stem* erect, 1—3 feet high, branched, roundish, somewhat angular, glabrous or with a few scattered hairs. *Leaves* alternate, on channelled thick petioles dilated at the base, rotundo-cordate, acute, sinuato-dentate, veiny, bright shining green, glabrous. *Flowers* white. *Sepals* erect, lanceolate, obtuse, white. *Petals* twice the length of the sepals, claws erect, limb patent obovate entire white. *Siliqua* nearly erect, linear, long, tetragonous, nodulose, on very short thick horizontally spreading pedicels. *Seeds* large, oblong, striato-rugose. The longer filaments have a nectary or gland externally between them, whilst the shorter ones each arise from the summit of a larger one.

12. HESPERIS, *Linn.* Dame's Violet.

1. H. *matronalis*, Linn. *Common Dame's Violet.* Stem erect; leaves ovato-lanceolate, toothed; limb of the petals obovate; siliqua erect, torulose, margins not thickened. *E. Fl. v. iii. p.* 207. *Hook. Br. Fl. p.* 308. *H. inodora, Linn. E. Bot. t.* 731.

Hilly pastures; naturalized. *Fl.* May, June. ♃.

"At Caermaen near Aston (near Oswestry); but though it might readily be mistaken for wild by a passing Botanist, I remember a very neat cottage there, inhabited by an ingenious turner, long ago pulled down; as is yet indicated by snowdrops and daffodils, and 'still where many a garden-flower grows wild.'" *J. F. M. Dovaston, Esq.*

Stem 2-3 feet high, slightly branched, round, hairy. *Leaves* scattered, all nearly or quite sessile, except some of the lowermost, obtusely irregularly and glandularly toothed, hairy. *Flowers* large, handsome, pale purple or white, fragrant in the evening and in rainy weather.

13. CARDAMINE. *Linn.* Bitter-cress.

1. C. *amara*, Linn. *Bitter Lady's Smock.* Leaves pinnate, radical leaflets roundish entire, cauline ones elliptico-oblong dentato-angulate; sepals ovato-acuminate, unequal, longer than the claws of the erect petals; style oblique, shorter than the stamens; stigma minute, rather acute; stem rooting at the base. *E. Bot. t.* 1000. *E. Fl. v. iii. p.* 190. *Hook. Br. Fl. p.* 304.

In watery places, by the sides of rivers and brooks; not uncommon. *Fl.* April—June. ♃.

Lord's Meadows near Albrighton; *H. Bidwell, Esq.* Marshy field between Bradley and Wyke near Wenlock; *W. P. Brookes, Esq.* Banks of Severn near Severn House, Coalbrookdale; *Mr. F. Dickinson.* Stanton; *Mr. E. Elsmere, junr.* Near Yeaton; *T. C. Eyton, Esq.* Between Lilleshall Church and Abbey. Chetwynd Moors; *R. G. Higgins, Esq.* In great plenty by the Pool in the Arkoll Wood near Wellington; *E. Lees, Esq.* Buildwas; *Miss H. Moseley.* By the side of Cound Paper mill stream. In an osier bed near Tern bridge. By the side of Cantlop Mill stream; *Rev. E. Williams's MSS.* New Dale pool near Ketley; *Dr. Du Gard.*

Banks of river Severn between Preston Boats and Uffington ferry near Shrewsbury.

Root creeping. *Herbage* smooth, pale-green. *Stem* 1 foot or more high, succulent, somewhat angular and zigzag. *Leaves* alternate, pinnate, on dilated winged and channelled petioles; leaflets of the radical leaves rounded, obtuse, entire, those of the stem elliptico-oblong, dentato-angulate, unequal at the base on the upper side, tipped with a small mucro, margins especially of the uppermost ciliated. *Flowers* corymbose. *Sepals* ovato-acuminate, slightly membranous at the edges, longer than the claws of the petals, the two opposite the shorter stamens the largest. *Petals* erect, obovato-oblong, emarginate, with a tooth on the claw, white with pellucid veins. *Filaments* as long again as the style; *anthers* violet-purple.

2. C. *pratensis*, Linn. *Common Bitter-cress, Lady's Smock.* Leaves pinnate, radical leaflets roundish, dentate; cauline ones lanceolate, nearly entire; sepals oblong, obtuse, equal, shorter than the claws of the spreading petals; style straight, as long as the stamens, stigma large, capitate. *E. Bot. t.* 776. *E. Fl. v. iii. p.* 189. *Hook. Br. Fl. p.* 304.

Moist meadows and ditches; common. *Fl.* April, May. ♃.

Root tuberous, somewhat toothed. *Herbage* generally smooth, dark shining green. *Stem* 1-2 feet high, stout, almost woody, angular and zigzag. *Leaves* on long dilated winged channelled petioles, leaflets of the radical ones roundish, wavy, angular and toothed; those of the stem narrow, lanceolate, nearly entire, all tipped with a short stout mucro. *Flowers* corymbose. *Sepals* oblong, obtuse, membranous at the edges, shorter than the claws, all equal in size. *Petals* spreading, broadly obovato-oblong, emarginate, with a tooth on the claw, "silver-white" or bluish-coloured with darker veins. *Filaments* as long as the style; *anthers* pale-yellow.

A variety with double flowers sometimes occurs. Meadows about Ross Hall near Shrewsbury; *T. and D. Bot. Guide.*

Bomere pool.

3. C. *hirsuta*, Linn. *Hairy Bitter-cress.* Leaves pinnate, radical leaflets rotundo-ovate, irregularly sinuato-dentate, cauline leaves generally wanting; sepals linear-oblong, obtuse, nearly equal, longer than the claws of the erect petals; stamens 4; style narrower than the linear shortly acuminated siliqua. *Linn. Sp. Pl.* 915. *E. Fl. v. iii. p.* 188. *(excl. syn.)*. *Hook. Br. Fl. p.* 304. *(excl. syn.)* *With. Arr.* 3d *ed. p.* 578. *Reich. Fl. Excurs.* 4304.

Hedge banks and shady places; common. *Fl.* March—June. ⊙.

Root fibrous. *Stems* several, spreading, 3—6 inches high, angular, smooth, zigzag. *Leaves* spreading in a circle on the ground, pinnate; petiole winged, channelled above and ciliated towards the base; leaflets petiolate, rotundo-ovate, irregularly sinuato-dentate, terminal one broad and dilated, hairy on upper surface, pale and glabrous beneath. *Flowers* corymbose, white. *Sepals* linear-oblong, obtuse, margins membranous, half the length of the petals. *Petals* erect, small, narrow obovate with a long claw or spathulate. *Stamens* 4. *Siliqua* linear, 2 or 3 times the length of the tetragonous erecto-patent pedicels, compressed, 2-edged, edges obtuse, shortly acuminate into the short *style* which is half the width of the siliqua.

4. C. *sylvatica*, Link. *Wood Bitter-cress.* Leaves pinnate, radical leaflets roundish sinuato-dentate, cauline ones ovate or lanceolate, toothed; sepals linear-oblong, obtuse, nearly equal, longer than the claws of the erect petals; stamens 6; style as broad as the very narrow linear siliqua. *Koch.* 43. *Reich. Fl. Excurs.* 4303. *C. flexuosa, With. Arr.* 578. *(C. hirsuta E. Bot. t.* 492*)?*

Wet ditches, sides of rivers, &c.; frequent. *Fl.* May, June. ⊙ or ♂.

Root fibrous. *Stem* 12–18 inches high, erect, branched, zigzag, angular and furrowed, glabrous or clothed with horizontally spreading soft hairs. *Leaves* on long strong dilated or winged petioles, deeply channelled above, glabrous, ciliated at the margins of the dilated base. *Leaflets* petiolate, about 5 pairs and a petiolate terminal larger one, those of the lower leaves rounded, unequal at the base on the upper side, sinuato or angulato-dentate, the teeth obtusely mucronate; terminal leaflet larger, round, sinuato-dentate, equal at the base; leaflets of stem leaves ovate, angulato-dentate, teeth shortly mucronate, unequal at the base on the upper side; uppermost narrower and lanceolate, strongly toothed; all glabrous, except a few hairs scattered on the margins. *Flowers* in terminal and lateral corymbs, subsequently elongated into a raceme. *Pedicels* round, compressed, glabrous, about as long as the siliqua, erecto-patent. *Sepals* linear-oblong, obtuse, membranous and white at the margins and apex, half the length of the petals. *Petals* erect, obovate, with a long claw, white. *Stamens* always 6. *Siliqua* erect, very narrow, linear, compressed, edges thickened and obtuse, slightly undulate, crowned with the short thick *style* and *stigma* nearly equalling the breadth of the siliqua.

5. C. *impatiens*, Linn. *Narrow-leaved Bitter-cress.* Leaves pinnate, leaflets lanceolate, deeply cut, rarely entire; stipules ciliated; petals linear or none. *E. Bot. t.* 80. *E. Fl. v. iii. p.* 187. *Hook. Br. Fl. p.* 304.

Moist rocks; rare. *Fl.* May, June. ⊙.

Tinker's Hill near Ludlow; *Rev. A. Bloxam.* Side of the road immediately on leaving Church Stretton for Ludlow; *Mr. T. Bodenham.* Clee Hill near Bitterley; *Miss Mc Ghie.* On the left hand side of the lane leading from the Wrekin to Huntingdon in the parish of Little Wenlock; *Rev. E. Williams's MSS.*

Root small, tapering. *Herbage* pale green, smooth. *Stem* erect, slender, 1-1½ foot high, branched, angular and zigzag. *Leaves* of numerous, equal, more or less notched, rarely entire, lanceolate leaflets. *Stipules* lanceolate, acute, deflexed, amplexicaul. *Flowers* extremely small, white, occasionally destitute of petals. *Siliqua* slender, the valves highly elastic and revolute, discharging the seeds with a crackling noise on the slightest touch, whence its specific name.

14. ARABIS. *Linn.* Rock Cress.

1. A. *hirsuta*, Br. *Hairy Rock cress.* Leaves all hispid, dentate, radical ones obovate, tapering at the base, cauline ones ovato-oblong, semiamplexicaul; siliqua straight. *E. Fl. v. iii. p.* 213. *Hook. Br. Fl. p.* 305. *Turritis hirsuta, Linn. E. Bot. t.* 587.

Walls and rocks; rare. *Fl.* June. ♂.

Blodwel rocks; *Rev. T. Salwey!*

Stem about 12 inches high, erect, stiff, rough with spreading hairs chiefly simple, sometimes with forked ones intermixed. *Radical leaves* obovate tapering at the base, hispid with forked hairs, shallowly and distantly toothed; those of the stem. ovato-oblong, semiamplexicaul, somewhat sagittate at the base, hispid and dentate. *Flowers* white. *Pedicels* short, smooth. *Siliqua* erect, about an inch long, linear, glabrous, valves with a single rib or keel, undulated. *Stigma* sessile on the very apex of the valves. *Seeds* oval, compressed, smooth and shining.

Arabis hispida, Linn. (Arabis petræa, DC.) is said in Turner and Dilwyn's Botanist's Guide to have been found "by the first milestone from Shrewsbury to Welchpool;" but it is presumed that this is a mistake. Be this however as it may, the station no longer exists, the bank having been removed many years ago in improving the Holyhead road.

15. TURRITIS. *Linn.* Tower-Mustard.

1. T. *glabra.* Linn. *Long-podded Tower-Mustard.* Radical leaves lyrato-runcinate, hispid; cauline ones oblongo-lanceolate, sagittate and amplexicaul, entire, glabrous; siliqua very long, erect. *E. Bot. t.* 777. *E. Fl. v. iii. p.* 215. *Hook. Br. Fl. p.* 305.

Banks and road-sides; not very common. *Fl.* May, June. ⊙.

Neach Hill in Donnington parish near Shiffnal; *Dr. G. Lloyd.* Near Marton; *T. C. Eyton, Esq.* Near Shotton near Shrewsbury; *Dr. T. W. Wilson.* Ditch banks about Beckbury, Badger, Ruyton, Stockton, Worfield, &c. By the side of the road between Allfield and King Street turnpike near Berrington; *Rev. E. Williams's MSS.*

Road side near Leaton Knolls between Shrewsbury and Leaton Shelf.

Stem 3-4 feet high, simple, round, glabrous and glaucous except at the very base which is clothed with soft deflexed hairs. *Radical leaves* spreading, lyrato-runcinate, hispid with forked hairs, soon withering; cauline ones oblongo-lanceolate, acuminated into an obtuse point, entire, smooth and glaucous, sagittate and amplexicaul at the base. *Flowers* yellowish white. *Pedicels* erect, round, glabrous, about one-fifth as long as the siliqua. *Siliqua* about 2½ inches long, linear, compressed, smooth; *valves* single-ribbed, plane, margins obtuse, indented, crowned with the short stout *style* and obtuse spreading 2-lobed *stigma* which is nearly as broad as the siliqua. *Seeds* numerous, minute, obovate, transversely striated.

16. BARBAREA. *Br.* Winter-cress.

1. B. *vulgaris*, Br. *Winter-cress, Yellow Rocket.* Lower leaves lyrate, terminal lobe ovali-rotund, cordate at the base, sinuated, upper ones obovate, angulato-sinuate, often pinnatifid at the base; siliqua erect, pedicels spreading. *E. Fl. v. iii. p.* 198. *Hook. Br. Fl. p.* 306. *Erysimum Barbarea, Linn. E. Bot. t.* 443.

Waste places and hedge-banks; not unfrequent. *Fl.* May—August. ♃.

Middle and near Cockshutt; *W. W. Watkins, Esq.* Oakley park meadows, near Ludlow; *Mr. H. Spare.* Ludlow; *Miss Mc Ghie.* Side of river Severn, Coalbrookdale; *Mr. F. Dickinson.* Near the Mill, Madeley; *W. P. Brookes, Esq.*

Road sides and cultivated ground generally about Shrewsbury.

Root tapering, woody. *Stem* 12—18 inches high, erect, branched, angular and furrowed, smooth. *Leaves* alternate, lower ones lyrato-pinnatifid, terminal lobe oval rounded or cordate at the base, upper ones obovate, deeply angulato-sinuate or often pinnatifid at the base, all smooth and sessile by their sagittate bases. *Flowers* yellow, on very short pedicels. *Sepals* ovate, concave, 3-ribbed,

erect. *Petals* twice the length of the sepals, oblongo-obovate, emarginate, veiny, the claw erect, the limb spreading. *Siliqua* erect, linear, tetragonous, smooth, six times as long as the spreading pedicel, crowned with the narrow *style* which is about half as long as the pedicel. *Seeds* obovate, compressed, minutely and deeply pitted. *Glands* one between each stamen.

17. NASTURTIUM. *Br.* Cress.

1. N. *officinale*, Br. *Water-cress.* Leaves pinnate, leaflets ovate, obtuse, unequal at the base, sessile, sinuate, glabrous; stem rooting from the axils; petals white; siliqua cylindrical, curved upwards; pedicels reclinate. *E. Fl. v. iii. p.* 192. *Hook. Br. Fl. p.* 253. *Sisymbrium Nasturtium, Linn. E. Bot. t.* 855.

Brooks, rivulets and plashy springs; frequent. *Fl.* June, July. ♃.

Root of many long white fibres. *Stems* variable in length, spreading or floating, angular, furrowed and smooth, alternately branched from the axils of the leaves, the branches protruding both from their upper and under sides at the point of attachment with the stem long white fibres, which rooting in the soil become in all probability, on the decay of the main stem, so many individual plants. *Leaves* pinnate, alternate, on long dilated winged smooth petioles deeply channelled above, pinnæ about 4 pairs, distant, sessile or nearly so, ovate, obtuse, unequal at the base, the upper edge cut away, the lower one rounded or subcordate, with 3 principal ribs at the base, margins sinuate or wavy, glabrous, terminal leaflet similar but larger and petiolate. *Flowers* white. *Pedicels* of the fruit tetragonous, horizontal or even reclining, glabrous. *Siliqua* about an inch long, cylindrical, smooth, curved upwards.

A well known and wholesome salad.

2. N. *sylvestre*, Br. *Creeping Nasturtium.* Leaves pinnatifid, segments lanceolate, inciso-serrate; petals yellow, longer than the sepals; siliqua short, cylindrical, patent; pedicels patent. *E. Fl. v. iii. p.* 193. *Hook. Br. Fl. p.* 306. *Sisymbrium sylvestre, Linn. E. Bot. t.* 2324.

Banks of rivers, &c.; not common. *Fl.* July, August. ♃.

Common on the banks of Severn; *Rev. E. Williams's MSS.* Severn banks near Marn woods, Coalbrookdale; *Mr. F. Dickinson.*

Banks of river Severn near Shrewsbury; abundant.

Root creeping. *Stems* 18 inches—2 feet high, ascending, angular, slightly furrowed, glabrous, purplish in the lower part, alternately branched. *Leaves* alternate, petiolate, lanceolate in their circumscription, deeply pinnatifid, *segments* of the lower leaves lanceolate, acute, inciso-serrate, serratures callously tipped, deep-green, glabrous; those of the upper leaves becoming gradually narrower, linear-lanceolate, more or less inciso-serrate, whilst those of the uppermost are entire. *Flowers* in long racemes at the extremities of the branches. *Pedicels* divaricated, round, smooth. *Sepals* yellow, erect, ovato-oblong, concave, rather more than half the length of the petals. *Petals* obovate, erect, yellow. *Nectary* obtuse, green, one between each stamen. *Style* short, round; *stigma* large, capitate, 2-lobed.

3. N. *terrestre*, Br. *Marsh Nasturtium.* Leaves lyrato-pinnatifid, segments ovate or oblong, unequally toothed; petals yellow, shorter than the sepals; siliqua obliquely ovate, turgid, curved upwards; pedicels reclinate. *E. Fl. v. iii. p.* 193. *Hook. Br. Fl. p.* 306. *Sisymbrium terrestre. E. Bot. t.* 1747.

Muddy and watery places; not common. *Fl.* June—August. ⊙.

Ludlow; *Miss Mc Ghie.* Eaton Mascott pool. Banks of Severn; *Rev. E. Williams's MSS.*

Banks of River Severn at Shelton Rough near Shrewsbury. Banks of Severn about Shrewsbury.

Root of many strong fibres. *Stem* erect, 6—18 inches high, angular, furrowed, glabrous, with alternate erecto-patent branches. *Leaves* alternate, on dilated winged and channelled petioles, clasping the stem by their sagittate or auricled bases, lyrato-pinnatifid, segments ovate or oblong, unequally toothed or serrated, serratures obtuse, obtusely mucronate, glabrous except a few hairs on the margins. *Flowers* yellow, minute, *petals* as short or shorter than the *sepals*. *Pedicels* glabrous, declinate, very little longer than the siliqua. *Siliqua* obliquely ovate often approaching to oblong, smooth, turgid, somewhat curved upwards, upper edge straight or nearly so, lower one gibbous or curved. *Style* short, thick and narrow, curved upwards. *Stigma* 2-lobed. *Seeds* numerous, roundish-ovate, red-brown, minutely pitted.

4. N. *amphibium*, Br. *Amphibious Nasturtium.* Leaves oblong, serrated or pectinato-pinnatifid; petals yellow, longer than the sepals; siliqua elliptical, straight, erect; pedicels reclinate. *E. Fl. v. iii. p.* 195. *Hook. Br. Fl. p.* 307. *Sisymbrium amphibium, Linn. E. Bot. t.* 1840.

Watery places; rare. *Fl.* June—August. ♃.

Banks of the river Severn near Bewdley; *Mr. G. Jorden.* Banks of river Severn; *Rev. E. Williams's MSS.*

Root long and strong, not creeping, throwing out numerous fibres. *Stem* 2-3 feet high, branched, striated, smooth. *Lower leaves* on long petioles, deeply pectinated under water, otherwise elliptic-lanceolate, cut or serrated; *upper* ones sessile or clasping by their auricled or sagittate bases, oblong, pectinated, serrated or sinuate, serratures obtusely mucronate, smooth. *Flowers* small, bright yellow. *Petals* longer than the sepals. *Pedicels* glabrous, reclinate, 3 times as long as the siliqua. *Siliqua* elliptical, tapering at the base into a stalk, erect, straight, not curved upwards. *Style* about one-third the length of the siliqua, straight and narrow; *stigma* 2-lobed. *Seeds* numerous, oval, minutely dotted.

18. CHEIRANTHUS. *Linn.* Wall-flower.

1. C. *Cheiri*, Linn. *Common Wall-flower.* Leaves lanceolate, acute, entire, with longitudinally divaricated appressed hairs; siliqua linear, lobes of the stigma patent; stem shrubby. *E. Fl. v. iii. p.* 203. *Hook. Br. Fl. p.* 308. C. *fruticulosus, Linn. E. Bot. t.* 1934.

Old walls and ruins; common. *Fl.* April, May. ♃.

Stem shrubby, 8—12 inches high, branched, spreading, rigid, naked and glabrous, clothed in the younger shoot or upper portion as well as on both surfaces of the leaves, pedicels, calyx and siliqua with close pressed silvery hairs, fixed by their middle and divaricating longitudinally in opposite directions. *Leaves* crowded, imbricated, lanceolate, acute, entire, tapering into a short dilated petiole, somewhat shining above, hoary beneath. *Flowers* corymbose, fragrant. *Sepals* linear-oblong, obtuse, erect, two of them gibbous at the base, stained with brown, about as long or longer than the claw, membranous and yellow at the margins. *Petals* obovate, entire or slightly notched, the claw erect, the limb flat, spreading, bright yellow. The two shorter *filaments* spring from a large fleshy green *gland*. *Siliqua* erect, linear, compressed, two edged, the *valves* with an elevated central line. *Style* short, stout, the width of the siliqua, slightly tapering upwards, with a few scattered bristles, crowned by the cloven divergent *stigma*.

19. BRASSICA. *Linn.* Cabbage, Turnip.

1. B. *Napus,* Linn. *Wild Navew, Rape or Cole-seed.* Root caulescent, fusiform; leaves smooth, upper ones cordato-lanceolate, amplexicaul, lower ones lyrate, toothed. *E. Bot. t.* 2146. *E. Fl. v. iii. p.* 217. *Hook. Br. Fl. p.* 309.

Corn-fields; not uncommon. *Fl.* May, June. ♂.
In great plenty between Cockshutt and Ellesmere. About Little Stretton. St. Giles's near Shrewsbury; *Rev. E. Williams's MSS.*
Corn-fields about Shrewsbury.
Stem erect, branched, striated, 1-2 feet high. *Leaves* all smooth and glaucous, *radical* ones lyrate, disappearing before the plant blossoms, *cauline* ones lanceolate, crenate or sinuato-dentate, clasping the stem by their dilated rounded or cordate base, uppermost lanceolate and entire. *Petals* yellow, rather small. *Siliqua* torulose.

2. B. *nigra,* Koch. *Common Mustard.* Leaves petiolate, lower ones lyrate, dentate, terminal segment the largest, lobed, upper leaves lanceolate generally entire; calyx spreading horizontally; siliqua appressed. *Babington Prim. Fl. Sarn. p.* 8. *Drejer Fl. Hafn. n.* 679. *Sinapis nigra, Linn. E. Bot. t.* 969. *E. Fl. v. iii. p.* 222. *Hook. Br. Fl. p.* 310.

Waste places; not common. *Fl.* May, June. ⊙.
Corn-fields near Ludlow; *Mr. H. Spare.* Banks of river Severn near Bewdley; *Mr. G. Jorden.* Fields, Welbach near Shrewsbury; *Mr. T. Bodenham.* Banks of river Severn near Preston Boats ferry near Shrewsbury; *Mr. F. Dickinson.* Ditch banks and on rubbish but not common; *Rev. E. Williams's MSS.*
Banks of river Severn beneath Shrewsbury Castle Mount.
Root tapering. *Stem* 2 feet high, erect, branched, round, nearly smooth or with a few scattered bristles. *Leaves* alternate, petiolate, lyrate or somewhat lyrato-pinnatifid, variously lobed, toothed, hispid, upper ones linear-lanceolate, shortly petiolate, entire, toothed, pendulous or deflexed, teeth tipped with an obtuse mucro. *Flowers* very small. *Sepals* patent, linear-obtuse, concave, yellowish. *Petals* obovate, yellow. *Siliqua* on a pedicel about one-third its length, closely appressed, tetragonous, valves strongly keeled, glabrous, crowned by the short tetragonous tapering narrow style or beak. *Seeds* about 8, globose, brown, covered with minute impressed dots.

20. SINAPIS. *Linn.* Mustard.

1. S. *arvensis,* Linn. *Wild Mustard, Charlock, Ketlock.* Leaves ovate, sublyrate, scabrous, inferior ones petiolate, uppermost sessile; siliqua torulose, valves 3-nerved; beak subulate, tetragonous; calyx patent. *Babington Prim. Fl. Sarn. p.* 8.

Corn-fields; too frequent. *Fl.* May, June. ⊙.
Var. a. vera. Siliqua smooth. *Bab. l. c.*
Root tapering. *Stem* 1-2 feet high, erect, furrowed, rough with deflexed bristles, branched. *Leaves* alternate, petiolate, ovate, obtuse, somewhat lyrate, sinuate and obtusely denticulate, rough and hairy on both sides, uppermost sessile. *Sepals* patent, linear, obtuse, glabrous, yellowish. *Petals* erecto-patent, obovate, yellow turning white in decay. *Siliqua* angular, crowned with a tetragonous subulate furrowed smooth beak, not half the length of the siliqua. *Seeds* numerous, globose, brown.

Var. β. retro-hirsuta. Siliqua rough with reflexed bristles. "S. *retro-hirsuta, Besser,*" Reich. *Babington Prim. Fl. Sarn. p.* 9. *E. Fl. v. iii. p.* 221.

Corn-fields about Battlefield and other places in the neighbourhood of Shrewsbury.

2. S. *alba,* Linn. *White Mustard.* Leaves lyrato-pinnatifid, rough; siliqua nodulose, valves 3-nerved, beak ensiform, flattened, 2-edged; calyx patent. *E. Bot. t.* 1677. *E. Fl. v. iii. p.* 222. *Hook. Br. Fl. p.* 310.

Waste places; not common. *Fl.* July—October. ⊙.
Fields, Welbch near Shrewsbury; *Mr. T. Bodenham.* Corn-fields but not common; *Rev. E. Williams's MSS.*
Side of turnpike road near Bank Farm near Shrewsbury.
Root tapering. *Stem* 1-2 feet high, erect, furrowed, rough with deflexed bristles, branched. *Leaves* lyrato-pinnatifid, segments ovate, inciso-serrated, the serratures callously pointed, roughish. *Flowers* numerous, yellow. *Sepals* linear, obtuse, margins incurved, patent. *Petals* twice the length of the sepals, limb obovate, spreading, on a very long narrow erect claw. *Siliqua* spreading, on nearly horizontal angular and furrowed smooth pedicels, short, nodulose, densely clothed with numerous minute deflexed bristles and with longer spreading more conspicuous white ones, terminated by a flattened two-edged beak, much longer than the valves, rough with erect bristles, 3-ribbed, the ribs approximate, the middle one most conspicuous, the edge of the beak always directed towards the stem. *Seeds* generally 3 in each cell, roundish, dark brown, with minute impressed dots.

21. DIPLOTAXIS. *DC.* Wall Mustard.

1. D. *tenuifolia,* DC. *Fine-leaved Wall Mustard.* Siliqua erect, linear, glabrous, stipitate at the base, attenuated upwards into the narrow style; pedicels half the length of the siliqua; leaves glabrous, petiolate, unequally pinnatifid, pinnæ linear-lanceolate, spreading, obtuse, denticulate, decurrent; stem glabrous. *Sinapis tenuifolia, Br. E. Fl. v. iii. p.* 223. *Hook. Br. Fl. p.* 310. *Sisymbrium tenuifolium, Linn. E. Bot. t.* 525.

Old walls and ruins; rare. *Fl.* July, August. ♃.
Walls of Ludlow Castle, plentifully. Walls of Shrewsbury Abbey.
Stems erect, 1-2 feet high, branched and bushy, round, glaucous and glabrous. *Leaves* alternate, generally unequally pinnatifid, often linear-lanceolate nearly entire and denticulate, glaucous green, smooth. *Flowers* large, lemoncoloured. *Sepals* erecto-patent, linear, concave. *Petals* obtuse, twice as long as the sepals. *Stigma* 2-lobed. *Seeds* in two rows.

22. RAPHANUS. *Linn.* Radish.

1. R. *Raphanistrum,* Linn. *Wild Radish or jointed Charlock.* Siliqua of one cell, jointed, striated; leaves simply lyrate. *E. Bot. t.* 856. *E. Fl. v. iii. p.* 226. *Hook. Br. Fl. p.* 310.

Corn-fields; rare. *Fl.* June, July. ⊙.
About Albrightlee, near Shrewsbury. Davenport House, near Bridgnorth. Church Stretton; *Rev. E. Williams's MSS.* Caynton; *H. Bidwell, Esq.*
Root tapering, slender. *Herb* rough with minute deflexed bristles. *Stem* 1-2

feet high, glaucous, branched. *Leaves* simply lyrate, bluntly toothed, terminal lobe rounded; upper ones oblong or lanceolate, acute, entire, coarsely serrated. *Flowers* yellow, veined. *Siliqua* erect, moniliform, smooth, longitudinally striated, terminated by a long subulate slightly bristly beak. *Seeds* large, globular, solitary in each knob or apparent joint, minutely dotted.
Mr. Williams mentions a variety (I presume with white flowers strongly veined with purple) as found near Albrightlee, near Shrewsbury.

CLASS XVI.

MONADELPHIA.

"How various, Lord, thy works are found,
For which thy wisdom we adore!
The earth is with thy treasure crown'd,
'Till Nature's hand can grasp no more!"
PSALM CIV.

CLASS XVI.

MONADELPHIA. *Filaments combined in one set.*

ORDER I. PENTANDRIA. *5 perfect stamens.*

1. ERODIUM. *Style* 1. *Calyx* of 5 leaves. *Corolla* of 5 petals. *Glands* 5. Five alternate *stamens* imperfect. *Fruit* beaked, separating into 5 one-seeded *capsules*, each with a long spiral *awn*, bearded on the inside.—*Nat. Ord.* GERANIACEÆ, *Juss.*—Name from ερωδιος, a *heron*; the fruit resembling the beak of that bird.

(See *Linum* in CL. V. ORD. I.—*Geranium pusillum* in ORD. DECANDRIA.—*Oxalis* in CL. X.

ORDER II. DECANDRIA. 10 *stamens.*

2. GERANIUM. *Style* 1. *Calyx* of 5 leaves. *Corolla* of 5 regular petals. *Glands* 5. *Fruit* beaked, separating into 5 one-seeded *capsules*, each with a long naked *awn.*—*Nat. Ord.* GERANIACEÆ, *Juss.*—Name from γερανος, a *crane*; the fruit resembling the beak of a Crane.

ORDER III. POLYANDRIA. *Many stamens.*

3. MALVA. *Styles* numerous. *Calyx* double; *exterior* of three leaves. *Capsules* numerous, circularly arranged, 1-seeded.—*Nat. Ord.* MALVACEÆ, *Juss.*—Name from μαλακος, *soft*; in allusion to the emollient nature of the species.

4. ALTHÆA. *Styles* numerous. *Calyx* double; *exterior* of 6—9 leaves. *Capsules* numerous, circularly arranged, 1-seeded.—*Nat. Ord.* MALVACEÆ, *Juss.*—Name from αλθω, to *cure*; from its healing properties.

MONADELPHIA—PENTANDRIA.

1. ERODIUM. *L'Herit.* Stork's-bill.

1. E. *cicutarium*, Sm. *Hemlock Stork's-bill.* Peduncles many-flowered; leaves pinnate, leaflets sessile pinnatifid and cut; stem prostrate, hairy. *E. Bot. t.* 1768. *E. Fl. v. iii. p.* 229. *Hook. Br. Fl. p.* 312. *Geranium cicutarium, Linn.*

Sandy pastures, dry ditch banks, &c.; not uncommon. *Fl.* Summer months. ☉.

Common in the sandy parts of the county; *Rev. W. Corbett.* Lincoln's Hill, Coalbrookdale; *Mr. F. Dickinson.* Ruyton and Knockin Heath; *T. C. Eyton, Esq.* About Newport; *R. G. Higgins, Esq.* Hord's Park near Bridgnorth; *Purton's Midl. Flora.* Bromfield near Ludlow; *Mr. H. Spare.* Sleap and Tilley; *Dr. T. W. Wilson.*

Haughmond Hill; Queen Eleanor's Bower, Haughmond Hill; near Shrewsbury.

Root tapering. *Whole herbage* clothed with horizontally spreading glandular hairs, more or less fetid. *Stems* numerous, procumbent, branched, roundish. *Leaves* alternate below, opposite above, on channelled petioles, pinnate, pinnæ ovate sessile deeply pinnatifid, segments cut acute ciliated. *Stipules* ovate, acute, glabrous, pellucid and membranous, margins ciliated. *Peduncles* opposite to the alternate leaves, otherwise axillary, round, longer than the leaves, bearing on their summits a monophyllous membranous cut *bractea*, with ovate acuminate ciliated segments, from which spring several short single-flowered *pedicels.* *Sepals* 5, oblong, obtuse, concave, 3-ribbed, membranous at the margins, hairy at the base, the midrib elongated into a short obtuse awn bearing two white bristles at the apex; 2 of the sepals having the midrib clothed with long horizontally spreading glandular hairs and a line of erect appressed short white bristles on either side, the membranous margins ciliated; whilst the remaining three sepals have all the ribs with glandular hairs, a line of erect appressed white bristles on each side the midrib and a similar one on each membranous margin which however are not ciliated. *Petals* 5, about equal in length to the sepals, rose-coloured, obovate, 3-ribbed, with a short ciliated claw. *Stamens* 10; 5 sterile ones exterior to and half the length of the others, subulate, not dilated at the base; 5 fertile ones subulate, dilated at the base, with a large rounded brown *gland* on the exterior at the base, and a few ciliæ immediately above. *Sepals* of the *fruit* erect, converging over the capsule, covered with erect appressed bristles, the glandular hairs disappearing. *Capsules* 5, linear-obovate, keeled, with 2 depressions at the summit, clothed with close pressed bristles which spread laterally on each side from the keel, tapering above into a very long flattened *beak* with short erect close pressed bristles, at length spirally twisted and adhering at the top. *Seeds* linear-obovate, smooth.

2. E. *moschatum*, Sm. *Musky Stork's-bill.* Peduncles many-flowered; leaves pinnate, leaflets subpetiolate, ovate, unequally cut; perfect stamens toothed at the base; stems depressed, hairy. *E. Bot. t.* 902. *E. Fl. v. iii. p.* 230. *Hook. Br. Fl. p.* 312. *Geranium moschatum, Linn.*

Roadsides, waste places, &c.; rare. *Fl.* June, July. ☉.

In a lane leading from Wellington to the Wrekin; *E. Lees, Esq.* High-road between Shelton and Montford Bridge; *T. and D. Bot. Guide.* On the right hand side of the road as you ascend the bank from Bicton towards Montford Bridge, August 1796; *Rev. E. Williams's MSS.* Ellesmere bowling-green; *J. F. M. Dovaston, Esq.* Near Lilleshall; *Dr. Du Gard.*

Root tapering. *Whole herbage* copiously clothed with horizontally spreading glandular hairs, of a powerful musky fragrance. *Leaves* pinnate on longish petioles, pinnæ very shortly stalked, nearly sessile, ovate, deeply and unequally cut, segments acute, ciliated. *Stipules* large, broadly ovate, acute, often rounded, very thin and membranous, glabrous. *Peduncles* opposite and axillary, longer than the leaves, bearing on their summits a monophyllous deeply cut membranous *bractea*, with oval obtuse glabrous segments, from which spring many short single-flowered *pedicels.* *Sepals* 5, oblong, obtuse, concave, 7-ribbed, midrib elongated into a bristle-pointed obtuse mucro, densely glanduloso-pilose. *Sterile stamens* with very broad filaments; *fertile* ones toothed on each side near the base. *Claws* of the *petals* not ciliated. *Fruit* similar to the last but larger and the hairs on the capsules longer.

3. E. *maritimum*, Sm. *Sea Stork's-bill.* Peduncles 1-2 flowered; leaves simple, ovato-cordate, petiolate, lobed and crenate; stems depressed, slightly hairy. *E. Bot. t.* 646. *E. Fl. v. iii. p.* 231. *Hook. Br. Fl. p.* 312. *Geranium maritimum, Linn.*

Hilly places; rare. *Fl.* May—Sept. ♃.

At the foot of the Longmont; *Miss Mc Ghie!* On the Morf near Bridgnorth in several places, particularly by the side of the road up the hill near the Hermitage. South end of the Wrekin; *Rev. E. Williams's MSS.*

Stems spreading close to the ground, 3—9 inches long, branched, leafy, hairy. *Leaves* petiolate, roundish, cordate, obtusely lobed and notched, covered on both sides with minute close hairs. *Stipules* ovate, acute, membranous, glabrous. *Peduncles* generally shorter than the leaves, bearing on the summit about 2 minute ovate acute membranous *bracteas*, from which spring 1-2 single-flowered *pedicels.* *Sepals* elliptical, obtuse, mucronate, 3-ribbed, hairy. *Petals* pale red, very minute, often wanting. *Fruit* as in the preceding but smaller.

MONADELPHIA—DECANDRIA.

2. GERANIUM. *Linn.* Crane's-bill.

* *Peduncles one-flowered.*

1. G. *sanguineum*, Linn. *Bloody Crane's-bill.* Stem decumbent, branched, clothed with horizontal hairs; leaves opposite, orbicular, palmato-fid, lobes 7 deeply trifid and incised; peduncles axillary, many times longer than the leaves, 1-flowered; sepals broadly elliptical, hairy, 5-ribbed, spreading in flower, erect in fruit; petals obcordate, spreading, twice the length of the sepals, hairy at the base, 5-ribbed; stamens all perfect, recurved, ciliated; capsules oblong, keeled, minutely pubescent, transversely wrinkled above, crowned with a few very long bristles, beak very long, hairy; seeds oblongo-subreniform, minutely wrinkled and dotted. *E. Bot. t.* 272. *E. Fl. v. iii. p.* 242. *Hook. Br. Fl. p.* 312.

Rocks and limestone pastures; rare. *Fl.* July. ♃.

Downton Castle; *T. and D. Bot. Guide.* Bloodwell rocks. Llanymynech white rock; *J. F. M. Dovaston, Esq.* Dowle Wood, plentiful; *Mr. G. Jorden.* Rocks on the Shropshire side of Dowle Brook, Wyre Forest; *Mr. W. G. Perry.*

Root woody, creeping. *Stems* numerous, decumbent, somewhat angular, branched, zigzag, clothed with soft prominent horizontal hairs, swollen above the joints. *Leaves* opposite, on hairy petioles, orbicular, palmato-fid, lobes 7, each deeply trifid and incised, segments linear-lanceolate, tipped with a callous crimson tubercle, softly hairy on both sides. *Stipules* ovate, membranous, marcescent, ciliated. *Peduncles* axillary, solitary, swollen at the base, many times longer than the leaves, clothed with horizontal hairs, with a swollen articulation and a pair of lanceolate ciliated marcescent *bracteas* above the middle. *Sepals* all broadly elliptical, membranous at the edges, 5-ribbed, the midrib elongated into a short callous awn, soft and hairy externally. *Petals* very large, handsome, red purple, with 5 branched darker veins, hairy at the base. *Filaments* dilated at the base. *Stigma* 5-fid. *Capsule* large, oblong, keeled, minutely pubescent, transversely wrinkled and having a few long bristles in the upper part, with a tuft of long white hairs on each side of the base of the inner margin, beak very long clothed with soft erecto-patent hairs. *Seeds* oblongo-subreniform, minutely wrinkled and dotted.

** *Peduncles 2-flowered.*

2. G. *phæum*, Linn. *Dusky Crane's-bill.* Stem erect, branched, clothed with decurved pubescence; lower leaves opposite, orbiculari-reniform, palmato-fid, lobes 5 deeply and irregularly inciso-dentate;

peduncles opposite to and longer than the leaves, 2-flowered; pedicels erect in flower and fruit; sepals oblong, glandular and hairy, 5-ribbed, recurved in flower, erect in fruit, about as long as the capsules; petals rotundo-obovate, wavy, recurved, 5-ribbed, hairy at the base, about as long as the sepals; stamens all perfect, recurved, ciliated; capsules oblong, keeled, clothed with erect appressed hairs, transversely wrinkled above, beak glandular; seeds oblong, punctato-striate. *E. Bot. t.* 322. *E. Fl. v. iii. p.* 232. *Hook. Br. Fl. p.* 313.

Woods and thickets; rare. *Fl.* May, June. ♃.

Stottesden, rare; *Mr. G. Jorden.* In the neighbourhood of Hopton Castle, Stanton Lacy, and at the Hope Shortwood; *Miss Mc Ghie.* In a wood near Corfton in Corve Dale; *Rev. T. Salwey.* Hedges near Buildwas church; *Rev. E. Williams's MSS.*

Root woody. *Stems* several, 1½-2 feet high, erect, round, branched, panicled at top, swollen above the joints, clothed with soft silky spreading more or less decurved hairs interspersed with minute glands, dotted with red and with red blotches at the swollen joints. *Lower leaves* in pairs, opposite, on hairy glandular petioles, orbiculari-reniform, palmato-fid, lobes about 5, deeply and irregularly inciso-dentate, tipped with callous tubercles, downy on both sides, somewhat hairy above; *upper leaves* alternate, similar but nearly sessile. *Stipules* large, brown, lanceolate, marcescent, ciliated. *Peduncles* opposite to and much longer than the upper leaves, hairy and glandular, having above the middle a swollen articulation, from which spring two hairy and glandular pedicels, each bearing a single *flower* with two pair of opposite brown lanceolate ciliated *bracteas* at the base. *Sepals* oblong, very glandular and hairy on the exterior, 5-ribbed, midrib elongated into a short mucro. *Petals* rotundo-obovate, wavy, deep chocolate colour, pale and hairy at the base, with 5 branched veins. *Filaments* dilated and ciliated at the base, united by 5 large green *glands* placed on their outside. *Capsules* oblong, keeled, clothed with erect appressed hairs, transversely wrinkled in the upper part; *beak* long and glandular. *Seeds* oblong, punctato-striate.

3. G. *sylvaticum*, Linn. *Wood Crane's-bill.* Stem erect, dichotomously branched, clothed with deflexed hairs; leaves palmate, in 7 deep lanceolate acute incised and serrated lobes; peduncles 2-flowered, longer than the upper leaves; sepals nearly equal, 3-ribbed, glanduloso-pilose, with a short obtuse awn; petals slightly notched; stamens fringed; capsules ovate, keeled, hairy, not wrinkled; seeds dotted. *E. Bot. t.* 121. *E. Fl. v. iii. p.* 234. *Hook. Br. Fl. p.* 313.

Woods, thickets, &c.; rare. *Fl.* June, July. ♃.

Dowle Wood, plentiful; *Mr. G. Jorden.* Clee Hill near Bitterley; *Miss Mc Ghie.* Near Park brook and Dowles brook, Wyre forest; *Mr. W. G. Perry.* Dudmaston woods; *Purton's Midl. Flora.* Woods at Mawley, (Sir Edward Blount's) near Cleobury Mortimer; *Rev. W. Corbett.* Near Hales Owen; *T. and D. Bot. Guide.* In the hedge round Hopton Wafers churchyard and the hedges of the adjoining fields. In Kinsley wood opposite Knighton, 1796; *Rev. E. Williams's MSS.*

Root woody. *Stem* 18 inches—2 or 3 feet high, erect, dichotomously branched, clothed with minute deflexed hairs. *Leaves* palmate, 7-lobed, lobes lanceolate acute, coarsely incised and serrated, veiny, finely hairy on both sides, the lower ones on long round petioles with deflexed hairs, upper ones on shorter petioles or nearly sessile. *Stipules* lanceolate, acuminated, membranous, glabrous. *Peduncles* axillary, considerably longer than the upper leaves, round and

as well as the pedicels glanduloso-pilose, with longer hairs interspersed, bearing near the summit 2 pairs of opposite lanceolate pubescent *bracteas* from which spring two short erect pedicels. *Sepals* nearly equal, elliptical, obtuse, 3-ribbed, midrib elongated into an obtuse awn 4 or 5 times shorter than the sepal, clothed with minute erect hairs, ribs glanduloso-pilose, margins membranous and minutely ciliated. *Petals* deep blue, entire or slightly notched, hairy at the claw. *Stamens* all nearly equal, subulate, membranous at the edges, fringed more than half way up. *Capsule* ovate, keeled, even, not wrinkled, most hairy about the keel, marked at each side towards the top with a brown rib. *Seeds* dotted.

4. G. *pratense*, Linn. *Blue Meadow Crane's-bill.* Stem erect, dichotomously branched, clothed with deflexed pubescence; leaves opposite, orbiculari-pentangular, in 7—9 very deep lanceolate acute inciso-pinnatifid lobes; peduncles axillary, considerably longer than the upper leaves, 2-flowered; flowers patent, petals resting on the sepals; sepals unequal, 3 larger broadly oblongo-ovate 5-ribbed glanduloso-pilose, 2 smaller linear-oblong, margins membranous, midrib glanduloso-pilose, all awned, erect in fruit, twice as long as the capsule; petals obovate, emarginate and notched, nearly twice as long as the calyx, 7-ribbed; stamens all perfect, recurved, triangular and hairy at the base; capsules oblong, keeled, glanduloso-pilose, with a tuft of long white hairs on each side at the base on the inner face, beak glanduloso-pilose; seeds oblong, compressed, minutely reticulated. *E. Bot. t.* 404. *E. Fl. v. iii. p.* 235. *Hook. Br. Fl. p.* 313.

Moist meadows; not common. *Fl.* June, July. ♃

Dowle, plentiful; *Mr. G. Jorden.* Fields below Coalport; *Mr. F. Dickinson.* Common on the Worcestershire border of Shropshire, viz. Burford, &c.; *Rev. W. Corbett.* Quatford Ferry, near Bridgnorth; *W. P. Brookes, Esq.* Oakley Park meadows near Ludlow; *Mr. H. Spare.* Meadows about Bitterley, Ashford Carbonel, Burford and Burraston. Banks of the Severn opposite Apley, &c.; *Rev. E. Williams's MSS.*

Oakley park, near Ludlow. Banks of the Ludwyche between Ludlow and Caynham Camp.

Root fibrous. *Stem* 3 feet high, erect, dichotomously branched, swollen above the joints, clothed with minute dense deflexed pubescence. *Leaves* all opposite, on very long round petioles with deflexed pubescence, orbiculari-pentangular in their circumscription, in 7—9 very deep lanceolate acute lobes, lobes deeply and irregularly inciso-pinnatifid, segments acute with a callous tip, minutely hairy and veiny on both sides, margins ciliated. *Stipules* lanceolate, very much elongated and acuminated, ribbed, ribs and margins pubescent. *Peduncles* axillary, solitary, very considerably longer than the upper leaves, round and as well as the pedicel glanduloso-pilose, bearing near the summit 2 pairs of opposite lanceolate elongated and acuminated ciliated *bracteas*, from which spring two very short pedicels swollen upwards beneath the flower, erect in flower, in fruit divergent at right angles. *Sepals* unequal, 3 longer ones broadly oblongo-ovate, 5-ribbed, midrib elongated into an obtuse awn half the length of the sepal, clothed with erect bristles, ribs and margins glanduloso-pilose; 2 smaller ones linear-oblong, membranous and ciliated at the margins, 3-ribbed, midrib glanduloso-pilose, awned as the others, lateral ribs glabrous, patent in flower, erect in fruit. *Petals* large, deep blue, with 7 red pellucid branched veins, nearly twice the length of the sepals, broadly obovate, somewhat emarginate, irregularly notched, with a very minute white claw, densely ciliated above the base. *Filaments* dilated into a triangular

figure at the base, hairy and ciliated, all united by large green glands; *anthers* deep purple, recurved. *Capsules* oblong, keeled, glanduloso-pilose, with a tuft of white hairs on each side at the base on the inner face; *beak* long, glanduloso-pilose, upper part with erect bristles. *Seeds* oblong, compressed, minutely reticulated, interstices impressed.

5. G. *Pyrenaicum*, Linn. *Mountain Crane's-bill.* Stem erect, dichotomously branched, clothed with decurved pubescence and spreading hairs; leaves opposite, orbiculari-reniform, in 7 rounded obtuse lobes; peduncles axillary, longer than the upper leaves, 2-flowered; flowers patent, petals resting on the sepals; sepals oblongo-ovate, ciliæ equal, patent in fruit, twice as long as the capsules; petals obcordate, twice as long as the sepals, 5-ribbed; stamens all perfect, recurved, ciliated; capsules oblong, obliquely pointed at each end, keeled, clothed with erect appressed bristles, beak with horizontal pubescence and pedicellate glands; seeds oblong, smooth. *E. Bot. t.* 405. *E. Fl. v. iii. p.* 239. *Hook. Br. Fl. p.* 313. *Fries Nov. 2nd ed. p.* 213. *Drej. Fl. Hafn. n.* 688.

Road-sides; rare. *Fl.* June, July. ♃.

Upton, near Shiffnal; *H. Bidwell, Esq.* Buildwas; *Miss H. Moseley.* Road-side from Shiffnal to the Lizard; *Mr. F. Dickinson.* Paper mills by the Teme side, Ludlow; *Miss Mc. Ghie.* In a field adjoining Cound rectory garden; *Rev. E. Williams's MSS.* Lane near Cound church; *Mr. T. Bodenham!*

Right hand side of the Holyhead road between Shrewsbury and Montford bridge, extending from the Four Crosses Public-house to Bicton Grove.

Root tuberous. *Stem* 2-3 feet high, erect, dichotomously branched, spreading, round, swollen above the joints, clothed with dense decurved pubescence interspersed with longer spreading or somewhat deflexed soft hairs. *Leaves* in pairs, opposite, on long round hairy and pubescent petioles, orbiculari-reniform, 7-lobed, lobes rounded obtuse incised, segments rounded obtuse with about 3 notches, callously tipped, softly hairy and veiny, margins ciliated; *upper leaves* on shorter petioles, with narrower lobes and more acute segments. *Stipules* broadly ovate, very acuminate, jagged at the points, hairy and ciliated. *Peduncles* axillary, solitary, longer than the upper leaves from which they chiefly arise, round and as well as the pedicels covered with horizontal pubescence, interspersed with innumerable pedicellate glands, bearing about the middle 2 pairs of opposite subulate ciliated *bracteas*, from which spring 2 pedicels erect in flower, in fruit decurved at right angles. *Sepals* nearly equal, oblongo-ovate, with callous points, larger ones 5-, smaller ones 3-ribbed, pubescent and with sessile glands, margins membranous, ciliated, patent in flower and fruit. *Petals* deep purple with 5 pellucid branched veins, twice as long as and resting on the sepals, obcordate, with a minute pale ciliated claw. *Filaments* dilated and ciliated at the base, alternately united by green glands; *anthers* bluish-purple, recurved. *Capsules* oblong, obliquely pointed at each end, keeled, clothed with erect appressed minute white hairs; *beak* long, with horizontal pubescence and pedicellate glands. *Seeds* oblong, reddish-brown, not compressed, smooth.

Reichenbach (*Fl. Excurs.* 4881) describes *G. pyrenaicum* "carpidiis lævibus;" in our Shropshire specimens as well as in those of Cambridgeshire the capsules are clothed with erect appressed minute hairs, and agree with Fries's description "arillis pubescentibus."

6. G. *lucidum*. Linn. *Shining Crane's-bill.* Stem spreading, dichotomously branched, glabrous and shining; leaves opposite, orbiculari-reniform, lobes 5 ovate trifid ciliated; peduncles axillary,

longer than the leaves; pedicels erect in flower, decurved in fruit, 2-flowered; sepals unequal, ovato-lanceolate, pyramidal in fruit, 5-angular, dentato-tuberculate, twice the length of the capsules; petals narrowly obcordate, limb nearly flat and spreading, claw as long as the sepals, 3-ribbed; stamens all perfect, erect, glabrous; capsules oblongo-obovate, keeled, longitudinally rugose, beak smooth; seeds oblongo-obovate, smooth. *E. Bot. t.* 75. *E. Fl. v. iii. p.* 236. *Hook. Br. Fl. p.* 313.

Walls and hedge-banks; not common. *Fl.* June. July. ⊙.

Badger dingle; *H. Bidwell Esq.* Walls at Oswestry and on the top of Llanymynech hill; *Rev. T. Salwey.* On the rocks at Church Stretton and Little Stretton; *E. Lees, Esq.* On a wall at Church Aston by the turnpike road; *R. G. Higgins, Esq.* In a wood near the Severn about a mile from Stourport down the river; *W. P. Brookes, Esq!* Oakley park near Ludlow; *Mr. H. Spare.* Amongst the loose stones on the S.E. side of the Wrekin near the top. Ditch banks, walls and roofs of houses in Oswestry, Bitterley, Ludlow, Earls Ditton, Silvington, Burwarton and Habberley. Between Halston and Hardwick; *Rev. E. Williams's MSS.* Wrenthall; *Mr. T. Bodenham.*

Sides of the turnpike road from Hanwood to Pontesbury, between the fifth milestone and the Lea Cross. Ruins of Ludlow Castle. Hedge banks between Oswestry and Westfelton. Pontesford Hill, south side.

Root fibrous, slender. *Whole herbage* shining and nearly glabrous. *Stem* round, smooth except a line of minute curved hairs on one side especially visible in the upper part, bright red, dichotomously branched from the base, swollen above the joints, weak, brittle and spreading in every direction. *Leaves* opposite, in pairs, on round smooth reddish petioles, rotundo-reniform, 5-lobed, lobes ovate, 3-fid, segments rounded and blunt, often notched, tipped with a short stout blunt mucro, edges slightly ciliated and with a few scattered hairs on the upper surface, pale and glabrous beneath. *Stipules* small, red, ovato-acuminate, glabrous. *Peduncles* axillary, solitary, longer than the leaves, glabrous except a line of decurved hairs on one side, with an articulation above the middle and 2 pairs of opposite lanceolate *bracteas* from which arise 2 pedicels erect in flower, in fruit divergent nearly at right angles, hairy on one side and bearing a single flower. *Calyx* pyramidal when closed, pentangular; *sepals* 5 unequal, 3 larger and equal ovato-lanceolate membranous 3-ribbed, midrib prominently and sharply keeled elongated into a short stiff mucro or awn, armed towards the centre of the face with 4 elevated triangular transverse processes on each side, lateral ribs with very sharp and prominently winged keels considerably more elevated than the midrib, with 3 elevated triangular transverse processes on the edges of the sepals which are membranous and incurved; 2 smaller sepals equal and opposite, half as broad as the others, lanceolate, mucronate, membranous, 3-ribbed, midrib prominent and keeled, lateral ribs fainter, without any elevated processes on the faces or edges. *Petals* of a faint and delicate rose-colour, small, narrowly obcordate, 3-ribbed, with a very long linear paler claw the length of the calyx, limb nearly flat and spreading. *Stamens* 10; *filaments* subulate, dilated at the base, glabrous, alternate ones connected by green glandular processes; *anthers* bright yellow. *Capsules* oblongo-obovate, slightly compressed, keeled, with about three elevated longitudinal rugose processes parallel to the keel connected by irregular transverse wrinkles, and on each side of their summits clothed with pedicellate glands which are also continued along the inner edge of the capsule; *beak* very long, smooth except the upper portion which is clothed with minute erect bristles. *Seeds* oblongo-obovate, reddish, smooth, rounded at the back, compressed towards the face.

7. G. *robertianum*, Linn. *Stinking Crane's-bill or Herb-Robert.*

Stem spreading, dichotomously branched, clothed with long horizontal hairs; leaves opposite, irregularly 5-angular, ternato-pinnatifid, lobes ovato-lanceolate deeply pinnatifid and incised; peduncles axillary longer than the leaves, pedicels erect in flower and fruit, 2-flowered; sepals ovato-lanceolate, hairy and glandular, 10-angular, pyramidal in fruit, twice the length of the capsules; petals obovate, entire or emarginate, limb nearly flat and spreading, claw nearly as long as the sepals, 3-ribbed, lateral ribs bifid; stamens all perfect, erect, glabrous; capsules oblong, keeled, hairy, transversely wrinkled, beak smooth; seeds oblong, smooth. *E. Bot. t.* 1486. *E. Fl. v. iii. p.* 235. *Hook. Br. Fl. p.* 314.

Hedges, thickets, woods, stony and waste ground; common. *Fl.* Summer months. ⊙.

Root fibrous, tapering. *Herbage* of a fetid smell. *Stem* dichotomously branched from the base, branches spreading, red, brittle, clothed with long horizontal hairs. *Leaves* opposite, in pairs, on round hairy petioles, irregularly 5-angular in their outline, ternato-pinnatifid, lobes ovato-lanceolate deeply pinnatifid, pinnæ deeply and irregularly incised, segments bluntish, tipped with a short obtuse crimson mucro, pale green, with scattered hairs. *Stipules* ovate, obtuse, ciliated. *Peduncles* axillary, solitary, longer than the leaves, round, glabrous except a line of incurved pubescence on one side, with an articulation and two pairs of opposite ovate ciliated *bracteas* near the summit, from which arise two short pedicels hairy on one side, erect both in fruit and flower. *Calyx* with 10 angles when closed; *sepals* unequal, 3 larger ovato-lanceolate, membranous at the edges, with 3 prominent ribs clothed with rows of long silky hairs and short glandular ones interspersed, midrib elongated into a rather long crimson awn; the other two narrower. *Petals* pale red-purple with three white ribs, lateral ribs bifid, somewhat concave at the base, obovate, margin entire, often emarginate or crenulate, claw nearly as long as the calyx. *Capsule* oblong, keeled, hairy, transversely wrinkled; *beak* smooth, upper portion with erect bristles. *Seeds* oblong, reddish brown, smooth.

Var: β. *flore albo.* Under the garden wall of the school-house near Haughmond abbey; probably an escape from the garden.

8. G. *molle,* Linn. *Dove's-foot Crane's-bill.* Stem decumbent, branched, clothed with horizontal pubescence and glands; leaves alternate, orbiculari-reniform, in 7—9 deep narrowly wedge-shaped lobes; peduncles opposite to and as long as the leaves, 2-flowered; flowers patent, petals resting on the sepals; sepals elliptical, ciliæ unequal, converging in fruit and twice the length of the capsules; petals oblong, deeply bifid, scarcely longer than the calyx, 3-ribbed, lateral ribs bifid at the base; stamens all perfect, recurved, glabrous; capsules oval, keeled, transversely wrinkled, glabrous; beak with erect bristles and pedicellate glands; seeds oval, smooth. *E. Bot. t.* 778. *E. Fl. v. iii. p.* 237. *Hook. Br. Fl. p.* 314. *Bab. Prim. Fl. Sarn. p.* 21. *Reich. Fl. Excurs.* 4879. *Fries Nov. 2d ed. p.* 215. *Drej. Fl. Hafn. n.* 686.

Dry pastures, roadsides, waste places; common. *Fl.* April—August. ⊙.

Root tapering. *Stems* variable in length, numerous, decumbent and straggling, round, clothed with short horizontal pubescence and pedicellate glands and with long silky hairs interspersed, scarcely branched, slightly swollen above the

joints. *Leaves* alternate, the lowermost only opposite, on long round hairy petioles, orbiculari-reniform, deeply 7—9-lobed, lobes narrowly wedge-shaped, anterior margin not truncate as in *pusillum,* unequally trifid, the central segments longer, all more acute, not rounded or obtuse, tipped with a very short mucro, both surfaces clothed with dense soft hairs. *Stipules* ovato-lanceolate, acuminate, membranous, ciliated, more or less laciniated. *Peduncles* opposite to and about as long as the leaves except the lower ones which are shorter, round, with soft dense short horizontal pubescence interspersed with glands and very long silky hairs, having an articulation above the middle with two pairs of opposite narrow lanceolate hairy *bracteas,* from which arise 2 glandular pubescent and hairy pedicels each single-flowered, erect in flower, in fruit divergent at right angles. *Sepals* nearly equal, elliptical, 3-ribbed, midrib elongated into a very short mucro, hairy glandular and pubescent on the exterior, margins ciliated, ciliæ of various lengths, in fruit converging, twice the length of the capsule. *Petals* red-purple, scarcely longer than the calyx, oblong, with a minute white ciliated claw, deeply bifid, 3-ribbed, lateral ribs bifid at the base. *Filaments* subulate, dilated at the base, glabrous, recurved at the summit, alternately united by green glands; *anthers* bluish-purple. *Capsules* oval, keeled, transversely wrinkled; *beak* with erect bristles and pedicellate glands interspersed. *Seeds* oval, not compressed, reddish-brown, smooth.

9. G. *rotundifolium,* Linn. *Round-leaved Crane's-bill.* Stem procumbent, dichotomously branched, clothed with horizontal glandular pubescence; leaves opposite, orbiculari-reniform, lobes 7 broadly wedgeshaped truncate trifid, margins ciliato-glandulose; peduncles axillary, not half the length of the leaves, 2-flowered; flowers campanulate, patent, petals not resting on but arching over the sepals; sepals unequal, elliptical, glanduloso-pilose, erecto-patent in fruit, twice the length of the capsules; petals spathulate, entire or slightly retuse, longer than the sepals, 3-ribbed, central rib bifid; stamens all perfect, erect, ciliated; capsules oblong, keeled, clothed with spreading bristles, with a tuft of long white hairs on each side the base on the inner face, beak glanduloso-pilose; seeds broadly oval, reticulated. *E. Bot. t.* 157. *E. Fl. v. iii. p.* 240. *Hook. Br. Fl. p.* 314. *Bab. Prim. Fl. Sarn. p.* 21. *Reich. Fl. Excurs.* 4878. *Drej. Fl. Hafn. n.* 689. *G. viscidulum, Fries Nov. 2nd ed. p.* 216.

Pastures and waste ground. *Fl.* June, July. ⊙.

Inserted on the authority of *Turn. & Dilw. Bot. Guide* in which it is stated to be "common about Shrewsbury." After repeated and diligent searches I have never been able to detect it. It is completely naturalized in my garden near Shrewsbury, where it is now a troublesome weed, from seeds taken from a specimen from Bath.

Root strong and woody, tapering. *Whole herbage* softly glanduloso-pilose. *Stem* dichotomously branched from the base, branches very divergent, procumbent and straggling in all directions, round, reddish, swollen above the joints, with dense horizontal soft glandular pubescence. *Leaves* in pairs, opposite, on long round glanduloso-pilose petioles, orbicular, 7-lobed, lobes broadly wedgeshaped blunt and truncate trifid, segments obtuse, central one broadest trifid, lateral ones entire or with a single tooth on the exterior side, all tipped with a minute obtuse mucro, margins ciliato-glandulose, both surfaces glanduloso-pubescent, when young the sinuses are marked with a red spot. *Stipules* red, lanceolate, acuminate, ciliated. *Peduncles* axillary, solitary, not half the length of the petioles, round

and as well as the pedicels glanduloso-pilose, bearing about the middle 2 pairs of opposite subulate ciliated *bracteas* from which spring 2 pedicels erect in flower in fruit divergent at right angles. *Sepals* unequal, 3 larger ones elliptical, others narrower, all 3-ribbed, glanduloso-pilose, midrib elongated into a short obtuse mucro. *Petals* somewhat longer than the calyx, patent, pale pinkish purple, with 3 darker ribs, central rib bifid, spathulate, the margin entire or slightly retuse. *Filaments* subulate, dilated and ciliated at the base, alternately connected by green glands; *anthers* whitish. *Capsules* of a greenish-brown colour, oblong, keeled, scarcely compressed, with spreading bristles, and a tuft of long white hairs on each side the base on the inner face; *beak* very long, glanduloso-piloso below, upper portion with erect bristles. *Seeds* broadly oval, light brown, covered with white elevated irregularly hexagonal reticulations.

10. G. *pusillum,* Linn. *Small-flowered Crane's-bill.* Stem decumbent, scarcely branched, clothed with deflexed pubescence; leaves opposite, orbiculari-reniform, in 5-7 broadly wedgeshaped truncated lobes; peduncles axillary, considerably shorter than the leaves, 2-flowered; flowers campanulato-patent, petals not resting on but arching over the sepals; sepals ovate, ciliæ equal, patent in fruit, scarcely longer than the capsules; petals oblongo-obcordate, scarcely longer than the calyx, 3-ribbed; 5 stamens only perfect, erect, ciliated; capsules oblong, keeled, densely clothed with erect white bristles, beak with erect bristles and pedicellate glands; seeds oblong, compressed, smooth. *E. Bot. t.* 385. *E. Fl. v. iii. p.* 238. *Hook. Br. Fl. p.* 314. *Bab. Prim. Fl. Sarn. p.* 21. *Reich. Fl. Excurs,* 4877. *Drej. Fl. Hafn. n.* 687. *G. rotundifolium, Fries Nov. ed. ii. p.* 212.

Waste ground, road-sides, &c.; frequent. *Fl.* June—September. ⊙.

Root tapering. *Stems* numerous, variable in length, decumbent, round, branched, clothed with deflexed pubescence, slightly swollen above the joints. *Leaves* in pairs, opposite, on long pubescent petioles, orbiculari-reniform, 5-7 lobed, lobes broadly wedgeshaped, anterior margin blunt or truncated, nearly equally trifid, segments all about the same length, rounded or obtuse, often toothed, tipped with a red callous tubercle, softly hairy on both sides. *Stipules* ovato-lanceolate, acuminate, ciliated, often laciniate. *Peduncles* axillary, solitary, very considerably shorter than the leaves, with deflexed pubescence, bearing about the middle an articulation and 2 pairs of opposite subulate hairy *bracteas* from which arise 2 pubescent and glandular single-flowered pedicels erect in flower, in fruit divergent. *Sepals* nearly equal, ovate, hairy and glandular, margins ciliated, ciliæ uniform in length, 3-ribbed, lateral rib bipartite, midrib elongated into a very short stout rigid crimson mucro, in fruit patent, scarcely longer than the capsules, margins membranous. *Petals* scarcely longer than the calyx, oblongo-obcordate, with a minute white claw slightly ciliated, 3-ribbed, very pale bluish-purple. Five of the *stamens* only with bluish-purple *anthers.* *Filaments* erect, dilated and ciliated at the base, with green glandular processes on the exterior at the base; *five* alternate stamens shorter, narrower, without anthers and without glandular processes, keeled. *Capsules* oblong, keeled, densely clothed with erect white bristles, *beak* with erect bristles and pedicellate glands. *Seeds* oblong, rounded at the back, compressed towards the front, smooth.

11. G. *dissectum,* Linn. *Jagged-leaved Crane's-bill.* Stem straggling, ascending, branched, clothed with soft deflexed bristles; leaves opposite, orbiculari-reniform, palmato-fid, lobes 5—7 battle-

dore-shaped trifid; peduncles axillary, shorter than the leaves, 2-flowered; flowers erecto-patent, petals resting on the sepals; sepals ovato-lanceolate, hairy and glandular, spreading in fruit, 3 times the length of the capsules; petals obovate, deeply emarginate, shorter than the sepals, with three impressed pale ribs; stamens all perfect, erect, ciliated; capsules oval, keeled, clothed with erect white bristles, beak with long pedicellate glands and short erect bristles; seeds oval, reticulated. *E. Bot. t.* 753. *E. Fl. v. iii. p.* 241. *Hook. Br. Fl p.* 314.

Hedges and pastures, gravelly and waste places; not unfrequent. *Fl.* May, June. ⊙.

Root tapering. *Stems* several, variable in length, decumbent and straggling or ascending by support of the surrounding vegetation, angular, swollen above the joints, clothed with soft white deflexed bristles. *Leaves* opposite, in pairs, on round petioles, lower ones the longest with deflexed bristles, orbiculari-reniform, palmato-fid, lobes 5—7 battledore-shaped in outline trifid, segments deeply incised, ultimate segments acute, tipped with a short blunt mucro, clothed with scattered hairs above, paler and hairy on the ribs beneath, margins ciliated. *Stipules* subulate, very much accuminated, red, ciliated at the margins. *Peduncles* axillary, solitary, shorter than the petioles, round, clothed with deflexed soft bristles, bearing about the middle 2 pairs of opposite subulate hairy *bracteas* from which arise 2 glandular and hairy pedicels supporting each a single flower. *Sepals* unequal, spreading in fruit, 3 larger ones ovato-lanceolate 5-ribbed, 2 smaller ones narrower 3-ribbed, the midrib of all elongated into a hairy awn, clothed with soft hairs and pedicellate glands. *Petals* shorter than the sepals, deep rose-colour, with 3 impressed pale ribs and a minute ciliated claw, obovate, deeply emarginate or notched. *Stamens* 10; *filaments* subulate, dilated and ciliated at the base, connected by green glands; *anthers* bluish-purple. *Capsules* oval, keeled, covered with erect white bristles, *beak* long, clothed with pedicellate glands and shorter erect bristles interspersed, the uppermost short portion with erect bristles only. *Seeds* large, oval, reticulated.

12. G. *columbinum,* Linn. *Long-stalked Crane's-bill.* Stem straggling, decumbent, branched, clothed with minute deflexed bristles; leaves opposite, orbiculari-reniform, palmato-fid, lobes 5—7 lanceolate incised; peduncles slender, axillary, 3 times longer than the leaves, 2-flowered; flowers erecto-patent, petals resting on the sepals; sepals broadly ovate, sparingly hairy, erect in fruit, twice as long as the capsules; petals narrow, obovate, emarginate, with a short obtuse tooth in the sinus, longer than the sepals, 3-ribbed; stamens all perfect, erect, scarcely ciliated; capsules ovate, keeled, nearly glabrous, beak with erect appressed bristles; seeds oval, reticulated. *E. Bot. t.* 259. *E. Fl. v. iii. p.* 241. *Hook. Br. Fl. p.* 314.

Gravelly and limestone soils; not very common. *Fl.* June, July. ⊙.

Buildwas; *Miss H. Moseley.* Side of the road between the Wrekin and Little Wenlock; *E. Lees, Esq.* Roadside in Little Wenlock; *Mr. F. Dickinson.* Roadside between Shineton and Wenlock; *W. P. Brookes, Esq!* Near Ellesmere; *Rev. A. Bloxam.* Roadside near Welbach; *Mr. T. Bodenham.* Near Eyton; *T. C. Eyton, Esq.* Near Oswestry; *Rev. T. Salwey.*

Pulley; Haughmond Hill; near Berrington; Hill Top, Wenlock Edge; Red Hill near Welbach.

Root tapering. *Stems* several, 6—8 inches long, straggling, decumbent, dichotomously branched, swollen above the joints, clothed with scattered minute deflexed soft bristles. *Leaves* opposite, in pairs, on round petioles, lower ones the longest with deflexed bristles, orbiculari-reniform, palmato-fid, lobes 5—7 lanceolate incised, segments rather obtusely pointed, tipped with a red callous point, clothed with scattered bristles, especially on the ribs beneath, margins subciliated. *Stipules* subulate, acuminate, red, ciliated at the margins. *Peduncles* axillary, solitary, 3 times longer than the petioles, round, clothed with deflexed closepressed soft bristles, bearing below the middle 2 pairs of opposite subulate red glabrous ciliated *bracteas* from which arise 2 long divaricated *pedicels*, similarly clothed, each bearing a single flower. *Sepals* unequal, erect in fruit, 3 larger ones broadly ovate 5-ribbed, 2 smaller ones ovato-lanceolate 3-ribbed, margins membranous and ciliated, clothed, especially on the ribs, with erect appressed scattered bristles, the midrib elongated into a hairy awn. *Petals* longer than the sepals, purple, with 3 red ribs, claw ciliated above the base, narrow obovate, emarginate, with a short obtuse tooth in the sinus. *Stamens* 10; *filaments* subulate, dilated at the base, very slightly ciliated on the exterior, connected by pale glands; *anthers* bluish-purple. *Capsules* oval, keeled, nearly glabrous or with a few very minute scattered hairs, *beak* long, clothed with short erect appressed bristles, the uppermost portion very narrow, inner margins with a few long geniculate deflexed bristles in the upper part and a tuft of long white hairs on each side at the base. *Seeds* large, oval, covered with elevated irregularly hexagonal reticulations.

MONADELPHIA—POLYANDRIA.

3. MALVA. *Linn.* Mallow.

1. M. *sylvestris*, Linn. *Common Mallow.* Stem herbaceous, erect, with spreading hairs, branched; leaves orbiculari-reniform, in 7 deep obtuse irregularly crenato-serrated lobes; stipules ovate, acute; outer sepals lanceolate obtuse hairy, inner sepals ovate acute; petals oblongo-obcordate, spreading; tube of anthers with stellate pubescence; peduncles erect in fruit; fruit glabrous. *E. Bot. t.* 671. *E. Fl. v. iii. p.* 244. *Hook. Br. Fl. p.* 315.

Waste places, waysides, &c.; common. *Fl.* June—August. ♃.
Root tapering. *Stem* erect or in barren soil decumbent, 2—4 feet high, branched, round, rough with long spreading hairs. *Leaves* alternate, on very long petioles, rough with spreading hairs, slightly channelled above, orbiculari-reniform, in 7 deep obtuse irregularly crenato-serrated lobes, plaited, veiny, hairy on both sides. *Stipules* 2, ovate, acute, margins denticulate and ciliated. *Flowers* about 5, in axillary fascicles, on round peduncles unequal in length but much shorter than the leaves, clothed with spreading horizontal hairs, erect in flower and fruit. *Outer calyx* of 3 lanceolate rather obtuse hairy leaves, denticulate and ciliated at the margins; *inner calyx* monophyllous, 5-fid, hairy, segments ovate acute, margins denticulato-ciliate. *Petals* many times longer than the calyx, pale pinky purple with darker veins, oblongo-obcordate, spreading, claw ciliated at the base. *Tube* of the anthers with stellate pubescence, *anthers* purple, *pollen* globose, hispid. *Fruit* orbicular, depressed, glabrous, minutely granulated.

2. M. *vulgaris*, "Trag." *Dwarf Mallow.* Stem herbaceous, decumbent, with deflexed pubescence, scarcely branched; leaves orbiculari-reniform, in 7 shallow rounded irregularly and acutely serrated lobes; stipules ovato-acuminate; outer sepals linear acute

---

glabrous, inner sepals ovato-acuminate; petals oblong, deeply emarginate, erect, 2-3 times longer than the calyx; tube of stamens with deflexed pubescence; peduncles decurved in fruit; fruit pubescent. " *Trag. Hist. p.* 369." *Fries Nov. 2d ed. p.* 219. *Reich. Fl. Excurs.* 4836. *Drejer Fl. Hafn. n.* 699. M. *rotundifolia*, L. *E. Bot. t.* 1092. *a. E. Fl. v. iii. p.* 246. *a. Hook. Br. Fl. p.* 315.

Waste places and waysides; not very common. *Fl.* June, September. ♃.
Waste places near Westbury; *Mr. F. Dickinson.* Eyton and Walford; *T. C. Eyton, Esq.* Wenlock; *W. P. Brookes, Esq.* Near Newport; *R. G. Higgins, Esq.* Ludlow; *Miss Mc Ghie.*
Coleham, Shrewsbury.
Root tapering. *Stem* decumbent, 1-2 feet long, scarcely branched, round, roughish with short deflexed pubescence. *Leaves* alternate, on very long petioles roughish with erect pubescence, slightly channelled above, orbiculari-reniform, in 7 shallow rounded obtuse irregularly but acutely serrated lobes, plaited, veiny, pubescent on both sides, of a duller darker green than in the preceding, and paler beneath. *Stipules* ovato-acuminate, margins ciliated. *Flowers* about 5, in axillary fascicles, on round peduncles unequal in length but much shorter than the leaves clothed with erect pubescence, erect in flower decurved in fruit. *Outer calyx* of 3 linear acute glabrous leaves ciliated at the margins; *inner calyx* monophyllous 5-fid hairy, segments ovato-acuminate, margins denticulato-ciliate. *Petals* rather more than twice the length of the calyx, white tinged with pink, with darker purple veins, oblong, deeply emarginate or 2-lobed, erect, claw ciliated at the base. *Tube* of the *stamens* with deflexed pubescence; *anthers* yellow; *pollen* white, globose, hispid. *Fruit* orbicular, depressed, minutely granulated, pubescent.
Fries considers that the M. *rotundifolia*, Linn. *Fl. Suec.* comprises two distinct species, *rotundifolia*, L. and *vulgaris*, Trag. and that the true *rotundifolia*, L. is an *annual* plant, the M. *pusilla*, E. Bot. t. 241. M. *rotundifolia* β. E. Fl. l. c. M. *parviflora*, Huds. Angl. 307, non Linn.

3. M. *moschata*, Linn. *Musk Mallow.* Stem herbaceous, erect, rough with horizontal tuberculate hairs, scarcely branched; leaves orbiculari-reniform, in 5—7 pinnatifid lobes; stipules lanceolate, acuminate; outer sepals linear-lanceolate; inner sepals broadly ovate, acute; petals broadly cuneate, deeply emarginate, spreading; tube of stamens with deflexed pubescence; peduncles erect in fruit; fruit hairy. *E. Bot. t.* 754. *E. Fl. v. iii. p.* 247. *Hook. Br. Fl. p.* 315.

Pastures and road-sides; not unfrequent. *Fl.* July, August. ♃.
Muckley Cross between Wenlock and Bridgnorth; *W. P. Brookes, Esq.!* Haughmond Abbey; *Mr. F. Dickinson.* Near Walford; *T. C. Eyton, Esq.* Astley; *Mr. E. Elsmere, junr.* Near Newport; *R. G. Higgins, Esq.* Albrighton near Shiffnal; *Dr. G. Lloyd.* Ludlow; *Miss Mc Ghie.* Near Dowles Brook, Wyre Forest; *Mr. W. G. Perry.*
Root woody. *Stems* 12—18 inches high, numerous, round, slightly branched, copiously covered with long horizontally spreading rough simple hairs arising from tubercles. *Radical leaves* on long petioles hispid and channelled above, rounded or reniform, obtuse, crenate, frequently lobed, soon withering away. *Stem leaves* alternate, on short petioles, orbiculari-reniform, deeply 5—7-lobed, lobes pinnatifid, segments linear acute channelled, nearly smooth above, hairy beneath. *Stipules* 2, lanceolate, acuminate, pubescent, margins ciliated. *Flowers* on short hairy axillary single-flowered peduncles, crowded at the summit of the stem and branches, delicate rose-colour, large and elegant. *Outer calyx* of 3

---

linear-lanceolate acute denticulate and ciliated 3-ribbed leaves, half the length of the interior ones. *Interior calyx* monophyllous, 5-cleft, segments broadly ovate acute, hairy and ciliated. *Petals* 5, broadly cuneate, with a short claw having a row of long white hairs on each side, anterior margin truncate and emarginate, minutely jagged, 9-ribbed, about 3 times longer than the calyx. *Tube* of stamens with deflexed hairs. *Pollen* globose, hispid; *anthers* pink. *Fruit* orbicular, depressed, densely hairy.

4. ALTHÆA. *Linn.* Marsh-mallow.

1. A. *officinalis*, Linn. *Common Marsh-mallow.* Leaves soft and downy on both sides, cordate or ovate, entire or 5-lobed; peduncles axillary, many-flowered, much shorter than the leaves. *E. Bot. t.* 147. *E. Fl. v. iii. p.* 244. *Hook. Br. Fl. p.* 316.

Marshes, rare; doubtfully wild. *Fl.* August, September. ♃.
Near Lutwyche Hall, Wenlock Edge; *W. P. Brookes, Esq.!*
Herbage remarkable for the dense velvety and starry pubescence. *Stem* 2-3 feet high, erect, simple. *Leaves* ovate or cordate at the base, plaited, 5-ribbed, more or less deeply divided into 5 acute lobes, unequally serrated. *Flowers* in very short dense axillary panicles, large and of a beautiful rose-colour.

CLASS XVII.

DIADELPHIA.

—

" 'Twas that delightful season when the broom
Full-flowered and visible on every steep
Along the copses runs in veins of gold."
WORDSWORTH.

CLASS XVII.

DIADELPHIA. *Filaments combined in two sets:—(except in the first division of Ord. iii.)*

ORDER I. HEXANDRIA. 6 *Stamens.*

1. CORYDALIS. *Calyx* of 2 deciduous leaves. *Petals* 4, one of them gibbous or spurred at the base. *Stamens* in 2 fascicles. *Capsule* siliquæform, 2-valved, many-seeded.—*Nat. Ord.* FUMARI-ACEÆ, *DC.*—Name from κορυδαλις, the Greek name for the *Fumitory*, with which the present genus was till lately united.

2. FUMARIA. *Calyx* of 2 deciduous leaves. *Petals* 4, one of them gibbous or spurred at the base. *Filaments* 2, each with 3 *anthers.* *Fruit* indehiscent, 1-seeded, the *style* deciduous.—*Nat. Ord.* FUMARIACEÆ, *DC.*—Name from *fumus, smoke;* on account of the smell emitted by the bruised herbage.

ORDER II. OCTANDRIA. 8 *Stamens.*

3. POLYGALA. *Calyx* of 5 leaves, 2 of them wing-shaped and coloured. *Petals* combined by their claws with the filaments, the lower one keeled. *Capsule* compressed. *Seeds* downy, crested at the hilum.—*Nat. Ord.* POLYGALEÆ, *Juss.*—Name, πολυ, *much,* and γαλα, *milk;* from some fancied property in the plant.

ORDER III. DECANDRIA. 10 *Stamens.*

(All belonging to the *Nat. Ord.* LEGUMINOSÆ, *Juss;* having the fruit a *Legume,* and the flowers *papilionaceous,* with the leaves mostly compound.)

* *Filaments all connected at the base or monadelphous.*

4. ULEX. *Calyx* of 2 leaves, with a small scale or *bractea* on each side at the base. *Legume* turgid, scarcely longer than the calyx.—Name, from *ec* or *ac,* Celtic, *a sharp point.*

5. GENISTA. *Calyx* 2-lipped; upper lip with 2 deep segments, lower one with 3 teeth. *Standard* oblong. *Legume* flat or turgid, many-seeded.—Name from *gen,* Celtic, *a shrub.*

6. SAROTHAMNUS. *Calyx* 2-lipped, lips scariose, upper one bi-, lower tri-dentate. *Style* very long, circinato-convolute, thickened upwards, plane on the interior side. *Stigma* terminal, minute, capi-tate.—Name from σαρος, a *broom* or *besom,* and θαμνος, a *shrub;* allusive to the general habit and appearance of the plant.

7. ONONIS. *Calyx* 5-cleft, segments linear. *Standard* large, striated. *Legume* turgid, sessile, few-seeded.—Name from *ovos,* an *ass;* because the plant is eaten by that animal.

8. ANTHYLLIS. *Calyx* inflated, 5-toothed. *Petals* nearly equal in length. *Legume* oval, 1—3-seeded, enclosed in the permanent calyx.—Name, ανθος, a *flower,* and ιουλος, a *beard* or *down;* from the downy calyx.

** *Stamens diadelphous, 9 united and 1 free.*

† *Style downy beneath the stigma.*

9. OROBUS. *Calyx* 5-fid or 5-dentate. *Stamens* diadelphous, filaments subulate. *Style* linear or dilated upwards, plane on the upper side, hairy beneath the straight or decurved stigma. *Legume* many-seeded.—Leaves *without* tendrils, the petiole terminating in a herbaceous mucro.—Name from *ορω,* to *strengthen* or *invigorate,* and *βους,* an *ox;* from the plants yielding food to cattle.

10. LATHYRUS. *Calyx* with its mouth oblique, its upper segments shortest. *Style* plane above, broader upwards, hairy beneath the stigma. Leaves *with* tendrils.—Name from *λαθυρος,* a leguminose plant of Theophrastus.

11. VICIA. *Calyx* 5-fid or 5-dentate. *Stamens* diadelphous, filaments subulate. *Style* filiform, entirely hairy in the upper part, or bearded on the under side corresponding to the inferior suture. *Legume* 2- or many-seeded.—Name originally derived, according to Theis, from *Gwig,* Celtic; *Wicken,* German; βικιον, Greek; *Vesce,* French; *Vetch,* English.

†† *Style glabrous.*

⊹ *Legume of 2, more or less complete, longitudinal cells.*

12. ASTRAGALUS. *Keel* of the corolla obtuse. *Legume* 2-celled (more or less perfectly); *cells* formed by the inflexed margins of the lower suture.—Name from *αστραγαλος,* the *vertebra;* in allusion to the knotted root of that individual plant to which it was formerly applied.

⊹⊹ *Legume more or less jointed.*

13. ORNITHOPUS. *Legume* compressed, curved, of many close single-seeded joints, whose sides are equal. *Keel* very small.—Name from *ορνις, ορνιθος,* a *bird;* and *πους,* a *foot;* from the simi-larity of the legumes to a bird's-foot.

14. HIPPOCREPIS. *Legume* compressed, submembranaceous, of numerous joints which are curved like a horse-shoe, so that each

legume has many deep notches on one side.—Name from *ιππος,* a *horse,* and *κρηπις,* a *shoe;* from the form of the legume.

⊹ ⊹ ⊹ *Legume of one cell, one or many-seeded, (not formed of many joints.)*

15. ONOBRYCHIS. *Legume* sessile, of one indehiscent joint, compressed, coriaceous, prickly, crested or winged.—Name from *ovos,* an *ass,* and *βρυχω,* to *eat;* the plant affording a valuable fodder.

16. MELILOTUS. *Legume* 1- or few-seeded, indehiscent, longer than the calyx. *Petals* distinct, deciduous. *Flowers racemose.* Leaves *ternate.*—Name from *mel, honey,* and *Lotus,* the genus so called.

17. TRIFOLIUM. *Legume* 1- or more-seeded, indehiscent, shorter than the calyx by which it is enclosed. *Petals* mostly combined by their claws and persistent. *Flowers capitate.* Leaves *ternate.* Name from *tres, three,* and *folium,* a *leaf.*

18. LOTUS. *Legume* cylindrical, somewhat spongy within, and imperfectly many-celled. *Keel* acuminated.—Name, supposed to be one of the three kinds (the *herbaceous)* of the Λωτος of the Greeks.

19. MEDICAGO. *Legume* falcate or spirally twisted.—Name, the *μεδικη* of the Greeks, so called because introduced by the Medes.

DIADELPHIA—HEXANDRIA.

1. CORYDALIS. *DC.* Corydalis.

1. C. *claviculata, DC.* *White climbing Corydalis.* Stem much branched, climbing; leaves pinnate, pinnæ stalked ternate or pe-date, leaflets elliptical acute, petioles ending in tendrils; pedicels very short, scarcely so long as the minute bracteas. *Fumaria clavi-culata, Linn.* E. Bot. t. 103. E. Fl. v. iii. p. 253. Hook. Br. Fl. p. 319.

Shady and wooded sides of sandy hills; not uncommon. *Fl.* May—July. ☉. East side of the Wrekin; *A. Aikin, Esq.* Chesterton mill, Caynton wood; *H. Bidwell, Esq.* Lodge Hill near Frodesley; *W. P. Brookes, Esq.* Wood at the west end of Lyth Hill; *Mr. T. Bodenham.* Upon Frodesley Hill, and near Church Stretton upon the rock on the right hand entering the town from Shrews-bury; *Rev. W. Corbett.* On the Wrekin; *Mr. F. Dickinson.* Harmer Hill; *T. C. Eyton, Esq.* Abundant on Lawrence Hill an eastern satellite to the Wrekin; *E. Lees, Esq.* Longmynd; Treflach woods near Oswestry; *Rev. T. Salwey!* Amongst the loose stones on the S.E. side of the Wrekin near the top and at the N.E. end. Upon the little rock in the wood on Lyth Hill. Amongst the stones at the back of Frodesley Park. On Hawkstone Hills. In Chetwynd Park; *Rev. E. Williams's MSS.* Ketley; *Dr. Du Gard.* Pimhill, west side. Woody base of the Wrekin near the brook between that hill and Lawrence Hill. Craigforda near Oswestry.

Stems delicate and slender, branched, irregularly tetragonous, climbing by the branched *tendrils* which terminate the petioles of the pinnate *leaves*. *Leaflets* elliptical, mucronate, stalked, entire, pedate, of a glaucous hue. *Flowers* in stalked clusters opposite the leaves, dense, small, pale-yellow. *Pods* lanceolate, acute, undulated. *Seeds* 3-4, orbicular, compressed, black, shining.

2. C. *solida*, Hook. *Solid-rooted Corydalis.* Stem simple, erect, with a scale beneath the lower leaf; leaves 3-4, biternate, their leaflets cuneate or oblong, and as well as the bracteas cut; root solid. *Hook. Br. Fl. p.* 319. *Fumaria solida, Linn. MSS. E. Bot. t.* 1471. *E. Fl. v. iii. p.* 252. *Corydalis bulbosa, DC. Fumaria Halleri, Willd.*

Groves and thickets; doubtfully wild. *Fl.* April. May. ♃.

"In a small alder copse adjoining the cascade, about a ¼ of a mile above Cound stank bridge, 1796. The specimens in my herbarium and garden are from this place, but I could not find it there in 1799. Near Hanwood and between Wellington and Wrockwardine." *Rev. E. Williams's MSS.*

Root orbicular, depressed, of several fleshy coats, but not hollow. *Stem* solitary, a little zigzag, angular, a span high. *Leaves* scattered, on channelled petioles, glaucous. *Cluster* terminal, solitary, erect, of 10—15 variegated purplish *flowers*, each with a long ascending blunt spur and a slight pale prominence at the opposite side. *Bracteas* 5-cleft, glaucous, slightly longer than the pedicel. *Sepals* very minute, rounded. *Pod* short. *Seeds* several.

2. FUMARIA. *Linn.* Fumitory.

1. F. *capreolata*, Linn. *Ramping Fumitory.* Fruit globose, more or less apiculate, the length exceeding the breadth; sepals attached above the base, broadly oval, apiculate, toothed, as broad as the corolla and about half its length; pedicels erecto-patulous, more or less decurved; bracteas equal to or one-third shorter than the fructiferous pedicel. *E. Fl. v. iii. p.* 255. *Hook. Br. Fl. 4th ed. p.* 265. γ. *Anglica, Arn. in 3d Ann. Report Bot. Soc. Edin. p.* 106. (excl. some syn.)

Hedges, walls, &c.; frequent. *Fl.* June—August. ⊙.

Hedges about Frankwell, Shrewsbury; *A. Aikin, Esq.* Dorrington pool dam; *H. Bidwell, Esq.* Near Whittington; *Rev. A. Bloxam.* Not uncommon about Wrockwardine and Wellington; *E. Lees, Esq.* About Ludlow; *Miss Mc Ghie.*

About Shrewsbury generally. Bayston Hill.

Stem climbing, several feet in height, or sometimes only diffuse and spreading near the ground, angular, glabrous, somewhat glaucous, branched. *Leaves* alternate, petioles triquetrous, long and twisting, bi-tri-pinnate; leaflets broad, obovato-cuneate, deeply incised into 3 unequal lanceolate obtuse mucronate segments, pale green, with more or less of a glaucous hue. *Flowers* in erect many-flowered lax racemes, on long angular and furrowed peduncles, opposite to and about as long as the leaves. *Pedicels* short, round, erecto-patulous and decurved. *Bractea* minute, lanceolate, acuminate, equal to or about two-thirds the length of the pedicel. *Sepals* broadly oval, apiculate, margin irregularly and variously toothed especially at the base, more or less entire towards the point, half the length of the corolla and equalling it in width. *Corolla* pale pink, tipped with dark crimson, upper and lower petals with a green keel. *Fruit* globose, with a minute obtuse apiculus, the length exceeding the breadth, laterally compressed. *Nut* globose, somewhat truncate at top, laterally compressed, 2-edged, the edge distinct throughout, with a minute elevated point on each side the edge at the base and a round depression on each side at the top confluent in the middle, smooth, the breadth slightly exceeding the length.

A plant not unfrequently occurs around Shrewsbury, very similar in general appearance and habit to the above, but differing from it in the following characters:—Fruit globose, truncate or subretuse, the breadth slightly exceeding the length; sepals attached above the base, broadly oval, apiculate, generally but not invariably toothed to the apex, broader than and nearly two-thirds as long as the corolla; pedicels erecto-patulous, more or less decurved; bracteas about one-third shorter than the fructiferous pedicels; flowers very large, pale with dark crimson tips. This Dr. Arnott recognizes as his *F. capreolata* β *Reichenbachii.* (*l. c.*) But as nearly all Botanists concur in assigning to *F. capreolata* a globose, more or less apiculate fruit, I submit that it would be scarcely correct to refer our plant to that species, notwithstanding the approximation in the form and relative length of the sepals and corolla. Indeed it seems to be an intermediate form between *capreolata* and *officinalis*, approximating to the former in the size of its flowers and relative length of its sepals, and to the latter in the subretuse form of the fruit, from which however the relative length of the sepals and also the form of the fruit keep it sufficiently distinct. It corresponds with *F. capreolata* of *Reich. Icones fig.* 4456, although in the diagnosis of that species in his *Fl. Excurs.* he says "nuculis globosis," whereas his figure represents them as subretuse. It may possibly be the *F. capreolata* intended in Badarro's diagnosis of that species in *De Candolle's Ic. Gall.* 34, cited by Reichenbach in *Fl. Excurs. n.* 4456, and which he states does not agree with the figure in that work, the fructiferous pedicels being "recurvati bracteâ subbreviores, sepala ovata corollæ tubo latiora, toto margine denticulata," whilst in the figure the fructiferous pedicels are represented "arrecto-patentes, bracteâ duplo longiores." Perhaps also *F. capreolata, Koch. Syn. p.* 32. "siliculis subrotundis obtusissimis," and *F. capreolata, Babington Prim. Fl. Sarn. p.* 4, "sepalis latitudine corollam paulo excedentibus, cariopside globoso truncato" should be quoted to this. Mr. Babington, who has also attentively studied this genus, regards this plant as Arnott's β *Reichenbachii*, but not the plant of *Reich. Icones*, which has the pedicels figured as decidedly ascending, not reflexed; and which he considers to be identical with Arnott's γ. *Anglica*. The *capreolata* of his *Primitiæ*, he informs me, appears more nearly allied to Arnott's β than to his γ, being an erect plant, with rather narrow leaves, pedicels reflexed, and sepals serrated nearly to the point. Our plant at all events merits attention and further research.

2. F. *officinalis*, Linn. *Common Fumitory.* Fruit globose, depressed, retuse or subemarginate, the breadth exceeding the length; sepals ovato-lanceolate, acute, incised, 2-3 times shorter and one-half narrower than the corolla; pedicels erecto-patulous; bractea one-half or one-third shorter than the fructiferous pedicel. *Babington Prim. Fl. Sarn. p.* 4. *Reich. Icones, fig.* 4454. *Reich. Fl. Excurs. n.* 4454. *Koch. Syn. p.* 33. *Drej. Fl. Hafn. n.* 703. *E. Fl. v. iii. p.* 254. *Arnott in 3d Ann. Rep. Bot. Soc. Edin. p.* 107. *a. Fries Nov. ed.* 2. *p.* 220. *a. Hook. Br. Fl. 4th ed. p.* 266.

Corn fields and road-sides; common. *Fl.* through the summer. ⊙.

Root tapering. *Stem* a foot or more high, angular, hollow, zigzag, succulent, branched, glaucous green as in the whole herbage. *Leaves* mostly alternate, on triquetrous petioles, bi-tri-pinnate; leaflets very variable in width, obovato-cuneate, deeply incised into about 3 irregular linear or lanceolate mucronate segments. *Flowers* in erect many-flowered lax racemes, on long angular and furrowed peduncles opposite to and as long as the leaves. *Pedicels* short, roundish or obsoletely angular, dilated upwards, erecto-patulous in flower and fruit. *Bracteas* about one-third shorter than the pedicel in flower, about one-half or one-third shorter in fruit. *Sepals* ovato-lanceolate, acute, incised all over, especially at the base, about as wide as or rather wider than the pedicel, 2-3 times shorter and one-half narrower than the corolla. *Corolla* rose-coloured, with dark crimson tips, upper and lower petals with a green keel. *Fruit* globose, depressed, retuse or subemarginate, with a minu e obtuse apiculus, the breadth exceeding the length, laterally compressed. *Nut* subreniform, laterally compressed, 2-edged, the edge at the top indistinct with a round depression on each side confluent in the centre, and a point on each side the edge or ridge at the base, rough, the length one-third less than the breadth.

Dr. Arnott in seperating this species into 3 marked varieties, appears to me to carry his divisions much too far, inasmuch as all his varieties may not unfrequently, I think, be observed in the same piece of ground, varying according to situation, and passing by such insensible gradations into each other, as to render it nearly impracticable, if not impossible, to define with precision their several limitations. If indeed varieties can be distinguished at all, perhaps his varieties β. and γ., which run into each other so imperceptibly, may probably constitute one variety; whilst his variety α. *grandiflora*, which is apparently more distinct, may be found sufficiently constant as to form a second variety. I have great doubts however whether they ought to be seperated at all.

DIADELPHIA—OCTANDRIA.

3. POLYGALA. *Linn.* Milkwort.

1. P. *vulgaris*, Linn. *Common Milkwort.* Stems many, branching, ascending; leaves obtuse, lower ones oblong, upper ones linear-lanceolate; flowers crested, racemose; wings of the calyx elliptical, about as long as the petals; capsules orbiculari-oblong, sessile. *E. Bot. t.* 76. *E. Fl. v. iii. p.* 257. *Hook. Br. Fl. p.* 266.

Dry hilly pastures; not uncommon. *Fl.* June, July. ♃.

Longnor; *Rev. W. Corbett.* Fairyland, Westfelton; *J. F. M. Dovaston, Esq.* Coxwood, Coalbrookdale; *Mr. F. Dickinson.* Walford; *T. C. Eyton, Esq.* Newport; *R. G. Higgins, Esq.* Ludlow; *Miss Mc Ghie.* Near Dowles brook, Wyre forest; *Mr. W. G. Perry.* Near Oswestry; *Rev. T. Salwey.* Whitecliff near Ludlow; *Mr. H. Spare.*

Sharpstones hill. Ebury, Bayston hill. Lyth hill. Haughmond hill.

Root tough and woody. *Stems* numerous, procumbent and ascending, branched from the base, 6—8 inches long, angular, covered with minute incurved hairs. *Leaves* copious, alternate, somewhat fleshy, bright green, clothed on both sides, especially the under one, with minute crystalline dots; lower ones oblong, upper ones linear-lanceolate, obtusely pointed, tapering at the base into a very short footstalk articulated at its connection with the stem, midrib prominent on lower surface, dilated towards the base. *Racemes* terminal, simple, solitary, many-flowered, lax. *Flowers* drooping, blue, pink or white, on short recurved coloured pedicels, with 3 *bracteas* at the base; intermediate bractea oblong, navicular, blunt, about as long as the pedicel, coloured, keel green, margins membranous, falling before the other two which are about half its size and acute, and leaving an elevated tubercle from which the pedicel arises. *Calyx* of 5 leaves, unequal, persistent, coloured; 2 inner ones or *wings* large, elliptical or often broader and more inclined to obovate, obtusely mucronate, attenuated at the base into a short claw, with 3 branching ribs; 3 other sepals about half the length, lanceolate, concave, obtusely pointed, upper one rather the larger, two lower ones united at the base, of the same colour as the corolla, with dark green keel, all distinctly 3-ribbed, the lateral ribs branched. *Petals* 3, about as long as or a little longer than the wings, closely united at the base so as to resemble a single deeply 3-fid one rolled together into a tube; lateral ones linear-oblong, blunt, with a single short rib which branches dichotomously, and a longitudinal line of white hairs on the inside extending as far as to the branching of the rib; intermediate one dissimilar in form, cucullate, geniculate, the edges converging green, margins ciliated, enclosing the anthers, the extremity crested with about 8 linear-obovate deeply bifid obtuse processes folded together. *Germens* obovate, compressed. *Style* compressed, dilated upwards, about as long as the germen; *stigma* 2-lipped, lower lip concave, purple, straight, terminating in a triangular jagged acute point, upper lip glanduliform, green, reflexed, not half the length of the lower. *Capsule* orbiculari-oblong, tapering at the base, compressed, deeply notched at the apex, surrounded by a broadish membranous margin, wavy at the edges, 2-celled. *Seeds* oblong, solitary, pendulous, clothed with soft hairs, with a succulent 3-partite oblong obtuse *arillus* at the base.

DIADELPHIA—DECANDRIA.

4. ULEX. *Linn.* Furze.

1. U. *Europæus*, Linn. *Common Furze, Whin, or Gorse.* Primary spines long, strong, polygonal, deeply furrowed, minutely scabrous in the upper part, leafy to about half their length; leaves all decurved, shaggy beneath; flowers from the primary and secondary spines; calyx shaggy, teeth converging and cohering; bracteas ovate, lax. *E. Bot. t.* 742. *E. Fl. v. iii. p.* 265. *Hook. Br. Fl. p.* 321.

Heathy places. *Fl.* early in spring and throughout the summer. ♄.

Stem 3—5 or 7 feet high, deeply furrowed, hairy or shaggy, branched; branches erect, ending in spines. *Leaves* subulate, mucronate, keeled, fleshy, somewhat rigid, decurved, shaggy on the under surface, bearing in the axil a spinous furrowed branch (primary spine) 2 inches or more in length, spreading at nearly right angles and more or less decurved, shaggy in the lower part as far as the leaves and spines extend, minutely scabrous above, bearing similar alternate decurved spiny leaves, each having in its axil a short angular furrowed rigid spreading sharp spine, (secondary spine) which is either simple or frequently bears a little above the base 2 nearly opposite, or sometimes 3 alternate spinous leaves each with a simple axillary spine. The leaves in the lower portion of the branches become broader and almost ovate and scariose. *Flowers* axillary, springing both from the primary and secondary spines, on a short solitary pubescent peduncle, having at its base an ovate keeled pointed minute externally pubescent *bractea*, and bearing two similar *bracteas* at its summit immediately under and alternate with the sepals. *Sepals* 2, ovate, compressed, tawny-yellow, covered with brown hairs, upper one or that subtending the standard 11-ribbed, with 2 cohering teeth at the apex; lower one with 3 cohering teeth. *Corolla* bright yellow. *Legumes* obliquely oblong, compressed, densely clothed with erect appressed silky hairs. *Seeds* 10, of which about 4 only are perfected, smooth.

It is recorded of Linnæus that during his visit to England he, for the first time in his life, beheld the *Ulex Europæus* in blossom, and immediately, enraptured at the sight, fell on his knees and offered his grateful homage to the Deity who had permitted him this gratifying spectacle. Connected with this plant and the above fact, the following singular and almost ominous coincidence is related in the "Memoirs and Correspondence of the late Sir James Edward Smith," the Linnæus of Britain;—"I became," says Sir James, "at the age of eighteen, desirous to study botany as a science. The only book I could then procure was 'Berkenhout,' 'Hudson's Flora' having become extremely scarce. I received 'Berkenhout' on the 9th of January, 1778, and on the 11th began, with infinite delight, to examine the *Ulex Europæus*, the only plant then in flower. I then first comprehended the nature of systematic arrangement and the Linnæan principles; little aware, that, *at that instant*, the world was losing the great genius who was to be my future guide; for Linnæus died on the night of January 11th, 1778."

2. U. *nanus*, Forst. *Dwarf Furze*. Primary spines short, slender, terete, striated, glabrous in the upper part, leafy at the base only; primary leaves horizontally spreading, curved upwards, glabrous beneath, with long silky hairs on the margin; flowers from the primary spines only; calyx finely pubescent, teeth diverging; bracteas minute, appressed. *E. Bot. t.* 743. *E. Fl. v. iii. p.* 266. *Hook. Br. Fl. p.* 321.

Heathy places; often with the preceding. *Fl.* mostly in autumn. ♄.

Near Whixall moss; *Mr. F. Dickinson.* About Ludlow; *Miss Mc Ghie.* Near Lee bridge; *Dr. T. W. Wilson.*

Haughmond hill. Sharpstones hill. Bickley coppice near Shrewsbury. Shelvock bank and Westfelton near Oswestry.

Smaller than the last in every part and very different in general appearance. *Stem* 1—3 feet high, deeply furrowed, hairy or shaggy, branched; branches erect or procumbent, spreading, ending in spines. *Leaves* triangulari-subulate, mucronate, keeled, fleshy, slightly rigid, spreading nearly horizontally and curved upwards, glabrous beneath, with long silky hairs on the margins, bearing in the axil a short spinous terete striated branch (primary spine) about an inch long, spreading at nearly right angles and decurved, shaggy in the lower part as far as the leaves extend, glabrous above, bearing at its base similar alternate somewhat decurved spiny leaves, each having in its axil a short angular striated rigid sharp spine, (secondary spine) which bears at the base 2 opposite spinous spreading leaves each with a simple axillary spine, and sometimes but not constantly one two or more erecto-patent spiny leaves above without axillary spines, which are also observable on the primary spines. *Flowers* axillary, springing from the primary spines only, on a short solitary peduncle with appressed pubescence, having at its base a minute appressed pubescent *bractea*, and bearing 2 similar *bracteas* at its summit immediately under and alternate with the sepals. *Sepals* 2, ovate, compressed, yellow approaching to the colour of the petals, covered with short appressed pubescence, upper one or that subtending the standard 11-ribbed, with 2 diverging teeth at the apex; lower one with 3 diverging teeth. *Corolla* pale yellow. *Legumes* shorter than in the last, obliquely oblong, compressed, clothed with erect appressed silky hairs. *Seeds* 5, seldom all perfected, smooth.

5. GENISTA. *Linn.* Green-weed.

1. G. *tinctoria*, Linn. *Dyer's Green-weed, Woad-Waxen.* Stems depressed, unarmed; leaves elliptico-lanceolate, acute, ciliated; peduncles and upper part of calyx hairy; calycine teeth equal, 3 lower ones connivent, finally decurved; corolla glabrous; legumes nearly cylindrical, glabrous. *E. Bot. t.* 44. *E. Fl. v. iii. p.* 262. *Hook. Br. Fl. p.* 321.

Pastures and borders of fields; not uncommon. *Fl.* July, August. ♄.

About Wem; *A. Aikin, Esq.* In a field at Round Acton mill abundantly. In a large field at Rowley near Harley. Frequent round Wenlock; *W. P. Brookes, Esq.* Near Wrentnall; *Mr. T. Bodenham.* Buckley farm near Oswestry; *J. F. M. Dovaston, Esq.* Fields near Broseley hill; *Mr. F. Dickinson.* Walford; *T. C. Eyton, Esq.* Benthall Edge; *Dr. G. Lloyd.* Ludlow; *Miss Mc*

Ghie. Whitecliff near Ludlow; *Mr. H. Spare.* Brandwood near Burlton; *W. W. Watkins, Esq.*

Sundorn. Cardiston. Base of Middleton hill. Hedge-banks near Longnor. Banks of a pit between Battlefield and Albright Hussee.

Root woody, creeping. *Stems* 1—2 feet high, depressed, with numerous ascending branches, straight, furrowed, smooth or more or less hairy especially when young. *Leaves* scattered, nearly sessile, elliptic-lanceolate, acute, the midrib and edges hairy, both surfaces covered with minute crystalline points. *Stipules* very small, setaceous. *Flowers* solitary, axillary, at the extremities of the branches, on short hairy *peduncles* bearing a pair of small subulate ciliated *bracteas* immediately below the flower. *Calyx* tubular, angular, with erect hairs especially on the teeth: teeth about equal in length, 3 lower ones subulate, connivent, finally decurved; 2 upper ones larger, subulate from a triangular base, straight. *Corolla* bright yellow; *standard* ovate, obtuse, margins incurved, erect; *wings* and *keel* finally deflexed. Above the base of the keel on each side is a gibbous protuberance. *Legume* nearly cylindrical, glabrous. *Seeds* many.

The whole plant yields a good yellow dye.

2. G. *Anglica*, Linn. *Needle Green-weed or Petty Whin.* Stems ascending, spiny; leaves ovato-lanceolate, mucronate, glabrous; peduncles and calyx glabrous; upper calycine teeth shorter, 3 lower ones straight; corolla glabrous; legumes oblong, turgid, glabrous. *E. Bot. t.* 132. *E. Fl. v. iii. p.* 263. *Hook. Br. Fl. p.* 322.

Moist heaths and moory ground; not common. *Fl.* June. ♄.

Whitecliff near Ludlow; *Mr. H. Spare.* Faintree; *Purton's Midl. Flora.* Tinkers hill near Ludlow; *Miss Mc Ghie.* Bridges, western end of Longmynd; *Rev. S. P. Mansel.* Borders of Wyre forest; *E. Lees, Esq.* Shawbury heath; *Mr. E. Elsmere, jun.* Base of the Stiperstones hill; *Mr. F. Dickinson.* Twyford Vownog near Westfelton; *J. F. M. Dovaston, Esq.* Caynton; *H. Bidwell, Esq.* In a pasture adjoining the west side of Berrington. On Harmer moss. In Pitchford Park. About the old coal-pits between Pitchford park and Frodesley; *Rev. E. Williams's MSS.*

Boggy ground north of Bomere pool.

Root woody, creeping. *Stem* 1 foot high, ascending, round, spiny, with short branches. *Leaves* scattered, ovato-lanceolate, entire, uppermost broader, mucronate, smooth, shortly petiolate, deciduous. *Flowers* solitary, axillary in the upper leaves, on short smooth erect *peduncles*, bright lemon-colour. *Calyx* smooth, 2-lipped; *upper lip* rather shorter than the lower one, of 2 deeply incised ovate acute segments; *lower lip* of 3 shallow triangular acute segments, all straight keeled and ciliated. *Standard* elliptical, slightly emarginate, revolute; *wings* inflated and concave, about as long as the standard; *keel* half as long again. *Legumes* oblong, turgid, smooth, crowned with the subulate recurved base of the *style.*

6. SAROTHAMNUS. *Wimm.* Broom.

1. S. *scoparius*, "Wimm." *Common Broom.* Branches angular, glabrous; leaves ternate, petiolate, upper ones simple, leaflets obovate; flowers axillary, shortly pedicellate; legumes hairy at the margin. "*Wimm. Fl. v. Schles. p.* 278." *Koch. Syn. p.* 152. *Cytisus scoparius, DC. Hook. Br. Fl. p.* 322. *Spartium scoparium, Linn. E. Bot. t.* 1339. *E. Fl. v. iii. p.* 260. *Genista scoparia, Lam.*

Dry hills and bushy places; frequent. *Fl.* May, June. ♄.

Shrub 3—6 feet or more high. *Branches* long, straight, green. *Leaves* deciduous, *leaflets* obovate, acute, silky when young. *Stipules* at the base of the young branches, obovate. *Flowers* solitary or in pairs, large, handsome, bright golden yellow; *keel* broad; *standard* and *wings* much spreading. *Legumes* large, compressed, dark brown, smooth at the sides, with long silky hairs along the sutures. *Seeds* several, rotundo-reniform, crested.

A decoction of the young green and very bitter tops is said to be powerfully purgative and diuretic.

7. ONONIS. *Linn.* Rest-harrow.

1. O. *arvensis*, Linn. *Common Rest-harrow.* Stems procumbent, glanduloso-pilose; lower leaves trifoliate, leaflets oblong, truncate, serrated in the upper part, subcuneate and entire below; flowers solitary, axillary; legumes shorter than the calyx, 2-seeded. *E. Bot. Suppl. t.* 2659. *O. arvensis a. Sm. E. Fl. v. iii. p.* 266. *O. repens, Linn. Reich. Fl. Excurs.* 3329. (except that the leaflets are not subrotund as there described.) *O. repens L. β. Koch. Syn.*

Barren pastures and borders of fields; common. *Fl.* June—August. ♃.

Root woody. *Stems* procumbent, ascending at the extremities, 8—16 inches long, more or less branched, round, one side clothed with short glandular hairs arising from tubercular bases, and the other or opposite side with longer soft silky spreading and decurved hairs destitute of glands, the hairy and glandular sides alternating between each node. *Leaves* alternate, nearly sessile, lower ones with 3 *leaflets*, upper ones with one, oblong, truncate at the apex, somewhat cuneate at the base, acutely serrated in the upper part, nearly entire in the lower, the apex with 3—5 teeth or serratures, of which the central one terminating the midrib is usually shorter than the rest and frequently slightly decurved, clothed on both sides with soft glandular hairs arising from tubercles. *Stipules* semiamplexicaul, adnate to the short petiole, combined, broadly ovate, acute, glanduloso-pilose, acutely serrated. *Flowers* solitary, axillary, on very short glanduloso-pilose peduncles. *Calyx* about half the length of the standard, tubular, striated, deeply 5-cleft below the middle, segments linear-lanceolate acute slightly curved upwards, clothed with short glandular pubescence and longer silky spreading hairs; 4 superior ones subtending the standard, of which the 2 central ones are close together and the 2 exterior ones somewhat divergent; the inferior or 5th segment subtending the keel. *Standard* much longer than the wings and keel, subrotund, entire, rose-coloured with darker streaks, clothed on the exterior with minute glandular pubescence; *wings* white, glabrous, oblongo-obovate, obtuse, claw unequally bifid, with a subulate acute process proceeding transversely from the interior side of the upper or shorter segments of the claw inserted into a corresponding fold on each side of the keel; *keel* rostrate, acuminate, white tinged with rose-colour, glabrous. *Legume* rather shorter than the calyx, obliquely oval, compressed, with a decurved apex, clothed with glandular pubescence and longer silky hairs. *Seeds* 2, orbiculari-reniform, granulated.

Var. β. spinosa. Branches spinous. *O. arvensis γ. E. Fl. l. c. O. repens a. Koch. l. c.*

Pasture between the New Inn and river Severn, near Shrewsbury. Pastures between Abery Wood and Shawbury heath; *Rev. E. Williams's MSS.*

8. ANTHYLLIS. *Linn.* Kidney-vetch.

1. A. *vulneraria*, Linn. *Common Kidney-vetch, or Lady's fingers.* Silky; leaves pinnate, leaflets unequal; legume semi-orbicular, on a

long peduncle, superior suture arched outwards, 1-seeded. *E. Bot. t.* 104. *E. Fl. v. iii. p.* 269. *Hook. Br. Fl. p.* 322.

Limestone pastures. *Fl.* June—August. ♃.

On all the limestone soils, especially Wenlock Edge; *A. Aikin, Esq.* Gleeton hill near Wenlock. In a field close to the Standhill coppice, Wenlock; *W. P. Brookes, Esq.* About Ellesmere; *Rev. A. Bloxam.* Llanymynech hill; *J. F. M. Dovaston, Esq.* Benthall Edge; *Mr. F. Dickinson.* Ludlow; *Miss Mc Ghie.* Pastures upon the Hill between Westhope, parish of Diddlebury, and Strefford parish of Wistanstow. Pastures above Stoke St. Milburgh; *Rev. E. Williams's MSS.*

Pastures at the Hill Top, Wenlock Edge.

Root somewhat woody, fibrous. *Stem* 1 foot high, several, ascending, round, hollow, clothed with erect close-pressed silky hairs, alternately branched. *Leaves* alternate, on a channelled silky petiole dilated at the base, unequally pinnate; *leaflets* about 8 pairs with a larger terminal one, gradually decreasing in size towards the base of the petiole, elliptical or lanceolate, submucronate, thickened and ciliated at the margin, glabrous above, with silky close-pressed hairs beneath. *Stipules* none. *Inflorescence* in terminal pairs of dense roundish many-flowered heads, subtended by a large sessile palmato-fid leaf. *Peduncles* very short, hairy, with a minute coloured *bractea* at the base. *Calyx* inflated, white below, purplish above, with erect silky hairs; 2 upper teeth very large, longer than the rest, converging, the sinus very small; 2 lateral ones triangular, acuminate; lowermost subulate. *Corolla* yellow, a little longer than the calyx. *Legume* on a long peduncle, semiorbicular, compressed, veiny, superior suture arched outwards, 1-seeded.

9. OROBUS. *Linn.* Bitter-vetch.

1. O. *tuberosus*, Linn. *Tuberous Bitter-vetch.* Stem simple, erect, winged, glabrous; leaflets 2-3 pairs, elliptical-oblong, mucronate, glabrous; stipules semisagittate, toothed at the base, glabrous; calyx tubular, angular, gibbous, glabrous, teeth unequal, 2 upper ones short triangular acute converging, sinus rounded, the rest longer triangular subulate diverging; legumes cylindrical. *E. Bot. t.* 1153. *E. Fl. v. iii. p.* 271. *Hook. Br. Fl. p.* 323.

Hilly pastures, thickets and woods; not uncommon. *Fl.* May, June. ♃.

Badger Quarry walk; *H. Bidwell, Esq.* In Shinewood, near Wenlock; *W. P. Brookes, Esq.!* Cox wood, Coalbrookdale; *Mr. F. Dickinson.* Leaton shelf and near Walford; *T. C. Eyton, Esq.* Astley; *Mr. E. Elsmere, junr.!* Wyre forest, abundant; *Mr. G. Jorden.* Ludlow; *Miss Mc Ghie.* Sundorn, Belswardine, &c.; *Rev. E. Williams's MSS.*

Somer wood. Sharpstones hill. Bomere pool. Bickley coppice. Under Cross hill near Shrewsbury. Bagley coppice near Battlefield. Whitecliff coppice near Ludlow.

Root creeping, with oblong tubers. *Stems* 1 foot high, trigonous, the 2 opposite angles winged. *Leaves* alternate, the dilated petiole projecting beyond the leaflets into a setaceous appendage; *leaflets* about 2 pairs, elliptical-oblong, opposite, sessile, glabrous, mucronate, with 3 longitudinal ribs. *Flowers* in long-stalked axillary racemes, purplish, elegantly veined. *Calyx* tubular, angular, glabrous, dark purple, somewhat gibbous on the upper side at the base; 2 upper teeth short, triangular, acute, somewhat converging, the sinus between them rounded; 3 lower ones much longer, triangulari-subulate, diverging. *Standard* obcordate, reflexed. *Legume* long, pendulous, cylindrical, black.

Var. β. tenuifolius, differs only in the leaflets being narrow and linear. *O. tenuifolius, Roth. D. Don.* Sharpstones hill near Shrewsbury, sparingly.

10. LATHYRUS. *Linn.* Vetchling and Everlasting Pea.

1. L. *pratensis*, Linn. *Meadow Vetchling.* Stem tetragonous, angles acute; leaflets lanceolate, mucronate, silky on the margins; tendril simple or divided; stipules as large as the leaflets, unequally sagittate, acuminate; peduncle tetragonous, many-flowered; calyx striated, 2 upper teeth shorter subulate acuminate crossing in convergence, the rest rather longer similar and straight; legumes nearly erect; seeds 7-8, orbicular. *E. Bot. t.* 670. *E. Fl. v. iii. p.* 276. *Hook. Br. Fl. p.* 324. *Drej. Fl. Hafn.* 747. *Koch. Syn. p.* 201. *Peterm. Fl. Lips. L. sepium, Scop. Reich. Fl. Excurs.* 3454.

Moist meadows and pastures; common. *Fl.* July, August. ♃.

Stem procumbent or climbing by its tendrils, branched, tetragonous, angles acute, smooth or pubescent. *Leaves* alternate, on a triquetrous petiole channelled above, bearing at its summit two opposite sessile lanceolate mucronate *leaflets*, prominently 3-ribbed, silky on the margins and ribs. *Tendril* simple, sometimes divided. *Stipules* very large, unequally sagittate, acuminate. *Peduncles* axillary, very long, tetragonous, pubescent. *Flowers* racemose, drooping to one side, *pedicels* short and silky, with a setaceous *bractea* at the base. *Calyx* tubular, angular, striated, with erect silky hairs on the angles, in other respects glabrous; 2 upper teeth shorter than the others, subulate, acuminate, crossing in convergence, the sinus rounded; 3 lower teeth rather longer, similar and straight. *Corolla* bright yellow; *standard* ovate, obtuse, entire, reflexed at the margin. *Legume* nearly erect, pubescent, compressed. *Seeds* 7-8, orbicular, smooth and shining, tawny with purple spots.

2. L. *sylvestris*, Linn. *Narrow-leaved Everlasting Pea.* Stem tetragonous, opposite angles broadly winged; leaflets linear-lanceolate, acuminate, glabrous; tendril branched; stipules very narrow, semisagittate, acuminate, incurved; peduncles pentangular, many-flowered; calyx 5-ribbed, 2 upper teeth very short triangular acuminate converging, the rest twice as long triangulari-subulate diverging; legumes long, drooping; seeds numerous, transversely oval. *E. Bot. t.* 805. *E. Fl. v. iii. p.* 276. *Hook. Br. Fl. p.* 324.

Thickets and waste places; not common. *Fl.* July, August. ♃.

Huck's barn, near Ludlow; and by the side of the old turnpike road between Ludlow and Tenbury, 1 mile from Ludlow; *Rev. T. Salwey.* Ludlow; *Miss Mc Ghie.* Cause castle near Westbury; *Mr. F. Dickinson.*

Plantations on banks of Severn and Canal between Shrewsbury and Uffington. Banks of Severn near White-horse fields, Shrewsbury. Shelton rough. Banks of Severn between Preston-boats and Uffington ferry.

Root creeping. *Stem* zigzag, branched, climbing to the height of 5 or 6 feet, tetragonous, the two opposite angles with broad dilated wings scabrous at the edges, contracted at the joints, hollow, glabrous. *Leaves* alternate, on a trigonous petiole, margin winged below the insertion of the leaflets, scabrous at the edges, elongated beyond into a triquetrous process from the extremity of which proceed three simple or branched divaricating *tendrils*. *Leaflets* a single opposite pair, tuberculately articulated below the middle of the petiole, spreading nearly at right angles, linear-lanceolate, acuminate, glabrous, deep green, somewhat glaucous, veiny and paler beneath, with 3 principal prominent ribs at the base. *Stipules* very narrow, half sagittate, acuminate and curved inwards. *Peduncles* laterally axillary, considerably longer than the leaves, pentangular, bearing at the extremity about

six drooping secund *flowers*. *Pedicels* very short, round, decurved, with a setaceous *bractea* about their own length at the base. *Calyx* short, tubular, angular, 5-ribbed, teeth unequal; 2 upper ones very short triangular acuminate converging, the sinus rounded; 3 lower ones twice as long, triangulari-subulate diverging, the sinus large and rounded. *Standard* variegated with pale rose-colour and yellowish-green, with darker red veins and a red stripe round the margin, reflexed, transversely dilated, notched; *wings* of a deep red violet, *keel* pale yellowish-green. *Legume* long, compressed, reticulated with veins, ventral suture winged, with a short stout decurved point at the apex. *Seeds* numerous, transversely oval, compressed, smooth.

Miss Mc Ghie informs me that *Lathyrus latifolius, L.* grows in a field near the road between Ludford and Ashford Bowdler, but not having seen specimens, I do not venture to insert it, it being in all probability only a broad-leaved variety of *sylvestris.*

11. VICIA. *Linn.* Vetch and Tare.

* *Peduncles elongated, few-flowered. Calyx not gibbous at the base on the upper side.*

1. V. *hirsuta*, Koch. *Hairy Tare.* Leaflets linear-oblong, truncate, mucronate; stipules 2-lobed, outer lobe 3-fid; peduncles many-flowered; standard reflexed, entire; calyx teeth equal, linear-subulate, 2 upper ones converging; legume broad, oblong, hairy; style short, curved upwards; seeds 2, orbicular, compressed, shining. *Koch Syn. p.* 191. *Drej. Fl. Hafn. n.* 740. *Bab. Prim. Fl. Sarn. p.* 30. *Ervum hirsutum, Linn. E. Bot. t.* 971. *E. Fl. v. iii. p.* 289. *Hook. Br. Fl. p.* 326.

Corn-fields and hedges; frequent. *Fl.* June. ⊙.

Root fibrous. *Stem* 2-3 feet long, tetragonous, winged and furrowed, slightly pubescent or with a few scattered stellate hairs, weak, straggling and climbing, branched. *Leaves* alternate, on a channelled hairy petiole terminating in a branched *tendril*, pinnate; *leaflets* about 7-8 pairs, alternate, on very short hairy stalks, linear-oblong, truncate, the midrib running out into a strong and stout acute mucro, deflexed, smooth or nearly so, dark green and glandularly mealy above, paler beneath. *Stipules* 2-lobed, one lobe lanceolate acute hairy beneath, the other lobe in 3 setaceous divaricating glabrous segments. *Peduncles* axillary, angular and furrowed, hairy, rather shorter than the leaves. *Flowers* in 2 pairs and an odd terminal one on short decurved pubescent *pedicels*, the peduncle extending beyond the terminal flower into a short mucro. *Calyx* with erect appressed hairs, segments equal linear-subulate acute, 2 upper ones converging, more than half the length of the corolla. *Corolla* very small, pale bluish-white, standard reflexed entire, with 2 dark blue spots on the keels. *Legume* short and broad, oblong, hairy, tipped with the short *style* which proceeds from the straightish upper suture and is curved upwards. *Seeds* 2, orbicular, compressed, bright shining red, perfectly smooth.

2. V. *tetrasperma*, Koch. *Smooth Tare.* Leaflets linear-oblong, obtuse, mucronate; stipules half-arrowshaped; peduncles 2-flowered; standard collapsed, emarginate; calyx-teeth unequal, triangulari-subulate, 2 upper ones shorter diverging; legume linear-oblong, glabrous, stipitate; style long, deflexed; seeds 4, globose, dull. *Koch Syn. p.* 191. *Drej. Fl. Hafn. n.* 741. *Bab. Prim. Fl. Sarn. p.* 30. *Ervum tetraspermum, Linn. E. Bot. t.* 1223. *E. Fl. v. iii. p.* 288. *Hook. Br. Fl. p.* 327.

Borders of fields, hedges, &c; not uncommon. *Fl.* June. ⊙.

On a bank at Patton by the roadside from Wenlock to Munslow. Hedge-bank near Wenlock on the road to Burton!; *W. P. Brookes, Esq.* Near Rushbury; *Mr. F. Dickinson.* Lane between Buildwas and Little Wenlock; *Dr. G. Lloyd.* About Wellington; *E. Lees, Esq.* Ludlow; *Miss Mc Ghie.* Whitecliff near Ludlow; *Mr. H. Spare.*

Hedges of fields near Berrington pool. Woods beyond Haughmond hill. Bomere. Lane leading from Hanwood road to Ascott. Near the confluence of rivers Corve and Teme, Ludlow.

Root fibrous. *Stem* 2 feet long, tetragonous, winged, compressed, with incurved erect hairs, weak, slender and straggling, branched from the base. *Leaves* alternate, on a channelled hairy petiole terminating in a branched *tendril*, pinnate; *leaflets* 5-6 pairs, alternate, on very short hairy stalks, linear-oblong, obtuse, not truncated, the midrib running out into a strong and stout acute mucro, spreading or incurved, dark green and glandularly mealy above, paler and hairy beneath. *Stipules* half arrowshaped, hairy. *Peduncle* axillary, very slender, angular and furrowed, with scattered hairs, rather shorter than the leaves. *Flowers* 2 not in a pair, on short decurved pubescent *pedicels*. *Calyx* with erect appressed hairs, teeth unequal, subulate from a triangular base, acute; 2 upper ones broader and shorter, diverging, one-third the length of the corolla. *Corolla* pale bluish-white, standard emarginate collapsed, with blue streaks and 2 dark blue spots on the keel. *Legume* linear-oblong, glabrous, porrect or stipitate at the base, tipped with the long deflexed *style* which proceeds from the middle of the apex of the legume. *Seeds* 4 rarely 5, globose, dull olive brown, scarcely rough.

** *Peduncles elongated, many-flowered. Calyx gibbous at the base on the upper side.*

3. V. *sylvatica*, Linn. *Wood Vetch.* Leaflets about 8 pairs, elliptic or elliptico-oblong, slightly concave, emarginate, mucronate, glabrous; stipules lunate, revolute, 2-lobed, deeply toothed at the base; tendril branched; peduncles many-flowered, longer than the leaves; calyx glabrous, teeth triangulari-subulate, 2 upper ones shorter, nearly straight, the rest somewhat divergent; legume rough, ventral suture curved upwards. *E. Bot. t.* 79. *E. Fl. v. iii. p.* 279. *Hook. Br. Fl. p.* 325.

Shady and bushy places; not common. *Fl.* July, August. ♃.

Whitecliff coppice, woods above Oakley park, and other woods near Ludlow, always on the side facing the north; *Rev. T. Salwey.* Faintree; *Purton's Middl. Flora.* Buildwas; *Miss H. Moseley.* In a beautiful wooded glen called the Jiggers Bank, Coalbrookdale. In the Limekiln woods near Wellington; *E. Lees, Esq.* Dowle, Wyre forest, plentiful; *Mr. G. Jorden.* Near Lutwyche hall and Longville hill, Wenlock Edge; *Mr. F. Dickinson.* Berwick wood near Shrewsbury; *T. and D. Bot. Guide.* Near Bridgnorth; *H. Bidwell, Esq.* In Shortwood adjoining the Arcoll hill at the north end of the Wrekin; *Rev. E. Williams's MSS.*

Shelton Rough near Shrewsbury. Hedges near Stokesay.

Root creeping. *Stem* zigzag, copiously branched, spreading widely and climbing to the height of 5—6 feet or more, round, compressed, the opposite sides narrowly winged, the intermediate faces deeply furrowed, with 3 prominent ribs, green stained with purple, glabrous except a slight pubescence about the joints. *Leaves* alternate, deflexed, pinnate, petiole pentangular deeply channelled above; *leaflets* about 8 pairs, alternate, on very minute petioles, elliptical or elliptico-oblong, slightly concave, the margins thickened and curved upwards, apex

emarginate, the midrib elongated into a short acute deflexed mucro, dark green. *Flowers* cream-coloured, streaked with bluish veins.

Decidedly one of the most elegant of our indigenous plants and richly deserving a place in our gardens, but unfortunately morbidly impatient of removal from its native habit.

4. V. *Orobus*, DC. *Wood Bitter-vetch.* Stem branched, recumbent, angular, hairy; leaflets 7—10 pairs, ovato-oblong or ovato-lanceolate, mucronate, hairy; stipules semisagittate, hairy; calyx tubular, teeth much shorter than the tube hairy, 2 upper ones short triangular converging, the rest longer subulate straight; legume ovato-oblong. *Koch. Syn. p.* 193. *Drej. Fl. Hafn. n.* 742. *Orobus sylvaticus, Linn. E. Bot. t.* 518. *E. Fl. v. iii. p.* 271. *Hook. Br. Fl. p.* 322.

Woods in the limestone dictrict; rare. *Fl.* May, June. ♃.

Whitecliff, near Ludlow; *Mr. H. Spare.*

Root creeping, woody. *Stems* numerous, spreading or recumbent, 1-2 feet long, branched, angular, with silky hairs. *Leaves* of 7—10 pairs of ovato-oblong or ovato-lanceolate *leaflets* mucronate, glabrous above, thickened and ciliated at the margins, veiny beneath, frequently with silky hairs, which are also scattered on the upper surface, articulated by very short petioles to a common stalk which is channelled above and hairy and elongated into a short straight process. *Stipules* semisagittate, hairy. *Clusters* of many secund flowers, crowded at the extremity of a hairy peduncle proceeding from the axil of and as long as the leaves; *pedicels* drooping, short, hairy, with a minute *bractea* at the base. *Calyx* tubular, teeth much shorter than the tube, more or less hairy; 2 upper teeth shorter, triangular, converging; 3 lower ones subulate, straight. *Corolla* cream-coloured, streaked and tipped with purple. *Legume* ovato-oblong, smooth, compressed. *Seeds* 1-3, smooth.

5. V. *Cracca*, Linn. *Tufted Vetch.* Leaflets about 10 pairs lanceolate, mucronate, silky; stipules semisagittate, silky; flowers imbricated; calyx tubular, 2 upper teeth minute triangular nearly glabrous, sinus broad and rounded, 3 lower ones considerably longer densely silky and ciliated, 2 lateral ones triangulari-subulate, central one longest narrower subulate; legume linear-oblong. *E. Bot. t.* 1168. *E. Fl. v. iii. p.* 280. *Hook. Br. Fl. p.* 325.

Moist hedges, &c.; frequent. *Fl.* July, August. ♃.

Stem weak, straggling, decumbent, scarcely branched, climbing by its tendrils to 3 or 4 feet or more in height, angular, furrowed, clothed with minute decurved pubescence. *Leaves* alternate, of about 10 pairs of lanceolate, mucronate, silky *leaflets* shortly petiolate on a channelled common stalk, which is elongated into a branched tendril. *Stipules* half arrowshaped, silky. *Flowers* beautifully variegated with crimson and purple, in long imbricated clusters, shortly pedicellate on an angular pubescent axillary *peduncle*. *Standard* bilobed at the apex, lobes rounded. *Calyx* with silky appressed hairs, tubular, purple, 5-cleft, 2 upper teeth minute, triangular, nearly glabrous, the sinus very broad and rounded, 3 lower ones very considerably longer, of which the 2 lateral ones are triangulari-subulate, the central one narrower, subulate, and rather the longest, densely silky and ciliated. *Legume* smooth.

*** *Flowers axillary, mostly subsessile.*

† *Calyx gibbous at the base on the upper side.*

6. V. *sativa*, Linn. *Common Vetch.* Leaflets elliptic-oblong

or obcordato-elliptical, truncated and retuse, mucronate; flowers solitary or in pairs, nearly sessile; calyx angular and tubular, gibbous at the base on the upper side, hairy, tube and teeth very long, nearly equal; teeth subulate, equal and straight; stipules half-arrowshaped, toothed, with a depressed spot at the back; legume erect, linear-lanceolate, silky, style decurved from the ventral suture; seeds smooth. *E. Bot. t.* 234.

Hedge banks and borders of fields; common. *Fl.* June. ⊙.

Root tapering, fibrous. *Stem* 1-2 feet high, straggling and climbing by the tendrils, branched from the very base, tetragonous and winged, clothed with short silky deflexed hairs. *Leaves* alternate, petiole deeply channelled above, clothed as are the leaves with copious silky hairs, considerably elongated at its extremity beyond the last leaflets, and sending forth from its apex from the very same point 3 *tendrils* which are either simple or branched and callously tipped. *Leaflets* 3—5 or 6 pairs, opposite or alternate, elliptico-oblong or obcordato-elliptical, the lower ones broader and more obcordate, retuse, the midrib elongated into a decurved subulate mucro. *Stipules* half-arrowshaped or rather 2-lobed, the outer or lower lobe the larger, strongly but variously acutely toothed, the upper lobe entire, bristle-pointed, ciliated, marked at the base with a dark depression. *Flowers* solitary or in pairs, axillary, on very short hairy peduncles, variegated with purplish-crimson, blue and white. *Calyx* tubular, pentangular, having 5 strong ribs and intermediate ones usually much branched, clothed with erect scattered silky hairs, tube long gibbous at the base on the upper or standard side, teeth subulate tapering acute nearly as long as the tube, all equal and straight, margins ciliated. *Legume* erect, linear-lanceolate, 1¼ inch long, flattish, downy; *style* long, decurved from the ventral suture. *Seeds* about 10, orbicular, rather compressed, smooth.

Varieties occur in which the leaves become narrower and almost linear, but still in other respects adhering to the above characters.

Var. β. angustifolia. Leaflets linear-lanceolate, truncate and retuse, mucronate, lower ones obcordate; flowers solitary or in pairs, nearly sessile. *E. Bot. Suppl. t.* 2614.

Hedge banks; frequent. *Fl.* June. ⊙.

Neach Hill, near Shiffnal; *Dr. G. Lloyd.* Ashford Carbonel; *Miss Mc Ghie.* Upper Berwick; Leaton Knolls; Cross Hill; Reabrook; Sutton; and generally near Shrewsbury.

Var. β. Bobartii. Leaflets linear, truncate and retuse, mucronate, lower ones obcordate; flowers solitary, nearly sessile. *E. Bot. Suppl. t,* 2708.

Dry gravelly pastures; not common. *Fl.* June. ⊙.

Ebury, Bayston Hill, near Shrewsbury.

A comparison of many specimens of what appeared to me to be intended by the above plants, shews them to be connected with each other and with the common form of *sativa* by such intermediate and insensible gradations, that I cannot withhold, whether correctly or not, from considering them as modifications of the same species. *Var. β.* usually grows on hedge-banks and road-sides, and appears to owe the slenderness of its habit, and the narrowness of its leaves, to its being drawn up by the denseness of the surrounding herbage. *Var. γ.* I have observed only in one locality, on the dry gravelly slopes of the ditches of the ancient encampment on Bayston Hill, where the pasture is short and close. It here assumes a short and straggling mode of growth, being about a span in length, and scarcely rising above the herbage, the leaves are narrower, the lower ones quite obcordate, the stipules less copiously toothed, and the flowers uniformly solitary, but a specimen or two may be occasionally found, which growing amid

the dense herbage usually congregated around a bramble or other "shrogge," supports itself thereon and becomes more elongated and taller in growth, the upper leaves becoming broader and more approaching to *var. β.* and the flowers occasionally appearing in pairs. In all other respects, all the forms agree with the characters described at length under *var. α.* Our *var. γ.* differs from the figure in *E. Bot. Suppl. t.* 2708. in having the hairs on the stem *deflexed* and not *erect*, as represented in that work. The proportional length of the teeth and tube of the calyx is slightly variable even on the same specimen, the teeth being sometimes somewhat shorter than the tube, though usually equal. The intermediate ribs are generally branched and the upper side of the tube is invariably gibbous at the base. These appearances accompany all the varieties. I do not presume to meddle with the synonymy.

7. V. *sepium,* Linn. *Bush Vetch.* Leaflets ovato-oblong, obtuse, emarginate, mucronate, gradually smaller upwards on the petiole; clusters 4—6-flowered, very much shorter than the leaves; calyx angular hairy, teeth short unequal triangular, 2 upper ones curved at the point converging, the rest straight acute; legume erect, glabrous. *E. Bot. t.* 79. *E. Fl. v. iii. p.* 286. *Hook. Br. Fl. p.* 326.

Woods and shady places; frequent. *Fl.* June, July. ♃.

Stem 1½-2 feet high, tetragonous, winged, smooth. *Leaves* alternate, petioles channelled and hairy above, elongated into a branched *tendril*, pinnate; *leaflets* about 5—7 pairs on very short stalks, nearly sessile, alternate, ovato-oblong, obtuse, emarginate, the midrib elongated into an acute decurved mucro in the sinus, hairy on both sides and with double rows of ciliæ on the margins, dark green above, paler and veiny beneath, gradually smaller upwards on the petiole; the leaflets of the lowermost leaves are short, subrotund, deeply emarginate at the apex, with an acute mucro in the sinus and on very short partial stalks. *Stipules* half arrowshaped, undivided or two-lobed, the outer lobe acutely toothed, the inner one marked with a round pale impression. *Flowers* in very short axillary 4—6-flowered secund clusters; *peduncle* with a few scattered hairs; *pedicels* alternate, short, clothed with erect hairs, drooping. *Calyx* pentangular, hairy, except in the upper part, especially on the angles; teeth short, unequal, 2 upper ones falcate converging, the others triangular straight very acutely pointed. *Standard* rounded, emarginate, with a mucro in the sinus, reflexed, bluish-purple, beautifully pencilled with darker veins. *Keel* bluish-pink at the apex.

†† *Calyx not gibbous at the base on the upper side.*

8. V. *lathyroides,* Linn. *Spring Vetch.* Leaflets elliptical-obovate or elliptical-lanceolate, retuse, mucronate; flowers solitary, nearly sessile; calyx campanulate, tube and teeth nearly equal, teeth subulate equal and straight; stipules half-halbertshaped, entire, not impressed with a dark spot; legume erect, linear, glabrous, with elevated dots, style decurved from the central suture; seeds nearly cubical, tuberculate. *E. Bot. t.* 30. *E. Fl. v. iii. p.* 283. *Hook. Br. Fl. p.* 325.

Limestone districts; rare. *Fl.* April, May. ⊙.

Whitecliffe near Ludlow; *Mr. H. Spare.*

Root fibrous. *Stems* several, 3 or 4 inches long, branched from the base, procumbent and straggling, angular, clothed with short silky horizontal pubescence. *Leaves* alternate, petiole deeply channelled above, clothed as are the leaflets with short silky hairs, elongated at its extremity into a very short simple

tendril. *Leaflets* generally 3 pairs, opposite, those of the lower leaves obcordate, of the upper leaves elliptical-obovate or elliptical lanceolate, all retuse, with a short stout triangular mucro in the sinus. *Stipules* small, half halbert-shaped, entire, without any teeth or any discoloured depression beneath. *Flowers* small, solitary, axillary, on a very short hairy peduncle, light-bluish purple. *Calyx* small, campanulate, with 5 strong prominent ribs and 5 finer unbranched intermediate ones, clothed with erect appressed long silky hairs, tube short; teeth subulate, tapering, acute, nearly as long as the tube, all equal and straight, margins ciliated. *Legume* erect, linear, destitute of pubescence, but covered with minute scattered elevated dots. *Seeds* 9, small, dark brown, nearly cubical, prominently tubercular or granulated.

Very similar in general appearance to *V. sativa γ. Bobartii*, but the small campanulate calyx which is destitute of all gibbosity on the upper side, the elevato-punctate legume and granulated cubical seeds completely distinguish it from that and every other variety of *V. sativa*, in every form of which the calyx is invariably large and tubular with a gibbous or inflated appearance on the upper side of the base of the tube, the legume pubescent, and the seeds orbicular compressed and smooth.

12. ASTRAGALUS. *Linn.* Milk-vetch.

1. A. *glycyphyllus,* Linn. *Sweet Milk-vetch.* Stem prostrate; leaflets 4-5 pairs, oval, obtusely pointed; stipules ovato-lanceolate, glabrous; peduncles shorter than the leaves; calyx glabrous, gibbous on the upper side, teeth unequal, 2 upper ones very short triangular converging, sinus very broad, the rest twice as long subulate straight; legumes long, subtriquetrous, arcuate, glabrous. *E. Bot. t.* 203. *E. Fl. v. iii. p.* 294. *Hook. Br. Fl. p.* 327.

Thickets and waste places; rare. *Fl.* July. ♃.

Banks of the Canal near Ellesmere; *Rev. A. Bloxam.!* Buck's orchard, Hord's park, near Bridgnorth; *Purton's Midl. Flora.* By the side of the turnpike road in the hollow way below the Fox farm between Shrewsbury and Cound. In a gravel pit at Round Hill near Coton Hill, Shrewsbury; *Rev. E. Williams's MSS.*

Root deeply creeping. *Stems* 2-3 feet long, prostrate, angular, zigzag, scarcely branched, nearly smooth. *Leaves* alternate, of 4-5 pairs of shortly petiolate oval obtusely pointed *leaflets* and an odd terminal one, bright green and smooth above, paler and with minute close-pressed hairs beneath. *Stipules* ovato-lanceolate, entire, glabrous. *Peduncles* axillary, angular, furrowed, slightly hairy, about half as long as the leaves, bearing at its summit an ovate *spike* of pale sulphur-coloured *flowers*, each on a very short hairy *pedicel*, subtended by a subulate ciliated *bractea* shorter than the calyx. *Calyx* short and stout, tubular, gibbous on the upper side, 12-ribbed, glabrous, 2 upper teeth very short triangular somewhat converging, the sinus very broad and roundish, 3 lower teeth twice the length subulate straight, all slightly ciliated. *Standard* oblong, notched. *Legume* full an inch long, nearly cylindrical, with a slight longitudinal furrow, curved upwards, pointed, smooth and even. *Seeds* 7-8, yellowish.

13. ORNITHOPUS. *Linn.* Bird's-foot.

1. O. *perpusillus,* Linn. *Common Bird's-foot.* Stems prostrate, hairy; leaflets elliptical, acute, hairy; stipules minute, subulate; flowers 3-4, nearly sessile; corolla about twice the length of the calyx; calyx hairy, teeth nearly equal, 2 upper ones slightly divergent; legumes curved, hairy and longitudinally wrinkled; seeds

elliptical, smooth. *E. Bot. t.* 369. *E. Fl. v. iii. p.* 290. *Hook. Br. Fl. p.* 328.

Sandy and dry gravelly soils; frequent. *Fl.* June. ⊙.

Ruckley; *H. Bidwell, Esq.* Beaumont Hill, Quatford; *W. P. Brookes, Esq.* Foot of the High Rock, Bridgnorth; *Rev. W. R. Crotch.* Hopton Hill; *J. F. M. Dovaston, Esq.* Near Whitchurch; *Mr. F. Dickinson.* Neach Hill near Shiffnall; *Dr. G. Lloyd.* Shelderton near Ludlow; *Mr. H. Spare.* Morf near Bridgnorth; *Purton's Midl. Flora.* Near Oswestry; *Rev. T. Salwey.* Shotton near Shrewsbury; *W. W. Watkins, Esq.* Market Drayton; *Dr. T. W. Wilson.*

Sharpstones Hill. Pulley Common. Haughmond Hill. Sandford, near Oswestry. Sandy heath near Wolf's Head Turnpike beyond Nesscliffe.

Root fibrous. *Stems* several, prostrate, 3—8 or 12 inches long, round, with silky hairs, somewhat zigzag, alternately branched. *Leaves* alternate, lowermost petiolate, upper ones sessile, pinnate, *leaflets* 5—10 or 12 pairs with an odd terminal one, elliptical, acute, small, hairy, the lowermost pair distant and at the base. *Stipules* minute, subulate. *Flowers* small, 3 or 4, nearly sessile at the extremities of the stem and branches in the axils of the leaf with 2 or 3 leaflets, the very short pedicels each subtended by a very minute subulate *bractea*. *Calyx* tubular, with erect silky hairs; teeth 5 nearly equal, 2 upper ones triangular slightly divergent, 3 lower ones triangular acuminate. *Corolla* about twice the length of the calyx, cream-colour; *standard* erect, obovate, slightly notched, beautifully streaked with crimson veins; *wings* white; *keel* greenish. *Legume* somewhat curved, compressed, jointed, hairy, with a short beak surmounted by a minute hook; joints 3—7, elliptical, longitudinally wrinkled when dry. *Seeds* elliptical, smooth.

14. HIPPOCREPIS. *Linn.* Horse-shoe Vetch.

1. H. *comosa,* Linn. *Tufted Horse-shoe Vetch.* Stems procumbent, smooth; leaflets obovato-elliptical, retuse, obtusely mucronate; stipules ovate, entire; peduncles very much longer than the leaves; flowers crowded, decurved; calyx short, glabrous, teeth nearly equal, 2 upper ones slightly converging; legumes arcuate, exterior margin repand, rough with prominent points; seeds arcuate, smooth. *E. Bot. t.* 31. *E. Fl. v. iii. p.* 291. *Hook. Br. Fl. p.* 328.

Banks and pastures; rare. *Fl.* July. ♃.

Near Ellesmere; *Rev. A. Bloxam.*

Root woody. *Stems* numerous, 6—8 inches long, branched at the base, angular, smooth, procumbent. *Leaves* alternate, of 4—6 pairs of obovato-elliptical leaflets, with an odd terminal one retuse obtusely mucronate in the sinus. *Stipules* ovate, entire. *Peduncles* terminal and axillary, long, stout, smooth, naked. *Flowers* 6 or more in terminal umbels, decurved, pedicels very short. *Calyx* short, teeth triangular acute, 2 upper ones rather shorter and stouter slightly converging. *Standard* deep yellow, striated; *wings* and *keel* paler. *Legumes* singularly curved, brown, rough with minute prominent points; joints 3, crescent-shaped, exterior margin repand. *Seeds* arcuate, smooth.

15. ONOBRYCHIS. *Tourn.* Saint-foin.

1. O. *sativa,* Lam. *Common Saint-foin.* Leaflets elliptic-oblong, mucronate, nearly glabrous; stipules ovate, acuminate, hairy; peduncles very much longer than the leaves; flowers in dense spikes; calyx tube short silky, teeth very long subulate acuminate straight, 4 uppermost nearly equal, lowermost shorter; legumes erect, semiorbicular, hairy, margins and ribs toothed.

Hedysarum Onobrychis, Linn. E. Bot. t. 96. *E. Fl. v. iii. p.* 292. *Hook. Br. Fl. p.* 328.

Hilly pastures in the limestone districts. *Fl.* June, July. ♃.

Llanymynech; *J. F. M. Dovaston, Esq.* Poughmill near Caynham Camp; *Miss Mc Ghie.* Pastures about Lutwych and Easthope. On the bank between Newman Hall and Eaton Brook near Berrington (probably cultivated there); *Rev. E. Williams's MSS.*

Root woody. *Stems* several, recumbent, 2 feet long, round, furrowed, hairy, scarcely branched. *Leaves* of about 10 pairs and an odd terminal one, elliptic-oblong, mucronate, entire *leaflets*, glabrous above, hairy on the midrib beneath. *Stipules* ovate, accuminate, hairy. *Peduncles* axillary, very much longer than the leaves, ascending, hairy. *Flowers* in dense terminal tapering spikes on short hairy pedicels with a narrow membranous *bractea* at the base. *Calyx* tube very short silky, teeth very long subulate acuminate, 4 uppermost equal, lowermost shorter, all straight. *Corolla* crimson, variegated with white and darker veins. *Legumes* erect, semiorbicular, bordered with sharp flat teeth, hairy at the sides and strongly reticulated with prominent partly spinous ribs.

16. MELILOTUS. *Tourn.* Melilot.

1. M. *officinalis*, Linn. *Common Yellow Melilot.* Stem erect; leaflets linear-obovate, truncate, strongly serrato-dentate; stipules subulate, entire; racemes lax; corolla thrice as long as the calyx, petals about equal in length; 2 upper calycine teeth longer than the rest; legumes obliquely ovate, acute at both ends, compressed, 2-edged, hairy and transversely wrinkled; seeds 2, oblong, gibbous, smooth. *Hook. Br. Fl. p.* 328. *Fries Nov. 2d ed. p.* 233. *Reich. Fl. Excurs. n.* 3200. *Trifolium Melilotus-officinalis, Linn. Trifolium officinale, E. Bot. t.* 1340. *E. Fl. v. iii. p.* 297.

Way sides; not common. *Fl.* June, July. ⊙.

Near Astley; *Mr. E. Elsmere, junr.!* Near Newport; *R. G. Higgins, Esq.* Banks of river Severn near Marn Wood, Coalbrookdale; *Mr. F. Dickinson.* Dowle on the river bank; *Mr. G. Jorden.* Whitecliffe near Ludlow; *Mr. H. Spare.* About Cound and Battlefield; *Rev. E. Williams's MSS.*

Banks of Severn near Buildwas bridge.

Root fibrous. *Stem* 2-3 feet high, erect, angular, furrowed, smooth, with diffuse branches. *Leaves* alternate, shortly petiolate, ternate; *leaflets* linear-obovate, truncate, margins thickened, strongly serrato-dentate, teeth submucronate, entire at the base, glabrous above, slightly hairy beneath, 2 lateral leaflets nearly sessile, terminal one on a longer stalk, somewhat incurved. *Stipules* subulate entire, the lower ones with one or more minute teeth next the base. *Peduncles* axillary, solitary, angular and furrowed, smooth below, hairy above, erect, considerably longer than the leaves. *Flowers* in lax many-flowered unilateral racemes on short decurved hairy pedicels each subtended by a minute setaceous *bractea.* *Calyx* campanulate, short, about one-third the length of the corolla, slightly hairy, 5-ribbed, with 5 subulate acute teeth of which the two upper ones are the longer. *Corolla* yellow, *standard* folded and keeled, notched, scarcely reflexed, about equal to the wings and keel. *Legume* small, drooping, obliquely oval, acute at both ends, compressed, 2-edged, terminated by the long filiform style, with erect hairs, transversely wrinkled. *Seeds* 2, oblong with a gibbous protuberance on one side, smooth.

17. TRIFOLIUM. *Linn.* Trefoil.

* *Legumes with several seeds.*

1. T. *repens*, Linn. *White Trefoil or Dutch Clover.* Stems

prostrate; leaflets rotundo-obovate, scarcely emarginate, mucronate, acutely denticulato-serrate, entire at the base, glabrous; stipules lanceolato-acuminate, glabrous; heads dense, globose, depressed, on very long axillary ascending pubescent peduncles, pedicels short decurved after flowering; calyx glabrous, teeth triangulari-subulate, acute, 2 upper ones largest and longest nearly parallel, sinus less deeply cleft; standard of corolla narrow, oblong, entire; legume short, linear, terete, smooth, 4-seeded. *E. Bot. t.* 1769. *E. Fl. v. iii. p.* 299. *Hook. Br. Fl.* 329.

Meadows and pastures; common. *Fl.* through the summer. ♃.

Root creeping. *Stems* numerous, prostrate, round, smooth, solid. *Leaves* alternate, on long smooth petioles channelled above; *leaflets* rotundo-obovate, scarcely emarginate, with a very minute mucro, acutely denticulato-serrate, entire at the base, glabrous, often marked with a white lunulate spot, on very short pubescent partial stalks. *Stipules* adnate to the petioles, lanceolato-acuminate, ribbed, glabrous. *Flowers* in globose depressed dense many-flowered heads, on very long axillary ascending angular furrowed pubescent peduncles. *Pedicels* spreading, short, round, slightly pubescent on the upper side, decurved in fruit, with a minute membranous lanceolate *bractea* at the base of each. *Calyx* 10-ribbed, glabrous, teeth unequal triangulari-subulate acute, 2 upper ones the largest and longest, nearly parallel, their sinus less deeply cleft. *Corolla* white or rather cream-coloured; *standard* narrow oblong entire erect, margin more or less deflexed. *Legume* short, linear, terete, smooth, ridged from the seeds, obtuse at the apex, from the middle of which proceeds the long straight style curved upwards at the extremity, invested by the withered brown corolla.

** *Legumes 1- or 2-seeded. Standard deciduous or unaltered. Calyx not inflated mostly hairy.*

2. T. *pratense*, Linn. *Common Purple Trefoil.* Stems ascending; leaflets oval, obovate or obcordate, entire; heads dense, ovate, sessile, subtended by two opposite leaves; calyx quite sessile, tube and teeth equal in length, hairy, 2 uppermost teeth slightly longer than the intermediate ones, lower one one-third longer, all straight and setaceous; standard of corolla linear-oblong, obtuse, emarginate; stipules ovate, abruptly bristle-pointed, subglabrous; legume obovate, 1-seeded. *E. Bot. t.* 1770. *E. Fl. v. iii.* 302. *Hook. Br. Fl. p.* 330. *Reich. Fl. Excurs. n.* 3166.

Meadows and pastures; common. *Fl.* Summer months. ♃

Root woody, fibrous. *Stems* numerous, ascending, 10—12 inches high, furrowed, clothed with long spreading silky hairs below, which in the upper portion under the heads of flowers are closely appressed. *Leaves* alternate, petioles channelled above, with spreading silky hairs, ternate; *leaflets* on very short stalks, oval or obovate, lower ones obcordate, entire, scarcely notched, tipped with a very minute obtuse mucro, with silky hairs on both surfaces, the upper one often marked with a pale whitish lunulate spot, two uppermost leaves mostly opposite close under the flowers. *Stipules* adnate to the base of the petiole, ovate, suddenly terminating in a bristle-shaped point, pale and membranous, marked by prominent purple ribs interbranching near the margins, nearly smooth. *Heads of Flowers* terminal, large, dense, ovate, obtuse, sessile. *Calyx* quite sessile, pale with 10 purple prominent ribs, tube and teeth about equal in length, tube with long erect silky hairs, teeth 5 narrow setaceous obtuse purple, 2 uppermost slightly longer than the intermediate ones, lower one one-third longer, all straight and clothed with distant

long spreading silky hairs, mouth with an internal thickened ring densely clothed with erect silky hairs. *Corolla* light purple, fragrant; *standard* linear-oblong, obtuse, erect, emarginate, marked with darker parallel veins; *wings* shorter than the standard; *tube* very much exserted, twice the length of the calyx; petals united to each other and to the stamens. *Legume* obovate, 1-seeded. *Seed* yellowish.

A variety with white flowers was found by the late Rev. E. Williams, in a pasture by the side of Berrington Pool and in a clover-field adjoining Cound church-yard, 1798.

3. T. *medium*, Linn. *Zigzag Trefoil.* Stems ascending, zigzag; leaflets elliptical or lanceolate, entire, mucronate; heads dense, ovate, shortly pedunculate, subtended by 2 opposite leaves; calyx nearly sessile, tube and teeth equal, tube glabrous, teeth hairy, 2 uppermost teeth slightly shorter than the intermediate ones, lower one one-third longer, all straight and setaceous; standard of corolla ovate, submucronate, entire; stipules linear-lanceolate, acuminate, hairy; legume oblong, 2-seeded. *E. Bot. t.* 190. *E. Fl. v. iii. p.* 302. *Hook. Br. Fl. p.* 330.

Pastures and borders of fields; not very common. *Fl.* July. ♃.

Near Astley; *Mr. E. Elsmere, junr.!*

Pulley Common. Shelton Rough. Plentiful by the side of the bridle-road leading from Battlefield to Albright Hussee. Right hand side of road up Wenlock Edge beyond Longville in the Dale.

Root woody, creeping. *Stems* numerous, decumbent at the base, then ascending, 1 foot high, zigzag, roundish, furrowed, clothed with long spreading silky hairs in the lower part and above with close-pressed ones. *Leaves* alternate, petioles channelled above, with spreading silky hairs, ternate; *leaflets* on very short stalks, elliptical or lanceolate, obtuse, entire, mucronate, glabrous and of a bright velvety green above, paler and with long silky hairs beneath, margins ciliated, 2 uppermost leaves mostly opposite under the flower. *Stipules* adnate to the petiole, linear-lanceolate, acuminate, pale and membranous, marked by purple ribs interbranching near the margins, clothed with long silky hairs. *Heads of Flowers* terminal, large, dense, subglobose, obtuse, shortly pedunculate, solitary. *Calyx* nearly sessile, pale, with 10 purple prominent ribs, tube and teeth about equal in length, tube glabrous, teeth 5 narrow setaceous obtuse purple, 2 uppermost ones slightly shorter than the intermediate ones, lower one about one-third longer, all straight and clothed with distant long spreading silky hairs, mouth with an internal thickened ring densely clothed with erect silky hairs. *Corolla* dark red-purple, fragrant; *standard* ovate, erect, pointed, submucronate, marked with darker parallel veins; *wings* nearly equal in length to the standard; *tube* very much exserted, twice the length of the calyx; petals united to each other and to the stamens. *Legume* oblong, 2-seeded.

4. T. *arvense*, Linn. *Hare's-foot Trefoil.* Silky, erect, alternately branched; leaflets lineari-elliptical, subretuse, mucronate, hairy; stipules ovate or lanceolate, acuminate, membranous; heads dense, cylindrical, obtuse, softly hairy; calyx with erect hairs; teeth twice the length of the tube, longer than the corolla, subulato-setaceous, nearly equal, spreading, plumose; standard of corolla ovate, obtuse, apex jagged; legume minute, membranous, 1-seeded. *E. Bot. t.* 944. *E. Fl. v. iii. p.* 305. *Hook. Br. Fl. p.* 331.

Dry pastures; not very common. *Fl.* July, August, ⊙.

Ruckley; *H. Bidwell, Esq.* On a hill above Quatford; *W. P. Brookes.!* Near Bridgnorth; *Mr. T. Bodenham.* Queen's-head near Oswestry. Dovaston; *J. F. M. Dovaston, Esq.* Ness; *T. C. Eyton, Esq.* Near Newport; *R. G. Higgins, Esq.* Near Ludlow; *Mr. H. Spare.* Walls near Ludlow; *Rev. T. Salwey.* Dodington near Whitchurch; *W. W. Watkins, Esq.* Ketley; *Dr. Du Gard.*

Haughmond Abbey. Dovaston Heath.

Whole Plant clothed with soft silky close-pressed hairs. *Root* fibrous. *Stems* 6—12 inches high, round, zigzag, alternately branched, branches divergenti-patent. *Leaves* alternate, on very short petioles, ternate; *leaflets* lineari-elliptical, somewhat retuse at the apex, with a sharp reflexed mucro in the sinus, hairy on both sides, margins ciliated. *Stipules* ovate or lanceolate, acuminate, membranous, strongly ribbed. *Flowers* on axillary peduncles, in terminal cylindrical obtuse dense soft hairy heads. *Calyx* tubular, 10-ribbed, clothed with erect hairs; teeth twice the length of the tube and exceeding the corolla in length, subulato setaceous, spreading, nearly equal, clothed with long soft spreading silky hairs which viewed under a powerful lens are found to be minutely denticulate. *Corolla* shorter than the calyx, cream-coloured; *standard* ovate, obtuse, the apex minutely jagged; *keel* with 2 crimson spots on the interior. *Legumes* minute, membranous, single-seeded.

5. T. *striatum*, Linn. *Soft-headed Trefoil.* Silky; stems procumbent or erect; leaflets obcordate or obovate, sinuato-denticulate above; stipules ovate, bristle-pointed, membranous, and ribbed; heads terminal and axillary, ovate or subglobose, subsolitary, sessile, subtended by 1-2 sessile leaves; calyx ventricose in fruit, 10-ribbed, hairy, teeth scarcely half the length of the tube, subulate, rigid, unequal, straight, ciliated; corolla exserted. *E. Bot. t.* 1843. *E. Fl. v. iii. p.* 307. *Hook. Br. Fl. p.* 332.

Dry gravelly pastures; not common. *Fl.* June. ⊙.

About Eaton Marscot, Berrington, Edgmond, Cound, on the Wrekin, Haughmond Hill, &c.; *Rev. E. Williams's MSS.*

Sharpstones Hill.

Whole Plant clothed with long white soft silky hairs. *Root* slender. *Stems* several, procumbent, 4—6 inches long, round, furrowed, spreading from the root, branched. *Leaves* alternate, distant, ternate, on short petioles; *leaflets* sessile, those of the lower leaves obcordate, of the upper ones obovate, sinuato-denticulate in the upper part. *Stipules* large, ovate, bristle-pointed, membranous, ribbed, adnate to the petiole and clasping the stem. *Heads of Flowers* ovate or subglobose, dense, axillary and terminal, nearly sessile, on short peduncles subtended by two stipulate ternate nearly sessile leaves. *Calyx* ventricose, elliptical, 10-ribbed and furrowed, with erect silky hairs, 5-toothed, teeth scarcely half the length of the tube unequal subulate spiny rigid straight ciliated, mouth contracted with an internal thickened ring bearing a row of erect hairs. *Corolla* pale rose-coloured, about as long as the calyx, standard lanceolate, wings and keel equal. *Legume* membranous, with rudiments of 2 seeds in an early state. *Seed* 1, oval, smooth.

Var. β. erectum. Stems 12 inches or more high, copiously branched from the base, erect or ascending. *Leaves* very distant, lower ones on rather long petioles, leaflets obcordato-cuneate, upper ones on shorter petioles or nearly sessile, leaflets obovato-elliptical, all sinuato-denticulate in the upper portion. *Heads of Flowers* much longer, ovate, subconical.

Hedge-banks near Cheney Longville, and Bayston Hill near Shrewsbury. *Fl. June.* ⊙.

Very different in general appearance and habit to *var. α*, even in similar habitats, though I cannot detect any character by which they can be specifically distinguished, unless such exist in the fruit which I have not seen. I have received an identical specimen from Barmouth, N. Wales; and Mr. C. C. Babington possesses one exactly similar from Naples, sent to him by Professor Gasparini, marked as *T. striatum var. erectum.*

*** *Calyx remarkably inflated after flowering and arched above. Standard of the corolla deciduous.*

6. T. *fragiferum*, Linn. *Strawberry-headed Trefoil.* Stems prostrate, creeping, leaflets obcordate, mucronato-serrate, glabrous; stipules ovate, very acuminate, glabrous; heads globose, dense, involucred, on long axillary peduncles; calyx of fruit unequally 2-lipped, decurved, upper lip inflated membranous reticulato-venose hairy with 2 subulate decurved teeth, lower lip half as long with 3 subulate straight teeth; standard of corolla narrow, oblong, emarginate, erect, subtended by upper lip of calyx; legume suborbiculari-oblong, smooth, 1-seeded. *E. Bot. t.* 1050. *E. Fl. v. iii. p.* 208. *Hook. Br. Fl. p.* 332.

Limestone Districts; rare. *Fl.* July, August. ♃.

By the side of the road up Wenlock Edge above Harley. Near the eleventh mile-stone on the left-hand side of the road as you descend the bank to Muckley Cross from Wenlock; *Rev. E. Williams's MSS.*

Stem prostrate, creeping by fibrous radicles, glabrous. *Leaves* on long spreading petioles; *leaflets* obcordate, mucronate in the sinus, with numerous prominent penniform nerves, margins mucronato-sinuato-serrate. *Stipules* large, whitish, ovate, very much acuminated. *Peduncles* axillary, solitary, erect, angular, glabrous, considerably longer than the leaves. *Heads* globose, solitary, erect, subtended by a multipartite glabrous involucre, small when in flower, subsequently very much enlarged, more or less coloured, and resembling a strawberry. *Calyx* erect *in flower*, nearly equally 2-lipped, upper lip with copious silky hairs, lower lip sparingly hairy, all the teeth subulate, the two of the upper lip rather the longest. *Calyx in fruit* decurved, unequally 2-lipped, contracted at the mouth; upper lip very large and inflated, membranous, reticulato-venose, hairy, with 2 decurved teeth; lower lip about half as long as the upper one, hairy, with 3 straight subulate teeth. *Flowers* small, purplish-red; *standard* emarginate, subtended by the bidentate upper lip of the calyx, erect. *Legume* concealed in the bottom of the inflated calyx, small, suborbiculari-oblong, laterally compressed, with a long lateral beak somewhat curved upwards, glabrous. *Seeds* 2, one only perfected, orbicular, compressed, shining.

**** *Standard of the corolla persistent, deflexed, dry, enveloping the fruit. (Flowers yellow.)*

7. T. *procumbens*, Linn. *Hop Trefoil.* Stems procumbent, spreading, ascending at the extremities; leaflets obovate, deeply emarginate, sinuato-denticulate above; stipules half ovate, acute, margins entire reflexed; peduncles longer than the leaves; heads broadly oval, of numerous densely imbricated flowers; standard obovate, deeply emarginate, at length deflexed, furrowed; legume elliptical, pointed at each end, 1-seeded. *T. procumbens, Linn. E. Bot. t.* 945. *E. Fl. v. iii. p.* 309. *a. Hook. Br. Fl. p.* 332.

Dry pastures, borders of fields &c.; frequent. *Fl.* June, July. ⊙.

Root strong, fibrous. *Stems* numerous, 12 inches long, branched, spreading

and decumbent, ascending at the extremity, round below, angular above, with scattered appressed hairs. *Leaves* alternate, distant, on a short hairy petiole channelled above, ternate; *leaflets* obovate, deeply emarginate, with a stout mucro in the sinus, entire in the lower half, sinuato-denticulate in the upper, the lateral leaflets on very short stalks, the central one on a longer one which is always considerably shorter than the common petiole, somewhat inflexed, glabrous above, paler and glaucous beneath, the midrib slightly hairy. *Stipules* adnate to the petiole, half ovate, acute, margins entire reflexed. *Peduncles* axillary, longer than the leaves, round, clothed with erect appressed silky hairs. *Heads* terminal, solitary, broadly oval; *flowers* numerous, densely imbricated, pale yellow, becoming tawny and hoplike after flowering. *Calyx* glabrous, 2 upper teeth very short triangular, the sinus between them large and rounded; 3 lower teeth twice the length of the upper ones, subulate, tipped with a single hair, the intermediate ones curved upwards, the rest straight. *Pedicels* very short, erect in flower, deflexed in fruit. *Standard* of *corolla* obovate, deeply emarginate, with a short mucro in the sinus, margins serrulate in the upper half incurved, furrowed, erect, at length deflexed and tawny, dry and membranous, deeply furrowed, 3-nerved at the base, the central one dividing into three immediately above the base, the central secondary nerve branching above the claw into 2 or 3 others and the secondary lateral ones branching above the claw into 4; the primary lateral nerves uniting with the exterior lateral secondary nerves above the claw, branching into 2, the subdivisions being continued parallel and penniform to the margin; wings longer than the keel. *Legume* small, elliptical, pointed at each end. *Seed* solitary, reniform.

8. T. *filiforme*, Linn. *Lesser Yellow Trefoil.* Stems procumbent; leaflets obcordate, denticulate above; stipules ovate acute, margins revolute ciliated; peduncles very much longer than the leaves; heads roundish of few lax flowers; standard narrow, oblongo-obovate, emarginate, at length reflexed, obscurely striated; legume oblong, obtuse, tapering at the base, stipitate, 1-seeded. *T. filiforme, a. major, Hook. Br. Fl. p.* 333. *T. minus, Relh. E. Bot. t.* 1256. *E. Fl. v. iii. p.* 310.

Dry pastures and roadsides; frequent. *Fl.* June, July. ⊙.

Root fibrous. *Stems* numerous, procumbent, copiously branched and straggling, angular, with spreading hairs in the lower part, appressed ones above. *Leaves* alternate, petiole very short channelled above with erect appressed hairs; *leaflets* obcordate, notched in the upper half, entire in the lower to the base, glabrous, except the midrib beneath, which terminates in an inconspicuous mucro in the sinus; the central leaflet on a short stalk always shorter than the common petiole, the rest nearly sessile. *Stipules* adnate to the petiole, ovate, acute, margins revolute ciliated. *Peduncles* axillary, solitary, very much longer than the leaves, slender, erect, clothed with long silky erect appressed hairs. *Heads* roundish, of 12—15 rather lax *flowers*, on very short pedicels, erect in flower, reflexed in fruit. *Calyx* glabrous, tube shorter than the teeth, teeth unequal, 3 lower ones subulate, obtuse, with a few long hairs at the apex; 2 upper ones triangular, obtuse, about one-third the length of the others, straight. *Corolla* yellow, persistent, becoming tawny as the seed ripens; *standard* narrow, oblongo-obovate, obscurely striated, emarginate, the sinus minutely denticulate, reflexed. *Legume* small, oblong, obtuse, tapering at the base, stipitate, glabrous, crowned from the middle of the apex with the straight style and curved stigma. *Seed* one, smooth and shining.

In a dry soil or on the close-cropped pasture of commons, the heads have fewer flowers, 6—8 or fewer, and the whole plant is smaller but still adhering to the above characters.

18. LOTUS. *Linn.* Bird's-foot Trefoil.

1 L. *corniculatus*, Linn. *Common Bird's-foot Trefoil.* Stems decumbent, solid; heads depressed, umbellate, 5—10-flowered; claw of standard transversely vaulted; calycine teeth connivent in bud, 2 upper ones converging in flower; stipules ovate, unequal; legume with a long deflexed rostrum from the middle of the apex. *Babington in Ann. Nat. Hist. v. ii. p.* 260. *E. Fl. v. iii. p.* 312. *Hook. Br. Fl. p.* 333.

Var. a. vulgaris, Koch. Nearly glabrous or with scattered hairs; stems ascending; leaflets obovate; stipules ovate, unequal. *E. Bot. t.* 2090.

Pastures, &c.; frequent. *Fl.* July, August. ♃.

Root woody. *Stems* solid, numerous, spreading, branched, procumbent or ascending, tetragonous, furrowed, clothed with minute erect appressed pubescence. *Leaves* alternate, on very short dilated channelled hairy petioles; *leaflets* obovate, scarcely emarginate, glabrous above, more or less hairy and glaucous beneath, margins ciliated. *Stipules* very shortly petiolate, ovate, obtuse, unequal and dilated at the base on the side farthest from the leaf, glabrous, margins ciliated. *Peduncles* very long, tetragonous, with erect appressed pubescence. *Clusters* terminal, of 5—10 *flowers*, depressed in the same plane, on very short erecto-patent glabrous pedicels which become decurved in fruit, subtended by a single nearly sessile terminal leaf. *Calyx-tube* and teeth about equal, nearly glabrous or with a few scattered hairs and ciliæ; tube angular, teeth connivent in the bud, triangulari-subulate, obtuse, the 2 upper ones converging, sinus rounded. *Corolla* yellow, *standard* striped with red at the base, obovate, reflexed, the claw much dilated and vaulted transversely. *Legume* linear, terete, straight, with a long setaceous deflexed rostrum springing exactly from the middle of the apex. *Seeds* numerous, oval, compressed, smooth.

2. L. *major*, Scop. *Narrow-leaved Bird's-foot Trefoil.* Stems ascending or procumbent; heads depressed, umbellate, 8—12-flowered; claw of standard linear, longitudinally vaulted; calycine teeth divergent or stellate in bud, 2 upper ones diverging in flower; stipules ovato-rotund, unequal; legume with a long setaceous straight rostrum from the upper suture. *Babington in Ann. Nat. Hist. v. ii. p.* 260. *E. Fl. v. iii. p.* 313. *Hook. Br. Fl. p.* 334.

Sides of ditches and moist places; frequent. *Fl.* July, August. ♃.

Benthall Edge; *Mr. F. Dickinson.* Eyton and Walford; *T. C. Eyton, Esq.* Near Newport; *R. G. Higgins, Esq.* Woods at the base of the Wrekin; *E. Lees, Esq.*

Var. a. vulgaris, Bab. Hairy, stems nearly erect. *E. Bot. t.* 2091.

Bomere Pool. Lane leading from Longden to the Oaks Hall near Pontesbury. Astley.

Root strong. *Stem* ascending, 1—3 feet high, round, striated, clothed with long spreading hairs. *Leaves* alternate, distant, on short dilated petioles; *leaflets* ternate, obovate, with an obtuse very short mucro, covered on both surfaces with long scattered hairs. *Stipules* very shortly petiolate, ovato-rotund, unequal and dilated at the base on the side farthest from the leaf, very minutely serrated, hairy. *Peduncles* very long, round, striated. *Clusters* terminal, of 8—12 *flowers*, depressed in the same plane, on very short erecto-patent glabrous pedicels which become deflexed in fruit, subtended by a single nearly sessile ternate leaf. *Calyx-tube* and

teeth about equal, longer than the tube of the corolla, hairy on the ribs and teeth, tube angular, teeth divergent or spreading like a star in the bud, triangulari-subulate, acute, 2 upper ones diverging, the sinus forming an acute angle. *Corolla* yellow, *standard* striped with red at the base, ovate, reflexed, the claw linear and longitudinally vaulted. *Legume* linear, terete, straight, having a long setaceous straight rostrum, springing from the upper surface. *Seeds* numerous, minute.

Var. β. glabriusculus, Bab. Glabrous, margins and nerves of the leaflets stipules and sepals with long ciliæ; stems erect or procumbent.

Bickley Coppice. Road-side Cross Hill, near Shrewsbury.

19. MEDICAGO. *Linn.* Medick.

1. M. *sativa*, Linn. *Purple Medick or Lucerne.* Erect, glabrous; leaflets obovato-oblong, serrato-dentate above, apex truncate with 3—5 teeth central one longest; stipules lanceolate, acute; flowers in lax racemes; calycine teeth nearly equal, subulate; legumes spirally twisted, many-seeded. *E. Bot. t.* 1749. *E. Fl. v. iii. p.* 317. *Hook. Br. Fl. p.* 335.

Corn-fields; not wild. *Fl.* June, July. ♃.

Corn-fields at Pulley near Shrewsbury; sparingly.

Stem 2-feet high, erect, angular, nearly smooth, alternately branched. *Leaves* alternate, petiolate; *leaflets* obovato-oblong, serrato-dentate in the upper part, entire below, apex truncate, with 3 or 5 teeth of which the central one is the longest, glabrous above, pubescent beneath; lateral leaflets nearly sessile, terminal leaflets shortly petiolate. *Stipules* lanceolate, acute. *Peduncles* axillary and terminal, solitary, longer than the leaves. *Flowers* in lax many-flowered erect racemes, on short pedicels with a small setaceous *bractea* under each. *Calyx* tubular, nearly smooth, 10-ribbed, teeth nearly equal subulate. *Corolla* bluish-purple. *Legume* spirally twisted, with 2 or 3 turns, veiny and silky. *Seeds* several, flattish.

2. M. *lupulina*, Linn. *Black Medick or Nonsuch.* Procumbent; leaflets rotundo-obovate, cuneate at the base, denticulate above, apex truncate, with a stout triangular mucro; stipules obliquely ovate, acuminate, denticulate; flowers capitato-spicate; calycine teeth unequal, subulate, curved upwaards, 2 upper ones shorter; legume orbiculari-falcate, scarcely spiral, ramoso-striato-rugose. *E. Bot. t.* 971. *E. Fl. v. iii. p.* 318. *Hook. Br. Fl. p.* 335.

Waste ground, roadsides, cultivated fields, &c.; common. *Fl.* May—Aug. ⊙.

Root fibrous. *Stems* numerous, spreading and ascending, alternately branched, tetragonous, furrowed, smooth below, softly hairy above. *Leaves* alternate, the lower ones on very long hairy petioles channelled above, upper ones on shorter petioles, uppermost nearly sessile; *leaflets* rotundo-obovate, cuneate at the base, margins denticulate in the upper half, apex truncate with a stout triangular acute mucro, silky and pubescent on both surfaces. *Stipules* obliquely ovate, acuminate, denticulate, pubescent, adnate to the petioles. *Peduncles* axillary, solitary, much longer than the leaves, pubescent, round and slender. *Flowers* yellow, numerous, in roundish terminal heads, becoming elongated and spicate as the fruit proceeds to maturity. *Pedicels* very short, each with a minute setaceous *bractea* at the base and as well as the calyx with erect appressed hairs. *Calyx teeth* unequal, subulate, curved upwards, 2 upper ones shorter, the sinus very large and rounded. *Standard* inflated at the base. *Legume* orbiculari-falcate, scarcely spiral, compressed, with scattered hairs and a few stipitate glands, ramoso-striato-rugose.

CLASS XVIII.

POLYADELPHIA.

———

"Hypericum, all bloom, so thick a swarm
Of flowers, like flies, clothing its slender rods
That scarce a leaf appears."

COWPER.

CLASS XVIII.

POLYADELPHIA. *Filaments combined in more than two sets.*

ORDER I. POLYANDRIA. *Many stamens.*

1. HYPERICUM. *Calyx* 5-partite or 5-leaved, inferior. *Petals* 5. *Filaments* united at the base into 3 or 5 sets. *Capsule* many-seeded. *Nat. Ord.* HYPERICINEÆ, *Juss.*—Name, the ὑπερικον of Dioscorides.

POLYADELPHIA—POLYANDRIA.

1. HYPERICUM. *Linn.* Saint John's-wort.

* *Styles* 3. *Sepals entire at the margins.*

1. H. *Androsæmum*, Linn. *Tutsan.* Stems shrubby, compressed, 2-edged, smooth; leaves cordato-ovate, obtusely pointed, sessile, pellucido-punctate; cymes trichotomous, few-flowered; sepals unequal, subcordato-ovate, obtuse, glabrous, glandular beneath; petals oval, obtuse, longer than the sepals; styles 3; capsule ovali-rotund, finally pulpy; seeds oblong, reticulato-striate. *E. Bot. t.* 1225. *E. Fl. v. iii. p.* 323. *Hook. Br. Fl. p.* 281.—*Androsæmum officinale, All.*

Thickets; not common. *Fl.* July. ♃.
Shinewood Coppice near Whitwell about a mile from Wenlock; Woodlane, Cressage! *W. P. Brookes, Esq.* In a thicket at Hayes near Oswestry; *T. and D. Bot. Guide.* Shortwood near Ludlow; *Mr. J. S. Baly.* Near Coreley and near Smethcote; *Rev. W. Corbett.* Coxwood and Farley Dingle; *Mr. F. Dickinson.* Dowle wood, not plentiful; *Mr. G. Jorden.* Benthal Edge; *Dr. G. Lloyd.* Near a spring at the base of the Wrekin, immediately below the camp on the northern side, very luxuriant; *E. Lees, Esq.* Buildwas; *Miss H. Moseley.* Rocks on the Shropshire side of Dowles brook, Wyre forest; *Mr. W. G. Perry.* On the side of the turnpike road between Bridgnorth and Faintree; *Purton's Midl. Flora.* Near Shawbury heath; *Dr. T. W. Wilson.* In the wood at the back of Acton Burnell park near the cold bath. A few plants upon Cound Moor on the bank above the brook near the stone quarry. Woods and thickets above Buildwas church. Hedges about West Coppice; *Rev. E. Williams's MSS.*
Lyd Hole near Pontesford Hill.
Stems several, 12—18 inches high, (2-3 feet, *Smith*) roundish, compressed, 2-edged, smooth, straggling, nearly simple. *Leaves* in opposite pairs, sessile, widely spreading, cordato-ovate, obtusely pointed, entire, firm and coriaceous, glabrous, of a bright light green above often tinged with red, paler beneath, with strong pellucid ribs and numerous pellucid dots. *Flowers* in a short terminal subtrichotomous few-flowered panicle in the axils of the uppermost pair of leaves; *peduncles* smooth, bright red, with small lanceolate glabrous *bracteas* at the ramifications. *Sepals* unequal, subcordato-ovate, obtuse, 3 larger, 2 smaller, 7-ribbed, red, with red tubercular glands on the under surface. *Petals* about as long as the sepals, yellow. *Styles* 3. *Capsule* ovali-rotund, finally pulpy, purplish black. *Seeds* oblong, minutely reticulato-striate.

2. H. *tetrapterum*, Fries. *Four-winged St. John's-wort.* Stem 4-winged; leaves ovali-oblong or elliptical, pellucido-punctate, with black glands on the margins beneath; flowers cymose; sepals erect, lanceolate, acuminate, with pellucid streaks; petals lanceolate, pale yellow; styles half the length of the capsule; capsule ovate, conical. *Fries Nov. 2d ed. p.* 236. *Reich. Fl. Excurs.* 5179. *Bab. Prim. Fl. Sarn. p.* 19. *Koch Syn. p.* 134. *Drej. Fl. Hafn. p.* 248. *H. quadrangulum, Sm. E. Bot. t.* 370. *E. Fl. v. iii. p.* 324. *Hook. Br. Fl.* 3d ed. p. 337, 4th ed. p. 281.

Moist places, sides of pools and rivulets; not very common. *Fl.* July. ♃.

Sides of the turnpike road from Harley to Wenlock near the lime kilns; *W. P. Brookes, Esq.!* Banks of river Severn near Coalport; *Mr. F. Dickinson.* Oakley park near Ludlow; *Mr. H. Spare.* About Oswestry; *Rev. T. Salwey.*

Abbots Betton pool near Shrewsbury. Golden, near Pitchford. Canal side near Blackmere, near Ellesmere.

Root creeping, woody. *Stems* several, 1-2 feet high, erect, acutely quadrangular and winged, interstices convex, smooth, reddish, with numerous opposite axillary short leafy branches. *Leaves* sessile, in opposite pairs, ovali-oblong or elliptical, entire, obtuse, many-ribbed, veiny, with numerous pellucid dots, and a row of black ones on the margins beneath. *Flowers* in terminal, forked, many-flowered cymes. *Sepals* lanceolate, acuminate, entire, glabrous, 5-ribbed, with pellucid streaks, half as long as the petals, erect. *Petals* lanceolate, pale yellow or lemon-colour, sometimes marked with purple dots and streaks. *Anthers* tipped with dark purple glands.

3. H. *quadrangulum*, Linn. *Square-stalked St. John's-wort.* Stem 4-winged; leaves elliptical-ovate, obtuse, somewhat retuse, mucronate, with black glands on the margins beneath, almost destitute of pellucid dots; flowers cymose; sepals reflexed, ovato-lanceolate, obtuse, denticulate, mucronate, with pellucid streaks; petals marked beneath with purple streaks and dots; styles half the length of the capsule; capsule ovate, conical, triquetrous, valves rounded at the back; seeds oblongo-cylindrical, reticulato-punctate. *Reich. Fl. Excurs.* 5178, (*excl. syn.*) *Fries Nov. 2d ed. p.* 237. (*excl. syn.*) *H. delphinense, Vill. Reich. Fl. Exsicc.* 1500.

Moist pastures, sides of ditches, rivers, &c.; frequent. *Fl.* July. ♃.

Roadside between Wenlock and Harley; *W. P. Brookes, Esq.!* Near Ludlow and Coalbrookdale; *Mr. F. Dickinson.* Lane between Buildwas and Little Wenlock; *Dr. G. Lloyd.* About Oswestry; *Rev. T. Salwey.* Eaton Mascott, Albrightlee, &c.; *Rev. E. Williams's MSS.*

Moist meadows between Canal and Sundorne Castle near Pimley. Haughmond Hill. Berrington. Sutton Spa. Shelton Rough. Sharpstones Hill. Banks of Severn at Dorset's Barn. Meole Brace, all near Shrewsbury. Westfelton. Banks of Canal between Blackmere and Colemere meres.

Root creeping. *Stem* 2 feet high, erect, branched, with 4 raised wings, discoloured with black glandular streaks and dots, the interstices rounded, perfectly glabrous. *Leaves* in pairs, opposite, sessile, elliptical-ovate, obtuse, somewhat but slightly retuse, the midrib terminated by a short obtuse mucro, margins pellucid, slightly scabrous especially towards the apex and with a single row of round black glandular dots imbedded on the under surface, opaque green above, paler and somewhat glaucous beneath, glabrous and nearly destitute of pellucid dots, (a very few large ones with one or two black dots only occurring) with 5 principal ribs at the base. *Flowers* in forked leafy terminal cymes. *Sepals* ovato-lanceolate, obtuse,

---

margin denticulate in the upper part, the midrib terminating in a short acute mucro, 5-ribbed, with short pellucid streaks or dots, reflexed. *Petals* elliptical, obtuse, bright yellow, marked on the under surface with long purple streaks and dots. *Capsule* ovate, conical, triquetrous, valves rounded at the back. *Styles* 3, about half the length of the capsule, spreading in flower, erect in fruit. *Stamens* in 3 fascicles; *anthers* yellow, with a black globular gland. *Seeds* oblongo-cylindrical, obtusely pointed at both ends, reticulato-punctate.

Mr. Babington in his *Prim. Fl. Sarn. p.* 19 informs us, that according to Wahlenberg and Fries, "the *H. dubium*, Leers is the true *H. quadrangulum, Linn.*" an opinion which he confirms by an examination of the Linnæan Herbarium, in which *H. dubium*, Smith has the name of *H. quadrangulum* appended to it by Linnæus, and also the authenticating number. Our plant agrees with the description of *H. quadrangulum, Linn.* in *Reich. Fl. Excurs.* but differs entirely both in general aspect and details from a plant gathered at Arran, Scotland, by Mr. G. Mc. Nab, and sent to me by the Edinburgh Botanical Society as the "*true H. dubium, Leers*," which corresponds well with the descriptions of that species in *Leers Fl. Herborn.* and *English Flora*, and of *H. quadrangulum* in *Pollich Fl. Palat.* n. 716, which latter however Fries designates as "descriptio optima" of *H. quadrangulum, Linn.* The Arran plant has the leaves truly elliptical, obtuse, very slightly retuse, and obtusely mucronate, destitute of pellucid dots, but with a row of black dots on the margins on the under surface; sepals very broad, elliptical, obtuse, perfectly entire, copiously marked on the exterior with black dots; petals large, elliptical and with copious black dots. Mr. Babington, to whom I communicated specimens of the two plants with my doubts of their identity, says, "The specimens of '*quadrangulum*' distributed by Reichenbach himself (!) agree with your Arran plant in all points; they are No. 1397 of his '*Flora Germ. Exsiccata*' and come from Dresden. Another also from Reichenbach (No. 1500, of the same collection) named '*H. delphinense, Vill.* an *H. quadrangulum, Linn.*' from the Alps near Bex, is exactly my Welsh *dubium* (alluding to specimens gathered near Llanberis, Caernarvonshire, identical with the Shropshire plant.) Fries, in his description of *quadrangulum*, (*Nov.* 237,) says, 'Sepala elliptica l. *lanceolato-elliptica*, obtusissima, *reflexa*,' which agrees well with *my* quadrangulum and dubium. I think, therefore, that our plant is undoubtedly the *quadrangulum* of *Linn.* and *Fries*, and the *delphinense, Vill.* and therefore part of *quadrangulum, Reich.*; not *dubium, Sm.* or *Leers*, nor *quadrangulum, Wimm. et Grab. Fl. Silec.* 3. 82. which is, I think, your Arran plant, and probably a distinct species for which Gmelin's name of *H. Leersii* (*Fl. Bad.* 3. 352.) probably ought to be adopted." The *H. quadrangulum* of *E. Fl.* and *E. Bot.* is *H. tetrapterum*, Fries.

4. H. *perforatum*, Linn. *Common perforated St. John's-wort.* Flowers cymose; stem 2-edged; leaves elliptical-oblong, obtuse, with numerous pellucid dots and black glands on the margins beneath; sepals lanceolate, acuminate, serrulate at the apex; petals obliquely oblong, crenulate, with black streaks and dots beneath; capsule ovate, conical, triquetrous, valves keeled, diagonally wrinkled; styles as long as the capsule. *E. Bot. t.* 295. *E. Fl. v. iii. p.* 325. *Hook. Br. Fl. p.* 337.

Hedge-banks, thickets, &c.; common. *Fl.* July. ♃.

Root somewhat creeping, woody. *Stem* 1-2 feet high, erect, rounded, with 2 opposite edges or wings on alternate sides between each joint proceeding from the midribs of the leaves, with opposite and decussate axillary branches, glabrous, with scattered black dots. *Leaves* opposite, in pairs, decussate, elliptical-oblong, obtuse, sessile, with 5—7 pellucid veins and numerous pellucid dots, and a few black ones chiefly on the margins at the back. *Flowers* cymose, bright yellow. *Sepals* lanceolate, acuminate, somewhat serrulate at the apex, 5-ribbed, the inter-

---

stices pellucid. *Petals* irregularly and obliquely oblong, the margins crenulate, with black dots in the sinuses. *Anthers* with a black dot on the summit between the lobes. *Capsule* ovate, conical, crowned with 3 divergent styles, the valves somewhat keeled, the faces diagonally wrinkled. *Seeds* oblong. *Styles* as long as the capsule.

This species is gathered with superstitious awe as a "plant of power," by youthful lovers in our county and North Wales, on Midsummer night, "the night of St. John," and by its fresh or withered state on the ensuing morn, "the voiceless flower" is deemed to prognosticate their future fortune in matrimony. If fresh, it was saved "to deck the young bride in her bridal hour;" but if withered, it seemed to say,—"more meet for a burial than bridal day."

    "Thou silver glow-worm, O, lend me thy light,
    I must gather the mystic St. John's-wort to-night,
    The wonderful herb whose leaf will decide
    If the coming year will make me a bride."

** *Styles* 3. *Sepals with glandular serratures on the margins.*

5. H. *humifusum*, Linn. *Trailing St. John's-wort.* Stems prostrate, somewhat 2-edged, glabrous; leaves ovali-oblong, obtuse, very slightly emarginate, mucronate, sessile, glabrous, minutely pellucido-punctate, margins with black dots beneath; flowers terminal, subcymose, few; sepals unequal, 3 larger ones oblong obtuse mucronate, 2 narrower lanceolate, glanduloso-serrate; capsule conical, triquetrous, valves rounded; seeds oblong, reticulato-punctate. *E. Bot. t.* 1226. *E. Fl. v. iii. p.* 326. *Hook. Br. Fl. p.* 337. *Bab. Prim. Fl. Sarn. p.* 20.

Gravelly and heathy pastures; not very common. *Fl.* July. ♃.

About Oswestry; *Rev. T. Salwey.* Near Hardwick and Lee Bridge; *W. W. Watkins, Esq.* Near Preston Brockhurst; *Dr. T. W. Wilson.* Whitecliff, Ludlow; *Mr. H. Spare.* Neach Hill near Shiffnall; *Dr. G. Lloyd.* Wyre forest; *Mr. G. Jorden.* Bomere Woods; *Mr. F. Dickinson.* Foot of Pontesford Hill; *Mr. T. Bodenham.*

Fields between Haughmond Hill and Sundorn. Sharpstones Hill. Nesscliffe.

Root fibrous. *Stem* 3—6 inches long, prostrate or ascending, slender, somewhat 2-edged, glabrous as is the whole plant. *Leaves* in opposite pairs, sessile, ovali-oblong, obtuse, very slightly emarginate, with a minute obtuse mucro in the sinus, glabrous, bright green above, glaucous beneath, with a row of black glandular dots on the margin which is thickened and somewhat recurved, pellucid dots minute, numerous. *Flowers* terminal, subcymose, few, bright yellow. *Sepals* unequal; 3 larger ones oblong, obtuse, mucronate, serrated, the serratures tipped with black glands, with many pellucid nerves and a few black streaks beneath; 2 smaller ones narrower, lanceolate, glanduloso-serrate. *Petals* with a row of black glands on the margin beneath, about one-third longer than the sepals. *Capsule* conical, triquetrous, rounded at the angles. *Styles* very short. *Seeds* numerous, oblong, reticulato-punctate.

    "Far diffus'd
    And lowly creeping, modest and yet fair,
    Like virtue, thriving most where little seen."

6. H. *hirsutum*, Linn. *Hairy St. John's-wort.* Stem erect, round, hairy; leaves shortly petiolate, ovato-elliptical, obtuse, pellucido-punctate, pubescent; flowers in terminal panicles; sepals lanceolate, smooth, glanduloso-serrate; capsule ovato-acuminate,

---

trigonous, longitudinally striated; seeds oblongo-cylindrical, longitudinally papillose. *E. Bot. t.* 116. *E. Fl. v. iii. p.* 328. *Hook. Br. Fl. p.* 338.

Woods and thickets; not very common. *Fl.* July. ♃.

Woods near Buildwas; *Mr. F. Dickinson.* Wenlock; *W. P. Brookes, Esq.!* Wyre forest; *Mr. G. Jorden.* Ludlow; *Miss Mc. Ghie.* Balderton near Middle; *W. W. Watkins, Esq.*

Hedges and fields Welbach. Sutton Spa near Shrewsbury. Hedges near Stokesay.

Root fibrous. *Stem* 2 feet high, erect, simple or scarcely branched, round, clothed with horizontal and decurved soft hairs. *Leaves* opposite, in alternate pairs, on very short petioles, ovato-elliptical, obtuse, pubescent, under surface paler, somewhat glaucous, margins recurved, ciliated, with 7 pellucid ribs and numerous scattered pellucid dots. *Flowers* in axillary and terminal forked panicles, interspersed with small lanceolate serrated *bracteas*, serratures with black globose glands. *Sepals* lanceolate, smooth, 3-ribbed, the interstices pellucid, serrated, serratures tipped with black globose glands. *Petals* linear-oblong, pale yellow, veiny, tipped with a black pedicellate globose gland. *Capsule* ovato-acuminate, trigonous, trifid at the apex, faintly longitudinally striated. *Styles* 3, deciduous. *Seeds* oblongo-cylindrical, round at both ends, reddish, marked with minute papillæ in longitudinal rows.

7. H. *montanum*, Linn. *Mountain St. John's-wort.* Stem erect, round, glabrous; leaves sessile, amplexicaul, ovato-oblong, glabrous, pellucido-punctate, with black glandular dots near the ciliated margin; flowers paniculato-corymbose; sepals lanceolate, acute, glabrous, glanduloso-serrate; capsule ovate, triquetrous, angles rounded, glabrous; seeds with longitudinal crisped furrows. *E. Bot. t.* 371. *E. Fl. v. iii. p.* 327. *Hook Br. Fl. p.* 337.

Bushy hills in the limestone districts; rare. *Fl.* July. ♃.

Ashford; Ludford and Steventon, near Ludlow; *Miss Mc Ghie.* Standhill Coppice, Wenlock; *W. P. Brookes, Esq.!*

Root fibrous. *Stems* 2 feet high, erect, simple, round, glabrous. *Leaves* opposite, sessile, amplexicaul, ovato-oblong, somewhat acute, glabrous, with numerous pellucid dots and a row of black glandular dots near the margin visible on both surfaces, 7-ribbed at the base, paler on the under surface, the prominent veins more or less clothed with mealy pubescence, margins very minutely ciliated; upper leaves few, distant. *Flowers* in terminal dense corymbose panicles, each division or fork subtended by a pair of opposite lanceolate acute *bracteas* with strongly serrated margins, serratures tipped with large black globular glands. *Sepals* lanceolate, acute, glabrous, 3-ribbed, margins strongly serrated and tipped with black globular glands. *Petals* pale lemon-coloured, elliptical, entire, without glands or dots. *Anthers* with a black glandular dot. *Stigma* capitate, purple. *Capsule* ovate, glabrous, triquetrous, angles rounded. *Styles* 3, half the length of the capsule. *Seeds* numerous, oblongo-cylindrical, rounded or very obtusely pointed at each end, brown, with longitudinally crisped furrows.

Hooker (*Br. Fl.*) describes the "margins of the leaves," though I presume by mistake, as having black glandular "*serratures.*"

8. H. *pulchrum*, Linn. *Small upright St. John's-wort.* Stem erect, round, slender, smooth; leaves amplexicaul, cordate, obtuse, decurved at the margins, glabrous, pellucido-punctate; flowers in subdichotomous panicles; sepals broadly ovate, glabrous, glanduloso-

serrate; capsule ovate, glabrous. *E. Bot. t.* 1227. *E. Fl. v.* iii. p. 329. *Hook. Br. Fl.* p. 338.

Hedge banks, dry bushy heaths; not uncommon. *Fl.* July. ♃.

Hughley common, near Wenlock; *W. P. Brookes, Esq.!* Welbach and Lyth Hill; *Mr. T. Bodenham.* Westfelton; *J. F. M. Dovaston, Esq.* Stanley's Coppice, Coalbrookdare; *Mr. F. Dickinson.* Wyre Forest; *Mr. G. Jorden.* Whitecliff, near Ludlow; *Mr. H. Spare.* Near Oswestry; *Rev. T. Salwey.*

Sharpstones Hill. Nesscliffe. Lyd Hole, near Pontesford Hill. Lane leading from Longden to Oaks Hall, near Pontesford Hill. Banks of the Canal near the Tunnel at Oteley, near Ellesmere.

Root woody. *Stem* 12—18 inches high, erect, nearly simple, round, smooth, slender, with many short leafy axillary branches. *Leaves* in opposite pairs, sessile, cordate, obtuse, decurved at the margins, smooth on both surfaces, slightly glaucous, thick and coriaceous, with pellucid dots; those on the young axillary branches, smaller, oblong, obtuse. *Flowers* in opposite erect axillary and terminal subdichotomous panicles, with a pair of opposite ovate *bracteas* at the ramifications. *Calyx* deeply 5-cleft, segments broadly ovate, glabrous, margins serrated, serratures with black globose glands, with 3 principal pellucid ramified ribs.

9. H. *elodes,* Linn. *Marsh St. John's-wort.* Stem trailing, roundish, shaggy; leaves sessile, semi-amplexicaul, roundish-ovate, obtuse or notched, shaggy, minutely pellucido-punctate; panicles few-flowered; sepals ovate, obtuse, glabrous, glanduloso-serrate; capsule ovato-cylindrical, longitudinally ribbed, 1-celled; seeds elliptical, longitudinally furrowed. *E. Bot. t.* 109. *E. Fl. v.* iii. p. 330. *Hook. Br. Fl.* p. 338.

Spongy boggy margins of pools; not unfrequent. *Fl.* July, August. ♃.

Bog near Ellesmere; *Rev. A. Bloxam.* Titterstone Clee Hill; *Rev. W. Corbett.* Hancot pool; *Mr. F. Dickinson.* Moss, Walford. Bog at Knockin Heath. Boreatton park; *T. C. Eyton, Esq.* Near Ludlow; *Miss Mc Ghie.* Clun Dale, Longmynd; *Rev. S. P. Mansel.* Shomere and Knockin Heath pools. Pits by the side of the road on Shawbury Heath and on the wet parts of the heaths, plentifully; *Rev. E. Williams's MSS.*

Bomere pool. Berrington pool. Oxon pool.

Stem trailing or prostrate at the base, jointed, throwing out numerous fibres from the joints and long scaly runners, ascending in the upper or flowering portion, roundish or obsoletely angular, shaggy, simple or branched. *Leaves* opposite, sessile, semiamplexicaul, roundish-ovate, obtuse or notched at the apex, shaggy on both sides, minutely pellucido-punctate, 5—7-ribbed. *Flowers* in terminal or axillary panicles, peduncles long naked downy bifid and bracteated in the upper part, each branch bearing 3—6 flowers with a solitary one in the axil, on glabrous pedicels longer than the sepals, each subtended by a pair of ovate obtuse *bracteas* with serrated margins tipped with crimson globular glands. *Calyx* ½ 5-fid, *sepals* ovate, obtuse, glabrous, about ¼ the length of the petals, margins strongly serrated and tipped with crimson globular glands, 3-ribbed, lateral ribs branched immediately above the base, the interstices with longitudinal pellucid streaks. *Petals* pale yellow with green ribs. *Styles* nearly as long as the capsules. *Capsules* ovato-cylindrical, longitudinally ribbed, 1-celled. *Seeds* elliptical, longitudinally furrowed, pale brown.

CLASS XIX.

SYNGENESIA.

—

"Star of the mead! sweet daughter of the day,
Whose opening flower invites the morning ray,
From thy moist cheek and bosom's chilly fold
To kiss the tears of eve, the dew-drops cold!
Sweet daisy, flower of love! when birds are pair'd,
'Tis sweet to see thee, with thy bosom bar'd,
Smiling in virgin innocence serene,
Thy pearly crown above thy vest of green.
The lark, with sparkling eye and rustling wing,
Rejoins his widow'd mate in early spring,
And, as he prunes his plumes of russet hue,
Swears on thy maiden blossom to be true."

LEYDEN.

CLASS XIX.

SYNGENESIA. *Anthers united into a tube. Flowers compound.—Nat. Ord.* COMPOSITÆ, *Juss.*

"In this Class, the *flowerstalk* is enlarged at the summit into a *receptacle,* which bears a great numerb of distinct, but closely placed, small *flowers* or *florets,* surrounded by a many-leaved *involucre,* so that the whole looks like one flower. Each *floret* has an inferior *germen,* the upper part frequently expanding into a hairy or feathery *calyx* called a *pappus,* and becoming a 1-seeded *fruit* (*achenium*). The *corolla* is of one *petal,* either tubular, or ligulate. *Stamens* 5. *Style* single. *Stigma* bifid."—HOOKER.

ORDER I. ÆQUALIS. *All the flowers perfect.*

* *All the Corollas ligulate or strap-shaped.*

1. TRAGOPOGON. *Achenium* longitudinally striated, beaked. *Pappus* feathery. *Receptacle* naked. *Involucre* simple, of several bracteas.—Name from τραγος, *a goat,* and πωγων, *a beard;* from the bearded appearance of the pappus.

2. HELMINTHIA. *Achenium* transversely striated, beaked. *Pappus* feathery. *Receptacle* naked. *Involucre* double; inner one of 8 close bracteas, outer of 4 or 5 large lax leary ones.—Name from ἑλμινς, ἑλμινθος, *a worm* or *moth,* and θηκη, *a case;* from the form of the achenium.

3. PICRIS. *Achenium* transversely striated, without a beak. *Pappus* with the inner hairs feathery, sessile. *Receptacle* alveolate, lacinulate. *Involucre* of many compact upright unequal bracteas, with several small lax linear ones.—Name from πικρος, *bitter.*

4. APARGIA. *Achenium* beaked. *Pappus* feathery. *Receptacle* naked. *Involucre* unequally imbricated, with hirsute black bracteas. —Name of uncertain origin.

5. THRINCIA. *Achenium* tapering into a beak. *Pappus* of the florets within the leaves of the involucre forming a short scaly cup, of the rest long feathery. *Receptacle* naked. *Involucre* imbricated. —Name from θριγκος, the *battlements of a fortification;* in allusion to the scaly pappus.

6. HYPOCHŒRIS. *Achenium* striated, often beaked. *Pappus* feathery. *Receptacle* chaffy. *Involucre* oblong, imbricated.—Name from ὑπο, *for,* and χοιρος, *a hog,* the roots being eaten by that animal.

7. LACTUCA. *Achenium* with a long beak. *Pappus* pilose. *Receptacle* naked. *Involucre* imbricated, cylindrical, few-flowered; its bracteas with a membranous margin.—Name from *lac, milk,* the plant abounding with a milky juice.

8. CREPIS. *Achenium* narrower upwards, striated. *Pappus* pilose, copious, soft, mostly white, deciduous. *Receptacle* naked. *Involucre* with scaly bracteas at the base.—Name from κρηπις, *a shoe.*

9. SONCHUS. *Achenium* transversely wrinkled, without a beak. *Pappus* pilose. *Receptacle* naked. *Involucre* imbricated with 2 rows of unequal at length connivent bracteas, tumid at the base.—Name σογχος, the Greek name for the plant.

10. LEONTODON. *Achenium* with a very slender beak. *Pappus* pilose. *Receptacle* naked. *Involucre* imbricated with bracteas, of which the outermost are frequently lax and flaccid.—Name from λεων, λεϑντος, *a lion,* and οδους, *a tooth ;* from the tooth-like lobes of the leaves.

11. HIERACIUM. *Achenium* angular, furrowed, with an entire or toothed margin at the top. *Pappus* pilose, in one row, sessile, frequently brownish, persistent. *Receptacle* nearly naked, dotted. *Involucre* imbricated.—Name from ιεραξ, *a hawk ;* because birds of prey were supposed to employ this plant to strengthen their powers of vision.

12. LAPSANA. *Achenium* compressed, striated. *Pappus* 0. *Receptacle* naked. *Involucre* in a single row of erect bracteas, with small ones at the base.—Name from λαπαζω, *to purge ;* from its laxative qualities.

13. CICHORIUM. *Achenium* turbinate, striated. *Pappus* sessile, scaly, shorter than the achenium. *Receptacle* naked or slightly hairy. *Involucre* of 8 bracteas, surrounded by 5 smaller ones at the base. (Flowers *blue.*)—Name *chikoùryeh,* the Arabic name.

** *Corollas all tubular and generally spreading so as to form a hemisphærical head, style jointed upwards.*

14. ARCTIUM. *Achenium* 4-sided. *Pappus* short, pilose. *Receptacle* chaffy. *Involucre* globose, the bracteas with an incurved hook at the point.—Name from *arceo, to hold fast ;* from the pertinacious adherence with which the involucres stick to animals.

15. SERRATULA. *Diœcious. Achenium* obovate. *Pappus* in 3-4 rows, of which the interior is the longest. *Receptacle* bristly or chaffy. *Involucre* oblong, imbricated with unarmed bracteas. *Anthers* muticous.—Name *serrula, a little saw,* from the serratures of the leaves.

16. CARDUUS. *Pappus* pilose rough or feathery, united by a ring at the base and deciduous. *Receptacle* bristly. *Involucre* tumid, imbricated with spinous bracteas.—Name from *ard,* Celtic, a *point ;* in allusion to the spinous armature of the plant.

17. ONOPORDUM. *Achenium* 4-angled. *Pappus* pilose, rough, united into a ring at the base and deciduous. *Receptacle* honeycombed. *Involucre* tumid, imbricated ; the bracteas spreading and spinose—Name from *ovos,* an *ass,* and περδω, *pedere ;* such being the effect, according to Pliny, upon the ass who eats it.

18. CARLINA. *Pappus* feathery. *Receptacle* chaffy. *Involucre* imbricated, tumid, the outer bracteas with numerous spines, the inner coloured, spreading, resembling a ray.—Name contracted from *Carolina,* from a tradition that the plant was shown by an angel to *Charlemagne,* as a remedy for the plague.

(See *Centaurea* in ORD. FRUSTRANEA.)

*** *Corollas all tubular but parallel, erect and crowded, forming a level top, without a ray, except casually.*

19. BIDENS. *Pappus* of 2—5 persistent awns, which are rough with minute deflexed prickles. *Receptacle* chaffy. *Involucre* of many bracteas ; the outer ones often leafy. *(Corollas sometimes radiant).*—Name from *bis, double,* and *dens, a tooth ;* from the awns or teeth which crown the fruit.

20. EUPATORIUM. *Pappus* pilose and rough or feathery. *Receptacle* naked. *Involucre* imbricated, oblong. *Florets* few. *Styles* much exserted.—Name from *Eupator,* the surname of Mithridates, king of Pontus, who brought the plant into use.

21. TANACETUM. *Achenium* crowned with a membranous margin or *pappus. Receptacle* naked. *Involucre* hemisphærical, imbricated. *Florets* of the ray trifid, sometimes wanting.—Name altered from Athanasia ; *a, not,* and θανατος, *death ;* or that which does not quickly fade.

22. ARTEMISIA. *Pappus* 0. *Involucre* few-flowered, ovate or rounded, imbricated. *Florets* of the ray, if any, slender, awlshaped. —Name from *Artemis,* the Diana of the Greeks.

23. GNAPHALIUM. *Pappus* pilose, the hairs often thickened upwards. *Receptacle* naked. *Involucre* scarious, imbricated, often coloured. *Florets* of the circumference, filiform.—Sometimes *diœcious.*—Name from γναφαλον, *soft down* or *wool,* with which the leaves are clothed.

24. FILAGO. *Pappus* pilose, caducous. *Receptacle* chaffy in the circumference. *Involucre* imbricated, conical, of few acuminated scales. *Florets* 4-toothed, those of the circumference filiform.— Name from *filum, a thread ;* from the thread-like hairs with which the whole plant is covered.

25. PETASITES. Nearly *diœcious. Pappus* pilose. *Involucre* imbricated in two rows of lanceolate scales. *Scapes* many-flowered, appearing before the leaves. Name from πετασος, a *covering to the head* or *umbrella,* from the great size of the foliage.

(See some species of *Senecio,* in the following ORDER.)

ORD. II. SUPERFLUA.

Florets of the centre tubular, perfect (having anthers and pistils) ; those of the circumference with pistils only, (thus as it were superfluous) and ligulate, forming a ray, all bearing seed.

* *Pappus pilose.*

26. TUSSILAGO. *Pappus* pilose. *Florets* of the ray long, narrow, numerous ; of the disk few (both yellow.) *Receptacle* naked. *Involucre* formed of a single row of equal linear bracteas. (Scapes *single-flowered, appearing before the leaves.)*—Name from *tussis,* a *cough ;* in the cure of which the plant has been employed.

27. ERIGERON. *Pappus* pilose, rough. *Florets* of the ray numerous, in many rows, very narrow, (mostly of a different colour from the disk.) *Receptacle* naked. *Involucre* imbricated with linear bracteas.—Name from ερι, *early,* and γερων, *an old man ;* from the baid heads of the receptacles after the flowers and fruit have fallen.

28. SENECIO. *Pappus* pilose. *Receptacle* naked. *Involucre* cylindrical, its bracteas linear equal, with several smaller ones at the bases, their tips often brown.—*(Flowers,* in the British species, yellow, their ray sometimes wanting.)—Name from *senex, an old man,* from the same reason as *Erigeron.*

29. SOLIDAGO. *Pappus* pilose, rough, in one row. *Receptacle* naked. *Involucre* closely imbricated. *Florets* of the ray few, in one row, and as well as those of the disk, yellow.—Name from *solidari, to unite,* from the attributed vulnerary properties.

30. INULA. *Achenium* beaked. *Pappus* pilose, in one row. *Receptacle* naked. *Involucre* imbricated. (Flowers *yellow. Anthers* with bristles at their base.)—Name said to be the same as *Helenium,* having sprung from the tears of *Helen.*

31. PULICARIA. *Achenium* not beaked. *Pappus* double ; *outer one* cupshaped, membranous, toothed ; *inner* pilose, rough. *Receptacle* naked. *Involucre* hemisphærical, closely imbricated with numerous bracteas. *(Flowers* yellow. *Anthers* with bristles at their base.)— Name from *pulex, a flea,* which is supposed to be driven away by its powerful smell.

32. DORONICUM. *Pappus* pilose, wanting in the florets of the ray. *Receptacle* naked, or nearly so. *Involucre* with the bracteas equal, in a double row. (Flowers *yellow.)*—Name from δωρον, a *gift,* and νικη, *victory ;* because it is said to have been formerly used to destroy wild beasts.

** *Pappus chaffy or none.*

33. BELLIS. *Pappus* none. *Receptacle* naked, conical. *Involucre* hemisphærical, its bracteas obtuse, equal, in a single row. *(Scape* single-flowered.)—Name from *bellus, pretty.*

34. CHRYSANTHEMUM. *Pappus* 0. *Receptacle* naked. *Involucre* hemisphærical or nearly flat ; the bracteas imbricated, membranous at the margins.—Name from χρυσος, *gold,* and ανθος, a *flower ;* from the colour of the blossoms of some species.

35. PYRETHRUM. *Achenium* crowned with a membranous border. *Receptacle* naked. *Involucre* hemisphærical or nearly flat, the bracteas imbricated, membranous at their margins.—Name from its resemblance to the πυρεθρον of Dioscorides, so called from πυρ, *fire,* on account of its acrid roots.

36. MATRICARIA. *Pappus* 0. *Receptacle* naked. *Involucre* hemisphærical or nearly flat, the bracteas imbricated, obtuse, not membranaceous at their margins.—Name from its reputed medical virtues.

37. ANTHEMIS. *Pappus* a membranous border, or 0. *Receptacle* convex, chaffy. *Involucre* hemisphærical or nearly plane, the bracteas imbricated, membranaceous at their margins. Name from ανθεμον, a *flower ;* from the profusion of its blossoms.

38. ACHILLÆA. *Pappus* 0. *Receptacle* flat, chaffy. *Involucre* ovate, imbricated. *Florets* of the ray 5—10, roundish or obcordate. Name from *Achilles* who is said to have first discovered its healing qualities.

(See *Bidens, Artemisia, Tanacetum* in ORD. EQUALIS. DIV.***)

ORD. III. FRUSTRANEA.

Florets of the disk perfect and fertile ; those of the circumference neuter ; all tubular.

39. CENTAUREA. *Pappus* pilose or 0. *Receptacle* bristly. *Involucre* imbricated. *Florets* of the ray narrow, funnelshaped, irregular, longer than those of the disk, (sometimes wanting.)—Name, because with this plant it is said the *Centaur Chiron* cured himself of a wound received in the foot from Hercules.

(See *Anthemis Cotula* in ORD. SUPERFLUA.)

SYNGENESIA—ÆQUALIS.

1. TRAGOPOGON. *Linn.* Goat's-beard.

1. T. *minor,* Fries. *Small yellow Goat's-beard.* Involucral bracteas twice as long as the florets; peduncles equal, swollen immediately beneath the flowers; leaves dilated at the base, tapering upwards into a long acuminate point, complicate, finely serrulate, smooth. *Fries Nov. 2nd ed. p.* 241. *T. pratensis var. minor. Fries Nov.* 1*st ed. p.* 95. *T. pratensis β. Hook. Br. Fl.* 4*th ed. p.* 288. *(excl. syn.) T. pratensis β. undulatus, Drej. Fl. Hafn. n.* 789. *T. major, (not T. major, Jacq.) Johnston Fl. Berw. vol.* 2. *p.* 286. *Hook. Br. Fl.* 3*rd ed. p.* 344. (*excl. syn.*) *T. pratensis, Bab. Fl. Bathon. p.* 29. *Poll. Fl. Palat. vol.* 2. *p.* 365. *(in part.) Relh. Fl. Cantab. p.* 290. *(in part.) With. Bot. Arr.* 3*rd ed. p.* 672. *(in part.) T. minus, DC. Prod. vol.* 7. *p.* 113.

Meadows and pastures; frequent. *Fl.* June. ♂.

Meole Brace, Shelton Rough, Green Fields, Canal side, all near Shrewsbury.

Root tapering. *Whole Plant* smooth and of a glaucous hue. *Stem* 2 feet high, erect, branched above, round, smooth, striated. *Leaves* alternate, erect, dilated and clasping at the base, linear and tapering upwards into a very long acuminated obtuse callous point, strongly keeled and complicate, margins finely serrulate, both surfaces covered with minute crystalline dots. *Peduncles* angular, very much swollen and hollow immediately beneath the flowers. *Flower* solitary, yellow, expanding only in the early morning. *Outer Florets* about half as long as the involucral bracteas, with a thick fleshy circle of erect yellow hairs at the orifice of the tube, the inner florets gradually shorter, the ray truncate, with 5 shallow obtuse teeth thickened and glandular at the apices. *Pollen* yellow, globular, covered with minute papillæ. *Involucral bracteas* 8—9, in two rows, connected at the base, lanceolato-subulate, obtusely pointed, strongly keeled, glabrous on the exterior, subpubescent on the interior surface, purplish at the smooth margins, twice the length of the outer florets, increasing in size as the fruit advances towards maturity when they become equal to the fruit in length and converge over it in a conical form, but are finally reflexed. *Achenium* linear-lanceolate, somewhat curved, angular and deeply longitudinally furrowed, the interstices covered with obtuse raised points, the angles with erect teeth, tapering into a round beak longer than the achenium bearing at its summit a circular appendage of crisped and curled white hairs immediately above which arise the long rays of the plumose *pappus.* *Receptacle* with very short thick ovate acute fleshy scales.

This plant appears to have been incorrectly confounded with the true *T. pratensis, Linn.* and is in all probability the commoner species, and generally diffused throughout England, although hitherto overlooked. The character from the proportional length of the florets and involucral bracteas is very constant. Possibly it may be the "*Tragopogon minus angustifolium, G. E.* 735. 3." mentioned in *How's Phytologia* and *Merrett's Pinax,* and noted in the *Indic. Pl. dub. in Raii Syn. 3rd ed.* as occurring "a mile on this side Epping, in the Forest." Certainly it is not *T. major, Jacq. Austr. t.* 29, which is referred by Fries (*l. c.*) as a yellow variety of *T. porrifolius,* inasmuch as that plant, though corresponding with ours in the relative length of the involucral bracteas and florets, and in the peduncles thickened upwards, is invariably described as having the rays of the florets "apice rotundatis," whilst in our plant they are truncate. On communicating to Professor Don a specimen of our plant with my reasons for suspecting it to be the *T. minor, Fries,* he replies:—"You are quite right with respect to *T. minor,* which

I consider a good species, and distinct from *major* and *pratensis.*" The following Shropshire localities for *T. pratensis, Linn.* have been communicated to me, some or all of which may on further research be ascertained to belong to *T. minor*:— Westfelton; *J. F. M. Dovaston, Esq.* Cox Wood, Coalbrookdale; *Mr. F. Dickinson.* Astley; *Mr. E. Elsmere, junr.* Newport; *R. G. Higgins, Esq.* Oakley Park near Ludlow; *Mr. H. Spare.* The Lodge near Ludlow; *Rev. T. Salwey.* The *true T. pratensis, Linn.* I have never myself observed in Shropshire, nor am I aware that its claims to a native of our county have been satisfactorily determined. By way of contrasting the specific differences of the two plants, and of attracting attention to the point, I subjoin the characters of the *true T. pratensis, Linn.* Involucral bracteas equalling the florets; peduncles equal; leaves linear, entire, keeled, subundulate, dilated at the base. "*Linn. Suec. p.* 648." *Fries Nov. ed.* 1. *A. Reich. Fl. Excurs. n.* 1850. *Poll. Fl. Palat. l. c.* (*in part.*) *Leers Fl. Herborn. n.* 601. *E. Bot. t.* 434. *Fl. Br. p.* 812. *E. Fl. v. iii. p.* 337. *Hook. Br. Fl.* 3*rd ed. p.* 344. *Lindl. Syn. p.* 161. *Huds. Fl. Angl. p.* 335. *With.* 3*rd ed. p.* 672. (*in part.*) *Hook. Fl. Scot. p.* 226. *Mackay Fl. Hib. p.* 165. *Purt. Midl. Fl. p.* 364. *Relh. Fl. Cantab. p.* 290. (*in part.*) *Sibth. Fl. Oxon. p.* 236. *T. pratensis, var. α. Hook. Br. Fl.* 4*th ed. p.* 288. *Drej. Fl. Hafn. n.* 789. *T. pratensis, var. α. subvar. b. Fries Nov.* 2*nd ed. p.* 240. *T. luteum, Raii Syn.* 2*nd ed. p.* 76. 3*rd ed. p.* 171. *T. calycibus corollæ radium æquantibus, &c. Hill Fl. Angl. p.* 392.

2. HELMINTHIA. *Juss.* Ox-tongue.

1. H. *echioides,* Gærtn. *Bristly Ox-tongue. Hook. Br. Fl. p.* 345. *Picris echioides, Linn. E. Bot. t.* 972. *E. Fl. v. iii. p.* 339.

Borders of fields; rare. *Fl.* June, July. ♃.

Stanton Lacy; *Mr. H. Spare.*

Root tapering. *Whole plant* clothed with rigid spines trifid and recurved at the apex, arising from tubercular bases. *Stem* 2-3 feet high, erect, round, furrowed, solid, branched. *Lower leaves* lanceolate, tapering at the base, which is cordately auricled and semiamplexicaul; *upper ones* sessile, cordate and amplexicaul. *Flowers* on solitary short peduncles, bright yellow. *Outer involucral bracteas* 5, broad, cordate, acuminate, spiny, loose, spreading, leafy, as long as the inner ones, which are 8, erect, lanceolate, membranous, equal, acute, and downy at the apex, the midrib bristly and separating at a short distance below the apex into a strong fimbriated bristly awn. At the base of the inner involucral bracteas alternate with them and about one-third their length, are smaller membranous nearly glabrous scaly bracteas. *Achenium* transversely wrinkled, crowned with a long beak. *Pappus* plumose. *Receptacle* naked.

3. PICRIS. *Linn.* Picris.

1. P. *hieracioides.* Linn. *Hawk-weed Picris.* Whole plant hispid with forked bristles; leaves lanceolate, sinuato-dentate, wavy, decurved; flowers corymbose; peduncles with many bracteas; involucre slightly hoary, hispid with forked bristles. *E. Bot. t.* 196. *E. Fl. v. iii. p.* 339. *Hook. Br. Fl. p.* 345.

Borders of fields, waste places, &c.; not very common. *Fl.* July, Aug. ♂.

Cox wood, Coalbrookdale; *Mr. F. Dickinson.* Ditch banks about Eaton Marscot; *Rev. E. Williams's MSS.*

Haughmond Hill. White Horse fields Shrewsbury. Shelton Rough.

Root tapering, strong. *Whole plant* hispid with strong spreading bristles forked at their extremities. *Stem* 3 feet high, erect, branched in the upper part, somewhat angular, ribbed and deeply furrowed, solid. *Leaves* lanceolate, decurved,

sinuato-dentate, obtusely pointed, tipped with a callous tubercle, hispid on both sides, margins wavy; radical ones tapering into long winged petioles; upper ones sessile, gradually decreasing in size upwards. *Flowers* large, in terminal bracteated corymbs at the extremities of the branches, bright yellow. *Peduncles* with several scattered bracteas and a larger one at the base, single-flowered. *Involucre* hemisphærical, imbricated, sparingly clothed with scattered minute white cottony pubescence. *Involucral bracteas* lanceolate, acute, outer ones short loose and spreading downy at the apex; inner ones long, erect, ciliated and downy at the margins, strongly keeled, all with the keel and lower portion armed with strong spreading bristles forked at the extremities. *Florets* yellow, 5-toothed, teeth much thickened and glandular at the back; orifice of the tube clothed with long yellow hairs. *Achenium* linear-oblong, angular, transversely wrinkled, and denticulate. *Pappus* plumose, sessile. *Receptacle* pitted, the interstices raised into a jaggedly toothed margin.

4. APARGIA. *Schreb.* Hawkbit.

1. A. *hispida,* Willd. *Rough Hawkbit.* Scape simple, single-flowered, peduncles swollen upwards, hispid; leaves oblongo-lanceolate, runcinate, hispid with forked hairs; involucre ovate, hispid. *Hook. Br. Fl.* 4*th ed. p.* 289. *E. Fl. v. iii. p.* 351. *Hedypnois hispida, Huds. E. Bot. t.* 554. *Leontodon hispidus, Linn.*

Meadows, pastures, &c.; frequent. *Fl.* June, July. ♃.

Root woody. *Leaves* all radical, oblongo-lanceolate, acute, runcinate, tapering at the base, hispid on both surfaces with prominent forked or 3-cleft hairs. *Peduncles* 1 foot high, erect, simple, single-flowered, round below, furrowed upwards, swollen beneath the involucre, hispid with forked horizontal spreading hairs. *Involucre* ovate, with dense long white spreading simple and forked hairs; inner bracteas linear-lanceolate, prominently keeled, the margins densely ciliated near the obtuse apex; outer bracteas smaller, irregularly placed. *Flowers* large, bright yellow. *Florets* with long yellow erect hairs about the exterior of the summit of the tube and with a mass of brown glands on the back of each tooth of the apex. *Achenium* slender, linear, transversely wrinkled. *Pappus* sessile, of 11 long rays with intermediate shorter exterior ones, feathery. *Receptacle* deeply pitted, intermediate margins raised and irregularly toothed acuminated into longish bristles.

2. A. *autumnalis,* Willd. *Autumnal Hawkbit.* Scape branched, scaly upwards, peduncles swollen upwards; leaves linear-lanceolate, toothed or pinnatifid, nearly glabrous; involucre cylindrical, pubescent. *E. Fl. v. iii. p.* 353. *Hook. Br. Fl.* 4*th ed. p.* 289. *Hedypnois autumnalis, Huds. E. Bot. t.* 830. *Leontodon autumnalis, Linn.*

Meadows and pastures; frequent. *Fl.* August. ♃.

Root premorse, with long stout lateral fibres. *Leaves* chiefly radical, linear-lanceolate, obtuse, tapering gradually into a long footstalk, very variable in their division, sometimes nearly entire, sometimes shallowly dentate, sometimes sinuato-dentate, and sometimes pectinato-pinnatifid, nearly glabrous or with a few scattered bristles both on the upper surface and on the stout midrib beneath. *Stems* 1 foot high, several, spreading and ascending, deeply furrowed and ribbed, with minute cottony pubescence in the furrows, dichotomously branched about midway, with a linear entire glabrous bractea at the base of the divisions, the lowermost one very much acuminate, hollow internally and containing a loose white cottony substance. *Peduncles* round, deeply furrowed and ribbed, with slight cottony pubescence in the furrows, bearing scattered scales which become numerous immediately beneath the involucre into which the peduncle gradually swells and becomes hairy. *Involucre* cylindrical, imbricated, covered sparingly with loose white cottony pubes-

cence; *outer involucral bracteas* short, erect, appressed; *inner ones* very long, linear, acuminate, the apex downy at the margins, with a strong keel armed with dark brown tawny spreading hairs. *Flowers* moderately large, yellow, orifice of tube hairy, 5-toothed, teeth deeply cut lanceolate acute. *Achenium* linear, transversely striated and denticulate. *Pappus* sessile, plumose. *Receptacle* naked, pitted.

5. THRINCIA. *Roth.* Thrincia.

1. T. *hirta,* Roth. *Hairy Thrincia.* Leaves lanceolate, sinuato-dentate, somewhat hispid with frequently forked hairs; scapes single-flowered, ascending, glabrous as well as the involucre. *Hook. Br. Fl.* 4*th ed. p.* 289. *Apargia hirta, Hoffm. E. Fl. v. iii. p.* 352. *Hedypnois hirta, E. Bot. t.* 555. *Leontodon hirtum, Linn.*

Gravelly pastures; not uncommon. *Fl.* July, August. ♃.

Near Madeley Lane pits; *Mr. F. Dickinson.* Oakley park, near Ludlow; *Mr. H. Spare.*

Bickley Coppice, near Montford Bridge.

Root premorse. *Leaves* all radical, spreading, lanceolate, obtuse, tapering at the base into a narrow winged petiole, sinuato-dentate, rough on both surfaces with long white hairs bifid or trifid at the apex. *Scapes* several, very considerably longer than the leaves, hairy below, smooth and striated above, naked. *Involucral bracteas* erect, glabrous, *outer ones* subulate, very short; *inner ones* lanceolate, membranous at the edges, downy on the margins at the apex. *Florets* yellow, hairy at the orifice of the tube, deeply toothed. *Achenium* linear, tapering at both ends, that of the *outer florets* smooth, longitudinally striated, crowned with a very short cup of scales inciso-dentate at the apex. *Achenium* of the *inner florets* longitudinally striated, transversely marked with raised dots or murications, which are largest towards the apex. *Pappus* sessile, plumose, dilated or with membranous margins at the base.

6. HYPOCHŒRIS. *Linn.* Cat's-ear.

1. H. *glabra,* Linn. *Smooth Cat's-ear.* Stem branched, leafless, glabrous; peduncles with small scales; leaves oblongo-lingulate, sinuato-runcinate or sinuato-dentate, slightly hispid; involucre conico-acuminate, glabrous.

Var. a. vera. Bab. Achenia of the marginal florets destitute of a beak, those of the disk with a long beak. *Bab. Prim. Fl. Sarn. p.* 56. *Hypochœris glabra, E. Bot. t.* 575. *E. Fl. v. iii. p.* 375. *Hook. Br. Fl.* 4*th ed. p.* 290. *Koch. Syn. p.* 427.

Hills and dry banky pastures; not very common. *Fl.* July. ⊙.

On a bank by the roadside at Dawley between Wellington and Ironbridge; *E. Lees, Esq.* Abundantly near Wellington; *Watson's New Bot. Guide.* About Eaton and Sundorn. Harmer Hill. Wrekin. Knockin Heath; *Rev. E. Williams's MSS.*

On new red sandstone, Hill of Hopton cliff near Nesscliffe.

Root tapering, fibrous. *Leaves* radical, spreading in a circle on the ground, oblongo-lingulate, obtuse, variable in their divisions, either sinuato-runcinate, the lobes subrhomboid or angular, or sinuato-dentate, the teeth acute, callously tipped, nearly glabrous, except a few scattered bristly hairs on the upper surface and on the midrib beneath, margins ciliated with distant bristly hairs, deep green above, paler beneath. *Stems* numerous, 4—6 inches high, spreading and ascending, branched at about one-third of their height from the base, round, smooth, hollow, with a minute entire ovate somewhat acuminate obtuse erect appressed fleshy scale at the base of the branches, and several similar ones alternately scattered on the peduncles above the branch, very slightly thickened immediately beneath the terminal solitary flower. *Flower* small, bright yellow, florets ligulate, acutely 5-toothed, 6-ribbed, the 4 inner ribs dividing below the sinus and uniting in the teeth; orifice of the tube hairy; apices somewhat thickened and glandular at the back. *Involucre* conico-acuminate, glabrous; *involucral bracteas* regularly imbricated, very variable in size, interior ones linear or linear-lanceolate with an obtuse discoloured point, keel strong green, margins pale and membranous,minutely, serrulate towards the ciliated apex, erect, finally reflexed. *Achenium* linear-elliptical, transversely wrinkled and longitudinally ribbed, the ribs armed with minute erect denticulations; that of the florets of the disk tapering upwards into a long beak twice the length of the seed, entirely scabrous with minute erect denticulations, that of the marginal florets without a beak or with a very short one. *Scales* of the *receptacle* pale, membranous, linear-lanceolate, very much acuminate, as long as the pappus and involucral bracteas. *Pappus* plumose.

Var. β. Balbisii, Bab. All the achenia with long beaks. *Bab. l. c. H. Balbisii, Lois. Reich. Fl. Excurs.* 1782. *Koch. Syn. p.* 427.

On the denuded surfaces of trap rocks on the middle portion of the eastern side of Pontesford Hill; sparingly.

My friend Babington in his *Primitiæ Floræ Sarnicæ*, frequently quoted in these pages, a work for botanical acumen and accuracy worthy of the highest praise, states that he has found "the character taken from the rostrated fruit, which is the only difference between *H. glabra, Linn.* and *H. Balbisii, Lois.* to be very variable; in some heads all the seeds being provided with equally long rostra, and in others upon the same plant, the outer row being nearly deprived of that elongation." In our Shropshire specimens of *var. β.* the seeds were equally rostrate in all the heads; whilst in the plants of *var α.* from the sandstone of Hopton Cliff the seeds of the heads were as uniformly constant to the characters of that variety. Reichenbach *(l. c.)* says of *H. Balbisii*, "Species certe distinctissima, e semine constantissima."

3. H. *radicata*, Linn. *Long-eared Cat's-ear.* Stem branched, leafless, glabrous; peduncles with small scales; leaves oblong, obtuse, runcinate, scabrous; involucral bracteas hairy on the keel. *E. Bot. t.* 831. *E. Fl. v. iii. p.* 377. *Hook. Br. Fl. p.* 353.

Meadows, pastures, way-sides, &c.; frequent. *Fl.* July. ♃.

Root strong, tapering, running deep into the ground. *Stems* several, 1 foot or more high, branched, spreading, angular, smooth, somewhat glaucous, leafless, with small scattered lanceolate scales. *Leaves* all radical, spreading in a circle on the ground, oblong, obtuse, runcinate, the segments and sinuses rounded, rough with long white simple rigid hairs. *Peduncles* thickened upwards, hollow, clothed with small scattered appressed subulate *bracteas*. *Flowers* rather large, bright yellow. *Involucral bracteas* imbricated, unequal, outer ones short, interior ones longer, lanceolate, acuminate, strongly keeled, glabrous except the keel which is rough, with rather long white rigid hairs and a few shorter black ones towards the apex; margins of the upper half minutely ciliated, densely so at the apex. *Florets* with a tuft of yellow hairs at the orifice of the tube, the ray deeply and acutely 5-toothed, glandular at the back, the glands on the exterior teeth frequently assuming the appearance of bristles. *Achenium* striated, tawny. *Pappus* stalked, plumose. *Scales of the receptacle* thin, membranous, lanceolate, acuminate, keeled, distantly ciliated in the upper part.

7. LACTUCA. *Linn.* Lettuce.

1. L. *virosa*, Linn. *Strong-scented Lettuce.* Leaves horizontal, oblongo-lanceolate, denticulate, obtuse, sagittate and amplexicaul at the base, their midrib prickly; flowers panicled. *E. Bot. t.* 1957. *E. Fl. v. iii. p.* 345. *Hook. Br. Fl. p.* 346.

Banks and waysides; rare. *Fl.* August. ♂.

Stanton Lacy; *Mr. H. Spare.*

Root tapshaped. *Stem* 2-3 feet high, solitary, erect, round, smooth, scarcely branched, prickly in the lower part. *Leaves* horizontal, nearly smooth, finely toothed and scabrous or prickly on the margins; radical ones numerous, obovate, undivided, depressed; cauline ones smaller, distant, frequently lobed, sagittate and amplexicaul at the base, midrib prickly. *Flowers* numerous, panicled, the divisions subtended by small cordate acute *bracteas*, which are also scattered on the pedicels. *Involucral bracteas* imbricated, erect, lanceolate, glabrous, downy at the apex, outer ones smaller, lax. *Florets* small, yellow. *Achenium* elliptical, compressed, tapering upwards into a long slender beak, longitudinally ribbed, transversely striated, black. *Pappus* short, rough. *Receptacle* naked.

2. L. *muralis*, Less. *Ivy-leaved Lettuce.* Leaves lyrato-runcinate, angled and toothed, terminal lobe triangular lobed and toothed; panicle with divaricated branches; florets 5. *Hook. Br. Fl. 4th ed. p.* 290. *Prenanthes, Linn. E. Bot. t.* 457. *E. Fl. v. iii. p.* 348.

On old walls and in woods; not very common. *Fl.* July. ♃.

Ludlow; *Mr. H. Spare.* Near Shawbury Heath; *Dr. T. W. Wilson.* Whitecliff near Ludlow; *Miss Mc Ghie.* Lilleshall Abbey. Chetwynd Park wall; *R. G. Higgins, Esq.* Walls at Hord's Park near Bridgnorth; *Purton's Midl. Flora.* Fairyland, Westfelton; *J. F. M. Dovaston, Esq.* Cox Wood, Coalbrookdale; *Mr. F. Dickinson.*

Haughmond Hill and Abbey.

Root woody. *Herb* slender, smooth. *Stem* 2 feet high, round, hollow, branched above, somewhat zigzag. *Leaves* lyrato-runcinate, lobes angled and toothed, dilated and clasping at the base, terminal lobe very large triangular lobed and toothed acute resembling an ivy-leaf, upper surface green, under surface pale and glaucous with more or less of a purplish hue. *Panicle* much branched and divaricated. *Bracteas* small, ovate, acute, amplexicaul. *Flowers* small, erect, light yellow. *Involucre* cylindrical, smooth, *outer bracteas* about 3, small and unequal; *inner ones* 5, equal, linear-lanceolate, obtuse, keeled, membranous at the margins, ciliated at the apex, purplish-green. *Florets* 5, in a single row, tube hairy about the orifice, teeth 5 acute large somewhat converging. *Achenium* obovate, striated, black, tapering into a short *beak* bearing a circular disk downy around the margin from which arises the plumose *pappus*.

8. CREPIS. Linn. *Hawk's-beard.*

1. C. *virens*, Linn. *Smooth Hawk's-beard.* Leaves glabrous, lanceolato-runcinate or remotely dentate, cauline ones pinnatifid and nearly entire, sagittate at the base; panicle subcorymbose; achenium oblong, shorter than the pappus, which latter nearly equals the involucre. *Bab. in Linn. Trans. v. xiii. p.* 455. *Hook. Br. Fl. 4th ed. p.* 291. *C. tectorum, E. Bot. t.* 1111. *E. Fl. iii. p.* 372.

Meadows, pastures, hedge-banks, &c.; common. *Fl.* July, ☉.

Stem furrowed, smooth, purplish, erect, branched above, sometimes branched at the base, diffuse and prostrate. *Leaves* glabrous, very variable in size and form; radical ones tapering below into a winged petiole, lanceolate or linear-lanceolate, obtuse, simply toothed, runcinate, runcinato-dentate or sinuato-dentate; cauline ones lanceolate or linear-lanceolate or linear, sagittate, mostly rounded at the end, sometimes acute, more or less entire, sinuate, runcinate or deeply pinnatifido-run-

cinate; uppermost leaves linear, entire or nearly so, sagittate; all glabrous. *Involucre* oval in bud, afterwards dilated or swollen at the base, conical, equalling the pappus; *outer bracteas* few, appressed, short, subulate; *inner* ones linear-lanceolate, acute, somewhat membranous at the edges, strongly keeled, the keel armed with short black bristles tipped with globose glands. *Florets* yellow, ligulate, orifice of the tube with long erect white hairs, acutely 5-toothed. *Achenium* oblong, glabrous, longitudinally ribbed. *Pappus* rough. *Receptacle* naked.

2. C. *paludôsa*, Mœnch. *Marsh Hawk's-beard.* Glabrous; stem erect, branched upwards and subcorymbose; radical leaves ovato-oblong, runcinato-dentate, attenuated into a footstalk, cauline ones lanceolate, toothed, heartshaped at the base and amplexicaul, much acuminated; involucre glanduloso-pilose; achenium striated, scarcely narrower upwards, about as long as the pappus. *Hook. Br. Fl. 4th ed. p.* 291. *Hieracium paludosum, Linn. E. Bot. t.* 1094. *E. Fl. v. iii. p.* 363.

Moist woods and meadows; rare. *Fl.* August. ♃.

Between Apley near Wellington and Preston on the Wildmoors; *E. Lees, Esq.* Meadows at Maesbury near Oswestry; *Rev. T. Salwey!* In the boggy part of the copse at the top of the hill opposite Mr. Kynaston's house at Hardwicke; *Rev. E. Williams's MSS.*

Root fibrous. *Stem* 18 inches to 2 feet high, erect, hollow, angular, striated and furrowed, branched upwards. *Radical leaves* ovato-lanceolate, acute, tapering at the base gradually into a rather long slightly winged petiole, glabrous on both sides, paler and veiny beneath, margin more or less deeply sinuato-dentate, the teeth or lobes directed towards the base, particularly in the lower portion of the leaf so as to become runcinato-dentate. *Stem leaves* several, amplexicaul by their dilated roundly auricled or cordate bases, sinuato-dentate, the teeth in the lower part pointing downwards. *Flowers* paniculato-corymbose, with minute subulate *bracteas* at the divisions. *Inner involucral bracteas* erect, equal, linear-lanceolate, with an obtuse downy apex, glabrous except the keel which is clothed with long black glandular hairs; *outer bracteas* similarly hairy, unequal and very much shorter, erect, appressed. *Florets* yellow, with 5 obtuse glandularly thickened teeth. *Achenium* linear, longitudinally ribbed. *Pappus* sessile, rough.

9. SONCHUS. *Linn.* Sow-thistle.

1. S. *palustris*, Linn. *Tall Marsh Sow-thistle.* Flower-stalks corymbose and involucre glanduloso-hispid; leaves denticulate, runcinato-pinnatifid, with few segments, arrowshaped at the base, upper ones simply sagittate. *E. Bot. t.* 933. *E. Fl. v. iii. p.* 341. *Hook. Br. Fl. p.* 346.

In a wild lane near Wellington; *E. Lees, Esq.*

Root fleshy, not creeping. *Stem* 6—8 feet high, cylindrical, angular and furrowed, scarcely branched, leafy. *Leaves* large, smooth, margins copiously ciliated with short rigid bristly hairs, lobes narrow acute chiefly directed downwards; uppermost leaves simply sagittate, sessile, linear, serrated. *Flowers* large, yellow. *Flower-stalks* and *involucre* rough with black bristly glandular hairs. *Achenium* furrowed. *Pappus* simple, smooth.

2. S. *arvensis*, Linn. *Corn Sow-thistle.* Root creeping; peduncles subumbellate and with the involucre glanduloso-hispid; leaves sinuato-runcinate, cordate at the base, denticulato-spinulose;

achenium with longitudinal denticulate ribs. *E. Bot. t.* 674. *E. Fl. v. iii. p.* 342. *Hook. Br. Fl. p.* 346.

Corn-fields; frequent. *Fl.* August. ♃.

Root fleshy, creeping. *Stem* 3 feet high, erect, scarcely branched, round, somewhat angular, hollow, glabrous except in the upper portion which as well as the peduncles is clothed with horizontally spreading yellow glandular bristles. *Leaves* alternate, linear-lanceolate, acute, sinuato-runcinate, the margins slightly revolute and sinuato-denticulate, teeth unequal mucronate, dilated and clasping by the rounded auricled base, bright green and glabrous above, glaucous green beneath; upper ones narrower, nearly entire, sinuato-denticulate. *Flowers* imperfectly umbellate, very large and conspicuous, bright yellow. *Involucre* imbricated, ventricose at the base; *outer bracteas* unequal, erect; *inner ones* linear-lanceolate, membranous at the edges, all strongly keeled and armed with yellow glandular bristles. *Achenium* with longitudinal denticulate ribs. *Pappus* sessile, nearly smooth, very slightly rough. *Receptacle* naked.

3. S. *oleraceus*, Linn. *Common Sow-thistle.* Leaves undivided or pinnatifid, toothed, clasping the stem with spreading sagittate auricles, lowest leaves stalked; achenia linear-oblong, longitudinally striated, striæ reflexo-dentate, interstices transversely rugose. *E. Bot. t.* 843. *α. et β. E. Fl. v. iii. p.* 343. *α. Hook. Br. Fl. 4th ed. p.* 293.

Waste places and cultivated ground; frequent. *Fl.* June, August. ☉.

Root fibrous. *Stem* 1-2½ feet high, erect, branched, angular, striated, more or less glaucous and tinged with red. *Leaves* alternate, lower ones oblongo-lanceolate, deeply pinnatifid, segments mucronate unequally lobed and sinuato-dentate, the teeth spinous nearly in the plane of the leaf, more or less glaucous on both sides, especially beneath, on a winged and toothed petiole dilated and clasping at the base; upper leaves more or less pinnatifid, sometimes nearly entire and sinuato-dentate, dilated and clasping at the base by their straight sagittate spreading auricles. *Flowers* subumbellate, yellow, with minute subulate sometimes toothed *bracteas* at the base of the pedicels, which are clothed more or less with a deciduous cottony down. *Involucral bracteas* imbricated; *inner ones* long, linear, prominently keeled, taper-pointed, obtuse, ciliated at the apex; *outer bracteas* shorter and narrower. *Tube* of the *florets* with long silky erecto-patent hairs on the exterior, teeth glandular at the back. *Involucre* converging and conical in *fruit.* *Achenium* linear-oblong, longitudinally striated, the striæ armed with reflexed teeth, transversely wrinkled. *Pappus* rough.

4. S. *asper*, Hoffm. *Rough Sow-thistle.* Leaves undivided or pinnatifid, sharply toothed, clasping the stem with rounded auricles, none on distinct stalks; achenia elliptical, longitudinally striated, striæ reflexo-dentate, interstices smooth. *E. Bot. Suppl. t.* 2765. *et* 2766. *S. oleraceus, γ. et. δ. E. Fl. v. iii. p.* 343. *β. Hook. Br. Fl. 4th ed. p.* 292.

Waste places and cultivated ground; frequent. *Fl.* June—August. ☉.

Root fibrous. *Stem* 1-2½ feet high, erect, branched, angular, striated, more or less glaucous and tinged with red. *Leaves* alternate, oblongo-lanceolate, deeply pinnatifid, segments runcinate unequally lobed and sinuato-dentate, margins crisped and wavy and consequently the spinous teeth turned in various directions, generally shining above, glaucous beneath, occasionally so above, none on distinct petioles, dilated and clasping at the base by their large and rounded auricles. *Flowers* subumbellate, yellow, with minute subulate often toothed *bracteas* at the base of the pedicels. *Tube* silky, and florets with glandular tips, as in the last,

Achenium broader and shorter than in *S. oleraceus*, elliptical, longitudinally striated. the striæ armed with deflexed teeth, smooth. *Pappus* rough.

10. LEONTODON. *Linn.* Dandelion.

1. L. *Taraxacum*, Linn. *Common Dandelion.* Leaves runcinate, glabrous, toothed; achenium longitudinally striated, transversely muricated above, rostrate.

Var a. officinale. Leaves broad, oblong; outer involucral bracteas lax, spreading and recurved. *E. Bot. t.* 510. *E. Fl. v. iii. p.* 349. *Hook. Br. Fl. 4th ed. p.* 292.

Meadows and pastures; common. *Fl.* All summer. ♃.

Root tapshaped. *Leaves* all radical, numerous, spreading, bright shining green, smooth or with scattered hairs, oblong, runcinate, lobes pointing downwards unequally toothed along the upper margins. *Scape* longer than the leaves, erect, smooth, hollow, naked. *Outer involucral bracteas* numerous, lanceolate, obtuse, with a gibbous protuberance at the back immediately below the apex, lax, spreading and recurved, glabrous; *inner ones* erect, very obtuse and somewhat thickened at the back of the apex, cohering in the lower part by their thin pale membranous incurved margins. *Florets* bright yellow, truncate, with 5 minute obtuse thickened and glandular teeth, mouth of tube naked. *Achenium* linear-oblong, compressed, longitudinally striated, transversely muricated in the upper portion, crowned with a long slender beak, bearing the rough *pappus*. *Receptacle* naked.

No plant can be more variable in the breadth and division of the leaves according to situation, whence many doubtful species have been formed. The following extreme forms have been observed:—

Var. β. palustre. Involucral bracteas all erect, appressed. *Hook. Br. Fl. 4th ed. p.* 292. *L. palustre, Sm. E. Bot. t.* 553. *E. Fl. v. iii. p.* 350.

Wet open pastures and moors; not very common. *Fl.* Summer. ♃.

Cound Moor and Abery wood; *Rev. E. Williams's MSS.*

Haughmond Park. Almond Park. Vownog near Westfelton.

Leaves generally narrower than those of *var. a.*, sinuato-dentate, nearly glabrous, frequently nearly entire. *Scape* glabrous or very slightly cottony beneath the involucre. I can perceive no difference in the fruit.

Var. γ. lævigatum. Leaves runcinato-pinnatipartite, lobes unequal acute; involucral bracteas with or without a callous subcorniculate apex, exterior ones lanceolate, appressed or somewhat spreading. *Taraxacum officinale, β. lævigatum, Babington Prim. Fl. Sarn. p.* 57.

Walls and dry hilly pastures. *Fl.* all Summer. ♃.

Sharpstones Hill. Old walls at Pulley near Shrewsbury, and probably common in similar situations throughout the county.

Smaller and stunted owing to situation. *Leaves* very slender, deeply pinnatifid down to the midrib or nearly so, the lobes pointing downwards, elongated or lineari-triangular, acute, distant, with smaller teeth between, more or less clothed with white cottony entangled down, which in some instances is very abundant about the axils of the leaves and on the scapes. *Involucral bracteas* glabrous with a callous corniculate appendage on the back of the apex, outer ones spreading, slightly reflexed.

My specimens were submitted to my friend Babington, who has paid particular attention to the extremely variable forms of this polymorphous Genus, and whose opinion being consequently valuable I here subjoin:—"I have taken some trouble to determine the real name of your Pulley and Sharpstones plant, but

find it quite impossible to settle the plant of De Candolle to which it ought to be referred. It is probably not his *T. lævigatum*, for he says "achenis pallidis." He does not notice the form of the involucrum. Neither is it the plant given under that name by Peterman, (*Fl. Lips.*), for that has a wholly exserted pappus, which ours has not. It is *Taraxacum officinale, β. lævigatum* of my Primitiæ."

11. HIERACIUM. *Linn.* Hawkweed.

* *Scape leafless, single-flowered.*

1. H. *Pilosella*, Linn. *Common Mouse-ear Hawkweed.* Scyons creeping; scape leafless, one-flowered; leaves elliptico-lanceolate, entire, hairy, hoary beneath with dense stellate pubescence. *E. Bot. t.* 1093. *E. Fl. v. iii. p.* 356. *Hook. Br. Fl. p.* 394.

Hedge-banks, dry pastures, &c.; common. *Fl.* May—July. ♃.

Root woody, throwing out many long creeping leafy scyons. *Leaves* all radical, spreading on the ground, elliptic-lanceolate, tapering at the base, the apex slightly recurved and tipped with a callous tubercle, margins entire except a few distant tubercles apparently the rudiments of teeth, upper surface deep green, smooth, covered with long white rigid hairs with tubercular bases, the hairs rough or minutely toothed when viewed under a powerful lens, under surface hoary with dense close-pressed stellate pubescence with longer rigid minutely toothed rigid hairs interspersed; petiole short and dilated, fringed with very long toothed hairs. *Scape* about 6 inches high, erect, round, simple, destitute of leaves and bracteas, clothed with dense close-pressed stellate pubescence with long white rigid minutely toothed horizontally spreading hairs interspersed and in the upper half especially beneath the flower with numerous shorter stout blackish stipitate glands. *Involucral bracteas* in two rows, the outer ones short, the inner ones longer, lanceolate, acuminate, membranous at the margins, clothed with dense close-pressed white stellate pubescence with numerous stout stipitate black glands, thickened at the base and interspersed with longer white minutely toothed spreading hairs similarly thickened and black at the base. *Flowers* solitary, terminal. *Florets* pale lemon-colour with a red central stripe at the back, tube clothed about the orifice with long crisped and curled soft white hairs, limb with 5 obtuse teeth, the exterior ones more deeply cleft, glandular at the back. *Pappus* rough. *Receptacle* with short stout acute scales.

When growing on the exposed and elevated parts of dry hills, the whole plant becomes smaller and the scyons very much abbreviated, or entirely wanting. In 1837, I gathered from the walls of Ludlow Castle a single specimen much exceeding the size of the ordinary state, with very long leafy scyons, the leaves much more elongated, linear-lanceolate, the margins distantly but distinctly and shallowly toothed and clothed with longer and more copious silky hairs, approaching in general appearance the variety *Peleterianum* which occurs on Craig Breidden in Montgomeryshire, but not so copiously hairy, and the involucre destitute of the copious silky hairs so conspicuous in specimens from that station. After inspecting this specimen, Mr. Babington says: "This is certainly *peleterianum*, differing only in its longer scyons. My Jersey plant (*Prim. Fl. Sarn. p.* 58.) takes the first step from the Breidden plant and this takes the second. I am convinced now that the plant cannot be distinguished." I should remark that I have had specimens of the Breidden plant gathered in 1834, in company with Mr. Babington, under cultivation ever since, and that they put forth in the first year after removal, leafy scyons as long or even longer than those of the Ludlow plant and still continue to do so, although in other respects they retain their peculiarities.

** *Stem with a single leaf, many-flowered.*

2. H. *murorum*, Linn. *Wall Hawkweed.* Stem corymbose,

bearing a single leaf, the upper part peduncles and involucres hoary with stellate pubescence and with very black glandular hairs; leaves green, hairy beneath and on the margins, ovate, subcordate, dentate, the basal teeth deeper and retroverse, radical leaves many petiolate persistent, cauline one shortly petiolate or subsessile. "*Linn. Suec. n.* 701. *ex optima parte*" *sec. Fries. Drej. Fl. Hafn. n.* 775. *E. Bot. t.* 2082. *Hook. Br. Fl. 3rd ed. p.* 350. *H. murorum, a. Fries Nov. 2nd ed. p.* 256, *et Mant. alt. p.* 44. *Koch. Syn. p.* 457. *Poll. Pal.* 2. 392. *Reich. Fl. Excurs. n.* 1768. *Hook. Br. Fl. 4th ed. p.* 294.

Walls and rocks, not uncommon. *Fl.* August. ♃.

Root with strong fibres. *Stem* 1-2 feet high, erect, simple, round, striated, hollow, clothed more or less with minute close-pressed stellate pubescence and long white denticulate hairs, in the upper part with more copious stellate pubescence and numerous black glandular hairs, bearing about the middle a single ovate *leaf*, deeply sinuato-dentate especially at the base, glabrous or nearly so on the upper surface, hairy beneath, on a short winged very hairy petiole. *Radical leaves* several, spreading, on rather long winged copiously hairy petioles dilated and clasping at the base, ovate, acute, subcordate at the base, sinuato-dentate, the teeth glandularly tipped, shallow towards the apex, deeper and longer towards the base, radiating in various directions, the lowermost frequently elongated and invariably pointing downwards or backwards, upper surface with few distantly scattered short hairs, bright green, frequently yellowish, under surface of a paler and somewhat glaucous hue, copiously clothed with long white denticulate hairs, especially on the midrb, margins ciliated. *Flowers* large, yellow, in a terminal corymb, the branches peduncles bracteas and involucres hoary with dense minute stellate pubescence, and with numerous black glandular hairs, the principal divisions subtended by subulate acuminate *bracteas*. *Inner involucral bracteas* nearly equal, linear-lanceolate, acuminate, *outer ones* few, smaller, irregularly imbricated, all erect and appressed, of a blackish-green on the keel, paler and submembranous on the margins, clothed with minute white dense stellate pubescence and black glandular hairs, the very apex with dense white simple hairs. *Achenium* black or blackish-brown linear, longitudinally ribbed and furrowed, when viewed under a powerful lens the ribs have very minute erect denticulations, the interstices transversely rough. *Pappus* more than twice the length of the achenium, minutely denticulate, sessile. *Receptacle* with minute elevated teeth on the interstices.

A very variable plant but recognizable amid all its modifications by the form of the leaves, their retroverse basal teeth, and the solitary leaf on the stem which although frequently very small and narrow and even reduced to a scale is never entirely absent.

Fries considers the *H. murorum* of *Smith* and of recent authors as identical with his *var. β. sylvaticum* which has broader leaves, cordate at the base, deeply and retroversely dentate, and is the *H. murorum, β. sylvaticum, Linn. Sp. Pl.* 2. *p.* 1128.

*** *Stem with many leaves, many-flowered.*

3. H. *sylvaticum*, Sm. *Wood Hawkweed.* Stem corymbose, bearing 3 or more leaves, the upper part peduncles and involucres hoary with stellate pubescence and with very black glandular hairs; leaves green hairy beneath and on the margins, ovato-lanceolate, attenuate at both extremities, dentate, basal teeth deeper pointing forwards, radical leaves few, persistent, and with the inferior cauline ones petiolate, superior subsessile. *E. Fl. v. iii. p.* 361. *E. Bot. t.* 2031. *a. Hook. Br. Fl. 3rd ed. p.* 351. *4th ed. p.* 294. "*H.*

murorum, a. Linn. ex parte—plene γ. Sp. Pl. 2. *p.* 1129. *Fl. Suec.*" *sec. Fries. H. murorum, β. Reich. Fl. Excurs. n.* 1768. *H. murorum, γ. caule magis folioso. Poll. Pal.* 2. 392. *H. vulgatum, Fries Nov. ed.* 2. *p.* 258. *et Mant. alt. p.* 48. *Koch. Syn. p.* 455. *Drej. Fl. Hafn. n.* 776.

Mountain woods, rocks, walls and banks; frequent. *Fl.* August. ♃.

Root with strong fibres. *Stem* 1-2 feet high, erect, simple, round, striated, solid, clothed with scattered long white denticulate hairs especially in the lower part, with scarcely any stellate close-pressed pubescence which becomes more copious in the upper part and is intermixed with a few scattered black glandular hairs, bearing at distant intervals several alternate ovato-lanceolate, somewhat acuminate *leaves*, deeply sinuato-dentate, teeth pointing forwards, elongated, acute and glandularly tipped, glabrous or nearly so on the upper surface, hairy beneath, the lower ones tapering at the base into a short winged very hairy petiole, the upper ones gradually diminishing in size upwards,tapering at the base and nearly sessile. *Radical leaves* properly speaking few, on rather long winged copiously hairy petioles dilated and clasping at the base, ovato-lanceolate, somewhat acuminate, tapering at the base into the petiole, sinuato-dentate, the teeth glandularly tipped, shallow in the upper part, deeper longer and more elongated towards the base, the lowermost frequently elongated, all pointing more or less forwards or towards the apex, bright green frequently yellowish, upper surface with few scattered short hairs, under surface paler, copiously clothed with long white denticulate hairs especially on the midrib, margins ciliated. *Flowers* large, yellow, in a terminal corymb, the branches peduncles bracteas and involucres hoary with dense minute stellate pubescence and with numerous black glandular hairs intermixed, the principal divisions subtended by subulate acuminate *bracteas*. *Inner involucral bracteas* nearly equal, linear-lanceolate, acuminate, *outer ones* few, smaller, irregularly imbricate, all erect and appressed, of a blackish-green on the keel, pale and membranous on the margins, clothed with minute dense white stellate pubescence and black glandular hairs intermixed. *Achenium* black or blackish-brown, linear, longitudinally ribbed and furrowed, when viewed under a powerful lens the ribs have very minute erect denticulations, the interstices transversely rough. *Pappus* more than twice the length of the achenium, minutely denticulate, sessile. *Receptacle* with minute elevated teeth on the intestices.

Equally variable with the preceding and in all probability mutually passing into each other by insensible gradations; an approximation towards which I observed in specimens of both species growing intermixed on the rocks at the Lyd Hole near Pontesford Hill. Recognizable generally by the form of the leaves, the forward direction of their teeth, and the leafy stem.

Mr. Babington has favoured me with the following remarks on these difficult and variable species. "I have paid some attention to this division of *Hieracium*, I mean the plants named *murorum* and *sylvaticum* by Smith, and have not been able to detect any other species in Britain. I am more and more drawn, although against my will, to suspect that even *murorum* and *sylvaticum* are not really distinguishable specifically. The number of leaves on the stem seems very inconstant and although the extremely leafy form of *sylvaticum*, (the true plant of *Smith*), appears very different from the form in which the stem bears only *one* leaf (which is, I suspect, *Schmidtii, Koch*), yet I cannot find any place at which the line can be drawn without violating nature, on account of the numerous intermediate states which occur. Again, if we look to the basal toothing of the leaves, many specimens occur in which it is difficult to say in which direction the large basal teeth point; they are sometimes tending upwards and then the leaf is attenuated below and the plant becomes *sylvaticum* or *Schmidtii*; at other times they spread horizontally, that is, at right angles to the midrib of the leaf, which is then either attenuated below or slightly cordate (this appears to be an intermediate state between the two above named species), and lastly, they are decidedly directed downwards in the true *murorum*. In *all* our plants I find glandular hairs upon the involucrum either in greater or less number, although they are sometimes much hidden by the shortness of their stalks and the length of the simple hairs."

4. H. *boreale*, Fries. *Northern Hawkweed.* Stem rigid, copiously leafy, hispid with denticulate hairs, branched above, branches subcorymbose; peduncles hoary with stellate pubescence; involucres slightly pubescent; leaves ovato-lanceolate or lanceolate, dentate, inferior ones attenuated at the base into a short petiole, superior ones sessile or nearly so, (but not amplexicaul), radical ones none; involucral bracteas appressed, uniform in colour, nearly black when dried. *Fries Nov. ed.* 1. *p.* 77. *ed.* 2. *p.* 261. *et Mant. alt. p.* 49. *Koch Syn. p.* 460. H. *subaudum (according to Koch) of Linn. Suec.* 274. *not of Sp. Pl.* 1131. *Sm. E. Fl. v.* iii. *p.* 367. *Hook. Br. Fl.* 4th *ed. p.* 295. E. *Bot. t.* 349. H. *sylvestre, Tausch. Reich. Fl. Excurs.* n. 1776.

Coppices, thickets, &c.; not common. *Fl.* August, September. ♃.
Cox wood, Coalbrookdale; *Mr. F. Dickinson.* Astley; *Mr. E. Elsmere, junr.* Near Coalport; *Dr. G. Lloyd.* Whitecliff, near Ludlow; *Mr. H. Spare.* Near Oswestry; *Rev. T. Salwey.*
Snailbeach works, near Minsterley. Bickley Coppice. Shomere Moss. Near Oxon.
Root fibrous. *Stem* 2 feet or more high, erect, simple, branched and corymbose above, round, striated and furrowed, hispid in the lower part with long denticulate or rough spreading hairs arising from tubercular bases, the upper portion hoary with dense minute close-pressed stellate hairs. *Leaves* alternate, ovato-lanceolate or lanceolate, acute; lower ones tapering at the base into a short winged petiole; upper ones sessile or nearly so, margins thickened and slightly revolute, rough with numerous minute short stiff bristles interspersed with longer denticulate hairs, distantly dentate, the teeth with callous tubercles, upper surface dark green with scattered denticulate hairs, under surface paler and prominently veined, very copiously clothed with denticulate hairs arising from thickened or tubercular bases. *Flowers* corymbose, with narrow lanceolate *bracteas* at the divisions and similar smaller ones scattered immediately beneath the involucre. *Florets* of a full palish yellow, 5-toothed, tube hairy. *Involucre* imbricated, *inner* bracteas nearly equal, erect, lanceolate, margins minutely serrulate, glabrous, except a slight pubescence on the keel; *outer* ones gradually diminishing in size as they approach the base, slightly pubescent on the keel and margin, very slightly spreading, not reflexed, of an uniform livid green when fresh, and nearly black when dried.
Fries states that "the true *H. subaudum* of Linn. *Sp. Pl.* well figured in *Allioni Fl. Ped. t.* 27. *f.* 2. is much more robust, all the leaves ovate, amplexicaul, hairy, unequally and doubly dentate."
"The subessile not amplexicaul leaves, pale margined (erect when young) involucral scales appear to distinguish *H. lævigatum, Willd.* and *Koch.* from its allies *H. boreale* and *H. subaudum.* The former of these (*boreale*) has the scales folded over the young florets and not forming an elevated crown round them as in *lævigatum*, its upper leaves are imperfectly sessile as in that plant and not amplexicaul as in the true *subaudum*; its involucral scales also are always of an uniform colour being of a rather livid green when fresh and nearly black when dried, whereas in *subaudum* they are always, even when dry, pale-margined." *Babington in lit.*

5. H. *umbellatum*, Linn. *Narrow-leaved Hawkweed.* Stem

---

rigid, hairy or glabrous, branched upwards, ultimate branches subumbellate; peduncles hoary; leaves lanceolate and linear, dentate or entire, inferior ones tapering into a short petiole, upper ones subsessile, radical ones none; involucral bracteas glabrous, recurved at the apex. E. *Bot. t.* 1771. E. *Fl. v.* iii. *p.* 370. *Hook. Br. Fl. p.* 295.

Groves or stony and rocky places; rare. *Fl.* August, September. ♃.
In some abundance between Wellington and Preston on the Wildmoors; *E. Lees, Esq.* Hord's park near Bridgnorth; *Purton's Midl. Flora.* In the lane between Badger and Stapleford. On the right hand side of the road between Atcham and Uffington. At the east corner of Bomere wood; *Rev. E. Williams's* MSS.
Stem 2-3 feet high, erect, round. *Leaves* distantly toothed or entire, roughish at the margins and slightly hairy, sometimes quite smooth, bright green above, paler beneath. *Flowers* bright yellow. *Bracteas* linear, few and small. *Achenium* angular, brown, finely dotted. *Pappus* rough. *Receptacle* slightly cellular.
Of this plant I have not seen any Shropshire specimens.

12. LAPSANA. *Linn.* Nipple-wort.

1. L. *communis*, Linn. *Common Nipple-wort.* Stem panicled, peduncles slender; radical leaves lyrate, cauline ones ovate or ovato-lanceolate, angulato-dentate, shortly petiolate; involucre of the fruit angular. E. *Bot. t.* 844. E. *Fl. v.* iii. *p.* 378. *Hook. Br. Fl. p.* 353.

Waste and cultivated ground; common. *Fl.* July, August. ☉.
Root fibrous. *Stem* 1—4 feet high, erect, branched, angular, striated, hispid with decurved glandular hairs. *Leaves* alternate, lyrate, the terminal lobe very large broadly ovate almost cordate, angulato-dentate, the teeth tipped with callous tubercles, rough on both surfaces, tapering into a deeply channelled rough petiole, the upper ones nearly reduced to the terminal lobe shortly petiolate and almost sessile. *Flowers* small, yellow, in terminal and axillary panicles on smooth angular peduncles with small subulate ciliated *bracteas* interspersed among the divisions. *Inner involucral bracteas* 8, equal, erect, linear-lanceolate, prominently keeled, glabrous, obtuse and ciliated at the apex, with 5 minute ovate acute glabrous denticulate *outer bracteas* at their base. *Tube* of the *florets* minutely hairy externally. Central tooth of the limb somewhat shorter than the others which bear on their backs a mass of yellow glands from which the central one is free. *Involucre* of *fruit* erect, angular. *Achenium* linear-obovate, compressed, striated. *Receptacle* naked.

13. CICHORIUM. *Linn.* Succory.

1. C. *Intybus*, Linn. *Wild Succory.* Flowers sessile, axillary, in pairs; leaves runcinate. E. *Bot. t.* 539. E. *Fl. v.* iii. *p.* 380. *Hook. Br. Fl. p.* 354.

Borders of fields; rare. *Fl.* July, August. ♃.
Dowle, sparingly; *Mr. G. Jorden.* Fields at Welbach near Shrewsbury; *Mr. T. Bodenham.*
Root fusiform. *Stem* 1—3 feet high, erect, branched, angular and furrowed, rough with bristly hairs. *Radical leaves* numerous, spreading, lanceolate, acuminate, runcinato-dentate, on dilated petioles, rough; *cauline ones* smaller, amplexicaul; *uppermost* cordato-lanceolate, entire. *Flowers* large, handsome, brilliant blue. *Outer involucral bracteas* few, unequal, oblong, lax, short, glanduloso-

---

ciliate; *inner* ones longer, linear, converging, finally recurved. *Achenium* clavate, obsoletely tetragonous, truncate. *Pappus* of very minute erect chaffy bristles. *Receptacle* slightly chaffy with slender scales shorter than the achenium.

14. ARCTIUM. *Linn.* Burdock.

1. A. *Lappa*, Linn. *Common Burdock.* Flowers subcorymbose; involucral bracteas nearly glabrous, all of the same colour, inner ones lineari-lanceolate, gradually attenuated into a mucronate point, margins scariose, longer than the florets. *Babington Ann. Nat. Hist. v.* 4. *p.* 253. E. *Fl. v.* iii. *p.* 380. *a. Hook. Br. Fl.* 4th *ed. p.* 296.

Waste places, road-sides, &c.; common. *Fl.* July, August. ♂.
Involucral bracteas longer than the florets, all yellowish-green, rigid, subulate and hooked except a very few of the innermost, which are linear-lanceolate, contracting gradually into a slightly curved rigid point, all of them in their lower part and some of the innermost throughout their whole length, with a very narrow scariose white minutely ciliated not serrulated margin. *Leaves* cordate, margins nearly flat and fringed with numerous rigid prickles formed by the excurrent points of the nerves.

2. A. *Bardana*, Willd. *Woolly Burdock.* Flowers racemose; involucral bracteas with a cobweb-like down, inner ones coloured, lineari-lanceolate, abruptly mucronate, shorter than the florets. *Babington Ann. Nat. Hist. v.* 4. *p.* 253. E. *Fl. v.* iii. *p.* 381. A. *Lappa*, β. *Hook. Br. Fl.* 4th *ed. p.* 296.

Waste places, road-sides, &c.; common. *Fl.* July, August. ♂.
Involucral bracteas shorter than the florets, nearly all purple-coloured, rigid, subulate and hooked, inner ones linear-lanceolate, contracted rather suddenly into an almost straight rigid point, more numerous in proportion than in A. *Lappa*: outer ones with a very narrow scariose white minutely ciliated margin; margin of the inner ones minutely serrulate throughout. *Leaves* much smaller, less wavy at their margins, fewer of the nerves excurrent.

15. SERRATULA. *Linn.* Saw-wort.

1. S. *tinctoria*, Linn. *Common Saw-wort.* Leaves entire or pinnatifid, finely serrated; outer involucral bracteas ovate, appressed, inner ones linear, coloured. E. *Bot. t.* 38. E. *Fl. v.* iii. *p.* 382. *Hook. Br. Fl. p.* 354.

Borders of woods, thickets, &c.; not common. *Fl.* August. ♃.
Wyre forest, plentifully; *Mr. G. Jorden.* Whitecliff near Ludlow; *Mr. H. Spare.* Astley; *Mr. E. Elsmere, junr.* Chesterton Roman Camp; *Dr. G. Lloyd!*
Root woody. *Stem* 2-3 feet high, erect, angular, striated, branched above. *Leaves* lyrato-pinnatifid, segments linear-lanceolate, acute, regularly and copiously mucronato-serrated, terminal lobe very much elongated, central lateral ones longer than the rest which gradually decrease towards both extremities, petiole dilated and amplexicaul at the base. *Flowers* corymbose. *Involucre* oblongo-cylindrical, regularly imbricated, *outer bracteas* ovate or ovato-lanceolate, acute, gradually lengthened upwards, the *inner* ones lanceolate acute, all rigid, erect, and more or less downy especially on the margins. *Corolla* purplish. *Achenium* lineari-obovate, compressed, somewhat angular, smooth. *Pappus* sessile, rough with rather long denticulations, yellowish. *Receptacle* with chaffy hairs as long or longer than the achenium.

---

16. CARDUUS. *Linn.* Thistle.

* Pappus rough.
† Leaves decurrent.

1. C. *nutans*, Linn. *Musk Thistle.* Leaves decurrent, spinous; flowers solitary, nodding; involucral bracteas lanceolate, cottony, outer ones deflexed. E. *Bot. t.* 1112. E. *Fl. v.* iii. *p.* 385. *Hook. Br. Fl. p.* 355.

Pastures, waste ground, &c.; frequent. *Fl.* July, August. ♂.
Stem 2 feet high, erect, angular, furrowed, solid, clothed with spreading tomentose hairs, angles with narrow sinuated wavy spinous wings. *Leaves* lanceolate, acute, pinnatifid, segments 3-lobed, wavy, terminated by a strong rigid spine, margins spinulose, hairy on both surfaces. *Flowers* terminal, solitary, large, handsome, crimson-purple, nodding. *Peduncles* clothed with appressed cottony down, unarmed. *Involucre* subglobose, imbricated, with more or less cottony web. *Bracteas* lanceolate, spinous, strongly keeled, *outer* ones deflexed, *middle* ones erecto-patent, *innermost* membranous, unarmed, erect. *Florets* tubular, limb divided in 5 narrow linear erect acute segments thickened and concave at the apex. *Filaments* hairy. *Anthers* apiculate, bristly at the base. *Achenium* smooth, polished, shining. *Pappus* very long, rough. *Receptacle* flat, with chaffy hairs or scales about half the length of the corolla. *Style* with a circular appendage of erecto-patent hairs immediately below the stigma.
The most elegant of our Thistles. Dr. Johnston in his *Flora of Berwick-upon-Tweed, Vol.* I. *p.* 178, considers the *Carduus Marianus* as the national flower of Scotland. "Proud Thistle! emblem dear to Scotland's sons!" On this interesting but doubtful subject Mr. Dovaston has communicated to me the following particulars:—" With regard to what peculiar Thistle is the floral badge of Scotland, it was an object of most particular inquiry with me in my long and laborious tour in that interesting country, in company with our eminently botanical friend Bowman. At Dumfries, around the grave of the Poet Burns, was planted profusely the *Onopordum Acanthium*, which in the rich sepulchral soil grew very large, and much resembled the Scotch Thistle in emblematic engravings: we were, however, told by an intelligent gentleman in the Hebrides, Donald Mc Lean, a young chieftain, that what he shewed us, the *Carduus Eriophorus*, was the Scotch Thistle. At Inverness, Sir James Grant said the Scotch Thistle was the only one that drooped, *Carduus nutans.* After many such remarks, we were at length told by a very intellectual Gardener at Roslin, and by Sir William Drummond at Hawthornden, that no particular Thistle, but any Thistle the poet or painter chose, was the national flower of Scotland; and this opinion we heard repeated in Edinburgh, at the tables of several learned and hospitable gentlemen. Though generally emblematical of the whole nation, it is in particular the badge of the clan Stewart. On the wet sides of some hills we, not unfrequently, found the *C. heterophyllus*, or gentle thistle, which was much and justly admired, and by some (erroneously) thought peculiar to Scotland: this, however, could not be the national emblem, as, being destitute of thorns, it would ill accord with their formidable Latin motto."

2. C. *acanthoides*, Linn. *Welted Thistle.* Leaves decurrent, spinous, somewhat cottony beneath; heads nearly globose, sessile or shortly stalked, solitary or clustered; involucral bracteas linear-subulate, terminating in a long spinous point. E. *Bot. t.* 973. E. *Fl. v.* iii. *p.* 386. *Hook. Br. Fl. p.* 355.

Way-sides, waste places, &c.; common. *Fl.* June, July. ☉.
Root tapering. *Stem* 3-4 feet high, erect, branched, angular, decurrently and

uninterruptedly winged, hairy and cottony, wings undulate, very spinous. *Leaves* alternate, oblongo-lanceolate, deeply sinuato-pinnatifid, lobes irregularly sinuato-dentate, the lower segments bent upwards and the anterior ones downwards, margins undulate with innumerable sharp spines of various sizes, upper surface deep green, hairy, lower surface paler, the prominent nerves and midrib with long white hairs, the interstices downy and cottony. *Flowers* at the extremities of the branches aggregated, often single, nodding, nearly sessile or on very short peduncles. *Involucres* globose, *outer bracteas* green, erecto-patent, linear subulate, single-ribbed, spiny, slightly downy or cottony, margins ciliated with spreading hairs, *inner ones* linear-lanceolate, erect, 3-ribbed, purple at the slightly recurved apex, which is acute but not rigid and spiny, margins very minutely ciliated with erect pubescence. *Florets* with a white tube the length of the pappus, the summit swollen where the stamens are inserted, limb deeply cleft into 5 erect linear acute segments. *Filaments* hairy at the base. *Achenium* oblong, smooth and shining. *Pappus* rough.

Smith lays much stress upon the want of the white cottony down on the under surface of the leaves as distinguishing his *C. acanthoides* from *C. crispus*, but in all the specimens of the former which have occurred to my notice there is invariably more or less of it; neither should any importance be attached to the inner coloured involucral bracteas as a distinctive mark of these plants.

A variety with white flowers has been found near Eaton Mascott.

3. C. *tenuiflorus*, Curt. *Slender-flowered Thistle.* Leaves decurrent, spinous, somewhat cottony beneath; heads nearly cylindrical, clustered, sessile; involucral bracteas lanceolate, attenuated upwards into a spinous point. *E. Bot. t.* 412. *E. Fl. v. iii. p.* 386. *Hook. Br. Fl. p.* 355.

Ditch banks; rare. *Fl.* June, July. ⊙.

Ditch bank around the old limestone quarry near Harnage; *Rev. E. Williams's MSS.*

"*Root* tapering, small. *Herbage* all white with cottony down. *Stem* erect, straight, slightly branched, 3 or 4 feet high, angular, with broad deeply lobed strongly spinous leafy wings. *Leaves* broadish, pinnatifid and sinuated, most cottony beneath, with strong yellowish spines, less numerous than in the last. *Flowers* sessile at the tops of the branches, several together, pale rose-coloured, with much fewer *florets*, and consequently a narrower more oblong *involvcre*. *Involucral bracteas* dilated and ovate, rather membranous, at the base; the upper part finally spreading and tipped with a yellowish spine. *Achenium* grey, compressed, shining. *Pappus* minutely rough." *Smith.*

†† *Leaves sessile.*

4. C. *Marianus*, Linn. *Milk Thistle.* Leaves amplexicaul, waved, spinous, the radical ones pinnatifid; involucral bracteas subfoliaceous, recurved, spinous at the margin. *E. Bot. t.* 976. *E. Fl. v. iii. p.* 387. *Hook. Br. Fl. p.* 355. *Silybum marianum Gært.*

Occasionally in cultivated ground but probably introduced. *Fl.* August, ♂.

Shineton; *W. P. Brookes, Esq.!* On Ford hill between Longnor and Preston. In Shineton churchyard; *Rev. E. Wllliams's MSS.*

Stem 3-4 feet high, erect, round, ribbed, furrowed, slightly pubescent, hollow, branched in the upper part. *Leaves* very large, oblongo-lanceolate, cordate and semiamplexicaul at the base, pinnatifid, margins and lobes strongly waved and undulate, armed with strong very sharp prickles, of a bright shining green, variegated along the veins with white on the upper surface, paler beneath and with a few scattered hairs along the prominent nerves; upper leaves much more entire. *Flowers* terminal and solitary on the extremity of the branches. *Iuvolucre* globose, more or less clothed with minute downy pubescence, *outer bracteas* few, close-pressed, erect, ovate, acute, bristle-pointed, tapering at the base, margins fringed with sharp bristles, *inner bracteas* ovate, erect, appressed, margins unarmed, contracted and terminated by an ovato-acuminate bristly pointed concave portion with bristly margins, curved back and spreading outwards; innermost linear, acuminate, erect, unarmed. *Florets* bright purple, tube very long and white, limb ventricose at the base, 5-cleft. *Filaments* downy. *Achenium* smooth and shining. *Pappus* rough. *Receptacle* with chaffy hairs.

On examining with a powerful lens the upper surface of the leaf, the green portion appears covered at intervals with very minute tubercles, and the white part entirely with minute irregular hexagonal reticulations, the interstices being hollow or depressed. This appearance seems caused by the epidermis in the white portion being raised and hollow beneath, whilst in the green portion it is closely pressed down on the parenchyma. The juice which is somewhat turbid, appears to contain innumerable pellucid globules of various sizes floating in it, between the smaller ones of which I imagined I perceived a mutual attraction and repulsion, but perhaps this appearance may have been caused by slight currents produced by vibrations consequent on persons moving about the room or touching the furniture.

** *Pappus feathery.*

† *Leaves decurrent.*

5. C. *lanceolatus*, Linn. *Spear Plume-thistle.* Leaves decurrent, pinnatifid, hispid, segments 2-lobed divaricate spinous; involucres ovate arachnoid, bracteas lanceolate spinous spreading. *E. Bot. t.* 107. *Cnicus lanceolatus, Willd. E. Fl. v. iii. p.* 288. *Hook. Br. Fl. p.* 355.

Waysides, pastures, &c.; frequent. *Fl.* July, August. ♂.

Root thick, branching, fibrous. *Stem* 3-4 feet high, erect, angular, furrowed, hairy or downy, with strong spiny wings, branched. *Leaves* alternate, sessile and decurrent at the base, lanceolate, pinnatifid, elongated at the apex into a lanceolate lobe, segments 2-lobed, the upper lobe somewhat lobed at the base, rigid, divaricating, all terminated by a strong yellow very sharp spine, margins somewhat recurved, minutely spinuloso-denticulate, hispid on the upper surface, paler and downy beneath. *Flowers* large, generally solitary on the extremities of the stem and branches, purple. *Involucre* ovate; *bracteas* lanceolate, dark green, glabrous, strongly keeled, scaly at the base, elongated into a rigid narrow spreading spine, margins ciliated about the middle with cottony down which gives to the involucre a cobweb-like appearance; points of the inner scales erect and more appressed. *Florets* tubular, tube twice the length of the limb, white, very slender, limb purple, dilated, deeply 5-cleft, segments linear, concave, thickened and obtuse at the apex. *Filaments* free, pubescent; *anthers* apiculate, bristly at the base. *Achenium* obovate, purplish, smooth and polished, obsoletely tetragonous, crowned with a raised margin and a conical obtuse appendage. *Pappus* sessile, plumose. *Receptacle* with long hairs.

6. C. *palustris*, Linn. *Marsh Plume-thistle.* Leaves decurrent, pinnatifid, scabrous, segments 4-lobed, spinous; involucres ovate, clustered, bracteas lanceolate, mucronate, appressed. *E. Bot. t.* 976. *Cnicus palustris, Willd. E. Fl. v. iii. p.* 389. *Hook. Br. Fl. p.* 356.

Moist meadows and shady places; frequent. *Fl.* July. ♂.

Root fibrous. *Stem* 3—6 feet high, erect, angular, furrowed, clothed with long white jointed hairs, angles with decurrent crisped and wavy wings armed with strong spines, alternately branched in the upper part, branches with spinous wings. *Leaves* alternate, lanceolate, radical ones petiolate, stem ones sessile and

decurrent, deeply pinnatifid, segments elongated, 4-lobed at the base, the lowermost lobe longer and narrower than the rest, all terminated by a strong spine, margins crisped, spinoso-serrated, roughish on both sides with long jointed hairs, especially on the strong prominent midrib beneath. *Flowers* clustered at the extremities of the stem and branches, deep crimson, frequently white, nearly sessile. *Involucre* ovate, a little broader at the base, subglobose in fruit, *bracteas* imbricated, lanceolate, appressed and keeled, *outer* ones shorter, terminated by a short subrecurved or spreading spine, downy at the margins, those of the middle portion with recurved harmless points and of the innermost with erect membranous ones. *Florets* tubular, tube shorter than the limb, limb dilated upwards, deeply 5-cleft, segments linear, acute, slightly concave. *Filaments* free, pubescent in the upper part; *anthers* violet, apiculate, bristly at the base. *Achenium* linear, obsoletely tetragonous, smooth, pale and shining, crowned with a margin within which arises a short obtuse cone. *Pappus* sessile, delicately plumose. *Receptacle* with long white hairs.

†† *Leaves sessile or nearly so.*

7. C. *arvensis*, Curt. *Creeping Plume-thistle.* Root creeping; stem panicled; leaves sessile, pinnatifid, spinous, segments oblong, remote, lobed at the base; involucre ovate, glabrous, outer bracteas broadly lanceolate, spinous. *E. Bot. t.* 975. *Cnicus arvensis, Hoffm. E. Fl. v. iii. p.* 390. *Hook. Br. Fl. p.* 356. *Serratula arvensis, Linn.*

Fields and by waysides; common. *Fl.* July. ♃.

Root extensively creeping, fleshy. *Stem* 2-3 feet high, round, angular, solid, glabrous, not winged, except in a very partial manner for a little way below the leaves. *Leaves* alternate, sessile, lanceolate, pinnatifid, segments oblong, remote, lobed especially at the base, all terminated with long strong pale spines, margins wavy and crisped, armed with strong rigid spines. *Flowers* panicled. *Peduncles* downy, each bearing a solitary flower. *Involucre* ovate, elongated and almost cylindrical after flowering, densely imbricated, glabrous; *outer bracteas* broadly lanceolate, terminated by a spreading spine, downy at the edges, dark green; *inner ones* longer and narrower, terminated by a purplish subrecurved soft point. *Florets* purple, tube very long, slender, white, limb in 5 erecto-patent linear concave segments, obtuse, thickened and incurved at the apex. *Filaments* white, adnate with the tube to a very short distance of their summits; *anthers* dark brown, submucronate at the base, each terminated by a pale apiculus. *Achenium* oblong, pale and shining, obsoletely tetragonous, crowned by a raised circular border within which rises a conical obtuse appendage. *Pappus* very long, sessile, delicately plumose. *Receptacle* with long white shining hairs.

8. C. *eriophorus*, Linn. *Woolly-headed Plume-thistle.* Upper leaves sessile, pinnatifid, hispid, segments 2-lobed, alternate, divaricate, alternately pointing upwards and downwards, spinous; involucres orbicular, depressed, densely arachnoid, bracteas linear-lanceolate, acuminate, erecto-patent. *E. Bot. t.* 386. *Cnicus eriophorus, Willd. E. Fl. v. iii. p.* 391. *Hook. Br. Fl. p.* 356.

Waste ground, in the limestone districts; not common. *Fl.* July. ♂.

Shadewell, Standhill Coppice, Wenlock; *W. P. Brookes, Esq.* New Inn between Ludlow and Shrewsbury; *T. and D. Bot. Guide.* Cause Castle near Westbury; *J. F. M. Dovaston, Esq.* Castle Meadow, parish of Burford. Near Tenbury. Near to Little Millechope, parish of Munslow; *Rev. W. Corbett.* Between Aston and the Longmynd; *E. Lees, Esq.* Caynham Camp near Ludlow; *Rev. T. Salwey.* About Buildwas mill. By the side of the road between Wilderhope and Stanway. About Farley between Buildwas and Wenlock; *Rev. E. Williams's MSS.*

Benthal Edge, sparingly.

Root tapshaped. *Stem* 3 feet high or more, angular, stoutly furrowed, softly hairy, alternately and copiously branched. *Radical leaves* very long, linear-lanceolate, on a long triquetrous somewhat winged petiole; *upper ones* alternate, sessile, not decurrent, all deeply pinnatifid, segments 2-lobed, alternate, divaricate, alternately pointing upwards and downwards, the anterior lobe with 2 smaller lobes at the base, terminated by long rigid sharp spines, upper surface green covered with close bristles, lower one white with soft cottony down, margin fringed with bristly prickles, the terminal lobe acuminated, *Flowers* solitary, sessile on the extremities of the branches, very large and handsome, subtended by a smaller leaf or two. *Involucre* very large, orbicular, depressed; *bracteas* linear-lanceolate, very much acuminated, erecto-patent, margins serrulate, their long spiny points projecting from a dense mass of white cobweblike down which envelopes the whole involucre. *Florets* purple. *Achenium* obovate, shining. *Pappus* plumose.

9. C. *pratensis*, Huds. *Meadow Plume-thistle.* Upper leaves sessile, lanceolate, sinuato-dentate, spinous, ciliated, pubescent above, cottony beneath; flowers mostly solitary; involucre globose, arachnoid, bracteas lanceolate, acuminate, erect. *E. Bot. t.* 177. *Cnicus pratensis, Willd. E. Fl. v. iii. p.* 394. *Hook. Br. Fl. p.* 357.

Wet pastures; rare. *Fl.* July. ♃.

Boggy field a little beyond Blackmere, by the Canal side; *J. E. Bowman, Esq.* Ellesmere; *Watson's New Bot. Guide.*

Shawbury Heath, sparingly.

Root fibrous. *Stem* 18 inches to 2 feet high, erect, simple, round, furrowed, clothed with greyish loose cottony down. *Leaves* lanceolate, acute, unequally sinuato-dentate, the teeth armed with bristly spines, the intermediate sinuations copiously fringed with smaller prickles, hairy and minutely downy on the upper surface, with copious grey loose cotton beneath; lower leaves tapering at the base into a winged petiole, upper ones gradually decreasing in size upwards, sessile and amplexicaul. *Flowers* solitary, terminal, pale purple. Sometimes a branch proceeds from the axil of the uppermost leaf bearing a second flower. *Involucre* globose, *bracteas* rigid, erect, lanceolate, acuminated, with a spinous point, pale green at the base, purplish at the apex, margins slightly membranous and serrated, with long loose spreading cobweblike down; innermost narrower, very much acuminated, and unarmed. *Filaments* downy. *Pappus* sessile, plumose, naked or minutely denticulate at the apex.

17. ONOPORDUM. *Linn.* Cotton-thistle.

1. O. *Acanthium*, Linn. *Common Cotton-thistle.* Leaves ovato-oblong, sinuato-dentate, spinous, decurrent, woolly on both sides; involucre globose, outer bracteas subulato-lanceolate, recurved and spreading, spinous. *E. Bot. t.* 977. *E. Fl. v. iii. p.* 396. *Hook. Br. Fl. p.* 357.

Waste ground, roadsides, &c.; rare. *Fl.* August. ♂.

Near Wenlock; *W. P. Brookes, Esq.* Wroxeter; *J. F. M. Dovaston, Esq.* Between Uckington Heath and Walcot. By the Flash, near Shrewsbury; *Rev. E. Williams's MSS.*

Roadside beyond Atcham, near Norton.

Whole plant covered with white loose cottony pubescence. *Stem* 4-5 feet high, erect, branched, angular, with broad leafy wings proceeding from the bases of the decurrent leaves, veiny and strengthened at intervals with strong spines which proceed from the stem through the wings, margins undulate. *Radical leaves* very large, upper ones alternate, ovato-oblong, terminated by a sharp spine, margins sinuato-dentate, spiny, decurrent. *Flowers* solitary, terminating the branches. *Involucre* globose, imbricated, *outer bracteas* subulato-lanceolate, recurved and spreading, terminated by a rigid spine, strongly keeled, margins serrated, *inner ones* unarmed and erect. *Florets* tubular, limb bluish-purple, 5-fid. *Achenium* tetragonous, obtuse. *Pappus* very rough, united at the base by an annular appendage.

18. CARLINA. *Linn.* Carline-thistle.

1. C. *vulgaris*, Linn. *Common Carline-thistle.* Stem corymbose, many-flowered, cottony; leaves lanceolate, sinuated, copiously spiny, cottony beneath; outer involucral bracteas lanceolate, with simple and branched spines. *E. Bot. t.* 1114. *E. Fl. v. iii. p.* 398. *Hook. Br. Fl. p.* 358.

Dry hilly pastures; not very common. *Fl.* June. ♂.

Oreton, plentiful; *Mr. G. Jorden.* Between Church Aston and Lilleshall; *R. G. Higgins, Esq.* Lilleshall lime kilns; *T. C. Eyton, Esq.* Near Lutwyche Hall and Wenlock; *Mr. F. Dickinson.* Longmynd; *Rev. W. Corbett.* Shipley Common, near Rudge; *H. Bidwell, Esq.*

Caradoc Hill. Haughmond Hill. Benthal Edge. Limestone quarries, near the Wrekin. Benthal Edge. Lawley Hill. Haughmond Hill.

Stem 12—18 inches high, erect, round, simple, cottony. *Leaves* alternate, lanceolate, acute, amplexicaul, rigid, margins sinuated, wavy, copiously spiny, glabrous above, cottony beneath. *Flowers* solitary and terminal, or corymbose and many-flowered. *Involucre* imbricated, more or less cottony; *outer bracteas* lanceolate, acute, lax, margins armed with simple and branched spines; *inner ones* linear, unarmed, entire, membranous, yellow, forming an horizontal ray to the purple flower.

19. BIDENS. *Linn.* Bur-marigold.

1. B. *cernua*, Linn. *Nodding Bur-marigold.* Leaves sessile, amplexicaul, undivided, lanceolate, serrated; outer involucral bracteas nearly equal, entire, twice as long as the flower; achenium linear, cuneate; pappus of 4 bristles. *E. Bot. t.* 1114. *E. Fl. v. iii. p.* 400. *Hook. Br. Fl. p.* 358.

Sides of rivulets, ditches and lakes; not very common. *Fl.* June—Aug. ⊙.

Marsh pool near Wenlock; *W. P. Brookes, Esq.!* Whitchurch, Marbury, and Ellesmere meres; *Mr. F. Dickinson.* Pit by Hadnall wood; *Mr. E. Elsmere, junr.!* Newport; *R. G. Higgins, Esq.* Tong Lodge lake; *Dr. G. Lloyd.*

Shelton Rough near the banks of Severn. Vownog near Westfelton.

Root fibrous. *Stem* 2-3 feet high, erect, angular, roughish, with opposite axillary branches. *Leaves* opposite, sessile and amplexicaul, undivided, lanceolate, acuminate, strongly coarsely and distantly but shallowly serrated, tapering at the base, margins denticulate with minute erect bristles, dark green and smooth above, paler beneath. *Flowers* large, solitary, terminal, brownish yellow, drooping. *Outer involucral bracteas* about 8, nearly equal, lanceolate, acute, entire, margins subciliated with erect bristles, about twice as long as the flower. *Inner bracteas* erect, ovato-lanceolate, obtuse, yellowish-brown with black striæ, margins subciliated. *Florets* tubular, dilated upwards, 5-cleft, segments acute, spreading or recurved. *Achenium* linear, cuneate, tetragonous, very much compressed, the angles with deflexed bristles. *Pappus* of 4 erect stout bristles, armed with smaller deflexed ones, nearly as long as the florets, arising from the angles, the 2 intermediate bristles rather the shorter. *Receptacle* nearly flat, covered with lanceolate

acute chaffy scales as long as the florets, tinged with yellow at the apex, paler below and marked with many dark-brown or blackish striæ or streaks at the back.

Var. β. *minima.* Leaves lanceolate, sessile; flowers erect. γ. *E. Fl. l. c.* *B. tripartita*, β. *minima*, *Huds. ed.* 2. 355.

This variety according to the Rev. E. Williams's MSS. is common in Shropshire. Smith considers it only a starved plant growing out of water in dry places where water has stagnated.

2. B. *tripartita*, Linn. *Trifid Bur-marigold.* Leaves petiolate, in 3—5 deep lanceolate serrated segments; outer involucral bracteas unequal, entire or serrated, many times longer than the flower; achenium obovato-cuneate; pappus of 2-3 bristles. *E. Bot. t.* 113. *E. Fl. v. iii. p.* 399. *Hook. Br. Fl. p.* 358.

Marshy places, sides of ponds and lakes; not very common. *Fl.* July. ⊙.

Tong Lodge lake; *Dr. G. Lloyd.* Oakley Park near Ludlow; *Mr. H. Spare* Newport; *R. G. Higgins, Esq.* Whitchurch and Marbury meres and pools; *Mr. F. Dickinson.* Rowley farm, Westfelton; *J. F. M. Dovaston, Esq.* Roadside near the bridge, Shineton; *W. P. Brookes, Esq.!*

Canal between Shrewsbury and Uffington. Bomere pool.

Root tapering, fibrous. *Stem* 2-3 feet high, erect, somewhat angular, solid, smooth, stained with purple, branches opposite, axillary. *Leaves* opposite on dilated winged connate ciliated petioles, in 3 sometimes 5 deep lanceolate acute strongly serrated segments, serratures submucronate, margins denticulate with minute erect bristles, dark green and smooth above, paler beneath. *Flowers* solitary, terminal, brownish-yellow. *Outer involucral bracteas* about 8, unequal, lanceolate, acute, serrated or entire, margins ciliated with erect bristles, many times longer than the flower. *Inner bracteas* erect, ovato-lanceolate, purplish and downy at the obtuse apex, margins membranous, marked with greenish-brown parallel streaks. *Florets* tubular, dilated upwards, 4 or 5-cleft, segments acute, spreading and recurved. *Achenium* obovato-cuneate, tetragonous, very much compressed, the angles with deflexed bristles. *Pappus* of 2 erect stout bristles armed with smaller deflexed ones, nearly as long as the floret, arising from the lateral angles. A third shorter bristle is sometimes present, arising from the angle or rib of the interior face. *Receptacle* nearly flat, covered with linear-lanceolate acute chaffy scales, as long as the florets.

Specimens not unfrequently occur in which the leaves are all undivided, but attention to their being petiolate, and to the outer involucral bracteas being unequal serrated and many times longer than the flowers will at once obviate any doubt, which may by possibility arise, as to which species the plant ought to be referred.

20. EUPATORIUM. *Linn.* Hemp-agrimony.

1. E. *cannabinum*, Linn. *Common Hemp-agrimony.* Leaves opposite, subpetiolate, 3—5-partite, segments lanceolate, serrated, central one longest; flowers corymbose. *E. Bot. t.* 428. *E. Fl. v. iii. p.* 401. *Hook. Br. Fl. p.* 358.

Banks of rivers and watery places; frequent. *Fl.* July, August. ♃.

Root tufted, creeping. *Stem* 3 feet high, erect, branched, round, with erect incurved pubescence, of a reddish hue. *Leaves* opposite, on short petioles, 3—5-lobed, lobes unequal; the central one longest, lanceolate, acute, strongly serrated, entire towards the point, deep green and roughish on the upper surface with minute hairs, paler beneath and finely hairy on the prominent ribs, covered with innumerable shining imbedded round globules. *Flowers* in dense terminal corymbs, pale pinkish; pedicels bracteated. *Involucre* narrow, oblong, imbricated; *bracteas*

erect, appressed, unequal, the outer ones shorter, green, downy, slightly membranous at the edges; inner ones lanceolate, obtuse, often truncate, apex downy, pale pink. *Florets* 5 or 6, tubular, gradually swelling upwards, with 5 acute erecto-patent segments. Style and bifid stigma elongated and exserted. *Achenium* linear, with about 6 angles, angles ribbed, covered with minute shining globules, surmounted by a small globular protuberance concave at the summit. *Pappus* equal to the florets, rough. *Receptacle* small, naked. Herbage slightly aromatic and the flowers emitting a seminal odour.

SYNGENESIA—SUPERFLUA.

21. TANACETUM. *Linn.* Tansy.

1. T. *vulgare*, Linn. *Common Tansy.* Leaves bipinnatifid, inciso-serrate; flowers corymbose. *E. Bot. t.* 1229. *E. Fl. v. iii. p.* 406. *Hook. Br. Fl. p.* 359.

Borders of fields, river sides; frequent. *Fl.* August. ♃.

Root woody, moderately creeping. *Stem* 3 feet high, erect, angular, furrowed, simple, more or less covered with glandular mealiness and a few scattered hairs. *Leaves* alternate, amplexicaul, bipinnatifid, segments lanceolate, acute, acutely toothed or cut, with smaller intermediate segments between the principal ones, dark green above, pale and slightly downy beneath, covered on both sides with innumerable minute glands. *Flowers* numerous, golden yellow, in dense terminal corymbs, the peduncles with minute scattered bracteas. *Involucre* hemispherical, imbricated; *bracteas* unequal, lanceolate, obtuse, slightly downy about the margins which are membranous laciniate and somewhat scariose. *Florets* tubular, 5-fid. *Achenium* tetragonous.

22. ARTEMISIA. *Linn.* Wormwood, Southernwood, Mugwort.

1. A. *Absinthium*, Linn. *Common Wormwood.* Leaves deeply bipinnatifid, hoary, silky and glandular; segments lanceolate, obtuse, subtrifid, thickened at the margins; flowers racemose, clusters hemisphærical, secund, pedicellate, silky. *E. Bot. t.* 1230. *E. Fl. v. iii. p* 409. *Hook. Br. Fl. p.* 359.

Waste places about villages; not very common. *Fl.* August. ♃.

Brocton, Near Wenlock; *W. P. Brookes, Esq.!* Haughmond Hill; *Mr. F. Dickinson.* Astley village; *Mr. E. Elsmere, junr.!* Lilleshall Abbey; *R. G. Higgins, Esq.!* Oreton, very abundant; *Mr. G. Jorden.* At Eudon and other places near Bridgnorth; very common; *Purton's Midl. Flora.* Road between Shotton and Wem; *Dr. T. W. Wilson.*

Astley churchyard. Lea Hall, near Shrewsbury. Pontesford Bridge works. Great and Little Ness, abundant. Kinton, Wilcot and places in that neighbourhood, plentifully. Bomere pool. Roadsides iu Upper Cound.

Root woody. *Whole herb* hoary with close-pressed silky hairs fixed by their centre and divaricating both ways at right angles, intensely bitter, emitting a peculiar powerful and pleasant aromatic odour. *Stems* numerous, bushy, 12—18 inches high, branched, angular and furrowed. *Leaves* alternate, on winged petioles channelled above, deeply bipinnatifid, segments lanceolate, obtuse, subtrifid, thickened at the margin, green on the upper surface, more or less hoary and with numerous opaque white glands imbedded in depressions, hoary, silky and shining beneath. *Flowers* in axillary and terminal bracteated racemes; clusters secund, hemispherical, on short decurved pedicels bearing above their middle one or two small bracteas and a larger lanceolate obtuse one at the base; those at the base

of the raceme trifid. *Involucre* imbricated, *outer bracteas* linear, obtuse, *inner ones* rotund, with broad membranous margins bearing long ciliæ. *Florets* pale yellow, tubular, gradually swelling upwards, acutely 5-cleft, segments recurved. *Anthers* apiculate, exserted. *Style* deeply cloven; *stigma* a transversely dilated ciliated disk. *Pappus* rough. *Achenium* smooth. *Receptacle* convex, with long silky hairs.

There is a tradition that this plant was extensively employed medicinally during the direful ravages of the plague; certain it is, that it now occurs in great abundance about the villages and hamlets in the neighbourhood of Oswestry and on the Welsh border where that frightful disease is known to have been rife. It has been noticed to grow frequently about pigeon-houses, and our beloved Shakspeare, whose works are ever "cramm'd with observation" of Nature, alludes to the circumstance in the scene in Romeo and Juliet where the garrulous old Nurse is twaddling with unstinted tongue about the age of her "lady-bird" Juliet, with all the tedious circumstance of "time's doting chronicles":—

> "'Tis since the earthquake now eleven years;
> And she was wean'd—I never shall forget it—
> Of all the days of the year, upon that day,
> For I had then laid WORMWOOD to my dug,
> Sitting in the sun under the dove-house wall."

2. A. *vulgaris*, Linn. *Mugwort.* Leaves pinnatifido-laciniate, tomentose beneath, segments lanceolate, acute, revolute at the margins; flowers racemose; clusters 3—7, ovate, sessile, tomentose. *E. Bot. t.* 978. *E. Fl. v. iii. p.* 410. *Hook. Br. Fl. p.* 359.

Hedges and waste places; common. *Fl.* August. ♃.

Stem 3-4 feet high, erect, branched, deeply furrowed, dark purple, more or less loosely tomentose, especially in the upper portion. *Leaves* alternate, on dilated petioles, with 3 or more small lanceolate acute segments, decreasing downwards at the semiamplexicaul base, pinnatifido-laciniate, the segments lanceolate acute deeply incised, glabrous and deep green above, white and tomentose beneath, margins entire, revolute. *Flowers* in axillary and terminal subsecund clusters 3, 5, or 7 together, sessile in the axil of a linear-lanceolate *bractea*, lower ones remote. *Involucre* ovate, *bracteas* imbricate, *outer ones* subulate, green, tomentose; *inner ones* obtuse, somewhat jagged and scariose at the apex, converging, membranous, with a glabrous green rib and clothed with long tomentose hairs. *Florets* all tubular, *exterior* ones few, female, 3-cleft, *stigma* linear, bifid, *interior* ones purple, 5-cleft, segments recurved, *stamens* apinulate, exserted, *style* dilated upwards, bifid, each segment crowned with a flat ciliated stigmatic disk. *Achenium* glabrous. *Receptacle* naked.

23. GNAPHALIUM. *Linn.* Cudweed.

* *Flowers diœcious.*

1. G. *dioicum*, Linn. *Mountain Cudweed.* Stems simple, with prostrate runners; radical leaves spathulate, woolly beneath, cauline ones lanceolate appressed; flowers diœcious, corymbose; inner involucral bracteas lanceolate, obtuse, dilated upwards, coloured. *E. Bot. t.* 267. *E. Fl. v. iii. p.* 414. *Hook. Br. Fl. p.* 360.

Mountain-heaths; rare. *Fl.* June, July. ♃.

Road from Trebrodind to Clun; *T. & D. Bot. Guide.*

Root woody, with many long simple fibres, sending forth from the crown numerous prostrate leafy runners which spread in a radiate manner and increase the plant. *Stem* 4—8 inches high, erect, simple, covered with dense white cottony down, leafy. *Radical leaves* tufted, obovate, apiculate, tapering at the base into a winged petiole, bright green and naked above, with dense close-pressed white cottony down beneath. *Stem leaves* erect, appressed, lanceolate, acute, cottony. *Flowers* 4 or 5 in terminal erect corymbs, dioecious. *Involucral bracteas* lanceolate, obtuse, *outer ones* short, green, cottony, *inner ones* dilated upwards, membranous, smooth, shining, margins serrulate, white, tinged with pale rose-colour. *Florets* tubular, 5-toothed. *Achenium* small, glabrous. *Pappus* sessile, very rough.

2. G. *margaritaceum*, Linn. *American Cudweed, Pearly Everlasting.* Herbaceous; stem erect, branched above; leaves linear-lanceolate, acute, alternate, cottony especially beneath; flowers corymbose, level-topped. *E. Bot. t.* 2018. *E. Fl. v. iii. p.* 413. *Hook. Br. Fl. p.* 360.

Moist meadows; rare. *Fl.* August. ♃.

Wyre Forest, (but uncertain whether in the Shropshire portion.) *Sm. E. Flora.* Coal-pit banks, Ketley; *Dr. Du Gard.*

Root somewhat creeping. *Stem* erect, 2 feet high, solid, densely cottony. *Leaves* alternate, upper surface green, even, covered with a loosely attached cottony web, with 3 parallel ribs, under surface densely white and cottony. *Peduncles* densely cottony. *Involucral bracteas* numerous, of a pure opaque pearly whiteness, obovate. *Florets* of the disk tubular, 5-cleft, yellowish. *Pappus* rough, dilated at the extremity. *Receptacle* tuberculated.

** *Flowers perfect.*

3. G. *sylvaticum*, Linn. *Highland Cudweed.* Stem simple, nearly erect, downy; leaves linear-lanceolate, downy; flowers axillary, forming an interrupted leafy spike.

Var. a. fuscum. Leaves woolly on both sides. *Hook. Br. Fl. p.* 360. *G. sylvaticum, E. Bot. t.* 913. *E. Fl. v. iii. p.* 415.

Groves, thickets and pastures; rare. *Fl.* August. ♃.

About Ludlow; *Miss Mc Ghie.*

Stem 3—12 inches high, solitary, simple, covered with cottony close-pressed down. *Leaves* alternate, lanceolate, acute, erect, tapering at the base into short petioles, cottony on both sides. *Flowers* in a dense leafy spike, slightly compound below. *Involucre* cylindrical, *bracteas* erect, oblong, obtuse, membranous, shining, with a brown border. *Florets* yellowish. *Achenium* papillose. *Pappus* rough. *Receptacle* minutely cellular.

Var. β. rectum. Leaves nearly glabrous above; spike longer, more interrupted. *Hook. Br. Fl. p.* 360. *G. rectum, Huds. E. Bot. t.* 124. *E. Fl. v. iii. p.* 415.

Sandy Heath, a mile from Shiffnal on the road to Wolverhampton; *Withering.* Ruckley Heath; *Dr. G. Lloyd.* In Cound paper-mill copse. In sandy roads about Berrington; *Rev. E. Williams's MSS.* Red lake, Ketley; *Dr. Du Gard.*

4. G. *uliginosum*, Linn. *Marsh Cudweed.* Stem very much branched, diffuse, woolly; leaves linear-lanceolate, woolly; flowers in terminal crowded clusters shorter than the leaves. *E. Bot. t.* 1194. *E. Fl. v. iii. p.* 417. *Hook. Br. Fl. p.* 360.

Wet places where water occasionally stagnates, fields, &c.; frequent. *Fl.* August, September. ⊙.

Root fibrous. *Whole plant* clothed with close-pressed cottony down especially in the upper part immediately beneath the heads of flowers. *Stem* 4—10 inches high, ascending or prostrate, copiously and alternately branched from the very base. *Leaves* alternate, linear-lanceolate, obtuse, submucronate, tapering gradually at the base, slightly cottony and greenish above, more densely so beneath. *Clusters of flowers* globose, 6 or 7 or more aggregated together and sessile amid the cottony down and leaves on the extremities of the branches. *Involucre* imbricated, *outer bracteas* few, ovate, *inner ones* equal, lanceolate, smooth, brownish-green, shining. *Florets* yellow, all fertile, of the disk tubular, 5-cleft, of the circumference filiform. *Achenium* smooth. *Pappus* sessile, rough. *Receptacle* naked.

24. FILAGO. *Linn.* Filago.

1. F. *minima*, Fries. *Least Filago.* Stem erect, subdichotomously branched; leaves linear-lanceolate, acute, appressed, cottony; flowers in lateral and terminal clusters longer than the leaves; involucre conical, naked at the apex. *Fries Nov. 2nd ed. p.* 268. *Hook. Br. Fl. 4th ed. p.* 302. *Drej. Fl. Hafn. n.* 815. *Gnaphalium minimum, Sm. E. Bot. t.* 1157. *E. Fl. v. iii. p.* 418. *Hook. Br. Fl. 3rd ed. p.* 361. *Gnaphalium arvense, Reich. Fl. Excurs. n.* 1389.

Dry and gravelly places; not unfrequent. *Fl.* July, August. ⊙.

Four miles from Whitchurch on the road to Ellesmere; *Mr. F. Dickinson.*

Caer Caradoc. Rocks on middle portion of south side of Pontesford Hill. Cold Hatton Heath. Sharpstones Hill. Dovaston Heath. Sweeney quarries near Oswestry.

Root small and slender. *Herb* all over cottony, of a greyish green hue. *Stems* one or more, 2—6 inches high, slender, dichotomously branched above. *Leaves* small, sessile, erect, appressed to the stem, cottony on both sides. *Flowers* 3 or more together, sessile; *florets* yellowish, tubular, 4-cleft. *Involucre* pentagonous, ovate at the base, conical upwards; *bracteas* lanceolate, convex, acute, green, downy, with narrow membranous edges, inner ones entirely membranous. *Achenium* of the florets between the outer involucral bracteas glabrous, destitute of *pappus*; that of the disk covered with erect papillæ with rough *pappus.*

2. F. *Germanica*, Linn. *Common Filago.* Stem erect, proliferous at the summit; leaves lanceolate, acute, wavy, erect, cottony; flowers in axillary and terminal globoso-capitate clusters; involucral bracteas nearly equal. *Hook. Br. Fl. 4th ed. p.* 302. *Drej. Fl. Hafn. n.* 817. *Gnaphalium Germanicum, Huds. E. Bot. t.* 1946. *E. Fl. v. iii. p.* 419. *Hook. Br. Fl. 3rd ed. p.* 361. *Reich. Fl. Excurs. n.* 1392.

Sandy and gravelly places and dry pastures; common. *Fl.* June, July. ⊙.

Stem one or several, 6—12 inches high, erect, straight, covered as is the whole herbage with dense grey loose cottony hairs, flowering at the extremity and sending forth from below the head of flowers 2 or more ascending branches, which are in like manner generally proliferous. This singular mode of growth caused the old botanists to designate it "herba impia" from the undutiful behaviour of the offspring in thus exalting themselves above their parents. *Leaves* numerous, lanceolate, acute, wavy, erect, amplexicaul. *Flowers* in terminal, dense, globose, sessile heads. *Florets* yellow, those of the centre of the disk 4-cleft; fertile florets also appear between the outer involucral bracteas. *Outer involucral bracteas* lanceolate, acuminate, concave, collapsed at the point so as to appear spiny, with a green keel, pale membranous margins and brown scariose points, serrulate

towards the apex, clothed on the exterior with long loose cottony hairs; *inner bracteas* similar but broader, glabrous and shining. *Achenium* with minute papillæ. *Pappus* pilose. *Receptacle* papillose.

25. PETASITES. *Desf.* Butter-bur.

1. P. *vulgaris*, Desf. *Common Butter-bur.* Thyrsus dense, oblong; leaves rotundo-cordate, unequally toothed, downy beneath, the lobes approximate. *Hook. Br. Fl. p.* 362.

Var. a. sterilis. Flowers sterile, bearing anthers, rarely seed. *Hook. l. c. T. Petasites, Linn. a. E. Fl. v. iii. p.* 426. *E. Bot. t.* 431.

Wet meadows and borders of brooks, rivers, &c.; not uncommon. *Fl.* April, May. ♃.

Buildwas; *Mr. F. Dickinson.* Between Church Aston and Lilleshall; *R. G. Higgins, Esq.* On the banks of the Rea near Bewdley, abundant; *Mr. G. Jorden.* Sides of the rivers Corve and Teme, Ludlow; *Miss Mc Ghie.* Oakley Park, near Ludlow; *Mr. H. Spare.* Near Oswestry; *Rev. T. Salwey.* In a meadow on the west-side of Wellington; *Withering.*

Sides of Meole brook, at Sutton Spa, near Shrewsbury. Morda pool near Oswestry, abundant.

Root thick, fleshy and extensively creeping, with many long fibres, aromatic, bitter. *Leaves* when full grown, which occurs long after the flowering, perhaps the largest of any British plant, petioles erect, thick. *Flower stalks* erect, stout, downy, clothed with concave tumid dilated membranous petioles, lower ones with rudimentary leaves, upper ones gradually becoming lanceolate *bracteas.* *Flowers* very numerous, in a dense ovate thyrsus, flesh-coloured, all flosculous, tubular, regularly 5-cleft. *Stigmas* of outer florets thickened and imperfect, central florets destitute of *anthers* and alone perfecting seeds. *Pedicels* slightly downy. *Involucre* and *bracteas* smooth.

Var. β. hybrida. Flowers fertile, bearing seed, rarely stamens. *Hook. l. c. T. Petasites, Linn. β. E. Fl. l. c. T. hybrida, Linn. E. Bot. t.* 430.

In similar situations but rarer. *Fl.* April, May. ♃.

In a moist meadow at the side of Walker Street, Wellington; *E. Lees, Esq.* On the banks of the river Mordda near Tre Vawr Clawd near Oswestry.

Leaves similar. *Flowers* in a long lax thyrsus, on downy bracteated pedicels. *Florets* all tubular, 4-5-cleft, mostly destitute of stamens except 1 or 2 in the centre of the disk. *Achenium* glabrous. *Pappus* sessile, very long, rough.

26. TUSSILAGO. *Linn.* Colt's-foot.

1. T. *Farfara*, Linn. *Common Colt's-foot.* Scape single-flowered, imbricated with scaly bracteas; leaves cordate, angular, dentate, downy beneath. *E. Bot. t.* 429. *E. Fl. v. iii. p.* 426. *Hook. Br. Fl. p.* 362.

Moist and clayey soils; common. *Fl.* March, April. ♃.

Root extensively creeping, very difficult of extirpation. *Flowers* appearing before the leaves, erect in blossom and seed, drooping before and after flowering, bright yellow, solitary, terminal. *Scapes* solitary or in clusters from the joints of the underground stem, 4—6 inches high, round, striated, more or less cottony, with numerous scattered reddish smooth oblong obtuse scaly *bracteas*, cottony on the margins. *Involucral bracteas* linear-oblong, obtuse, membranous, glabrous.

Florets of the circumference numerous, narrow, ligulate, spreading, generally fertile; those of the disk few, tubular, swelling upwards, 5-cleft, generally barren. *Achenium* glabrous. *Pappus* sessile, rough. *Leaves* erect, on a stout channelled petiole, rotundo-cordate, slightly lobed, copiously sinuato-dentate, teeth tipped with callous discoloured tubercles, smooth and somewhat glaucous above, white and densely cottony except on the prominent veins beneath.

27. ERIGERON. *Linn.* Flea-bane.

1. E. *acris*, Linn. *Blue Flea-bane.* Hispid; peduncles alternate, corymbose, single-flowered; lower leaves obovate, tapering into the petiole; stem leaves lingulate, acute, sessile; pappus as long as the florets of the ray. *E. Bot. t.* 158. *E. Fl. v. iii. p.* 423. *Hook. Br. Fl. p.* 362.

Dry places, wastes, &c.; in the limestone districts. *Fl.* August. ♃.

Old walls at Ludlow; Blodwell rocks; *Rev. T. Salwey.* On a wall at Hord's park and at the side of the turnpike road, opposite to Faintree house near Bridgnorth; *Purton's Midl. Flora.* In rough stony places at the base of the Arkol hill near Wellington; *E. Lees, Esq.* Ruckley Grange; *Dr. G. Lloyd.* Oreton, plentiful; *Mr. G. Jorden.* Lilleshall Abbey and Chetwynd park wall; *R. G. Higgins, Esq.!* Lilleshall lime kilns; *T. C. Eyton, Esq.* Lincoln's hill near Coalbrookdale; *Mr. F. Dickinson.*

Wenlock Abbey. Limestone quarries near the Wrekin. Benthal Edge.

Root fibrous. *Stem* 12—18 inches high, erect, simple, angular, striated, rough with short hairs, purplish, with several alternate axillary simple branches in the upper portion. *Lower leaves* obovate, tapering at the base into winged petioles, entire, rough on both sides with hairs; *stem-leaves* alternate, narrow, lingulate, acute, hairy on both sides, sessile. *Flowers* solitary on the extremities of the branches which are corymbose, peduncles striated and rough with hairs. *Involucral bracteas* unequal, subulate, acute, rough with white hairs. *Florets* of the ray narrow, ligulate, blue, nearly erect; those of the disk tubular, 5-cleft, yellow. *Achenium* with erect hairs. *Pappus* sessile, rough, tawny. *Receptacle* naked.

28. SENECIO. *Linn.* Groundsel.

* *Flowers without rays.*

1. S. *vulgaris*, Linn. *Common Groundsel.* Leaves semiamplexicaul, pinnatifid, toothed; flowers in clustered corymbs, destitute of a ray. *E. Bot. t.* 747. *E. Fl. v. iii. p.* 429. *Hook. Br. Fl. p.* 363.

Waste and cultivated ground, fields, &c.; common. *Fl.* all the year. ⊙.

Root fibrous. *Stem* 6—12 inches high, erect, more or less branched, round, furrowed, more or less downy especially in the axils of the leaves. *Leaves* alternate, pinnatifid, toothed and cut, obtuse, dilated, sessile or semiamplexicaul by their rounded toothed auricled bases, the lower ones mostly obovate, jagged, tapering into petioles. *Flowers* corymbose, on terminal and axillary leafy branches. *Involucre* oblongo-conical, glabrous, *bracteas* erect, outer ones small, subulate, scattered and appressed at the base of the inner ones, their apices discoloured; *inner ones* as long as the florets, linear-lanceolate, striated, erect, with a small black apex. *Florets* all tubular, 5-cleft, yellow. *Achenium* linear, ribbed and furrowed, clothed with erect appressed silky hairs. *Pappus* sessile, rough. *Receptacle* naked.

*** Flowers rayed, with the ray rolled back.*

2. S. *sylvaticus*, Linn. *Mountain Groundsel.* Ray revolute; leaves pinnatifid, denticulate, pubescent; outer involucral bracteas appressed, very much shorter than the inner ones; stem erect, straight, branched; flowers corymbose; achenium cano-pubescent. *E. Bot. t.* 748. *E. Fl. v. iii. p.* 431. *a. Hook. Br. Fl. p.* 363.

Dry sandy and gravelly hilly places; not uncommon. *Fl.* July. ⊙.

Wrekin; *Mr. F. Dickinson.* Hardwicke; *Mr. E. Elsmere, junr.*

Haughmond hill. Grinshill. Shawbury heath. Sharpstones Hill. Pimhill. Nesscliff. Sunderton Camp, near Haughmond. Sandford, near Oswestry.

Root fibrous. *Stem* erect, simple or branched, 10 inches to 3 feet high, round, furrowed, hairy, leafy. *Leaves* alternate, deeply pinnatifid, lobed, lobes denticulate, the teeth with an obtuse callous tip, margins revolute, more or less dilated and amplexicaul at the base, upper surface dark green, midrib deeply furrowed, slightly pubescent, under surface paler, somewhat cottony or tomentose with long straggling jointed hairs. *Flowers* axillary and terminal, corymbose, with subulate acuminate hairy *bracteas* at the base of the ramifications and with a few smaller ones scattered on the ultimate branches or pedicels. *Involucre* conical, swollen or dilated at the base, pubescent. *Outer involucral bracteas* very few, scattered, subulate, acute, pubescent, apex stained with brown, very short, scarcely one-fourth the length of the inner ones. *Inner involucral bracteas* long, subulate, acuminate, strongly keeled, margins membranous, apex obtuse, stained with dark brown and ciliated. *Florets* of the ray small, revolute, apex with 3 obtuse shallow teeth; those of the disk tubular, 5-cleft, segments erecto-patent, their apices with a large cluster of glands. *Achenium* longitudinally ribbed and furrowed, clothed with erect appressed short silky hairs. *Pappus* rough. *Whole plant* with a disagreeable aromatic smell when bruised, hairy or pubescent but not glandular.

"Whitecliffe, near Ludlow," and "about Pimhill," have been communicated as habitats for *S. viscosus, Linn.* but having never met with any specimens in Shropshire, and particularly in the latter locality, which could be referred to any other species than *sylvaticus*, I conclude that there is some mistake and accordingly refrain from inserting *S. viscosus* as a native of our County. It differs from *sylvaticus* in the achenia being quite glabrous or at most with a few rows of *very minute* distant hairs running up them; the leaves, involucre, &c., very glandularly pubescent and viscid; the outer involucral bracteas much longer, often equalling half the length of the involucre.

**** Flowers with patent rays. Leaves pinnatifid.*

3. S. *erucæfolius*, Huds. *Hoary Ragwort.* Ray spreading; lower leaves petiolate, oblongo-obovate, acute, deeply incised and toothed, auricled at the base, upper ones sessile and amplexicaul by their auricled bases, deeply pinnatifid, segments linear, acute, toothed, margins somewhat revolute, pubescent especially beneath; achenium longitudinally ribbed, silky; stem erect, corymbose. *Huds. Fl. Angl. p.* 366. "*Linn. Suec.* 750." *Fries Nov. 2nd ed. p.* 266. *Koch Syn. p.* 387. *S. tenuifolius, Sm. Fl. Br. p.* 884. *E. Fl. v. iii. p.* 433. *E. Bot. t.* 574. *Hook. Br. Fl. 3rd ed. p.* 364. *4th ed. p.* 305.

Hedges, pastures, &c.; not very common. *Fl.* July, August. ♃.

Hedge banks near Minsterley; *Mr. F. Dickinson.* Grinshill; *Mr. E. Elsmere, jun.*

Canal side between Shrewsbury and Uffington.

Root creeping. *Stem* 2 feet high, erect, simple, alternately branched in a corymbose manner, angular and furrowed, clothed with soft downy hairs. *Leaves* alternate, lower ones petiolate, oblongo-obovate, acute, deeply incised and toothed, auricled at the base, upper leaves sessile and amplexicaul by their auricled bases, deeply pinnatifid, segments linear, acute, toothed, sometimes pinnatifid, margins somewhat revolute, pubescent on both surfaces, especially the under surface which is paler and frequently cottony. *Flowers* corymbose, at the extremities of the stem and branches. *Peduncles* and *involucre* cottony. *Outer involucral bracteas* few, long, linear-lanceolate, pale or discoloured at the incurved apex, resembling the bracteas immediately below them. *Inner involucral bracteas* erect, appressed, broader, keeled, margins membranous, ciliated towards the apex which is brown and downy. *Florets* of the ray bright yellow, ligulate, linear-oblong, with 3 minute obtuse converging teeth, the central one the shorter, somewhat revolute, tube hairy; those of the disk orange-yellow, tubular, dilated upwards, 5-cleft, segments acute, erecto-patent. *Achenium* longitudinally ribbed, clothed with erect silky bristles. *Pappus* sessile, rough. *Receptacle* flat, with very minute erect points or teeth.

4. S. *Jacobæa*, Linn. *Common Ragwort.* Ray spreading; lower leaves lyrato-pinnatifid, upper ones bipinnatifid, sinuato-denticulate; achenium of the disk hairy, of the ray glabrous; stem erect, corymbose. *E. Bot. t.* 1130. *E. Fl. v. iii. p.* 434. *Hook. Br. Fl. p.* 364.

Hedge-banks, waysides, neglected pastures; common. *Fl.* July, Aug. ♃.

Root fleshy. *Stem* 2-3 feet high, erect, round, striated, glabrous or with a few scattered hairs, branched in the upper part. *Leaves* alternate, lower ones petiolate, lyrato-pinnatifid, denticulate, slightly pubescent and veiny above, paler and pubescent on the ribs beneath, upper leaves sessile, bipinnatifid, sinuato-denticulate, the segments divaricate, the pinnæ crowded at the semiamplexicaul base. *Flowers* bright yellow terminal and axillary, corymbose, the divisions subtended by small, more or less pinnatifid bracteas. *Peduncles* slightly and loosely downy or cottony. *Involucre* nearly hemispherical, glabrous; *outer bracteas* few, scattered, lax, subulate, acute, the apex stained with black, about one-third shorter than the inner ones, very slightly downy or cottony; *inner bracteas* lanceolate, acute, glabrous, strongly keeled, 3-ribbed, somewhat membranous at the margins, the apex stained with dark brown, and ciliated. *Florets* of the ray, narrow, ligulate, with 3 shallow obtuse teeth, and a few short glandular hairs at the summit of the tube; those of the disk tubular, 5-cleft, segments erecto-patent glandular on the back of the apex. *Achenium* of the ray glabrous, reticulato-striate; of the disk with erect appressed, short, silky bristles. *Pappus* rough.

5. S. *erraticus*, Bert. *Straggling Ragwort.* Leaves lyrate, inferior ones petiolate, inciso-dentate, the terminal lobe very large ovate frequently cordate at the base, lateral lobes obovato-oblong or spathulate, dentate, lobes of the upper leaves cuneate; achenium glabrous; stem round, divaricately branched upwards. *Bab. Prim. Fl. Sarn. p.* 53. *Koch Syn. p.* 388. *Reich. Fl. Excurs. n.* 1576.

Wet places, river sides, &c.; frequent. *Fl.* July, August. ♃.

Lane leading from Longden to Oaks Hall, near Pontesford; Banks of river Severn about Shrewsbury; plentifully.

Root fibrous. *Stem* 3-4 feet high, erect, round, furrowed, glabrous, purplish in the lower part, with wide-spreading branches in the upper portion. *Leaves* alternate on rather long dilated winged petioles, lyrate, the terminal lobe very large, frequently 3 inches long and 2 broad, ovate, or oblong, obtuse, subcordate

at the base, irregularly inciso-dentate, the lateral lobes few, very small in comparison, decreasing in size as they approach the stem, obovato-oblong, or somewhat spathulate, coarsely and irregularly dentate; the upper leaves with narrower and more copious divisions, which are continued down to the very base which thence becomes dilated and semiamplexicaul, all glabrous veiny and deep green. *Flowers* corymbose at the extremities of the stem and branches, yellow, large. *Peduncles* striated, very slightly cottony. *Involucre* hemispherical, *outer bracteas* very few, scattered, lax, subulate, with a discoloured point; *inner ones* nearly equal, linear-lanceolate, glabrous, strongly keeled, often cohering by their membranous serrulate margins, point discoloured ciliated. *Florets* of the ray linear-elliptical, with 3 minute acute teeth, orifice of the tube glabrous; of the disk tubular, 5-cleft, segments erect. *Achenium* of both ray and disk glabrous. *Pappus* rough.

The following localities have been sent to me for *S. aquaticus* which it is probable are referable to the present species, as the plant prevalent on the banks of the Severn about Shrewsbury and in such other parts of the County which I have examined is *S. erraticus.* Newport; *R. G. Higgins, Esq.* Ludlow; *Miss Mc Ghie.* Severn side near Buildwas; *Mr. F. Dickinson.* I have never met with any specimens which I could decidedly pronounce identical with *S. aquaticus, Huds.* of which the following are the characters in Mr. Babington's work before quoted. "Lower leaves petiolate, crenato-dentate, undivided or lyrate, the terminal lobe obovate, obtuse, lateral ones oblongo-linear; upper leaves pinnati-partite, lobes linear; achenium glabrous, submuricate; stem round, branched, corymbose."

***** Flowers rayed. Leaves undivided.*

6 S. *Saracenicus*, Linn. *Broad-leaved Groundsel.* Ray syreading, nearly entire; leaves lanceolate sessile, minutely glanduloso-serrate, glabrous; stem erect, solid, glabrous; corymbs terminal, of rather few flowers; bracteas linear-setaceous. *E. Bot. t.* 2211. *E. Fl. v. iii. p.* 436. *Hook. Br. Fl. p.* 364.

Moist meadows and pastures; rare. *Fl.* July, August. ♃.

In a watery lane in the hamlet of Hintz, in the Parish of Coreley. *Rev. W Corbett.*

Stem 3—5 feet high, solid, angular, smooth, scarcely branched. *Leaves* scattered, sessile, broadly lanceolate, acute, glabrous. *Flowers* corymbose, bright yellow. *Involucre* somewhat downy; *involucral bracteas* lanceolate, erect, acute, downy and blackened at the apex, outer ones smaller. *Achenium* smooth. *Pappus* rough.

29. SOLIDAGO. *Linn.* Golden-rod.

1. S. *Virgaurea*, Linn. *Common Golden-rod.* Stem erect, round, pubescent; lower leaves elliptical, acute, serrated, petiolate, pubescent, cauline ones lanceolate, acute, serrated or entire, subsessile. *E. Bot. t.* 301. *E. Fl. v. iii. p.* 439. *Hook, Br. Fl. p.* 365.

Woods, thickets and hilly places; not unfrequent. *Fl.* July, September. ♃.

Near the Knowle lime kilns. Titterstone Clee hill. Between Pulverbatch and Habberley; *Rev. W. Corbett.* High rock near Bridgnorth; *Rev. W. R. Crotch.* Westfelton; *J. F. M. Dovaston, Esq.* Coxwood, Coalbrookdale; *Mr. F. Dickinson.* Rock at Cheswell in the parish of Longford; *R. G. Higgins, Esq.* Wyre forest, plentifully; *Mr. G. Jorden.* Chesterton; *Dr. G. Lloyd.* Linley near Bishop's Castle; *W. P. Brookes, Esq.* Whitecliff near Ludlow; *Mr. H. Spare.* Craigforda near Oswestry; *Rev. T. Salwey.* Near Middle; *W. W. Watkins, Esq.* On the rocks of Harmer hill. Ruyton near Baschurch. Banky

pastures about Drayton. On the rock and walls of Red Castle near Hawkstone; *Rev. E. Williams's MSS.*

Ludlow Castle. Haughmond hill. Grinshill. Pontesford hill. Limestone quarries near the Wrekin. Lyd Hole near Pontesford hill. Nesscliffe.

Root woody, fibrous. *Stem* 1—3 feet high, erect, simple, angular, furrowed, reddish or green, more or less downy. *Lower leaves* elliptical, acute, serrated, tapering at the base into a winged petiole, margins ciliated, more or less downy on both sides especially beneath; *upper leaves* lanceolate, acute, serrated or entire, nearly sessile. *Flowers* in an erect simple or compound panicled raceme, pedicels pubescent bracteated. *Involucral bracteas* erect, appressed, imbricated, unequal, linear-lanceolate, acute, pubescent, membranous at the margins and ciliated towards the apex. *Florets* of the ray few, 6—10, ligulate, elliptic-oblong, spreading, with 3 shallow obtuse teeth, yellow, tube externally pubescent; of the disk tubular, deeply 5-cleft, segments ovate acute yellow. *Achenium* linear, longitudinally striated, clothed with short erect hairs. *Pappus* sessile, rough. *Receptacle* naked.

Very variable in size, the compactness or laxity of its inflorescence and serratures of the leaves. Koch remarks of it "Planta polymorpha, sub qua fortasse plures species latitant, sed collatis multis speciminibus, limites, quibus certé distinguantur, reperire non potui."

30. INULA. *Linn.* Elecampane.

1. I. *Helenium*, Linn. *Common Elecampane.* Leaves amplexicaul, ovate, rugose, downy beneath; involucral bracteas ovate, downy. *E. Bot. t.* 1546. *E. Fl. v. iii. p.* 441. *Hook. Br. Fl. p.* 365.

Hedge-banks, moist pastures, &c.; not common. *Fl. July, Aug.* ♃.

Near Marton pool; *Dr. T. W. Wilson.* Oakley park near Ludlow; *Mr. H. Spare.* London road near Burford House. Hopton castle and village; *Miss Mc Ghie.* In a wild woody spot close to the Severn between Bridgnorth and Upper Areley; *E. Lees, Esq.* Road to Bitterley about 2 miles from Ludlow; *Dr. G. Lloyd.* Lowe, plentiful; *Mr. G. Jorden.* Field near Buildwas; *Mr. F. Dickinson.* Westfelton. Llwyn-y-groes. Llwyntidman. Melverley; *J. F. M. Dovaston, Esq.* Near Llanymynech. Two miles from Bishop's Castle on the Montgomery road. Marlow near Ludlow; *T. and D. Bot. Guide.* Between Harley and Wenlock; *W. P. Brookes, Esq.!* By the side of the road between Buildwas Abbey and the mill and below the mill. In the corner of a field by the side of the road between Broseley and Linley. Waste ground about Alberbury Priory, in great plenty. Pastures adjoining Wigwig. By the side of Cantlop mill stream. Side of road in Bourton near Wenlock. Meadow between Buildwas Bridge and Coalbrookdale; *Rev. E. Williams's MSS.*

Near Middlehope on Wenlock Edge.

Stem 3—5 feet high, round, furrowed, solid, branched, downy. *Leaves* large, ovate, serrated, veiny, downy and hoary beneath, petiolate, upper ones amplexicaul. *Flowers* very large, terminal, solitary. *Involucral bracteas* ovate, broad, recurved, downy on both surfaces. *Florets* of the ray numerous, long, narrow, with 3 unequal teeth, yellow. *Achenium* tetragonous, smooth. *Pappus* rough.

2. I. *Conyza*, D.C. *Ploughman's Spikenard.* Leaves ovato-lanceolate, crenato-serrate, pubescent, upper ones entire; stem herbaceous, corymbose; involucral bracteas recurved, leafy. *Hook. Br. Fl. 4th ed. p.* 306. *Conyza squarrosa, Linn. E. Bot. t.* 1195. *E. Fl. v. iii. p.* 421.

Roadsides and hedge-banks; not common. *Fl.* September, October. ♂.

Near Bridgnorth; *Rev. A. Bloxam.* Lincoln's Hill and Coalbrookdale, Cundor Hill; *Mr. F. Dickinson.* Lilleshall Abbey; *R. G. Higgins, Esq.* Oreton, plentiful; *Mr. G. Jorden.* Chesterton; *Dr. G. Lloyd.* On the limestone in the vicinity of Wellington; *E. Lees, Esq.* Common by the sides of the roads about Ludlow; near Oswestry; *Rev. T. Salwey.* Ditch banks about Eaton Mascot, Wenlock, between Stoke Castle and Onibury, &c.; *Rev. E. Williams's MSS.* Near Frankwell, Shrewsbury; *Mr. J. S. Baly.*

Limestone quarries near the Wrekin. Roadsides at Red Hill and beyond Welbach. Benthal Edge, abundant. Woods near Bomere pool. Roadsides near the Black Barn near Eaton Mascott.

Root tapering, fleshy. *Stem* 2-3 feet high, erect, angular, purplish, clothed with soft hairs and mealy glandular particles. *Leaves* ovato-lanceolate, petiolate, crenato-serrate, soft and hairy on the upper surface, more densely hairy and paler with glandular mealiness beneath; upper leaves entire and nearly sessile. *Panicle* terminal, copiously branched, corymbose, leafy. *Involucre* nearly ovate, imbricated, *bracteas* linear-lanceolate, acute, erect, purple, hairy on the exterior, margins ciliated, the outer ones broader and with recurved or squarrose green points, covered with yellow glandular meal on the inner side. *Florets* yellow, outer ones female, tubular, with a very short 3 toothed ray; those of the disk hermaphrodite, tubular, 5-cleft. *Anthers* with plumose bristles at the base. *Achenium* longitudinally furrowed, with erect bristles in the upper part. *Pappus* plumose. *Receptacle* naked.

31. PULICARIA. *Gærtn.* Flea-bane.

1. P. *dysenterica,* Cass. *Common Flea-bane.* Leaves cordato-oblong, amplexicaul, obsoletely dentate, rugose, hoary; stem woolly, corymbose; involucral bracteas setaceous. *Hook. Br. Fl. p.* 365. *Inula dysenterica, Linn. E. Bot. t.* 1115. *E. Fl. v. iii. p.* 441.

Moist meadows and watery places; not uncommon. *Fl.* August. ♃.

Smethcott; *Dr. T. W. Wilson.* Oakley park, near Ludlow; *Mr. H. Spare.* About Donnington, near Shiffnal; *Dr. G. Lloyd.* Coalport; *Mr. F. Dickinson.* Roadside Harley to Wenlock, near the Limekilns; *W. P. Brookes, Esq.*

Pit between Battlefield and Albright Hussee. Canal between Shrewsbury and Uffington. Canal-side near Queen's Head turnpike gate and near Oswestry.

Root creeping. *Herbage* more or less hoary with loose white cottony down, emitting an aromatic acrid odour when bruised. *Stem* 12—18 inches high, erect, round, branched at the summit in a corymbose manner, the branches not unfrequently overtopping the stem. *Leaves* alternate, cordato-oblong, acute, amplexicaul by their dilated rounded auricles, margins somewhat recurved wavy and with minute bristles, obsoletely dentate, upper surface veiny and rugose. *Involucre* hemispherical, flattened, imbricated, *bracteas* setaceous, discoloured and recurved at the apex, cottony. *Florets* of the ray numerous, narrow, 3-toothed, pale yellow, tube glabrous. *Disk* very broad, hemispherical, flattened, florets numerous, deep orange-yellow, tubular, dilated upwards, limb with 5 acute erect segments, thickened and glandular on the exterior of their margins and apex. *Anthers* with minute bristles at the base. *Achenium* longitudinally striated, bristly in the upper part. *Pappus* double, outer very short,compressed, membranous, irregularly toothed; inner about as long as or longer than the tube, pilose, rough, sessile. *Receptacle* flat, rough with short erect projecting teeth or points.

32. DORONICUM. *Linn.* Leopard's-bane.

1. D. *Pardalianches,* Linn. *Great Leopard's-bane.* Leaves cordate, toothed, the lowermost on long naked petioles, the intermediate with the petioles dilated into 2 broad semiamplexicaul ears at the base, the uppermost sessile and amplexicaul. *E. Bot. Suppl. t.* 2654. *E. Fl. v. iii. p.* 447. *Hook. Br. Fl. p.* 366.

Roadsides, wastes, &c.; doubtfully wild. *Fl.* June, July. ♃.

Banks of Severn below Bridgnorth; *Withering.* In a hedge by the road from Much Wenlock to the Ironbridge; *Smith's English Flora.* On the left hand side of the road about 200 yards beyond Cardington on the road to Chatwall, 1794; *Rev. E. Williams's MSS.*

Root creeping, with compressed transversely sulcate woolly tubers, which throw out beneath coarse fibres and from the sides white fleshy scaly horizontal threads which produce other tubers. *Stem* 3 feet high, rising solitary and erect from the tubers, hollow, round, furrowed, hairy, glanduloso-pilose in the upper part. *Leaves* soft and pliant, hairy on both sides, somewhat wavy, irregularly toothed, obtuse, except the uppermost which is rather acuminate; radical ones large, on channelled petioles, rounded, cordate, lobes of the sinus overlapping each other; stem leaves 6—9, lower one petiolate, their petioles except the lowest winged and auricled, auricles in the higher leaves confluent and in the upper ones quite lost. *Involucral bracteas* lanceolato-linear, taper-pointed, about half as long as the ray, glanduloso-pilose. *Flowers* bright yellow, solitary, terminal, the earliest overtopped by succeeding ones. *Florets* of the ray numerous, in a single row, linear-oblong, sometimes entire, usually irregularly 3-toothed. *Achenium* oblong, furrowed, that of the ray naked, destitute of *pappus* or bearing only a bristle or two; that of the disk hairy and with a sesile crown of simple roughish bristles.

33. BELLIS. *Linn.* Daisy.

1. B. *perennis,* Linn. *Common Daisy.* Scape single-flowered; leaves spathulate, obovate, crenate. *E. Bot. t.* 424. *E. Fl. v. iii. p.* 448. *Hook. Br. Fl. p.* 366.

Pastures; everywhere. *Fl.* from early spring till the end of autumn. ♃.

34. CHRYSANTHEMUM. *Linn.* Ox-eye.

1. C. *Leucanthemum,* Linn. *Great white Ox-eye.* Radical leaves obovate petiolate, cauline ones oblongo-obovate variously pinnatifid and incised at the amplexicaul base; involucral bracteas lanceolate, obtuse, with a tawny narrow membranous margin. *E. Bot. t.* 601. *E. Fl. v. iii. p.* 450. *Hook. Br. Fl. p.* 367.

Dry pastures; frequent. *Fl.* June, July. ♃.

Root woody, fibrous. *Stem* 1-2 feet high, erect, simple or branched, angular, furrowed, more or less hairy. *Lower leaves* obovate, serrated or deeply cut, tapering into a winged petiole dilated at the base; *upper ones* narrower, but similar in outline, variously pinnatifid, especially at the amplexicaul dilated base, deep green, nearly smooth. *Flowers* solitary, terminal, large, disk yellow broad and flat. *Florets* of the ray white, elliptic-oblong, with 3-obtuse teeth, marked underneath with 2 prominent longitudinal elevations terminating in the sinuses; tube glabrous. *Involucral bracteas* imbricated, unequal, lanceolate, obtuse, with a green keel, tawny towards the margins which have a shining white jagged membrane.

2. C. *segetum,* Linn. *Corn Marigold, yellow Ox-eye.* Leaves oblong, inciso-serrated above, toothed at the amplexicaul base, glaucous; involucral bracteas oblong, obtuse, with a broad membranous laciniate border. *E. Bot. t.* 540. *E. Fl. v. iii. p.* 450. *Hook. Br. Fl. p.* 367.

Corn-fields; not unfrequent. *Fl.* June, August. ⊙.

Near Churton; *Mr. T. Bodenham.* Occasionally about Longnor; *Rev. W. Corbett.* Whitchurch; *Mr. F. Dickinson.* Stanton; *Mr. E. Elsmere, junr.* Newport; *R. G. Higgins, Esq.* Fields about Albrighton near Shiffnal. Tong! *Dr. G. Lloyd.* Hopton village. Felton. Bromfield, near Ludlow; *Miss Mc*

Ghie. Oakley park, near Ludlow; *Mr. H. Spare.* Near Marton; *W. W. Watkins, Esq.* Plentifully about Harley, Church Stretton, St. Giles's Fields, and other gravelly soils; *Rev. E. Williams's MSS.*

Root tapering. *Stem* 1 foot high, erect, alternately branched, angular, furrowed, smooth. *Leaves* alternate, oblong, inciso-serrated above, toothed at the base, amplexicaul, succulent, smooth. *Flowers* solitary, terminal, large, deep yellow, peduncles swelling upwards. *Involucre* hemispherical, imbricated, *bracteas* oblong, obtuse, green with a broad membranous laciniate border, outer ones smaller. *Florets* of the ray oblong, abrupt, with 2 large obtuse teeth, and often a third very small obtuse one in the deep sinus. *Achenium* compressed, smooth, longitudinally ribbed and furrowed.

35. PYRETHRUM. *Hall.* Feverfew.

1. P. *Parthenium,* Sm. *Common Feverfew.* Leaves petiolate, pinnate or deeply pinnatifid, segments ovate or ovato-oblong, pinnatifid or deeply cut; flowers corymbose, disk flat; involucral bracteas linear, obtuse; achenium longitudinally ribbed and furrowed, crowned with a short jagged membrane. *E. Bot. t.* 1231. *E. Fl. v. iii. p.* 452. *Hook. Br. Fl. p.* 367. *Matricaria Parthenium, Linn.*

Waste places, about villages and in hedges; not very common. *Fl.* July. ♃.

Shotton; *Dr. T. W. Wilson.* Oakley park, near Ludlow; *Mr. H. Spare.* Newport; *R. G. Higgins, Esq.* Wenlock Abbey; *W. P. Brookes, Esq.!* Munslow and Diddlebury; *Mr. F. Dickinson.*

Ludlow Castle. Roadside between Hadnall and Hardwick, abundantly. Pontesford Bridge works. Near Shrewsbury.

Root tapering. *Stem* 18 inches to 2 feet high, erect, round, furrowed, pubescent, branched above. *Leaves* alternate, petiolate, pinnate or rather deeply pinnatifid, leaflets ovate or ovato-oblong, pinnatifid or deeply and variously cut, clothed with a hoary pubescence. *Flowers* corymbose, peduncles furrowed, hoary, swelling upwards. *Involucre* hemispherical, downy. *Involucral bracteas* imbricated, linear, obtuse, with a prominent green keel and a broad membranous laciniated margin, downy. *Disk* yellow, flat; *ray* white, oval, 3-toothed, teeth obtuse, central one shortest, 5-ribbed, the ribs uniting below the apex and not continued into the teeth. *Achenium* tetragonous, longitudinally ribbed and furrowed, the furrows with a row of crystalline papillæ, at least in a young state, crowned with a short jagged membrane. *Plant* strong-scented, bitter and tonic.

2. P. *inodorum,* Sm. *Corn Feverfew or scentless Mayweed.* Leaves sessile, bipinnatifid, segments subulate, mucronate; flowers solitary, disk conical; involucral bracteas lanceolate, obtuse; achenium transversely rugose on the exterior, with two round glandular dots immediately below the elevated entire border. *E. Bot. t.* 676. *E. Fl. v. iii. p.* 453. *Hook. Br. Fl. p.* 367. *Chrysanthemum inodorum, Linn.*

Fields and waysides; not uncommon. *Fl.* August, October. ⊙.

Madeley wood; *Mr. F. Dickinson.* Hadnall; *Mr. E. Elsmere, junr.* Ludlow; *Miss Mc Ghie.* Frodesley; *T. A. Gleadowe, Esq.*

Near Astley Church. Cross Hill, near Shrewsbury. Old Heath, near Shrewsbury. Cross Houses near Berrington.

Stems several, erect, 10—12 inches high, glabrous, tetragonous, deeply furrowed, branched above. *Leaves* alternate, dilated and sessile at the base, 3-ribbed and furrowed beneath, bipinnatifid, segments subulate, mucronate, glabrous, furrowed beneath, round and flattened above, covered as are all the species in this genus with minute whitish glandular dots. *Flowers* solitary, terminal, large; disk conical, yellow; rays white, oblong, 3-toothed, the exterior teeth somewhat

rounded, the central one acute, 4-ribbed, the ribs continued into the teeth and meeting a little below their apices. *Involucre* hemispherical, *bracteas* lanceolate, obtuse, with a prominent green keel, pale membranous margin, tawny and scarious at the outer edge. *Achenium* oblong, abrupt, subtetragonous, compressed, the interior surface with a prominent rib smooth, the exterior ones rounded transversely rugose, and 2 large round glandular greenish dots immediately beneath the elevated entire border. *Receptacle* conical, naked.

36. MATRICARIA. *Linn.* Wild Chamomile.

1. M. *Chamomilla,* Linn. *Wild Chamomile.* Leaves bi-tri-pinnatifid, segments subulate, mucronate; flowers solitary, disk conical; involucre nearly plane, bracteas linear obtuse; achenium angular, destitute of crown or border. *E. Bot. t.* 1232. *E. Fl. v. iii. p.* 455. *Hook. Br. Fl. p.* 367.

Corn-fields, waste ground, roadsides, &c.; common. *Fl.* August. ⊙.

Root fibrous. *Stem* 12—18 inches high, erect, simple or branched chiefly from the base, angular, striated, smooth, solid. *Leaves* alternate, bi-tri-pinnatifid, 3 or 4 of the primary pinnæ crowded on each side near the dilated semiamplexicaul base, segments subulate mucronate flattened, plane on the upper surface, under surface with a central and 2 marginal ribs, the interstices deeply furrowed, fleshy, covered with minute glandular dots. *Peduncles* naked, smooth and furrowed. *Involucre* flat or nearly so; *bracteas* imbricated, linear, obtuse, with a prominent green keel, and a pale membranous margin which is dilated towards the extremity so as to approach an oblongo-obovate form, the membrane at the obtuse apex being very copious and scariose. *Florets* of the ray white, reflexed at night, oblong, very obtuse, truncated, with 3 obtuse teeth, the two outer teeth rounded, the central one more pointed generally a little shorter, 4-ribbed, the ribs converging and uniting below the apices of the teeth. *Disk* yellow, conical. *Receptacle* conical, hollow, smooth.

37. ANTHEMIS. *Linn.* Chamomile.

1. A. *nobilis,* Linn. *Common Chamomile.* Scales of the receptacle oblong, obtuse, scariose at the margins and apex; achenium subtrigonous, smooth, crowned with an obsolete margin. *E. Bot. t.* 980. *E. Fl. v. iii. p.* 457. *Hook. Br. Fl. p.* 368.

Dry gravelly pastures and waste places; rare. *Fl.* August. ♃.

Oreton, common; *Mr. G. Jorden.* At the north-east corner of Cound moor by the side of a small rill near the finger post, plentifully. Upon Bayston hill and Bicton heath. By the side of the road as you ascend the hill from Lidbury (the new way) to Bishop's Castle; *Rev. E. Williams's MSS.*

Stem 1 foot long, prostrate and branched. *Leaves* bipinnate, segments linear-subulate, somewhat downy. *Flowers* solitary, terminating the branches, disk yellow elongato-conical, ray white. *Involucre* hairy, *exterior bracteas* elongated, laciniate and membranous, *interior ones* acute, much shorter. *Whole plant* intensely bitter, highly aromatic and used medicinally.

2. A. *arvensis,* Linn. *Corn Chamomile.* Scales of the receptacle lanceolate, acuminated into a rigid mucro; achenium obsoletely tetrangular, equally furrowed, that of the exterior florets crowned with a plicato-rugose tumid ring, that of the interior ones crowned with an acute margin. *E. Bot. t.* 602. *E. Fl. v. iii. p.* 458. *Hook. Br. Fl. p.* 368.

Corn-fields; not common. *Fl.* July. ♂.

Oakley park near Ludlow; *Mr. H. Spare.*

Corn-fields near Battlefield. Hancott pool.

Stem 12—18 inches high, erect, much branched, hoary with fine shaggy hairs. *Leaves* bipinnatifid, segments linear-lanceolate, acute, mucronate, pubescent. *Flowers* solitary, on long terminal furrowed peduncles downy at the summit. *Disk* yellow, elongato-conical, *ray* broad and white. *Involucre* pubescent, *inner bracteas* lanceolate, with a membranous laciniate border, *outer ones* ovate, acute, shorter.

3. A. *Cotula*, Linn. *Stinking Chamomile.* Scales of the receptacle lineari-setaceous; achenium nearly terete, tuberculato-striated, crowned with a crenulate margin surrounding a subconvex disk. *E. Bot. t.* 1772. *E. Fl. v. iii. p.* 459. *Hook. Br. Fl. p.* 368.

Waste places, corn-fields, roadsides; not very common. *Fl.* July, Aug. ☉.
Hadnall; *Mr. E. Elsmere, junr.* Corn-fields opposite Coalport; *Mr. F. Dickinson.*

Stem 1 foot or more high, erect, branched, bushy, angular, furrowed, smooth. *Leaves* bipinnatifid, segments subulate, submucronate, glabrous or nearly so. *Flowers* solitary, on terminal striated downy peduncles. *Disk* elongato-conical, pale yellow; *ray* white. *Involucre* pubescent, *bracteas* nearly equal, lanceolate, obtuse, margins membranous and jagged.

28. ACHILLÆA. *Linn.* Yarrow.

1. A. *Ptarmica*, Linn. *Sneeze-wort Yarrow.* Leaves linearlanceolate, acutely serrated, serratures mucronate, minutely scabrous at the margins; flowers lax. *E. Bot. t.* 757. *E. Fl. v. iii. p.* 461. *Hook. Br. Fl. p.* 368.

Moist meadows and pastures in hilly districts. *Fl.* July, August. ♃.
Banks of Severn, Coalbrookdale; *Mr. F. Dickinson.* Rowley near Wenlock; *W. P. Brookes, Esq.* Newport; *R. G. Higgins, Esq.* Wyre Forest; *Mr. G. Jorden.* Ludlow; *Miss Mc Ghie.* Oakley Park near Ludlow; *Mr. H. Spare.* Near Oswestry; *Rev. T. Salwey.* Near Shawbury heath; *Dr. T. W. Wilson.* Fields foot of the Lawley. Uffington. Sundorn. Shelton. Sharpstones hill. *Root* creeping. *Stem* 2 feet high, erect, angular, smooth. *Leaves* alternate, sessile, semiamplexical, linear-lanceolate, acute, acutely serrated ·ratures mucronate, minutely scabrous at the margins, deep shining green,▴. .ib. *Flowers* corymbose; peduncles hoary, single-flowered. *Florets* of the ray ɔout 8, female, milk white, broad, oval, with 3 obtuse teeth, tube short; those of .he disk tubular, 5-cleft, segments recurved. *Involucre* hemispherical, imbricated, *bracteas* appressed, unequal, somewhat pubescent, green, with a membranous brown margin. *Achenium* compressed, smooth. *Receptacle* with large membranous chaffy scales.

2. A. *Millefolium*, Linn. *Common Yarrow or Milfoil.* Leaves bipinnatifid, segments linear, bi-tri-fid, mucronate; flowers crowded. *E. Bot. t.* 758. *E. Fl. v. iii. p.* 463. *Hook. Br. Fl. p.* 369.

Pastures and waysides; common. *Fl.* all summer. ♃.
Root woody, somewhat creeping. *Stem* 1-2 feet high, erect, slightly branched, roundish, subangulate, cottony. *Leaves* alternate, semiamplexical, linear-lanceolate, bipinnatifid, segments narrow, linear, bi-tri-fid, mucronate, scabrous above, hairy beneath, petiole channelled and hairy. *Flowers* corymbose, with a leaf or leaflets at the divisions of the cottony striated peduncles. *Involucre* oblong, *bracteas* imbricated, unequal, scariose at the margins, silky. *Florets* of the ray 5, with pistils only, claw long very narrow erect, ray spreading horizontally round tridentate, the central tooth shorter and smaller; those of the disk tubular, hermaphrodite, 5-cleft, segments recurved. *Achenium* oblong, slightly compressed, glabrous.

A variety with pale rose-coloured flowers frequently occurs, which differs only in the rays being reflexed and their teeth more equal in size.

SYNGENESIA—FRUSTRANEA.

39. CENTAUREA. *Linn.* Knapweed, Blue-bottle.

1. C. *nigra*, Linn. *Black Knapweed.* Leaves petiolate, lanceolate, remotely sinuato-serrate; involucre subglobose; central bracteas lanceolate, contracted above the middle, dilated above into an ovatoacuminate pectinated appendage; pappus very short, squamiform, denticulate. *E. Bot. t.* 278. *E. Fl. v. iii. p.* 466. *Hook. Br. Fl. p.* 369.

Meadows and pastures; frequent; *Fl.* June, August. ♃.
Root strong, woody. *Stems* several, spreading and ascending, 18 inches to 2 feet high, furrowed, clothed sparingly with loose cottony pubescence and rough with short rigid hairs, branched in a corymbose manner, solid, except immediately beneath the flowers where it is swollen upwards and hollow. *Lower leaves* lanceolate, acute, tapering at the base into a very long narrow winged petiole, roughish on both surfaces with short hairs, dark green above, pale and veiny beneath, margins remotely sinuato-serrate, ciliated, serratures tuberculate; *upper leaves* narrower, subpectinato-pinnatifid, chiefly so towards the base, the terminal lobe very narrow and elongated, sessile; *uppermost* nearly entire, continued up to the very base of the involucre. *Flowers* solitary on the extremity of the branches, light purple. *Involucre* subglobose, *bracteas* closely and regularly imbricated, erect, uppermost recurved, lowermost ones small, ovate, dusky brown, margins pectinated; those of the middle lanceolate, of a shining green in the lower part with loose white cottony pubescence above the middle where they become narrower and contracted and dilated above into an ovato-acuminate dusky brown appendage with a pectinated margin; teeth very long, divaricate and spreading, fringed with minute erectopatent bristles or ciliæ; uppermost bracteas similar but longer in the green portion, the terminal appendage larger more rotund of a paler brown membranous at the margins which are not pectinated but only irregularly laciniate or jagged, recurved. In all the terminal appendages only are visible with the small cottony portion immediately beneath, thus giving to the involucre an opaque dusky brown appearance. *Florets* apparently all fertile and tubular, gradually dilated upwards, with a deeply 5-cleft limb, segments erect linear acute; *filaments* hairy a little below the anthers, glabrous in the lower portion, exserted; *pistil* exserted beyond the anther-tube, surrounded with a small tuft of spreading hairs immediately at the base of the papillose stigma. *Ray* wanting. *Achenium* hairy, crowned with the short squamiform denticulate *pappus*. *Receptacle* densely clothed with long narrow linear acute white membranous chaff, about half the length of the narrow lower portion of the florets.

Var. β. radiata. Flowers radiant. *Hook. Br. Fl. p.* 370.
In similar situations; not common.
The late Rev. E. Williams found a variety with *white* flowers on a ditch bank on the right hand side of the road between Uffington and Sundorn; and by the side of the footpath between Eaton Mascott and Berrington. His MS. states that all the plants raised from the seed from the former locality produced *purple* flowers.

2. C. *Scabiosa*, Linn. *Greater Knapweed.* Leaves petiolate, lyrato-pinnatifid, segments ovato-lanceolate, irregularly inciso-serrate; involucre large, globose, central bracteas ovato-lanceolate, acute or with a short spine, pectinated, teeth increasing in size upwards, divaricate, ciliated; pappus long, bifariously bristly. *E. Bot. t.* 56. *E. Fl. v. iii. p.* 468. *Hook. Br. Fl. p.* 370.

Barren pastures, roadsides, &c.; frequent. *Fl.* July, August. ♃.
Stem 2-3 feet high, erect, branched in a corymbose manner, pentangular and furrowed, rough with scattered hairs, solid. *Leaves* lyrato-pinnatifid, segments ovato-lanceolate, deeply and irregularly inciso-serrate, submucronate, terminal lobe trifid, hispid on both surfaces, pale green, margins ciliated, lower ones on a long dilated hispid petiole, upper ones sessile, decreasing in size upwards, the segments entire or nearly so. *Peduncles* deeply furrowed and angular, gradually swollen upwards and with a few scattered acute entire or subpectinated bracteas immediately below the involucre. *Flowers* solitary on the extremity of the branches, large, handsome, bright purple. *Involucre* large, globose, *bracteas* densely and regularly imbricated, erect, uppermost recurved; lowermost ones ovate; central ones ovato-lanceolate, acute, frequently with a short acute spine, dark green and slightly hoary with white cottony pubescence, the apex and a narrow strip of the margins from about the middle scariose and dusky brown, pectinated, the teeth gradually increasing in size upwards as they approach the apex, divaricate and spreading, fringed with minute bristles or ciliæ; uppermost bracteas similar in form but larger, the apex dilated into a large subovate scariose dusky brown paler membranous recurved appendage. *Florets* of the ray tubular, very long and spreading in a radiate manner, limb deeply 5-cleft, segments linear-lanceolate, without stamens or pistils; those of the disk tubular, gradually dilated upwards, 5-angular, with 5 prominent darker-coloured ribs, the interstices furrowed, limb 5-cleft, segments linear, erecto-patent. *Anthers* exserted, dark purple; *filaments* hairy immediately below the anther-tube, glabrous in the lower part; *pistil* exserted beyond the anther-tube, with a circular tuft of spreading hairs immediately below the papillose stigma. *Achenium* sparingly hairy. *Pappus* tawny, about half the length of the narrow portion of the tube of the florets, most beautifully and minutely bifariously bristly. *Receptacle* densely clothed with long white linear acuminate membranous scales or chaff.

3. C. *Cyanus*, Linn. *Corn Blue-bottle.* Leaves sessile, linearlanceolate, acute, entire, lowermost toothed or pinnatifid; involucre ovate, central bracteas ovato-lanceolate or lanceolate, margins membranous, deeply incised at the apex, segments ciliated; pappus longer than the silky achenium, paleaceous, bristly. *E. Bot. t.* 56. *E. Fl. v. iii. p.* 467. *Hook. Br. Fl. p.* 370.

Corn fields; not unfrequent. *Fl.* July, August. ☉.
Oakley park near Ludlow; *Mr. H. Spare.* Fields by side of road from Shrewsbury to Ludlow; *Mr. J. S. Baly.*
Fields near Bomere pool. Westfelton Moor.
Root fibrous. *Stem* 2 feet high or more, erect, branched, angular and furrowed, whitish with scattered cottony pubescence. *Leaves* alternate, narrow, linearlanceolate, acute, entire, sessile, 3-ribbed, lowermost ones broader, toothed or pinnatifid. *Inflorescence* solitary, terminal. *Involucre* ovate, smooth or slightly cottony, *outer bracteas* ovate, very acute, pale and membranous, not discoloured, inciso-laciniate at the edges, segments very acute, margins ciliated; *central bracteas* ovatolanceolate and lanceolate, erect, green with a membranous margin, scariose, discoloured and dull brown, deeply incised at the apex, segments ciliated; *innermost bracteas* longer and narrower, green, the upper portion pink, with an obtuse laciniate brown scariose apex. *Flowers* of a rich bright blue. *Florets* of the ray tubular, with a large dilated spreading irregularly many-cleft limb, abortive; those of the disk smaller, tubular, with a short regularly 5-cleft limb, segments linear obtuse, purplish. *Filaments* with a circular appendage of hairs immediately below the dark purple *anthers*. *Achenium* with erect appressed silky hairs. *Pappus* short, longer than the fruit, scarcely half as long as the tube, paleaceous, bristly. *Receptacle* flat, with very narrow linear acute chaffy scales shorter than the tube.

CLASS XX.

GYNANDRIA.

"The book of Nature is before us,—that noblest of volumes,— where we are ever called to wonder and to admire, even when we cannot understand."

Sir WALTER SCOTT.

CLASS XX.

GYNANDRIA.　*Stamens situated upon the style or column,*
above the germen.

The plants in this Class belong to the *Nat. Ord.* ORCHIDEÆ, *Juss.*—*Roots* often tuberous ; *stems* herbaceous ; *leaves* striated, sheathing at the base. *Flowers* in six divisions, of which the 3 outer ones are the *calyx* though coloured, the 3 inner the *corolla* ; lower *petal* (so situated by the twisting of the inferior *germen*) mostly larger, differently shaped and termed the *lip*. *Style* a column more or less elongated bearing the *stigma* on which is fixed the *anther*. *Pollen* either pulverulent, loosely collected into a mass ; or of grains elastically cohering, fixed to a *stalk* ; or of a definite number of waxy masses. Between the bases of the anther-cells is a *rostellated process*, which in the genera *Orchis* and *Gymnadenia* is a linear projection from the upper edge of the stigma and extending for a short distance upwards between the lower parts of the cells of the anthers. In the second division of the family, that is, in the genera *Neottia*, *Listera*, *Neottidium* and *Epipactis*, this part becomes a separate organ projecting from the upper edge of the stigma and generally bearing at the top an appendage usually of a flat expanded form but sometimes globular ; this appendage attaches itself to the glutinous part of the pollen-mass and ultimately falls off from the rostellum with the pollen-mass. It appears to be intended as a means of conveying the pollen-mass into such a situation that the pollenic cord may find its way to the stigma.

To the kindness of my excellent friend Mr. C. C. Babington who has studied the plants of this singular tribe with ardent perseverance and success, I am deeply indebted for the whole of the generic and specific characters, and numerous observations. To those who may be desirous of investigating the Order I would be permitted to recommend the beautiful and elaborate lithographs in Esenbeck's Genera Plantarum Floræ Germanicæ, as essential aids towards an accurate knowledge of the peculiar structure.

ORDER I. MONANDRIA.　*One Stamen.*

A. *Pollen-masses in divisible lobes, elevated on a stalk with a gland*
at its base.

* *Cells of the Anther with a rostellated process between their bases.*

1. ORCHIS. *Perianth* ringent, hooded. *Lip* 3-lobed, spurred. *Glands* of the stalks of the pollen-masses in a common pouch, and a *rostellated process* between the bases of the cells of the anther.—Name, an ancient appellation of the plant.

2. GYMNADENIA. *Perianth* ringent, hooded. *Lip* 3-lobed, spurred. *Glands* of the stalks of the pollen-masses without a pouch, the cells closed below but at length open in front, and a *rostellated process* between their bases.—Name from γυμνος, *naked*, and αδην, a *gland* ; one of the essential characters of this genus.

** *Cells of the Anther without a rostellated process between their bases.*

3. HABENARIA. *Perianth* ringent, hooded. *Lip* 3-lobed or entire, spurred. *Glands* of the stalks of the pollen-masses exserted,

---

naked, *rostellated process* wanting.—Name from *habena*, a *thong* or *lash*, which the spur sometimes resembles.

4. OPHRYS. *Perianth* patent. *Lip* variously lobed, without any spur. *Glands* of the stalks of the pollen-masses each in a distinct pouch, *rostellated process* wanting.—Name from οφρυς, the *eyebrow*, which Pliny says the plant was used to blacken. The flowers are beautiful and curious, resembling certain insects after which the species are named.

B. *Pollen-masses granular, not stalked.*

* *Cells of the Anther with a rostellated process between their bases.*

5. NEOTTIA. *Perianth* ringent. *Lip* channelled, unguiculate at the base, fringed. *Stigma* roundish. *Rostellum* straight, bifid, with an elongate linear appendage between its points. *Germen* not twisted.—Name from νεοττια, a *bird's-nest*, formerly applied to our *Neottidium Nidus-avis*, on account of its densely tufted fibres.

6. LISTERA. *Perianth* ringent. *Lip* deflexed, 2-lobed. *Stigma* transverse. *Rostellum* elongated, acute, with a minute globose appendage at its somewhat reflexed apex. *Column* very short. *Germen* not twisted.—Name in honour of *Dr. Martin Lister*, an eminent British Naturalist.

7. NEOTTIDIUM. *Perianth* hooded. *Lip* deflexed, 2-lobed, saccate at the base. *Stigma* transverse. *Rostellum* flat, broad, prominent, without an appendage. *Column* elongated. *Germen* not twisted.—Name, a diminutive of *Neottia*.

8. EPIPACTIS. *Perianth* patent. *Lip* interrupted, the basal lobe concave, the terminal lobe larger with two projecting plates at its base above. *Stigma* nearly square. *Rostellum* short, terminated by a globose appendage. *Anthers* terminal, erect, sessile, 2-celled, cells without septa. *Column* short. *Germen* straight, upon a twisted stalk.—Name from επι, *upon*, and πακτος, *fixed* or *jointed* ; from the appearance of the lip.

** *Cells of the Anther without a rostellated process between their bases.*

9. CEPHALANTHERA. *Perianth* converging, (in *rubra* spreading.) *Lip* interrupted, the basal lobe saccate and articulated with the recurved terminal lobe. *Stigma* transverse. *Rostellum* wanting. *Anther* terminal, erect, moveable, shortly and thickly stalked, 2-celled, the cells with imperfect septa. *Column* elongated. *Germen* sessile, twisted.—Name from κεφαλη, a *head*, and ανθηρα, *flowery*.

GYNANDRIA—MONANDRIA.

1. ORCHIS. *Linn.* Orchis.

A. *Glands of the stalks of the pollen-masses separate ; lip in*
æstivation erect. (ORCHIS VERA.)

† *Bracteas* 1-*nerved ; tubers* 2, *undivided.*

* *Lip* 3-*lobed, lobes broad, short.*

1. O. *Morio*, Linn. *Green-winged Meadow Orchis.* Lip 3-lobed, crenulate, the middle lobe truncate-emarginate ; spur ascending, subclavate, rather shorter than the germen ; sepals and petals obtuse, connivent ; anther obovate, rather acute. *E. Bot. t.* 2059. *E. Fl. v. iv.* p. 11. *Hook. Br. Fl.* p. 373.

Meadows and pastures ; not unfrequent. *Fl.* June. 4.

Kenley Common, near Hughley ; *W. P. Brookes, Esq.* Welbach and Lythwood, near Shrewsbury ; *Mr. T. Bodenham.* Pasture fields under the Lawley ; *Rev. W. Corbett.* Coalbrookdale ; *Mr. F. Dickinson.* Hadnall ; *Mr. E. Elsmere, junr.* Oakley Park, near Ludlow ; *Mr. H. Spare.* Near Oswestry ; *Rev. T. Salwey.* Shotton near Shrewsbury ; *W. W. Watkins, Esq.*

Almond pool banks. Bomere woods.

Stem from a span to a foot high. *Leaves* lanceolate, inferior ones obtuse spreading, upper ones acute sheathing appressed. *Flowers* few, in a lax *spike.* *Calyx* purplish-green, forming a sort of helmet over the rest of the flower. *Lip* purple, pale in the middle, with purple spots. *Sepals* and *petals* marked with green veins.

2. O. *mascula*, Linn. *Early purple Orchis.* Lip 3-lobed, crenate, the middle lobe emarginate ; spur ascending, cylindrical, rather longer than the germen ; sepals acute, the 2 outer ones reflexed upwards ; petals convergent ; anther obcordate, apiculate. *E. Bot. t.* 631. *E. Fl. v. iv.* p. 11. *Hook. Br. Fl.* p. 373.

Woods and pastures ; not unfrequent. *Fl.* June. 4.

Oakley park, near Ludlow ; *Mr. H. Spare.* Pasture fields under the Lawley ; *Rev. W. Corbett.* Newport ; *R. G. Higgins, Esq.* Near Walford ; *T. C. Eyton, Esq.* Welbach and Lythwood, near Shrewsbury ; *Mr. T. Bodenham.* Abundant around Wenlock ; *W. P. Brookes, Esq.*

Haughmond Hill and especially about Shrewsbury.

Stem ½ foot high. *Leaves* chiefly radical, elliptic-lanceolate, shining, generally marked with dark purple spots. *Flowers* in a lax oblong *spike*, purple ; the centre of the *lip* whitish at the base and spotted, sometimes altogether white. *Sepals* and *petals* not veined.

** *Lip* 3-*lobed, central lobe dilated bifid and often having an intermediate tooth.*

3. O. *ustulata*, Linn. *Dwarf dark-winged Orchis.* Lip 3-partite, lobes linear, the middle one bifid ; spur one-third as long as the germen, decurved ; sepals ovate acute connivent, covering the linear obtuse petals ; anther rather obovate, acuminate ; bractea shorter than the germen. *E. Bot. t.* 18. *E. Fl. v. iv.* p. 12. *Hook. Br. Fl.* p. 373.

Pastures in the limestone districts ; rare. *Fl.* June. 4.

At the Woodlands, near Bridgnorth; *Purton's Midl. Flora.* The Lodge, near Ludlow; *Rev. T. Salwey.* Pasture field near the Sallow Coppice, near Millechop, Parish of Eaton; *Rev. W. Corbett.* On the side of the Hill opposite Downton Castle; *Plymley Agric. Surv.*

Smaller than the preceding. *Stem* 4-5 inches high. *Leaves* lanceolate, acute, spreading. *Spike* oblong, dense, obtuse; *flowers* small, numerous. *Sepals* forming a sharp helmet-like covering, within which are the 2 small linear lateral *petals*, and with them of a dark dingy purple. *Lip* white, with purple raised spots.

†† *Bracteas many-nerved; tubers 2, palmate.*

4. O. *maculata*, Linn. *Spotted palmate Orchis.* Lip flat, 3-lobed, crenate; spur subulate, shorter than the germen; sepals 3 patent; petals connivent; stem solid. *E. Bot. t.* 632. *E. Fl. v. iv. p.* 22. *Hook. Br. Fl. p.* 375.

Pastures; not unfrequent. *Fl.* June, July. ♃.

Frequent round Wenlock; *W. P. Brookes, Esq.* Very common about Ludlow; *Mr. J. S. Baly.* Welbach and Lythwood, near Shrewsbury; *Mr. T. Bodenham.* Cox Wood, Coalbrookdale; *Mr. F. Dickinson.* Near Walford; *T. C. Eyton, Esq.* Newport; *R. G. Higgins, Esq.* Extremely common and of great size in Coalbrookdale; *E. Lees, Esq.* Oakley Park, near Ludlow; *Mr. H. Spare.* Shotton, near Shrewsbury; *W. W. Watkins, Esq.*

Haughmond Hill, Bomere Pool, and generally around Shrewsbury.

Stem a foot high, slender. *Leaves* lanceolate, remote, spotted with purple. *Spike* short, dense, oblong. *Flowers* pale purple, more or less spotted and streaked, especially the lip. *Lip* deeply lobed, exterior lobes rounded, central one longer and ovate. Intermediate *sepal* with an acute flat point. Lower *bracteas* longer than the germen, upper ones equal to it.

5. O. *latifolia*, Linn. *Marsh Orchis.* Lip obscurely 3-lobed, its sides reflexed, crenate; spur subulate, shorter than the germen; 2 lateral sepals patent; petals connivent; stem hollow. *E. Bot. t.* 2308. *E. Fl. v. iv. p.* 21. *Hook. Br. Fl. p.* 375.

Marshes and moist meadows; not unfrequent. *Fl.* June. ♃.

Oakley Park, near Ludlow; *Mr. H. Spare.* Marshy spots about the base of the Arcoll Hill, near Wellington; *E. Lees, Esq.* Hadnall and near Shawbury Heath; *Dr. T. W. Wilson.* Field near Marn wood, Coalbrookdale; *Mr. F. Dickinson.* Whixall moss; *Rev. A. Bloxam.* Buildwas; *Miss H. Moseley.*

Albrightlee; Shawbury Heath; Hancott pool; Shelton Rough; and generally around Shrewsbury.

Intermediate *sepals* usually hooded at the point. Lower *bracteas* longer than the flower.

Fries (*Nov. Mant.* 1. *p.* 16. *et Mant. alt. p.* 53.) and Koch (*Syn. p.* 687.) consider that under *O. latifolia, Linn.* (*Suec. n.* 802.) three species are included, the distinguishing characters of which are here inserted in order to direct attention to them, as all of them may possibly be recognized in Britain. Whether their differences be sufficient to separate them as species seems very doubtful, since they seem such as we should expect to arise from variation of soil and situation, and as consequently intermediate states are likely to exist, it may ultimately be found impossible to draw the line between them:—

1. O. *latifolia.* Lateral petals reflexed upwards; inferior and intermediate bracteas longer than the flowers, all 3-nerved and veiny; leaves spreading, inferior ones oval or oblong obtuse, superior ones smaller lanceolate acuminate. *Koch Syn. p.* 687. *Fries Nov. Mant.* 1. *p.* 16. *O. majalis, Reich. Icon. f.* 770.

Robust, the leaves often spotted. Considered by Fries (*l. c.*) as the *true O latifolia, Linn.*

2. O. *angustifolia.* Lateral petals patent, afterwards reflexed upwards; all the bracteas longer than the flowers, 3-nerved and veiny; leaves erect, parallel to the stem, elongato-lanceolate, attenuated, cucullato-contracted at the apex, uppermost exceeding the base of the spike, inferior one shorter, spreading. *Koch. l. c. O. latifolia, Reich. Icon. f.* 769. *O. angustifolia, Wimm. et Grab. Fl. Sil. v. iii. p.* 252.

Taller, the leaves without spots.

3. O. *incarnata.* Lateral petals reflexed upwards; bracteas 3-nerved and veiny, inferior ones equalling, superior ones shorter than the flowers; inferior leaves lanceolate, erecto-patent, superior ones linear erect plane at the apex subcanaliculate. *Koch. l. c. p.* 688. sub O. Trausteineri, *Saut.* O. *incarnata, Linn. certe. Fries Nov. Mant.* 1. *p.* 16. *Hartm. Scan. ed.* 3. *O. angustifolia, Reich. Icon. f.* 1140. forma foliis angustissimis.

Slenderer and with nearly the habit of small specimens of *O. maculata*; leaves without spots.

Mr. Babington, who inclines to consider them as varieties, says—"Since my letter containing an account of the varieties of this plant, I have examined the Linnæan Herbarium, and find that his *O. latifolia* does not exactly agree with either of my varieties and Koch's species. It has the leaves erecto-patent and cucullate at the tip, and the *lower* bracteas longer than the flowers; the form of the lower leaves I could not determine. I should not recommend you to separate the varieties, as I am inclined to think that we shall ultimately find it impossible to draw the line between them. Your plants from Hancott pool and Albrightlee I consider as *majalis*, Reich! which is *latifolia*, Sm.! and perhaps Linnæus; but then, they both appear to have the leaves cucullate."—*Babington in lit.*

B. *Glands of the stalks of the pollen-masses united, lip in æstivation erect.* (Anacamptis, *Rich.*)

6. O. *pyramidalis*, Linn. *Pyramidal Orchis.* Lip with 3 equal lobes and 2 tubercles at the base; spur filiform, longer than the germen; lateral sepals ovato-lanceolate, acute, spreading; bracteas 3-nerved; tubers undivided. *E. Bot. t.* 110. *E. Fl. v. iv. p.* 10. *Hook. Br. Fl. p.* 375.

Pastures and waste ground in the limestone districts; not unfrequent. *Fl.* July. ♃.

Shadwell, a part of the Standhill Coppice close to Wenlock; *W. P. Brookes, Esq.* Benthal Edge; *Mr. T. Bodenham.* Coreley; *Rev. W. Corbett.* About the High Rock near Bridgnorth; *Rev. W. R. Crotch.* Near Cox Wood and Marn Wood, near Coalbrookdale; *Mr. F. Dickinson.* Oreton; *Mr. G. Jorden.* Whole fields covered with it on the southern side of the Limekiln Woods, near Wellington; *E. Lees, Esq.* Near the fourth milestone from Ludlow to Wenlock; *Rev. T. Salwey.* Buildwas; *Miss H. Moseley.* Old pastures about Stevens Hill near Cound. Much Wenlock. Earls Ditton; *Rev. E. Williams's MSS.*

Hill Top, Wenlock Edge. Limestone quarries near the Wrekin. Benthal Edge.

Stem 12—18 inches high. *Leaves* linear-lanceolate, acute, uppermost shorter. *Spike* broad, ovate, dense, at first pyramidal; *flowers* of a delicate rose-purple, spirally arranged.

2. GYMNADENIA. *Br.* Gymnadenia.

1. G. *conopsea*, Br. *Fragrant Gymnadenia.* Lip 3-lobed, lobes equal entire obtuse; lateral sepals spreading; spur filiform twice as long as the germen; tubers palmate. *Hook. Br. Fl. p.* 375. *Orchis conopsea, Linn. E. Bot. t.* 10. *E. Fl. v. iv. p.* 23.

Dry hilly pastures, chiefly in the limestone districts. *Fl.* June—August. ♃.

Oakley park, near Ludlow; *Mr. H. Spare.* Near Park brook, Wyre Forest; *Mr. W. G. Perry.* Moors near Shortwood, Hope Bagot, Felton moors, &c.; *Miss Mc Ghie.* Near Bridgnorth; *Purton's Midl. Flora.* Wyre Forest; *Mr. G. Jorden.* Benthal Edge; *Mr. F. Dickinson.* Meadows near Coreley parsonage, and under the Lawley near Longnor; and under the Longmynd; *Rev. W. Corbett.* Shadwell, Standhill Coppice, Wenlock; *W. P. Brookes, Esq.* Between Pitchford and Golden. Golden and Cound moor. Between Longner Green and Lidley-Hayes; *Rev. E. Williams's MSS.*

Between Middlehope and Westhope, Wenlock Edge.

Stem 12—18 inches high. *Tubers* palmate. *Leaves* linear-lanceolate, keeled. *Spikes* cylindrical, variable in density, *flowers* rose-purple. The 2 lateral *sepals* spreading, their margins revolute. The 2 lateral *petals* connivent. *Spur* pointing downwards, curved. The 2 *cells* of the *anther* are perforated at the base, through which the naked large and oblong *glands* of the *stalks* of the *pollen-masses* appear.

2. G. *albida*, Scop. *Small white Gymnadenia.* Lip 3-lobed, lobes entire acute, the central one longest; sepals and lateral petals connivent; spur much shorter than the germen; tubers clustered. *Scop. Carn. ed.* 2. *v. ii. p.* 201. *Orchis albida, Sm. E. Fl. v. iv. p.* 18. *Satyrium albidum, Linn. E. Bot. t.* 305. *Habenaria albida, Br. Hook. Br. Fl. p.* 376.

Mountain pastures; very rare. *Fl.* June, July. ♃.

Benthal Edge; *Dr. G. Lloyd.*

Stem about a span high. Lower *leaves* obovate, superior ones lanceolate, uppermost narrower. *Spike* cylindrical, dense, of many small white fragrant *flowers*. *Bracteas* ovato-lanceolate, obtuse, about as long as the germen. *Lip* scarcely longer than the sepals, deflexed.

"The structure of this plant is certainly that of *Gymnadenia* not *Habenaria.* I am far from being certain that this genus and *Aceras* ought to be retained separate from *Orchis. G. conopsea* comes so very near to *O. pyramidalis* in habit and appearance, as often to be hardly distinguishable without examination, and the structure of the anther is scarcely sufficiently different to constitute a genus." *Babington.*

3. HABENARIA. *Br.* Habenaria.

1. H. *viridis*, Br. *Green Habenaria.* Spur very short, 2-lobed; lip linear, flat, 3-pointed, intermediate point the shortest. *Hook. Br. Fl. p.* 376. *Orchis viridis, Sm. E. Fl. v. iv. p.* 20. *Satyrium viride, Linn. E. Bot. t.* 94.

Dry hilly pastures, not unfrequent. *Fl.* June, July. ♃.

Large uncultivated field at Rowley near Wenlock; *W. P. Brookes, Esq.!* Meadow near Ludlow; *T. and D. Bot. Guide.* Woodside near Donnington, (Shiffnal); *H. Bidwell, Esq.* Near Cox Wood, Coalbrookdale, in a field opposite Comer; *Mr. F. Dickinson.* Hadnall; *Mr. E. Elsmere, junr.* Oreton; *Mr. G. Jorden.* Caynham Camp near Ludlow; *J. Walcot, Esq.* Near Oswestry; *Rev. T. Salwey.* Near Shawbury Heath; *Dr. T. W. Wilson.* Buildwas; *Miss H. Moseley.* Meadows about Eaton Mascot, and Battlefield; *Rev. E. Williams's MSS.*

Bomere Woods. Meadows between Westhope and Middlehope. Wenlock Edge.

Stem 6—12 inches high. *Tubers* palmate. *Leaves* ovate or elliptical obtuse, upper ones lanceolate acute. *Spikes* lax. *Inferior bracteas* longer than the flowers. *Sepals* ovate, obtuse, and with the narrow linear lateral *petals* which are about equal to them in length connivent and forming a helmet, green. *Lip* half

as long again as the sepals, linear, ending in 2 sharp lobes with a smaller central one, greenish-brown.

"In this genus the rostellated process which exists between the bases of the anther-cells in *Orchis* is wanting, or perhaps may be considered as represented in *H. chlorantha* by the slight point in the centre of the upper edge of the stigma, and possibly by the transverse elevated line between the glands of the pollen-masses in *H. viridis.* In *H. bifolia* I do not detect it in any form."—*Babington.*

2. H. *chlorantha*, Bab. *Yellow Butterfly Habenaria.* Spur twice as long as the germen; lip linear, entire; petals connivent, obtuse; anther truncate, twice as broad at its base as its top, with the pollen-cells ascending obliquely and converging upwards. *Babington in Linn. Trans. v. xvii. p.* 463. *Orchis bifolia, a. E. Bot. t.* 22. *E. Fl. v. iv. p.* 9. *Habenaria bifolia et chlorantha, Hook. Br. Fl. 3d ed. p.* 376. *H. bifolia β. Hook. Br. Fl. 4th ed. p.* 315.

Pastures in the limestone districts; not unfrequent. *Fl.* June. ♃.

Buildwas; *Miss. H. Moseley!* Benthal Edge; *J. E. Bowman, Esq.*

Pastures between Westhope and Middlehope on Wenlock Edge, plentifully, growing with *H. bifolia.* Bomere wood.

Stem 12—18 inches high. *Leaves* usually two, sometimes 3, elliptical, spreading, shining. *Bracteas* lanceolate, about as long as the germen, one to each flower, besides a few larger ones scattered along the stem. *Flowers* numerous, in a loose spike, pure white, except a greenish tinge on the lip and spur. *Upper lateral petals* about one-third longer than the anther, obtuse. *Lip* linear-oblong; *spur* about twice as long as the germen, thickened towards the end. *Anther* very large, truncate, the bases of the cells being twice as far apart as their tops, giving to the whole anther a somewhat semicircular character, the longitudinal central line between the cells in front elevated into a prominent ridge and forming a groove or furrow on the back. *Stigma* very broad at its top and slightly pointed in the middle, curved into a semicircular form.

3. H. *bifolia*, Bab. *Smaller Butterfly Orchis.* Spur twice as long as the germen, lip linear entire; petals connivent, obtuse; anthers oblong, truncate, with the pollen-cells parallel.. *Babington in Linn. Trans. v.* 17. *p.* 463. *E. Bot. Suppl. t.* 2806. *H. b. var. a. Hook. Br. Fl. 4th ed. p.* 315. *Orchis bifolia, β. E. Fl. v.* 4. *p.* 9.

Pastures in the limestone districts; not unfrequent. *Fl.* June. ♃.

Near Oswestry; *Rev. T. Salwey.* Oakley park, near Ludlow; *Mr. H. Spare.* Near Dowles Brook, Wyre forest; *Mr. W. G. Perry.* Woods about Apley, near Bridgnorth; *E. Lees, Esq.* Wyre forest; *Mr. G. Jorden.* Near Aiksmore Coppice, Astley; *Mr. E. Elsmere, junr.!* Longnor, Coreley, Smethcot parishes, &c.; *Rev. W. Corbett.* Common round Wenlock; *W. P. Brookes, Esq.* White-cliff Coppice, near Ludlow; *Mr. J. S. Baly.* Buildwas; *Miss H. Moseley.* Old pastures about Berrington, Pitchford, Lee near Ellesmere. Plentifully in pastures adjoining Netchwood. Common at Monks Hopton; *Rev. E. Williams's MSS.*

Burwood, Wenlock Edge. Bomere wood. Meadows between Westhope and Middlehope, Wenlock Edge, plentiful with *H. chlorantha.*

Much smaller than the last. *Leaves* 2, rarely 3, oblong, spreading, very bright and shining. *Bracteas* lanceolate, about as long as the germen, one to each flower and several larger ones scattered on the stem. *Flowers* in a loose spike, white with a greenish tinge, fewer and much smaller than in *H. chlorantha.* *Anther* truncate or slightly emarginate, rarely somewhat rounded at the top, its cells nearly parallel throughout their whole length; the longitudinal central line between the cells being a deep furrow in front and a keel behind. *Stigma* rather broad, truncate, folded so as to leave a channel between its pointed lobes, the middle emarginate.

4. OPHRYS. *Linn.* Ophrys.

1. O. *apifera*, Huds. *Bee Ophrys.* Lip tumid 5-lobed, 2 lower lobes prominent and with a hairy base, 2 intermediate ones reflexed truncate, terminal one acute elongated reflexed; anthers with a hooded point; petals oblong, bluntish, downy. *E. Bot. t.* 383. *E. Fl. v. iv. p.* 30. *Hook. Br. Fl. p.* 377. *O. insectifera, ι. Linn.*

Limestone districts; not unfrequent. *Fl.* July. ♃.

Abundant between Eaton and Upper Millechope; *Rev. W. Corbett.* In a field at Rowley; also near the Hill Top, close to the Limekilns on the right hand side of the road from Wenlock to Lutwyche; *W. P. Brookes, Esq.* Benthal Edge; *T. and D. Bot. Guide.* Lilleshall lime kilns; *T. C. Eyton, Esq.* Limekiln wood near Wellington; *E. Lees, Esq.* Buildwas; *Miss H. Moseley.* About Hughley, and lime kilns at the north end of the Wrekin; *Rev. E. Williams's MSS.* Near the Steeraway lime kilns. Madeley. Near Lincoln's Hill; *Dr. Du Gard.*

Limestone quarries near the Wrekin. Wenlock Edge. Lime quarries near the Five Chimnies.

Stem about a foot high. *Flowers* large, few, remote, sessile, each subtended by a lanceolate *bractea*, longer than the germen. *Sepals* widely spreading, rose-coloured with green nerves, elliptical, obtuse. *Upper petals* linear, obtuse, spreading, hairy on the interior, shorter than the sepals. *Lip* large, prominent, tumid, velvety, of a rich brown, variegated with angular or curved yellow lines or spots; *lower lobes* prominent, hairy, short, oblong; *central lobe* with a thin rounded lobe on each side, and elongated at the extremity into a subulate acute reflexed hairy point.

2. O. *aranifera*, Huds. *Spider Ophrys.* Lip tumid obscurely 3-lobed, the middle lobe large emarginate without an appendage; anther acute; petals linear, glabrous. *E. Bot. t.* 65. *E. Fl. v. iv. p.* 32. *Hook. Br. Fl. p.* 378.

Limestone districts; very rare. *Fl.* April, May. ♃.

Priors Halton Farm, near Ludlow; *Mr. H. Spare.*

Stem about a span high. *Flowers* very few. *Sepals* green. *Upper petals* linear, obtuse, glabrous. *Lip* shorter and broader than in *O. apifera*, deep brown, with 2 paler parallel shining smooth lines connected by a cross-bar, not unfrequently resembling the Greek letter π.

3. O. *muscifera*, Huds. *Fly Ophrys.* Lip oblong 3-fid, the middle segment elongated bifid; anther short, obtuse; petals filiform. *E. Bot. t.* 64. *E. Fl. v. iv. p.* 29. *Hook. Br. Fl. p.* 378.

Limestone districts; very rare. *Fl.* June. ♃.

In a few spots in a field at Rowley, and on lime rocks at Farley near Wenlock; *W. P. Brookes, Esq.*

Stem about a foot high. *Flowers* few, distant, each subtended by a lanceolate *bractea* longer than the germen. *Sepals* ovato-lanceolate, obtuse, widely spreading, green. *Lateral petals* ascending, linear-filiform, brown, downy, resembling the antennæ of an insect. *Lip* oblong dependent, disk convex smooth brown, with a broad pale bluish spot in its centre; lateral lobes spreading downy, central one oblong elongated bifid.

5. NEOTTIA. *Jacq. et Swarz.* Lady's Tresses.

1. N. *spiralis*, Sw. *Fragrant Lady's Tresses.* Tubers ovato-

oblong, thick; radical leaves ovato-oblong, stem-leaves like bracteas; spike dense. *E. Fl. v. iv. p.* 36. *Hook. Br. Fl. p.* 378. *Ophrys spiralis, Linn. E. Bot. t.* 541. *Spiranthes autumnalis, Rich.*

Limestone districts; not common. *Fl.* August—September. ♃.

Buildwas; *Miss H. Moseley.* Priors Halton farm, near Ludlow; *Mr. H. Spare.* Meadow on the bank of the Corve, near Delbury; *Rev. T. Salwey.* Caynham Camp, near Ludlow; *Miss Mc Ghie.* Near Detton; *Mr. G. Jorden.* In a field at a farm called the Springs, near Harley; *W. P. Brookes, Esq.!*

Tubers ovato-oblong, 3-4. *Stem* 4—6 inches high, rather bracteated than leafy. *Flowers* singularly spiral on the stalk, greenish-white. *Upper sepal* and 2 *inner petals* combined. *Lip* longer than the rest of the flower, oblong, broader and crenate at the apex. *Stigma* and *anther* both acuminate.

6. LISTERA. *Br.* Tway-blade.

1. L. *ovata*, Br. *Common Tway-blade.* Leaves 2, opposite, ovate; lip bifid; column of fructification with a crest which includes the anther. *E. Fl. v. iv. p.* 38. *Hook. Br. Fl. p.* 379. *Ophrys ovata, Linn. E. Bot. t.* 1548.

Woods and moist pastures; frequent. *Fl.* June. ♃.

Wenlock; *W. P. Brookes, Esq.* Near Hanwood; *Mr. T. Bodenham.* Westfelton; *J. F. M. Dovaston, Esq.* Cox wood, Coalbrookdale; *Mr. F. Dickinson.* Newport; *R. G. Higgins, Esq.* Wyre Forest; *Mr. G. Jorden.* Benthal Edge; *Dr. G. Lloyd.* Buildwas; *Mss H. Moseley.* Ludlow; *Miss Mc Ghie.* Oakley park, near Ludlow; *Mr. H. Spare.* The Lodge, near Ludlow; *Rev. T. Salwey.* Shotton, near Shrewsbury; *W. W. Watkins, Esq.* Long Dale, near Cound; *Rev. E. Williams's MSS.*

Stem 1 foot high. *Leaves* striated. *Flowers* distant upon the *spike*, yellowish-green. *Sepals* ovate; two lateral *petals* linear-oblong; *lip* long, bifid, without any teeth at the base. *Bracteas* very short.

2. L. *cordata*, Br. *Heart-leaved Tway-blade.* Leaves 2, opposite, cordate; lip 4-lobed; "column without any crest." *E. Fl. v. iv. p.* 39. *Hook. Br. Fl. p.* 379. *Ophrys cordata, Linn. E. Bot. t.* 358.

Sides of mountains in heathy spots; very rare. *Fl.* July, August. ♃.

Among the heath on the east side of the Stiperstones Hill immediately below the Devil's Arm-chair.

Root a few long fleshy fibres. *Stems* 3—5 inches high. *Flowers* few, very small, spiked, greenish-brown. *Leaves* of the *perianth* somewhat spreading, those of the *calyx* ovate. *Lateral petals* linear-oblong; *lip* pendent, linear.

"In *L. ovata* we have a curious crest which includes the anther, curving forwards over it. Hooker says that *L. cordata* has no crest. I have been unable to satisfy myself upon this point from my dried specimens, but think that he is in error, and if so, the crest at the apex of the column ought to be inserted in the generic character, as it does not exist in *Neottidium.*" *Babington.*

7. NEOTTIDIUM. *Schltd.* Bird's-nest.

1. N. *Nidus-Avis*, Schl. *Common Bird's-nest.* Leaves none, stem with sheathing scales; lip linear-oblong, with 2 spreading lobes. *Listera Nidus-Avis, Hook. E. Fl. v. iv. p.* 39. *Hook. Br. Fl. p.* 379. *Ophrys Nidus-Avis, Linn. E. Bot. t.* 48. *Neottia Nidus-Avis, Rich.*

Shady woods; not common. *Fl.* May, June. ♃

Wood below the Blodwell Rocks, near Oswestry! Coppice below the Fir Plantation on the top of Delbury Common; *Rev. T. Salwey.* Whitecliff, near Ludlow; *Mr. H. Spare.* Woods at Coalbrookdale; *Purton's Midl. Flora.* Plowden woods, near Bishop's Castle; *Miss Mc. Ghie.* Buildwas; *Miss H. Moseley.* In the deepest and most obscure parts of the Limekiln woods, near Wellington; *E. Lees, Esq.* Benthal Edge, and Farley woods; *Mr. F. Dickinson!* Ash Coppice adjoining the Downes Farm, near Wenlock; *W. P. Brookes, Esq.* Woods under the Wrekin; *Rev. E. Williams's MSS.*

Root of many short thick densely aggregated fleshy fibres. *Stem* 1 foot high. *Flowers* spiked, of a dingy brown. *Sepals* and lateral *petals* oblong-oval, nearly equal. Lobes of the *lip* spreading.

The Hon. and Rev. W. H. Herbert has investigated the economy of this plant, which has been considered as parasitical on the roots of trees. He found several dead flower-stalks which had grown out of bundles of fleshy fibres, diverging every way, but the fibres were falling apart, and the plant appeared to have died as an annual after flowering. On stirring the ground further, at a short distance, he discovered a live bundle of similar fibres, with a very strong white shoot or eye like the dormant shoot of a perennial herbaceous plant, which was evidently to produce a flowering-stem in the next summer. Pursuing his researches he soon discovered similar bundles of different sizes, which were clearly immature and not ready to sprout in the following spring. On examination of the smallest, he found that it grew from the end of a half-dead fibre; and recurring to the dead plant which he had first taken up, he perceived that its several fibres or at least many of them, though dead at the base, *were alive at the other end* and beginning to bristle or protrude young fibres near the extremity. By further research, he clearly ascertained that the plant *dies after flowering*, but is capable of *reproducing a new plant* from the point of each of its fibres after they have fallen apart, the extreme point becoming the eye or shoot, which increases in size till its maturity, and the lateral bristles becoming the fibres by which the plant is to be nourished and afterwards propagated. The young roots continue thus to increase in bulk under ground till they come to the flowering age, when they push up vigorously, die and spawn again in the same extraordinary manner.

8. EPIPACTIS. *Rich. et Sw.* Helleborine.

1. E. *latifolia*, Sw. *Broad-leaved Helleborine.* Leaves broadly ovate, amplexicaul; perianth connivent; lower bracteas longer than the drooping flowers; lip 3-lobed, middle-lobe roundish, shortly acuminated. *E. Fl. v. iv. p.* 41. *Hook. Br. Fl. p.* 379. *Serapias latifolia, Linn. E. Bot. t.* 269.

Woods; not unfrequent. *Fl.* July, August. ♃.

Mog Forest, Lutwyche. Buildwas Park; *W. P. Brookes, Esq.* Longnor, Coreley, and Burford parishes; *Rev. W. Corbett.* Cox Wood, and other woods near Coalbrookdale; *Mr. F. Dickinson.* In a new plantation near Walford, "not know there before 1835." Limekiln woods near the Wrekin; *T. C. Eyton, Esq.* Dowle Wood, not plentiful; *Mr. G. Jorden.* Benthal Edge; *Dr. G. Lloyd.* Buildwas; *Miss H. Moseley.* In great profusion and beauty on the borders of a horse-tract through Whitecliffe woods, near Ludlow; *E. Lees, Esq.!* By the Pool at the Lodge, near Ludlow; *Rev. T. Salwey.* Oakley Park shrubbery, near Ludlow; *Mr. H. Spare.*

Benthal Edge. Mount Sion, near Oswestry. Woods on west side of Bomere Pool. Near Steeraway lime kilns, near Wellington. Copse near Sharpstones Hill.

Mr. Babington has directed my attention to a plant which we gathered in 1835 in the woods on the west side of Bomere pool, and which we supposed at

the time to be E. *latifolia*, but which he has recently determined to be E. *viridiflora, Reich.* He thus characterizes the two:

1. E. *latifolia*, All. Leaves broadly ovate, longer than the internodes; lower bracteas longer than the flowers; terminal division of the lip roundish-cordate obtuse with a small recurved point, shorter than the broadly ovate sepals and petals. E. *Bot. t.* 269.

Leaves ovate, very broad, the upper ones ovato-oblong; the lower *bracteas* foliaceous, lanceolate, attenuated. *Flowers* green, with the lip purple, sometimes nearly all purple; *peduncles* shorter than the downy *germen.* Lobe of the *lip* broader than long, crenate.

2. E. *viridiflora*, Reich. Leaves ovato-oblong, the upper ones lanceolate acute; lower bracteas longer than the flowers; the terminal division of the lip triangular-cordate acute, as long as the lanceolate petals and sepals. *Reich. Icon. f.* 1142. *Reich. Fl. Excurs. n.* 891. *Petermann Fl. Lips.* 641.

Narrower and more elongated in all its parts than *E. latifolia*, only the lowest *leaves* ovate, the intermediate ones lanceolate, and the upper ones lanceolato-attenuated and merging gradually into the linear-lanceolate *bracteas. Flowers* "green, tinged with purple;" *peduncle* shorter than the downy *germen.* Lobe of the *lip* longer than broad, crenate.

Woods at Bomere pool, Salop; and Luton, Kent.

This is considered by some to be a variety of the last, but as I consider incorrectly, few authors notice it at all, describing only the former. Mr. Babington subsequently writes: "I much fear that we can hardly distinguish the *viridiflora* from the *latifolia* without more study from fresh specimens. I consider your Oswestry and Bomere plants as the former, and the Sharpstones and Whitecliff specimens as *true latifolia.* There is something remarkably different in their general appearance and habit and I think the characters drawn from the lip will ultimately prove them to be different species. There are specimens of *viridiflora* in Smith's Herbarium from Matlock; and "Abberley, [Worcestershire] Mr. Butt." Of *latifolia vera*, Smith's herbarium has one specimen from the north of England, which is probably it and no other. The plants so named in the Linnæan Herbarium are certainly only *E. palustris*, as determined by Smith."

Fries *(Nov. Mant. alt. p.* 54.*)* considers *E. viridiflora, Reich.* as a variety "floribus viridibus" of his own *E. media* which he thus describes:—

"*E. media.* Foliis ovato-lanceolatis æqualiter acuminatis internodio longioribus; pedicellis ex erecto arcuato-cernuis ovarium pubens æquantibus; lamina attenuato-acuminata perigonii lacinias æquante, superne carina plicato-crenata.

a. floribus albis. *Serapias microphylla, Bot. Dan. non Ehrh. S. latifolia, γ. albens, Wahl. Suec. p.* 589. [hæc forma evidentissime a E. latifolia differt.]

b. floribus viridibus. *Ser. viridiflora, Reich. Ic. f.* 1142. *sec. Koch.*

c. floribus roseo-rubris. *Wahl, l. c. β. Fl. Dan. t.* 1938. E. *atrorubens, Reich. Icon. f.* 1141. *(atrorubentes* modo in E. latifolia vidi.*)*

Si ad colores floris attendatur cum E. *latifolia* necesse confundetur; hæc enim analoga ratione coloribus ludit, modo semper magis squalidis et obscuris, nec unquam ut in præsente puris: albidis, roseis, &c.—prorsus dubium est, cui potissimum pertineat *S. viridiflora, atrorubens*, Hoffm.—nec igitur mirum plantas ipsas frequenter confundi: certe e colore denominari nequeunt. Est quædam gentilitas et gracilitas in singulis partibus, quibus semper dignoscitur a *Epipactis latifoliæ*, vulgo longe majoris, formis pumilis et obesis. Vaginæ inferiores aphyllæ, apertæ, infundibuliformes. Folia e basi amplexicauli ovato-lanceolata, in apicem æqualiter attenuata, nec abrupte acuminata ut in E. *latifoliæ.* Racemus gracilis, bracteis flore brevioribus. Flores minores, labello in acumen æqualiter producto, carina plicato-crenata, quo certissime differt a E. *latifolia*, in qua labelli lamina subrotunda cum brevi involuto apiculo et carina non plicato-crenata. Germen pubens demum glabrescit. Media est non modo versus E. *palustrem*, sed etiam E. *rubrum*, &c."

2. E. *palustris*, Sw. *Marsh Helleborine.* Leaves lanceolate; bracteas shorter than the slightly drooping flowers; upper division of the lip roundish obtuse crenate as long as the perianth. *E. Fl. v. iv. p.* 43. *Hook. Br. Fl. p.* 380. *Serapias palustris, Scop. E. Bot. t.* 270. *S. longifolia, Linn.*

Marshy places, in the limestone districts; not common. *Fl.* July. ♃.
In a large uncultivated field near Harley; *W. P. Brookes, Esq.!* Llwyn-y-groes wood near Llanymynech, and Oakley Wood near Bishop's Castle; *T. and D. Bot. Guide.* Canton Rough, near Bridgnorth; *Dr. G. Lloyd.* Moors, Felton farm, near Ludlow; *Miss Mc Ghie.* In a boggy field between Golden Covers and on Cound Moor; *Rev. E. Williams's MSS.*
Stem 1 foot high, purplish above. *Calyx* purple-green, lateral *petals* and *lips* white with rose-coloured streaks at the base.

—

9. CEPHALANTHERA. *Rich.* White Helleborine.

1. C. *ensifolia*, Rich. *Narrow-leaved White Helleborine.* Leaves lanceolate, pointed; bracteas much shorter than the glabrous germen; lip obtuse, included. *Epipactis ensifolia, Sw. E. Fl. v. iv. p.* 45. *Hook. Br. Fl. p.* 380. *Serapias ensifolia, Linn. E. Bot. t.* 494.

Hilly woods; rare. *Fl.* May, June. ♃.
Plantation near Ruckley; *H. Bidwell, Esq.* Wyre forest most abundantly where a recent fall of wood has taken place or on the borders of the old rides; *E. Lees, Esq.!*
Stem 12—18 inches high. *Flowers* pure white, with a yellow protuberance on the *lip* whose disk has several white elevated ribs.

CLASS XXI.

MONŒCIA.

—

" last
Rose, as in dance, the stately trees, and spread
Their branches hung with copious fruit, or gemm'd
Their blossoms : with high woods the hills were crown'd ;
With tufts the valleys and each fountain side,
With borders long the rivers : that earth now
Seem'd like to heav'n, a seat where Gods might dwell,
Or wander with delight, and love to haunt
Her sacred shades."

MILTON.

" Much can they praise the trees so straight and hy,
The sayling pine ; the cedar proud and tall ;
The vine-propp elme ; the poplar never dry ;
The builder oake, sole king of forrests all ;
The aspine good for staves ; the cypresse funerall ;

The laurel, meed of mightie conquerours
And poets sage ; the firre that weepeth still ;
The willow, worne of forlorne paramours ;
The eugh, obedient to the benders will ;
The birch for shaftes ; the sallow for the mill ;
The mirrhe sweete-bleeding in the bitter wound ;
The warlike beech ; the ash for nothing ill ;
The fruitful olive ; and the platane round ;
The carver holme ; the mapple, seeldom inward sound."

SPENSER.

CLASS XXI.

MONŒCIA. *Stamens and Pistils in separate flowers on the same plant.*

ORDER I. MONANDRIA. 1 *Stamen.*

1. EUPHORBIA. *Involucre monophyllous, campanulate or turbinate ; limb* 8—10-*fid, exterior segments glandular, coloured, entire or often* 2-*horned, enclosing* 12 *or more male flowers with a single female one in the centre. Male flower.—Perianth a minute palaceous scale, entire or cut. Stamen single. Female flower.—Perianth none or rarely a very minute one. Pistil single, central, pedicellate, at length exserted. Germen subglobose,* 3-*celled, cells* 1-*seeded. Styles* 3, *connate at the base, stigmas bifid.—Nat. Ord.* EUPHORBIACEÆ, *Juss.*—Named from *Euphorbus,* physician to Juba, King of Mauritania, who brought the plant into use.

2. CALLITRICHE. *Male flower.—Perianth of* 2 *opposite fleshy falcate leaflets, (sometimes wanting). Stamen single. Anther* 1-*celled. Female flower.—Perianth similar. Germen* 4-*lobed, lobes laterally compressed, indehiscent,* 4-*celled, cells* 1-*seeded. Styles* 2, *erect or recurved ; stigma acute. (Flowers axillary, sometimes hermaphrodite).—Nat. Ord.* HALORAGEÆ, *Br.*—Name from καλος, *beautiful,* and θριξ, *hair ;* in allusion to the slender stem or long capillary roots.

3. ZANNICHELLIA. *Male flower.—Perianth a membranaceous interpetiolar stipule. Stamen single, anther* 2—4-*celled. Female flower.—Perianth membranaceous, short, cupuliform. Germens* 4, *oblong, compressed,* 1-*seeded. Style* 1 ; *stigma peltate. (Flowers axillary, sometimes hermaphrodite).—Nat. Ord.* NAIADES, *Juss.*—Named in honour of *John Jerome Zannichelli,* a Venetian apothecary and Botanist.

ORDER II. DIANDRIA. 2 *Stamens.*

(See *Callitriche* in ORD. I. *Carex* in ORD III.)

ORDER III. TRIANDRIA. 3. *Stamens.*

4. TYPHA. *Flowers in very dense cylindrical spikes or catkins, the male ones terminal, the female ones below the male. Male flower.—Perianth of* 3 *elongated bristles. Stamens* 3, *united below into a filament, anthers* 4-*celled. Female flower.—Perianth of many hairs. Germen at first sessile at length elevated on a long setiform pedicel,* 1-*celled,* 1-*seeded. Style simple, elongated. Stigma unilateral.—Nat. Ord.* AROIDEÆ, *Juss.*—Name from τυφος, a *marsh,* where the plant grows.

5. SPARGANIUM. *Flowers* in dense sphærical heads, the male ones superior. *Male flower.—Perianth* of 3—6 membranaceous scales. *Stamens* numerous, *anther* 2-celled. *Female flower.—Perianth* similar. *Germens* 2, 1-celled, 1-seeded. *Style* simple. *Stigma* unilateral. *Drupe* dry, 1-seeded.—*Nat. Ord.* AROIDEÆ, *Juss.*—Name from σπαργανον, a *little band;* in allusion to the long and narrow leaves.

6. CAREX. *Flowers* in an imbricated *spike. Male flower.—Calyx* (as it is usually called) a scale or glume. *Corolla* 0. *Stamens* 3, *anthers* 2-celled. *Female flower.—Calyx* a scale or glume. *Corolla (Perigynium)* of 1 piece, urceolate, swollen. *Stigmas* 2-3. *Nut* triquetrous, included within the persistent corolla (which is thus considered to form part of the *fruit.*)—*Nat. Ord.* CYPERACEÆ, *Juss.* —Name from κειρω, to *shear* or *cut;* in allusion to the sharp leaves and stems.

ORDER IV. TETRANDRIA. 4 *Stamens.*

7. LITTORELLA. *Male flower.—Calyx* of 4 leaves. *Corolla* funnelshaped, 4-fid. *Stamens* 4, very long. *Female flower.—Calyx* of 4 leaves. *Corolla* urceolate, contracted and 4-toothed at the mouth. *Style* very long. *Capsule* 1-seeded.—*Nat. Ord.* PLANTAGINEÆ, *Juss.* —Name from *littus* a *shore,* from its place of growth.

8. ALNUS. *Flowers* in imbricated *catkins. Male flower.—Scale* of the *catkin* 3-lobed, with 3-*flowers. Perianth* single, 4-partite. *Female flower.—Scale* of the *catkin* single; those of the *perianth* 4, adnate. *Germens* 2, 2-celled. *Styles* 2. *Nut* compressed.—*Nat. Ord.* AMENTACEÆ, *Juss.*—Name from the Celtic, *al, near,* and *lan, the river-bank.*

9. BUXUS. *Flowers* clustered, axillary. *Male flower.—Perianth* single, of 4 unequal leaves, 2 opposite ones smaller, with one *bractea* at the base. Rudiment of a *germen. Female flower.—Perianth* similar, with 3 *bracteas* at the base. *Styles* 3. *Capsule* with 3 beaks, 3-celled, *cells* 2-seeded.—*Nat. Ord.* EUPHORBIACEÆ; *Juss.*— Name, altered from πυξος, the Greek name for this tree.

10. URTICA. *Flowers* in spikes or heads. *Male flower.—Perianth* single, 4-partite, containing the cup-shaped rudiments of a *pistil. Female flower.—Perianth* single, of 4-leaves. *Pericarp* 1-seeded, shining. *Nat. Ord.* URTICEÆ, *Juss.*—Name from *uro,* to *burn;* in allusion to its stinging property.

(See *Myrica* in CL. XXII.)

ORDER V. PENTANDRIA. 5 *Stamens.*

11. BRYONIA. *Male flower.—Calyx* 5-toothed. *Corolla* 5-cleft.

Filaments 3. *Anthers* 5. *Female flower.—Calyx* 5-dentate. *Corolla* 5-cleft. *Style* 3-fid. *Berry* inferior, globose, many-seeded.—*Nat. Ord.* CUCURBITACEÆ, *Juss.*—Name from βρυω, to *shoot* or *grow rapidly;* in allusion to its quick growth.

ORDER VI. POLYANDRIA. *Many Stamens.*

12. CERATOPHYLLUM. *Flowers* axillary. *Male flower.—Perianth* multipartite. *Stamens* 12—20. *Female flower.—Perianth* multipartite. *Germen* single, 1-celled, 1-seeded. *Style* straight; *stigma* filiform, curved. *Nut* superior, 1-seeded.—*Nat. Ord.* CERATOPHYLLEÆ, *Gray.*—Name from κερας, a *horn,* and φυλλον, a *leaf;* in allusion to the form of the leaves.

13. MYRIOPHYLLUM. *Flowers* axillary or spicate. *Male flower. Calyx* 4-partite. *Corolla* of 4 deciduous petals. *Stamens* 8. *Female flower.—Calyx* of 4 leaves. *Corolla* 0. *Germens* 4, coalescent, 4-celled, cells 1-seeded. *Stigmas* 4, sessile. *Nuts* 4, sessile, subglobose, 1-seeded.—*Nat. Ord.* HALORAGEÆ, *Br.*—Name from μυριος, a *myriad,* and φυλλον, a *leaf;* in allusion to its numerous leaves.

14. SAGITTARIA. *Male flower.—Calyx* of 3 persistent leaves. *Petals* 3, deciduous. *Stamens* numerous. *Female flower.—Calyx* and *corolla* similar. *Germens* numerous, collected into a head. *Style* short. *Stigma* simple, obtuse. *Pericarps* 1-seeded, compressed, margined.—*Nat. Ord.* ALISMACEÆ, *Bich.*—Name from *sagitta,* an *arrow,* from the form of the leaves.

15. ARUM. *Spatha* of one leaf, convolute at the base. *Perianth* 0. *Spadix* clavate and naked at the apex, bearing numerous sessile *stamens* near the middle and *germens* at the base. *Berry* 1-celled, many-seeded.—*Nat. Ord.* AROIDEÆ, *Juss.*—Name formerly written *Aron,* and supposed to be an Egyptian name of some plant of this tribe.

16. POTERIUM. *Flowers* in a head, with 3 *bracteas* at the base of each; female flowers superior. *Male flower.—Perianth* 4-partite, persistent, tube short, contracted at the mouth with an annuliform torus. *Stamens* 20—30, *filaments* long, pendulous. *Female flower.— Perianth* as in male. *Germens* 2, included in the tube of the perianth. *Styles* 2. *Stigmas* tufted. *Pericarps* 2, 1-seeded, invested with the hardened 4-angled tube of the perianth.—*Nat. Ord.* ROSACEÆ, *Juss.* Name from *poterium,* a *drinking-cup,* the plant having been used in the preparation of a drink called a " cool tankard."

17. QUERCUS. *Male flowers* in a lax slender pendulous *catkin* or spike. *Perianth* single, 6—8-partite. *Stamens* 8 and more. *Female flowers* sessile on a peduncle more or less elongated. *Invo-*

---

lucre of many little scales united into a *cup. Perianth* single, closely investing the germen, 6-toothed. *Germen* 3-celled. *Style* 1. *Stigma* 3-lobed. *Nut* or *acorn* 1-celled, 1-seeded, invested with the persistent enlarged perianth and surrounded at the base by the enlarged cupshaped scaly involucre.—*Nat. Ord.* AMENTACEÆ, *Juss.*—Name from the Celtic, *quer, beautiful,* and *cuez,* a *tree.*

18. FAGUS. *Male flowers* in a globose pedunculate pendulous *catkin. Perianth* single, monophyllous, campanulate, 5-6-fid. *Stamens* 8—12, exserted. *Fertile flowers* in a subglobose erect *catkin. Involucre* 4-lobed, prickly. *Perianth* single, urceolate, with 4-5 minute lobes. *Germen* incorporated with the perianth, 3-celled, 2 becoming abortive. *Styles* 3. *Nuts* 1-seeded, invested with enlarged involucre.—*Nat. Ord.* AMENTACEÆ, *Juss.*—Name from φαγω, to *eat,* on account of the nutritive qualities of the fruit.

19. CASTANEA. *Male flowers* in a very long cylindrical *catkin. Perianth* single, of 1 leaf, 6-cleft. *Stamens* 10—15, exserted. *Female flowers* in a sessile subglobose *catkin. Involucre* 4-lobed, thickly muricated. *Perianth* single, urceolate, 5, 6, 7 or 8-fid, having the rudiments of 12 *stamens. Germen* invested with the perianth, 5—8-celled, each cell 2-seeded. *Styles* 5—8. *Nut* 2-3-seeded, invested with the enlarged involucre.—*Nat. Ord.* AMENTACEÆ, *Juss.*—Name from *Castanea* in Thessaly, which produced magnificent chesnut trees.

20. BETULA. *Male flowers* in a long cylindrical terminal *catkin. Bracteas* squamiform, peltate, with a smaller scale on either side. *Perianth* of 3 entire scales. *Stamens* 4. *Female flowers* in a dense cylindrical lateral *catkin. Bracteas* squamiform, 3-lobed, pubescent, 3-flowered. *Perianth* 0. *Styles* 2. *Germen* compressed, 2-celled, (1 abortive). *Nuts* compressed, with a membranous margin, 1-seeded. *Nat. Ord.* AMENTACEÆ, *Juss.*—Name from *betu,* Celtic, *birch.*

21. CARPINUS. *Male flowers* in a lateral cylindrical *catkin. Bracteas* roundish, ciliated at the base. *Stamens* 12 and more. *Female flowers* in a terminal lax *catkin. Bracteas* large, foliaceous, 3-lobed, 1-flowered. *Perianth* monophyllous, urceolate, 6-dentate, incorporated with the 2-celled *germen,* of which 1 cell is abortive. *Styles* 2. *Nut* ovate, striated, 1-seeded.—*Nat. Ord.* AMENTACEÆ, *Juss.*—Name from Celtic, *car, wood,* and *pin,* a *head;* the wood having been employed to make yokes for oxen.

22. CORYLUS. *Male flowers* in a cylindrical *catkin. Bracteas* simple. *Perianth* of 2 adnate scales. *Stamens* 8. *Anthers* 2-celled. *Female flowers* gemmaceous. *Bracteas* ovate, entire. *Perianth* obsolete. *Germens* several, surrounded by a scaly involucre. *Stigmas* 2. *Nut* 1-seeded, invested at the base by the enlarged united coriaceous

---

scales of the involucre.—*Nat. Ord.* AMENTACEÆ, *Juss.*—Name from κορυς, a *casque* or *cap;* in allusion to the appearance of the fruit.

MONŒCIA—MONANDRIA.

1. EUPHORBIA. *Linn.* Spurge.

* Glands of the involucre rounded on the outside.

1. E. *helioscopia,* Linn. *Sun Spurge.* Leaves broadly spathulate, emarginate, serrated, smooth; umbels of 5 principal trichotomous branches; bracteas obovate, denticulate; capsule globose, depressed, triquetrous, angles rounded, glabrous; seeds rounded-obovate, reticulato-rugose. *E. Bot. t.* 883. *E. Fl. v. iv. p.* 63. Hook. *Br. Fl. p.* 387.

Waste and cultivated ground; common. Fl. July—September. ☉.
Root fibrous. *Stem* 8—12 inches high, erect, simple or with opposite branches from the base, round, smooth or with a few scattered horizontal hairs. *Leaves* alternate, broadly spathulate, emarginate, serrated, smooth. *Umbels* of 5 principal branches arising from a whorl of 5 nearly obcordate serrated leaves. *Branches* with horizontal hairs, trichotomously divided, with a solitary flower in the axil, each division subtended by 3 unequal *bracteas,* the two outer larger irregularly oval or obovate, the inner one smaller round, all denticulate. *Scales* of the *involucre* obtuse, notched, converging. *Glands* transversely oval, green, disk with impressed dots. *Capsule* globose, depressed, triquetrous, the angles rounded. *Seeds* rotundo-obovate, reticulato-rugose, dark brown.

** Glands of the involucre pointed or angular.

2. E. *Cyparissias,* Linn. *Cypress Spurge.* Leaves linear, obtuse, entire; umbels of many principal dichotomous branches and several scattered peduncles below; bracteas subrotundo-cordate, entire; capsule scabrous when young, finally smooth; seeds obovate, smooth. *E. Bot. t.* 840. *E. Fl. v. iv. p.* 66. Hook. *Br. Fl. p.* 388.

Thickets and a weed in cultivated ground; doubtfully wild. Fl. June, July. ♃. Whitecliffe near Ludlow; *Mr. H. Spare.*
Root woody, creeping. *Stem* 8—12 inches high, erect, round, smooth, branched, the branches without flowers overtopping the central umbel. *Leaves* linear, obtuse, entire, sessile, reflexed at the margins, bright green above, somewhat glaucous beneath, those of the stem distant, of the branches densely crowded, reflexo-patent. *Branches* of the umbel 10—15, twice forked, subtended by as many linear leaves. *Bracteas* of 2 subrotundo-cordate sessile entire leaves, becoming scarlet in the autumn. *Glands* 4, yellow, lunate. *Involucre* hairy internally at the mouth. *Capsules* scabrous when young, afterwards smooth and even. *Seeds* smooth.

3. E. *exigua,* Linn. *Dwarf Spurge.* Leaves linear, obtuse, submucronate; umbel of 3 principal dichotomous branches; bracteas linear, acute, dilated, unequal and subcordate at the base; capsule triquetrous smooth, valves with a thickened tuberculate evanescent border; seeds oval, quadrangular, rugose. *E. Bot. t.* 1336. *E. Fl. v. iv. p.* 60. Hook. *Br. Fl. p.* 389.

Corn-fields; not unfrequent. Fl. July. ☉.

Fields opposite Coalport; *Mr. F. Dickinson.* Astley; *Mr. E. Elsmere, junr.!* Newport; *R. G. Higgins, Esq.* Sutton near Ludlow; *Dr. G. Lloyd.*

Fields near Bomere and Berrington pools. Westfelton. Limestone quarries near the Wrekin.

Root fibrous. *Stem* 6—8 inches high, erect, oppositely branched from the base, round, smooth and glaucous. *Leaves* alternate, trifarious, sessile, linear, obtuse, submucronate, glaucous. *Umbel* with 3 principal branches arising from a whorl of 3 larger leaves dilated at the base. *Branches* dichotomously divided, with a solitary flower in the axil, subtended by 2 linear acute *bracteas* dilated and unequal at the base sometimes approaching to cordate and often with a single tooth on one side. *Segments* of the *involucre* minute, converging. *Glands* yellow, rounded, with a pair of filiform divergent distinct horns arising from beneath. *Capsule* triquetrous, smooth, valves with a swollen and thickened tuberculate border which nearly disappears when ripe. *Seeds* oval, quadrangular, wrinkled.

4. E. *Peplus*, Linn. *Petty Spurge.* Leaves broadly ovate, subemarginate, petiolate; umbels of 3 principal dichotomous branches; bracteas ovate, obtuse, mucronate, petiolate; capsule glabrous, valves with thickened rugose keels; seeds oval, foveolate. *E. Bot. t.* 959. *E. Fl. v. iv. p.* 60. *Hook. Br. Fl. p.* 389.

Cultivated and waste ground; common. *Fl.* July, August. ⊙.

Root fibrous. *Stem* 6—10 inches high, erect, branched, round, glabrous, tinged with red. *Leaves* alternate, broadly obovate, more or less emarginate, tapering into a short petiole, glabrous. *Umbel* of 3 principal branches, with a pedicellate flower in the axil, arising from a whorl of 3 obovate petiolate leaves. *Branches* spreading, compressed, dichotomous, with a central flower, the divisions subtended by 2 ovate obtuse mucronate petiolate *bracteas*. *Scales* of the *involucre* ciliated, converging. *Glands* lunate, horns very long. *Germen* winged with a swollen and thickened rugose border. *Seeds* oval, honeycombed, invested with a minutely granulated white pellicle.

Reichenbach (*Fl. Excurs.* 4773.) states that the seeds of this species have 22 depressions thus distributed, 4 on each of the dorsal faces with 3 smaller ones interposed on either side, then three alternate ones on each side and an oblong one on each anterior face. In our specimens the depressions do not preserve this regularity, being often confluent, variable in number and not so numerous; nevertheless they do not correspond with the arrangement in the allied species, *E. peploides, Gouan.* (*Reich.* 4774.) which has subrotund entire leaves and bracteas, and 14 depressions on the seeds, thus disposed on each half:—3, 1, 2 and an oblong one.

5. E. *Lathyris*, Linn. *Caper Spurge.* Leaves oblongo-lanceolate, cordate at the base, mucronate, 4-farious; umbel of 4 principal dichotomous branches; bracteas cordato-acuminate, mucronate; capsule globose, depressed, triquetrous, angles rounded, valves incurved at the margins and dorsally marked with a longitudinal indented line; seeds oval, smooth. *E. Bot. t.* 2255. *E. Fl. v. iv. p.* 61. *Hook. Br. Fl. p.* 389.

Cultivated ground; doubtfully wild. *Fl.* June, July. ♂.

Lilleshall Abbey; *R. G. Higgins, Esq.*

Cultivated ground about Shrewsbury.

Root fibrous. *Stem* 4-5 feet high, erect, simple, round, glaucous, reddish below. *Leaves* sessile, 4-farious, oblongo-lanceolate, cordate at the base, mucronate, wavy, dark green above, paler beneath, glabrous. *Umbel* of 4 principaln branches arising from a whorl of 4 leaves with a sessile flower in the axil. *Branches*

dichotomously divided, the divisions subtended by 2 cordato-acuminate mucronate *bracteas*, with a sessile flower in the axil. *Segments* of the *involucre* ovate, obtuse, the apex minutely toothed, converging. *Glands* thick, fleshy, deep yellow, lunate, the horns obtuse and rounded. *Capsule* globose, depressed, triquetrous, the angles rounded, valves incurved at the margins, dorsally marked with a longitudinal indented line. *Seeds* large, oval, black, smooth, invested with a whitish slightly wrinkled pellicle.

On submitting to the microscope some of the milky juice with which all the Euphorbiæ abound, innumerable minute needle-shaped transparent bodies appear floating in it. These are called *raphides* and according to Raspail (*Nouv. Syst. de Chem. Organ.*) are found exclusively in the interstitial passages of the cellular tissue, though other physiologists contend that they are constantly formed inside the bladders of the cellular tissue. Raspail states them to be crystals of phosphate of lime, forming six-sided prisms, terminated at each end by a pyramid with the same base. Their functions in the vegetable economy are unknown.

6. E. *amygdaloides*, Linn. *Wood Spurge.* Leaves obovato-lanceolate, obtusely mucronate, attenuated into a petiole, hairy beneath; umbel of 5-6 principal simply forked branches, with several scattered axillary peduncles below; bracteas perfoliate, rounded, mucronate, cloven on each side; capsule elliptical, glabrous; seeds smooth. *E. Bot. t.* 256. *E. Fl. v. iv. p.* 68. *Hook. Br. Fl. p.* 389. *E. sylvatica, Linn.—Jacq.*

Woods and thickets; not unfrequent. *Fl.* March, April. ♃.

New Mills, river Perry, near Baschurch; *Dr. T. W. Wilson.* Wyre Forest, very abundant; *Mr. G. Jorden.* Farley woods; *Mr. F. Dickinson.* Welbach Coppice, near Shrewsbury; *Mr. T. Bodenham.* Coppices near Homer; *W. P. Brookes, Esq.!* Woods about the Stiperstones. About Farley, near Wenlock; *Rev. E. Williams's MSS.*

Whitecliff Coppice, near Ludlow. Lyth Hill. Almond Park wood. Woods beyond Haughmond Hill.

Root woody. *Stems* several, 12—18 inches high, rather shrubby, simple, round, generally naked and red below. *Leaves* numerous, crowded towards the extremity whence arises the peduncle, obovato-lanceolate, tapering at the base into a short petiole, entire, more or less obtusely pointed and tipped with a short stout obtuse mucro, deep green and nearly glabrous or with a few scattered hairs above, paler or not unfrequently purplish beneath and clothed with soft hairs, margins somewhat thickened, softly ciliated, lower leaves larger, diminishing gradually upwards, the upper or floral leaves becoming truly obovate obtuse mucronate and more remote. *Peduncle* striated, hairy, bearing a principal terminal umbel of 5-6 simply forked branches, subtended by 5 rotundo-obovate obtuse mucronate sessile leaves, with several similar forked branches below from the axils of the floral leaves. *Bracteas* of 2 yellowish pale combined rounded mucronate smooth leaves, cloven half-way down on each side. *Glands* 4, lunate, acute, fleshy, yellow. *Involucre* with white hairs on the interior. *Capsule* elliptical, glabrous. *Seeds* smooth, when viewed under a powerful lens minutely dotted.

2. CALLITRICHE. *Linn.* Water-Starwort.

1. C. *verna*, Linn. *Vernal Water-Starwort.* Leaves variable; fructiferous peduncles very short; bracteas arcuate; styles always erect; carpels tetragonous, laterally compressed, obtusely keeled at the back, in parallel pairs. *E. Fl. v. i. p.* 10. *Babington Prim. Fl. Sarn. p.* 37. *Hook. Br. Fl. p.* 390. *Reich. Fl. Excurs. n.*

4746. *Bluff and Fing. Comp. Fl. Germ.* 2*d ed. t.* 1. *pt.* 1. *p.* 3. *C. aquatica, E. Bot. t.* 722.

Ditches, pools, and slow streams; frequent. *Fl.* April, May. ⊙.

2. C. *platycarpa*, Kuntz. *Broad-fruited Water-Starwort.* Leaves ovali-spathulate, slightly emarginate; fructiferous peduncles short; bracteas arcuate; styles arcuate and reflexed; carpels tetragonous, laterally compressed, obtusely keeled at the back, in parallel pairs. *Babington Prim. Fl. Sarn. p.* 36. *Reich. Fl. Excurs. n.* 4748. *Bluff and Fing. Comp. Fl. Germ.* 2*d ed. t. i. pt.* 1. *p.* 4.

Muddy ditches, &c.; not unfrequent. *Fl.* June, July. ⊙.

Near Preston Gobald churchyard; near Oaks Hall near Pontesbury; Sharpstones Hill near Shrewsbury.

All the *leaves* ovali-spathulate, slightly emarginate, opposite and connate at the base, with a single rib which at the summit of the narrow portion of the leaf branches into 3, of which the central one proceeds straight to the apex, and the 2 lateral ones curve upwards and unite with the mid-rib just below the apex. Each of the lateral ribs sends off on its exterior side from a point below the middle of the leaf a secondary rib, which curving upwards again unites with the primary lateral ribs below the point of their own union with the mid-rib. *Bracteas* curved, persistent, although minute when the fruit is enlarged. *Filament* moderately long. *Styles* recurved, even in their early state. *Fruit* on a very short peduncle, nearly orbicular, tetragonous, laterally compressed; carpels oblongo-reniform, laterally compressed, obtusely margined or keeled, placed paralled to each other.

3. C. *pedunculata*, DC. *Pedunculated Water-Starwort.* Leaves emarginate; fructiferous peduncles elongated or very short; bracteas none; styles divaricate in flower, finally reflexed and appressed; carpels tetragonous, laterally compressed, obtusely keeled at the back, in parallel pairs. *E. Bot. Suppl. t.* 2606. *Bab. Prim. Fl. Sarn. p.* 37. *Hook. Br. Fl. p.* 390.

Ponds, &c.; not very common. *Fl.* June. ⊙.

Var: a. vera, Bab. Lower leaves linear emarginate, upper ones broader in the middle; carpels pedunculate.

Pits on Grinshill near Clive church.

Lower leaves linear or linear-spathulate, obtuse, emarginate, 3-ribbed; *upper ones* oblongo-spathulate, 3-ribbed, emarginate, opposite and connate. *Bracteas* none. *Styles* recurved. *Peduncle* of *fruit* elongated. *Fruit* nearly orbicular, tetragonous, laterally compressed; *carpels* oblongo-reniform, laterally compressed, obtusely margined or keeled, placed parallel to each other.

Var: β. sessilis, Bab. Lower leaves linear emarginate, upper ones obovato-spathulate truncate at the apex; carpels sessile.

Pit in a field south of the Sharpstones Hill near Shrewsbury. Golden Pool, near Pitchford.

Lower leaves linear, slightly narrow upwards, emarginate, single-ribbed; *upper ones* obovato-spathulate, truncate, 3-ribbed. *Bracteas* none. *Peduncle* of *fruit* very short, scarcely visible. *Fruit* as in the preceding variety.

3. ZANNICHELLIA. *Linn.* Horned-Pondweed.

1. Z. *palustris*, Linn? *Common Horned-Pondweed.* Style equal-

ling or exceeding in length half the mature carpel. *E. Bot. t.* 1844.? *E. Fl.*?

Pools and stagnant waters; not common. *Fl.* August. ⊙.

Shrewsbury Canal, near Uffington; *Rev. E. Williams's MSS.* North-west margin of Ellesmere Mere below the House of Industry; *J. E. Bowman, Esq.!*

Pit in an old brick-yard, near the second Canal bridge on the left hand side of the road leading from Castle foregate to the Old Heath, Shrewsbury.

Roots proceeding from the joints, fibrous. *Stem* filiform, slender, branched, floating. *Leaves* 2 or 3 together, very narrow and acute. *Inflorescence* axillary. *Stamen* simple, 2-celled, obtusely apiculate. *Stigma* large, peltate, slightly concave, irregularly crenulate or toothed. *Umbel* of *fruit* very shortly pedunculate. *Carpels* generally 3 sometimes 4 or 5, on pedicels longer than the peduncle and about one-third the length of the carpel, oblong, incurved, laterally compressed, with a denticulate membrane on the dorsal margin, crowned with the persistent *style* about half as long as the carpels.

M. Ad. Steinheil in his "Observations sur la Spécification des Zannichellia, &c.," in *Ann. des Sciences Nat. n. s. vol.* 9. *p.* 87. considers that Linnæus incorrectly united the two species of Micheli, and that Koch is equally incorrect in regarding all the forms described by recent continental botanists as mere varieties of *Z. palustris, Linn*; and proposes to separate them afresh into the original species named by Willdenow, *dentata* and *palustris*. He rejects as of very uncertain character all distinctions derived from the leaves, peduncles and pedicels, stipules, filaments, and number of carpels, and founds his specific differences on the comparative length of the style and fruit, the number of anther-cells and the entire or toothed stigma. The characters he assigns are:—*Z. dentata, Willd.* anthers 2-celled, uniapiculate; stigmas crenulate, papillose; style shorter than half the mature fruit. *Z. palustris, Willd.* anthers 4-celled, biapiculate; stigmas perfectly entire, not papillose; style nearly equalling in length the mature fruit. He moreover states that *palustris* is almost always found near the coast whilst *dentata* grows in inland situations. Under these two species he arranges as subspecies or varieties nearly all the numerous forms of the continental botanists. Assuming Steinheil's characters to be constant our Shropshire plant would appear identical with the *var: δ*, of his *subspecies repens* of *Z. dentata, Willd.* to which he quotes *Z. pedunculata, α. stagnalis, Reich. Fl. Excurs.* 9. An examination however of specimens collected by Mr. J. E. Bowman, at Gresford, Denbighshire, with 2-celled anthers; by Mr. C. C. Babington in Needwood forest, Staffordshire, with 2-3 and 4-celled anthers on the same plant; by myself at Cambridge, with 4-celled anthers; as well as those from the before mentioned localities in Shropshire with 2-celled anthers; shows that whatever be the number of the anther-cells, in all the stigma is irregularly crenulate or toothed and the style about half the length of the carpels. In these latter characters a specimen from Dover in which the anthers could not be detected likewise corresponded. In all the carpels were shortly pedicellate and the dorsal margin armed with 1 or 3 denticulate membranes.

MONŒCIA—TRIANDRIA.

4. TYPHA. *Linn.* Cat's-tail or Reed-mace.

1. T. *latifolia*, Linn. *Great Reed-mace.* Leaves linear, nearly plane; sterile and fertile catkins continuous. *E. Bot. t.* 1455. *E. Fl. v. iv. p.* 71. *Hook. Br. Fl. p.* 391.

Borders of ponds and lakes; not uncommon. *Fl.* July, August. ♃.

Benthal Edge; *Mr. T. Bodenham.* Llynclys, near Oswestry; *J. F. M. Dovaston, Esq.* Pool at Horsehay, near Shiffnal; *Mr. F. Dickinson.* Newport; *R. G. Higgins, Esq.* Oakley park, near Ludlow; *Mr. H. Spare.*

Hancott pool. Sutton, near Shrewsbury. Canal between Shrewsbury and Uffington. Morda pool, near Oswestry.

Root thick, creeping, fibrous. *Herbage* bright green and smooth. *Stem* 6-7 feet high, erect, round, striated, simple, leafy, solid. *Leaves* alternate, linear, dilated and flat in the upper part and terminating in an acute point, narrowed at the base, convex beneath, concave above, entirely clasping the stem by their dilated sheaths, membranous at the edges. *Inflorescence* terminal, erect, about 1 foot long. *Male spike* in the upper portion, cylindrical, uninterrupted, composed of several spikes superimposed one on the other and each subtended by a deciduous leafy *bractea*. *Perianth* of 3 elongated hairs surrounding the stamens. *Anthers* 3 or more, on a common simple *filament*, yellow. *Female spike* below the male one, entire and uninterrupted, dense, of a dark brown colour. *Perianth* of many long soft hairs. *Germen* oblong, elevated on a long setiform peduncle. *Style* elongated, simple. *Stigma* unilateral, persistent. *Receptacle* scarcely chaffy.

2. T. **angustifolia**, Linn. *Lesser Reed-mace.* Leaves linear, grooved below; sterile and fertile catkins a little distant from each other. *E. Bot. t.* 1456. *E. Fl. v. iv. p.* 72. *Hook. Br. Fl. p.* 391.

Borders of lakes and pools; not common. *Fl.* July. ♃.

Pond on the right hand side of the road between Bromfield and Downton Castle, near Ludlow; *Rev. T. Salwey.* Astley; *Mr. E. Elsmere, junr.* Oakley Park, near Ludlow; *Mr. H. Spare.* Near Kingswood; *H. Bidwell, Esq.* Shomere pool. Betton pools, near Shrewsbury; *Rev. E. Williams's MSS.*

Bomere, Hancot, Berrington, and Almond Park pools, all near Shrewsbury. Golden Pool near Pitchford.

Very similar but slenderer than the last. *Leaves* very narrow, perfectly linear, acuminate, convex or nearly semicylindrical in the lower part beneath, concave above, sheathing and amplexicaul at the base. *Male and Female spikes* separate, with about ½—1 inch naked space between them. *Female spike* slender, pale brown. *Receptacle* with chaffy scales. *Hairs* of the perianth somewhat thickened at the apex.

5. SPARGANIUM. *Linn.* Bur-reed.

1. S. **ramosum**, Huds. *Branched Bur-reed.* Radical leaves triangular at the base, their sides concave beneath; flowers in axillary and terminal spikes; male clusters numerous; female clusters 1-2, sessile; stigma linear. *E. Bot. t.* 744. *E. Fl. v. iv. p.* 74. *Hook. Br. Fl. p.* 391. *S. erectum,* Linn.

Ditches and stagnant waters; frequent. *Fl.* July, August. ♃.

Oakley Park, near Ludlow; *Mr. H. Spare.* Tong Lodge Lake; *Dr. G. Lloyd.* Westbury; *Mr. F. Dickinson.* Marsh Pool, near Wenlock; *W. P. Brookes.*

Bomere and Hancott Pools. Haughmond Hill. Mare Pool. Canal between Shrewsbury and Uffington. Westfelton Moor.

Stem 2 feet high, erect, simple, somewhat angular above, terminating in several flowering branches. *Leaves* alternate, ensiform, elongated, erect, obtuse, notched, radical ones triangular at the base, the sides concave beneath, upper or floral leaves dilated concave and semiamplexicaul at the base. *Flowers* in axillary and terminal spikes, each subtended by a leafy *bractea* at the base; peduncle nearly erect to the first female cluster, then diverging and incurved. *Male flowers* in numerous globular dense clusters, alternate and sessile in the upper part of the peduncles, white, *anthers* yellow. *Female flowers* in a single, rarely 2, dense globose cluster sessile on the lower portion of the peduncle. *Sepals* 3—6, membranaceous, brown, spathulate. *Stigma* elongated, unilateral, concave, downy. *Fruit* ovato-acuminate, angular from compression. Whole plant glabrous.

2. S. **simplex**, Huds. *Unbranched upright Bur-reed.* Radical leaves triangular at the base, their sides plane beneath; flowers in a terminal simple spike; male clusters numerous; female clusters 3-4, lowermost pedicellate; stigma linear. *E. Bot. t.* 745. *E. Fl. v. iv. p.* 75. *Hook. Br. Fl. p.* 392. *S. erectum,* β. Linn.

Ditches and stagnant waters; not so common as the last species. *Fl.* July, August. ♃.

Marsh Pool, near Wenlock; *W. P. Brookes, Esq.* Lincroft Pool, near Plealey; *Mr. T. Bodenham.* Near the Wrekin; *Mr. F. Dickinson.* Dowle; *Mr. G. Jorden.* Tong Lodge Lake; *Dr. G. Lloyd.* Near Ludlow; *Miss Mc Ghie.* Near Dowle Brook, Wyre Forest; *Mr. W. G. Perry.*

Benthal Edge. Canal between Shrewsbury and Uffington. Hancott Pool.

Stem 12—18 inches high, erect, simple, zigzag, round, compressed, unbranched. *Leaves* alternate, ensiform, elongated, erect, acute, radical ones triangular at the base, the sides flat beneath, upper or floral leaves dilated concave and semiamplexicaul at the base. *Flowers* in a terminal simple spike. *Male flowers* in small dense globular heads, sessile and approximate in the upper portion of the spike, *anthers* deep yellow. *Female flowers* in 3-4 globular dense heads placed alternately and remotely on the lower portion of the spike, erect, sessile by reason of their pedicels which arise from the axils of the floral leaves or bracteas being closely agglutinated to the main stem, the lowermost head or sometimes the 2 lowermost heads only being elevated on a partially free pedicel. *Sepals* 3, membranous, pale green, spathulate. *Stigma* elongated, unilateral, downy. *Fruit* ovato-acuminate, angular from compression. Whole plant glabrous.

3. S. **natans**, Linn. *Floating Bur-reed.* Leaves floating, plane; flowers in a terminal sessile spike; male clusters mostly solitary; female clusters 2-3, lowermost pedicellate; stigma ovate, very short. *E. Bot. t.* 273. *E. Fl. v. iv. p.* 75. *Hook. Br. Fl. p.* 392.

Lakes, ditches, and stagnant waters; not common. *Fl.* July. ♃.

In the ditches upon a large bog belonging to Mr. Lloyd of Aston, near Belmont, 4 miles from Oswestry; *T. and D. Bot. Guide.* North-west margin of Colemere Mere; *J. E. Bowman, Esq.* In a ditch on Shomere Moss, near Shrewsbury. In the Canal between the Queen's-head Turnpike and Woodhouse, near Oswestry; *Rev. E. Williams's MSS.*

Hancot Pool. Ditches north of Bomere Pool.

Root fibrous, throwing out scaly runners. *Stem* erect, somewhat flexuose above, round, striated, glaucous, leafy. *Radical leaves* numerous, very long, tufted, those of the stem alternate, gradually diminishing upwards, somewhat sheathing and amplexicaul at the base, linear, obtuse, entire, flat or slightly channelled, flaccid, floating, pale-green, pellucid. *Flowers* in a terminal sessile spike. *Male flowers* in a small dense mostly solitary terminal head, *anthers* yellow. *Female flowers* in 2 or 3 globular dense green heads, placed alternately and mostly together below the male head, sessile, except the lowermost which is elevated on a short pedicel, each subtended by a floral leaf or bractea. *Sepals* 3, lanceolate, pale green. *Stigma* solitary, very short, ovate, peltate, oblique, on a short thick style. *Fruit* ovato-acuminate, angular from compression.

6. CAREX. *Linn.* Carex or Sedge.

* Spike simple. Stigmas 2.

† *Diœcious.*

1. C. **dioica**, Linn. *Creeping separate-headed Carex.* Spike simple, diœcious; fruit ascending, at length divaricate; perigynium

substipitate, ovate, shortly acuminated, ribbed, serrulate at the margins upwards, orifice bifid; nut ovali-subrotund, slightly attenuated at the base, obsoletely triquetrous; glumes ovate, acute; leaves and stem smoothish; root creeping. *E. Bot. t.* 544. *E. Fl. v. iv. p.* 77. *Hook. Br. Fl. 4th ed. p.* 330.

Spongy bogs; rare. *Fl.* May, June. ♃.

On Cound Moor; *Rev. E. Williams's MSS.* In a field at Rowley, near Much Wenlock; *W. P. Brookes, Esq.!*

Leaves setaceous, subtriquetrous, slightly channelled above, erect, smooth, shorter than the stem. *Stem* erect, slender, triquetrous, smooth. *Spikes* oblong. *Bractea* ovate, mucronate. *Glumes* ovate, more or less acute, brown, with a green keel and white edges. *Fruit* when young erect, subsequently spreading nearly horizontally, longitudinally ribbed. *Nut* ovali-subrotund, compressed, obsoletely triquetrous, very minutely dotted.

†† *Androgynous.*

2. C. **pulicaris**, Linn. *Flea Carex.* Spike simple, upper half with barren flowers; fruit lax, finally reflexed; perigynium oblong, attenuated at both ends, compressed, smooth, orifice bifid; nut oblongo-obovate, compressed; glumes lanceolate. *E. Bot. t.* 1051. *E. Fl. v. iv. p.* 78. *Hook. Br. Fl. 4th ed. p.* 331.

Bogs; not uncommon. *Fl.* May, June. ♃.

On Cound Moor. Haughmond Hill. Under the Wrekin. About the Stiperstones; *Rev. E. Williams's MSS.* Caynton; *H. Bidwell, Esq.* In a field at Rowley, near Much Wenlock; *W. P. Brookes, Esq.!* Bog near Ellesmere; *Rev. A. Bloxam.* Field by footpath from Madeley to Shiffnal, about half-way near a wood; and near Madeley; *Mr. F. Dickinson.* Hadnall; *Mr. E. Elsmere, junr.!* Near Park brook, Wyre Forest; *Mr. W. G. Perry.* Below the Blodwell rocks; *Rev. T. Salwey.!*

Haughmond Hill; Stiperstones Hill. Boggy ground north of Bomere pool.

Root tufted, fibrous. *Stem* erect, slender, 6—10 inches high, angular, smooth, naked. *Leaves* all on the lower portion of the stem, smooth, setaceous or filiform, semicylindrical, channelled above, contracted above the middle into a long acuminated triangular scabrous point, shorter than the stem, sheathing and ribbed at the base. *Spikes* oblong. *Bracteas* none. *Glumes* as long and broad as and forced off by the reflexed fruit. *Fruit* dark brown, shining, at first erect, finally reflexed. *Perigynium* with 2 strong marginal ribs, otherwise nerveless, the orifice slightly membranous entire and truncate bifid on one side only. *Nut* minutely dotted, about two-thirds as long as the perigynium, crowned by the long persistent style.

** Spike compound, androgynous. Styles 2.

† Spikelets alternate, sterile at the base.

3. C. **stellulata**, Gooden. *Little prickly Carex.* Spikelets 3-4, sterile at their base, roundish, somewhat distant; fruit squarroso-patent; glumes broadly ovate, acute; perigynium nearly sessile, broadly ovate, attenuated, plano-convex, margins serrulate, orifice bifid; nut stipitate, ovate, obtuse, compressed. *E. Bot. t.* 806. *E. Fl. v. iv. p.* 80. *Hook. Br. Fl. 4th ed. p.* 332.

Marshes and heathy places; not uncommon. *Fl.* May, June. ♃.

Boggy grounds about Eaton Mascot; *Rev. E. Williams's MSS.* Whixall Moss; *Rev. A. Bloxam.*

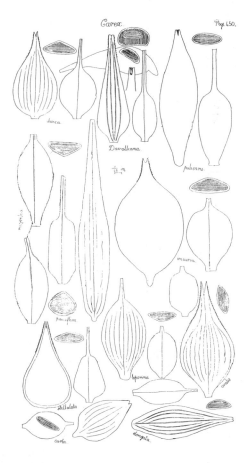

Carex.   Pag. 450.

Bomere Pool, west side. Hancott Pool. Haughmond Hill. Stiperstones Hill. Shawbury Heath. Boggy ground base of Caradoc Hill. Morda Pool, near Oswestry.

Root tufted. *Herbage* of a grass-green. *Stem* 6—9 inches high, triquetrous, edges nearly smooth. *Leaves* sheathing at the base, channelled above, linear, tapering into an acute triquetrous point, keels and edges rough. *Spike* erect, of 3-4 small roundish distant sessile few-flowered spikelets, lowermost subtended by a bractea which is frequently but by no means constantly elongated into a setaceous leaflet. *Glumes* membranous, greenish-brown. *Perigynium* 7-ribbed on the convex side, numerously obscurely striated on the plane one. *Nut* smooth, about half the length of the perigynium.

4. **C. ovalis**, Gooden. *Oval-spiked Carex.* Spikelets 3—6, sterile at the base, oval, approximate; fruit erect; glumes lanceolate, acute, as long as the fruit; perigynium ovato-acuminate, tapering at the base, compressed, plano-convex, striated, with a broad membranous margin, serrulate at the edges, orifice bifid; nut oval, compressed. *E. Bot. t.* 806. *E. Fl. v. iv. p.* 82. *Hook. Br. Fl.* 4*th ed. p.* 332.

Bogs, marshes, and margins of ponds; not common. *Fl.* June. ♃.

Meadows and pastures about Eaton Mascot, Sundorn, &c.; *Rev. E. Williams's MSS.* Hadnall; *Mr. E. Elsmere, junr.!* Oakley Park, near Ludlow; *Miss Mc Ghie.* Oswestry; *Rev. T. Salwey.!*

Bomere Pool. Hancott Pool. Canal between Shrewsbury and Uffington. Welbach, near Shrewsbury. Ellesmere and Colemere Meres.

Root creeping, scaly. *Stem* 1 foot high, triangular, smooth, roughish above. *Leaves* chiefly on the lower portion of the stem, deep-green, short, linear, acute, rough at the edges and midrib, with membranous sheaths at the base. *Spikelets* 3—6, alternate, crowded, erect, brownish-green, shining. *Bracteas* small, shorter than the spikelets, that of the lower one ovate membranous, the nerve extending into a rough mucro about as long as the bractea, those of the upper ones resembling the glumes, ovato-lanceolate, membranous at the edges with a green nerve. *Glumes* concealing the fruit, lanceolate or ovato-lanceolate, brownish, with a membranous margin and green nerve, as long and broad as the fruit. *Perigynium* obscurely ribbed or striated on the convex side, 5-ribbed at the base on the plane side. *Nut* smooth.

5. **C. curta**, Gooden. *White Carex.* Spikelets 5-6, alternate, approximate, elliptical, sterile at the base, lowermost slightly distant, bracteated; fruit erect; glumes ovate, cuspidate, shorter than the fruit; perigynium stipitate, ovate or subelliptical, acute, plano-convex, faintly striated, margins rough, orifice bifid; nut elliptical, tapering at the base, compressed. *E. Bot. t.* 1386. *E. Fl. v. iv. p.* 81. *Hook. Br. Fl.* 4*th ed. p.* 332.

Boggy margins of pools; not common. *Fl.* June. ♃.

In a boggy field on the south side of Berrington. Round Hancott Pool. Pool between Nesscliffe and Knockin. Almond Park Moss; *Rev. E. Williams's MSS.* Whixall Moss; *Rev. A. Bloxam.*

Bomere pool.

Root creeping. *Stem* 1 foot high, triangular, rough above. *Leaves* narrow, linear, acute, striated, rough at the edges and keels below the taper point. *Spikelets* of a short tumid elliptical shape and a pale silvery green hue, upper ones crowded, lowermost very slightly distant, subtended by a very short acute *bractea.* *Glumes* membranous and whitish with a green keel, lower ones suddenly cuspidate, upper ones broadly ovate acuminato-cuspidate. *Nut* smooth.

6. **C. elongata**, Linn. *Elongated Carex.* Spikelets numerous, oblong, lax, rather distant, sterile at the base, lowermost bracteated; fruit patent; glumes rotundo-ovate, obtuse, shorter than the fruit; perigynium sessile, oblongo-acuminate, plano-convex, strongly ribbed, margins rough, orifice scarcely bifid; nut linear-oblong, tapering at the base, compressed, obsoletely subtriquetrous, smooth. *E. Bot. t.* 1920. *E. Fl. v. iv. p.* 82. *Hook. Br. Fl.* 4*th ed. p.* 332.

Marshy margins of lakes; rare. *Fl.* June. ♃.

Colemere mere, below the point where the Ellesmere Canal leaves the direction of the mere; *J. E. Bowman, Esq!*

Root tufted, with long stout fibres. *Stem* 18 inches to 2 feet high, triangular, rough above. *Leaves* verry narrow, linear, acute, striated, rough at the edges and keel below the point. *Spikelets* 9-10 or more, rusty brown, oblong, narrow, scarcely tumid, crowded above, lax and distant below, the lowermost subtended by a *bractea* more or less leafy. *Glumes* rotundo-ovate, more or less obtuse, dark brown with a green keel and whitish membranous edges. *Perigynium* 7-ribbed on convex side, about 5-ribbed on plane one.

7. **C. remota**, Linn. *Distant-spiked Carex.* Spikelets several, small, sessile, sterile at the base, 3 lowermost very distant, subtended by very long leafy bracteas, the lowermost reaching beyond the spike, uppermost spikelets crowded, inconspicuously bracteated; fruit erect; glumes oblong, acute, single-nerved, rather shorter than the fruit; perigynium oblongo-ovate, shortly acuminate, attenuate at the base, plano-convex, ribbed, margins rough, orifice bifid; nut elliptical, compressed, very minutely elevato-punctate. *E. Bot. t.* 832. *E. Fl. v. iv. p.* 84. *Hook. Br. Fl.* 4*th ed. p.* 332.

Moist shady places; not common. *Fl.* June. ♃.

Moist ditch banks and moist woods about Eaton Mascott and Battlefield; *Rev. E. Williams's MSS.* Cox wood, Coalbrookdale; *Mr. F. Dickinson.*

Canal between Shrewsbury and Uffington. Lane leading from Longden to Oaks Hall, near Pontesbury. Morda pool, near Oswestry.

Root tufted, fibrous. *Stem* 1-2 feet high, slender, smooth and roundish below, triangular and rough above. *Leaves* very narrow, linear, acute, edges incurved, rough. *Spikes* flexuose or zigzag. *Spikelets* numerous, ovate, pale, sessile, 3 lower ones remote, each subtended by a very long narrow leafy bristle-pointed *bractea*, that of the lowermost spikelet extending beyond the extremity of the stem; upper spikelets small, more or less crowded, with small inconspicuous not leafy *bracteas*. *Glumes* broadly oblong, acute, membranous and whitish, the single pellucid nerve terminating below the apex and with a stripe of green on each side, rather shorter than the fruit. *Perigynium* obscurely 5-ribbed at the base on the plane side; the convex side with a cordlike mass of ribs which proceed downwards from the bifid orifice to a little above the middle, where it diverges into five ribs, which again converge and unite a little above the base.

C. *axillaris* seems to be chiefly distinguished from *remota* by the lowermost spikelet being compound and subtended by a long leafy bractea extending beyond the spike, the bracteas of the other spikelets being shorter; whilst in *remota* the bracteas of the 3 lowermost spikelets which are simple all extend beyond the spike. The character from the number of nerves of the glumes seems variable. In *axillaris* the perigynium is elliptical, attenuated at both ends, with similar ribbing to that of *remota.* The form of the nut in both is nearly alike.

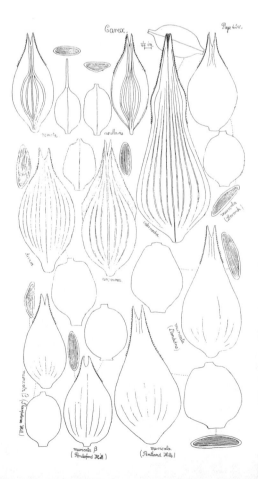

Carex. Page 452.

renota *axillaris*

C. leporina

C. ovalis

C. curta

C. muricata (β) (*Malvern*)

C. muricata (*Llandaw*)

C. muricata β (*Pontesford Hill*) *C. muricata* (*Pentland Hills*)

†† *Spikelets alternate, sterile at their extremity.*

8. C. *intermedia*, Gooden.   *Soft brown Carex.*   " Inferior and terminal spikelets fertile, all crowded into an oblong interrupted head, the intermediate ones sterile ; fruit acutely margined, longer the glumes ; bracteas membranaceous, the lower ones somewhat leafy ; stem triangular, leaves plane." *Hook. Br. Fl. 4th ed. p.* 333. *E. Bot. t.* 2042.   *E. Fl. v. iv. p.* 86.

Wet meadows ; rare.   *Fl.* June. ♃.

Meadows under Cronkhill. On the banks of the river Tern. Between Halston and Hardwick ; *Rev. E. Williams's MSS.*

Root deeply creeping. *Stems* 1-1½ foot high, erect, with 3 rough unequal angles. *Leaves* chiefly in the lower part of the stem, sheathing, scarcely reaching to its summit. *Spike* oblong, bluntish, of a rusty brown, moderately compressed, lowermost one subtended by an ovate *bractea* terminating upwards in a setaceous leaflet, the rest with an ovate leafless bractea. *Fruit* large, not so distinctly winged as gradually flattened towards the margin, beak serrulate upwards, deeply bifid, striated. In all the specimens which I have ever seen of this species the nut has been invariably abortive.

9. C. *muricata*, Linn.   *Greater prickly Carex.*   Spike oblong, dense or interrupted at the base, spikelets compact or approximate ; fruit spreading ; glumes ovate, submucronate ; perigynium tapering at the base or substipitate, ovato-acuminate, plano-convex ; margins rough, orifice bifid ; nut broadly oblong, plano-convex, minutely granulated. *E. Bot. t.* 1097.   *E. Fl. v. iv. p.* 88. *Hook. Br. Fl. 4th ed. p.* 334. *a. Koch. Syn. p.* 751.

Hedge-banks and gravelly pastures ; not uncommon.   *Fl.* June. ♃.

Pastures about Sundorn, and Eaton Mascott ; *Rev. E. Williams's MSS.* Welbach, near Shrewsbury ; *Mr. T. Bodenham.* Cox Wood, Coalbrookdale ; *Mr. F. Dickinson.* Hadnall ; *Mr. E. Elsmere, junr.!* -

Hedge-banks near Lower Berwick on the turnpike road-side. Near Berrington. Lane leading from Longden to the Oaks Hall, near Pontesbury. Hedge-banks between Shrewsbury New Race Course and Monkmoor.

Root tufted, fibrous. *Herbage* of a bright grass-green. *Stem* erect, 1-2 feet high, slender, triangular, smooth, striated and roughish above. *Leaves* chiefly in the lower portion of the stem, narrow, linear, acute, rough at the edges and keel. *Spikelets* 8—10, of a rusty green, upper ones dense, 1 or 2 of the lower ones only slightly remote. *Bractea* of the lowest spikelet small setaceous dilated and membranous at the base, those of the upper ones broadly ovate membranous with a long rough mucro, somewhat similar to the glumes. *Glumes* membranous, single-nerved. *Fruit* yellow-brown.

In specimens from Llandidno, North Wales, (communicated by Mr. W. Wilson) and from the Pentland Hills, Scotland (sent by Rev. T. B. Bell) the spikelets are approximate and only 6 in number, the perigynium much larger and full one-fourth longer, the beak being much more elongated and acuminated, and the nut broadly ovate obtuse compressed subplano-convex. In size and other characters they apparently correspond with our Shropshire *muricata.* Of the Llandidno plant Mr. Wilson remarks : " rather different from any elsewhere seen." The size of the perigynium and form of the nut of the Llandidno specimen are identical with those parts in *Vignea muricata.* No. 409. *Reich. Fl. Germ. Exsicc. !* Do not these differences indicate the probability of two distinct species being comprehended in that usually named *muricata ?*

Var. β. E. Fl. l. c. differs only in having smaller rounder simple spikelets and a smoother stem.

Gravelly ditch-banks about Eaton Mascott. Cound. On the rocky bank of Haughmond Hill. Wrekin. Ditch-banks about Ruyton, near Beckbury ; *Rev. E. Williams's MSS.*

Queen Eleanor's Bower, Haughmond hill. Sharpstones hill. Pontesford hill.

10. C. *divulsa*, Gooden.   *Grey Carex.*   Spike elongated, lax, lower spikelets remote, uppermost approximate ; fruit suberect ; glumes pale, membranous, acute, with a green dorsal stripe ; perigynium ovate, acute, plano-convex, margins slightly rough, orifice bifid ; nut ovate, obtuse, plano-convex. *Hook. Br. Fl. 4th ed. p.* 334. *E. Bot. t.* 629. *(young.) E. Fl. v. iv. p.* 89. *Koch. Syn. p.* 751.

Moist shady pastures ; rare.   *Fl.* May, June. ♃.

Ditch-bank on the south-east side of the Cloud Coppice, near Berrington.

Very different in appearance from the preceding, of a paler colour, and slenderer habit. The elongated lax spikes of which the lowermost spikelets are widely separated from each other, the smaller spikelets with fewer fertile flowers, the form of the perigynium and nut keep it quite distinct.

Root fibrous, tufted. *Stems* 1-2 feet high, weak, partially reclining, triquetrous, angles rough upwards. *Leaves* of a bright grass-green, sheathing the stem, narrow, as long or longer than the stem, margins and keel rough. *Spike* elongated, lax, interrupted ; *spikelets* small, sessile, the lower ones widely separated, the upper ones approximate, generally simple, except one or two of the lowermost which are not unfrequently lengthened into a short branch bearing 2, 3 or 4 sessile subremote spikelets and then probably constitutes Smith's var. *β.* and that represented in *Micheli's Gen.* 69. *t.* 33. *f.* 10. *Bracteas* small, ovate, close, membranous, more or less elongated into a setaceous hispid point ; the lowermost bractea sometimes becoming completely foliaceous and longer than the whole spike. But all these variations both of the spikelets and bracteas are observable in the same tuft of specimens. *Glumes* ovate, pale, membranous, acute, mucronate, with a bright green dorsal stripe or keel, shorter than the fruit. *Perigynium* quite smooth, except a slight roughness on the margins of the beak.

††† *Spikelets compound.*

11. C. *vulpina*, Linn.   *Great Carex.*   Spikelets compound, collected into a cylindrical crowded spike ; fruit squarroso-patent ; glumes ovate, mucronate ; perigynium stipitate, ovato-acuminate, plano-convex, obscurely ribbed, margins serrulato-scabrous, orifice bifid ; nut broadly oval, tapering at the base, compressed, minutely elevato-punctate, crowned with the somewhat thickened base of the style ; stem very acutely triangular ; leaves broad. *E. Bot. t.* 307. *E. Fl. v. iv. p.* 90. *Hook. Br. Fl. 4th ed. p.* 334.

Wet shady places, margins of ponds and rivers ; common.   *Fl.* June. ♃.

Sides of pools and ditches, and canals ; *Rev. E. Williams's. MSS.* Malt House Pool, Coalbrookdale ; *Mr. F. Dickinson.* Oakley park, near Ludlow ; *Miss Mc Ghie.* Near Oswestry ; *Rev. T. Salwey.!*

Canal between Shrewsbury and Uffington. Benthal Edge. Sutton Spa. Eyton on the Wildmoors, &c. Ditches between Battlefield and Albright Hussee.

Root tufted, fibrous. *Stem* 2 feet high, erect, leafy in the lower part only, angles very rough. *Leaves* deep green, broad, sharply acuminate, rough at the edges and keel. *Spikes* large, dense, very compound, greenish. *Bracteas* very rough at the edges and keels, dilated at the base, frequently tapering into a long

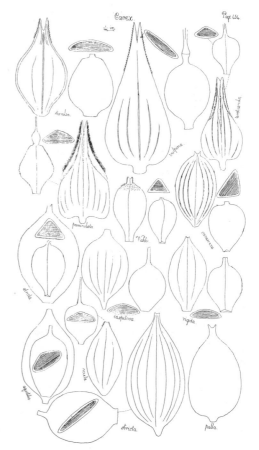

setaceous leafy extremity, very conspicuous when in flower, but becoming obsolete in fruit. *Glumes* with a rough mucro, single-nerved, brownish-green, as broad as but shorter than the fruit. *Perigynium* obscurely 5-ribbed on the convex side, with 6-7 short ribs at the base on the plane one, the 2 ribs on the margin strong and prominent.

Judging from the specimen *no.* 411. *Reich. Fl. Germ. Exsicc ! C. nemorosa, Rebent.* appears to be only a variety of *vulpina*, with a shorter denser less interrupted spike, and of a paler hue, probably from growing in shade. The form of the perigynium and nut are quite identical with those of *vulpina*.

12.　C. *teretiuscula*, Gooden. *Lesser panicled Carex.* Spike compound, oblong, consisting of ovate compact spikelets ; glumes broadly ovate, acute, membranous, slightly shorter than the fruit ; fruit erect ; perigynium stipitate, ovato-gibbous, acuminated into a winged serrulate bidentate beak, subplano-convex, with 3-4 central nerves on the convex surface ; nut turbinate, triquetrous, angles rounded, very minutely rough ; base of the style not thickened. *E. Bot. t.* 1065. *E. Fl. v. iv. p.* 91. *Hook. Br. Fl. 4th ed. p.* 334.

Wet boggy margins of pools ; rare. *Fl.* June. ♃.

By the sides of Shomere and Hancott pools. By the side of a ditch between Adeney and Batterey, near Edgmond. Ditches at the east end of Colemere mere ; *Rev. E. Williams's MSS.* Ellesmere and Colemere meres ; *J. E. Bowman, Esq.* Near Oswestry ; *Rev. T. Salwey.*

Bomere pool. Colemere mere.

Root fibrous, slightly creeping into scattered simple not dense tufts. *Stems* 1-2 feet high, erect, slender, with 3 sharp rough angles, the intermediate spaces convex with a prominent longitudinal rib, which renders it roundish. *Leaves* sheathing the lower part of the stem, erect, about as tall as the stem, deep green, rough on the edges and keel, very narrow, acute. *Glumes* membranous, whitish, with a strong dark brown keel. *Perigynium* brown, smooth and shining, of an ovate gibbous form, tapering gradually into the beak, the convex side with 3-4 raised central ribs and a strong membranous elevated wing extending along the centre of the beak downwards frequently to half the length of the perigynium ; on either side of the central ribs is a short single rib at the base, the plane side quite smooth.

13.　C. *paniculata*, Linn. *Great panicled Carex.* Spike panicled, consisting of ovate spikelets arranged on the elongated diverging branches of a common axis ; glumes ovate, membranous, acute, about as long as the fruit ; fruit spreading ; perigynium stipitate, broadly ovato-gibbous, acuminated into a winged subciliato-serrulate bidentate beak, gibboso-plano-convex, obscurely many-nerved ; nut ovate, obtuse, attenuate at the base, compresso-triquetrous, minutely dotted ; base of the style thickened or swollen. *E. Bot. t.* 1064. *E. Fl. v. iv. p.* 92. *Hook. Br. Fl. 4th ed. p.* 335.

Spongy bogs ; rare. *Fl.* June. ♃.

Ditches about Shineton, Halston, and Hardwick. Under the Stiperstones. Between Beckbury and Ruyton. Sandford Pool, near Westfelton. Croesmere Mere ; *Rev. E. Williams's MSS.* Colemere, near Ellesmere ; *Rev. A. Bloxam.* Twyford Vownog, near Westfelton.

Root of many strong fibres, densely tufted, not creeping. *Stem* 2-3 feet high, with 3 rough acute angles, intermediate spaces flat, striated. *Leaves* broad, rough at the edges and keel. *Spikes* 2—4 inches long. The beak of the perigynium on the convex surface has a winged membrane as in the preceding species.

The form of the fruit and especially of the nut, independently of its habit of growth in large dense tufts and of other characters widely separates this from the preceding species.

*** *Terminal spike barren solitary (or in* 15 *and* 16, *sometimes more than* 1*); the rest fertile.*

† *Stigmas* 2.

14.　C. *cæspitosa*, Linn. *Tufted Bog Carex.* Barren spike solitary, rarely 2 ; fertile spikelets 2-3, erect, cylindrical or oblong, obtuse, rarely with barren flowers at the extremity, sessile, lower one shortly pedunculate ; glumes elliptical or oblong, obtuse, shorter than the fruit ; perigynium more or less stipitate, ellipsoid or oblong, subplano-convex, with a very short entire point, obscurely ribbed at the base ; nut elliptico-subrotund. *E. Bot. t.* 1507. *Hook. Br. Fl. 4th ed. p.* 336.

Marshes, wet and boggy pastures, &c. ; frequent. *Fl.* May, June. ♃.

Moist meadows, about Eaton Mascott, &c. ; *Rev. E. Williams's MSS.* Boggy meadows behind Marn Wood, near Coalbrookdale ; *Mr. F. Dickinson.*

Hancott Pool. Haughmond Hill. Bomere Pool. Stiperstones Hill. Shawbury Heath. Golden Pool, near Pitchford. Morda Pool, near Oswestry. Colemere and Ellesmere Meres.

Root creeping, in small loose tufts. *Stem* 8 inches to 1 foot high, erect, weak, triangular, smooth and striated, angles rough upwards. *Leaves* radical, erect, narrow, linear, flattish, channelled, edges and keels rough. *Spikelets* erect, short, dense, obtuse, subtended by sheathless *bracteas* of which the lower one is subfoliaceous, about as long as the spike and with small dark roundish auricles. *Stigmas* very long, spreading and flexuose, covered with coarse pubescence. *Glumes* about two-thirds as long as the erect fruit, oblong, obtuse, scarcely pointed, equalling the fruit in breadth, dark purple brown with a central green streak. *Perigynium* more or less distinctly stipitate, about 5-ribbed on the plane side, 5—7-ribbed on the convex side, the ribs not extended to the apex.

15.　C. *acuta*, Linn. *Slender-spiked Carex.* Sheaths none, bracteas long, foliaceous ; fertile spikelets long, cylindrical, acuminate, slender, erect when in fruit ; perigynium oval or ovali-elliptical, plano-convex, subacuminate, entire at the point ; nut rotundo-obovate, elevato-punctate. *E. Bot. t.* 580. *Hook. Br. Fl. 4th ed. p.* 337.

Sides of pools and rivers ; frequent. *Fl.* May. ♃.

Ponds and sides of rivers ; River Tern, &c. ; *Rev. E. Williams's MSS.* Malthouse Pool, Coalbrookdale ; *Mr. F. Dickinson.*

Almond park pool, near Shrewsbury.

Root creeping extensively. *Stem* 2-3 feet high, acutely angular, scabrous. *Leaves* broad, scarcely glaucous, shorter than the stem, rough at the edges and keel. *Fertile spikelets* 3-4, with a few barren flowers at the extremity. *Glumes* lanceolate, acute, black, with a pale rib. *Stigmas* always 2.

16.　C. *stricta*, Gooden. *Straight-leaved Carex.* Barren spike 1 or more ; fertile spikelets 2—4, erect, cylindrical, elongated, often acuminated with barren flowers at the extremity, nearly sessile, lower one shortly pedunculate ; glumes lanceolate, acute, as long as the fruit ; perigynium stipitate, oblongo-elliptical, compressed, with a

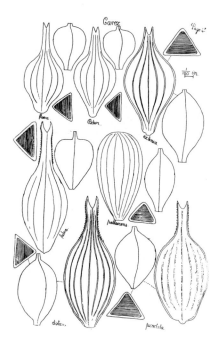

short entire beak, ribbed; nut oblongo-elliptical. *E. Bot. t.* 914. *Hook. Br. Fl.* 4th ed. p. 337.

Marshes and boggy margins of pools; frequent. *Fl.* April, May. ♃.

Pits, ponds and ditches about Eaton Mascott, Pitchford, &c.; *Rev. E. Williams's MSS.* Meadow near Marn Wood, Coalbrookdale; *Mr. F. Dickinson.* Near Oswestry; *Rev. T. Salwey.*

Bomere Pool. Hancott Pool. Shawbury Heath. Morda Pool, near Oswestry. Golden Pool, near Pitchford. Blackmere, Colemere and Ellesmere Meres.

Root fibrous, growing in large dense solid tufts. *Stem* 2-3 or more feet high, erect, strong, triangular, with acute rough angles. *Leaves* long, broadish, rough, with a filamentous reticulation at their sheathing bases. *Bracteas* with sheaths, lowermost subfoliaceous, shorter than the spike, with large oblong brown auricles. *Stigmas* spreading and flexuose, pubescent. *Glumes* as long or longer than the fruit, lanceolate, acute, about half as broad as the fruit, dark purplish-brown, with a single green pellucid nerve. *Perigynium* with about 5—7 ribs on one side and 3—5 on the other, some of them vanishing or obliterated above and the others extended to the apex.

†† *Stigmas* 3.

+ *Fruit glabrous.*

a. Fertile spikelets abbreviated, subsessile.

17. C. *flava*, Linn. *Yellow Carex.* Sterile spikelet solitary, sessile or stalked; fertile spikelets 2—4, subrotundo-ovate, lowermost on a peduncle longer than the sheath, frequently remote, uppermost subapproximate, subsessile; bracteas long, leafy, sheathing, finally spreading or reflexed; glumes ovate, acuminate, single-ribbed, shorter than the fruit; fruit spreading; perigynium obovate, inflated, ribbed, smoothish, acuminated into a more or less deflexed longer or shorter rough-edged beak, bifid at the point; nut turbinate, trigonous, minutely reticulated. *E. Bot. t.* 1294. *E. Fl. v. iv. p.* 106. *Hook. Br. Fl.* 4th ed. p. 337.

Boggy meadows; not unfrequent. *Fl.* May, June. ♃.

Sides of pools, bogs, and wet commons; *Rev. E. Williams's MSS.* Titterstone Clee Hills; *Mr. J. S. Baly!* Longmynd and Caradoc Hills; *Mr. T. Bodenham.* Wrekin Bog, west side; *Mr. F. Dickinson.* Near Oswestry; *Rev. T. Salwey.!*

Hancott Pool. Haughmond Hill. Stiperstones Hill. Shawbury Heath. Ellesmere Mere. Golden Pool, near Pitchford. Morda Pool, near Oswestry. Bomere Pool.

Root fibrous, tufted. Whole plant of a stiffish appearance and yellowish hue. *Stem* erect, 6—8 inches or a foot high, bluntly trigonous, smooth, leafy at the base. *Leaves* bright grass-green, erect, various in length, flat, broad, ribbed, channelled above, acutely keeled beneath, contracted and acuminated into a triangular point, scabrous at the edges and keel. *Bracteas* very foliaceous, lowermost resembling the broad acuminated leaves, overtopping the spikes, uppermost shorter, deflexed. *Fertile spikes* 2 or 4. *Glumes* ovate, acuminate, tawny, with a green pellucid nerve, scabrous at the point and keel. *Perigynium* green, at length yellow, smooth, 5-ribbed on the one side, 7-ribbed on the other.

In vain have I sought to detect some constant characters whereby to distinguish *C. flava* and *Œderi.* The sterile spikelet is in both sessile or stalked; the approximation of the fertile spikelets is very variable in both; the lowermost pedunculate and remote one on a peduncle exceeding in length the sheath of the bractea occurs in each; the form and number of ribs of the perigynium is similar

and the beak equally very variable in length; whilst the shape of the nut is identical. The only character which would seem to approach towards anything like constancy is the direction of the beak of the perigynium, and yet in *flava* this varies by such intermediate gradations from a decidedly decurved state to one scarcely or very slightly decurved, that too much reliance should not perhaps be placed on it as a discriminating mark, particularly as a very evident and marked decurvature in the beak occurs in some specimens gathered near the Menai Bridge and communicated to me by Mr. W. Wilson, as " *C. Œderi var.?* extremely variable and very closely approaching *C. flava.*" On the whole, therefore, I feel myself impelled to regard them both as modifications of the same species.

Var. β. polystachya, Koch. Fertile spikelets 5-6. *Koch. Syn.* p. 764.

Golden Pool, near Pitchford, June 1840, intermixed with the usual states.

18. C. *extensa*, Gooden. *Long bracteated Carex.* Sheaths very short, scarcely any; bracteas extremely long, foliaceous; sterile spikelet sessile, fertile spikelets nearly sessile, oblong, approximate; glumes rotundo-ovate, mucronate, 3-ribbed; fruit erecto-patent; perigynium obovate, attenuate at the base, turgid, ribbed, with a short accuminated beak, smooth on the margin, orifice bifid; nut oblongo-elliptical, attenuate at both ends, trigonous. *E. Bot. t.* 833. *E. Fl. v. iv. p.* 108. *Hook. Br. Fl.* 4th ed. p. 338.

Whixall Moss; *Mr. F. Dickinson.* Fl. June. ♃.

Root black, fibrous, slightly creeping. *Stem* 6—12 inches high, bluntly triangular, smooth, leafy in the lower part, generally more or less curved. *Leaves* long and narrow, convolute, rough-edged at the extremity. *Bracteas* very long, foliaceous, spreading widely and decurved. *Fertile spikelets* usually 3, approximate, occasionally with a remote one, on very short peduncles. *Glumes* membranous, tawny, mucronate, strongly 3-ribbed, shorter than the fruit. *Perigynium* 4-ribbed on triquetrous side, 5-ribbed on plane one. *Nut* dark-brown, minutely reticulato-dotted.

The form of the nut independently of any other differences, essentially distinguishes this from every modification of *flava.*

b. Fertile spikes stalked, erect.

19. C. *fulva*. Gooden. *Tawny Carex.* Sheaths elongated, shorter than the peduncles; bracteas foliaceous; glumes membranous at the edges, acute; fertile spikelets oblongo-ovate, distant; fruit ascending; perigynium ovato-ellipsoid, plano-triquetrous, acuminated into a straight beak, margins rough, orifice bicuspidate; nut turbinate, trigonous, elevato-punctate. *E. Bot. t.* 1295. *E. Fl. v. iv. p.* 107. *Hook. Br. Fl.* 4th ed. p. 338.

Boggy meadows; not frequent. *Fl.* May, June. ♃.

Boggy margin of Golden Pool. Boggy field called Mosterley, adjoining Cound Moor. In an Alder copse at the south end of the Wrekin. About Hardwick, near Ellesmere; *Rev. E. Williams's MSS.* Near Eaton; *T. and D. Bot. Guide.* Whixall Moss; *Rev. A. Bloxam.* Wrekin Bog, west side; *Mr. F. Dickinson.*

Haughmond Hill. Hancott Pool. Stiperstones Hill. Shawbury Heath. Boggy ground north of Bomere Pool.

Root creeping. *Stem* ascending, 8—15 inches high, slender, acutely triangular, rough-edged. *Leaves* chiefly in the lower part of and not half the length of

the stem, broad, short, acute, rough at the edges and keel, channelled above, keeled beneath, edges somewhat recurved. *Bracteas* not rising above the stem. *Peduncle* of the lowermost spikelet half as long again as its sheath. *Fertile spikelets* 2 rarely 3, erect, short, lax, pale-coloured and few-flowered. *Glumes* ovate, acute, never mucronate, brownish-green, with a green pellucid nerve and a white membranous border, shorter than the fruit. *Perigynium* 5-ribbed on the triquetrous side, 7-ribbed on the plane one, the 2 ribs on the margins and also the 2 lateral ribs frequently thickened.

The form of the perigynium, direction of the fruit and habit of growth keep this sufficiently distinct from *flava*, while the shape of the nut readily separates it from *extensa* and *distans* with its allies.

20. C. *pallescens*, Linn. *Pale Carex.* Sheaths scarcely any; fertile spikelets about 3, pedunculated, oblongo-cylindrical, scarcely pendulous; glumes oblong, cuspidato-mucronate, as long as the fruit; fruit erect; perigynium ovato-elliptical, tumid, striated, obtuse, glabrous; nut lineari-elliptical, attenuated at both ends, trigonous. *E. Bot. t.* 2185. *E. Fl. v. iv. p.* 105. *Hook. Br. Fl.* 4th ed. p. 338.

Marshy places; not very common. *Fl.* June. ♃.

Moist meadows above Pitchford. Battlefield. Cantlop and Bomer woods; *Rev. E. Williams's MSS.* Field on a side of a hill near Madeley; *Mr. F. Dickinson.* Hadnall; *Mr. E. Elsmere, junr.!* Measbury and Porkington! near Oswestry. *Rev. T. Salwey.*

Golden, near Pitchford. Bog north of Bomere pool.

Root tufted, fibrous. *Stem* 12—15 inches high, sharply triangular, rough above. *Leaves* erect, flat, striated, slightly hairy, with pubescent sheaths. *Bracteas* leafy, one to each spikelet, lowermost longest reaching above the stem. *Spikes* obtuse, pale-green.

21. C. *distans*, Linn. *Loose Carex.* Barren spikelet solitary, with obtuse glumes; fertile 2-3, remote, erect, oblong, stalked, the barren stalks longer than the sheathing bracteas; glumes mucronate; fruit ovate or rather ellipsoid, triquetrous, equally 7-ribbed, smooth or rough at the upper margins and at the edges of the narrow short bifid beak; nut oblongo-elliptical, attenuated at both ends, trigonous, reticulato-dotted. *Hook. Br. Fl.* 4th ed. p. 339. *E. Bot. t.* 1234. *E. Fl. v. iv. p.* 109.

Moist meadows about Eaton Mascott; Pitchford and Golden; under the Wrekin. Shawbury Heath; *Rev. E. Williams's MSS.* Fl. June. ♃.

Stem 8 inches to 1 or 1½ foot high, slender. *Spikes* very distantly placed, their rather long *peduncles* entirely concealed by the sheathing bases of the *bracteas.* *Glumes* rather pale brown. *Fruit* green, inclining to brown when ripe. *Perigynium* with 7 equal ribs on either side, and 2 very prominent ribs on the margins.

Of this species I have never seen Shropshire specimens, nor have I been able to verify Mr. Williams's habitats, and as all the recorded localities in Britain are either on the sea-coasts or where the sea has recently been or still occasionally flows, I should have hesitated to have inserted it as a Shropshire plant had I not in so many instances ascertained the uniform accuracy of the localities mentioned by Mr. Williams in his MS. Catalogue, more especially as it is quite evident from Mr. Williams's communications to Sir J. E. Smith and the Bishop of Carlisle, that he was well acquainted with the species of this difficult genus.

The form of the nut in the 4 closely allied species *punctata, distans, binervis* and *lævigata* affords an admirable character for correctly distinguishing them. In *punctata*, the nut is ovato-rhomboidal, attenuated at both ends; in *distans*, oblongo-elliptical, attenuated at both ends; in *binervis*, obovate, attenuate below; and in *lævigata*, pyriform, attenuated below.

22.   C. *binervis*, Sm.   *Green-ribbed Carex.*   Barren spikelet solitary, with obtuse glumes; fertile 3, erect, remote, cylindrical, lowermost most remote, on peduncles exserted from and longer than the sheathing foliaceous bracteas; glumes ovate, obtuse, with a short scabrous mucro; fruit erecto-patent; perigynium ovate or rather ellipsoid, triquetrous, with a straight beak, serrulato-scabrous at the margin, bifid at the orifice, the outer or keeled surface with about 7 obscure ribs and 2 distant stronger prominent submarginal ones; nut obovate, attenuated below, triangular, minutely elevato-punctate. *E. Bot. t.* 1387. *E. Fl. v. iv. p.* 122. *Hook. Br. Fl.* 4*th ed. p.* 339.
Dry heaths and moors; not common. *Fl.* June. ♃.
On Shawbury Heath, and in a piece of boggy ground on the east side of the Stiperstones.
Root tufted, with long stout fibres. *Stem* 2-3 feet high, triangular, striated, smooth, leafy. *Leaves* bright green, broad, acuminated, half the height of the stem, flat, striated and deeply channelled on the upper surface, acutely keeled beneath, edges and keels rough, with long striated smooth sheaths. *Bracteas* narrow, rough at the edges and keel, with smooth sheaths shorter than the peduncles, except the uppermost which is equal to the peduncle. *Sterile spikelet* rarely more than 1. *Fertile spikelets* 3 sometimes 4, on compressed smooth peduncles, the 2 lowermost remote and on longish peduncles, the rest subapproximate and on very short peduncles, of a dark or blackish-brown hue. *Glumes* ovate, membranous, dark brown, with a green keel and rough mucro, shorter but as broad as the fruit. *Fruit* erect. *Perigynium* ovato-triquetrous, somewhat compressed, with a rather short broad beak, slightly rough, sometimes smooth at the margin, orifice bifid, the plane or inner surface obscurely many-ribbed, the outer or keeled surface with 5—7 obscure ribs about the central prominent angle or keel and 2 very distinct strong elevated smooth ribs towards the margin, though totally distinct from it.

23.   C. *panicea*, Linn.   *Pink-leaved Carex.*   Barren spikelets solitary; fertile spikelets 2, remote, subcylindrical, with lax flowers, on exserted peduncles; sheaths elongated, shorter than the peduncles; bracteas leafy; glumes oblong, acute, shorter than the fruit; fruit erecto-patent; perigynium ovate-subglobose, obsoletely triquetrous, inflated, minutely granulated, with 2 lateral ribs and a short round somewhat decurved beak, orifice oblique emarginate; nut obovato-globose, bluntly trigonous. *E. Bot. t.* 1505. *E. Fl. v. iv. p.* 114. *Hook. Br. Fl.* 4*th ed. p.* 340.
Marshy places and bogs; common. *Fl.* June. ♃.
Boggy meadow near Marn wood, Coalbrookdale; *Mr. F. Dickinson.* Hadnall; *Mr. E. Elsmere, junr.!*
Nobold, near Shrewsbury. Hancott pool. Bomere pool. Shawbury heath. Haughmond hill.   Colemere and Ellesmere meres.   Golden, near Pitchford. Morda pool, near Oswestry; and generally throughout the County.
Similar in its glaucous foliage to C. *recurva*, but widely different in character. *Root* creeping. *Stem* 1-1½ foot high, erect, striated, smooth, obtusely trigonous. *Leaves* chiefly radical, broad, glaucous and spreading, ribbed, rough at the edges and keel towards the top. *Bracteas* erect, leafy, with pale sheaths, lowermost

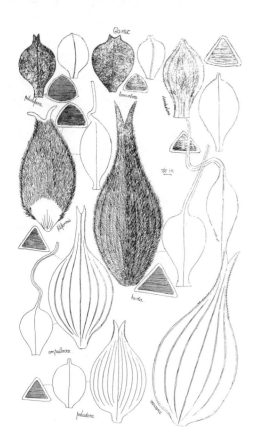

about as long as the adjacent spikelet, uppermost shorter. *Fertile spikelets* 2, remote, often with barren flowers at their summits, the lowermost on a long slender triquetrous stalk, generally twice the length of the sheath, its lower flowers lax and distantly placed. *Glumes* oblong, acute, dark chocolate with a green keel and pellucid nerve, slightly membranous at the edges, as broad as the fruit. *Perigynium* greenish-yellow, inflated, with a prominent rib on each margin.

c. Fertile spikelets stalked, drooping.

a. Fertile spikelets abbreviated.

24. C. *limosa*, Linn. *Mud Carex.* Sheaths extremely short, scarcely any; fertile spikelets oblongo-ovate, pendulous; bracteas subsetaceous; fruit erect; glumes rotundo-ovate, mucronate or sub-apiculate, as long or longer than the fruit; perigynium substipitate, elliptico-rotundate, compressed, ribbed, with a very short sub-emarginate point; nut broadly obovato-elliptical, bluntly triquetrous, smooth. *E. Bot. t.* 2043. *E. Fl. v. iv. p.* 102. *Hook. Br. Fl. 4th ed. p.* 340.

Bogs and marshes; rare. *Fl.* May, June. ♃.

Bog below Weeping Cross turnpike, near Shrewsbury. Bogs about Lee, near Ellesmere; *Rev. E. Williams's MSS.* Bog near turnpike road to Wenlock, from Salop 1½ mile; *Mr. F. Dickinson.*

Bomere pool.

Root creeping deeply and extensively, with densely downy fibres, ascending obliquely. *Stems* 8—10 inches high, erect, slender, triangular, striated, rough-edged above, leafy below. *Leaves* very narrow, linear, acuminated, more or less complicated and channelled, striated, rough at the edges and at the strong keel below the very acuminated point, slightly glaucous, enveloped at the base with brown scales. *Bracteas* erect, taper, with a pair of small membranous brown rounded auricles at the base, the lowermost one foliaceous, upper ones setaceous short. *Fertile spikelets* 2, occasionally 1, rarely 3, on long slender semicylindrical smooth pendulous peduncles thickened immediately below the spikelet. *Glumes* dark brown, with a greenish gold tinge, keel broad green, strongly and prominently single-ribbed, the point acutely mucronate, as long as or frequently when the point and mucro are more acuminated, longer than the fruit. *Fruit* glaucous green, compressed, with 2 strong ribs on the margins and 4-5 obscure ones towards the base on the faces, the very short point emarginate.

β. Fertile spikelets elongated.

25. C. *strigosa*, Huds. *Loose pendulous Carex.* Sheaths elongated, equal to the peduncles; fertile spikelets 4, remote, filiform, slender, nearly erect; glumes ovato-lanceolate, mucronate, one-third shorter than the fruit; fruit slightly recurved; perigynium lanceolate, acutely triquetrous, with an obliquely truncate entire point, ribbed; nut lineari-oblong, attenuated at both ends, triquetrous, minutely granulate. *E. Bot. t.* 994. *E. Fl. v. iv. p.* 95. *Hook. Br. Fl. 4th ed. p.* 341.

Groves and thickets; rare. *Fl.* May, June. ♃.

In the Pool Tail Coppice above Pitchford House, plentifully; *Rev. E. Williams's MSS.*

Root fibrous. *Stem* 18 inches to 2 feet high, erect, bluntly triangular, smooth, striated. *Leaves* broadly linear, acuminate, edges and ribs rough. *Bracteas* leafy, sheathing the peduncles almost entirely. *Fertile spikelets* about 4, loose and slender. *Glumes* ovato-lanceolate, with a rough mucro, membranous, pale, with a central green stripe and prominent midrib, about one-third shorter than the fruit. *Perigynium* lanceolate, acutely triquetrous, the faces with 4-5 prominent ribs, green, without a beak, the orifice entire.

26 C. *sylvatica*, Huds. *Pendulous Wood Carex.* Sheaths one-third to one-half as long as the peduncles; fertile spikelets 4, remote, filiform, pedunculate, rather slender, slightly drooping; glumes ovate, acuminate, submucronate, nearly equalling the fruit; fruit erect; perigynium elliptical, obtusely triquetrous, tapering into a linear bifid beak, glabrous; nut obovato-elliptical, triquetrous, elevato-punctate. *E. Bot. t.* 995. *E. Fl. v. iv. p.* 96. *Hook. Br. Fl.* 4th ed. p. 341.

Moist woods; not unfrequent. *Fl.* May, June. ♃.
Welbach Coppice; *Mr. T. Bodenham.* Captain's Coppice, Coalbrookdale; *Mr. F. Dickinson.* Hadnall; *Mr. E. Elsmere, junr.!* Near Oswestry; *Rev. E. Salwey.!*
Shelton Wood, near Shrewsbury. Golden, near Pitchford. Cloud Coppice, near Berrington.

Root tufted, with stout fibres. *Stem* 2 feet high, erect, triangular, smooth, striated. *Leaves* narrow, linear, acute, rough on the upper surface, edges, keel and ribs. *Bracteas* leafy, sheathing the peduncle about one-third upwards. *Fertile spikelets* 4, loose and slender. *Glumes* ovate, acuminate, submucronate, brownish, membranous, with a strong rib, nearly as long as the fruit. *Perigynium* elliptical, obtusely triquetrous, acuminated into a long linear beak slightly rough at the edge, deeply cleft, smooth and without ribs except the 2 prominent marginal ones.

27. C. *pendula*, Huds, *Great pendulous Carex.* Sheaths elongated, nearly equal to the peduncles; fertile spikelets about 6, cylindrical, very long, pendulous; glumes ovate, mucronate, shorter than the fruit; fruit spreading and recurved at the beak; perigynium substipitate, ellipsoid, subtriquetrous, tumid, ribbed, acuminated into a short bifid decurved beak; nut elliptical, attenuated at both ends, triquetrous, minutely granulated. *E. Bot. t.* 2315. *E. Fl. v. iv. p.* 95. *Hook. Br. Fl.* 4th ed. p. 341.

Moist wooded and shady places; not common. *Fl.* May, June. ♃.
In the Birch Coppice near Buildwas. By the side of several rills between Leighton and Buildwas. Near Coalport Turnpike; *Rev. E. Williams's MSS.* Wood between Buildwas Inn and the Birches, Coalbrookdale; *T. and D. Bot, Guide.* Old Brook Wood, near Coalbrookdale; *Mr. F. Dickinson.*
Benthal Edge. Lyd Hole, near Pontesford Hill. Woods by the roadside between Much Wenlock and Buildwas Bridge.

Root tufted, with many long stout fibres. *Stem* 5-6 feet high, erect, gracefully drooping in the upper part, triquetrous, smooth, striated. *Leaves* very long, recurved, linear, acute, deeply longitudinally channelled above, half of the limb recurved on each side of the midrib, roughish, bright green and shining above, paler and glaucous beneath, midrib rough, margins minutely serrulate, with long close sheaths at the base, which in the upper leaves are about half the length of the leaves. *Barren spikelet* solitary, terminal. *Fertile spikelets* about 6, on triquetrous rough peduncles a little longer than the sheaths by which they are enclosed, 3-4 inches long, cylindrical, clavate, gradually thickening towards the extremity, lax at the base, gracefully and elegantly pendulous. *Glumes* shorter than the fruit, ovate, mucronate, somewhat membranous at the margins, of a deep chocolate colour, with a broad green keel. *Fruit* spreading and decurved at the beak.

Perigynium ellipsoid, somewhat triquetrous, tumid, with a short bifid beak whose lobes are minutely notched, having 3 ribs on the upper tumid side and one central rib on the inner or plane surface, with 2 ribs on the margins. Nut elliptical, slightly tapering at the base and apex, triquetrous, minutely granulated.
The largest and by far the most elegant of our native Carices.

28. C. *Pseudo-Cyperus*, Linn. *Cyperus-like Carex.* Sheaths scarcely any; fertile spikelets on long slender peduncles, cylindrical, pendulous; bracteas large, very foliaceous; glumes setaceous, dilated and membranous at the base, 3-nerved, serrulate at the edges, as long as the fruit; fruit spreading horizontally; perigynium stipitate, ovato-lanceolate, very much acuminated into a bicuspidate smooth beak, slightly rough-edged, pale, smooth and ribbed; nut narrowly elliptical, triquetrous, minutely granulated. *E. Bot. t.* 242. *E. Fl. v. iv. p.* 101. *Hook. Br. Fl.* 4th ed. p. 341.

Margins of rivers and ponds; not common. *Fl.* June. ♃.
Ponds and ditches about Golden and Sundorn. Mare pool, near Shrewsbury. Between Battlefield and Shrewsbury; Hancott pool; *Rev. E. Williams's MSS.* Snowdon pool; *H. Bidwell, Esq.* Hampton's Bank, near Ellesmere, in low ground by the Canal; *Mr. F. Dickinson.* Hadnall; *Mr. E. Elsmere, junr.* Oakley Park, near Ludlow; *Miss Mc. Ghie.* Maesbury marsh, near Oswestry; *Rev. T. Salwey.!*
Canal between Shrewsbury and Uffington. Bomere pool. Hancott pool. Battlefield. Pit between Battlefield and Albright Hussee. Mare pool.

Well distinguished by its size, bright green colour and elegantly pendulous spikelets. *Root* tufted, fibrous. *Stem* 2 feet high, acutely triangular, rough at the edges above, leafy. *Leaves* very broad, acute, scabrous above and at the edges, smooth beneath, with a smooth acute keel. *Bracteas* similar and nearly as large as the leaves. *Fertile spikelets* 3-4, long, with numerous flowers, elegantly pendulous, on rough trigonous slender peduncles.

29. C. *recurva*, Huds. *Glaucous Heath Carex.* Sheaths short, scarcely any; bracteas leafy, auricled at the base; fertile spikelets cylindrical, scarcely drooping, densely imbricated, on long slender peduncles; glumes ovate, acute, nearly equal to the fruit; fruit erecto-patent; perigynium obovato-globose, slightly downy, entire at the small point; nut rotundo-obovate, triquetrous, minutely elevato-punctate. *E. Bot. t.* 1506. *E. Fl. v. iv. p.* 114. *Hook. Br. Fl.* 4th ed. p. 342.

Moist meadows, clayey pastures, and wet barren heathy ground; common. *Fl.* May, June. ♃.
Root creeping, sheathed with purplish-brown scales. *Stems* about 1 foot high, obsoletely trigonous. *Leaves* chiefly radical, partially recurved, very glaucous, much resembling the foliage of pinks. *Fertile spikelets* 2, often 3. *Barren spikelet* generally solitary, sometimes accompanied by a smaller one and the upper portion of the upper fertile spikelets frequently consisting of barren florets. *Bracteas* leafy, lowermost very long, their sheaths very short, crowned with rounded brown auricles. *Fruit* closely placed, erecto-patent. *Perigynium* brownish when ripe, minutely dotted or tuberculated and with minute short bristly hairs or pubescence. *Glumes* ovate, about as broad and long as the fruit, acute, of a chocolate colour with a green rib.
Varieties occur in which the sheath of the lower fertile spikelet is more elongated and the peduncle very long; the barren spikelet either solitary and the

fertile spikelets entirely fertile; or, with 4 barren spikelets and the upper portion of the fertile spikelets consisting also of barren florets. In all these variations there does not appear any available difference in the fruit, which in all is of the same shape and downy as in the common state.

++ *Fruit downy.*

Fertile spikelets sessile.

30. C. *præcox*, Jacq. *Vernal Carex.* Sheaths short, scarcely any, equal to the peduncles; fertile spikelets 2, oblong, approximate; glumes broadly ovate, acuminate, as long as the fruit; fruit erect; perigynium rhomboideo-obovate, compressed, subtriquetrous, downy, without ribs, orifice abrupt entire spinulose; nut obovate, attenuated below, triquetrous, minutely dotted. *E. Bot. t.* 1099. *E. Fl. v. iv. p.* 111. *Hook. Br. Fl.* 4th ed. p. 342.

Dry pastures and heaths; common, *Fl.* April, May. ♃.
Root creeping. *Stem* 4—8 inches high, erect, triangular, smooth, leafy at the base only. *Leaves* forming close tufts, short, spreading or recurved, acute, flattish, rough on the edges points and ribs. *Upper bracteas* very short; lower one as long or longer than its spikelet, erect, dilated and 3-nerved at the base, acuminated with a setaceous leafy point, minutely serrulate at the margins, arising from short and abrupt sheaths dilated upwards, and about equal to the short peduncles. *Fertile spikelets* generally 2, approximate, erect, elliptico-oblong, dense, but not many-flowered. *Glumes* broadly ovate, acuminate, often mucronate, brown and membranous, with a green rib. *Barren spikelet* solitary. *Anthers* numerous, bright yellow, rendering it conspicuous at an early season of the year. *Perigynium* without ribs except the 2 obscure marginal ones. *Nut* with a circular protuberance or rugosity around the base of the style corresponding with the position of the orifice of the perigynium.

31. C. *pilulifera*, Linn. *Round-headed Carex.* Sheaths none; bracteas small, subfoliaceous; fertile spikelets 2-3, sessile, roundish, approximate; glumes broadly ovate, strongly mucronate, as long as the fruit; fruit erecto-patent; perigynium stipitate, obovato-globose, pubescent, with 2 marginal ribs and a short bifid beak; nut obovato-globose, scarcely triquetrous, very minutely granulated. *E. Bot. t.* 885. *E. Fl. v. iv. p.* 112. *Hook. Br. Fl.* 4th ed. p. 342. *C. montana, Linn.*

Hills and moory grounds; not uncommon. *Fl.* June. ♃.
By the side of Shomere pool and under Harmer Hill. Battlefield woods; *Rev. E. Williams's MSS.*
Haughmond Hill. Pimhill. Shawbury Heath.
Readily distinguished by the almost spherical fertile spikelets and pubescent fruit. *Root* shaggy, fibrous, tufted. *Stems* 6—12 inches high, acutely trigonous, slender, weak, curved, mostly recumbent, naked, striated, roughish. *Leaves* chiefly radical, tufted, pliant, grassy, long and narrow, but shorter than the stem, striated, channelled and scabrous above and at the edges, smooth and keeled beneath. *Bracteas* subulate, scabrous, the lower one only rising above its spikelet, all quite destitute of sheaths. *Perigynium* green, with a brown beak. *Glumes* brownish, with a green nerve. *Nut* brown with prominent paler ribs on the obscure angles.

**** *Terminal spikelets barren, 2 or more; the rest fertile. Stigmas 3.*

† *Fruit hairy.*

32. C. *filiformis*, Linn. *Slender-leaved Carex.* Glabrous; sheaths scarcely any; bracteas long, very narrow; fertile spikelets 2, shortly pedunculate, oblongo-cylindrical; glumes oblongo-lanceolate, cuspidate and ciliated at the point; fruit erect; perigynium substipitate, elliptical, obsoletely triquetrous, with a short bicuspidate beak, lanato-pilose; nut narrowly elliptical, pointed at both extremities, stipitate, triquetrous, minutely reticulated. *E. Bot. t.* 904. *E. Fl. v. iv. p.* 128. *Hook. Br. Fl.* 4th ed. p. 343.

Bogs; not common. *Fl.* May, June. ♃.
Shomere pools. Bogs about Ellesmere. Ellesmere Mere. White Mere; *Rev. E. Williams's MSS.* Near Eaton; *T. and D. Bot. Guide.* Colemere and Blackmere, near Ellesmere; *J. E. Bowman, Esq.* Betton Pool; *Mr. F. Dickinson.* Berrington and Bomere Pools. Colemere Mere.

Whole plant smooth, rushlike, shining. *Root* creeping. *Stem* 1-2 feet high, erect, slender, obscurely triangular, scabrous in the upper part. *Leaves* chiefly in the lower part of the stem, slender, narrow, erect, semicylindrical, involute and rough at the margins, with membranous filamentous sheaths at the base. *Bracteas* leaflike, sheathing and auricled, lowermost long rising above the stem. *Fertile spikelets* 2, somewhat remote; peduncles rather longer than the sheaths. *Perigynium* covered with dense tawny shining hairs. *Glumes* as long as the fruit, dark brown with a green midrib. Nut tipped with the curved base of the style. *Barren spikelets* 2, rarely solitary, very long and narrow.

33. C. *hirta*, Linn. *Hairy Carex.* Hairy; sheaths elongated, shorter than the peduncles; bracteas long, foliaceous; fertile spikelets 2-3, pedunculate, cylindrical, short, distant; glumes lanceolate, aristate; fruit erect; perigynium ovato-elliptical, tumid, scarcely triquetrous, attenuated into a long bicuspidate serrulate beak, ribbed, hairy; nut obovate, attenuated below, triquetrous, very minutely granulated. *E. Bot. t.* 685. *E. Fl. v. iv. p.* 125. *Hook. Br. Fl.* 4th ed. p. 343.

Margins of pools, not common. *Fl.* May, June. ♃.
Sides of pools, and moist meadows, and sandy lanes; *Rev. E. Williams's MSS.* Pools dam, Wenlock Abbey; *W. P. Brookes, Esq.* Pastures in Coalbrookdale; *Mr. F. Dickinson.* Hadnall; *Mr. E. Elsmere, junr.!* Near Oswestry; *Rev. T. Salwey.!*
Hancott Pool. Ellesmere and Colemere Meres. Bomere Pool. Golden Pool, near Pitchford. Morda Pool, near Oswestry.

Root extensively creeping, with scaly runners and shaggy fibres. Whole herbage more or less clothed with soft shaggy white hairs, which in wet situations (as in our specimens) are confined chiefly to the sheaths of the leaves which are densely hairy on the inner side or that opposite to the limb of the leaf and with denser tufts at the summit. *Stems* about 18 inches high, erect, leafy, triquetrous, rough-edged. *Leaves* flat, erect, acuminate, rough-edged. *Bracteas* large, foliaceous, their hairy sheaths about half the length of the peduncles though not unfrequently they include nearly the whole of them. *Barren spikelets* 2-3, light brown, their *glumes* densely hairy. *Fertile spikelets* 2-3, distant, pedunculate, erect. *Glumes* lanceolate, as long as the fruit, pale, membranous, with a few scattered hairs and a green central stripe and strong midrib elongated into a strong though slender serrulate awn, about one third the length of the glume. *Perigynium* with numerous ribs and clothed with white hairs.

Var. β. Fertile spikelets compound. *E. Fl. l. c.*
Golden pool, near Pitchford ; June, 1840.

†† *Fruit glabrous.*

34. C. **ampullacea**, Gooden. *Slender-beaked Bottle Carex.* Sheaths none ; bracteas foliaceous, long ; fertile spikelets 2-3, cylindrical, long, nearly erect ; glumes lanceolate, acute, shorter than the fruit ; fruit crowded, spreading ; perigynium stipitate, subglobose, inflated, with a long narrow beak, bifid at the point, smooth, obscurely ribbed ; nut obovate, triquetrous, minutely reticulated. *E. Bot. t.* 780. *E. Fl. v. iv. p.* 124. *Hook. Br. Fl. 4th ed. p.* 343.

Bogs, marshes, and margins of pools ; not uncommon. *Fl.* May, June. ♃.
Berrington, Cound, Shomere, Shrawardine, Hancott, and Oxon pools ; *Rev. E. Williams's MSS.*
Snowdon pool. Lodge Lake, near Tong ; *H. Bidwell, Esq.* Ellesmere ; *Rev. A. Bloxam.* Betton pool ; *Mr. F. Dickinson.* Astley ; *Mr. E. Elsmere. junr.!* Near Oswestry ; *Rev. T. Salwey.!*
Bomere pool. Hancott pool. Mare pool. Ellesmere and Colemere meres. Morda pool, near Oswestry. Golden pool, near Pitchford.

Readily distinguished by the shape of the perigynium which resembles an antique bottle or flask, (*ampulla.*) *Root* creeping. *Stem* 1-2 feet high, erect, smooth, rough upwards, obsoletely triangular. *Leaves* long and narrow, acute, glaucous, channelled, rough at the edges and keel near the extremity, the radical ones frequently filamentous at the base. *Bracteas* long and narrow, lowermost over-topping the spike. *Barren spikelets* 2-3. *Fertile spikelets* 2-3, on short smooth triquetrous peduncles, slightly drooping. *Fruit* longer than the glumes, moderately spreading, arranged in 8 or 9 rows. *Perigynium* smooth, yellowish-brown, with 2 strong marginal ribs and 5—7 obscure ribs on one side and 4-5 on the other. *Glumes* narrow, lanceolate, acute, brownish, with a strong midrib.

35. C. **vesicaria**, Linn. *Short-spiked Bladder Carex.* Sheaths none ; bracteas foliaceous, long ; fertile spikelets 2-3, oblongo-cylindrical, slightly drooping ; glumes lanceolate, acuminate, shorter than the fruit ; fruit lax, spreading ; perigynium substipitate, ovato-oblong, inflated, with a conical beak bifid at the point, smooth, ribbed ; nut elliptical, triquetrous, minutely granulated. *E. Bot. t.* 779. *E. Fl. v. iv. p.* 123. *Hook. Br. Fl. 4th ed. p.* 343.

Bogs and marshes ; not uncommon. *Fl.* May, June. ♃.
Snowdon pool ; *H. Bidwell, Esq.* Astley ; *Mr. E. Elsmere, junr.!*
Bomere pool. Hancott pool. Mare pool. Shawbury heath. Golden pool, near Pitchford.

Root creeping. *Stem* 2 feet high, erect, with 3 very sharp rough angles. *Leaves* erect, narrow, acuminate, rough on the edges and keel beneath. *Bracteas* foliaceous, very narrow, the lowermost rising above the stem. *Barren spikelets* 2-3. *Fertile spikelets* 2-3 or 4, on very short smooth triquetrous peduncles, slightly drooping, pale yellow. *Glumes* membranous, brown, with a prominent green midrib. *Perigynium* with 2 strong marginal ribs and 5 others on each face, margins of the beak rough. *Nut* small, not near filling the cavity of the inflated perigynium, crowned with the long permanent style.

36. C. **paludosa**, Gooden. *Lesser Common Carex.* Sheaths

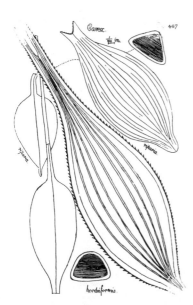

none ; bracteas very long, foliaceous ; glumes of the sterile spikelets obtuse ; fertile spikelets 3, cylindrical, obtuse, erect, shortly pedunculate ; glumes lanceolate, cuspidate, as long as the fruit ; fruit erecto-patent ; perigynium stipitate, oblongo-ovate, with a short bifid beak, compressed, triquetrous, many-ribbed ; nut rotundo-obovate, triquetrous, minutely granulated. *E. Bot. t.* 807. *E. Fl. v. iv. p.* 120. *Hook. Br. Fl. 4th ed. p.* 344. "*C. acuta, Curt.*"

Banks of rivers and ditches. *Fl.* May. ♃.
Old bed of River Severn below Cross Hill, near Shrewsbury. Twyford Vownog, near Westfelton.

Root creeping. *Stem* 2 feet high, erect, unequally triquetrous, angles rough. *Leaves* rather broad, erect, rough at the edges and keel. *Bracteas* without sheaths, scarcely auricled, lowermost large overtopping the stem. *Barren spikelets* 2 or 3. *Fertile spikelets* 3, frequently with barren flowers at the extremities, slightly pedunculate. *Perigynium* green, smooth, with 2 marginal ribs and 7-8 other prominent ones on each face. *Glumes* brown with a green midrib.

37. C. **riparia**, Curt. *Great Common Carex.* Sheaths none ; bracteas very long, foliaceous ; glumes of the sterile spikelets acuminated ; fertile spikelets 3—5, scarcely pedunculated, broadly cylindrical, acute ; glumes cuspidate ; perigynium stipitate, oblongo-ovate, subtriquetrous, tumid, finely ribbed, acuminated into a short broad cloven beak, orifice spinulose ; nut pyriform, attenuated at the base, trigonous, elevato-punctate. *E. Bot. t.* 579. *E. Fl. v. iv. p.* 121. *Hook. Br. Fl. 4th ed. p.* 344.

Sides of rivers and ditches. *Fl.* May. ♃.
Weston, near Oswestry. Maesbury Marsh and between the Holyhead Road and Maesbury Marsh, by the side of the Canal near Oswestry ; *Rev. T. Salwey.!*

Root extensively creeping. *Stem* 3 feet high, erect, with 3 sharp very rough angles. *Leaves* very broad, erect, acuminate, rough on the edges and keel, deep green. *Bracteas* foliaceous, very large, rising above the stem, with short sheaths and pale round membranous auricles. *Barren spikelets* 3—5. *Fertile spikelets* 3-4, erect, on short peduncles, cylindrical, very long and thick. *Perigynium* with numerous fine ribs, 11 on the plane side and 11—13 on the triquetrous one. *Nut* pale yellow.

MONŒCIA—TETRANDRIA.

7. LITTORELLA. *Linn.* Shore-weed.

1. L. **lacustris**, Linn. *Plantain Shore-weed.* *E. Bot. t.* 468. *E. Fl. v. iv. p.* 130. *Hook. Br. Fl. p.* 406. *Plantago uniflora, Linn.*

Margins of pools, &c. ; not common. *Fl.* June. ♃.
Ellesmere mere below Otley ; *Rev. A. Bloxam.* Bomere and Betton pools. A pool on the Perthy common, near Ellesmere. Croesmere. Whitemere ; *Rev. E. Williams's MSS.*
Colemere and Ellesmere meres.

Root fleshy, fibrous, with numerous horizontal runners. *Stem* none. *Leaves* all radical, linear, ascending, entire, fleshy, semicylindrical, dilated very considerably at the base with a membranous border. *Scapes* several. *Male flower* solitary on a simple erect scape 2-3 inches long, with a membranous *bractea* above the base, arising from the axils of the leaves. *Calyx* persistent, 4-partite ; segments equal, lanceolate, acute, about as long as the tube of the corolla. *Corolla* funnel-shaped, with a long cylindrical tube, limb in 4 equal acute segments. *Stamens* 4 inserted in the base of the calyx, *filaments* elongated very much exserted, *anthers* incumbent. *Female flowers* in pairs, sessile in the axils of the leaves, at the base of the scapes of the male flowers, bracteated. *Calyx* persistent, 4-partite; segments lanceolate, acute, erect, slightly shorter than the corolla. *Corolla* tubular, contracted at the apex which has 4 minute acute teeth. *Style* simple, filiform, elongated; *stigma* pubescent. *Fruit* an oblong 1-celled nut, slightly compressed, 2-edged, deeply and transversely rugose.

Var. β. hirsutus. Leaves hairy, chiefly on the upper surface. *E. Fl. l. c.*

On the east shore of Bomere pool, near the watering place.

Whole plant smaller, with much narrower leaves, considerably more dilated and membranous at the base in proportion to the size.

8. ALNUS. *Tourn.* Alder.

1. A. *glutinosa,* Gærtn. *Common Alder.* Leaves rotundo-cuneiform, shallowly lobed, obtusely and irregularly serrated, somewhat glutinous, downy in the axils of the nerves beneath. *E. Fl. v. iv. p.* 131. *Hook. Br. Fl. p.* 407. *Betula Alnus, Linn. E. Bot. t.* 1508.

Moist meadows, and banks of rivers and pools; frequent. *Fl.* March, April. ♄.

Stiperstones Hill. Shomere Bog. Banks of Severn about Shrewsbury. Morda Pool, near Oswestry. Twyford Vownog, near Westfelton. Golden Pool, near Pitchford. Cound Brook.

A middle-sized *tree*, not very ornamental though of agreeable verdure and dense shade. *Bark* nearly black and full of clefts. *Wood* pale flesh-colour, soft, tender and without much tenacity, of great durability under water, used for turning and cabinet-making, very subject to be perforated by a small beetle. *Branches* crooked, spreading, round and smooth, glutinous when young. *Leaves* rotundo-cuneiform, margin shallowly lobed, obtusely and irregularly serrated, glutinous, of a deep shining green, paler and downy in the axils of the transverse parallel nerves beneath. *Inflorescence* in a compound raceme of male and female catkins proceeding from the extremities of the branches. *Male catkin* elongated, cylindrical, *bracteas* stipitate and peltate, 3-flowered. *Perianth* 4-partite. *Anthers* 4, filaments very short. *Female catkins* short, ovate or oblong, *bracteas* cuneiform, 2-flowered. *Perianth* of 4 adnate scales. *Germens* 2, 2-celled. *Fruit* a strobilus; *nuts* small, ovate, compressed, angular, 1-seeded by abortion.

This tree is rather common on the banks of the River Severn near Shrewsbury, and in very early times must have been very abundant, since the ancient British name of that town, Pengwern, (*pen, head* or *hill*, and *gwern*, an *alder*,) was, according to Leland, assigned to it from the circumstance: "Accipiens patria sibi lingua nomen ab *alnis*."

9. BUXUS. *Linn.* Box.

1. B. *sempervirens,* Linn. *Common Box-tree.* Leaves oval-oblong, retuse, convex, coriaceous, shining above, opaque beneath, their petioles slightly hairy; anthers ovato-sagittate. *E. Bot. t.* 1341. *E. Fl. v. iv. p.* 133. *Hook. Br. Fl. p.* 407.

Limestone districts; rare. *Fl.* April. ♄.

Woods and thickets on Tinker's Hill, near Ludlow; *Miss Mc Ghie.*

A small shrubby *tree* of slow growth; *wood* yellowish, hard and compact, highly valuable for carving and engraving upon. *Bark* pale, longitudinally cracked. *Leaves* opposite, oval-oblong, retuse, convex, coriaceous, dark shining

ever-green above, paler and minutely dotted beneath, consisting of two lamina free in the middle and united at the margins, on short petioles. *Inflorescence* in axillary clusters, yellow, several male flowers surrounding a solitary terminal female one. *Male perianth* of 4 roundish obtuse concave unequal leaflets, 2 interior and 2 exterior ones opposite, subtended by a single *bractea*. *Stamens* 4. *Female perianth* similar, but subtended by 3 *bracteas*. *Germen* 3-celled. *Styles* 3. *Capsule* globose, beaked with the 3 persistent styles, coriaceous. *Seeds* 2 in each cell, oblong, subtriquetrous.

10. URTICA. *Linn.* Nettle.

1. U. *pilulifera,* Linn. *Roman Nettle.* Leaves opposite, broadly ovate or cordate, acuminate, coarsely dentate; stipules oblongo-ovate; fertile flowers in globose pedunculate clusters; seeds punctato-tuberculate. *Babington in Ann. Nat. Hist. v. i. E. Bot. t.* 148. *E. Fl. v. iv. p.* 134. *Hook. Br. Fl. p.* 407. "*Linn. Sp. Pl.* 1395. *Wallr. Sched. Crit.* 488. *Reich. Fl. Excurs.* 1105. *Koch. Syn.* 635."

Waste and cultivated ground; rare. *Fl.* June, July. ⊙.

Garden at Eaton Mascott, but very rare; *Rev. E. Williams's MSS.*

Stem erect, bluntly quadrangular. *Leaves* broadly ovate, usually cordate at the base, the margins deeply cut into large divaricated teeth, on long thick petioles. *Stipules* oblongo-ovate. *Peduncles* very short, scarcely exceeding the diameter of the large globular heads. *Seed* dark-brown with numerous darker prominent points, rather opaque.

I have not seen Shropshire specimens of *U. pilulifera.*

2. U. *urens,* Linn. *Small Nettle.* Leaves opposite, elliptical, with about 5 nearly parallel ribs; stipules linear, reflexed; clusters of flowers oblong, subsimple. *E. Bot. t.* 1236. *E. Fl. v. iv. p.* 134. *Hook. Br. Fl. p.* 407.

Waste and cultivated ground; not uncommon. *Fl.* June—October. ⊙.

Ludlow; *Miss Mc Ghie.* Eyton and Walford; *T. C. Eyton, Esq.* Astley; *Mr. E. Elsmere, junr.*

Uffington, Shrewsbury, &c.

3. U. *dioica,* Linn. *Great Nettle.* Leaves opposite, ovato-acuminate, cordate, coarsely deeply and acutely serrated; stipules lanceolate, ciliated; clusters of flowers much branched, in pairs, mostly diœcious. *E. Bot. t.* 1750. *E. Fl. v. iv. p.* 135. *Hook. Br. Fl. p.* 407.

Waste places, &c.; frequent. *Fl.* July, August. ♃.

Root fleshy and creeping. *Stems* 1-2 feet high, erect, simple, tetragonous, angles round, interstices furrowed, hispid with short decurved bristles interspersed with longer perforated ones arising from tubercles containing an irritating venomous liquid, which upon pressure is transmitted through the tubular bristle, and causes a stinging sensation on that part of the skin touched. *Leaves* in pairs, opposite, ovato-acuminate, cordate at the base, coarsely deeply and acutely serrated, hairy and bristly; petioles much shorter than the leaves, channelled, armed with venomous bristles and short erect hairs. *Stipules* lanceolate, ciliated. *Flowers* in long alternately branched pubescent clusters, two arising from the axil of each leaf, those of the male plant erect or horizontally spreading, of the female plant reflexed. *Male flowers* either single or 2 or 3 together, on short pubescent pedicels, with a minute ovate toothed pubescent bractea at the base. *Sepals* 4, roundish, mem-

branous at the obtuse apex, hairy. *Filaments* opposite the sepals, transversely plicate, at first inflexed, in the fully expanded blossom becoming spreading and reflexed by an elastic jerk, which bursts the anthers and scatters the globose pollen in a copious cloud. *Rudimentary pistil* globose, shortly pedicellate, perforated. *Female flowers* more densely clustered on the branches of the raceme, sessile, sometimes with a minute *bractea* at the base. *Sepals* 4, unequal, erect, hairy, 2 opposite ones larger round enveloping the fruit, 2 smaller, oblong. *Fruit* ovate, compressed, minutely dotted.

Mr. G. Jorden writes "on the summit of the Titterstone Clee Hill, (1730 feet) I found the nettle to flourish most luxuriantly at an altitude where few plants will grow." The highest station in which I have noticed it is on the summit of Pontesford Hill (950 feet.)

MONŒCIA—PENTANDRIA.

11. BRYONIA. *Linn.* Bryony.

1. B. *dioica,* Jacq. *Red-berried Bryony.* Leaves palmato-quinquelobed, rough on both sides; flowers dioecious, berries red. *E. Bot. t.* 439. *E. Fl. v. iv. p.* 138. *Hook. Br. Fl. p.* 408.

Thickets and hedges; frequent. *Fl.* May—September. ♃.

Root very large, fusiform, fleshy. *Male plant*:—*Stem* herbaceous, annual, rough, angular and furrowed, more or less branched, climbing by tendrils to the height of many feet. *Leaves* alternate, on a short rounded petiole, palmato-quinquefid, central lobe elongated mucronate, lobes angulato-dentate, 5-ribbed at the base, bright pale green, roughish on both sides and on the margins with small curved bristles arising from tubercular bases. *Tendrils* opposite to the flowers, very long and simple, spirally curved after contact. *Flowers* in axillary panicles; peduncles longer than the leaves, pubescent, pedicels short. *Calyx* monophyllous, adnate to the corolla, 5-fid, segments triangulari-subulate. *Corolla* yellowish-white, reticulated with green veins, 5-fid; segments ovato-lanceolate, obtuse, spreading, pubescent, 5-ribbed. *Filaments* 3, short, hairy. *Anthers* 5, 2 combined on 2 of the filaments, the 5th solitary, their line of dehiscence bordered on each side with a row of semipellucid green globules surmounted by a short erect bristle. *Female plant*:—*Stem* smooth, angular and furrowed. *Leaves* alternate, petiolate, cordato-quinque-lobed, terminal lobe cuspidate, all mucronate, margins subdentate, clothed on both sides with curved bristles or tubercles, but not rough as in the male plant, 5-ribbed at the base. *Flowers* axillary, nearly sessile, on a very short peduncle bearing 3 or 4 longer single-flowered pedicels, shorter than the leaves. *Calyx* pedunculate, on the summit of the berry, monophyllous, adnate to the corolla, 5-fid, segments subulate acute recurved. *Corolla* yellowish-white, with 3 parallel green veins, 5-fid, segments lanceolate acute recurved pubescent. *Style* glabrous, *stigma* 2-3-cleft, laciniato-papillose. *Berry* globose, glabrous, scarlet.

MONŒCIA—POLYANDRIA.

12. CERATOPHYLLUM. *Linn.* Hornwort.

1. C. *demersum,* Linn. *Common Hornwort.* Fruit armed with 2 spines near the base and terminated by the curved subulate style. *E. Bot. t.* 947. *E. Fl. v. iv. p.* 141. *Hook. Br. Fl. p.* 409.

Ponds, ditches and pools! rare. *Fl.* July. ♃.

Pond at Pitchford; *Rev. E. Williams's MSS.*

Shrawardine Pool. Ellesmere mere.

Herb floating entirely under water, dark green, long, slender, copiously

branched, densely clothed with whorled setaceous 2 or 3 times forked *leaves*; segments fine linear acute, the ultimate ones distantly dentato-serrate on the inferior margin. *Flowers* axillary, small, whorled, sessile in the axils of the leaves, pale green. *Anthers* sessile, crowded, spotted, 2-beaked, 2-celled. *Fruit* armed with 2 spreading lateral spines and a terminal one from the elongated style, all variable in length.

13. MYRIOPHYLLUM. *Linn.* Water-Milfoil.

1. M. *spicatum,* Linn. *Spiked Water-Milfoil.* Leaves whorled, pectinato-pinnate, segments capillary, nearly opposite; male flowers in a long interrupted leafless many-whorled erect spike. *E. Bot. t.* 83. *E. Fl. v. iv. p.* 143. *Hook. Br. Fl. p.* 409.

Ponds and ditches, frequent. *Fl.* July, August. ♃.

Newport; *R. G. Higgins, Esq.* Oakley Park, near Ludlow; *Mr. H. Spare.* River Severn, near Shrewsbury.

Herb smooth, floating under water, with branching round stems, dark green. *Leaves* 4 in a whorl, large. *Inferior bracteas* incised, equalling or slightly exceeding the whorl, all the rest entire shorter than the whorl.

2. M. *alterniflorum,* DC. *Alternate-leaved Water-Milfoil.* Leaves whorled, pinnati-partite; segments capillary, distant, alternate or opposite; male flowers in a naked spike about 6-flowered, nodding before flowering; female flowers about 3 together, arranged in axillary whorls at the base of the male spike. *Babington Prim. Fl. Sarn. p.* 36. *Reich. Fl. Excurs. n.* 4073. *Koch Syn.* 244. "*D. C. Bot. Gall.* 190."

Ponds and ditches. *Fl.* July. ♃.

Discovered by the *Rev. A. Bloxam*, in a pond by the side of the Canal near Whixall Moss about 6 miles from Ellesmere, and near Colemere.

Berrington Pool, near Shrewsbury. Plentifully in the ditches on the Twyford Vownog, near Westfelton. Morda Pool, near Oswestry. Colemere and Ellesmere Meres.

Much slenderer and more copiously branched in the upper portion, and with much finer more delicate and smaller *leaves* than the preceding. *Inferior floral leaves* pinnately divided and about as large as the other leaves; *upper ones* entire, shorter than the flowers. *Female flowers* rarely alternate.

"This species grows very abundantly in the Isle of Wight. I cannot find in the few authors who have described this plant (which appears to have been well known even on the continent only within these few years) any allusion to a most essential distinguishing character, that of having only 3 instead of 4 leaves in a whorl, sometimes they are opposite or binary, never, so far as I have observed, quaternary; neither De Candolle, Koch nor Loiseleur say a word on the subject. That *M. alterniflorum* was not unknown to our old English herbalists I think I am fully prepared to prove, by reference to Petiver's British Herbal designed to illustrate the plants of Ray, where in tab. vi. fig. 6. is a rude but not inexpressive portrait of our plant under the name of "*Fine feather Pondweed*," and in which the ternary arrangement of the leaves is partially but clearly expressed, though most of the leaves are drawn opposite or in pairs, but the appearance the small naked spike so often assumes, the male flowers having fallen away leaving only a solitary bractea, and the single or sometimes double whorl of fertile flowers at its base remaining, is very characteristically depicted. The *M. spicatum*, β. of *Eng. Flora* is, I suspect, our *alterniflorum* also, and I am inclined to think that a broad-leaved form of the same plant found in little pools near Yarmouth, Isle of Wight, in which the segments of the leaves are broad fleshy and a little obtuse, not capillary, and of a much brighter green colour than in the normal state, is that represented in Morison's figure referred to by the same author. *M. alterniflorum* seems restricted to certain parts of the continent only; it is common according to the French Botanists in the west of France, but in Germany has as yet been detected only along the course of the Rhine and a few of its tributaries."—*Dr. Bromfield in lit.*

3. M. *verticillatum,* Linn. *Whorled Water-Milfoil.* Leaves whorled, pinnatipartite, segments setaceous; flowers in axillary whorls. *E. Bot. t.* 218. *E. Fl. v. iv. p.* 143. *Hook. Br. Fl. p.* 409.

Ponds and ditches; not common. *Fl.*

Canal, near Ludlow; *Miss Mc Ghie.* Ditches below Eyton on the Wildmoors. In a pond between Golden and the Watling Street Road. At the west end of Blackmere, near Ellesmere; *Rev. E. Williams's MSS.*

14. SAGITTARIA. *Linn.* Arrow-head.

1. S. *sagittifolia,* Linn. *Common Arrow-head.* Leaves sagittate, lobes acuminate; scapes simple. *E. Bot. t.* 84. *E. Fl. v. iv. p.* 144. *Hook. Br. Fl. p.* 409.

Ditches and margins of rivers, &c.; not common. *Fl.* July, August. ♃.

Wildmoors, near Eyton; *T. C. Eyton, Esq.* Ludlow; *Miss Mc Ghie.*

Canal between Shrewsbury and Uffington, and pit at Uffington.

Root tuberous, with long fibres and bulbiferous runners. *Herb* milky and glabrous. *Leaves* all radical, on long triquetrous striated internally cellular petioles tapering upwards, submersed ones linear obtuse, the rest sagittate, lobes nearly equal and very much acuminate, 2 of them pointing downwards and one upwards, with parallel veins connected by transverse diagonal reticulations. *Scape* a foot or more high, erect, simple, angular, furrowed, internally cellular. *Flowers* in whorls of 3, in the upper part of the scape on short pedicels, those of the male flowers much the longest, subtended by 3 ovate acute membranous combined *bracteas*. *Male flowers* superior. *Calyx* of 3 ovate, obtuse, concave, membranous, persistent sepals. *Petals* 3, alternate with and larger than the sepals, round, spreading, somewhat decurved, deciduous, white with a short violet purple claw. *Filaments* short; *anthers* violet, subsagittate. *Pistils* 1—5. *Female flowers* inferior, with a similar calyx and corolla. *Germens* numerous, in a globose head. *Style* short; *stigma* simple, obtuse, papillose. *Receptacle* subglobose, fleshy. *Fruit* obovate, compressed, surrounded with a broad dilated gibbous membranous margin and crowned with a short recurved beak, one-seeded.

15. ARUM. *Linn.* Cuckoo-pint.

1. A. *maculatum,* Linn. *Cuckoo-pint or Wake-robin.* Leaves all radical, hastato-sagittate, lobes deflexed; spadix club-shaped, obtuse, shorter than the spatha. *E. Bot. t.* 1298. *E. Fl. v. iv. p.* 148. *Hook. Br. Fl. p.* 411.

Hedge-banks, thickets, &c.; frequent. *Fl.* April, May. ♃.

Root tuberous. *Leaves* petiolate, hastato-sagittate, acute, erect, shining green, often spotted with black. *Spatha* large, erect, oblong, convoluted at the base, pale green. *Spadix* clavate, naked, dark purple, marcescent, bearing at the base a ring of pointed spreading filaments (apparently imperfect *germens.*) Below these is a ring or circle of 2-celled sessile *anthers*, and below these the sessile *germens*. *Berries* scarlet, crowded into an oblong spike, remaining through the winter after the rest of the plant has decayed.

16. POTERIUM. *Linn.* Salad-Burnet.

1. P. *Sanguisorba,* Linn. *Common Salad-Burnet.* Spines none; stem somewhat angular. *E. Bot. t.* 860. *E. Fl. v. iv. p.* 147. *Hook. Br. Fl. p.* 410.

Limestone districts; frequent. *Fl.* July. ♃.

Oreton, plentifully; *Mr. G. Jorden.* Ludlow; *Miss Mc Ghie.* Benthal Edge; *Dr. G. Lloyd.* Gleaton Hill, near Wenlock; *W. P. Brookes, Esq.* By the side of the turnpike road about Presthope, near Wenlock. About Farley, near Wenlock; *Rev. E. Williams's MSS.*

Llanymynech Hill. Blodwell Rocks.

Stem 1 foot high, erect, angular, smooth, branched. *Leaves* pinnate; *leaflets* petiolate, ovate or rounded, deeply serrated, the terminal tooth much shorter and smaller than the rest, glaucous green and glabrous above, paler beneath, the nerves hairy, having the smell and taste of cucumbers. *Flowers* dull-purple, segments of the perianth with a minute downy tuft at the apex. *Stigmas* bright crimson, very elegant.

17. QUERCUS. *Linn.* Oak.

I have great pleasure in acknowledging my obligations to Professors Don and Graham for the determination of our three forms of Oak. Both these eminent Botanists are inclined to adopt the opinion of those who regard our British Oaks as varieties only; inasmuch as the leaves vary without the least reference to the absence or presence or relative length of the peduncle, but they consider that a series of careful observations and experiments are requisite to settle this question in a satisfactory manner. In the meantime as the three forms have been long distinguished and the claim of each to be regarded as a distinct species is equal, Professor Don has drawn up the following characters by which they will be readily recognized whether as species or varieties.

1. Q. *Robur,* Linn. *Common British Oak.* "Young branches glabrous; leaves on short footstalks, cuneately oblong, pinnatifid, slightly pubescent beneath, lobes oblong rounded, with deep narrow somewhat acute sinuses, bases biauriculate equal; female catkins on long peduncules; fruit oblong." *Prof. D. Don. E. Bot. t.* 1342. *Martyn Fl. Rust. t.* 10. *E. Fl. v. iv. p.* 148. *Hook. Br. Fl. p.* 410. *Hook. Scot. p.* 237. *Mack. Fl. Hib. p.* 255. *Lindl. Syn. 2nd ed. p.* 240. *Johnst. Fl. Berw. vol.* 1. *p.* 206. *a. Huds. Fl. Angl. 2nd ed. p.* 421. *Q. latifolia, Raii Syn. 2nd ed. p.* 286. *3rd. ed. p.* 400. *Q. femina, With. 3rd ed. p.* 387. *Loch Lomond Oak No.* 1. *of Dr. Graham in 1st Ann. Rep. Edin. Bot. Soc. p.* 35. *Q. pedunculata, Willd. and foreign Authors.*

Woods and hedges. *Fl.* April, May. ♄.

Kemberton and Shiffnal, abundant; *Mr. F. Dickinson.* Ludlow; *Miss Mc Ghie.* Wyre Forest; *Loudon Arbor. Brit.*

Nesscliff hill in extensive masses, certainly wild.

Scales of the *cupula* ovate, obtuse, glabrous, minutely ciliated at the margins.

2. Q. *intermedia,* Don. *Intermediate Oak.* "Young branches glabrous; leaves on long footstalks, cuneately oblong, slightly pinnatifid, glaucous and copiously clothed with fine starry pubescence beneath, lobes short rounded, sinuses shallow spreading obtuse; base

obtuse unequal; female catkins on very short peduncles; fruit oblong." *Profr. D. Don. Norwood Oak, Martyn Fl. Rust. t.* 11. *Q. latifolia mas, quæ brevi pediculo est. Raii Syn. 2d ed. p.* 286. *3d ed. p.* 440. *Q. Robur, β. Huds. Fl. Angl. p.* 421. *Q. Robur, With. 3d ed. p.* 387. *Q. sessiliflora, β. Sm. E. Fl. v. iv. p.* 150. (excl. syn.) *Quercus "near to Robur." Smith. Herb. Loch Lomond Oak No.* 2. *of Dr. Graham, in 1st Ann. Rep. Edin. Bot. Soc. p.* 35. (according to a specimen! from Dr. Graham.)

Hilly woods. *Fl.* April, May. ♄.

On the western summit and base of Nesscliff hill in considerable masses, decidedly wild. Treflach wood near Oswestry, in an isolated copse at the west base of Sweeney Mountain, immediately below the tower.

With us a small stunted *tree. Leaves* alternate, petiolate, obtuse, the lobes larger towards the middle, more numerous and smaller towards the apex which as well as the base is somewhat tapered, deep shining green and glabrous above, the under surface of a pale hoary green arising from dense short stellate pubescence, the sides of the midrib and principal ribs are also clothed with stellate clusters of pale hairs. In the young leaves when examined in May, the petiole is very much longer and the ribs on the upper surface are covered with stellate pubescence which disappears in a more advanced state. *Fruit* in axillary clusters of 3-4 small seldom perfected acorns. *Scales* of the *cupula* rhomboid, with erect appressed pubescence and scariose ciliated margins.

"I fear the confusion about the British forms of *Quercus* is increased by Smith. Martyn seems to have had tolerably precise notions regarding three forms; which you will understand by my using the names *pedunculata, intermedia,* and *sessiliflora*. The first and third you will not mistake my meaning in referring to:—the first having a largely developed peduncle, and the third, according to Martyn, absolutely no peduncle; the second has one short distorted, and is I feel sure your Nesscliff Oak and the specimen I send to you from Loch Lomond. It is the Norwood oak of Martyn's *Fl. Rust. t.* 11. and not his Durmast, *t.* 12, which is the name given according to him in the New Forest to the last of the three (*Q. sessiliflora.*) Smith in *Eng. Bot.* considered the second and third identical, and in *E. Fl.* he strangely considered the second the type of *sessiliflora,* and the third the *β.* of that form, though he thus made *a.* to be the *intermedia,* and *β.* the plant to which alone the name could apply." *Prof. Graham in lit.* Professor Don thus writes: "There is no doubt that your Oak from Nesscliff is *Martyn's tab.* 11, and not *t.* 12. I cannot account for Smith's mistake in quoting the Durmast as *β,* instead of *a. sessiliflora,* as the specimen marked "*sessiliflora*" in his herbarium is correct. There are likewise specimens of *intermedia,* or the Norwood oak of Martyn, in the same herbarium, with the words in Smith's hand-writing "near to Robur" on the paper. These specimens are from Holkham, in Norfolk."

For my first acquaintance with this oak I am indebted to my friend Dovaston, whose acuteness of observation detected it in very early life among the other oaks at Nesscliff.

3. Q. *sessiliflora,* Salisb. *Sessile-fruited Oak.* "Young branches pubescent; leaves on long footstalks, oblong, pinnatifid, glabrous beneath, lobes ovate-oblong obtuse, sinuses rather deep forming a somewhat acute angle, base unequal obtuse or frequently more or less attenuated; female catkins sessile; fruit ovate." *Prof. D. Don. E. Bot. t.* 1845. *Durmast Oak, Martyn Fl. Rust. t.* 12. *Hook. Scot. p.* 273. *Lindl. Syn. 2nd ed. p.* 240. *Johnst. Fl. Berw. p.* 207. *Loch Lomond Oak No.* 3. *of Dr. Graham in 1st Ann. Rep.*

Edin. Bot. Soc. p. 35. *Q. sessiliflora, a. Sm. E. Fl. v. iv. p.* 150. (excl. syn.) *"Q. sessiliflora," Smith. Herb. Q. Robur, Willd. and most foreign authors.*

Hilly woods. *Fl.* April, May. ♄.

Coalbrookdale woods, abundant; *Mr. F. Dickinson.* Ludlow; *Miss Mc Ghie.* About the Bwlthy, Bauseley, and slopes of hills on the Welsh border; *J. F. M. Dovaston, Esq.* Wyre Forest; *Loudon Arbor. Brit.*

Mr. W. Atkinson in "Observations on the quality of the Oak timber produced in Great Britain," (*Hort. Trans. n. s. vol.* 1. *p.* 336.) considers *Q. pedunculata* (*Q. Robur, Linn.*) and *Q. sessiliflora* to be the only species native to Great Britain and Ireland. The Durmast or Dunmast Oak of the New Forest (of Martyn) he concludes from an examination of numerous specimens and travel through a portion of the forest itself to be only an accidental variety, with dark or dun-coloured acorns, arising probably from some peculiarity in the soil, particularly as he has found such acorns on trees both of *pedunculata* and *sessiliflora,* and has invariably raised those species from such acorns. *Q. pedunculata* has acorns on long stalks and leaves with very short footstalks or none. It contains a great quantity of silver grain and consequently splits clean and easy and is adapted for split paling and lath. It is also a stiffer wood, and though it may be broken with a less weight than that of *Q. sessiliflora,* yet it requires a much greater weight to bend it, and is therefore best calculated for beams or to bear the greatest weight without bending. *Q. sessiliflora* has acorns which sit close to the branch having hardly any stalks and leaves with footstalks ½ to 1 inch long. It contains so small a portion of silver grain, that wood of this species from old buildings, as for instance the roof of Westminster Hall, and the oldest buildings about London generally, has been mistaken for sweet Chestnut (*Castanea vesca*) but on examination he has found it to be in reality, *Q. sessiliflora.* The black oak of our bogs is also of this species: and he hence concludes that the chief part of our natural woods was of this species, some centuries ago, though now *Q. pedunculata* occurs chiefly in the woods in the south, whilst *Q. sessiliflora* is more abundant in the north of England. The wood of *Q. sessiliflora* bends from a weight much sooner than that of *Q. pedunculata,* but requires a much greater weight to break it, and from its toughness is best calculated for ship timber. Both species grow equally well, but *sessiliflora* is a handsomer tree in its foliage, and probably the wood is more durable.

18. FAGUS. *Linn.* Beech.

1. F. *sylvatica,* Linn. *Common Beech.* Leaves ovate, glabrous, obsoletely dentate, their margins ciliated. *E. Bot. t.* 1846. *E. Fl. v. iv. p.* 152. *Hook. Br. Fl. p.* 411.

Woods; doubtful if wild. *Fl.* April, May. ♄.

Haughmond hill, doubtfully wild.

A large and noble *tree* of combined majesty and elegance; *branches* widely spreading, umbrageous; *bark* pale and smooth; *wood* white and brittle, of considerable durability under water, but soon falling to decay when exposed. *Leaves* on short petioles, alternate, ovate, obsoletely dentate, glabrous and shining above, paler beneath, the nerves and their axils downy; when young delicately thin and semipellucid, the margins and also the nerves on both surfaces fringed with long silky hairs:—"the silken green of the young beech leaves in May." *Male flowers* in globose racemose catkins pendulous on long silky peduncles. *Perianth* campanulate, 5-6-fid, silky. *Stamens* 8—12, on long exserted filaments. *Female flowers* above the male in subglobose catkins on shorter stout erect silky peduncles. *Involucre* urceolate, 4-lobed, clothed with numerous unequal linear downy *bracteas,* enclosing two triquetrous 3-celled *germens,* each invested with a silky *perianth* which is elongated upwards into a laciniate limb. *Styles* 3. *Fruit* formed of the 4-cleft somewhat woody involucre, externally clothed with simple pliant prickles. *Nuts* 2, acutely triquetrous, crowned with the silky segments of the perianth, 1-seeded. *Pericarp* coriaceous, smooth, internally hairy. The seedlings with their pale thick fleshy cotyledons are very singular in appearance.

19. CASTANEA. *Tourn.* Chestnut.

1. C. *vulgaris*, Lam. *Spanish Chestnut.* Leaves oblongo-lanceolate, acuminate, mucronato-serrate; glabrous on each side. *Hook. Br. Fl. p.* 411. *Fagus Castanea*, Linn. *E. Bot. t.* 886. *E. Fl. v. iv. p.* 151.

Woods, apparently wild. *Fl.* July, August. ♄.
About Alveley; *Purton's Midl. Flora.* Old trees about Ludlow; *Miss Mc Ghie.* About Hardwick near Ellesmere, and in Pitchford Park; *Rev. E. Williams's MSS.*

A stately and majestic *tree*; *bark* with remarkably deep clefts; wood of extreme durability said to have been formerly used extensively in buildings; (see however observations under *Quercus sessiliflora*) branches widely spreading, round and smooth when young. *Leaves* oblongo-lanceolate, acuminate, mucronato-serrate, glabrous, rich shining green above, paler beneath. *Male flowers* in axillary solitary yellow elongated pendulous racemose catkins of numerous globose sessile bracteated clusters. *Female flowers* in small subglobose catkins, either subsessile at the base of the male catkins or solitary in the axils of the leaves. *Involucre* campanulate, 4-partite, clothed with numerous unequal bracteas, enclosing the 3 *germens*, each of which is invested with the downy 6-8-fid *perianth*. *Fruit* formed of the 4-cleft somewhat woody involucre, externally covered with complicated sharp prickles. *Nuts* large, 2-3, broadly ovate, convex on one side, plane on the other, attached to the involucre by a broad basal scar, invested with the perianth.

20. BETULA. *Linn.* Birch.

1. B. *alba*, Linn. *Common Birch.* Leaves ovato-deltoid, acuminate, strongly unequally and doubly serrated, glabrous. *E. Bot. t.* 2198. *E. Fl. v. iv. p.* 153. *Hook. Br. Fl. p.* 411.

Hilly woods and in boggy soils; frequent. *Fl.* April, May. ♄.
Tick wood; *Mr. F. Dickinson.* Ludlow; *Miss Mc Ghie.*
Nesscliff. Bomere and Shomere Mosses. Bogs near Wolf's Head Turnpike on Shrewsbury and Oswestry road. Twyford Vownog, near Westfelton.

A very hardy middle-sized *tree*, adapting itself to almost every kind of soil situation and climate, of a peculiar light and airy appearance, with a white papyraceous *epidermis* which peels off; when old with deep-black shaggy longitudinal clefts in the *bark*. *Branches* alternate, repeatedly subdivided, erect or spreading, wiery, pliant, flexible and springy, with a reddish-brown smooth *bark*, more or less resinous. *Leaves* alternate, bright green, smooth, reticulato-venose, petiolate, very variable in size and shape, margins thickened, coarsely duplicato-serrate. *Male catkins* terminal, solitary, cylindrical, lax, imbricated, pendulous. *Bracteas* squamiform, peltate, with a smaller scale on either side. *Perianth* of 3 entire scales, with 4 *stamens* at their base. *Female catkins* lateral, cylindrical, dense, imbricated, pendulous. *Bracteas* squamiform, 3-lobed, pubescent, 3-flowered. *Perianth* 0. *Stigmas* 2. *Nut* minute, oblong, compressed, with a broad membranous wing on each side.

The trunk is subject to the occasional production of large knots of a reddish tinge, marbled, light and solid, much sought after by turners; and the branches (especially when the tree grows in a boggy soil) to larger masses of intertwined twigs, caused possibly by the extravasation of sap incident on the puncture of

some insect, and which, as Withering very aptly remarks, appear at a distance similar to rooks' nests. The *wood* is hard, tough, white shaded with red, with a grain intermediate between coarse and fine, of moderate durability, but very subject to the attacks of worms, chiefly employed in cooperage turnery and lighter kinds of work, especially pattens and heels of women's shoes. The white epidermis is proverbial for durability and as a defence against humidity, and was used by the ancients for their MSS. previous to the introduction of paper. The oil distilled from the bark imparts that powerful fragrance peculiar to Russia leather. The sap abounds in saccharine matter and is formed into a pleasant pungent wine. The whole tree diffuses an agreeable odour and is noticed by Burns as the "fragrant birk."

This "most beautiful of forest trees—the Lady of the woods"—intimately connected as it is with the literary history and ceremonials, civil and religious, of earlier times, is still more forcibly associated in our memories with the bright happy and buoyant days of youth,—with scenes of exquisite but evanescent mingled pains and pleasures, when to our minds "life and its thousand joys seemed but as one long summer's day:"—that we may indeed exclaim with the writer in the *Nouveau du Hamel*, "the sight of a birch-tree offers a vast subject of interesting meditation; and happy the man to whom its flexible pendant branches do not recall to mind, that to him, they were formerly instruments of punishment." To the application of the twigs in the formation of the well-known instruments of castigation, our Shropshire poet, the elegant-minded Shenstone, feelingly alludes in his "Schoolmistress:"—

> "And all in sight doth rise a birchen tree,
> Which learning near her little dome did stowe;
> Whilom a twig of small regard to see,
> Tho' now so wide its waving branches flow;
> And work the simple vassals mickle woe;
> For not a wind might curl the leaves that blew,
> But their limbs shudder'd, and their pulse beat low;
> And as they look'd they found their horror grow,
> And shap'd it into rods, and tingled at the view."

Where the trees of the forest have been fancifully compared to the heathen gods, as the Ash to Apollo, the Birch has with peculiar felicity been given to Venus.

Of its varieties there are very numerous, the following ones are the principal.

Var. β. pendula. Leaves ovato-rhomboid, acuminate, strongly unequally and doubly serrated, entire at the base, glabrous. *B. alba, β. pendula*, E. Fl. l. c.

This variety is usually known as the *weeping* birch, and attains a much larger size than the common form. Its branches are also straighter and more erect, and the lateral ones become very much elongated, wiery and elegantly pendulous at the extremities, "arching like a fountain shower," or as Wordsworth beautifully has it:—

> "Light Birch, aloft upon the horizon's edge,
> Transparent texture, framing in the east
> A veil of glory for the ascending Moon."

The upper surface of the leaves is more or less marked with roundish white resinous spots.

Var. γ. salax. Leaves rhombeo-triangular, acuminate, slightly cordate at the base, strongly unequally and doubly serrated, resinous.

In this variety, "profuse of nursing sap," both surfaces of the leaves, but especially the upper one, are abundantly covered with pale resinous spots and the branches with minute crystalline spiculæ interspersed with innumerable pale fragrant highly inflammable resinous globules, whence it has derived the name of the *salty birch.*

Var. δ. pubescens. Leaves cordato-ovate, acute, coarsely and irregularly serrated, clothed as are the petioles and branches with dense soft pale velvety pubescence.

When this variety has attained to considerable age its "velvet leaves" become very small and obtuse.

21. CARPINUS. *Linn.* Hornbeam.

1. C. *Betulus*, Linn. *Common Hornbeam.* Scales or bracteas of the fruit oblong, serrated, with 2 smaller lateral lobes. *E. Bot. t.* 2032. *E. Fl. v. iv. p.* 155. *Hook. Br. Fl. p.* 412.

Woods and hedges. *Fl.* May. ♄.
Near the Wrekin; *Mr. F. Dickinson.*

A small and not unhandsome *tree*; *bark* smooth; *wood* white, tough and hard; *branches* rigid. *Leaves* alternate, petiolate, ovate, doubly and acutely serrated, acute, plaited when young, smooth, the transverse ribs and axils hairy beneath. *Stipules* oblong, obtuse, smooth, reddish, deciduous. *Male flowers* in lateral sessile cylindrical imbricated catkins. *Female flowers* in a lax terminal catkin.

22. CORYLUS. *Linn.* Hasel-nut.

1. C. *Avellana*, Linn. *Common Hasel-nut.* Nut ovoid; involucre of the fruit campanulate, spreading, margins lacerato-dentate; leaves subrotund, cordate, acute. *E. Bot. t.* 723. *E. Fl. v. iv. p.* 157. *Hook. Br. Fl. p.* 412.

Hedges and copses; common. *Fl.* March, April. ♄.
A small shrubby *tree*, with hairy or glandular branches when young. *Leaves* light green, downy, especially beneath. *Male flowers* in terminal clustered long pendulous catkins. *Female flowers* in ovate scaly buds; *stigmas* bright crimson. *Involucre* of the *fruit* very variable in size and laciniation, and in all probability diligent search would bring to light many curious varieties.

CLASS XXII.

DIŒCIA.

> "The willow wreath weare I, since my love did fleet;
>   O willow, willow, willow!
> A garland for lovers forsaken most meete.
>   O willow, willow, willow!
>   O willow, willow, willow!
> Sing, O the greene willow shall be my garlànd!"
>   OLD BALLAD.

CLASS XXII.

DIŒCIA.   *Stamens and pistils in separate flowers and on different plants.*

(MONANDRIA. 1 *Stamen.* For some *Salices* see ORD. II.)

ORDER I. DIANDRIA. *Stamens 1—5, mostly 2.*

1. SALIX. *Male flower.*—Scales of the *catkin* single-flowered, imbricated, with a nectariferous gland. *Perianth* 0. *Stamens* 1—5. *Female flower.*—Scales of the *catkin* single-flowered, imbricated, with a nectariferous gland. *Perianth* 0. *Stigmas* 2, often cleft. *Capsule* 1-celled, 2-valved, many-seeded. *Seeds* comose.—*Nat. Ord.* AMEN-TACEÆ, *Juss.*—Name from the Celtic, *sal, near*, and *lis, water.*

ORDER II. TRIANDRIA. *3 Stamens.*

2. EMPETRUM. *Male flower.*—Calyx of 3 equal convex sepals, with several imbricated scaly bracteas at the base. *Petals* 3, equal, coloured, inserted into the receptacle, marcescent and deciduous. *Stamens* 3; filaments capillary, exserted. *Rudiment* of a pistil. *Female flower.*—Calyx and corolla as in male. *Stamens* 0. *Germen* subglobose, 6—9-celled. *Style* short. *Stigmas* 6, erect, peltate, radiate. *Berry* superior, globose, 6—9-seeded.—*Nat. Ord.* EMPE-TREÆ, *Nutt.*—Name from εν, *in*, and πετρος, a *stone*; from growing in rocky places.

(See *Valeriana dioica* in CL. III. Some *Salices* in ORD. I.)

ORDER III. TETRANDRIA. *4 Stamens.*

3. VISCUM. *Male flower.*—Calyx obsolete. *Petals* 4—8, ovate, fleshy, united at the base and bearing each a single *anther* adnate with the upper surface. *Female flower.*—Calyx an obscure margin, superior. *Petals* 4, sometimes 3, erect, ovate, very minute. *Stigma* sessile. *Berry* inferior, 1-seeded, with 1-2 embryos.—*Nat. Ord.* LORANTHEÆ, *Juss.*—Name from ιξος, the Greek name for the plant; or ισχυς, *power*, on account of the potent and mysterious virtues formerly attributed to it.

4. MYRICA. *Male flower.*—Scales of the *catkin* concave. *Perianth* 0. *Female flower.*—Scales of the *catkin* concave. *Perianth* of 2 (2—4; *Esenbeck)* minute scales adhering to the germen. *Styles* 2. *Drupe* 1-celled, 1-seeded.—*Nat. Ord.* MYRICEÆ, *Rich.*—Name, μυρικη, in Greek, synonymous with the Tamarix.

(See *Rhamnus* in CL. V. *Urtica* in CL. XXI.)

ORDER IV. PENTANDRIA. *5 Stamens.*

5. HUMULUS. *Male flower.*—Perianth single, of 5 leaves. *Anthers* 5, longitudinally dehiscing. *Female flower.*—Perianth of a

single scale infolded at the base and enclosing the germen, persistent, subsequently enlarged. *Styles* 2. *Seed* 1.—*Nat. Ord.* URTICACEÆ, *Juss.*—Name from *humus, rich soil* or *mould*, in which the plant flourishes.

(See *Ribes* in CL. V. *Bryonia* in CL. XXI. *Salix* in ORD. I.)

ORDER V. HEXANDRIA. *6 Stamens.*

6. TAMUS. *Male flower.*—Perianth single, in 6 deep segments. *Female flower.*—Perianth single, in 6 deep segments, contracted at the neck, superior. *Stigmas* 3. *Berry* 3-celled.—*Nat. Ord.* SMILA-CEÆ, *Juss.*—Name supposed to be the *Uva Taminia* of Pliny, or Black Bryony.

(See *Rumex* in CL. VI.)

ORDER VI. OCTANDRIA. *8 Stamens.*

7. POPULUS. *Male flower.*—Scales of the *catkins* incised. *Perianth* a single, cupshaped, oblique, entire scale. *Anthers* 8—12. *Female Flower.*—Scales of the *catkin* incised, ciliated. *Perianth* a cupshaped scale. *Stigmas* 4 or 8. *Capsule* superior, 1-celled, 2-valved, many-seeded. *Seeds* comose.—*Nat. Ord.* AMENTACEÆ, *Juss.*—Name *populus*, or the tree of the *people*, as it was esteemed to be in the time of the Romans, and of the French revolution.

ORDER VII. ENNEANDRIA. *9 Stamens.*

8. MERCURIALIS. *Male flower.*—Perianth single, 3-partite. *Stamens* 9—12. *Anthers* of 2 globose lobes. *Female flower.*—Perianth single, 3-partite. *Stigmas* 2. *Capsule* 2-celled; *cells* 1-seeded, superior. *Stigmas* 3.—*Nat. Ord.* EUPHORBIACEÆ, *Juss.*—Named in honour of *Mercury* who is said to have discovered the virtues of the plant.

9. HYDROCHARIS. *Flowers* spathaceous. *Male flower.*—Calyx in 3 deep segments. *Corolla* of 3 petals. *Stamens* 9, in 3 rows, within which are 3 imperfect *styles.* *Female flower.*—Calyx-tube adnate to the germen, limb in 3 deep segments. *Corolla* of 3 petals. *Staminodia* 3, filiform. *Nectary* of 3 fleshy scales. *Styles* 6, each with 2 *stigmas.* *Capsule* coriaceous, roundish, 6-celled, many-seeded. —*Nat. Ord.* HYDROCHARIDEÆ, *Juss.*—Name from υδωρ, *water*, and χαιρω, *to rejoice*; being aquatic plants.

(ORD. DECANDRIA. See *Silene* and *Lychnis* in CL. X.—ORD. ICO-SANDRIA. See *Rubus* and *Fragraria* in CL. XII.—ORD. POLYAN-DRIA. See *Stratiotes* in CL. XXI. See *Populus* in ORD. VI.)

ORDER VIII. MONADELPHIA. *Stamens combined.*

10. JUNIPERUS. *Male flower.*—Catkin ovate, destitute of scaly bracteas. *Stamens* inserted in the axis of the catkin, imbricated; *filaments* dilated into a scale bearing the *anthers* on the margin at the

base; *anthers* 3—6, globose, 1-celled. *Female flower.*—Catkin ovate, gemmæform. *Bracteas* scaly, imbricated, those at the base sterile, upper ones surrounding 3 fleshy squamæform ovaries. *Berry* consisting of the enlarged and united ovaries, scaly at the base, 3-seeded. *Seeds* subtriquetrous, with 5 glanduliferous cells at the base.—*Nat. Ord.* CONIFERÆ, *Juss.*—Name from *jeneprus*, Celtic, *rude* or *rough.*

11. TAXUS. *Male flower.*—Catkin subglobose, formed of imbricated simple scaly *bracteas*, exterior ones shorter, interior elongated. *Calyx* and *Corolla* 0. *Filaments* numerous, united into a column, exserted. *Anthers* peltate, with 4—6, rarely more, cells, adnate and longitudinally dehiscing beneath. *Female flower.*—Catkin gemmæ-form, short, single-flowered, formed of imbricated scaly *bracteas.* *Drupe* fleshy, open at the apex and enclosing an oval brown *nut* unconnected with the fleshy portion.—*Nat. Ord.* CONIFERÆ. *Juss.*—Name from ταξις, *order* or *arrangement*, in allusion to the distichous arrangement of the leaves.

DIŒCIA—DIANDRIA.

1. SALIX. *Linn.* Willow, Sallow, and Osier.

The Shropshire species of this difficult genus were determined by Mr. Borrer, whose descriptions in Hooker's British Flora, 4th ed. are inserted, and to the localities from which specimens were submitted to him the letter (B) is appended.

I. Monandræ. *Borr.* Filament 1, *with a double anther. Trees of low stature, or shrubs, with twiggy branches and more or less lanceolate and serrated leaves often broader upwards. Catkins very compact.*

1. S. *Helix, Linn. Rose Willow.* Monandrous, erect; leaves lanceolate, broadest upwards, attenuated below, serrated, glabrous; germens oblongo-ovate, very pubescent, sessile, style short, stigmas almost linear emarginate. *E. Bot. t.* 1343. *E. Fl. v. iv. p.* 188. *Hook. Br. Fl. 4th ed. p.* 354.

Osier holts, hedges, banks of rivers, &c. *Fl.* March, April. ♄.
Hedges about Eaton Mascott; *Rev. E. Williams's MSS.* Brook near Westbury; *Mr. F. Dickinson.*
Almond park, near Shrewsbury. (B.)

2. S. *Lambertiana, Sm. Boyton Willow.* Monandrous, erect; leaves lanceolate, broadest upwards, serrated, glabrous; germens shortly ovate, very pubescent, sessile; stigmas ovate, emarginate. *E. Bot. t.* 1359. *E. Fl. v. iv. p.* 189. *Hook. Br. Fl. 4th ed. p.* 354.

Hedges, banks of rivers, &c. *Fl.* April. ♄.
In a hedge at the bottom of a field called the Hop-yard, on the estate of E. Williams, Esq., Eaton Mascott; *Rev. E. Williams's MSS.*
Very nearly allied to the last, but distinguishable by its leaves, which are generally broader at the base, and the purplish glaucous hue of the young shoots.

3. S. *Forbyana*, Sm. *Fine Basket Osier.* Monandrous, erect ; leaves with small downy stipules, lanceolato-oblong, serrated, glabrous ; style equal in length to the linear divided stigmas. *E. Bot. t.* 1344. *E. Fl. v. iv. p.* 191. *Hook. Br. Fl. 4th ed. p.* 355.

Meadows and wet places. *Fl.* April. ♄.
Eyton on the Wildmoors. (B.)
Stems yellowish-green, glossy. Allied to *S. Helix,* especially in the fructification, but differing in foliage. This species is much esteemed by basket-makers for the finer sorts of wicker-work.

II. Triandræ. *Borr. Stamens* 3. *Leaves lanceolate, approaching to ovate, with evident deciduous stipules, serrated, glabrous. Catkins lax. Germens stalked, mostly glabrous. Most of the species constitute excellent osiers, and become trees if left to themselves.*

4. S. *undulata*, Ehrh. *Sharp-leaved triandrous Willow.* Triandrous ; leaves lanceolate, acuminated, serrated, glabrous ; germens stalked, ovato-acuminate, style as long as the linear bifid stigmas ; scales very villous. *Hook. Br. Fl. 4th ed. p.* 356. *S. lanceolata, Sm. E. Bot. t.* 1436. *E. Fl. v. iv. p.* 168.

Banks of brooks and wet places. *Fl.* April, May. ♄.
Large tree over Meole Brace bridge. (B.)
A small tree, which casts its bark annually. Used by the basket-makers, but not so well calculated for the finer sorts of work as *S. triandra.*

5. S. *triandra*, Linn. *Long-leaved triandrous Willow.* Triandrous ; leaves oblongo-lanceolate, acute, serrated, glabrous ; germens stalked, oblongo-ovate, as well as the retuse scale ; stigmas sessile, retuse. *E. Bot. t.* 1435. *E. Fl. v. iv. p.* 166. *Hook. Br. Fl. 4th ed. p.* 356.

Sides of brooks, edges, &c. *Fl.* May—August. ♄.
By the side of Condover, Cantlop, and Eaton brooks. Hedges about Wheatley, near Sundorn ; *Rev. E. Williams's MSS.*
Cold Hatton heath. (B.)
This, if left to itself, becomes a tall tree, 20—30 feet high, casting its bark in autumn. It is abundantly cultivated, and reckoned among the most valuable of the osiers.

6. S. *Hoffmanniana*, Sm. *Short-leaved triandrous Willow.* Triandrous ; leaves shortly and broadly lanceolate, acute, slightly rounded at the base, serrated, glabrous ; germens stalked, ovate, compressed, glabrous ; stigmas nearly sessile. *E. Bot. Suppl. t.* 2620. *E. Fl. v. iv. p.* 168. *Hook. Br. Fl. 4th ed. p.* 357.

Rivers sides. *Fl.* May. ♄.
Banks of river Severn, near Coalbrookdale ; *Mr. F. Dickinson.*
A much branched *shrub* or crooked tree, scarcely exceeding 12 feet. *Bark* of the stem and large *branches* deciduous, as in the other triandrous Willows. The humbler growth, the short flat lanceolate *leaves* more rounded at the base, with larger rounded earshaped *stipules,* distinguish this plant from *S. triandra,* with which it is said to agree in the fertile *flower* as it does in wanting the deep furrows of the young twigs, so remarkable in *S. amygdalina.*

7. S. *amygdalina,* Linn. *Almond-leaved Willow.* Triandrous ;

---

leaves oblongo-ovate, acute, rounded at the base, serrated, glabrous ; germens much stalked, ovate, glabrous ; stigmas sessile, bifid ; young branches furrowed. *E. Bot. t.* 1636. *E. Fl. v. iii. p.* 169. *Hook. Br. Fl. 4th ed. p.* 357.

Banks of rivers. *Fl.* April, May—August. ♄.
Banks of river Severn, near Atcham ; *Mr. F. Dickinson.*
A *tree,* with much furrowed yellowish young *branches,* considered inferior as an osier to *S. triandra,* which it approaches in botanical character.

III. Pentandræ. *Borr. Stamens more than 3, usually 5, in each catkin, so numerous and long as to render the flowers, which too are in perfection at the same time with the foliage, quite handsome ; whilst the tree itself is the most ornamental of the whole genus. Germens glabrous. Moderately-sized trees, with ample glossy fragrant foliage, exuding a resin from the glandular serratures of the leaves.*

8. S. *pentandra*, Linn. *Sweet Bay-leaved Willow.* Stamens 5 ; leaves elliptical-lanceolate, acuminated, glanduloso-serrate, glabrous, with several glands at the base ; germens lanceolate, glabrous, nearly sessile, style scarcely any, stigmas bifid. *E. Bot. t.* 1805. *E. Fl. v. iv. p.* 171. *Hook. Br. Fl. 4th ed. p.* 357.

Banks of rivers, and watery places. *Fl.* May, June. ♄.
Babin's wood, between Whittington and Oswestry ; *J. F. M. Dovaston, Esq.*
Eyton on the Wildmoors. (B.)
18—20 feet high. Its large and copious shining foliage almost gives this plant the appearance of an evergreen. *Sterile catkins* broad, fragrant, as well as the leaves. This species is much sought after by the Irish harvest-men, who call it "the black willow," and cut it for their *shillelahs.*

IV. Fragilis. *Borr. Stamens 2, (as in the following groupes.) Trees of considerable size, with lanceolate glabrous serrated stipulated leaves, and very lax catkins with elongated more or less stalked glabrous germens.*

9. S. *decipiens*, Hoffm. *White Welsh or Varnished Willow.* Leaves lanceolate, pointed, serrated, very glabrous, floral ones partly obovate and recurved, footstalks somewhat glandular ; germens tapering, stalked, glabrous ; style longer than the cloven stigmas ; branches smooth, highly polished. *E. Bot. t.* 1937. *E. Fl. v. iv. p.* 183. *Hook. Br. Fl. 4th ed. p.* 358.

Low meadows, moist hedges &c. *Fl.* May. ♄.
Banks of the river Severn near Marn wood, Coalbrookdale ; *Mr. F. Dickinson.*

10. S. *fragilis*, Linn. *Crack Willow.* Leaves ovato-lanceolate, acute, serrated, glabrous ; germens shortly pedicellate, oblongo-ovate, glabrous, style short, stigmas bifid ; scales pubescent and much ciliated. *E. Bot. t.* 1807. *E. Fl. v. iv. p.* 184. *Hook. Br. Fl. 4th ed. p.* 358.

Hedges, banks of rivers, &c. *Fl.* April, May. ♄.
Hedges and sides of pools ; *Rev. E. Williams's MSS.*
Fields near Meole Bridge. (B.) Banks of Severn, Coleham, Shrewsbury. (B.)
(Much like *S. ambigua,* Pursh, *Borrer.)*
Wood of little or no value.

---

11. S. *Russelliana*, Sm. *Bedford Willow.* Leaves lanceolate, tapering at each extremity, strongly serrated, glabrous, very pale beneath ; germens stalked, lanceolate, acuminate, glabrous, style as long as the bifid stigmas ; scales narrow-lanceolate, slightly ciliated or pubescent. *E. Bot. t.* 1808. *E. Fl. v. iv. p.* 186. *Hook. Br. Fl. 4th ed. p.* 358.

Banks of rivers, marshy woods, &c. *Fl.* April, May. ♄.
Banks of river Corve, Ludlow ; *Mr. F. Dickinson.* Pitmoor pool ; *Rev. M. Berkley, in Henslow's Herbarium, Cambridge.*
Large trees by roadside near Nobold, near Shrewsbury. (B.) Banks of river Severn, Coleham, Shrewsbury. (B.) Sutton Spa. (B.)
An extremely valuable tree and the most profitable for cultivation of any of the species, for the value of its timber as well as bark, the rapidity of its growth, and the handsome aspect of the tree. The bark contains the tanning principle in a superior degree to that of the oak, and is moreover a valuable substitute for the Cinchona in medical practice.

V. Albæ. *Borr. Trees of considerable elevation, having lanceolate serrated leaves, with long silky hairs beneath, especially in a young state, which give to the foliage a light or whitish hue : the serratures glandular. Catkins lax : germens glabrous.*

12. S. *alba*, Linn. *Common White Willow.* Leaves ellipticallanceolate, regularly glanduloso-serrate, acute, silky beneath, often so above ; germens ovato-acuminate, nearly sessile, glabrous ; stigmas nearly sessile, short, recurved, bifid ; scales short, pubescent at the margin. *E. Bot. t.* 2430. *E. Fl. v. iv. p.* 231. *Hook. Br. Fl. 4th ed. p.* 359.

River-sides, moist woods, &c. *Fl.* May. ♄.
On the bank of Meole brook, near Meole bridge. Common about Ludlow, Burford, and Millichope. Newport ; *Rev. E. Williams's MSS.* Banks of river Corve, Ludlow ; *Mr. F. Dickinson.*
Trees in field near Meole bridge. (B.)
A well-known tree of considerable size.

Var. *β. cærulea.* Under side of the leaves less silky, often quite glabrous. *Sm. and Hook. l. c. S. cærulea, (blue Willow) E. Bot. t.* 2431.

Shrewsbury castle mound. (B.)
Of such exceedingly rapid growth, that it is still by many deemed a distinct species. Both varieties are alike valuable for their bark and timber, and are both amply deserving of cultivation.

13. S. *vitellina*, Linn. *Yellow Willow or Golden Osier.* Leaves lanceolate, with glandular serratures, acuminate, more or less silky beneath, often so above ; germens lanceolate, sessile, glabrous ; style short, stigmas bipartite ; scales lanceolate. *E. Bot. t.* 2430. *E. Fl. v. iv. p.* 182. *Hook. Br. Fl. 4th ed. p.* 359.

Hedges and banks of rivers, &c. *Fl.* May. ♄.
"Hedges ;" *Rev. E. Williams's MSS.*
Near Marn wood, Coalbrookdale ; *Mr. F. Dickinson.*
This is rendered striking by the bright yellow colour of its branches, of which tint the leaves often partake.

VI. Fuscæ. *Borr. Small shrubs, with generally procumbent stems and leaves between elliptical and lanceolate, mostly silky beneath, nearly entire. Catkins ovate or cylindrical. Germens silky, stalked.* [*The habit of* S. fusca *rather approaches the* Monandræ *group.*]

14. S. *fusca*, Linn. *Dwarf silky Willow.* Leaves elliptical or elliptic-lanceolate, acute, entire, or with minute glandular serratures, somewhat downy, glaucous and generally very silky beneath ; germens upon a long stalk, lanceolate, very silky, stigmas bifid ; stems more or less procumbent. *Hook. Br. Fl. 4th ed. p.* 331. *S. repens, Hook. Scot.* 1. *p.* 284.

Var. *a.* Stem much branched, upright, decumbent below ; leaves elliptical-lanceolate. *S. fusca, E. Bot. t.* 1960. *E. Fl. v. iv. p.* 210.

Var. *β.* Stem depressed, with short upright branches ; leaves elliptic-lanceolate. *S. repens, E. Bot. t.* 183. *(with young leaves only.) E. Fl. v. iv. p.* 209.

Var. *γ.* Stem prostrate, with elongated straight branches ; leaves elliptic-oblong. *S. prostrata, E. Bot. t.* 1959. *E. Fl. v. iv. p.* 211.

Var. *δ.* Stem recumbent ; leaves elliptical. *S. fœtida, E. Fl. v. iv. p.* 208.—*S. adscendens, E. Bot. t.* 1962.—*Subvar.*leaves smaller. *S. fœtida, β. E. Fl. v. iv. p.* 208. *S. parvifolia, E. Bot. t.* 1961.

Var. *ε.* Stem procumbent ; leaves elliptic-lanceolate. *S. incubacea,* Linn. *E. Fl. v. iv. p.* 212. *(excl. of all the other syns.? Borr.) E. Bot. Suppl. t.* 2600.

Var. *ζ.* Stem erect or spreading ; leaves elliptical with a recurved point, very silvery beneath. *S. argentea, E. Bot. t.* 1364. *E. Fl. v. iv. p.* 206.

Moist and dry heaths, moors, and sandy situations. *Fl.* April, May. ♄.
β. On Shawbury heath. Stocket moor, near Cockshut. Meadows between Whittington and Halston. Bog between the Cloud coppice and the Gouter, near Berrington ; *Rev. E. Williams's MSS.* Shawbury heath. (B.)
δ. Shawbury heath. (B.)
ζ. " Bog by the side of the Canal near the Queen's Head turnpike, between Westfelton and Oswestry, as Dr. Smith informs me ; " *Rev. E. Williams's MSS.*
Usually a small procumbent *shrub,* with rather long straight *branches,* but varying exceedingly according to situation and other circumstances, as do the *leaves* also, which are more or less glabrous above, and more or less silky beneath where the nerves are prominent.
Although the whole of the above varieties have not been discovered in our county, I have thought it better to insert descriptions of all, since the plant being so liable to variations from various causes, it seems not improbable that most of them may be yet detected.

VII. Viminales. *Borr. Trees of a more or less considerable size, with long pliant branches and lanceolate leaves. Germens nearly sessile, hairy or silky ; their styles elongated, their stigmas linear, mostly entire.*

15. S. *viminalis*, Linn. *Common Osier.* Leaves linear-lanceolate, obscurely crenate, white and silky beneath; stipules very small, sublanceolate; branches straight and twiggy; germens upon very short stalks, lanceolato-subulate, style elongated, stigmas long, linear, mostly entire. *E. Bot. t.* 1898. *E. Fl. v. iv. p.* 228. *Hook. Br. Fl. 4th ed. p.* 363.

Banks of rivers, brooks, &c. *Fl.* April, May. ♄.
"Banks of rivers and brooks, common;" *Rev. E. Williams's MSS.*
Banks of river Severn, Shrewsbury. (B.) Sutton Spa, near Shrewsbury. (B.)
This is held in great esteem for basket-work.

16. S. *stipularis*, Sm. *Auricled Osier.* Leaves lanceolate, very indistinctly crenate, white and downy beneath; stipules large, semicordate, acute, often with a tooth or lobe at the base; germens stalked, lanceolate, very downy, style elongated, stigmas linear, undivided; scales very shaggy. *E. Bot. t.* 1214. *E. Fl. v. iv. p.* 230. *Hook. Br. Fl.* 4*th ed. p.* 363.

Hedges, woods, wet places, &c. *Fl.* March. ♄.
By the side of Cound paper mill-stream. By the side of the turnpike road between Uckington and Haygate. By the side of the Canal at the bridge leading to Longnor. Hedges about Haughton, near Sundorn. On the south side of Shawbury heath. Wheatley, near Sundorn. About Preston Gobalds. Middle. Cockshut. Ellesmere; *Rev. E. Williams's MSS.*
Allied to the preceding in *fructification*: differing in its larger and coarser *leaves*, less white beneath, and with large very remarkable stipules.

17. S. *Smithiana*, Willd. *Silky-leaved Osier.* Leaves lanceolate, obscuely crenate, white and covered with starry pubescence beneath; stipules very small narrow acute; germens lanceolato-subulate, very silky, shortly stalked; style elongated; stigmas long, linear, mostly entire. *Hook. Br. Fl. p. 4th ed. p.* 363. *E. Fl. v. iv. p.* 229. *S. mollissima. E. Bot. t.* 1509. *(not Ehrh.)*

Meadows and osier grounds. *Fl.* April, May. ♄.
Cold Hatton Heath. (B.)

18. S. *ferruginea*, And. MSS. *Ferruginous Willow.* Leaves thin, lanceolate, with wavy crenatures and small teeth, minutely hairy on both sides, paler beneath; stipules small, half ovate; scales oblongo-lanceolate; germen silky, stalked, style about as long as the oblong stigmas. *E. Bot. Suppl. t.* 2665. *Hook. Br. Fl. 4th ed. p.* 364.

Banks of streams. *Fl.* April, May. ♄.
Banks of river Severn, Shrewsbury. (B.) *("ferruginea*, I think, but with catkins more silky, scale more rounded, stigmas more divided than in my plants." *Borrer.)*—Near the Sharpstones hill. (B.)

19. S. *holosericea*, Willd. *Soft shaggy-flowered Willow.* Leaves lanceolate, acuminate, serrated, glabrous above, pale downy and strongly veined beneath; catkins cylindrical; germens stalked, densely clothed with silky wool; stigmas ovate, sessile; scales (black) very shaggy. *Hook. Br. Fl. 4th ed. p.* 364.

Hedges and marshy places. *Fl.* April, May. ♄.

Cold Hatton heath. (B.) Eyton on the Wildmoors; (B.) ("Willdenow's differs a little."—*Borrer.*) Hedges, Oswestry; (B.)

20. S. *acuminata*, Sm. *Long-leaved Willow.* Leaves lanceolato-oblong, pointed, wavy, finely toothed, glaucous and downy beneath; stipules half-ovate, then kidney-shaped; catkins cylindrical; germen stalked, ovate, hairy, style as long as the undivided stigmas. *E. Bot. t.* 1434. *E. Fl. v. iv. p.* 227. *Hook. Br. Fl. 4th ed. p.* 364.

Moist woods and hedges. *Fl.* April. ♄.
"Hedges, common;" *Rev. E. Williams's MSS.*

VIII. Cinereæ. *Borr. Trees or low shrubs, with downy branches, and mostly obovate grey hoary toothed more or less wrinkled and stipuled leaves, very veiny beneath. Germens sericeo-tomentose.—(This group is usually denominated the Sallows.)*

21. S. *cinerea*, Linn. *Grey Sallow.* Leaves obovato-elliptical, sometimes approaching to lanceolate, more or less glaucous above, beneath pubescent and reticulated with veins, the margins slightly recurved; stipules semicordate; germens stalked, lanceolato-subulate, silky; styles short; stigmas mostly entire. *E. Bot. t.* 1897. *E. Fl. v. iv. p.* 215. *Hook. Br. Fl. p.* 364.

Banks of rivers, and moist woods. *Fl.* April. ♄.
Four miles from Ludlow, on Wenlock road; *Mr. F. Dickinson.*
Shomere bog, near Shrewsbury. (B.)
A tree, 20—30 feet high, of no beauty and little use.

22. S. *aquatica*, Sm. *Water Sallow.* Stem and branches erect; leaves slightly serrated, obovato-elliptical, minutely downy, flat, rather glaucous beneath; stipules rounded, toothed; germens silky, stalked; stigmas nearly sessile. *E. Bot. t.* 1437. *E. Fl. v. iv. p.* 218. *Hook. Br. Fl. 4th ed. p.* 365.

Wet hedge-rows, swampy places, &c. *Fl.* April. ♃.
"Hedges;" *Rev. E. Williams's MSS.* Rushbury; *Mr. F. Dickinson.*
Banks of river Severn, near Shrewsbury. (B.) Shomere moss, near Shrewsbury. (B.) Meole Brace. (B.)

23. S. *oleifolia*, Sm. *Olive-leaved Sallow.* Stem erect; branches straight, spreading; leaves obovato-lanceolate, flat, rather rigid, minutely toothed, acute, glaucous, reticulated and finely hairy beneath; stipules small, notched, rounded; catkins oval, nearly half as broad as long. *E. Bot. t.* 1402. *E. Fl. v. iv. p.* 219. *Hook. Br. Fl. 4th ed. p.* 365.

Marshy places and hedges. *Fl.* March. ♄.
Shomere moss, near Shrewsbury. (B.) Eyton on the Wildmoors. (B.) Middleton, Oswestry. (B.) The Nant, Oswestry. (B.) Oswestry. (B.) Cold Hatton heath. (B.)

24. S. *aurita*, Linn. *Round-eared Sallow.* Leaves obovate, repando-dentate, wrinkled with veins, more or less pubescent, very downy beneath, tipped with a small bent point, recurved at the

margins; stipules roundish, semicordate; germens lanceolato-subulate, stalked, silky; style very short; stigmas generally entire. *E. Bot. t.* 1487. *E. Fl. v. iv. p.* 216. *Hook. Br. Fl. 4th ed. p.* 365.

Moist woods, and thickets. *Fl.* May. ♄.
"Hedges;" *Rev. E. Williams's MSS.*
Shomere moss, near Shrewsbury. (B.) Sharpstones hill. (B.)
A small bushy *tree* with straggling *branches.* One of the least equivocal species, although its *leaves* vary in length and in roundness. They are usually much wrinkled and vaulted, the *stipules* large and stalked.

25. S. *caprea*, Linn. *Great round-leaved Sallow.* Leaves ovato-elliptical, acute, serrated and waved at the margin, downy beneath; stipules semicordate; germens pedicellate, lanceolato-subulate, silky; stigmas sessile, undivided. *E. Bot. t.* 1488. *E. Fl. v. iv. p.* 225. *Hook. Br. Fl. 4th. ed. p.* 365.

Woods, hedges, &c. *Fl.* April, May. ♄.
"Hedges;" *Rev. E. Williams's MSS.* Captain's Coppice, Coalbrookdale; *Mr. F. Dickinson.*
Oswestry. (B.) Sutton Spa, near Shrewsbury. (B.) Shomere moss, near Shrewsbury. (B.)
A small *tree*, distinguished by being in the spring loaded with handsome yellow *blossoms* before any of the *leaves* appear. The *catkins* of both kinds are broader and shorter than in most of the species with crowded *flowers.* The *wood* has been employed for various agricultural implements, and the *bark* in tanning and as a successful substitute for *Cinchona.*

26. S. *sphacelata*, Sm. *Withered-pointed Sallow.* Stem erect; leaves elliptico-obovate, even, veiny, entire or slightly serrated, downy on both sides, discoloured at the point; stipules half heart-shaped, toothed, erect; germens stalked, ovato-lanceolate, silky; stigmas notched, longer than the style. *E. Bot. t.* 2333. *E. Fl. v. iv. p.* 224. *Hook. Br. Fl. 4th ed. p.* 365.

Banks of streams, wet woods, &c. *Fl* April, May. ♄.
Bomere wood, near Shrewsbury. (B.) Banks of Severn, between Preston Boats & Uffington, near Shrewsbury. (B.) Meole Brace, near Shrewsbury. (B.)

DIŒCIA—TRIANDRIA.

2. EMPETRUM. *Linn.* Crow-berry.

1. E. *nigrum*, Linn. *Black Crow-berry or Crake-berry.* Procumbent; leaves linear-oblong. *E. Bot. t.* 526. *E. Fl. v. iv. p.* 234. *Hook. Br. Fl. p.* 438.

Heathy hills; not common. *Fl.* May. ♄.
On Selattyn mountain; *T. and. D. Bot. Guide.* Bog near Ellesmere; *Watson's New Bot. Guide.*
Stiperstones hill. Castle Ring hill, near Stiperstones. Shomere moss, near Shrewsbury; sparingly.
A small procumbent *shrub*, with numerous partly ascending branches covered with stipitate glands. *Leaves* scattered, crowded, linear-oblong, obtuse, shortly petiolate, margins with a dense downy fringe, revolute and meeting beneath, thus giving the appearance of a white longitudinal streak, the false margin with a single row of stipitate glands, upper surface smooth, under surface with scattered minute resinous dots or glands. *Flowers* axillary towards the summits of the

branches, small, solitary, nearly sessile, gemmæform, purplish. *Berries* purplish-black, globose, subtended at the base by the persistent calyx.

DIŒCIA—TETRANDRIA.

3. VISCUM. *Linn.* Misseltoe.

1. V. *album*, Linn. *Common Misseltoe.* Stems dichotomously branched; leaves oblongo-lanceolate, obtuse; flowers in axillary clusters. *E. Bot. t.* 1470. *E. Fl. v. iv. p.* 236. *Hook. Br. Fl. p.* 439.

Parasitic on various trees. *Fl.* December—April. ♄.
On Hawthorn, profusely; Loton park. On Hazel, rare; Fettis farm, Melverley. On American Poplar, when it becomes larger; Calcott hall, near Vyrniew. On Pyrus Aucuparia, rare; on an old tree in a hedge between the villages of Dovaston and Kynaston. On Crab, Apple, and Pear; common. *J. F. M. Dovaston, Esq.* Plentifully about Ludlow; *Miss Mc Ghie.* In an orchard, Muckley Cross; *Mr. F. Dickinson.* Walford and Eyton; *T. C. Eyton, Esq.*
On the Apple, Upper Edgbold; Nobold, &c., near Shrewsbury. On the Pear; near the Criften on Haughmond hill. On the Larch; plentifully, at Cold Weston, as I have been informed by an intelligent labourer.
A small, evergreen, parasitical, crowdedly branched, bushy *shrub*, of very slow growth, pendant from the trunks and branches of trees. *Root* woody, inserting itself into the bark of the tree on which it grows and propagating itself as shewn by Professor Henslow *(Loudon's Mag. Nat. Hist.* 6. 500.*)* by extending between the bark and young wood, green filamentary *scyons*, which at intervals give off at right angles other portions of a lighter colour, and these striking into the young wood of the tree throw up suckers which become individual plants. A longitudinal section of a branch on which it has grown for many years exhibits the root in a wedgeshaped form surrounded by the wood. The aspect of the male plant is of a yellowish hue, the branches short and rigid, and the leaves small, whilst that of the female plant is of a dark and bright green, the branches longer more luxuriant and less rigid, and the leaves frequently larger. The herbage emits a slight but agreeable aromatic scent. *Stem* rounded, with thickened joints, repeatedly dichotomously branched. *Leaves* coriaceous, in opposite pairs from the joints, oblongo-lanceolate, obtuse, tapering at the base, and bearing on the upper surface a thickened concave scale or protuberance, *(stipule?)* with 5 parallel ribs, glabrous, entire, slightly glaucous, covered on both surfaces when viewed under a powerful lens with innumerable pale-yellow glandular dots. *Male flowers* in axillary and terminal 3-flowered clusters, the central flower the larger, sessile on a tetragonous fleshy receptacle crowned with two triangular acute erect *bracteas* which subtend the lateral florets. *Calyx* none. *Corolla* generally of 4, (sometimes 5, 6, and even 8) ovate acute erecto-patent fleshy segments united at the base, each bearing a single oblong *anther* adnate with its upper surface, without lobes or filaments, but whose substance is broken up into a number of hollow oblong cavities containing the pale-yellow oval *pollen. Female flowers* in axillary and terminal whorls of 4 spreading distant 3-flowered clusters with an erect central one, each cluster sessile on a short rounded fleshy receptacle crowned with two erect triangular acute internally fringed *bracteas.* From this primary receptacle arises another smaller one crowned with similar bracteas which are alternate with the others and enclose the central floret, the two lateral florets being in general only partially developed. *Calyx* a thickened margin on the summit of the germen. *Corolla* of 4, (sometimes 3) minute, ovate, converging fleshy, deciduous *petals. Style* none. *Stigma* an obtuse point. *Berry* globular, smooth, white, pellucid, veiny, subtended by the secondary bracteas, consisting of a viscid mass, enveloping a single heartshaped compressed *seed* with one or two *embryos*, whose *radicle* is naked and superior.

The following account of the germination of the seeds is derived from *Loudon's Arboretum Britannicum* 2. 1024. "The first indication of germination is the appearance of one or more radicles which at first rise up and then bend over until they reach the body of the substance to which the seed has been attached. Having reached that substance, the point of the radicle swells out like the extremity of the sucker of the house-fly, and having fixed itself to the bark, if more than one have proceeded from a single seed, the embryos all separate, and each putting out leaves at its upper extremity becomes a separate plant."

Dutrochet in his *Memoir Sur la Motilité des Vegetaux*, p. 114, mentions that he had ascertained both in aquatic and terrestial plants that that portion of the stem which possessed a coloured parenchyma was directed upwards and became an ascending stem, whilst that which was colourless descended and became a root. The radicle of the embryo of the misseltoe which is green presented however a remarkable exception to this law, by invariably manifesting a tendency to avoid the light by growing in a directly opposite direction. He details many curious experiments which he made by causing misseltoe seeds to germinate on bricks, tiles, stones, trees, cannon balls, in glass tubes, and even on the glass panes of windows, and found that in all these situations and under all circumstances the radicle invariably grew towards the body to which the seed was attached, in a direction diametrically opposite to the light; and from thence deduces this conclusion :—that the radicle of the embryo of misseltoe is influenced by two different agents; one, a peculiar attraction to the body on which it germinates, the other, an unconquerable and powerful tendency to shun the light. He also states that the plumule remains undeveloped and in a rudimentary state between the cotyledons during the first year after germination, no portion of the ascending stem of the embryo being apparent except the part comprised between the insertion of the cotyledons and the origin of the radicle. In the spring of the second year the cotyledons become dried up and detached from the stem which then begins to develope its first leaves.

My friend Mr. Dovaston has succeeded in causing the seeds to germinate on "the Oak, several Pines, Cherry, Laurel, Portugal Laurel, Holly, Lime, Elms, Hornbeam, Birch, Sycamore, Ash, Chestnut, Hazel, Acacia, and its own friends the Crab, Apple, Hawthorn and Pear on which it grew well and still grows most luxuriantly. On the resinous it did germinate but took little or no hold : on the gummy, a little better ; on the others better still, but on all, except on its usual foster-nurses 'sickened and so died.' On one apple called 'the Lady,' its leaves become much larger and more vigorous ; as also I have found them on the Black Poplars, at Callcott Hall."

The Misseltoe particularly that which grew upon the Oak, as is universally known, was highly esteemed and gathered with many mystical rites by the Druids; and "the youths and maidens of our villagery," still—

"When the year its course hath rolled,
And brought blithe Christmas back again,
Forth to the woods with merrie hearts go
To gather in the misletoe :"—

which is destined to bear so important a part in the hearty mirth and uncontrolled delight which celebrate the Yuletide.

The oft-quoted lines of Virgil are so admirably characteristic of the appearance and habits of the plant that I cannot with propriety omit them :—

" liquidumque per aera lapsæ
Sedibus optatis geminæ super arbore sidunt:
Discolor unde auri per ramos aura refulsit.
Quale solet silvis brumali frigore viscum
Fronde virere nova, quod non sua seminat arbos,
Et croceo fœtu teretes circumdare truncos.
Talis erat species auri frondentis opaca
Ilice : sic leni crepitabat bractea vento."

4. MYRICA. *Linn.* Gale.

1. M. *Gale*, Linn. *Sweet Gale or Dutch Myrtle.* Leaves obovato-lanceolate, acute, distantly serrated in the upper part; stem shrubby. *E. Bot. t.* 582. *E. Fl. v. iv. p.* 239. *Hook. Br. Fl. p.* 439.

Bogs; not common. *Fl.* May. ♄.

Moss at Walford and Yestalls, near Walford ; *T. C. Eyton, Esq.* Marbury bog ; *Mr. F. Dickinson.* Bog near Ellesmere ; *Rev. A. Bloxam.* About Lee, near Ellesmere ; *Rev. E. Williams's MSS.*

Twyford Vownog near Westfelton, in great abundance.

Shrub 3-4 feet high, erect, bushy ; *branches* alternate, erect, brown, pubescent. *Leaves* alternate, very shortly petiolate, obovato-lanceolate, acute, distantly serrated in the upper part, margins thickened and somewhat recurved, hairy, upper surface dark green pubescent with bright pellucid resinous yellow globules, under surface paler and veiny, hairy, with innnmerable round mealy and resinous dull and bright particles with larger globules interspersed. *Catkins* numerous, sessile, lateral and terminal, on the extremities of the branches, formed during summer in the axils of the leaves and remaining through the winter, becoming full grown in the following March, and expanding in May. *Male Catkins* erect, simple, cylindrical, formed of imbricated bracteal scales, ovato-rhomboid, sinuated or cut away at the base, concave, pointed and keeled, margins ciliated, with a tuft of deflexed hairs at the base on each side, and a mass of bright yellow pellucid resinous globules, strongly bitter in the taste, yielding when bruised a most delightfully aromatic perfume, pale greenish-yellow, with a brown margin, having on the inside at the base the 4 short erect *stamens*. *Female catkins* erect, short, oval, formed of similar imbricated bracteal scales bearing the flower in their concavities. *Scales* of the *perianth* generally 2, (2—4 *Esenbeck*) minute, erect, fleshy, adhering to the germen. *Stigmas* filiform, elongated, generally 2 sometimes 3, when there is a third minute obtuse scale to the perianth. *Berry* very small, covered with resinous dots.

Mr. Dovaston has observed male and female catkins on the same plant. The resinous aromatic globules with which the plant abounds are probably similar in their properties to those found on the Hop, in which the bitter principle called Lupuline resides, and consequently have been substituted for hops in brewing by the poor in Sweden. Linnæus also says that the berries boiled in water yield wax like those of the *Myrica cerifera* or Candleberry Myrtle.

DIŒCIA—PENTANDRIA.

5. HUMULUS. *Linn.* Hop.

1. H. *Lupulus*, Linn, *Common Hop.* *E. Bot. t.* 427. *E. Fl. v. iv. p.* 240. *Hook. Br. Fl. p.* 440.

Hedges ; frequent, apparently wild. *Fl.* July. ♃.

Root creeping. *Stem* herbaceous, twining and climbing to a considerable height, branched, hexangular, hollow, rough as is the whole plant with very minute rigid scabrous points and decurved bristles. *Leaves* opposite, on scabrous angular petioles channelled above, united at the base by a pair of large cordato-ovate acuminate reflexed bifid *stipules*, rotundo-cordate, 5-lobed, acute, margins coarsely serrated, bristly serratures cuspidate and mucronate, dark green, wrinkled on the upper surface, paler green beneath and with innumerable yellow resinous globules. *Flowers* pale green, pendulous, in axillary panicles shorter than the leaves. *Male flowers* in compound axillary panicles, bracteated at the divisions. *Perianth* 5-partite, patent. *Stamens* 5. *Female flowers* either in panicles or on simple peduncles, bearing 4 ovate acuminate acute *bracteas* at the base of the opposite pedicels and several similar ones scattered above. *Catkin* ovate, densely imbri-

cated ; *bracteas* rotundo-ovate, shortly acuminate, pubescent. *Perianth* of a single membranous ovate acute scale infolded at the base and enclosing the germen. *Stigmas* 2, elongated, subulate, pubescent, connate at the base. *Fruit* strobiliform, ovate, formed of the enlarged imbricated scales of the perianth, resinous at the base and enfolding the subglobose erect small hard 1-seeded *nut*, covered with resinous aromatic globules, *seed* pendulous.

The yellow globules with which the plant is covered, when immersed in water emit according to Raspail a resinous oily fluid in which resides the bitter principle termed Lupuline, so importantly useful in brewing ale. Linnæus asks " what is that electrical murmur like very distant thunder when the hoppoles are shaken by the wind?" As the whole plant is rough with rigid scabrous points which are probably like the cuticle of grasses composed of silica, may not this murmuring noise arise from the attrition of the leaves, stems, &c., against each other?

DIŒCIA—HEXANDRIA.

6. TAMUS. *Linn.* Black Bryony.

1. T. *communis*, Linn. *Common Black Bryony.* Leaves cordato-acuminate, entire. *E. Bot. t.* 91. *E. Fl. v. iv. p.* 241. *Hook. Br. Fl. p.* 440.

Hedges and thickets ; frequent. *Fl.* June, July. ♃.

Root large and fleshy, black externally, white within, acrid. *Whole plant* smooth and shining. *Stem* round, twining and climbing to a great height and gracefully hanging from trees and bushes. *Leaves* alternate, on a thick succulent pentangular petiole channelled above swollen at the base, cordato-acuminate, entire, wavy or minutely crisped at the margins, veiny, 7-ribbed at the base, the 3 central ribs only reaching to the apex. *Stipules* in pairs, subulate, succulent, spreading at right angles. *Male inflorescence* greenish yellow, on a long pedunculated axillary raceme longer than the leaves. *Flowers* on short pedicels, with a linear lanceolate *bractea* at the base, 1-2 rarely 3 together ; when there are 2 flowers together the second pedicel arises from the primary one at its very base and has a pair of opposite minute bracteas a little below the flower. *Perianth* deeply 6-partite, campanulato-patent ; segments lanceolate, recurved, the alternate ones obtuse and membranous at the apex, the rest pointed, margins of all subrevolute. *Female inflorescence* greenish-yellow, on a short pedunculated axillary decurved raceme, much shorter than the petioles. *Flowers* on short pedicels, with a pair of minute setaceous *bracteas* either at the base or on the pedicel. *Perianth* deeply 6-partite ; segments ovate, obtuse, spreading. *Stamens* 6, abortive, at the base of the petals. *Style* tubular, 3-fid. *Stigma* dilated, bifid, reflexed. *Germen* ovate or oblong, 3-celled. *Berry* nearly globose, fleshy, scarlet, 3—6-seeded. *Seeds* globose, pendulous.

DIŒCIA—OCTANDRIA.

7. POPULUS. *Linn.* Poplar.

1. P. *alba*. Linn. *Great White Poplar or Abele.* Leaves rotundo-cordate, lobed, toothed, glabrous above, downy and very white beneath ; fertile catkins ovate ; stigmas 4. *E. Bot. t.* 1618. *E. Fl. v. iv. p.* 243. *Hook. Br. Fl. p.* 440.

Hilly woods. *Fl.* April. ♃.

Lincoln's hill, Coalbrookdale ; *Mr. F. Dickinson.*

A large *tree*, with creeping roots and plentiful suckers ; *bark* smoothish ; *wood* white, soft, tough, close-grained ; *branches* horizontally spreading, white and cot-

tony when young. *Leaves* generally with 3 principal lobes, dark green and glabrous above, white and densely downy beneath, shortly petiolate. *Male flowers* in long cylindrical pendulous catkins with brown fringed scales. *Anthers* violet. *Female flowers* in ovate catkins.

This tree is usually considered identical with the Shittim-wood of the Scriptures.

2. P. *canescens*, Sm. *Grey Poplar.* Leaves roundish, deeply waved, toothed, hoary and downy beneath ; fertile catkins cylindrical ; stigmas 8. *E. Bot. t.* 1619. *E. Fl. v. iv. p.* 243. *Hook. Br. Fl. p.* 440.

Woods. *Fl.* March. ♃.

Borders of Wyre forest ; *E. Lees, Esq.*

A tall handsome *tree* with creeping root, slower in growth and producing firmer and better *wood* than the other species, valuable for floors ; *bark* smooth, silvery grey ; *branches* upright and compact. *Leaves* large, conspicuously 3-ribbed, white and downy beneath, "the poplar that with silver lines his leaf." *Catkins* all cylindrical, pendulous.

3. P. *tremula*, Linn. *Aspen.* Leaves nearly orbicular, broadly toothed, glabrous on both sides ; petioles vertically compressed ; stigmas 4, erect, auricled at the base. *E. Bot. t.* 1909. *E. Fl. v. iv. p.* 244. *Hook. Br. Fl. p.* 440.

Woods. *Fl.* March, April. ♃.

Near Marn wood, Coalbrookdale ; *Mr. F. Dickinson.* Wyre forest ; very abundant ; *Mr. G. Jorden.* Astley ; *Mr. E. Elsmere, junr.*

A lofty round-headed *tree* with creeping roots ; *bark* smooth ; *wood* white, soft, light, fine-grained ; *young branches* brown, pubescent. "The many-twinkling leaves of aspen tall" are roundish, slightly pointed, wavy and bluntly toothed, smooth, on vertically compressed petioles which causes them to tremble obedient to the slightest breeze.

4. P. *nigra*, Linn. *Black Poplar.* Leaves deltoid, acute, serrated, glabrous on both sides ; fertile catkins cylindrical, lax ; stigmas 4. *E. Bot. t.* 1910. *E. Fl. v. iv. p.* 245. *Hook. Br. Fl. p.* 441.

Woods and river-banks. *Fl.* April. ♃.

Captain's Coppice, Coalbrookdale ; *Mr. F. Dickinson.* Banks of the Teme, between Ludlow and Tenbury ; *E. Lees, Esq.*

Limekiln woods, near the Wrekin. Grinshill.

A tall umbrageous *tree* without suckers, of quick growth ; *bark* blackish ; *wood* tough, close-grained ; *branches* smooth. *Leaves* deep green, paler beneath. *Catkins* long, lax, pendulous.

DIŒCIA—ENNEANDRIA.

8. MERCURIALIS. *Linn.* Mercury.

1. M. *perennis*, Linn. *Perennial or Dog's Mercury.* Stem perfectly simple ; leaves rough ; root creeping, perennial. *E. Bot. t.* 1872. *E. Fl. v. iv. p.* 248. *Hook. Br. Fl. p.* 441.

Banks, woods, and shady places ; very common. *Fl.* April, May. ♃.

Root creeping extensively. *Herbage* rough. *Stem* about 1 foot high, obsoletely tetragonous, simple, slightly winged on two opposite sides. *Leaves* chiefly in the upper part, ovate, acute, serrated, petiolate, with small acute *stipules.* *Flowers* green, in axillary lax spikes, those of the female plant short. Supposed *nectary* awlshaped. *Capsule* bristly. The plant is said to be very poisonous and in drying generally turns of a bluish-green.

2. M. *annua*, Linn. *Annual Mercury.* Stem with opposite branches; leaves glabrous; root fibrous, annual. *E. Bot. t.* 559. *E. Fl. v. iv. p.* 248. *Hook. Br. Fl. p.* 441.

Waste and cultivated ground; rare. *Fl.* August. ☉.
Gardens in Frankwell, Shrewsbury; *Rev. E. Williams's MSS.*
Root much branched, fibrous. *Herbage* smooth, of a bright shining green. *Stem* about 1 foot high, erect, bushy, with numerous opposite branches. *Leaves* ovate or lanceolate, serrated and ciliated, shortly petiolate, with small acute *stipules. Flowers* green, *barren* ones in small tufts ranged in interrupted *spikes; fertile* ones fewer, stalked, axillary. *Fruit* covered with stiff hairs from tubercular bases.

9. HYDROCHARIS. *Linn.* Frog-bit.

1. H. *Morsus Ranæ.* Linn. *Common Frog-bit. E. Bot. t.* 808. *E. Fl. v. iv. p.* 250. *Hook. Br. Fl. p.* 441.

Ditches, ponds, &c.; not uncommon. *Fl.* Fuly. ♃.
Ditch in Edgmond moors a few yards from the Canal; *R. G. Higgins, Esq.* Ellesmere mere; *Mr. F. Dickinson.* Hordley, Tetch-hill, Ellesmere; *J. F. M. Dovaston, Esq.* Ditches about Newport, Kynnersley, Ellesmere, Hardwick, Hordley. Pits about Home Barns, near Sundorn; *Rev. E. Williams's MSS.*
Canal between Shrewsbury and Uffington. Pit at base of Haughmond hill. Floating and sending down long *radicles* from the horizontal *stems. Leaves* petiolate, rotundo-reniform, entire, purplish beneath. *Male flowers* 3, on long peduncles arising from a two-leaved pellucid membranous *spatha. Female flowers* on a solitary peduncle from a single *spatha. Sepals* ovate-oblong, concave, mem-. branous at the edges. *Petals* larger, roundish, undulate, very delicate, white with a yellow central stain. *Seeds* small, ovate, attenuated at the base, with a lax somewhat fleshy covering formed of warts consisting of spiral cellules.

DIŒCIA—MONADELPHIA.

10. JUNIPERUS. *Linn.* Juniper.

1. J. *communis*, Linn. *Common Juniper.* Leaves three in a whorl, linear, mucronate, spreading, longer than the fruit. *E. Bot. t.* 1100. *E. Fl. v. iv. p.* 251. *Hook. Br. Fl. p.* 442.

Woods and heaths; rare. *Fl.* May. ♄.
Near Kingswood, truly wild; *Mr. G. Jorden.* Woods Burford, Ashford Carbonel and in the neighbourhood of Ludlow; *Miss Mc Ghie.* Wyre Forest; *Mr. W. G. Perry.*
A bushy *shrub*, extremely variable in size, more or less erect, smooth; *wood* reddish; *branches* very numerous, spreading, subdivided, leafy, quadrangular. *Leaves* 3 in a whorl, spreading, evergreen, linear, mucronate, channelled and glaucous above, convex keeled and dark green beneath. *Flowers* axillary, sessile, small. *Berry* globose, bluish-black.

11. TAXUS. *Linn.* Yew.

1. T. *baccata*, Linn. *Common Yew.* Leaves linear, distichous, crowded; flowers axillary, sessile. *E. Bot. t.* 746. *E. Fl. v. iv. p.* 253. *Hook. Br. Fl. p.* 442.

Hilly places. *Fl.* March. ♄.
In woods and hedges about Eudon Burnell, near Bridgnorth; *Purton's Midl. Flora.* Sides of the Wrekin. Limekiln woods near Wellington. Wyre forest, sparingly. Woods on the Shatterford basaltic ridge between Bridgnorth and Kidderminster; *E. Lees, Esq.* Hatton; *T. C. Eyton, Esq.* Glen north of Coalbrookdale. Hedges between Wenlock and Bridgnorth; *T. & D. Bot. Guide.* Wenlock Edge near Lutwych. By the side of Pitchford brook; *Rev. E. Williams's MSS.*
Below Pimley, near Sundorn. Haughmond Hill. Lyth Hill. Westfelton. Grinshill. Battlefield. Blodwell rocks.
A *tree* of low stature, and of proverbially slow growth, but with a straight trunk frequently of considerable diameter, variously channelled longitudinally; *bark* smooth, deciduous; *wood* hard, beautifully veined; *branches* horizontal, spreading in opposite directions. *Leaves* scattered, nearly sessile, distichous, crowded, linear, entire, tapering gradually at the apex into a mucronate point, very slightly revolute, persistent, deep green, smooth and shining above, paler beneath, the midrib prominent on both surfaces. *Flowers* axillary, solitary, very pale flesh-colour, from scaly imbricated buds. *Drupe* sweet, internally glutinous, scarlet, with a delicate glaucous bloom, esteemed poisonous.
See a valuable Paper "On the Longevity of the Yew," by J. E. Bowman, Esq., F.L.S., in *Loudon's Mag. Nat. Hist. n. s.* 1. *p.* 28. 85.

Var. β. *Dovastoniana.* Branches pensile; leaves linear, suddenly pointed. *Loudon Arbor. Brit. v. iv. p.* 2083 *fig.* 1990.

In the grounds of J. F. M. Dovaston, Esq., at Westfelton, obtained from a hedge-bank near Sutton, parish of Westfelton.
This very remarkable variety differs from the common form in the extraordinary pensility of its branches; in the leaves being broader, suddenly and not gradually brought to a point; and in the drupes being frequently in pairs. I have dedicated it to the memory of its rescuer and planter, the late Mr. John Dovaston, of Westfelton, a self-educated gentleman of great ingenuity, talent, and science, devotedly attached to planting and botany: a feeble offering of respect from one who, though unknown to him, has again and again contemplated with delight and astonishment the productions of his active and ingenious mind, amid the calm retirement of the groves of curious and beautiful trees with which he enriched his grounds. Of this—

> "pillar'd shade
> Beneath whose sable roof
> Of boughs, as if for festal purpose, deck'd
> With unrejoicing berries, ghostly shapes
> May meet at noontide."

I insert the following description kindly communicated by my friend J. F. M. Dovaston, Esq. The fact of this tree being monœcious was first observed on 27 October, 1822; though Mr. Dovaston had previously in 1821 noticed the occurrence of some very healthy large and perfect berries on a small *male* yew-tree (since destroyed) close to the north-west corner of his house. "It is about sixty years since my father John Dovaston, a man without education, but of unwearied industry and acute ingenuity, had with his own hands sunk and constructed a pump; and the soil being light, it continually fell in; he secured it with wooden bars, but foreseeing their speedy decay, he planted near to it, a Yewtree, which he bought off a poor cobbler for sixpence, who had plucked it up from a hedge-bank near Sutton; rightly judging that the fibrous and matting tendency of the yew-roots would hold up the soil. They did so; and independent of its utility, the yew (as you have to your great admiration witnessed) grew into a tree of the most striking and distinguished beauty; spreading horizontally all around to the diameter of 63 feet, with a single spiral leader to a great height; each branch in every direction dangling in tressy verdure down to the

very ground, pendulous and playful as the most graceful birch or willow, and visibly obedient to the feeblest breath of summer air. Its foliage, like that of the Asparagus, is admirably adapted for retaining the dew drops; and at sunrise it would seem that Titania and a bevy of her Fairies had been revelling the night around it, and left their lamps in capricious frolick, so glitteringly coruscant is every branch with its millions of every-coloured scintillations, as it were all a-blaze.—To descend, however, to prose:—this lovely tree has food for the mind of the philosopher, as well as for the eye of the poet: for, strange to tell, and what few unseeing believed, though a male, and smoking like furnace, or a very volcano, with farina to the blasts of February, it has one entire branch self-productive, and exuberantly profuse in Female berries, full, red, rich, and luscious; from which I have raised 17 plants, every one of which already markedly partakes largely of the parents' *pensility.* Of these seedlings several have been presented to the following friends who have planted them in the localities appended to their names, viz.—Thomas Bulkeley Owen, Esq., at Tedsmere Hall, near Westfelton, Shropshire; Mrs. Hayman, Gresford, Denbighshire; Mr. George Yates, merchant, Woolton, near Liverpool; Thomas Jeffreys, Esq. Wilcott, near Nesscliffe, Shropshire; Edward Williams, Esq., Lloran House, Oswestry; Rev. Thomas Archer, Whitchurch, near Aylesbury, Bucks; Rev. C. A. A. Lloyd, Whittington, near Oswestry; and two in the grounds of W. A. Leighton, Esq., near Shrewsbury. The remaining trees are still in my possession and are intended to be distributed to Societies or persons who will undertake to plant them in situations where they are likely to be preserved. Berries will also, at the proper season, be given with pleasure, to such persons who may be curious in these matters."

CLASS XXIII.

POLYGAMIA.

"O Nature! all sufficient! over all!
Enrich me with the knowledge of thy works!
Snatch me to heaven; thy rolling wonders there,
World beyond world, in infinite extent,
Profusely scatter'd o'er the blue immense,
Show me; their motions, periods and their laws
Give me to scan; through the disclosing deep
Light my blind way: the mineral strata there;
Thrust, blooming, thence the vegetable world;
O'er that the rising system, more complex,
Of animals; and higher still, the mind,
The varied scene of quick-compounded thought,
And where the mixing passions endless shift;
These ever open to my ravish'd eye;
A search, the flight of time can ne'er exhaust!
But if to that unequal; if the blood,
In sluggish streams about my heart, forbid
That best ambition; under closing shades,
Inglorious, lay me by the lowly brook,
And whisper to my dreams. From Thee begin,
Dwell all on Thee, with Thee conclude my song;
And let me never, never stray from Thee!"

THOMSON.

CLASS XXIII.

POLYGAMIA. *Stamens and pistils separate or united, on the same or on different plants, and having two different kinds of perianth.*

ORDER I. MONŒCIA. *Flowers different on the same plant.*

1. **ATRIPLEX.** *Sterile flower and united flower.*—*Perianth* single, 5-partite, inferior. *Stamens* 5. *Style* bipartite. *Pistilliferous flower.*—*Perianth* single, of 2 valves. *Stamens* 0. *Fruit* 1-seeded, covered by the persistent enlarged *perianth.*—*Nat. Ord.* CHENOPO-DEÆ, *Juss.*—Name from *a*, *not*, and τριπλαξ, *triple.*

POLYGAMIA—MONŒCIA.

1. ATRIPLEX. *Linn.* Orache.

1. **A. angustifolia,** Sm. *Spreading narrow-leaved Orache.* Stem erect or prostrate; leaves lanceolate, entire, the lower ones with 2 ascending lobes from a wedgeshaped base; fruit-bearing calyces rhomboidal, with ascending prominent acute lateral angles, acute, entire, with the back smooth, longer than the fruit, and collected into nearly simple interrupted spikes; seeds smooth and shining. *Bab. Prim. Fl. Sarn. p.* 82. *Smith. Herb. E. Bot. t.* 1774. *E. Fl. v. iv. p.* 258. *Hook. Br. Fl. 4th ed. p.* 379. *Drej. Fl. Hafn. n.* 299.
Waste and cultivated ground; common. *Fl.* June. ☉.
Spikes wandlike, with distant clusters of *flowers.* Calyx-valves reticulated. Lateral angles of the *calyx* of the fruit acute and prominent; the valves suddenly and remarkably contracted above the lateral angles. When growing on cultivated land it becomes highly rampant, every part being increased and reduplicated in a remarkable manner; its smooth and shining *fruit* however keep it distinct from *patula*, even if the shape of the calyx-valves was not sufficient.

2. **A. patula,** Linn. *Spreading Halberd-leaved Orache.* Stems mostly erect, with ascending branches; lower leaves ovate-hastate, with 2 horizontally spreading lobes, denticulate; upper leaves lanceolate, nearly entire; fruit-bearing calyces triangular-rhomboidal, nearly entire, slightly muricated on the back, longer than the fruit, and collected into nearly simple interrupted spikes; seeds opaque, rough. *Babington Prim. Fl. Sarn. p.* 83. *Linn. Herb. E. Bot. t.* 936. *Hook. Br. Fl. 4th ed. p.* 379. *a. E. Fl. v. iv. p.* 257. *A. latifolia, β. Wahl.* 660.
Waste and cultivated ground; common. *Fl.* July. ☉.
Lateral angles of the *calyx* of the fruit obtuse. The wand-like nearly simple spikes with separated clusters of flowers, the deltoid leaves with spreading or even slightly deflexed lateral lobes, the form of the calyx-valves, and opake wrinkled seeds separate this both from *angustifolia* and *deltoidea.*

3. **A. deltoidea,** Bab. *Deltoid-leaved Orache.* Stem erect, with

ascending branches; leaves opposite, all hastato-triangular, with 2 descending lobes, unequally dentate or sinuato-dentate; fruit-bearing calyces ovato-triangular, dentate, muricated on the back, rather longer than the fruit, collected into a branched many-flowered panicle; seeds smooth, shining. *Babington Prim. Fl. Sarn. p.* 83.
Waste ground. *Fl.* September. ☉.
Old Heath near Shrewsbury, sparingly.
Root fibrous. *Stem* erect, striated, quadrangular, branched, the branches ascending, often reddish, 1-2 feet high. *Leaves* mostly opposite, all triangular-hastate, truncate at the base, with descending lobes, irregularly sinuato-dentate, acute-angled at the apex, dull green above, mealy beneath. *Spikes* terminal and axillary, mostly confined to the upper part of the stem and branches, shortly branched, many-flowered. *Flowers* in small round dense tufts placed near together, and thereby differing totally from *A. patula.* *Calyx* of the fruit ovate-triangular, sometimes almost cordate at the base, acute, dentate, strongly muricated, a little longer than the fruit, thickly covered with a mealy coat. *Seed* black, smooth and shining, not half so large as that of *A. patula*, reddish when immature. The whole upper part of the plant is covered with a minutely crystalline afterwards mealy coat.

END OF THE PHÆNOGAMOUS OR FLOWERING PLANTS.

ADDITIONS AND CORRECTIONS.

Page 7.—*Ligustrum vulgare.*
Buildwas abbey walls. In a hedge on right hand side of the road between Edgmond and Chetwynd. In the hedge of a coppice on the left hand side of the turnpike road between Battlefield and Hadnall; *Rev. E. Williams's MSS.*
On the north-west side of Colemere mere, probably planted.

Page 9.—*Veronica montana.* Add as syn. *Fries Nov. 2nd ed. p.* 1. *Reich. Fl. Excurs. n.* 2501. *Bluff and Fing. Comp. 2nd ed. t.* 1. *pt.* 1. *p.* 23. *Drej. Fl. Hafn. n.* 25.
Ditch-banks about Madeley; *Rev. E. Williams's MSS.*
On a rough bank opposite the foot-bridge leading over the river Morda to Tre Vawr Clawd, near Oswestry. Dunn's Dingle, Leaton Shelf, near Shrewbury. On Leaton Shelf.

Page 10.—*Veronica agrestis.* Add as syn. *Koch Syn. p.* 530. *Reich. Fl. Excurs. n.* 2490. *Reich. Fl. Germ. Exsicc. No.* 251! *Bluff and Fing. Comp. Fl. Germ. t.* 1. *pt.* 1. *p.* 28. *V. versicolor, Fries, Drej. Fl. Hafn. n.* 18. *a. Fries Nov. 2nd ed. p.* 2.
"There are two varieties of this plant; one with the capsules downy all over, and the other named 'ciliata' by my cousin Churchill Babington, junr., in which the keel is ciliated and the fruit otherwise glabrous. I make the hairy form the type, because Fries quotes *Reichenbach Iconogr. t.* 277. (' pro more nitidissime,') and that is this form. In *E. Bot. Suppl.* the capsule is represented as hairy. Bertoloni also (*Fl. Ital.* l. 100.) describes them as ' parce pilosa, pilis longiusculis.' " *Babington in lit.* Since receiving the above I have examined numerous living specimens and find that the keel is ciliated with glandular hairs, and that the uppermost exposed portion of the lobes which is not in contact with the sepals is covered with similar hairs, whilst the lower part or that concealed by and in contact with the sepals is perfectly glabrous. The hairy portion consequently varies in size according to the greater or lesser expansion of the sepals, whilst the quantity is very variable, even on the same specimen, from tolerable copiousness to a few scattered hairs; the keels however in all remaining uniformly ciliated. I have never seen the capsule *entirely* hairy. In these respects our specimens agree with *V. agrestis, Reich. Fl. Germ. Exsicc. No.* 251! Fries describes the sepals as "enerviis," but in our specimens and in that of Reichenbach they have 3 nerves. The hairs on the edges of the sepals I have never observed to be glandulose. The stamens are inserted, as remarked by Koch, in the lower margin of the tube.

Page 10.—*Veronica polita.* Add as syn. *Fries Nov. 2nd ed. p.* 1. *Reich. Fl. Excurs. n.* 2488. *Reich. Fl. Germ. Exsicc. n.* 248! *Drej. Fl. Hafn. n.* 17. *Bluff and Fing. Comp. p.* 28. *V. didyma,* "Ten." *Koch. Syn. p.* 531.
The hairs on the capsule are patent or spreading and tipped with glands; the lobes globoso-convex, not compressed at the keel; and the stamens are inserted in the inferior margin of the tube.

Page 11. 12.—*Veronica Buxbaumii,* Ten. *Buxbaum's Speedwell.* Stem procumbent; leaves petiolate, cordato-ovate, inciso-serrate, shorter than the flower-stalk; calyx-segments lanceolate or ovato-lanceolate, nearly acute; capsule divaricato-obcordate, transversely dilated, twice as broad as long, of 2 compressed sharply keeled divaricated ciliated lobes; seeds 7-8 in each cell. *V. Buxbaumii,* "*Ten. Fl. Neap. v.* 1. *p.* 7." *E. Bot. Suppl. t.* 2769. *Hook. Br. Fl. ed.* 3. *p.* 8. *ed.* 4. *p.* 7. *Koch, Syn. p.* 531. *Reich. Fl. Excurs. n.* 2409. *Bluff and Fing. Comp. Fl. Germ.* 2nd *ed. t.* 1. *pt.* 1. *p.* 29. *V. filiformis, (not of Vahl.) Johnst. Fl. Berw. v. i. p.* 225. *Bab. Fl. Bathon. p.* 37. *Hook. Br. Fl. ed.* 1. *p.* 6. *V. agrestis, β. Hook. Br. Fl. ed.* 2. *p.* 6. *V. persica, Fries, Nov.* 2nd *ed. p.* 4. ☉.

Corn field at Pen-y-lan, near Oswestry; *Rev. T. Salwey! Fl.* Summer and Autumn. ☉.

Root fibrous. *Stem* procumbent, filiform, divided immediately above the root, spreading to the length of 2 feet or more and throwing down roots from the lower sides of the joints and branches upwards, clothed on opposite sides with dense short pubescence intermixed with longer hairs. *Leaves* lowermost opposite, upper ones alternate, all petiolate, cordato-ovate, coarsely inciso-serrate, with a few scattered hairs on both surfaces and 3 principal ribs. *Flowers* large, bright blue; *peduncles* pubescent, longer than the leaves, erect, fractured at the apex in fruit. *Calyx* longer than the fruit, segments lanceolate or ovato-lanceolate, nearly acute, with 3 principal prominent ribs from which smaller ones ramify, margins ciliated. *Capsule* divaricato-obcordate, transversely dilated, the breadth twice the length, of 2 large compressed sharply keeled divaricated lobes, ciliated on the keels, the sides transversely reticulated with prominent veins when dry, crowned with the style which is about two-thirds the length of the capsule. *Seeds* 7-8 in each cell, similar in form and marking to those of *V. agrestis* and *polita.*

Page 11.—*Pinguicula vulgaris.*
Cound moor, and a piece of ground adjoining called Mosterley. Adjoining the Fox Farm, near Shrewsbury. Moors between Tern Hill and Moreton Say. In a boggy copse at the south end of the Wrekin; *Rev. E. Williams's MSS.*

Page 11.—*Utricularia vulgaris.*
In Sir Edward Smyth's pool, near Eaton Mascott. Ditches about Halston. Pits about Battlefield; *Rev. E. Williams's MSS.*
Ditches on Twyford Vownog, near Westfelton.

Page 12.—*Utricularia minor.*
Ditches between the Queen's-head turnpike and Woodhouse near Oswestry. Ditch on Shomere moss, near Condover. Pool by the side of the road on Knockin heath; *Rev. E. Williams's MSS.*

Page 12.—*Lycopus Europæus.*
Hadnall church-yard. Colemere mere. Morda pool, near Oswestry.

Page 13.—*Salvia verbenaca.*
Correct the description of the calyx thus :—*Calyx* tubular, with 13 ribs, unequally 2-lipped, ciliated, permanent; upper lip broadly obovate, with 3 minute triangular teeth, the exterior ones acuminato-mucronate converging and the middle one straight terminating the central rib; lower lip of 2 large ovato-acuminate mucronate segments.

Page 13.—2. *Salvia pratensis,* Linn. *Meadow Clary or Sage.* Lower leaves cordato-oblong, acute, irregularly duplicato-crenate, petiolate, upper ones lanceolate, acute, sessile, amplexicaul; bracteas

cordate, sharply acuminate; corolla large, many times longer than the calyx, glandular and viscid at the summit. *E. Bot. t.* 153. *E. Fl. v. i. p.* 34. *Hook. Br. Fl. p.* 11.
Dry meadows and hedge-banks; rare. *Fl.* July. ♃.
Road-side near the Fox Farm, near Shrewsbury; *W. P. Brookes, Esq!* Oakley park, near Ludlow; *Mr. H. Spare.*
Root woody. *Stem* erect, 4-sided, furrowed, clothed with decurved and spreading hairs. *Herbage* not very aromatic. *Leaves* dark-green, lower ones cordato-oblong, acute, duplicato-crenate, somewhat lobed, petiolate; upper ones lanceolate, acute, subcordate, doubly crenate, nearly sessile; uppermost linear-lanceolate, acute, sessile, semiamplexicaul; very much wrinkled and veiny, both surfaces covered with minute glandular mealiness, hairy chiefly on the ribs and margins beneath. *Bracteas* 2 under each whorl, cordato-acuminate, entire, ciliated and hairy with jointed hairs. *Calyx* tubular, with 13 glanduloso-pilose ribs, margins ciliated, permanent, unequally 2-lipped; upper lip broadly obovate, with 3 minute triangular teeth, the 2 exterior ones acuminato-mucronate converging, and the middle one shorter straight terminating the central rib; lower lip of 2 large ovate acuminate mucronate segments, reflexed in fruit. *Flowers* whorled at regular intervals. *Corolla* very large, exserted, many times longer than the calyx, dark purple; upper lip galeate, recurved and arched, linear-oblong, obtuse, notched, compressed, glanduloso-pilose on the exterior; lower lip 3-lobed, middle lobe largest, cloven; throat inflated in front. *Seeds* 4, oval, giving out a brown mucilage in water.

Page 14.—*Lemna trisulca.*
Ditches on Twyford Vownog, near Westfelton. Ditches on the common at the south end of Ellesmere mere.

Page 14.—*Lemna polyrrhiza.*
Ditches on the common at the south end of Ellesmere mere.

Page 18.—Insert after ERIOPHORUM.

9. * BLYSMUS. *Stems* simple, leafy. *Spikes* terminal, compound, distichous, compressed, bracteated. *Glumes* imbricated in 4 rows, upper ones smaller, all fertile. *Perigynium* of 6 persistent, more or less scabrous bristles. *Stigmas* 2. *Fruit* (Caryopsis) plano-convex, obsoletely trigonous, tapering upwards into the acuminated style.—*Nat. Ord.* CYPERACEÆ, *Juss.*—Name from βλυσμος, a *source* or *spring*; in allusion to the usual places of growth.

Page 26.—*Valeriana dioica.*
Twyford Vownog, near Westfelton. Morda pool, near Oswestry.

Page 25.—*Valeriana officinalis.*
Morda pool, near Oswestry.

Page 27.—*Valerianella olitoria.* Add the following synonyms : *V. olitoria, Koch Syn. p.* 339. *Reich. Fl. Excurs. n.* 1184. *Bluff and Fing. t.* 1. *pt.* 1. *p.* 58. *Fedia olitoria, Babington Prim. Fl. Sarn. p.* 48.
The form of the fruit in our specimens of this species differs from the figure given in Mr. Woods' *"Observations on the Species of Fedia,"* in the Linn. Trans. *xvij. p.* 421. *fig.* 1. in the achenium being much more laterally compressed and consequently narrower, more divided or lobed, furrowed before and behind, and having a lateral rib on each of the abortive cells. Indeed the transverse section of our fruit approaches more nearly in general appearance to the transverse sec-

tion of the achenium of an allied species *V. gibbosa,* represented in *fig.* 3. *pl.* 3. of *De Candolle's Mem. sur la famille des Valerianées.* Assuredly it is quite different from the fruit of *V. olitoria,* given in *fig.* 2. of the same work. In our plant the dissepiment separating the abortive cells is incomplete, but nevertheless is always more or less present and the bracteas are dentato-ciliate, characters which seem chiefly to distinguish *V. olitoria* from *V. gibbosa.* The character of the fruit designated by Koch in the expression "lateribus bicostatis, costa altera tenuissima," and by Bluff and Fing. " utrinque bistriatum," it is presumed alludes to the single slender rib on the side of the abortive cells and also to the larger prominent thickened margin of the dorsal corky mass subtending the fertile cell, on the edge of the furrow which externally marks the limits of the fertile and abortive cells. With respect to the furrows of the fruit, Koch says "fructibus utrinque planiusculis, margine sulco exaratis," and Bluff and Fing. " unisulcatum :" by which phrase Koch seems to imply that the anterior and posterior faces are destitute of furrows, the only furrow being that on the sides between the sterile and fertile cells; whilst the expression of Bluff and Fing. agrees with De Candolle's figure of *olitoria* before mentioned, which has a furrow on the anterior face only, and also with Woods' fig. 1. which has one furrow on the posterior face only and none on the anterior: our plant however possesses a furrow invariably on both faces. In all these characters specimens from Berwickshire, Scotland, correspond with our Shropshire ones. Of course the above remarks are appliable only to the *mature* fruit.

Page 27.—*Valerianella carinata.* Add as synonyms :—*V. carinata, Koch Syn. p.* 339. *Reich Fl. Excurs. n.* 1183. *Bluff and Fing. Comp. t.* 1. *pt.* 1. *p.* 58. *F. carinata, Babington Prim. Fl. Sarn. p.* 48.
Hedge-banks under the quarries near Leigh Hall, one mile from Grimmer rocks; *J. E. Bowman, Esq!*

Page 27.—*Valerianella dentata.* For *V. dentata, DC.* read *V. Morisonii, DC.,* and add as synonyms : *V. Morisonii,* "*DC. Prodr.* 4. *p.* 627." *Koch Syn. p.* 340. *F. dentata, Bab. Prim. Fl. Sarn p.* 49. *Drej. Fl. Hafn. n.* 34. *Fries Nov.* 2nd *ed. p.* 5. *V. dentata, Reich. Fl. Excurs. n.* 1186. *Reich. Fl. Germ. Exsicc.* No. 182! *Bluff and Fing. Comp. Fl. Germ.* 2nd *ed. t.* 1. *pt.* 1. *p.* 59.
Neither our Shropshire plants-nor any specimens so named from other parts of Britain which have come under my observation are referable to the description of *V. dentata* of DC. and Koch, which is a totally different species, having the sterile and fertile cells nearly equal in size and consequently is included in a different section of the genus, the Platycœlæ of DC., whilst ours is comprehended in the section Psilocœlæ, in which the sterile cells are very small and filiform. Our plant is indeed the *F. dentata* of Woods' Observations. The crown of all our plants differs in being more or less open ; in some specimens it is closed by the convergence of the teeth, so as to resemble the appearance in Woods' fig. 4 and 5, *α*; whilst in others it is more open and expanded, resembling his figures 3, 6 and 7. In all, the abortive cells of the intermediate rib on the anterior face terminate in a thickened process somewhat below the crown, and the centre of the anterior face has invariably a tooth immediately above the conjunction of the 2 abortive cells. Mr. Woods states that he had never observed this central tooth although Reichenbach's figures frequently represent it, but on referring him to the circumstance in our plants he replies—" On recurring to my herbarium and examining the seeds in the way you have pointed out I observe that a central tooth does often occur, and is not unfrequent among the Psilocœlæ, but not so universally as it seems to be with you. Sometimes the 3 teeth on the anterior face are nearly equal, sometimes the middle one is much smaller, sometimes it

seems reduced to a mere bunch of the ciliæ with which these teeth are, I believe, always furnished. Now and then there is clearly and distinctly no central tooth, while at other times the 3 teeth seem to be united into an irregularly jagged membrane." *Woods in lit.*

The figure of the fruit of our β. *eriocarpa* agrees generally with the open-crowned form of Woods' *dentata*, except in being perhaps a little stouter; the crown and teeth also perfectly correspond, and it seems to differ chiefly in being hairy. The anterior tooth is here also visible. Upon this Mr. Woods observes— " If *F. eriocarpa* of DC. be a species, it must probably rather be determined by the habit of the plant, the panicle terminating in secund spikes from the repeated abortions of the branches and the upward thickening of the flowerstalks, than by the seeds. I have been sometimes inclined to doubt whether the variety of *dentata* with involute teeth and a very small and a very oblique mouth, might not be distinct from the rest. If so the name *dentata* must be preserved to the open-mouthed variety which is the most common, and which is no doubt what has been called *F. mixta*." *Woods in lit.*

Valerianella dentata. Reich Fl. Germ. Exsicc. No. 182 *!* is identical with our form with the open-mouthed calyx from Bomere and Sharpstones hill.

Page 29.—*Montia fontana.*
Boggy ground near the Wolf's-head turnpike gate, on the road from Nesscliff to Oswestry.

Page 30.—*Eleogiton fluitans.*
Ditch on east side of morass between the Queen's-head turnpike and Wood-house, near Oswestry. On the north side of the largest pool near Abbot's Betton. In abundance in a ditch on Shomere moss. In a ditch on the west side of Berrington pool. Ditches about Ellesmere and Birch; *Rev. E. Williams's MSS.*

Page 30.—*Eriophorum vaginatum.*
Shomere moss. Harmer moss, near Pimhill. Bog below the Weeping Cross turnpike, near Shrewsbury; *Rev. E. Williams's MSS.*

Page 31.—*Eriophorum polystachion.*
On a piece of boggy ground called Mosterley adjoining Cound moor. In a boggy copse adjoining Vessons Wood under the Stiperstones; *Rev. E. Williams's MSS.*

Page 32.—Insert before 10. SCIRPUS.

9.* BLYSMUS. *Panz.* Blysmus.

1. B. *compressus*, Panz. *Broad-leaved Blysmus.* Spikelets many-flowered; bristles 6, as long as the persistent style, with deflexed teeth; fruit elliptico-obovate, plano-convex, obsoletely trigonous, attenuated upwards into the very long filiform style, punctato-striate. *Hook. Br. Fl.* 4th ed. p. 27. *Schœnus compressus,* Linn. E. Bot. t. 791. *Scirpus caricinus,* Schrad. E. Fl. v. i. p. 59.
Moist meadows; rare. *Fl.* July. ♃.
In a field at the bottom of Wenlock Edge, by the side of the road from Lutwyche to Kenley; *Rev. E. Williams's MSS.*
Root somewhat creeping. *Stem* 6—8 or 12 inches high, simple, leafy. *Leaves* linear, acute, channelled, sheathing at the base. *Spikes* of a bright chestnut colour, terminal, compound, compressed, distichous; *spikelets* many-flowered, each subtended by a membranous ribbed *bractea,* the lowermost bractea generally foliaceous. *Glumes* 3-ribbed, with smaller intermediate ones. *Stigmas* 2, pubescent. *Fruit* greyish. Bristles 6, arranged on the outside of the stamens.

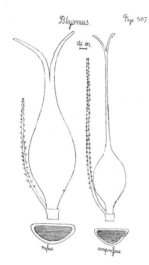

Blysmus. *Page* 507.

rufus. *compressus.*

Page 32.—*Scirpus lacustris.*
Ellesmere and Colemere meres. Golding pool, near Pitchford.

Page 32.—*Scirpus sylvaticus.*
At the upper end of the water at Sundorn. Side of the brook a little above Pitchford Forge. Betton pool. Above Woodhouse. Meadows about Duncote. Under Cronkhill; *Rev. E. Williams's MSS.*
Dunn's Dingle, Leaton Shelf, near Shrewsbury. In wet meadows through which the bridle road leads from the turnpike road near Sundorn to the Shrewsbury Canal, at Pimley, abundant. Morda pool, near Oswestry.

Page 33.—*Eleocharis palustris.*
Pit between Battlefield and Albright Hussee. Boggy part in meadows south side of Battlefield church. Morda pool, near Oswestry. Ellesmere and Colemere meres.

Page 34.—*Scirpidium aciculare.*
On the east side of the water at Sundorn. By the side of the river Roden at Rodington hall. On the south side of Betton upper pool. Sides of Colemere and Whitemere; *Rev. E. Williams's MSS.*

Page 34.—*Bæothryon Halleri.*
Ditch on Shomere moss and on the moss on the south-east side of Bomere pool. North side of Berrington pool. Upon a piece of ground called Mosterley, adjoining Cound moor. Betton upper pool; *Rev. E. Williams's MSS.*

Page 34.—*Bæothryon cæspitosum.*
Heath above Silvington; *Rev. E. Williams's MSS.*

Page 35.—*Cladium Mariscus.*
Round Colemere mere in abundance, and in a bog on the right hand side of the road between the Queen's-head turnpike and Woodhouse, near Oswestry; *Rev. E. Williams's MSS.* Colemere mere; *H. Bidwell, Esq!*

Page 36.—*Rhynchospora fusca.*
This plant was inserted on the authority of a MS. list of rare plants occurring at Bomere pool in the hand-writing of the Rev. E. Williams, communicated to the author by Dr. Du Gard; but as the species is not noticed as a Shropshire plant in Mr. Williams's MS. Catalogue it is probable that there is some mistake, and consequently it must be erased.

Page 37.—*Lolium temulentum.*
Corn-fields between Eaton and Pitchford, and about Shawbury heath; *Rev. E. Williams's MSS.*

Page 37.—*Lolium festucaceum.*
Moist meadows about Berrington and Pitchford; *Rev. E. Williams's MSS.*

Page 38.—*Agropyrum caninum.*
" Old hedges; " *Rev. E. Williams's MSS.*

Page 39.—*Hordeum pratense.*
Upon the lawn at Sundorn and adjoining pastures. In the church-yard at Battlefield, and in the meadow adjoining; *Rev. E. Williams's MSS.*

Page 42.—*Alopecurus geniculatus.*
Morda pool, near Oswestry.

Page 42.—*Alopecurus geniculatus. Var.* γ. *bulbosus.* Root with copious long fibres; lower joints of the culm oval and fleshy. *E. Fl. l. c.*
" Sandy roads; " *Rev. E. Williams's MSS.*

Page 43.—*Alopecurus agrestis.*
Corn-fields between Culmington and Onybury. About Upper Nash. By the side of the road in Priors Ditton. Below Harnage; *Rev. E. Williams's MSS.*

Page 44.—*Glyceria fluitans.*
Morda pool near Oswestry.

Page 45.—*Glyceria spectabilis.*
On the banks of the Roden and Tern. In a mill-pool by the side of the road between Muckley Cross and Morvill. In the pool at Lilleshall House; *Rev. E. Williams's MSS.*

Page 47.—*Sclerochloa rigida.*
Wenlock Abbey. Round the old limekilns on Stevens Hill near Cound; *Rev. E. Williams's MSS.*

Page 48.—*Baldingera arundinacea.*
Morda pool near Oswestry.

Page 49.—*Phragmites communis.*
Colemere mere.

Page 51.—*Festuca gigantea.*
Banks of the river Morda between Pen-y-lan Mill and the bridge above, near Oswestry; *Rev. T. Salwey !*

Page 52.—*Vulpia Myurus.*
Old walls Shrewsbury and Ludlow; *Rev. E. Williams's MSS.*

Page 52.—*Vulpia bromoides.*
By the side of turnpike road on Bayston Hill, and near Atcham bridge; *Rev. E. Williams's MSS.*
South side Pontesford Hill.

Page 53.—*Danthonia decumbens.*
Twyford Vownog near Westfelton.

Page 53.—*Bromus secalinus.*
Corn-field near Berrington; *Rev. E. Williams's MSS.*

Page 55.—*Bromus erectus.*
On the bank of the Severn by the Austin Friars below the Welsh bridge, Shrewsbury; *Rev. E. Williams's MSS.*

Page 58.—*Calamagrostis Epigejos.*
Locality of Aqualate mere, no part of which water is in Shropshire.

Page 58.—*Calamagrostis lanceolata.*
About Sundorn and Battlefield. Marton pool; *Rev. E. Williams's MSS.* The locality of Aqualate mere, which is in Staffordshire.

Page 60.—*Aira flexuosa.*
About Pulverbatch. Shelve. Moors about Ellesmere. Harmer hill. Pimhill; *Rev. E. Williams's MSS.*
Grinshill. Lyd hole near Pontesford hill. Nesscliffe hill.

Var. γ. *E. Fl. l. c.* Leaves longer; panicle less wavy and less regularly ternate; glumes much paler or greenish.
Haughmond hill; *Rev. E. Williams's MSS.*

Page 60.—*Aira caryophyllacea.*
Pontesford hill, (east side.) Blodwell rocks.

Page 61. *l.* 9.
For " coppiece" read " coppice."

Page 61.—*Aira precox.*
Pontesach hill. Cound moor, near the stone quarry.

Page 67.—Add after MŒNCHIA.

16. RADIOLA. *Calyx* of 4 leaves united up to their middle, and mostly 3-cleft. *Petals* 4. *Capsule* of 8 cells and 8 valves.—*Nat. Ord.* LINEÆ, *DC.*—Name from *radius,* a *ray;* probably from the ray-like segments of the calyx.

Page 67.—*Dipsacus pilosus.*
By the side of Eaton and Cound brooks. In Mr. Corbet's wood at Sundorn. Upper end of the water at Sundorn. About Alveley &c. By Buildwas mill. Under Red Castle, near Hawkstone; *Rev. E. Williams's MSS.* In the road between Lubstree Park and Lilleshall; *H. Bidwell, Esq.*

Page 69.—*Scabiosa columbaria.*
By the side of the turnpike road near Presthope; *Rev. E. Williams's MSS.* Llanymynech Hill. Blodwell rocks.

Page 69.—*Galium Mollugo.*
Hedges about Shiffnal and Coalbrookdale. Quat; *Rev. E. Williams's MSS.*

Page 70.—*Galium saxatile.*
Grinshill. Pontesford Hill. Llanymynech Hill. Sweeney mountain.

Page 72.—*Plantago media.*
About Much Wenlock, Buildwas and other limestone soils. Burford. Cleobury Mortimer. Neen Savage. Farlow. Nash; *Rev. E. Williams's MSS.*

Page 72.—*Plantago coronopus.*
At the junction of 3 roads between Venus Bank and the Paper mill in the parish of Cound. By the side of the road between Cardiston and Cross Gates. Side of the road near Dudmaston. Upon Haughmond Hill, near Downton. About Edgmond, &c. Harmer Hill and by the side of Whitemere, near Ellesmere. Upon Cound moor. The Morf, near Bridgnorth; *Rev. E. Williams's MSS.* Lea Hall, near Shrewsbury.

Page 72.—*Cornus sanguinea.*
Hedges of turnpike road between Weeping Cross gate and the Fox Farm, near Shrewsbury. Hedges near the brook near Cound moor. Penyvoel and Llwyntidman, near Llanymynech. Hedges of turnpike road 1 mile from Oswestry on Shrewsbury road.

Page 73.—*Alchemilla vulgaris.*
Eastern summit of Pontesford Hill. Twyford Vownog, near Westfelton.

Page 75.—*Potamogeton crispus.*
Morda pool near Oswestry. Ellesmere mere.

Page 75.—*Potamogeton pectinatus.*
Lilleshall Canal, near Lilleshall Abbey; *Rev. E. Williams's MSS.*

Page 76.—*Potamogeton pusillus.*
In the brook above Pitchford Hall. Ditches between Queen's Head turnpike and Woodhouse, near Oswestry; *Rev. E. Williams's MSS.*
β. *major.* Ditches below Mr. Kynaston's house at Hardwicke. Pond below Golding. Hancott pool, plentifully. Eaton Mascot pool; *Rev. E. Williams's MSS.*

Page 77.—*Potamogeton heterophyllus.*
Whitemere and Blackmere, near Ellesmere; *Rev. E. Williams's MSS.*

Page 77.—*Potamogeton oblongus.* Add as syn. *Bab. Prim. Fl. Sarn. p.* 99.
Ditches on Twyford Vownog, near Westfelton.

Page 78.—*Potamogeton rufescens.*
Pond on Hollyhurst common, near Longnor. Ditches by the side of the road between Melverley and Oswestry. Ditches about Hordley and Rednall Mill; *Rev. E. Williams's MSS.*
Boggy ground near Wolf's Head turnpike on the road from Shrewsbury to Oswestry.

Page 79.—*Sagina apetala.* Add as syn. *S. apetala,* α. *ciliata, Bab. Prim. Fl. Sarn. p.* 15.

Page 79.—*Mœnchia erecta.*
Knockin heath. Abery wood. Kenley common. Harmer hill; *Rev. E. Williams's MSS.*
Haughmond hill. Pontesford hill, (south side.)

Page 79.—Add after MŒNCHIA.

16. RADIOLA. *Gmel.* Flax-seed.

1. *R. millegrana,* Sm. *Thyme-leaved Flax-seed. E. Bot. t.* 890. *E. Fl. v. i. p.* 244. *Hook. Br. Fl. p.* 80. *R. linoides, Gmel. Linum Radiola, Linn.*
Moist gravelly and boggy soils; rare. *Fl.* July, August. ☉.
On the south-east side of Shawbury heath. On Uckington heath; *Rev. E. Williams's MSS.* Rudge heath; *H. Bidwell, Esq!*
A very minute plant, 1-2 inches high, repeatedly dichotomous. *Leaves* distant, ovate, entire, glabrous, dotted. *Flowers* axillary and terminal, stalked, solitary, on short peduncles. *Calyx-segments* united so as to form a monophyllous many-toothed calyx.

Page 88.—In the generic character of ULMUS, for " *Perianth single, superior,*" read " *Perianth single, inferior.*"

Page 95.—*Echium vulgare.*
Clover and corn-fields and pastures, in a gravelly soil about Eaton Mascot. Cound. Hardwick near Hadnall. Cardiston, &c.; *Rev. E. Williams's MSS.* Hopton hill. Fields near the Black Barn, near Eaton Mascot.

Page 96.—*Lithospermum officinale.*
On right hand side of the road up the hill between Wenlock and the Marsh. On the right hand side of the road up the hill from Coalport. On left hand side of the road above Harley turnpike. Walls in Cleobury North. Woods and thickets between Buildwas and Wenlock. Walls of Red Castle near Hawkstone; *Rev. E. Williams's MSS.*
Canal side near Blackmere, near Ellesmere.

Page 97.—For *Symphytum officinalis* read *Symphytum officinale.*
By the side of the brook in several places between Pitchford and Cound. Plentifully in meadows by the side of the brook at Shiffnal. Meadows adjoining Sutton Maddock. Side of Condover Brook. Side of the brook at Ruyton, near

Beckbury. (With purple blossoms.) By the side of the Tern at Duncot; *Rev. E. Williams's MSS.*
Westfelton.

Page 97.—*Lycopsis arvensis.*
Lane near Croeswylan, near Oswestry.

Page 98.—*Anchusa sempervirens.*
By the side of the road between Barrow and Willey. Croeswylan, near Oswestry; *Rev. E. Williams's MSS.*

Page 100. *l.* 15.
For " Hancott park" read " Hancott pool."

Page 101.—*Lysimachia vulgaris.*
Bog between the Cloud Coppice and the Gouter, plentifully; *Rev. E. Williams's MSS.*
Ellesmere and Colemere meres; abundant.

Page 102.—*Lysimachia Nummularia.*
Lanes and roadsides about Culmington, Seifton and Millichope, plentifully; *Rev. E. Williams's MSS.*

Page 103.—*Primula Veris, var.* γ. *elatior.*
Meadows at Warbrook, near Westfelton.

Page 105.—*Hottonia palustris.*
Pit in the field opposite the Inn at Uffington, near Shrewsbury. Ditches on Twyford Vownog, near Westfelton.

Page 105.—*Menyanthes trifoliata.*
Ellesmere and Colemere meres.

Page 105.—*Erythræa Centaurium.*
Var. with white blossoms about Eaton; *Rev. E. Williams's MSS.*

Page 106.—*Hyoscyamus niger.*
Near the village of Wyke; *J. E. Bowman, Esq.*
Astley church-yard.

Page 106.—*Atropa Belladonna.*
Among the loose stones below the old limekilns on Stevens Hill, near Cound; *Rev. E. Williams's MSS.*

Page 107.—*Solanum dulcamara.*
In immense abundance around Shrawardine pool.
Var. with white blossoms on Haughmond hill; *Rev. E. Williams's MSS.*

Page 108.—*Verbascum Blattaria.*
By the side of the road between Ludlow and Bromfield. In the village of Felshampton. By the side of the road between Cound moor and Cressage. Upon Haughmond hill, near Downton. By the side of the road between Upton (near Shiffnal) and Ivelith; *Rev. E. Williams's MSS.*
Var. with white blossoms near Bridgnorth; *Rev. E. Williams's MSS.*

Page 108.—5. *Verbascum nigrum,* Linn. *Dark Mullein.* Leaves crenate, nearly glabrous above, slightly tomentose beneath; lower cauline ones oblongo-ovate, cordate, on long petioles, upper ones ovato-oblong, subsessile; stem acutely angular upwards; flowers pedicellate, in subremote bracteated clusters, on an elongated raceme; pedicels twice the length of the calyx; filaments with purple woolly

hairs about their middle. *E. Bot. t.* 59. *E. Fl. v. i. p.* 312. *Hook. Br. Fl. p.* 112.
Banks and way-sides; rare. *Fl.* July, August. ♃.
Pastures between Longnor and Preston. At Bobaston Bank, near Edgmond; *Rev. E. Williams's MSS.*
Stem 2-3 feet high, erect, generally simple, sparingly clothed with stellate clusters of hairs. *Leaves* dark green, veiny, somewhat downy. *Flowers* in subremote clusters of 6—8, on stellately hoary bracteated pedicels. *Calyx* stellately hoary. *Corolla* rather large, yellow, with purple spots in the throat.

Page 109.—*Polemonium cæruleum.*
In a ditch bank on the right hand side of the turnpike road between the Golden Heart public-house and Burlton; *Rev. E. Williams's MSS.*

Page 109.—*Vinca minor.*
In the plantation adjoining Berwick laundry, near Shrewsbury; *Rev. E. Williams's MSS.*

Page 109.—*Vinca major.*
Hedges in Cound, Berrington, Acton Burnell, Eyton on the Wildmoors, Shelton, &c., &c. On the right-hand side of the road between Atcham and Tern bridge. Between the turnpike and Battlefield; *Rev. E. Williams's MSS.*

Page 110.—*Jasione montana.*
Road-side at Lea Hall, near Shrewsbury. Grinshill. Lyd hole, near Pontesford hill.

Page 110.—*Lobelia Dortmanna.*
Whitemere, near Ellesmere; *Rev. E. Williams's MSS.*

Page 110.—*Campanula patula.*
Ditch-banks about Eaton Mascot, Cound, Berrington, Montford, Shipton, &c., common; *Rev. E. Williams's MSS.*

Page 111.—*Campanula latifolia.*
By the side of the turnpike road about ⅓ of a mile beyond Muckley Cross towards Morvill. In a copse near Belswardine; *Rev. E. Williams's MSS.* Oswestry parish; *Rev. T. Salwey!*

Page 112.—*Campanula Trachelium.*
Hedges about Eaton Mascot, common; *Rev. E. Williams's MSS.* Oswestry parish; *Rev. T. Salwey!*

Page 112.—*Campanula glomerata.*
Ditch banks about Lutwych; *Rev. E. Williams's MSS.*

Page 112.—*Wahlenbergia hederacea.*
North side of Titterstone Clee Hill towards the bottom; *Rev. E. Williams's MSS.*

Page 113.—*Prismatocarpus hybridus.*
Corn-fields near the Gouter upon the Eaton Mascot Estate; *Rev. E. Williams's MSS.*

Page 114.—*Rhamnus catharticus.*
A tree on the right-hand side of turnpike road about 150 yards beyond Cound stank bridge towards Pitchford. Upon the bank below the limekilns on Stevens Hills, near Cound; *Rev. E. Williams's MSS.* In hedges between Westbury and Minsterley; *J. E. Bowman, Esq.*

Page 114.—*Rhamnus Frangula.*
Cound moor, and on an adjoining piece of ground called Mosterley. Wood

called Colon's Rough, near Sundorn. Copse at south end of the Wrekin; *Rev. E. Williams's MSS.*

Twyford Vownog, near Westfelton. Colemere mere.

Page 114.—*Euonymus Europæus.*

Hedges about Eaton Mascot; *Rev. E. Williams's MSS.* Craigforda, near Oswestry. Hedges of roadside near the Black Barn near Eaton Mascot. Hedges near the brook near Cound moor.

Page 115.—*Viola hirta.*

Llynymaen, once the residence of Edward Llhwyd, the Antiquary. Blodwell rocks.

Page 115.—*Viola odorata.*

Montford village. Lane leading from Shrawardine towards Shrawardine pool. Near Oswestry.

Page 116—*Viola odorata, var. γ. imberbis.*

"This comes very near to *V. odorata*, and is much more common than is generally believed. Should the beardless petals prove uniformly constant, it may be regarded as a distinct species, and your name *imberbis* retained for it. *V. suavis* of Bieberstein, although referred by De Candolle as a variety under *odorata*, is not the same form as your violet."—*Profr. D. Don in lit.*

Page 116.—*Viola palustris.*

Berrington pool and in bogs, not uncommon; *Rev. E. Williams's MSS.* Near the Steeraway lime works near the Wrekin; *H. Bidwell, Esq.*

Page 117.—*Viola canina, var. γ. flavicornis.* "Stem ascending, woody, somewhat angular, much branched; leaves heart-shaped, coriaceous, smooth and even; stipules and bracteas fringed; calyx-leaves lanceolate."—*Smith.* *V. flavicornis, Sm. E. Fl. v. i. p.* 805. *E. Bot. Suppl. t.* 2736. *V. canina, β. minor, Hook. Br. Fl. 3rd ed. p.* 121. *4th ed. p.* 105.

Twyford Vownog, near Westfelton.

Chiefly distinguished from *V. canina* by its smaller size, deeper colour of the corolla, the short blunt yellowish spur, and the short firm rigid very even cordate obtuse leaves on equally short petioles.

"This is certainly *V. flavicornis* of Smith, but I fear it is only a variety of *V. canina.*"—*Profr. D. Don in lit.*

Page 119.—*Viola lutea.*

Hills above Shelve, Bishop's Castle, Bettws, and all parts of the Hundred of Clun. On the right-hand side of the road between Nesscliff and Knockin heath; *Rev. E. Williams's MSS.* North end of Oswestry race-ground; *Rev. T. Salwey.* Var. with blue flowers, about Shelve and Knockin heath; *Rev. E. Williams's MSS.*

Var. with whitish flowers, near Knockin heath; *Rev. E. Williams's MSS.*

Profr. D. Don identifies our Stiperstones plant as *V. lutea, Huds.* and considers it "a good species, although chiefly distinguished from *V. tricolor* by its habit and perennial root."

Page 120.—*Ribes rubrum.*

Eaton Mascott. Coppice by the brook side, and by the side of Cound paper-mill stream and Cound brook; *Rev. E. Williams's MSS.*

Page 120.—*Ribes nigrum.*

Brook in the lane leading from Bayston hill to the Sharpstones hill.

Page 122.—*Chenopodium intermedium.* Add as syn: *Hook.*

Br. Fl. 4th ed. p. 124. *Bluff and Fing. Comp. Fl. Germ. 2d ed. t. i. pt.* 1. *p.* 449. *C. urbicum, β. intermedium, Koch Syn. p.* 605.

Koch says (*l. c.*) "Cultura, sæpe repetita, edoctus sum, *C. intermedium* nil nisi varietatem *C. urbici* efficere."

Rev. E. Williams in his MS. Catalogue inserts " *C. urbicum*; dunghills, not uncommon:" but as I have had no opportunity of identifying his specimens and the only Shropshire specimen which I have seen being *C. intermedium, M. and K.* which is assuredly the *C. urbicum, E. Fl.* it appears highly probable that Mr. Williams's plants are similar, at all events I do not feel myself justified in inserting the *true C. urbicum, Linn.* as a native of Shropshire.

Page 123.—*Chenopodium album.* Add to the specific character: seeds orbicular, depressed, smooth and shining, convex on both sides, obtusely but distinctly keeled on the margins.

Page 123.—4. *Chenopodium murale*, Linn. *Nettle-leaved Goose-foot.* Stem erect, much branched; leaves rhombeo-ovate, entire at the cuneate base, acutely sinuato-dentate above, teeth unequal acute directed towards the apex somewhat incurved, with 3 principal ribs at the base; flowers in subaxillary and terminal, divaricately branched, leafless cymes; seeds depressed, opaque, elevato-punctate, acutely keeled at the margins. *E. Bot. t.* 1722. *E. Fl. v. ii. p.* 11. *Hook. Br. Fl. p.* 142.

Waste places near towns and villages. *Fl.* August. ☉.

Dunghills and rubbish as you enter the Abbey Foregate, Shrewsbury; *Rev. E. Williams's MSS.*

Whole plant fetid, of a darkish slightly glaucous green, with purple streaks occasionally on the stem and branches. *Stem* much and alternately branched, angular and furrowed. *Leaves* alternate, rhombeo-ovate, acutely and copiously sinuato-dentate in the upper part, the cuneate base being entire, teeth irregular and unequal in size, very acute, all directed towards the apex and more or less incurved, the basal teeth rather more elongated and somewhat more spreading than the rest, with 3 principal ribs at the cuneate base which slightly tapers into the petiole, dark green and shining. *Cymes* subaxillary and mostly terminal, copiously divaricately branched, leafless. *Flowers* in numerous small round dense green or glaucous mealy clusters, sessile or rather remotely arranged on the spreading and slightly nodding branches. *Seeds* orbicular, horizontally depressed, convex on both sides, surrounded with an acutely keeled margin, black, opaque, rough with very minute elevated points.

5. *Chenopodium ficifolium*, Linn. *Fig-leaved Goose-foot.* Stem erect, branched; leaves oblong, subhastate, unequally 3-lobed, entire at the rhomboid or cuneate base, intermediate lobe elongated, sinuato-dentate, teeth unequal acute more or less spreading, with 3 principal ribs at the base; upper leaves lanceolate, entire; flowers in axillary and terminal, compound, erect, nearly leafless racemes; seeds depressed, shining, excavato-punctate, obtuse but not keeled at the margins. *E. Bot. t.* 1724. *E. Fl. v. ii. p.* 13. *Hook. Br. Fl. p.* 143.

Dunghills and waste ground. *Fl.* August, September. ☉.

" Dunghills and among potatoes, common ;" *Rev. E. Williams's MSS.*

Stem erect, alternately branched, 2-3 feet high, obtusely angular, striped with green and white. *Leaves* alternate, petiolate, oblong, unequally 3-lobed, some-

what hastate, the intermediate lobe elongated oblongo-lanceolate obtuse mucronate shallowly and remotely sinuato-dentate, the teeth irregular and unequal, more or less spreading, lateral lobes abbreviated obtuse entire more or less spreading, the base rhomboid or cuneate and entire tapering into the petiole, with 3 principal ribs at the base; uppermost leaves oblong or lanceolate, obtuse, mucronate, more or less entire; clothed on both surfaces as is the whole plant with minute mealiness. *Flowers* in terminal and axillary, compound, erect, nearly leafless racemes, the ramifications diverging more or less at right angles and bearing the numerous round dense crowded sessile clusters; the lower racemes elongated and interspersed with numerous petiolate narrow entire linear-lanceolate leaves, the upper racemes much more abbreviated denser and nearly leafless. *Fruit* black, shining, orbicular, depressed, somewhat flattened above, convex beneath, minutely excavato-punctate, margins rounded and obtuse not keeled.

C. album may be readily distinguished from this by its smooth fruit, convex on both sides, obtusely but distinctly keeled on the margins.

Page 123.—*Orthospermum Bonus Henricus.*

Preston Gobalds church-yard. Little Ness. Croeswylan, near Oswestry. Llanymynech.

Page 124.—*Ulmus montana.*

Very large and noble trees on Haughmond hill, near Queen Eleanor's bower, and one immense tree close to the south side of Haughmond Abbey.

Page 125.—*Gentiana Amarella.*

By the side of the road up Wenlock Edge above Harley. Pastures between Clee St. Margaret and Abdon. Between Westhope in the parish of Diddlebury and Strefford. As you ascend Wenlock Edge from Kenley to Lutwyche. Pastures above Buildwas and between Buildwas Abbey and Wenlock; *Rev. E. Williams's MSS.*

Page 126.—In *Torilis Anthriscus* the prickles of the fruit when examined under a lens will be seen to be clothed with minute erect bristly teeth, and in *T. infesta* with deflexed ones.

Page 126.—*Torilis nodosa.*

Pastures, &c., about Berrington and Condover; *Rev. E. Williams's MSS.*

Page 127.—*Heracleum Sphondylium, var. β.*

" Found by Dr. Bowle in Shropshire ;" *How's Phytologia Britannica,* 1650. *p.* 118.

Page 128—*Œnanthe fistulosa.*

Pit between Battlefield and Albright Hussee.

Page 128.—*Œnanthe crocata.*

Ditches near Pitchford forge horse bridge. Ditch by the side of the road below Cronkhill. Near Berwick Coal-wharf. Near Meole Bridge. About Minton, in the parish of Church Stretton. The Tern at Duncot; *Rev. E. Williams's MSS.*

Side of turnpike road 1 mile from Oswestry on Shrewsbury road. Morda pool, near Oswestry.

Page 128.—*Œnanthe Phellandrium.*

In a pool near Sidney, on the Wildmoors; *H. Bidwell, Esq.*

Page 130.—*Sium angustifolium.*

In the moat at Hadnal. Ditches between Halston and Whittington. On the Wildmoors. Lilleshall old mill-pool; *Rev. E. Williams's MSS.* Westfelton moor.

Page 130.—*Pimpinella magna.*

On the right-hand side of the road as you descend the Hill from the Red Hill turnpike towards Weston; *Rev. E. Williams's MSS.*

Page 131.—*Ægopodium Podagraria.*

In Madeley and Stirchley church-yards. By Cound bridge. Hedges in Upper Cound; *Rev. E. Williams's MSS.* The Dean on the new road between Broseley and Bridgnorth, 5 miles from the latter town; *J. E. Bowman, Esq !*

Page 131.—*Sison Amomum.*

Ditch-banks and pastures about Chelmarsh, Billingsley, Cainham, &c. Aston Aer. In Brocton near Worthen, and between Brocton and Binweston; *Rev. E. Williams's MSS.*

Page 131.—*Helosciadium repens* is stated to be found on Cound moor in Rev. E. Williams's MSS.; but I know not any characters by which it can be distinguished from *H. nodiflorum.*

Page 132.—*Helosciadium inundatum.*

By the side of Ellesmere mere. Ditches below Hardwicke House near Ellesmere. Uckington Heath. Abery wood. Berrington pool. Betton large pool. Pool by Knockin Heath. Eaton Mascot pool; *Rev. E. Williams's MSS.* Pit near Old Heath turnpike on left-hand side of the road from Shrewsbury to Battlefield.

Page 132.—*Apium graveolens.*

" Stank meadow near Bishop's Castle, according to Camden's Britannia, but I could not find it there ;" *Rev. E. Williams's MSS.*

Page 132.—*Cicuta virosa.*

Pool adjoining Cound Hall. Pit by a farm-house in the lane between Tern turnpike and Uckington heath. Upper Berrington pool. West end of the largest pool at Betton. Sandford pool, near Oswestry. Pits between Upton Magna and the Forge by the Ray farm. Croesmere mere. Pools by the Wolf's-head near Knockin heath. Ditches at the east end of Colemere; *Rev. E. Williams's MSS.*

Page 133.—For *Chærophyllum temulentum,* Linn. read *Chærophyllum temulum,* Linn.

Page 134.—*Myrrhis odorata.*

Ditch round Stokesay castle. In a field by the side of the wood in the village of Hopton Wafers; *Rev. E. Williams's MSS.*

Page 134.—*Smyrnium Olusatrum.*

Upon the rocks of Bridgnorth as you descend the steps from the High to the Low Town; *Rev. E. Williams's MSS.*

Page 136.—*Viburnum Opulus.*

Near Hanwood. Right-hand side of turnpike road between Fox farm and Chilton Grove near Shrewsbury.

Page 136.—*Sambucus Ebulus.*

Waste ground about Presthope. In a lane adjoining Culmington church-yard. In the village of Bitterley and Upper Ledwich. In Peeton, near Munslow; *Rev. E. Williams's MSS.*

Page 136.—*Sambucus nigra, var. β. fructu albo.* *E. Fl. v. ii. p.* 109.

Adjoining Cound parsonage; *Rev. E. Williams's MSS.*

Page 136.—*Parnassia palustris.*
Upon Cound moor and a piece of ground adjoining called Mosterley. About Baschurch. Moor between Tern Hill and Moreton Say, plentifully. By the side of a pool between Golding and Watling-street Road ; *Rev. E. Williams's MSS.*

Page 138.—*Drosera longifolia* is one of the three plants mentioned in our earliest British Catalogue, How's Phytologia Britannica, 1650, as occurring "In England, by Ellesmere, in Shropshire."

Page 144.—*Berberis vulgaris.*
Hedges about Eaton Mascot. By the side of the road opposite the Canal-bridge leading to Longnor. Hedges between Wistanstow and Ludlow. Between Buildwas Abbey and West Coppice ; *Rev. E. Williams's MSS.*

Page 145.—*Peplis Portula.*
Near Clive church-yard on Grinshill. Ellesmere mere.

Page 145.—*Galanthus nivalis.*
By the side of the brook between Cound and Cound-moor ; *Rev. E. Williams's MSS.*

Page 145.—*N arsissus poeticus*
In a small meadow by the side of the road at Frodesley lane, between Frodesley and Longnor ; *Rev. E. Williams's MSS.*

Page 145.—*Narcissus biflorus.*
Meadow adjoining Mr. Williams's house at Eaton Mascott ; *Rev. E. Williams's MSS.*

Page 145.—*Narcissus Pseudo-Narcissus.*
In a meadow between Cound bridge and the gravel pit adjoining the New Mill. Orchard near Longnor ; *Rev. E. Williams's MSS.* Abundantly in a meadow at Woodseaves, near Market Drayton ; *R. G. Higgins, Esq.!*

Page 146.—*Convallaria majalis.*
Erase "Abundantly in a meadow at Woodseaves near Market Drayton ;" *R. G. Higgins, Esq.!*

Page 146.—*Allium vineale.*
Plentifully in a small triangular meadow adjoining Cantlop bridge. Meadows under Pimley near Sundorn ; *Rev. E. Williams's MSS.*

Page 146.—*Allium ursinum.*
In great abundance by the side of the brook and millstream near Muckley Cross, between Wenlock and Morvill ; *Rev. E. Williams's MSS.*

Page 147.—*Ornithogalum umbellatum.*
In the meadow adjoining Mr. Williams's House at Eaton Mascot ; *Rev. E. Williams's MSS.*

Page 147. 2. *Ornithogalum Pyrenaicum,* Linn. *Spiked Star of Bethlehem.* Racemes elongated ; peduncles equal, spreading, erect in fruit ; filaments all dilated. *E. Bot. t.* 499. *E. Fl. v. ii. p.* 143. *Hook. Br. Fl. p.* 160.
Pastures ; rare. *Fl.* June, July. ♃.
In Mr. Whitmore's wood at Apley ; *Rev. E. Williams's MSS.*
Bulb ovate. Leaves all radical, long, linear, acuminate, channelled, spreading, appearing with the flowers. Scape 1½—2 feet high, erect, round, smooth and polished, bearing an elongated *raceme* of numerous greenish-white *flowers,* each on

a short stalk, bracteated at the base. *Segments of the perianth* linear-oblong, obtuse, spreading, with incurved margins. *Filaments* broad and dilated at the base, cuspidate into a slender point. *Capsule* ovate, with 3 furrows, erect.

Page 147.—*Narthecium ossifragum.*
Moors about Lee and Birch near Ellesmere. About Shelve ; *Rev. E. Williams's MSS.*

Page 151.—*Juncus squarrosus.*
Clive end of Grinshill.

Page 151.—*Luzula pilosa.*
Dunn's Dingle, Leaton Shelf, near Shrewsbury. Nesscliffe hill.

Page 152.—*Rumex Hydrolapathum.*
Ditches about Newport. Between the Cloud Coppice and the Gouter. Banks of the rivers Tern and Roden ; *Rev. E. Williams's MSS.*

Page 153.—*Rumex maritimus.*
Eaton Mascot pool ; *Rev. E. Williams's MSS.*

Page 155.—*Scheuchzeria palustris* does not occur in Mr. Williams's MSS. Catalogue.

Page 156.—*Triglochin palustre.*
By the side of the water at Sundorn. Cound moor. About Halston ; *Rev. E. Williams's MSS.*

Page 156.—*Colchicum autumnale.*
About Wenlock, Wellington, and Bitterley. By the side of the Severn near Tern bridge. Eaton Mascot ; *Rev. E. Williams's MSS.*
The following extract from a letter from Professor Christison of Edinburgh, affords such a valuable addition to the account of the vital economy of this plant, and opens such an interesting field of observation, that I cannot refrain from inserting it, in the hope that botanists favourably situated for observations may be stimulated to investigate the matter. "The department of the inquiry to which my attention was lately directed, is the formation of infant-bulbs from the parent during its second winter. Besides the principal bulb which begins to form, in mid-summer or a little later, almost between the bulb and root-proper, as you describe, little bulbs are produced upon the body of the parent,—some, I suspect, about the same time,—others during the winter and ensuing spring, when the original bulb, though plump and large, is quite watery, free of starch (so as to shrivel to a skin, if dried), and probably no longer of any use towards the nourishment of what I have called the principal young bulb (then pushing up its germen amidst its leaves), but merely serving the purpose of preparing infant-bulbs for propagation, which are speedily thrown off altogether from the parent, and may be found about the size of lemon-seeds entangled in its radicles. What I wish still to ascertain is, how long these infant-bulbs take to attain such maturity as to throw up flowers, and what changes take place in the interval. I suspect several years are necessary ; and it is not improbable that during the interval the little nursling grows, not by direct enlargement, but by feeding a new bulb every year, itself annually dying, exactly like the mature flowering bulb." *Prof. Christison in lit.*

Page 157.—*Actinocarpus Damasonium.*
On Abery wood ; *Rev. E. Williams's MSS.*

Page 158.—*Alisma Plantago, var. β. lanceolata, E. Fl. v. ii. p.* 203.
At the edge of the water at Sundorn ; *Rev. E. Williams's MSS.*

Page 158.—*Alisma natans.*
By the side of the water at Sundorn in the wood. Pits at the north-east corner of Abery wood. Blackmere and Whitemere. Pits by the side of the turnpike road across Shawbury heath, and between Shawbury heath and Shawbury. In the brook by the side of the road between Loppington and Woverley ; *Rev. E. Williams's MSS.*

Page 158.—*Alisma ranunculoides.*
Bomere pool. Ditches near Hordley, Newport, &c. Betton pool ; *Rev. E. Williams's MSS.*

Page 163.—*Chlora perfoliata.*
Pastures about Chelmarsh, Billingsley, Clee Town, &c. About the lime-kilns on Stevens hill in the parish of Cound. Forge-bank, Eaton ; *Rev. E. Williams's MSS.*

Page 164.—*Erica Tetralix.*
Var. with white blossoms. Near Ellesmere ; *Rev. E. Williams's MSS.*

Page 164.—*Erica cinerea.*
Var. with white flowers. Hopton hill.

Page 165.—*Calluna vulgaris, var. β. hirsuta.* Leaves densely hoary. *E. Fl. v. ii. p.* 225.
On Hodnet heath in abundance ; *Rev. E. Williams's MSS.* Bomere pool. *Var.* with white flowers. Near the Devil's Arm-chair, Stiperstones hill.

Page 165.—*Vaccinium Myrtillus.*
In the Fir coppice in Acton Burnell park. On the Wrekin and Caradoc hills. Woods about Birch near Ellesmere. Between Ruyton and the Platt Mills. Brown and Titterstone Clee hills. Harmer hill ; *Rev. E. Williams's MSS.* Grinshill.

Page 166.—For *Oxycoccus palustris,* Linn. read *O. palustris, Rich.* and add before the synonyms. *Lindl. Syn. p.* 134. *Vaccimum Oxycoccus,* Linn.

Page 168. *l.* 10. from the bottom ; for Hincham read Hineham.

Page 168.—*Daphne Laureola.*
On the left-hand side of the road between Harley mill and the turnpike. In Birch coppice between Buildwas and Coalbrookdale, and near the church. Hedges and lanes about Little Wenlock. Hedges between Cornhill common and Neen Solars. Hedges between Stanway and Wilderhope. Hedges and lanes about Stottesden ; *Rev. E. Williams's MSS.*

Page 169.—*Polygonum Bistorta.*
Meadows about Pitchford, Underdale, Cleobury North, Sidbury, and Stockton. Home-barns, near Sundorn ; *Rev. E. Williams's MSS.* Plentiful in the meadow adjoining the west side of Haughmond abbey. Meadow south side of Battlefield church.

Page 171.—For *Fagopyrum vulgaris,* Meisn. read *Fagopyrum esculentum,* Mœnch.

Page 172.—*Paris quadrifolia.*
In a field adjoining the upper end of Pitchford-pool tail coppice ; by the side of the coppice and near the brook or rill. In a copse upon the Vall hill opposite Hardwicke house, near Ellesmere ; *Rev. E. Williams's MSS.*

Page 173.—*Adoxa moschatellina.*
Dunn's Dingle, Leaton Shelf, near Shrewsbury. Banks of river Severn near Monkmoor, near Shrewsbury.

Page 177.—*Butomus umbellatus.*
Eaton Mascot pool ; *Rev. E. Williams's MSS.*

Page 184.—*Andromeda polifolia.*
Bogs about Lee and Birch near Ellesmere ; *Rev. E. Williams's MSS.*
This plant is mentioned in our earliest Catalogue of British Plants How's Phytologia Britannica, 1650, p. 106, as occurring "in great quantity at Birch in the moores of Ellesmere, in Shropshire," in which locality it is still found.

Page 185.—*Scleranthus annuus.*
Sandy heathy ground near Wolf's Head turnpike, on the road from Shrewsbury to Oswestry.

Page 185.—For *Arctostaphylos Uva-ursi,* Linn. read *A. Uva-ursi,* Spreng. And add before the other synonymes ; *Lindl. Syn. p.* 174. *Arbutus Uva-ursi,* Linn.

Page 186.—*Chrysosplenium oppositifolium.*
Eaton Mascot wood below Pitchford forge. By the side of the brook in woods above and below Pitchford ; *Rev. E. Williams's MSS.*

Page 187.—*Saponaria officinalis.*
Side of the road near Farley. Under the walls of Shipton church-yard. By the side of the road between Benthal and Wenlock. Hedges at Morton Corbet. Pastures in Upper Cound. Plealey ; *Rev. E. Williams's MSS.*

Page 188.—*Dianthus plumarius.* Add as syn. *Mackay Fl. Hib. pt.* 1. *p.* 40.

Page 188.—*Dianthus deltoides.*
Upon the lawn at Davenport House, near the dog-kennel. By the side of the road between Davenport House and Worvill. By the side of the road between Norbury and Linley. About the stone quarry at Downton, under Haughmond hill ; *Rev. E. Williams's MSS.*

Page 191.—*Stellaria graminea, var.*
The variety here mentioned is by no means uncommon in this county.

Page 191.—*Stellaria glauca.*
By the side of Marton pool, near Chirbury. At the east side of Colemere mere, near Ellesmere ; *Rev. E. Williams's MSS.*
At the north end of Ellesmere mere, abundant.

Page 192.—*Moehringia trinervis.*
Eaton Mascot ; *Rev. E. Williams's MSS.*
Montford village. Road-side between Weeping Cross turnpike and the Fox farm, near Shrewsbury.

Page 192.—*Lepigonum rubrum.*
Add as syn. *Lepigonum rubrum, Bluff and Fing. Comp. t.* 1. *pt.* 2. *p.* 93. *Alsine rubra,* "*Wahl.*" *Drej. Fl. Hafn. p.* 163. *Reich. Fl. Excurs. n.* 3660. *Arenaria rubra, Bab. Prim. Fl. Sarn. p.* 16.
On Uckington heath, plentifully. On Lyth hill ; *Rev. E. Williams's MSS.* Grinshill. Hopton hill.

Page 193.—*Cotyledon Umbilicus.*
Grinshill. Rocks on south and east sides of Pontesford hill. Hopton hill.

Page 193.—*Sedum Telephium.*
Rocks on south side of Pontesford hill. Lyd Hole, near Pontesford hill.

Page 194.—*Sedum reflexum.*
Marrington Dingle, near Chirbury; *J. E. Bowman, Esq.*
Chambre Wên, Llanymynech.

Page 194.—*Sedum dasyphyllum.*
Walls at Pitchford House. Old walls in Uppington; *Rev. E. Williams's MSS.*

Page 194.—*Sedum acre.*
Llanymynech hill. Blodwell rocks.

Page 194.—*Sedum Anglicum.*
North side of the Castle-ring hill, near Stiperstones.

Page 195.—*Sedum Forsterianum.*
Rocks on south and east sides of Pontesford hill.

Page 196.—*Oxalis Acetosella, var. β. purpurea.*
By the side of the lane between Habberley and Marsley; *Rev. E. Williams's MSS.*

Page 196.—*Lychnis Flos-Cuculi.*
For Chetwynd moss, read Chetwynd moors.

Page 196.—*Lychnis dioica, Linn.*
"I now consider *L. vespertina* and *diurna* of Sibthorp as distinct species and characterize them as follows:—

1. *L. vespertina*, Sibth. *Petals* ½ bifid, crowned; stem villose; leaves peduncles and calyxes hairy; leaves ovato-lanceolate; flowers dichotomously panicled, diœcious; teeth of the calyx linear-lanceolate, elongated; capsule conical, with its teeth erect. *E. Bot. t.* 1580.

White, sometimes reddish *flowers*. *Teeth* of the *calyx* twice as long as those of *dioica (diurna, Sibth.)*

2. *L. dioica*, Linn. (ex parte). Petals ½ bifid, crowned; stem, leaves, peduncles and calyxes villose; leaves ovate, acute; flowers dichotomously panicled, diœcious; teeth of the calyx ovate or somewhat triangular, short; capsule nearly globular, its teeth ultimately reflexed. *E. Bot. t.* 1579. *L. diurna, Sibth.*

Flowers red, very rarely nearly white."—*Babington in lit.*

Page 199.—*Spergula vulgaris.*
Sandy field base of Hopton hill. Horse-leap on Dovaston heath.

I fear the only distinguishing character, the absence or presence of white papillæ in *S. arvensis* and *vulgaris* is too inconstant to afford a specific discrimination: they must, therefore, be reduced to varieties as Koch has done in his Synopsis.

Page 199.—*Spergula nodosa.*
On the boggy parts of Haughmond hill; *Rev. E. Williams's MSS.*

Page 210.—*Prunus insititia.*
Hedges about Eaton, Cronkhill, Harnage, Sundorn, Downton, Shawbury, &c.; *Rev. E. Williams's MSS.*

Page 211.—*Cerasus Padus.*
In the lane between Lydham and Bishop's Castle. By the side of the brook between Ruyton and Dorrington; *Rev. E. Williams's MSS.*
Hedges about Oswestry, on the Shrewsbury road.

Page 211.—Erase the whole description of *Cerasus Avium* and substitute the following:—

2. *Cerasus Avium*, Mœnch. *Common Wild Cherry-tree.* Arborescent, branches divaricate; leaves drooping, oblongo-obovate, suddenly cuspidate, the point more or less elongated, inciso-serrate, downy beneath, with 2 glands on the petiole; outer scales of leaf-buds deflexed; umbels aggregate, lax, pedicels 3—5, drooping; calyx-tube contracted beneath the sepals; sepals oblongo-ovate, obtuse, entire, reflexed; petals oblongo-obovate, bifid, with a minute claw; fruit heart-shaped, black, lusciously bitter-sweet, on long stalks.—*Prunus Avium, Linn.*

Woods; frequent. *Fl.* May. ♄.

Ludlow; *Miss Mc Ghie.* Jigger's bank, Coalbrookdale; *E. Lees, Esq.* Walford; *T. C. Eyton, Esq.* Westbury; *Mr. F. Dickinson.* Wood at the back of Sundorn castle; *Rev. E. Williams's MSS.*

Orange-grove, Westfelton. Bomere woods. Coppices about Church Stretton. Haughmond hill. Almond park.

A *tree* 20—30 feet high, with smooth *bark* and spreading divaricate *branches*. *Leaves* on deeply channelled petioles, oblongo-obovate, suddenly cuspidate, the point more or less elongated, simply scarcely doubly inciso-serrate, serratures obtuse tipped with an obtuse red glandular mucro; with 2 large oval depressed crimson *glands* on the summit of the petiole a very little below the base of the leaf, pale opaque green above, channelled on the midrib and veins, glabrous, paler beneath with silky hairs chiefly but not entirely on the nerves and in their axils. *Stipules* a pair at the base of each petiole, linear-subulate, glanduloso-incised, those of the outer scales broader, the outer scales of the leaf-buds reflexed. *Flowers* in crowded lax umbels; *pedicels* 3—5, about 2 inches long, somewhat drooping, thickened upwards beneath the germen, sessile in the bud whose scales are hairy on the interior and glabrous on the exterior, ciliato-dentate on the margins, recurved and bent upwards at the apex, with linear-subulate incised *bracteas* at the base. *Calyx* 5-fid; *segments* oblongo-ovate, obtusely pointed, reflexed, 3-nerved, the lateral nerves exteriorly branched, midrib unbranched; *calyx-tube* campanulate, contracted beneath the sepals, glabrous, scarcely ribbed. *Petals* 5, spreading, oblongo-obovate, bifid, with a very minute claw, white, veiny. *Stamens* filiform, as long as the pistil, *anthers* yellow. *Pistil* subulate; *stigma* capitate, disciform, with a lateral fissure. *Germen* oblong, round, with a lateral longitudinal depression communicating upwards with the groove on the style and terminating in the lateral fissure of the stigma. *Ovule* 1, oval. *Fruit* heart-shaped, small, black, lusciously sweet with an almost almond bitter, on long stalks.

A variety with white fruit occurs about Westfelton.

My friend Dovaston, who first directed my attention to our wild cherries, considers that all our cultivated varieties have sprung from three original wild stocks, which he thus arranges:—

1. The parent of the Hearts, so called from the heart-shaped form of the fruit, which is black, red or white, on long stalks, and ripens second in time or intermediate between the Dukes and the Morellos. This he recognizes in our Orange-grove specimens, and regards as identical with No. 3. of *Ray's Syn.* 3rd *ed. p.* 463. "Cerasus sylvestris fructu minimo cordiformi;" and the var. *β.* of Smith *E. Fl. vol. ii. p.* 355.

2. The parent of the Morello, Kentish and short-stalked cherries, with small round red subacid fruit, which he identifies with the Hailstones specimens, [described below under *C. austera*] and Ray's No. 4. "Cerasus sylvestris septentrionalis fructu parvo serotino," and Smith's *var.* γ.

3. The parent of the Dukes, so called from their *leading* precocity in ripening, with heart-shaped fruit and of an upright fastigiate habit of growth. Of this he observes—"To what wild cherry to refer these I know not, though I am thoroughly persuaded they have a different parentage from the other two. I have, from very early years, attentively marked this very striking difference, obviously visible to any one having an eye in his head, sharp to botanical physiognomies: and have been always very much astonished more notice of these very marked characters has not been made in the books. The discrepancies are not easily to be described in words, or even drawn,—they must be seen and felt:—but that there is a most complete distinction, 'water will not wash, nor fire burn it out of my bones.'"—*Dovaston in lit.*

Mr. Borrer regards our two Shropshire Cherries as identical with two which he finds in Sussex, and of which he has kindly communicated specimens, from which it is evident they are altogether similar. He considers them to be the *Prunus Avium, Linn.* and *P. Cerasus, Linn.* and with respect to the synonyms inclines to refer Smith's var. *a.* to *P. Cerasus, Linn.* and all his other varieties to *P. Avium, Linn.*

Mr. Edward Forster has the two Sussex Cherries under cultivation in his garden at Woodford, Essex; and believes them to be *P. Avium* and *P. Cerasus, Linn.* regarding our Shropshire forms as identical. *P. Cerasus* does not occur in Essex, but *P. Avium* is very abundant. The specimen he communicates of *P. Avium* has the leaves of an ovato-lanceolate acuminate form different from the oblongo-obovate shape of those sent by Mr. Borrer, and also of our Shropshire specimens, but this may arise probably from cultivation. Mr. Forster remarks, "these trees seem to have been well enough known until Hudson absurdly put them together in his second edition, in which unfortunately he was followed by Smith, who probably had not paid sufficient attention to them." He adds the following as synonymous with our *C. Avium:*

P. avium, Linn. Sp. Pl. 680. (though Linnæus says the fruit is red.) *Huds. Fl. Angl. ed.* 1. (1762.) *p.* 187. *P. Cerasus, E. Bot. t.* 706. *P. Cerasus,* γ. *Huds. Fl. Angl. ed. ii. p.* 213. *P. Cerasus,* δ. *Sm. E. Fl. vol. ii. p.* 355. *Cerasus sylvestris fructu nigro. Raii Syn.* 463.

Dr. Bromfield who has studied our wild Cherries with the most minute attention and with constant reference to the works and specimens of the continental botanists, identifies our Shropshire Cherries with those detected by him in the Isle of Wight. He has kindly communicated specimens of his plants accompanied with detailed descriptions and synonymes, which so characteristically depict our Shropshire forms that I gladly avail myself of his liberal permission to insert them here:—

"*Prunus* (Cerasus) *Avium, Linn.* arborescent; leaves ovato-lanceolate, drooping, downy beneath; umbels sessile, lax, aggregate around the leaf-buds; sepals somewhat pointed; petals flaccid, a little connivent.

Very abundant in woods, thickets, and hedges in all parts of the Isle of Wight, and I should say decidedly an aboriginal here. With us sometimes a *tree* of 40 feet, usually about 20 feet. Differs from *P. Cerasus* in its single arborescent *stem*, which is not stoloniferous, in its thinner longer more veiny *leaves* which droop almost vertically when full grown, have a reddish tinge and are downy beneath. *Flower-stalks* not springing from leafy scales, forming lax *umbels* mostly clustered around the leaf-buds on the spurs and main branches, more rarely scattered. *Petals* thin, weak, but little spreading. *Sepals* more or less pointed." *Dr. Bromfield in lit.*

3. *Cerasus austera.* *Late Acid Cherry Tree.* Fruticose,

branches divaricate; leaves not drooping, subrotundo-obovate, somewhat cuspidate, obtuse, doubly crenato-serrate, glabrous, with 2 glands on the base of the leaf; outer scales of leaf-buds erect; umbels mostly scattered, pedicels 3, erect or patent; calyx-tube not contracted beneath the sepals; sepals oval, obtuse, concave, crenato-dentate, reflexed; petals rotundo-obovate, subemarginate, sessile, without a claw; fruit round, red, subacid, on short stalks.—*Prunus Cerasus, Linn. et Auctorum.*

Hedges; rare. *Fl.* May. ♄.

Braddws near the Hailstones, Westfelton.

An erect bushy *shrub*, 3—6 feet high, sending up numerous slender erect suckers, with very smooth *bark* and divaricating *branches*. *Leaves* on channelled petioles, elliptical ovate, frequently rounder and approaching to subrotundo-obovate, somewhat cuspidate, obtuse, duplicato-crenato-serrate, serratures tipped with an obtuse pale red glandular mucro, with 2 large crimson subglobose or oval depressed *glands* at the base of the leaf, often on the summit of the petiole, shining or varnished opaque green above, paler and veiny beneath with straggling silky hairs on the ribs and in their axils, and somewhat complicate, at least in a young state, finally quite glabrous, firm, erect or horizontal, never drooping. *Stipules* a pair at the base of the petiole, linear, dentate, glandular, those of the outer leaves or scales of the buds broader and oblong, glanduloso-dentate, erect. *Flowers* in scattered nearly sessile umbels; pedicels 3, about 1½ inch long, erect or spreading, gradually thickened upwards beneath the germen, with a minute submembranous oblong bractea deeply acutely incised at the base of each. *Calyx* 5-fid; *segments* oval, obtuse, concave, glanduloso-crenato-dentate, erect in bud and before the expansion of the flowers finally reflexed, 3-nerved, lateral nerves very much branched on the exterior side, midrib unbranched; *calyx tube* campanulate, furrowed and obscurely ribbed, not contracted beneath the sepals. *Petals* 5, spreading, rotundo-obovate, subemarginate, sessile or without any claw, white, veiny. *Stamens* subulate, *anthers* yellow. *Stigma* capitate, disciform, somewhat lobed, margins recurved, with a lateral cleft. *Germen* subovato-oblong, attenuated into the style, with a longitudinal elevation on one side which is continued upwards as a slight groove along the style and terminates in the lateral fissure of the stigma. *Ovules* 1 or 2, usually 1, oval. *Fruit* small, round, red, gratefully subacid, on short stalks, ripening late.

The following is Dr. Bromfield's beautifully characteristic description of his Isle of Wight cherry, the specimens of which leave not the slightest doubt of its identity with our Shropshire *C. austera:*—

"*Prunus Cerasus, Linn.*" Fruticose; leaves broadly ovato-lanceolate, firm, glabrous, not drooping; umbels few-flowered, mostly scattered, erect or patent; inner scales of the flower-buds leafy; petals firm, widely spreading; sepals obtuse. *P. Cerasus, Linn. et pler. Auct. Europ.* (excl syn. Script. Anglican.) *Willd. Sp. Plant. II. p.* 991. *Wahlenb. Upsal. p.* 164. *Lejeune Fl. de Spa. I. p.* 222. *Revue de la Fl. de Spa. p.* 92. *Koch Syn. p.* 206. (descr. bona.) *Mertens and Koch in Rohling's Deutschland's Flora* (aptiss. descript. plantæ nostræ) *IIIer. Band. s.* 408. *Pollich Palat. II. p.* 27. (bona.) *Leers Herborn. ed. alt. p.* 116. *Cerasus acida, Endlicher Fl. Poson. p.* 468. *Prunus acida vel austera, Ehrh.* (teste Wahlenb. et fide spec. Upsaliensis in herb. amic. Bentham cum nostrat. Vectianis compar.) *Prunus Caproniana, Gaudin Helv. III. p.* 307. (optima.) *Haller Stirps Helv. No.* 1083. (teste Gaudin et Leers, etiam bona.)

A bushy *shrub* 6 or 8 feet high, branching from the base and sending up from the root many slender *stems* covered with a reddish-brown *bark*, and having much of the habit and mode of growth of *P. spinosa* or some of its less branched varieties. *Leaves* appearing with the flowers, (with us at the close of April or early in May) at first very small, scarcely 2 inches long, bright green and varnished as it were with gum, without any of the red tinge so conspicuous in those of *P. Avium*, folded when young, and in that state have a slight hairiness beneath, which quite disappears subsequently, and long before they are full grown are perfectly glabrous on both sides. In this earlier state they are broadly ovate and shortly acuminate, finally enlarging to 3 or 4 inches and inclining to ovato-lanceolate with longer points, but in all stages of development the leaves are erect or horizontal, never pendant, as is so strikingly the case in *P. Avium*, besides possessing a firmness and opacity quite wanting to those of this last. There is mostly a small reddish gland or two at the base of each leaf of which the petioles themselves are destitute unless on their extreme upper part and that but seldom. *Umbels* lateral, subsessile, mostly scattered, solitary or two together, (at least in the really wild plant, for in the cultivated Morello Cherry of which Mr. Borrer and others consider this as the parent, the umbels are often crowded or aggregate as in *P. Avium*) few (2—4) flowered, with an occasional single flower-bud interspersed. *Peduncles* simple, erect, spreading or deflexed, but not drooping from their laxity as in *P. Avium*, springing from buds whose inner scales are leafy, which is considered by Koch as a character of the species. *Blossoms* large, white and sweet-scented, making a handsome appearance in contrast with the bright green of the young leaves. *Sepals* ovate, very blunt and rounded, strongly deflexed, with a few crenate serratures, scarcely one-third the length of the broad round firm and widely spreading *petals*. *Fruit* small, red and gratefully acid.

The synonymes of *P. Avium* and *P. Cerasus* seem never to have been understood by British Botanists, though clearly discriminated on the continent; forms of our common wild Cherry exhibited chiefly in the colour, form, and flavour of the fruit having been vaguely designated by these names in England. The Linnæan Herbarium, which I have inspected for the purpose of clearing up the difficulty, contains no specimens of its possessor's *P. Cerasus*, the greater part of the authentic examples of that genus being now either wholly or in part destroyed by time and insects, and the few remaining are bad and not characteristic of the species they represent. What Linnæus understood by *P. Cerasus* was one of the cultivated cherries of which our present plant was his *var. γ. austera;* that it could not have been our common wild cherry is evident from his calling *that* which is the only native Swedish species except *P. Padus, P. Avium*, thus making it a species apart; but the descriptions and references of the Swedish botanists themselves, as well as of Ehrhart who resided at Upsal, and from whose collection I have seen and compared a specimen of his plant with mine and find them agree perfectly, prove from their coincidence that these two cherries were well known and distinguished by them. The excellent descriptions of our plant by Pollich, Leers, Haller, and especially of Gaudin, show the universality of the opinion that it is the species Linnæus intended by his *P. Cerasus*, which included several of the cultivated sorts. I cannot, with any feeling of certainty, refer our Shropshire and Isle of Wight plants to any of the varieties or presumed species of Ray or Smith. Their notices of both are much too vague or concise to gather anything like distinct evidence that our present species was known to either. Ray's No. 1. (*Syn. ed.* 3.) is surely not the same, as he calls it "the common wild cherry," and this of all the four kinds mentioned by him is the only one that so far as the description goes, agrees at all with ours in its fruit, but that of *P. Avium* is occasionally both red and black, so that colour alone is no sufficient definition. Smith seems to have had no clear ideas of our two cherries, for he joins the synonymes of several old authors which refer to *our P. Cerasus* to the *P. Avium* of Linnæus (*Fl. Suec.*) and which by his own shewing and not giving any special habitat he rightly enough considers as the common wild cherry, though quite wrong in attaching the above synonyms referring to a totally different thing."—*Dr. Bromfield in lit.*

Mr. Edward Forster has furnished the following synonyms for our *C. austera:—Prunus Cerasus, Linn. Sp. Pl.* 679. *P. Cerasus, a. Sm. E. Fl. vol. ii. p.* 354. (I can hardly suppose Smith ought to have referred to *P. Avium, Fl. Suec.* Linnæus should know his own plant. Smith was probably led by the reference to

Bauhin, which I should say does belong to this plant.) To this I refer—*Cerasus sylvestris fructu rubro, J. Bauh.* 220. *Raii Syn.* 463. *Common Wild Cherry-tree* —considering as varieties:—*β. Cerasus sylvestris fructu minimo cordiformi, Raii Syn. l. c. γ. Cerasus sylvestris septentrionalis fructu parvo serotino, Raii Syn l. c.*

Mr. Borrer has observed our plant in Sussex, Cornwall, and Devon. Dr. T. Bell Salter has a specimen from Canford, Dorsetshire. Mr. Dovaston has noticed it growing about Llanrwst and Llangollen, North Wales, and near Keswick in Cumberland, but in all these localities sparingly. Neither he nor I have seen it in any other place in Shropshire than the one mentioned, the prevailing species in our woods and coppices being *C. Avium*. Neither Mr. Borrer nor Dr. Bromfield find it in the *interior* of their woods but always in hedge-rows or thickets in the enclosed country though remote from human habitation, and hence consider it as probably a naturalized species. Lejeune and Gaudin appear to entertain no doubt of its indigenous origin in Belgium and Switzerland; other continental authors regard it as naturalized.

Page 212.—*Cratægus Oxyacantha.*

Many old trees on the south side of Pontesford hill.

Page 212.—*Pyrus Malus.*

Two very distinct forms of the Common Crab-apple are observable in our hedges, and probably occur throughout England. Ray distinguished them, but considered that they differed from each other only as a cultivated form differs from a truly wild one. Later botanists have overlooked them, and one of them is now quite expunged from our books, though the continental botanists still retain both, and De Candolle considers them distinct species. Their difference of appearance is recognizable in their extreme states by the most cursory observer, and there is also a marked difference in the form of the leaf but too inconstant to be regarded as a character. The only constant characters reside in the presence or absence of pubescence on the young branches and on the peduncles and calyx-tube. This pubescence varies from perfect woolliness to slight pubescence in the one form, whilst the other is perfectly glabrous, and though specimens continually recur which at first sight may render it doubtful to which form they should be referred, an attention to the young branches *conjointly* with the young fruit resolves the difficulty. Whether as Ray asserts, one form be the effect of cultivation I know not, but both are now equally plentiful, growing indiscriminately in the same hedge and to all appearance equally wild, and certainly have as just claims to be mentioned in our books as the recorded varieties of many other species. I would characterize them thus:—

Pyrus Malus, Linn.

Var. a. sylvestris. Young branches glabrous; leaves glabrous on the under surface; peduncles and calyx-tube glabrous; *Pyrus Malus, Linn. E. Bot. t.* 179. *a. Huds. Fl. Angl. 2nd ed. p.* 216. *With. 3rd ed. p.* 462. *a. glabra, Bluff and Fing. Comp. Fl. Germ. 2nd ed. t. i. pt.* 2. *p.* 175. *Pyrus acerba, Reich. Fl. Excurs. n.* 4066. *Pyrus (sec. Malus) acerba, DC. Prod. v. ii. p.* 635. *DC. Bot. Gall.* 181. *a. austera, Wallr. Sched. Crit.* 215. *Fl. Siles. ii.* 20. *a. glabra, Koch Syn. p.* 235. *β. acerba, Peterm. Fl. Lips.* 355. *Malus sylvestris, Raii Syn. 3rd ed. p.* 452. *Hill Fl. Brit. p.* 255. *Malus acerba, Merat Fl. Par.* 187. *and DC. Fl. Fr. Suppl.* 530. *P. Malus sylvestris, Fl. Dan. t.* 1101.

Var. β. sativa. Young branches pubescent; leaves more or less pubescent or even woolly on the under surface; peduncles and calyx-tube more or less pubescent or woolly. *Pyrus Malus, Linn. E. Fl.*

v. ii. p. 363. (excl. syn.) *Reich. Fl. Excurs. n.* 4067. *Peterm. Fl. Lips.* 354. *β. Huds. Fl. Angl. 2nd ed. p.* 216. *Var.* 2. *With. 3rd ed. p.* 462. *β. tomentosa, Bluff and Fing. Comp. p.* 175. *P. (sec. Malus) Malus, Linn. DC. Prod. v. ii. p.* 635. *Bot. Gall.* 181. *β. mitis, Wallr. Sched. Crit.* 215. *Fl. Siles.* 2. 20. *β. tomentosa, Koch Syn. p.* 235. *Malus sativa, Raii Syn. 3rd ed. p.* 451. *Hill Fl. Brit. p.* 255.

Page 212.—*Pyrus torminalis.*

In the hedge on the left-hand side of the road between Cound and Cound moor. On the left-hand-side of the road between Uffington and the first canal-bridge towards Atcham. In the wood at Sundorn. On the left-hand side of the road above Harley mill. Hedges near Battlefield. Mr. Burton's wood, Longnor; *Rev. E. Williams's MSS.* Harlot's gate, near Westfelton; *J. F. M. Dovaston, Esq.*

Page 212.—*Pyrus aucuparia.*

Blodwell rocks.

Var. with leaflets doubly serrated.

Stunted trees on the summit of rock on north side of the Castle-ring hill, near Stiperstones.

Page 213.—*Spiræa salicifolia.*

A single bush on the bank of turnpike road at the Pissing-bank near Nesscliffe.

Page 213.—*Spiræa Filipendula.*

Round the limestone quarries at Farley, between Wenlock and Buildwas; *Rev. E. Williams's MSS.*

Page 215.—*Rosa spinosissima.*

By the side of the road between Snailbatch and Penally lead mines; *Rev. E. Williams's MSS.*

Page 216.—*Rosa tomentosa.*

"Woods and hedges, common;" *Rev. E. Williams's MSS.*

Page 216.—*Rosa rubiginosa.*

"Woods and hedges, common;" *Rev. E. Williams's MSS.* Broad-meadow fields, Lady hills and other places about Westfelton, frequent; *J.F.M.Dovaston,Esq.*

Hedges on road-side leading from Minsterley to Snailbeach.

Page 222.—*Rubus idæus.*

In the wood at the top of Acton Burnell park. By the side of the road between Pulverbatch and Westcote. In Chetwynd park. Around Marton pool. In a plantation near Halston gate; *Rev. E. Williams's MSS.*

Twyford Vownog near Westfelton, plentifully.

Page 225.—*Rubus fissus.*

Twyford Vownog near Westfelton.

Page 226.—*Rubus affinis, var. γ.*

Hedges on the road-side at the Gouter, near Eaton Mascot.

Page 239.—*Rubus chamæmorus.*

"Near Hodnet, as informed by the Rev. Reginald Heber, since Bishop of Calcutta;" *Rev. E. Williams's MSS.* Surely a mistake.

Page 240.—*Comarum palustre.*

Morda pool near Oswestry. Golding pool near Pitchford.

Page 240.—*Potentilla anserina.*

"A specimen with 4 leaves to the blossom and 8 to the calyx, found July, 1798, near Pitchford Forge;" *Rev. E. Williams's MSS.*

Page 240.—*Potentilla argentea.*

By the side of Cound park-walls; *Rev. E. Williams's MSS.*

Page 241.—*Potentilla verna.*

In the lane between Cound and Golding; *Rev. E. Williams's MSS.*

Page 243.—*Geum rivale.*

In a wood by the side of a pool between Golding and the Watling-street road. Ditch-banks in a meadow adjoining Bitterley church. In Cantlop wood. In a boggy copse at the south end of the Wrekin; *Rev. E. Williams's MSS.*

Page 251.—*Helianthemum vulgare.*

Queen's Bower, Haughmond hill. About Farlow lime-kilns; *Rev. E. Williams's MSS.*

Llanymynech hill. Blodwell rocks; plentifully.

Page 253.—*Helleborus fœtidus.*

Under Whittington castle garden near S.E. corner. Amongst the bushes by the side of the road at Farley between Wenlock and Buildwas; *Rev.E. Williams's MSS.*

Page 253.—*Helleborus viridis.*

Near Lilleshall abbey; *Rev. E. Williams's MSS.*

Page 253.—*Aconitum Napellus.*

By the side of the brook a few yards above Gossart bridge between Ludlow and Burford, in abundance; *Rev. E. Williams's MSS.*

Page 254.—*Aquilegia vulgaris.*

By the side of the road between Habberley and Minsterley. Between the Bull farm and Evenwood near Acton Burnell. Woods about Harley; *Rev. E. Williams's MSS.*

Page 255.—*Thalictrum flavum.*

On the banks of the Severn at the turn of the road between Cronkhill and Cronkhill turnpike. Near Underdale. On the banks of the Tern at Duncot; *Rev. E. Williams's MSS.*

Page 255.—*Clematis Vitalba.*

In a cottage garden hedge on the right-hand side of the road a little beyond Norley Green, leading to Willey; *Rev. E. Williams's MSS.*

Page 263.—*Trollius Europæus.*

Morda pool near Oswestry.

Page 273. *l.* 2. from the bottom; for Weston Lullingfields read Weston Lullingfield.

Page 277.—*Origanum vulgare.*

Blodwell rocks.

Page 289. *l.* 4. from the bottom; for Buckley read Ruckley.

Page 296.—*Pedicularis palustris.*

Morda pool near Oswestry.

Page 302.—*Verbena officinalis.*

Penyvoel near Llanymynech.

Page 302.—*Orobanche major.* Add as syn.—*Sutton in Linn.Trans. vol. iv. p.* 173. *O. rapum, Bluff and Fing. Comp. t. i. pt.* 2. *p.* 432.

Page 303.—*Orobanche elatior.*

"I find the lobes of the upper lip to be inflexed and usually having an elevated point between the lobes. Stamens usually glabrous in the upper half, but rarely with a few scattered hairs even as far as the summit of the filament. Style with a very few glandular hairs within. Stigma bilobed, yellow. I take these characters from a *fresh* specimen gathered by Mr. Borrer near Barton, Suffolk."—*Babington in lit.*

Page 310.—*Lepidium Smithii.* Add as syn.—*Bab. Prim. Fl. Sarn. p. 7.*
Penyvoel near Llanymynech. Grimmer Rocks. Hogstow mill.

Page 313.—*Sisymbrium Thalianum.*
Var. *flore pleno.* Rugeley, Salop ; *Rev. M. Berkley in Henslow's Herbarium, Cambridge.*

Page 316.—*Cardamine impatiens.*
Shelderton rocks in the south-west part of the county ; *Rev. T. Salwey.*

Page 318.—*Nasturtium terrestre.*
Wet place near Slate Row not far from the Donnington-wood Works; *H. Bidwell, Esq.!*
North end of Ellesmere mere.

Page 321.—*Sinapis alba.*
Penyvoel and Llwyntidman, near Llanymynech.

Page 325.—*Erodium cicutarium.*
Llanymynech hill.

Page 330.—*Geranium Pyrenaicum.*
On the road-side between Wellington and Wrockwardine ; *H. Bidwell, Esq.*

Page 331. *l.* 18. for Wrenthall read Wrentnall.

Page 331.—*Geranium lucidum.*
Blodwell rocks.

Page 335.—*Geranium columbinum.*
Near Colemere mere. Fields near the Black Barn, near Eaton Mascot.

Page 336.—*Malva vulgaris.*
The Mount, Frankwell, Shrewsbury. Nesscliffe. Colemere. Berrington.

Page 337.—*Malva moschata.*
Generally about Shrewsbury, and in other parts of the County.

Page 345.—*Fumaria capreolata.*
I find the plant with the truncate or subretuse fruit, mentioned under this species as probably distinct, to be very generally diffused on the hedge banks near Oswestry, together with true *F. capreolata,* but retaining all the characters I have assigned to it.

Page 347.—*Ulex Europæus.* Add as syn. *Babington in Ann. Nat. Hist. vol. v. p.* 301.

Page 348.—*Ulex nanus.* Add as syn. *Babington in Ann. Nat. Hist. vol. v. p.* 302.

Page 348.—*Genista tinctoria.*
Meadows near Golding pool, near Pitchford.

Page 363.—*Trifolium striatum, var. a.*
Colemere mere. Penyvoel, near Llanymynech.

Page 371.—*Hypericum Androsæmum.*
Hope coppice ; *J. E. Bowman, Esq.*

Page 374.—*Hypericum humifusum.*
Sweeney Mountain.

Page 375.—*Hypericum montanum.*
Above Wollerton on Wenlock Edge ; *Rev. T. Salwey !*

Page 384.—*Tragopogon minor.*
I have seen specimens of the Tragopogon from Westfelton and the Lodge near Ludlow, two of the localities mentioned in the text as communicated for *T. pratensis,* and they prove to be identical with our *T. minor.*

Page 386. *l.* 20, for *Leontodon hispidus* read *L. hispidum.*

Page 386. *l.* 28, for *Leontodon autumnalis,* read *L. autumnale.*

Page 389.—*Lactuca muralis.*
Blodwell rocks.

Page 445.—*Euphorbia amygdaloides.*
Grimmer rocks ; *J. E. Bowman, Esq.*

Page 404.—*Carlina vulgaris.*
Llanymynech hill.

Page 406.—*Artemisia Absinthium.*
Colemere mere.

Page 431.—*Habenaria bifolia.*
Woody slope north of Shelve Church ; *J. E. Bowman, Esq.*

Page 446. *l.* 4, for Kuntz read Kutz.

Page 449.—*Sparganium simplex.*
Canal near Lubstree park. In ditches on the Wildmoors ; *H. Bidwell, Esq.*

Page 459.—*Carex pallescens.*
Meadows at Acton Scott ; *Rev. T. Salwey.!* Near Kinnersley ; *H. Bidwell, Esq.*

Page 460.—*Carex panicea.*
Near Kinnersley ; *H. Bidwell, Esq.*

Page 466.—*Carex paludosa.*
Ditches under Cronkhill. About Ruyton, near Baschurch ; *Rev. E. Williams's MSS.*

Page 467.—*Carex riparia.*
Banks of the Tern near Duncott. Side of Marton pool near Chirbury. By Rednal mill. Near Hordley ; *Rev. E. Williams's MSS.*

Page 470.—*Ceratophyllum demersum.*
In the ditches on the Wildmoors near Sidney ; *H. Bidwell, Esq.*

Page 471.—*Myriophyllum verticillatum.*
In ditches on the Wildmoors near Sidney ; *H. Bidwell, Esq.!*

Page 472.—*Sagittaria sagittifolia.*
In ditches on the Wildmoors, near Crudgington ; *H. Bidwell, Esq.*

Page 491. *l.* 10, for Calcott Hall read Dwyffrwd.

Page 492. *l.* 38, for Calcott Hall read Dwyffrwd.

INDEX OF LOCALITIES.

Shrewsbury has been assumed as a centre, and the county divided into four portions, N.E. N.W. S.E. and S.W.; the distance and direction is next added; and then the geological formation from Murchison's Silurian System. (T.) Trap. (Q.) Quartz or altered Caradoc Sandstone. (Sl. Camb. R.) Slaty Cambrian Rocks. [(Ll. F.) Llandeilo Flags, and (C.S.) Caradoc Sandstone.] *Lower Silurian Rocks.* [(W.S.) Wenlock Shale. (W.L.) Wenlock Limestone. (L.L.R.) Lower Ludlow Rocks. (A.L.) Aymestry Limestone. (U.L.R.) Upper Ludlow Rocks.] *Upper Silurian Rocks.* (T.O.R.) Tilestone of Old Red Sandstone. (O.R.S.) Old Red Sandstone. (Carb. L.) Carboniferous Limestone. (M. G.) Millstone Grit. (L.C.M.) (U.C.M.) Lower and Upper Coal Measures. (L.N.R.S.) Lower New Red Sandstone. (M.L.) Magnesian Limestone. (L.R.M.) (U.R.M.) Lower and Upper Red Marl. (U.N.R.S.) Upper New Red Sandstone. (L.L.) Lower Lias. Parishes or Church-towns are printed in Capitals. The figures between parentheses indicate the elevation in feet above the sea. References are made to places within 10 miles of any locality, so as to exhibit the Flora of each district of the County.

Ellardine Moss, near Hawkstone, N.E. 9 m. s.e. Wem, (UNRS) 30, 31, 146, 185, 260.

Ellesmere, N.W. 17 m. n.n.w. Shrewsbury, (UNRS) 14, 29, 30, 31, 35, 102, 105, 131, 132, 150, 185, 252, 263, 274, 296, 297, 320, 326, 335, 351, 358, 359, 372, 376, 403, 450, 465, 466, 467, 488, 490, 493, 496, 507, 509, 520.

See *Birch. Blackmere. Cockshutt. Colemere. Croesmere Mere. Ellesmere Canal. Ellesmere Mere. Hardwick. Hordley. Lee. Otley. Perthy. Tetchhill. Vall Hill. Whitemere.*

Ellesmere Canal, N.W. 252.

Ellesmere Mere, N.W. 8, 34, 157, 158, 173, 404, 447, 451, 455, 456, 457, 460, 465, 466, 467, 470, 471, 496, 505, 508, 510, 512, 517, 521, 530.

Ellerton Hall, N.E. 4 m. n.n.w. Newport, (UNRS) 137.

Emstry, S.E. 2½ m. s.e. Shrewsbury, (LNRS) 97, 112.

Ensdon, N.W. 6 m. n.w. Shrewsbury, (UNRS) 276, 279.

Eudon Burnell, S.E. 3 m. s.s.w. Bridgnorth, (LCM) 210, 406, 497.

Evenwood Common, N.E. 1 m. e. Acton Burnell, (Car.S) 276, 529.

Exford's Green, S.W. 4 m. s.s.w. Shrewsbury, (Sl.Camb.R.) 136.

Eyton on the Wildmoors, N.E. 2 m. n. Wellington, (LNRS) 26, 29, 49, 67, 78, 97, 116, 168, 177, 250, 253, 335, 337, 366, 454, 469, 472, 484, 485, 489, 491, 513.

Flash, near Shrewsbury, 45, 131, 238, 403.

Fox Farm, S.E. 2¾ m. s.e. Shrewsbury, (LNRS) 358, 504, 505, 510, 517, 521.

Ford Hill, (between Longnor and Preston), 400.

Frankwell, n.w. suburb of Shrewsbury, 344, 416, 496.

Frodesley, S.E. 8 m. s.s.e. Shrewsbury, (Car.S.) 111, 278, 343, 349, 518.

See *Lodge Hill.*

Faintree, S.E. 4½ m. s.w. Bridgnorth, (ORS) 212, 349, 354, 371, 411.

Fairy-land, (Westfelton), N.W. 11, 101, 264, 346, 389.

Farley, S.E. 2½ m. n.e. Wenlock, (WL) 55, 403, 432, 434, 445, 473, 521, 528, 529.

See *Farley Dingle.*

Farley Dingle, S.E. 2½ n.e. Wenlock, (WL) 145, 371.

Farlow, 510, 529.

Felshampton, S.W. 4 m. s. Church Stretton, (WS) 512.

Felton Farm, near Ludlow, S.E. 30, 101, 137, 145, 417.

Felton Moors, near Ludlow, S.E. 430, 436.

Fennymere, N.W. 5 m. s.w. Wem, (UNRS) 96, 252, 296.

Fettis Farm, Melverley, N.W. 491.

Fitz, N.W. 5 m. n.n.e. Shrewsbury, (UNRS) 73.

See *Fitz Coppice.*

Fitz Coppice, N.W. 71.

Five Chimnies, (Wenlock Edge), S.E. 3 m. s.w. Wenlock, (WL) 59.

Gamester Lane, N.W. 1 m. s. Westfelton, (UNRS) 72, 119, 291.

Gleeton Hill, S.E. 1 m. n. Wenlock, (W.S.) 164, 213, 282, 351, 473.

Golding, S.E. 7 m. s.e. Shrewsbury, (LNRS) 372, 430, 446, 448, 456, 457, 458, 459, 460, 462, 463, 465, 466, 468, 472, 508, 510, 517, 528, 529, 531.

Gossart Bridge, S.E. 529.

Gouter, near Berrington, S.E. 5 m. s.e. Shrewsbury, (LNRS) 487, 512, 513, 519, 528.

Grimmer Rocks, N.W. 2 m. s.w. Minsterley, (LSR) 506, 530, 531.

Grinshill, N.E. (589) 7 m. n. Shrewsbury, (UNRS) 60, 193, 412, 415, 446, 495, 497, 509, 510, 513, 518, 519, 520, 521, 522.

See *Grinshill Church-yard.*

Grinshill Church-yard, 123.

Grafton, N.W. 5½ m. n.w. Shrewsbury, (UNRS) 186.

Habberley, S.W. 8 m. s.w. Shrewsbury, (Sl.Camb.R.) 263, 331, 414, 522, 529.

See *Vessons Wood.*

Hadnal, N.E. 5 m. n. Shrewsbury, (UNRS) 37, 43, 45, 51, 53, 54, 105, 131, 168, 273, 404, 418, 420, 427, 430, 450, 451, 453, 459, 460, 462, 463, 465, 503, 504, 516.

See *Birches Lane.*

Hailstones, (Westfelton), N.W. ½ m. e. Westfelton, (UNRS) 298, 525.

Hales-Owen, S.E. (LCM) 8, 157, 186, 328.

See *Broadmoor. Leasowes.*

Halford, S.W. ½ m. n.n.e. Craven Arms, (WS) 10.

Hall Hussee, (see Albright Hussey), 298.

Halston, N.W. ½ m. e. Whittington, (LNRS) 243, 264, 331, 453, 455, 487, 504, 516, 519, 528.

See *Maestermin Bridge.*

Halton, S.E. 2 m. w. Ludlow, (ORS) 97.

Halton, (Priors) S.E. 2 m. n. Ludlow, (ORS) 350, 433.

Hampton Bank, near Coalbrookdale, S.E. (UCM) 258, 463.

Hancott Pool, 2 m. n. Shrewsbury, (UNRS) 8, 11, 12, 30, 32, 33, 43, 45, 48, 49, 58, 100, 105, 128, 129, 132, 135, 152, 153, 154, 156, 158, 166, 168, 240, 249, 252, 258, 310, 376, 428, 448, 449, 450, 455, 456, 457, 458, 460, 463, 465, 466, 510.

Hanmer Hall, 14, 215, 511.

Hanwood, S.W. 4 m. s.w. Shrewsbury, (LNRS) 169, 263, 295, 297, 331, 344, 354, 433, 517.

Hardwicke, N.E. 5½ m. n.n.e. Shrewsbury, (UNRS) 60, 105, 374, 412, 418, 511.

Hardwick, N.W. 1½ m. s.w. Ellesmere, (UNRS) 297, 331, 390, 453, 455, 458, 476, 496, 510, 517.

See *Maestermin Bridge.*

Harley, S.E. 10½. m. s.e. Shrewsbury, (Car.S.) 26, 31, 112, 125, 156, 164, 251, 255, 288, 348, 364, 372, 415, 416, 418, 433, 511, 514, 520, 528, 529.

See *Springs. Wigwig.*

Harlescote, 2½ m. n. Shrewsbury, (UNRS) 109.

Harlot's Gate, near Westfelton, N.W. 528.

Harmer Hill, N.W. 6 m. n. Shrewsbury, (UNRS) 193, 194, 311. 343, 387, 414, 464, 509, 510, 520.

See *Harmer Moss.*

Harmer Moss, N.W. 349, 507.

Harnage, S.E. 7 m. s.e. Shrewsbury, (Car.S.) 145, 290, 400, 509, 522.

Harriet's Hayes, S.E. 1½ m. e.n.e. Albrighton, (UNRS) 232.

Hatton, S.E. 2 m. s.e. Shiffnal, (UNRS) 74, 132, 497.

See *Hatton Dingle.*

Hatton Dingle, S.E. 109, 146.

Hatton Hine Heath, N.E. 3 m. e.s.e. Lee Brockhurst, (UNRS) 138.

Hatton (Cold) Heath, N.E. 1½ m. n.n.w. Waters Upton (UNRS) 138, 145, 166, 409.

See *Hatton (Cold) Moss.*

Hatton (Cold) Moss, N.E. 185.

Haughmond Hill, N.E. (850) 4 m. n.e. Shrewsbury, (Sl.Camb.R.) 9, 29, 31, 38, 53, 55, 61, 70, 73, 100, 101, 108. 110, 114, 115, 135, 136, 146, 158, 173, 189, 192, 193, 211, 212, 226, 236. 238, 251, 259, 263, 290, 292, 311, 325, 335, 346, 348, 354, 359, 363, 372, 374, 385, 389, 392, 404, 406, 412, 415, 427, 428, 445, 448, 450, 451, 454, 456, 457, 458, 460, 464, 475, 491, 496, 497, 509, 510, 511, 512, 516, 521, 522, 523, 529.

See *Downton. Haughmond Abbey. Somer wood. Sunderton Camp.*

Haughmond Abbey, 4 m. n.e. Shrewsbury, 73, 106, 187, 188, 193, 194, 232, 235, 241, 274. 288, 332, 337, 363, 389, 516, 520.

Haughton, N.E. 5 m. n.e. Shrewsbury, (Sl.Camb.R.) 274, 488.

Hawkestone, N.E. (812) 4 m. n.e. Wem, (UNRS) (812), 9, 31, 79, 189, 193, 249, 281, 311, 343, 165, 29. 415.

See *Ellardine. Hawkstone heath. Red Castle. Weston.*

Hawkestone Heath, N.E. 29,

Hayes, N.W. 1 m. n.w. Oswestry, (MG) 127, 252, 263, 371.

Haygate, S.E. 10 m. e.s.e. Shrewsbury, (LNRS) 145, 488.

Heath, near Ludlow, S.E 240.

Heath, Old, (near Shrewsbury), N.E. 1 m. n.n.e. Shrewsbury, (UNRS) 106, 302, 418, 502, 517.

Helion, near Rushbury, S.E. 193.

Higford, S.E. 1 m. s.s.w. Beckbury, (UNRS) 135.

High Holborn, S.E. 1 m. n. Albrighton, (UNRS) 228.

High Rock, (Bridgnorth), S.E. 44, 96, 111, 112, 186, 359, 414, 429.

Hill Top, (Wenlock Edge), S.E. 4½ m. s.w. Wenlock, (WL) 101, 131, 187, 283, 335, 351, 429, 432.

Hineham, near Ludlow, S.E. 168.

Hintz, Parish Coreley, S.E. (ORS) 414.

Hogstow Mill, N.W. 1½ m. s. Minsterley, (L.Sil. R) 530.

Hodnet, N.E. 6 m. s.w. Market Drayton, (UNRS) 105, 520, 528.

Hollyhurst Common, near Longnor, 511.

Home Barns, near Sundorn, N.E. 5 m. n.e. Shrewsbury, (UNRS) 496, 520.

Homer, S.E. 1½ m. n. Wenlock, (WS) 213, 445.

Hope Bagot, S.E. 4½ m. e. Ludlow, (ORS) 430.

Hope, near Ludlow, S.E. 147, 186.

Hope Bowdler Hill, S.W. (1100) 2 m. e. Church Stretton, (T) 116, 303.

Hope Coppice, 531.

Hope Mead, near Bishop's Castle, S.W. 156.

Hopton Cliff, near Nesscliff, N.W. 72. 279.

See *Hopton hill.*

Hopton Hill, near Nesscliff, N.W. 101, 298, 359, 387, 511, 520, 521, 522.

Hopton, S.W. 240, 252, 417.

Hopton Castle, S.E. 7 m. n.e. Ludlow, (ORS) 106, 148, 274, 328, 415.

Hopton Monks, S.E. 4 m. s. Wenlock, (ORS) 272, 275.

Hopton Wafers, S.E. 2½ m. w.n.w. Cleobury Mortimer, (ORS) 328, 517.

Hordley, N.W. 3 m. s.w. Ellesmere, (UNRS) 496, 511, 520, 531.

Hord's Park, near Bridgnorth, S.E. 77, 157, 173, 295, 310, 325, 358, 389, 397, 411.

Horsehay, near Shiffnal, S.E. 168, 169, 447.

Hucklement, near Ludlow, S.E. 137.

Huck's Barn, near Ludlow, S.E. 37, 352.

Hughley, S.E. 3½ m. s.w. Wenlock, (WS) 136, 172, 252, 376.

Hungerford, S.E. 8 m. s.w. Wenlock, (ULR) 275.

Huntingdon, S.E. 1 m. n.n.e. Little Wenlock, (UCM) 316.

Hurst Coppice, (Ness) N.W. 186.

Ingerton, S.E. 222.

Ironbridge, S.E. 13½ m. s.e. Shrewsbury, (UCM) 52, 73, 128, 204, 417.

See *Benthal Edge. Coalbrookdale. Tike's Nest.*

Ivelith, S.E. 2 m. s. Shiffnal, (UNRS) 512.

Jigger's Bank, (Coalbrookdale) S.E. 211, 354, 523.

Kemberton, S.E. 2½ m. s.s.w. Shiffnal, (LNRS) 97, 473.

Kenelm's Chapel, S.E. (Staffordshire) 173, 185.

Kenley, S.E. 3½ m. w. Wenlock, (Car.S.) 193, 313, 427, 507, 511, 516.

Ketley, S.E. 1 m. s.e. Wellington, (UCM) 97, 188, 193, 315, 363, 408.

See *Red Lake.*

Kingsland, (near Shrewsbury) S.W. ½ m. w.s.w. Shrewsbury, (LNRS) 68, 302.

King Street, near Berrington, S.E. 3½ m. s.s.e. Shrewsbury, (LNRS) 317.

Kingswood, S.E. 2 m. s.e. Albrighton, (UNRS) 448, 496.

Kinlet, S.E. 4 m. n.e. Cleobury Mortimer, (T) 12, 73, 156.

Kinnersley, N.E. 3½ m. n.n.e. Wellington, (UNRS) 158, 496, 531.

See *Sidney.*

Kinsley Wood, (opposite Knighton) 328.

Kinton, N.W. 2 m. n.w. Nesscliffe, (UNRS) 119, 406.

Knockin, N.W. 6 m. s.s.e. Oswestry, (UNRS) 106, 187, 451.

See *Knockin Heath. Kynaston.*

Knockin Heath, N.W. 2 m. s.e. Knockin, (UNRS) 166, 325, 387, 30, 504, 511, 514, 517.

Knowle, parish Burford, S.E. 125, 414.

Kynaston, N.W. 2 m. s.e. Knockin, (UNRS) 491.

Larden, (Wenlock Edge) S.E. 2 m. s.s.e. Easthope, (ULR) 111, 169.

See *Larden Hall.*
Larden Hall, 146.
Lawley Hill, S.E. (900) 1½ m. s.e. Leebotwood, (T&Q) 52, 61, 70, 195, 404, 421, 427, 430.
Lawley's Cross, S.E. 1 m. s. Buildwas. (WS) 274.
Lawrence Hill, near Wrekin, S.E. 343.
Lea Farm House, (between Wenlock and Church Stretton) S.E. 279.
Lea Hall, N.W. 5 m. n. Shrewsbury, (UNRS) 510, 513.
Lea Cross, S.W. 5 m. w.s.w. Shrewsbury, (CM) 263, 331.
Leasowes, (Hales Owen) S.E. 71.
Leaton Shelf, N.W. 4 m. n.n.w. Shrewsbury, (UNRS) 46, 68, 111, 112, 186, 351, 503, 508, 519, 521.
Leaton Knolls, N.W. 3¼ m. n.n.w. Shrewsbury, (UNRS) 317, 356.
Ledwich, 283, 517.
Lee Bridge, N.E. ¼ m. s.w. Lee Brockhurst, (UNRS) 102, 348, 374.
Lee, near Ellesmere, N.W. 1¾ m. s. Ellesmere, (UNRS) 431, 461, 493, 519, 521.
Leebotwood, S.W. 3½ m. n.n.e. Church Stretton, (UCM) 186, 188.
See *Caer Caradoc. Causeway Wood. Lawley. Lidley Hayes.*
Leigh Hall, N.W. 2½ m. s.w. Minsterley, (L.Sil.R.) 506.
LEIGHTON, S.E. 3½ m. n. Wenlock, (Car.S.) 462.
Leegomery, N.E. 1 m. n.e. Wellington, (LNRS) 283.
Lidley-Hayes, S.W. ½ m. s.e. Leebotwood, (LNRS) 430.
Lightmoor, S.E. 1 m. n.e. Coalbrookdale, (UCM) 132.
LILLESHALL, N.E. 3 m. s.w. Newport, (UCM) 106, 129, 144, 158, 164, 275, 298, 302, 315, 326, 404, 410, 411, 432, 509, 510.
See *Lilleshall Mill-pool. Lilleshall Abbey. Lilleshall Hill.*
Lilleshall Mill-pool, 38, 47, 78, 254, 516.
Lilleshall Abbey, N.E. 1 m. s.e. Lilleshall, (LNRS) 68, 106, 144, 164, 190, 253, 276, 315, 389, 406, 411, 415, 444, 510, 529.
Lilleshall Hill, N.E. (350) ¼ m. n. Lilleshall, (T) 193, 311.
Limekiln Woods, (Wrekin), 125, 165, 172, 243, 416, 429, 432, 434, 444, 495, 497.
Linley, S.W. (1000) 3 m. n.n.e. Bishop's Castle, (Sl.Camb.R.) 222, 414, 415, 521.
Lincoln's Hill, Coalbrookdale, S.E. (479) ¼ m. s.e. Coalbrookdale, (UCM) 47, 50, 73, 115, 129, 136, 192, 255, 290, 291, 302, 325, 411, 415, 432, 494.
Lincroft Pool, near Pleale̲y, S.W. 6. m. s.w. Shrewsbury, (Sl.Camb.R.) 449.
Littlehales, near Newport, N.E. 258, 296.
Lizard, S.E. 2½ m. n.e. Shiffnal, (UNRS) 330.
See *Lizard Wood.*
Lizard Wood, S.E. 2½ m. n.e. Shiffnal, (UNRS) 165.
Llynclys, N.W. 3 m. s. Oswestry, (LNRS) 157, 447.
See *Llynclys Pool.*
Llynclys Pool, N.W. 3 m. s. Oswestry, (LNRS) 252.
Llanvorda, N.W. 1½ m. s.w. Oswestry, (MG) 166.
LLANYMYNECH, N.W. 5 m. s.s.w. Oswestry, (Carb.L.) 136, 156, 187, 243, 276, 283, 360, 415, 516.

See *Blodwell Rocks. Chambre Wên. Devolog. Dwyffrwd. Llanymynech Hill. Llwynygroes. Llwyntidman. Penyvoel. Wern Ddû.*
Llanymynech Hill, N.W. 6 m. s.s.w. Oswestry, (Carb.L.) 279, 327, 331, 351, 473, 510, 522, 529, 530, 531.
Llymymaen, N.W. 1 m. s.w. Oswestry, (LCM) 514.
Llwyn-y-groes, N.W. 1 m. e.s.e. Llanymynech, (LNRS) 415, 436.
See *Llyntidman.*
Llwyntidman, N.W. near Llwyn-y-groes, (LNRS) 415, 510, 530.
Lodge, near Ludlow, S.E. 186, 254, 295, 385, 428, 433, 434, 531.
Lodge Hill, near Frodesley, S.E. (Car.S.) 343.
Long Dale, near Cound, S.E. (Car.S.) 433.
LONGDEN, S.W. 5½ m. s.w. Shrewsbury, (CM) 300, 366, 376, 400, 452, 453.
LONGDEN UPON TERN, N.E. 3 m. n.n. w. Wellington, (UNRS) 145.
LONGFORD, N.E. 1½ m. s.w. Newport, (LNRS) 414.
See *Cheswell.*
Long Lane Quarry, (near Cheney Longville) S.W. ½ m. s.s.w. Cheney Longville, (Car.S.) 27, 52.
Longmont, S.W. (U.Sil.R.) 327.
Longmynd, S.W. (1200—1680) (Sl.Camb.R.) 110, 116, 119, 147, 150, 151, 195, 343, 349, 376, 402, 404, 430, 457.
See *Bridges. Clun Dale.*
Longner, S.E. 2½ m. e.s.e. Shrewsbury, (LNRS) 186, 400, 488, 512, 518, 528.
See *Ford Hill.*
LONGNOR, 8 m. s. Shrewsbury, (LNRS) 39, 97, 110, 188, 240, 258, 259, 274, 346, 349, 417, 430, 431, 434, 518.
See *Hollyhurst Common.*
Longville, (Cheney) S.W. 2 m. n.w. Craven Arms, (Car.S.) 44, 52, 127, 186, 310, 363.
See *Long-lane Quarry.*
Longville Hill, S.E. 354.
Longville in the Dale, S.E. 6 m. e. Church Stretton, (WS) 46, 112, 362.
LOPPINGTON, N.W. 2¼ m. w. Wem, (LRM) 520.
See *Woverley.*
Lord's Meadow, Albrighton, near Shiffnal, S.E. 74, 315.
Loton Park, near Alberbury, N.W. 9 m. w. Shrewsbury, (ML) 491.
Lowe, N.E. 1 m. n.w. Wem, (LRM) 415.
Lubstree Park, 243, 259, 276, 510, 531.
LUDFORD, S.E. 3¾ m. s.e. Ludlow, (ULR) 188, 189, 194, 353, 375.
LUDLOW, S.E. 29, m. s. Shrewsbury, (ULR) 9, 11, 12, 26, 28, 29, 31, 33, 37, 43, 44, 47, 51, 60, 68, 72, 97, 100, 101, 106, 108, 109, 112, 114, 115, 120, 121, 124, 126, 127, 128, 131, 136, 146, 151, 154, 157, 166, 168, 169, 172, 188, 189, 190, 192, 194, 195, 211, 212, 217, 243, 249, 250, 251, 254, 256, 263, 276, 278, 283, 286, 290, 291, 295, 296, 297, 299, 302, 303, 309, 311, 313, 316, 317, 319, 320, 330, 331, 337, 344, 346, 348, 351, 352, 354, 363, 372, 375, 376, 389, 402, 408, 410, 411, 416, 420, 422, 428, 429, 430, 433, 449, 469,

472, 473, 475, 476, 486, 489, 491, 495, 496, 509, 512, 518, 523, 529.
See *Ashford. Aston. Aston Botterell. Bitterley. Brook House. Bromfield. Burford. Burraston. Burway. Cainham. Caynham Camp. Clee St. Margaret. Cleobury Mortimer. Coreley. Corfeton. Corve dale. Corve river. Craven Arms. Culmington. Diddlebury. Downton Hall. Downton Castle. Downton Hall Woods. Felton Farm. Halton. Priors Halton. Heath. Hopton Castle. Hineham. Hope Bagot. Hope. Hucklement. Hucks Barn. Ledwich. Lodge. Ludford. Ludlow Castle. Ludwyche. Lynney. Marlow. Upper Nash. Oakley Park. Onibury. Peeton. Poles Farm. Saltmoor. Shelderton. Shortwood. Silvington. Stanton Lacy. Steventon. Stoke St. Milborough. Stoke Castle. Stokesay. Sutton. Teme River. Tinker's Hill. Titterstone Clee Hill. Brown Clee Hill. Vinalls. Cold Weston. Wheathill. Whitbach. Whitecliff.*
Ludwyche, (River), near Ludlow, S.E. 329.
Lumhole Pool, (near Coalbrookdale), S.E. 9, 96, 109.
Lutwyche, (Wenlock Edge), S.E. 1 m. s.w. Easthope, (LLR) 55, 112, 338, 354, 360, 404, 432, 434, 497, 507, 513, 516.
See *Mog Forest.*
LYDBURY, S.W. 3 m. s.e. Bishop's Castle, (U.Sil.R.) 286.
Lydham, S.W. 2 m. n.n.e. Bishop's Castle, (Sl.Camb.R.) 523.
Lyd Hole, S.W. n.e. Pontesford Hill, 371, 376, 395, 415, 462, 509, 513, 522.
Lynney, near Ludlow, S.E. 190.
Lyth Hill, S.W. 3¾ m. s.s.w. Shrewsbury, (Sl.Camb.R.) 29, 38, 52, 60, 61, 71, 72, 135, 136, 173, 193, 286, 343, 346, 376, 445, 497, 521.
See *Lythwood.*
Lythwood, S.W. 3 m. s.s.w. Shrewsbury, (Sl.Camb.R.) 164, 427, 428.
MADELEY, S.E. 15 m. s.e. Shrewsbury, (UCM) 31, 38, 48, 59, 97, 129, 130, 135, 144, 146, 157, 317, 432, 450, 459, 503, 517.
See *Madeley-lane Pits. Madeley Wood.*
Madeley-lane Pits, S.E. 387.
Madeley Wood, S.E. 418.
Maesbury, N.W. 5 m. s.s.e. Oswestry, (UNRS) 33, 137, 264, 300, 302, 390, 459, 463, 467.
Maestermin Bridge, (between Halston and Hardwick), N.W. 264.
Malt-house Pool, Coalbrookdale, S.E. 454, 456.
Marbury Mere, N.E. 3 m. n.n.e. Whitchurch, (not in Shropshire), (LRM) 152, 252, 258, 260, 404, 405, 493.
March Pool, (Wilcot) N.W. 105.
See *Marsh.*
Mare Pool, S.E. 1¾ m. s.e. Shrewsbury, (LNRS) 32, 128, 129, 258, 418, 463, 466.
Marlow, near Ludlow, S.E. 415.
Marn Wood, S.E. 1 m. w.s.w. Coalbrookdale, (WS) 33, 109, 128, 212, 318, 428, 456, 457, 460, 485, 486, 495.
Marrington Dingle, near Chirbury, S.W. 1 m. s.e. Chirbury, (L.Sil.R.) 522.
Marsh, S.E. 1¼ m. e. Much Wenlock, (ULR) 511.
See *Marsh Pool.*

Marsh Pool, S.E. 1 m. e. Wenlock, (ULR) 105, 404, 448, 449.
Marsley, S.W. 1 m. s.s.w. Habberley, (Sl.Camb.R.) 522.
Marton, near Baschurch, N.W. 5 m. s.w. Wem, (UNRS) 105, 240, 317, 418.
Marton Pool, near Baschurch, N.W. 5 m. s.w. Wem, (UNRS) 252, 258, 415, 509, 528.
Marton Pool, near Chirbury, S.W. 15½ m. s.w. Shrewsbury, (U.Sil.R.) 521, 531.
Maulbrook, near Walton, N.E. 243.
Mawley, S.E. 1½ s.e. Cleobury Mortimer, (ORS) 328.
MELVERLEY, N.W. 10 m. w.n.w. Shrewsbury, (UNRS) 415, 491, 511.
See *Fettis Farm.*
MEOLE BRACE, 1½ m. s. Shrewsbury, (CM) 61, 97, 135, 167, 169, 189, 251, 275, 286, 288, 309, 310, 372, 410, 484, 486, 489, 490, 516.
Merrington, N.W. 5½ m. n.n.w. Shrewsbury, (UNRS) 106.
MIDDLE, N.W. 7 m. n.n.w. Shrewsbury, (UNRS) 73, 97, 106, 193, 194, 276, 317, 414, 488.
See *Balderton.*
Middleton, Oswestry, 489.
Middleton Hill, N.W. 2 m. n.w. Woolaston, (U.Sil.R.) 60, 145, 283, 349.
Middlehope, S.E. 2 m. n.n.w. Diddlebury, (LLR) 136, 302, 415, 430, 431.
Mile House, N.W. 1½ m. s.e. Oswestry, (LNRS) 211.
Milford, near Baschurch, N.W. 7 m. n.w. Shrewsbury, (UNRS) 146, 147.
Millechope, S.E. 3 m. e.s.e. Eaton, (ULR) 216, 428, 432, 486, 512.
See *Little Millechope.*
Millechope, (Little) S.E. 2 m. e.s.e Eaton, (ULR) 145, 402.
Milloon, (Westfelton) N.W. 298.
Milson Wood, S.E. 3 m. s.w. Cleobury Mortimer, (ORS) 146.
MINSTERLEY, S.W. 9½ m. s.w. Shrewsbury, (Ll.F.L.Sil.) 290, 396, 412, 513, 528, 529.
See *Grimmer Rocks. Hogstow. Leigh Hall. Snailbeach.*
Minton, parish Church Stretton, S.W. 2½ m. s.s.w. Church Stretton, (Sl.Camb.R.) 516.
Moelydd, N.W. 3 m. s.w. Oswestry, (MG) 115.
Mog Forest, Lutwyche, S.E. 1 m. s.s.w. Easthope, (AL) 434.
Monkmoor, N.E. 1½ m. n.e. Shrewsbury, (LNRS) 112, 151, 286, 312, 453, 521.
MONKS HOPTON, S.E. 4½ m. s. Wenlock, (ORS) 272, 275, 431.
Montford Bridge, N.W. 4 m. n.w. Shrewsbury, (UNRS) 111, 326, 330, 387.
MONTFORD, N.W. 5 m. w.n.w. Shrewsbury, (UNRS) 263, 513, 514, 521.
Morda Pool, N.W. 1½ m. s. Oswestry, (LCM) 105, 410, 448, 451, 452, 456, 457, 460, 465, 466, 468, 471, 504, 505, 508, 509, 510, 516, 528, 529.
See *Morda River.*
Morda (River) N.W. 410, 503.
MORETON CORBET, N.E. 4½ m. s.e. Wem, (UNRS) 243, 521.
MORETON SAY, N.E. 3 m. w. Market Drayton, (URM) 504, 517.
Morfe, S.E. ½ m. e. Bridgnorth, (LNRS) 189, 279, 327, 510.
MORVILL, S.E. 3 m. w.n.w. Bridgnorth, (ORS) 276, 509, 513, 518.

INDEX OF GENERA, SPECIES, SYNONYMS, &c.

THE END.

THE

Filices, Lycopodiaceæ, Marsileaceæ,

AND

Equisetaceæ of Shropshire,

BY

WILLIAM PHILLIPS, F.L.S.

Reprinted from the Transactions of the Shropshire Archæological Society, 1877.

PRIVATELY PRINTED FOR THE SOCIETY

SHREWSBURY :

ADNITT AND NAUNTON.

OSWESTRY .

WOODALL AND VENABLES.

1877.

THE
FILICES, LYCOPODIACEÆ, MARSILEACEÆ,
AND
EQUISETACEÆ OF SHROPSHIRE.
BY WILLIAM PHILLIPS, F.L.S.

WHILE Shropshire can boast of one of the best local Floras ever published, comprising more than one half the species of indigenous flowering plants of Britain, up to the present time there exists no work devoted to the Cryptogamic plants of the county. It was the intention of the learned author of the "Flora of Shropshire" to have supplemented that work by one comprising the Cryptogamia, but although much work has been done by him towards the fulfilment of this intention, especially in the determination of the Lichens, I am not aware that there is any immediate prospect of such a work appearing. Mr. R. Anslow has published a List of the Mosses of the Wrekin which it is hoped may soon be followed by one embracing the species of the entire county. The list now given of the Ferns and their allies is intended to be a small contribution to the same general object of supplying hereafter a complete Salopian Cryptogamic Flora.

Shrewsbury, Sept., 1877.

The names of contributors to this list (to whom I tender my thanks) and the sources of information are abbreviated or given in full as below :—

Agriculture of Shropshire, by Archdeacon Plymley, the list of plants by Dr. Babington of Ludlow. 1803. ... ... *Plym.*
An unpublished list of plants of Shropshire by the late Rev. E. Williams. (From a copy kindly lent me by the Rev. W. A. Leighton.) ... ... *W.*
Turner and Dillwyn's Botanical Guide ... ... *Turn. & Dill.*
Moore's History of British Ferns... ... ... *Moore.*

Newman's History of British Ferns ... ... *Newman.*
A Botanical Guide to the Environs of Church Stretton, by G. H. Griffiths, M.D., Shrewsbury ... ... *G.*
A Guide to the Ferns and many of the Rarer Plants growing about Ludlow : A Marston. Ludlow, 1870 ... ... *M.*
Miss Brown, Bridgnorth, a Manuscript List of Ferns, &c., of the neighbourhood of Bridgnorth, confirmed by her own observation (kindly communicated to me by J. R. Jebb, Esq).... *Miss Brown.*
Mrs. Auden, Ford, Salop ... ... *Mrs. Auden.*
Rev. W. A. Leighton, B.A., Cam., F.L.S., F.B.S. ED. *Rev. W. A. Leighton.*
J. R. Jebb, Esq., Shrewsbury ... ... *J. R. Jebb.*
R. Anslow, Esq., Wellington, Salop ... ... *R. Anslow.*
Rev. T. H. Eyton, Bridgnorth ... ... *Rev. T. H. Eyton.*
Rev. W. T. Burges, Shrewsbury ... ... *Rev. W. T. Burges.*
Rev. J. H. E. Charter, Shrewsbury ... *Rev. J. H. E. Charter.*
Mr. R. M. Serjeantson, Acton Burnell, Salop. ... *R. M. Serjeantson.*
W. Beckwith, Esq., Eaton Constantine.

FILICES.

HYMENOPHYLLUM UNILATERALE, *Willd.*
Longmynd, near the Stiperstones : *G.* ; on the same mountain, Spout Valley, near Church Stretton : *Rev. J. F. Couch* ; Treflach Wood near Oswestry : *Moore.*

ADIANTUM CAPILLUS—VENERIS, *L.*
Mr. Newman in his "History of British Ferns" quotes a statement from the Phytologist (v. I, p. 579) that Mr. Westcott, sixteen years previous to the publication of the announcement, had found this plant on the Titterstone Clee Hill, but that on again visiting the spot he was unable to find it. Many botanists since then have carefully searched the rocks there, but I need scarcely add without success.

Mr. W. Beckwith sent me a frond of this fern in October, 1876, gathered by him on the S.W. end of the Wrekin " amongst loose stones ;" on subsequent enquiry, however, he learned that it had been planted there some three or four years previously by a person living in the neighbourhood.

PTERIS AQUILINA, *Linn.*
This species is very commonly distributed throughout the whole of Shropshire.

LOMARIA SPICANT, *Desv.*
This is a very common species on the mountains, hills, and heathy ground, throughout the county. Common in Whitcliff, and woods about Ludlow : *M.* Shawbury Heath, The Longmynd, Caer Caradoc, near Church Stretton, Stiperstones : *W.P. Miss Brown* gives the following localities near Bridgnorth ; Thatcher's Wood, Apley Drive, Shirlett, Darley Dingle, Burcott, Cantern Dingle, Cliff Wood, Bowman Hill, The Woodlands.

ASPLENIUM RUTA MURARIA, *Linn.*
Old walls, not uncommon : *W.* Henly, Ashford and Bromfield bridges, near Ludlow : *M.* Cardington : *G.* Llanymynech and

Much Wenlock : *Rev. W. A. Leighton. Mr. Anslow* reports it in the following places : On bridge over the Tern at Crudgington, walls in Baschurch village, walls at Much Wenlock and Easthope, Dothill Park wall, near Wellington. *Miss Brown* gives the following localities : Worfe Turnpike, High Rock, Cat-brain, Danesford, Davenport, Morville, Halton, Worfield, Farmcott. Acton Burnell church : *R. M. Serjeantson.* On garden wall, Quatford, and bridge at Quat. : *Rev. T. H. Eyton.*

ASPLENIUM TRICHOMANES, *Linn.*
Walls and rocks of Haughmond Hill : *W.* Caer Caradoc, near Church Stretton : *G.* Whitcliff : *M.* Walls about Ludlow, Wenlock, Clungunford, Llanymynech, on the Wrekin (but few plants only), old walls at Tern near High Ercall, walls in Much Wenlock : *R. Anslow.* Gaer ditches, near Bucknell : *W.P.* Crickheath Hill : *J. R. Jebb. Miss Brown* records it at Worfe Turnpike, High Rock, Bowman Hill, Thatcher's Wood, Burcott, Quatford, Shipton, Morville Hall, Farmcott, Apley Terrace.

ASPLENIUM ADIANTUM, *Linn.*
Rocks of Haughmond Hill, *W.* (not now to be found, *W.P.*) Cheney Longville : *G.* Norton Camp near Craven Arms : *M.* Hedge banks near Wrockwardine, Condover, Clungunford : *R. Anslow* gives the following : Easthope, Cluddeley, near the Wrekin, in 1862 (but now extinct there), Ringer's lane near Admaston, Cleobury North. *Miss Brown* reports it from High Rock and Hermitage Hill, Bowman Hill, Quatford, Burcott, Davenport, Apley Terrace. Lane near Golding, Gaer Ditches, near Bucknell : *W.P.* Near Little Ness and Crickheath Hill : *J. R. Jebb.*

ATHYRIUM FILIX—FŒMINA, *Bernh.*
This is a very generally distributed species, and abundant throughout the county.

CETERACH OFFICINARUM, *Willd.*
Walls of Ludlow : *Plym.* Old walls at Sweeney, Oswestry : *R. Anslow.* Crickheath Hill near Oswestry : *J. R. Jebb.* Law farm, Cleobury Mortimer : *Miss Brown.*

SCOLOPENDRIUM VULGARE, *Sm.*
Rocks, shady lanes, and mouth of wells : *W.* Rocks and walls about Church Stretton : *G.* Hawkstone Park : *Rev. W. A. Leighton.* In pit shafts at Steeraway near Wellington, Roadside near Upton Magna, old wells on Cold Hatton and Ellerdine Commons, Grindle near Shiffnal : *R. Anslow.* Canal side, Woodseams near Market Drayton : *J. R. Jebb.* Astbury, Tusley, Rindleford, Daniel's Mill, Darley Dingle, The Woodlands, Potseithing : *Miss Brown.*

CYSTOPTERIS FRAGILIS, *Bernh.*
    var : DENTATA.
Lee Bridge, sandstone rocks : *W.P.* Whitcliffe Rocks near Ludlow, Craig-y-rhiw near Oswestry : *J. R. Jebb.* Cistern at Shipton Hall, Broseley Churchyard : *Miss Brown.*

ASPIDIUM ACULEATUM, *Sw.*
    Shady lanes and hedges: *W.* Ragleth wood near Church
    Stretton, wood east end of Wrekin: *W.P.* Mary's Dingle near
    Leighton under Wrekin: *R. Anslow.* Near Baschurch: *J. R.
    Jebb.* Burcott rocks, Swancott, Davenport woods, Astley
    Abbots, Cantern Dingle: *Miss Brown.*
ASPIDIUM ACULEATUM.
        var. LOBATUM.
    Acton Scott: *G.* Hedge banks between Railway Station and
    Grinshill: *W.P.* Limekiln Woods near the Wrekin, Clee Hill,
    Natchwood: *R. Anslow.* Road between Burwarton and Charlcott,
    Stanley dingle, Footbridge on Morville road: *Miss Brown.*
ASPIDIUM ANGULARE, *Willd.*
    Steeraway, Limekiln wood near the brooks, Madoc's Hill, Prest-
    hope near Much Wenlock: *R. Anslow.* Ragleth wood, Church
    Stretton, Callow Hill dingle near Minsterley, Farley Dingle, near
    Buildwas: *W.P.* Hopton's woods, Cleobury Mortimer, Cliff wood:
    Miss Brown.
NEPHRODIUM FILIX-MAS, *Rich.*
    Generally distributed, and abundant throughout the county.
NEPHRODIUM SPINULOSUM, *Desv.*
    Shomere Moss, Boggy ground under Arkoll Hill, boggy ground
    under Titterstone Clee Hill above Acton Botterill; *R. Anslow.*
    Darley Dingle (Linley), Mawley, Cleobury Mortimer: *Miss
    Brown.*
NEPHRODIUM DILATATUM, *Desv.*
    In woods and heathy ground throughout the county.
        var. NANUM.
    Dudmaston Hill, Quatford, Shifnal road, Rindleford: *Miss
    Brown.*
NEPHRODIUM ÆMULUM, *Baker.*
    Near La mole pool, in Coalbrookdale: *Rev. J. Hayes.* *R.
    Anslow.*
NEPHRODIUM THELYPTERIS, *Desv.*
    Berrington upper pool: *W.* Whittington: *Rev. W. W. How.*
    Bomere pool and Shomere Moss: *R. Anslow.* Marton Pool,
    near Baschurch: *Rev. J. H. E. Charter.* Colemere Mere, near
    Ellesmere: *Rev. W. T. Burges.*
NEPHRODIUM OREOPTERIS, *Desv.*
    Shineton Common, Shawbury Heath, Cound Moor: *W.* Titter-
    stone Clee Hill: *M.* Woolstone: *G.* Longmynd, near Church
    Stretton: *W.P.* Lawrence's Hill and Arkoll Hill, on the Wrekin,
    roadside at foot of Wrekin, abundant, Ellardine Common, Shirlett:
    R. Anslow. Crickheath Hill: *J. R. Jebb.* Quatford, Bowman
    Hill, Linley, Darley Dingle, Burcott: *Miss Brown.*
POLYPODIUM VULGARE, *Linn.*
    On walls, trunks of trees, &c. Common throughout the county.
POLYPODIUM PHEGOPTERIS, *Linn.*
    Titterstone Clee Hill: *Newman.* Longville Common: *M.*

    Ragleth Wood, near Church Stretton: *W.P.* Darley Dingle,
    near Linley: *R. Anslow.* Craigforda, Oswestry: *Salwey.* Brown
    Clee Hill: *Miss Brown.*
POLYPODIUM DRYOPTERIS, *Linn.*
    Wrekin, foot of Wenlock Edge, between Lutwych and Kinley:
    W. Titterstone Clee Hill, Downton Castle: *Plym.* Canton Rough,
    about a quarter of a mile from Bridgnorth: *Purton.* Longville
    Common near Craven Arms: *M.* Spout Valley, Longmynd,
    near Church Stretton: *W.P.* Eastern declivity of Wrekin Hill:
    R. Anslow. Frodesley Hill: *Moore.* Hoar Edge, Whitcliffe
    Coppice, near Ludlow: *Newman.* Goose's Meadow, Chesterton,
    Burcott, Apley drive: *Miss Brown.*
POLYPODIUM ROBERTIANUM, *Hoffm.*
    Cheney Longville: *G.* (This is very doubtful. *W.P.*)
OSMUNDA REGALIS, *Linn.*
    Between Birch and Lee, near Ellesmere, Woodhouse near
    Oswestry, Shawbury Heath: *W.* Titterstone Clee Hill: *M.*
    Knockin Heath: *Turner & Dill.* Colemere and Sandford Heath
    near Westfelton: *Rev. W. A. Leighton.* Roddington Heath
    (1862), Ellerdine Common, near the Hazles Farm below Eller-
    dine, Hodnet Heath: *R. Anslow.* Between Black Mere and
    Kettle Mere, near Ellesmere: *Rev. W. T. Burges.* Hopton
    wood, Cleobury Mortimer, Stanley: *Miss Brown.*
OPHIOGLOSSUM VULGATUM, *Linn.*
    Meadows about Eaton, Leighton, Golden, Berrington, *W.*
    Berwick near Shrewsbury, Westfelton: *Rev. W. A. Leighton.*
    Ford: *Mrs. Auden.* Coppice near Onslow Hall: *W.P.* Steeraway
    in the wood, near the Arkol Hill, Fields under the Wrekin,
    Donnerville Lawn, near Wellington: *R Anslow.* Fields
    adjoining Eyton Park: *Rev. W. Houghton.* Oldbury, Harps-
    ford, Stanley, Cantern brookside, Hoards Park Farm, Shirlett:
    Miss Brown.
BOTRYCHIUM LUNARIA, *SW.*
    In a meadow near Ludlow: *Turner & Dill.* Whitcliffe: *M.* Sutton
    Spa: *late Revd. S. P. Mansell.* Craig-y-rhiw, near Oswestry:
    J. R. Jebb. "The late H. Bidwell Esqr found this plant in a
    large heathy field under the Wrekin." *R. Anslow.*

LYCOPODIACEÆ.

LYCOPODIUM CLAVATUM, *Linn.*
    Bedstone Hill, near Knighton, *Plym.* Hodnet Heath, Stiper-
    stones Heath, Bettws near Clun: *W.* Caradoc Hill, near Church
    Stretton, Corndon Hill, Brown Clee Hill: *Rev. W. A. Leighton.*
LYCOPODIUM ALPINUM, *Linn.*
    Stiperstones Heath, according to Dillenius: *W.*
LYCOPODIUM SELAGO, *Linn.*
    Stiperstones Heath, Brown Clee Hill, Titterstone Clee Hill: *W,*

    Longmynd Hills, abundant: *G.* Caradoc Hill, near Church
    Stretton, Corndon Hill: *Rev. W. A. Leighton.*
SELAGINELLA SELAGINOIDES, *Gray.*
    Longmynd Hills, *G.*
ISOETES LACUSTRIS, *Linn.*
    Eastern shore of Bomere Pool: *W.* South end of the lake, 1874.
    W.P. Pools, near Darnford: *G.*

MARSILEACEÆ.

PILULARIA GLOBULIFERA, *Linn.*
    "Bomere, Betton Pool, Pool between Knockin and Nescliffe: *W.*
    Although I have carefully searched the first two localities, given
    by Mr. Williams, I have failed to discover the plant." *W.P.*

EQUISETACEÆ.

EQUISETUM ARVENSE, *Linn.*
    This is generally distributed throughout the county, and too
    often existing in profusion, being a weed difficult of extinction.
EQUISETUM MAXIMUM, *Lam.*
    Golden Pool, and near Harley: *W.* (It exists in the former
    locality no longer, the pool having been drained.) Weir Coppice,
    near Shrewsbury. Leaton Knolls: *W.P.* Rodington Heath:
    R. Anslow. Holley Coppice, the Albyns: *Miss Brown.*
EQUISETUM SYLVATICUM, *Linn.*
    Wood between Sidbury and Bridgnorth, East end of Stiperstones
    Hill, Priors' Ditton, Brown Clee Hill: *W.* Ticklerton, Strefford:
    G. Benthall Edge, Wenlock's wood, near the Wrekin: *R.
    Anslow.* Gittins' Hey-Wood, near Westbury, wood near
    Steeraway, Wellington: *W.P.* Linley, Darley Dingle, Faintree,
    Shirlett: *Miss Brown.*
EQUISETUM PALUSTRE, *Linn.*
    Shallow ponds and boggy ground: *W.* Marsh Brook: *G.*
    Ellesmere Mere: *W.P.* Bog under the Wrekin, Preston on the
    Wildmoors, Eyton: *R. Anslow.* Brown Clee Hill: *Miss Brown.*
EQUISETUM LIMOSUM, *Linn.*
    Ponds and pools: *W.* Plowden: *G.* River Severn, near
    Shrewsbury: *Rev. W. A. Leighton.* Shawbury Heath: *W.P.*
    Worfe turnpike: *Miss Brown.*
EQUISETUM HYEMALE, *Linn.*
    East end of Cantlop wood: *W.* Dell at Bitterley, below the
    Clee Hill: *Moore.* Town's Mills pond: *Miss Brown.*